IFAC

International Federation of Automatic Control

CONTROL SCIENCE AND TECHNOLOGY FOR DEVELOPMENT (CSTD'85)

IFAC Proceedings Series, 1986. Number 7

IFAC PROCEEDINGS SERIES

Editor-in-Chief

JANOS GERTLER, Department of Computer and Electrical Engineering, George Mason University, Fairfax, Virginia, USA

GERTLER & KEVICZKY (*General Editors*): A Bridge Between Control Science & Technology (*Ninth Triennial World Congress, in 6 volumes*)

- Analysis and Synthesis of Control Systems (*1985, No. 1*)
- Identification, Adaptive and Stochastic Control (*1985, No. 2*)
- Large-scale Systems, Decision-making, Mathematics of Control (*1985, No. 3*)
- Process Industries, Power Systems (*1985, No. 4*)
- Manufacturing, Man–Machine Systems, Computers, Components, Traffic Control, Space Applications (*1985, No. 5*)
- Biomedical Applications, Water Resources, Environment, Energy Systems, Development, Social Effects, SWIIS, Education (*1985, No. 6*)

BARKER & YOUNG: Identification and System Parameter Estimation (1985) (*1985, No. 7*)

NORRIE & TURNER: Automation for Mineral Resource Development (*1986, No. 1*)

CHRETIEN: Automatic Control in Space (*1986, No. 2*)

DA CUNHA: Planning and Operation of Electric Energy Systems (*1986, No. 3*)

VALADARES TAVARES & EVARISTO DA SILVA: Systems Analysis Applied to Water and Related Land Resources (*1986, No. 4*)

LARSEN & HANSEN: Computer Aided Design in Control and Engineering Systems (*1986, No. 5*)

PAUL: Digital Computer Applications to Process Control (*1986, No. 6*)

YANG JIACHI: Control Science & Technology for Development (*1986, No. 7*)

MANCINI, JOHANNSEN & MARTENSSON: Analysis, Design and Evaluation of Man–Machine Systems (*1986, No. 8*)

GELLIE, FERRATE & BASANEZ: Robot Control "Syroco '85" (*1986, No. 9*)

JOHNSON: Modelling and Control of Biotechnological Processes (*1986, No. 10*)

IFAC Related Titles

BROADBENT & MASUBUCHI: Multilingual Glossary of Automatic Control Technology

EYKHOFF: Trends and Progress in System Identification

ISERMANN: System Identification Tutorials (*Automatica Special Issue*)

CONTROL SCIENCE AND TECHNOLOGY FOR DEVELOPMENT (CSTD'85)

Proceedings of the IFAC/IFORS Symposium
Beijing, People's Republic of China, 20–22 August 1985

Edited by

YANG JIACHI

Chinese Academy of Space Technology,
Beijing, People's Republic of China

Published for the

INTERNATIONAL FEDERATION OF AUTOMATIC CONTROL

by

PERGAMON PRESS

OXFORD · NEW YORK · BEIJING · FRANKFURT
SÃO PAULO · SYDNEY · TOKYO · TORONTO

U.K.	Pergamon Press, Headington Hill Hall, Oxford OX3 0BW, England
U.S.A.	Pergamon Press, Maxwell House, Fairview Park, Elmsford, New York 10523, U.S.A.
PEOPLE'S REPUBLIC OF CHINA	Pergamon Press, Qianmen Hotel, Beijing, People's Republic of China
FEDERAL REPUBLIC OF GERMANY	Pergamon Press, Hammerweg 6, D-6242 Kronberg, Federal Republic of Germany
BRAZIL	Pergamon Editora, Rua Eça de Queiros, 346, CEP 04011, São Paulo, Brazil
AUSTRALIA	Pergamon Press Australia, P.O. Box 544, Potts Point, N.S.W. 2011, Australia
JAPAN	Pergamon Press, 8th Floor, Matsuoka Central Building, 1-7-1 Nishishinjuku, Shinjuku-ku, Tokyo 160, Japan
CANADA	Pergamon Press Canada, Suite 104, 150 Consumers Road, Willowdale, Ontario M2J 1P9, Canada

First edition 1986

ISBN 0 08 033473 3

These proceedings were reproduced by means of the photo-offset process using the manuscripts supplied by the authors of the different papers. The manuscripts have been typed using different typewriters and typefaces. The lay-out, figures and tables of some papers did not agree completely with the standard requirements: consequently the reproduction does not display complete uniformity. To ensure rapid publication this discrepancy could not be changed: nor could the English be checked completely. Therefore, the readers are asked to excuse any deficiencies of this publication which may be due to the above mentioned reasons.

The Editor

Printed in Great Britain by A. Wheaton & Co. Ltd., Exeter

IFAC/IFORS SYMPOSIUM ON CONTROL SCIENCE AND TECHNOLOGY FOR DEVELOPMENT

Organized by
Chinese Association of Automation

Sponsored by
International Federation of Automatic Control (IFAC)
Sponsoring Technical Committees: Developing Countries (DECOM), Applications (APCOM), Economic and Management Systems (EMSCOM), Education (EDCOM), Social Effects of Automation (SOC.EFF.), Systems Engineering (SECOM)

Co-sponsored by
International Federation of Operational Research Societies (IFORS)

Supported by
China Association for Science and Technology (CAST)
United Nations Educational, Scientific and Cultural Organization (UNESCO)
United Nations Industrial Development Organization (UNIDO)

International Program Committee
M. Najim, Chairman, Morocco
Yan Xiaojun, Co-Chairman, PRC
A. Alonso, Mexico
A. Bhattacharyya, India
A. van Cauwenberghe, Belgium
R. Chaussard, France
N. Cohn, USA
A. R. Curran, USA
P. Eykhoff, Netherlands
G. Guardabassi, Italy
S. Kahne, USA
S. Abdel Kader, Egypt
E. Kaskurewicz, Brazil
Y. Kaya, Japan
U. Luoto, Finland
B. Tamm, USSR
M. Thoma, Germany
Y. Sawaragi, Japan
W. Schaufelberger, Switzerland
T. Vamos, Hungary
Cai Fuyuan, PRC
Chen Zhenyu, PRC
Huang Taiyi, PRC
Kwei Hsiangyuan, PRC
Zheng Weimin, PRC
Zheng Ying-Ping, PRC

National Organizing Committee
Yang Jiachi, Chairman
Fang Jun, Vice-Chairman
Hu Chiheng, Vice-Chairman
Jia Wenhua
Ling Weihou
Wang Hongbing
Wang Zhengzhong
Wu Jong-Ming

FOREWORD

Over the long history of mankind, the productive power of man has always been in the process of development. This process consists of periods of gradual evolution with certain waves of rapid revolutionary events which were usually initiated by a number of scientific discoveries and technical inventions. The first industrial revolution was probably the single, most momentous event in our history. It was produced by mankind's mastery over mechanical power as a replacement for human and animal muscle power. Now we are in the midst of a new technological revolution. Many disciplines of science and technology are affecting the course of this revolution. However, the explosive development of informatics, including communication, computers and control, exerts the most immediate and extensive influence on the course of this revolution. For today, the main limitation to the advance of technology is the very complexity of the machines and systems we have created. Control is the limiting factor, and modern informatics is the direct result of our need to control the complexity of the devices and systems that we have so ingeniously created.

Systems scientists of many countries have studied the dynamics of the development of technology over the centuries and found that the time period between technological innovations and practical applications in production has been shortened and that the world is more and more becoming a very complex system where technology and socio-economic factors are all interrelated state variables. Without control, this new technological revolution will inevitably follow along lines similar to the first industrial revolution and the result will be an unbalanced world with enormous economic power concentrating with a few nations. This development will not leave some developed countries with traditional industries uneffected, yet it is already a great challenge to the developing countries, especially those still not industrialized.

More than a hundred years ago, Karl Marx predicted that science will also be a productive power. In order to meet the challenge of the world-wide technological revolution, science and technology should be looked upon as a useful tool, vital to the economic development of developing countries. In the meantime, academic work in science and technology must closely integrate with the needs of production. The achievements of control science and technology will certainly give an opportunity to the developing countries if they are effectively applied to saving water, energy and natural resources and to the improvement of management. By using appropriate technology including proper level of automation, the technical gap between advanced and backward facfories can be narrowed, the quality of products improved, and the efficiency of the operation raised.

Scientists from several international professional societies and organizations of the United Nations have long concerned with the status quo of the world technical revolution and its effect on the future of mankind. Many conferences have been organized to discuss new trends of the technical revolution in general. The aim of this Conference is to provide an international forum for the presentation of the latest research results and experiences dealing with a broad scope of control science and technology for development.

According to the aims and scope of the Developing Countries Committee of IFAC, "DECOM is concerned with identifying and promoting the fulfilment of the diverse needs of countries seeking increased industrialization. ...... In the context of this Committee, the term industrialization is intended to pertain primarily to the application of control systems and automation to production processes, agriculture, housing, transportation, trade as well as health and education services."

We hope that this Conference will be able to fulfil the aims our committee members are dedicated to and also the intentions of our International Program Committee.

CONTENTS

SYSTEMS SCIENCE AND CHINA'S ECONOMIC REFORMS

Jian Song

Institute of Information and Control, Beijing, China

Widespread and far-reaching economic reforms are sweeping China in a planned manner. The courage and pragmatic spirit demonstrated by the Chinese government in these reforms has not only won enthusiastic support from the Chinese people but also attracted much attention of both statesmen and economists throughout the world. The opening of the fourspecial economic zones,Hainan Island, and 14 coastal cities to foreign investment, the smooth settlement of the Hongkong issue by means of "one country, two systems" have undoubtedly added new meaning and implications to China's economic reforms, thus giving them an even greater momentum. Some press people even contend that China's strategy of "one county, two systems" is the most daring and fascinating initiative ever attempted in the 20th century.

China is a big country and once had a glorious ancient civilization. But ever since the industrial revolution, it has been lagging behind. The history of economic development throughout the world with different civilizations has provided valuable experience for China's reforms. Summing up its own experience of the last 35 years with successes and failures, China has adopted a new economic policy. This new policy, in my view,has been formulated by absorbing the most splendid essentials of various civilizations of mankind and discarding what has been proven by practice to be useless. The life-and-death competition on the free market in the ear Adam Smith(1723-1790), the theory of Parson Malthus(1776-1834) on ruthless war and pestilence, the Utopian Socialism,the theory of moderate government intervention advocated by John Maynard Keynes(1883-1946), and the postwar resurgence of Japan have all served as philosophical references for China's economic reforms. However, the direct reason behind the current reforms has in fact been China's own bitter experience over the past thirty years with the straits caused by the rigid structure of centrally running the entire national economy. Yet according to the title of this paper, I am not going to discuss in detail those political, economic, and ethical motivations and implications associated with China's economic reforms, but only to brief on the important role that the modern systems science has played and is potentially capable of playing in the current reform of China.

Several decades ago, people referred to political economy as a dismal science. Economists eventually recognized that by the only means of emotional and conceptual descriptions it was difficult to reveal a nation's economic state and particularly the dynamics of economic progresses. It was even harder to provide means for unbiased evaluation of given economic policies and, moreover, to predict the outcome of such policies. Thus econometrics emerged as the times required. Not surprisingly, one could say that economitrics has saved the old economics from its impending misery. Because of the advancement made in statistics and particularly with the commercialization of computer technology, the application of systems science and system control theory to the study of economic problems, military affairs, managerial science, population issues as well as sociology has become a powerful trend of our times.

This trend has been further accelerated by the emergence of the general systems theory, the founder of which is widely believed to be Von Bertalanffi who started his carrer as a researcher in the feild of biology (Theoretische Biologie, 1932). Towards the end of 1960s, Professor Ilya Prigogine (1917-) who worked for years as the director of the center for Statistical Mechanics and Thermodynamics at University of Texas, US, developed a theory on disspasive structure which greatly promoted the development of systems science, for which he received 1977 Nobel Prize. The rapid development of system control theory and computer technology over the past twenty years has provided new and powerful theoretical methods and tools for the study of econometrics and other social phenomena.

Beginning in 1977, after ten years of internal upheaval, natural and social scientists in China quickly caught up with the pace of the world scientific advancement and started using systems science and control theory to study Chinese economics, demography, and other most pressing issues, and have obtained heartening results in every aspect. It can be said without exageration that current economic reforms in China have benefitted considerably from the application of systems science. Chronologically, the earlier achievements were made in the field of population system control which was subsequently realized in China's family planning program. Then a series of excellent investigations were conducted in relation of economic issues, especially in the study of energy and pricing systems. Recently, the method of system analysis is beginning to be used in the country's macroeconomic planning. In all these areas, the State Council can timely solicit opinions from scientists. At the same time, a large amount of data from system analysis are available for policy-making at the State Council on a fairly reliable scientific basis.

Having faced the entire history of mankind in the past, China today is fortunate enough to be able to draw on experience and lessons from the evolution of civilization in various countries and, in particular, from the achievements of modern science.

I. POPULATION SYSTEM CONTROL

Population size and its growth rates can greatly affect the develoment of a country's economy. For example, the amount of discovered main kinds of China's maieral reserves ranks third in the world, however, it recedes into eightieth if its per-capita share is considered. Having faced the huge size and high growth rate of population, since seventies the government of China has given a top priority to the study of population problem, analysis of its present state and trends of growth in the future.

As far back as 200 years ago, when Thomas Robert Malthus wrote his essay on population, the population size of China had already reached 200 million. 150 years after, according to China's first census in 1954, it went up to about 600 million. Beginning to feel the pressur from its population, the government of China realized that there was a need for controlling the rate of population growth. In 1957, Mao ze-dong said ironically:" In terms of child births, human beings seem to be least capable of controlling themselves and there dose exist a situation of anarchism. If this situatino was allowed to proceed unchecked,China's population would experience a ten-fold increase to reach 6 billion and mankind would take the road of early destruction and doom". Despite what he said, at that time the government did not take much effective measures to check the pace at which the population was growing. In 1964, seven years after Mao's remarks, the second census showed an increase of another 100 million, making a total of 700 million. Soon the "Cutural Revolution" came. By 1969, even before people could extricate themselves from chaos and agony, an increase of another 100 million people was recorded. By 1974, the total reached 900 million. During that time, people lived in confinement, yet they were completely free to indullge themselves in reproduction capability. Between 1964 and 1978, China recorded a population growth of 250 million, more than the population of US was added within 13 years, but hardly any growth in GNP.

After more than ten years isolation from the outside world, during a visit to Europe in 1978, I happened to learn about the application of system analysis theory by European scientists to the study of population problems with a great success. For instance, In a "Blueprint for Survival" published in 1972, British scientists contended that Britain's population of 56 million had greatly exceeded the sustaining capacity of ecosystem of the Kingdom. They argued Britain'S population should be gradually reduced to 30 million, namly, a reduction by nearly 50 percent; some Dutch scientists also believed that Netherlands' population of 13.5 million had far gone beyond the limit of what the country's 40,000-square-kilometre territory could possibly bear and should therefore be reduced at least by a half. I was extremely excited about these documents and determined to try the method of demography.

In China, the controvercy over the population issue has beengoing on for well over 100 years. Scholars, basing their arguments on their own philosophies and theories, have often come up with completely opposite views. Some statement once held that the ultimate weapon for China to resis foreign invasion was to maintain a large population. consequently, any theory that stood for controlling the rate of population growth was regarded as the reprint of Malthus and was therefore condemned. Nevertheless, those who advocated population control resisted furiously, what frequently led to firestorm of political conflicts. Up until late 1960s, the efforts to find a solution of this age-old proposition whether or a smaller population was good for China was still permeated with danger. History already made it clear that the often misused literal exposition alone was incapable of resolving the pending confussions, and only the language of nature science would be powerful enough to clarify the concept and problem-setting in China's population study excluding sentimental biases of the public. This time we decided to go another way, differdnt from that of the old generation of Chinese demographers who had been rounly abused in 1950s. In fact,data,equations, theorems and corresponding conclusions, especially the irrefutable mathematical logicare far vulnerable to criticism. Moteover, one could withdraw himself into the sanctuary with high prestige of natural science if necessary. Today we are gratified to note that all of these objectives have been attained. We indded have been enjoying the blessings of system science and mathematics.

More precisely, in the research on population systems, we have done the following; 1-3

1) We have elaborated two kinds of mathematical models of population evolution in accordance with the actual conditious in China. One is a system described by partial differential equation with closed feedback at the boundary. The other one is expressed by a system of differencial equations. The two models are equivalent in terms of physics, but each has its own advantages. The first is convenient for theoretical analyses while the latter is suited to caculation. The verification of the models by the historical data of cencus and survey statitics has shown that their logical structure and accuracy are quite satisfactory.

2) In 1979, for the first time in China, we accomplished long-term projections of the trend of China's population growth. Being published in 1980, it shocked the people throughout the country and also caught the attention of the government. The six curves of projections made it amply clear to the people that if the average fertility rate was maintained above the level of 3.0 of 1970s over a long period of time, China's population by the second half of the next century would go up to 4.5 billion equaling the total world population today, and it would continue to grow forever. If the people and government want to see China's population stabilized on a stationary level, the average number of child-bearing per woman must not exceed 2.0; if people would like to have China's population reduced below 700 million, which is admited the preferable size for China, during the next two decades each married couple should be encouraged to have only one child. Even then, it would take 70 years to achieve this goal. The predicted trends of population growth according to different total fertility rates β for a century period are shown in fig. 1.

3) We formulated the Lyapunov Stability Theory for Population Systems, and succeeded in proving a central theorem of instability. It is proved, that for any community there exists a critical total fertility rate β_{cr} , which is uniquely determined by age distribution, mortility rates and fertility pattern of the female population The population system is stable in the sense of Lyapunov if and only if the actual total fertility rate β is not greater than β_{cr}. In other word, if the total fertility rate is permanently greater than its critical value,

the population system is unstable, and any small positive perturbation would result in an unlimited increase in population size as time tends to infinity. In late 1970s and early 1980s, the value of critical fertility rate of China's population was estimated as 2.2. But 1982's census show was 2.45. Despite the spectacular success China's family planning program has already achieved, continuous effort must be made to reduce the total fertility rate below the level of 2.2, so as to ensure the stability of China's population system. Theoretically, this stability theorem is applicable to any country or nationality, it might be particularly useful for analysis of population in developing countries. The reason for rapid growth of the world population rests with the developing countries whose total fertility rate is three to four times greater than their critical fertility rates. It is surprising that today there are still some scholars who kick up a row against family-planning program instead of worrying about the fate of billions of our future descendants. This demonstrates how weak the voice of systems science is, for there are still quite many intellectuals who, starting off with biassed sentiment, go so far as to challenge the irrefutable logic of natural sciences.

4) In the aspect of mathematical study, we have thoroughly investigated the properties of so called population operators. We accomplished the study of the asymptotic behavior, spectral properties, contrallability, completeness of eigensolutions, and so forth of the population operator. In addition to its possible theoretical significance, the results of the study has also enhanced our sense of confidence and security.

5) Following the policy relaxation for China's rural areas, the peasants have been prospering day by day. Many of them now ask the government to permit them to have more children. Why not let each married couple have two children, one boy and one girl to make an ideal pairing? But our policy-making support software promptly reports that, if each married couple of the peasants, who represent 80 percent of the total population, was allowed to have more than two children, China would have to be prepared for accomodating two billion inhabitants by the end of the next century, doubling the current population size. To other questions related to the two billion, economists have been called upon to find answers. But unfortunately, not a single economist has ventured to satisfy the request of the peasants.

Theories have revealed to the people a conclusion that some years later the ideal arrangement will be two children per couple, one boy and one girl. But not today. At least before the end of this century, we must stick to the policy of one child per couple, so that the people of the 21th century will be able to enjoy family happiness. This is an expedient measure intended as a compensation for the erroneous policies of the past that fostered uncontrolled population growth in the three decades. It is a price our children must pay for the mistakes made by their grandparents.

II. A RATIONAL PRICE SYSTEM —— KEY FOR TRANSITION TO NEW SYSTEM

In October 1984, the government of China openly confessed that a rigid economic system has taken shape in China over the last 30 years due to both historical and political reasons. The main maladies inherent in such a system are: the government places excessive and extremely rigid control over the enterprises; the role of market mechanism is neglected; equalitarianism has prevailed over a long period of time in the distribution of social wealth, under which enterprises live off the "big pot" of the state whereas workers live off the "big pot" of firms. Such practices have severely stufled the enthusiasm and creativeness of the worker and inactivated the enterprises.

Roughly speaking, the current economic reform is mainly aimed at considerably developing commodity production and widening market activities so as to enable the market to play a more active role in advancing and regulating economic growth. The government of China realized that a fully developed commodity economy is an inescapable stage in attaining a modern social economy, and the centrally planned sector of the national economy must be narrowed by a wide margin. For a country as big as China, it is absolutely inconceivable to place the development of its national economy solely on the basis of centrally planned production and consumption. Therefore ought to be encouraged to complete with one another in the marketplace, thus permitting some firms and individuals to get better off earlier.

A distribution system based on absolute equalitarianism can only lead to common poverty instead of common prosperity. We must also draw on and absorb all advanced managerial experiences and practices of various countries in the world, including the highly developed capitalist countries. Finally, the most important prerequisite of the present economic reform is to relax the government control over pricing system, and enlarge the circulation sphers of commodities with prices left to respond to supply and demand situation of the market. In short, the ultimate goal of China's economic reform is to build an economic system full of vigor and vitality inspired by market activities and, at the same time, controlled by the government macro-planning. Thus, such system should be driven by a mixed mechanism of market flexibility with moderately centralized control.

Obviously, the economic reform must begin with changing the pricing system, gradually releasing the prices of commodities to be floated instead of the rigid practice of government hold throughout the country. The fate of the intended economic reform will be entirely dependent upon the success of pricing system reform. It is said that economic reforms are different and cold and they sometimes appear to be full of charm, glamor, and fascination. Whereas on other occasions they might be accompanied by danger and disaster. For if the social wealth failed to increase rapidly, the relaxation of price control and the spontaneous wage increases would give rise to unchecked inflation, that could adversely affect the living standards of hundreds of millions of people. As an agraring country, Chinese people have been hitherto enjoying the stable and low-priced agroproducts due to a huge sum of financial subsidies provided by the government. The first problem facing the present economic reform is to design a strategy for relaxing the government price control over agro-products.

Two years age, as intrusted by the government, we organized a team of system analysists to study the above mentioned problem. Our efforts resulted in the establishement of a model for the pricing dynamics of agricultural and related products, which covers 237 items of 45 categories and is consisted of 114 equations, among them 19 equations in time series of dynamics, 43 for state description, and 52 for equlibrium. There are defined 142 structural parameters, including 43 endogennous and 20 exogenous variables and three types of policy control variables: purchasing and retailing prices, rate of wage increases, and taxation. The inter-dependent relationships among different variables constitute a large-scale dynamic system. Having reached the level of stable operation, the system now is permanently resident in a large computor data base and ready for running at any time. A careful verification of the model with over 3000 items of statistical data from the last couple of years has shown an error less than 5 per cent for short-term projection. It has provided a convenient tool for pricing system. Taking into account the supply-demand balance, price index, increase of government revenue, cash flow, etc. as the most important parameters for policy evaluation, we studied many possible ways for transition from the existing pricing system to a new one, and submitted official report with a number of choices for government to make its decision. Some general remarks and conclusions contained in this report may be briefed as follows.

1) The average annual government subsidies have amounted to ¥20 billion in the past several years to hold down the prices of agricultural products. The 1983 figure of governmental compensation is well over 30 billion, about a quarter of the total government financial revenue. If the pricing system was to remain unchanged, the subsidies would probably grow at a rate of ¥10 billion each year. Then after three to five years, annual expenditure of government in the form of subsidies for price stabilization of only agricultural products would consititute about one third of the total government revenues, thus it would become unbearable burden on the government and then even harder to carry out the transition to the market-floating price system.

2) Over the last five years, China has experienced an average annual GNP growth of 9 per cent and a national income increase of 8 percent. In 1984, a 13 percent growth in GNP and 12 percent increase in national income were recorded. A careful study of the price system dynamics described by abovementioned model has shown that the favourable economic growth rates in recent years laid a sound foundation for the government to initiate the process of reform, reducing gradually subsidies for agricultural products and releasing their prices to be regulated by

market mechanism. Some practical schemes have been devised to complete the reform within three years. The related projections indicate, the increase of wages would cover the ascent of price-index and people's real income would not decrease nor would inflation occur.

3) The release of the centralized pricing system for agricultural product should be started up at the earliest possible time. If this could be done right now, the government revenues would go up considerably in the years to come.

4) The short-term predictions made with emphases on different economic targets provide a number of choices to the government to make its decision. One of conclusions has shown that, with a 18 percent increase in retailing prices, about 20 billion RMB would have sufficed to add to the total amount of wages for keeping the balance between demand and supply os daily necessities for the whole population, and a 8 percent annual growth rate of financial revenues for the government would be assured.

A detailed report on this issue had already been submitted to the government for its reference, and we are happy that it played a certain role in the government policy-making on the current economic reform.

III. MACRO-CONTROLLED AND MICRO-ENLIVENED ECONOMY

China's economic system has now started moving apart from the system of utmost centralized planning, and it will probably never turn to the other extreme-Adam Smith's Completely Free Market. For such a big country as China with many different economically developed provinces, the attempt to make a perfect and all-compassing economic plan is nothing short of bureauratic utopia. On the other hand, if things are allowed to drift alone without status analysis and control of the overall economic system, it would surely lead to disaster. Therefore, to establish a micro-enlivened and macro-controlled economic system is the strategic goal of the current economic system reform in China. For example, China today is terribly short of energy. The shortage of electric power is up to 100 billion kwh, which makes many factories operate only four days a week. China possesses huge hydro-power reserves but only less than five percent has been developed. The discovered deposits of fossil energy rank third in the world, but the exploitation capacity is insufficient. The improvement of communication and transportation system is urgently needed. Agriculture calls for modernization and investment. In all of these areas, long-term investment plans must be worked out in accordance with the limited fund accomulation rates estimated for five to ten years to come.

By the way, in the course of economic reform, the concept of capital has changed radically. A century ago, Carl Marx said:" Capital is dead labour, it lives only by sucking living labours; Capital came into the world, dripping from head to foot, from every pore, with blood and dirt." However, the contemporary bankers alleged that: " Capital is far more important that your mother." The Chinese economists are now to explore some new concept and definitions between these extremes that would be applicable to this country.

Facing the shotage of fund, the government must put most sectors of the economy under macro-control in order to optimize the benefit of investment with constraint of national capital accumulation and social wealth growth. The results of system analysis indicate that the insufficiency of supply is the principle feature for the nest decades. Thus, the growth rate of supply capacity will determine the speed of China's economic development. However, the delay and dynamic effect of investment existing in different sectors would considerably complicate the policy-making of resource allocation. For example, food industry needs 2-3 years for construction, power plant, 5-10 years and 10 years for building a railway, etc. It is completely inadequate to make decision only on the basis of observation and intuitive judgement in any dynamic system. In recent years, the efforts of Chinese system scientists and economists have afforded the government a solid scientific foundation and enriched possible alternatives for its economic policy formulation.

In 1976, 17 countries and BNL and KFA jointly developed a linear programing model (MARKAL: MARKEL ALLOCATION) for energy policy evaluation, which has given evident impact on China's energy planning. Several Chinese universities have recently finished, by joint effort, a detailed dynamic analysis and projection on current

China's energy supply and future development. The result indicates that by 1990 the yearly energy demand will reach 900 million ton standard coal (against 700 million tons in 1984), 600 billion kw-hour electro-power; by the end of this century the annual demand will be increased to 1.4 billion tons standard coal and 1200 billion kwh elctro-power. In order to meet the need, the total amount investment for the remaining years is estimated as 20 billion US dollars, which is meant 10 billion dollars a year. In the way, the system analysists simply conclude that the government has to invest at the average of 10 billion US dollars each year into energy sector in order to quadruple the nation's GNP till the end of this century.

The State Planning Commission developed a model for more reliable planning. Their considerations included analysis of nineteen industrial sectors. This model consists of 254 dynamic equations of endougenous variables and 675 equilibrium equationswith 122 exogenous variables. A verification of this model by historical statistics from 1952 to 1982 shows that the coefficient of correlation has reached 0.95-0.99 and accuracy of short-term extrapolation up to 85%. In this way, the system science has provided the State Council with significant reference for its long-term planning of investments of different sectors.

It is no doubt that the vitalization of China's economy eventually depends upon the hard working of Chinese people encouraged by the reformed economic policy and the opendoor strategy. However, the system science will provide an indispensable guideline for government in its policy-making.

REFERENCE

1. J.Song, et al., Theory on prediction of Population Evolution Precesses. Scientia Sinica, Vol. 24, No.3. (1981)
2. J.Song, Some Development in Mathematical Demography. Theoretical Population Biology, Vol. 22, No.3. (1982)
3. J.Song, J.Y.Yu, On Stability Theory of Population Systems and Critical Fertility Rates. Mathematical Modelling. Vol.2, No.2 (1981)
4. J.song, et al., Spectral Properties of Population Operator and Asymtotic Behaviour of Population Semigroup. Acta Mathematical Sinica, Vol.2, No.2 (1982)

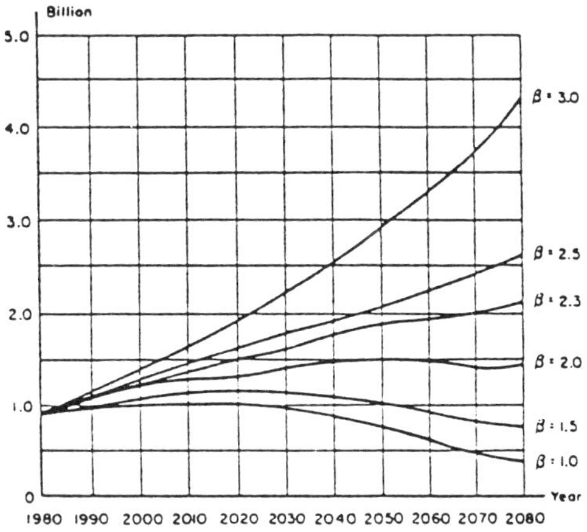

Fig. 1.

NEW TRENDS IN SYSTEMS APPROACH BY THE INTERACTIVE USE OF MICROCOMPUTERS

Y. Sawaragi* and K. Inoue**

**Faculty of Science, Kyoto Sangyo University, Kyoto 603, Japan*
***Faculty of Engineering, Kyoto University, Kyoto 606, Japan*

Abstracts. Placting stress on 1) the use of heuristics or experiences of an expert and 2) the interactive use of microcomputers or personal computers, some of the new trends are reviewed and discussed in systems approach especially in the field of interactive modeling of large scale systems, system failure diagnosis by use of knowledge engineering techniques, multiple criteria decision making and gaming system for complex problems.

Keywords. Systems approach; microcomputers; modelling; large-scale systems; failure diagnosis; decision theory; interactive programming; game theory; environment control; economics.

INTRODUCTION

A distinctive feature of a large-scale system is of its poor-definedness or ill-definedness, whereas systems we have dealt with so far are well-defined. Most real systems have this tendency. Social and economic systems are typical examples, which involve human beings as their important elements. Namely, in these systems, the factor of uncertainty plays a vital role, which turns out to be of great difficulty to deal with in mathematical formulations.

In discussing how an approach to ill-defined systems should be, we must consider the following two points.

The first point is how we modify various methodologies well established for well-defined systems. Indeed, methods and tools developed for a well-defined system are not necessarily useless for an ill-difined system, but eventually, what is able to bridge the gap between the real system and the theory must be heuristics that we human beings have. Therefore, an interactive systems approach in which human beings take part is strongly suggested.

The other point is that we have to pay attention to the capability of a computer which has been very quickly developed in this decade. In the past, much effort was concentrated on developing numerical and calculating capability, whereas in recent years we can expect an appearance of a computer which can not only manipulate symbolic notations but also process knowledge information by an inferential engine. An example of it is the on-going project of the fifth generation computer which was first proposed in Japan.

Moreover, as a direction of computer development, we have to point out the beautiful progress in micro-processors created by the development of micro-electronics technology. The smaller the cost/performance and the size of a computer becomes, the closer the distance between man and computer gets, and also the more user-friendly the use of the computer becomes.

Because of the reasons described above, the methodology is to be realized that supports decision making to resolve various problems such as structural identification, behavioral prediction and optimization of a large-scale system via taking mutual advantages of heuristics based on human expriences and the ability of a knowledge-based computer and via iterations of interactive processes between man and computer, especially microcomputer.

Based on this view, in this paper, we discuss the state of art in new trends in systems approach forcussing our attention on results of research groups headed by the authors, which include 1) interactive modeling of large scale systems, 2) system failure diagnosis by use of knowledge engineering techniques, 3) multiple criteria decision making and 4) gaming system for complex problems.

INTERACTIVE MODELING OF LARGE SCALE SYSTEMS

It is very difficult to formulate the practical model of a large complex system including human elements, for example, an environmental system, a traffic system, an economic system and other socio-technical systems. In this case, it is most important to combine the mathematical approach and heuristic one which is fostered through ripe experiences on field studies and expert knowledges of human being. A new modeling method, the interactive method of data handling (IMDH) is described, which aims to build the practical model of large scale systems by means of the communication between man and computer (Ryobu and Sawaragi, 1979, 1982).

IMDH is composed of two kinds of communication process. The first communication is qualitative and the second is quantitative. The outline of the modeling process is summarized in Fig.1:

Step-1 MAN arranges the information and classifies each relation among the variables into three types--"white", "gray" and "black". We suppose that a system is composed of variables $(x_1, x_2, \ldots x_n)$ and the model is expressed as the following equations;

$$x_i = f_i(x_1, x_2, \ldots, x_j, \ldots, x_n),\ i=1\sim n, j=1\sim n, i\neq j.$$

If all of the relations between x_i and each of $(x_1, \ldots, x_n)$ are apparent we call them "white". If some of them are not apparent, we call them "gray". If all of them are not apparent, we call them "black".

Step-2 COMPUTER makes the linear equations with the method of least squares for "white" and with self-learnging organization for "gray" and "black".

Step-3 COMPUTER shows all the relations of the variables by the tier-structural di-graph, from these equations.

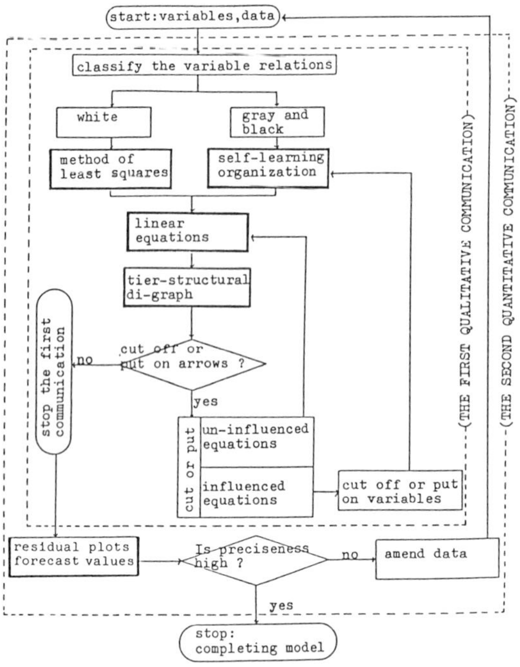

Fig. 1. Modeling process by IMDH

Step-4 MAN cuts off unnecessary arrows and puts on necessary ones in the di-graph. If there is nothing to do, MAN stops the first communication.

Step-5 Finding which variables are influenced on Step-4, MAN excepts the unnecessary variables or adds the necessary ones to the predictor variables. And then, return to Step-2.

Step-6 COMPUTER makes out the residual plots and the forecast values from the results (the equations) of the first communication.

Step-7 MAN picks up low-precise equations with reference to the residual plots or insufficient equations whose forecast values are much different from his experience, image and so on.

Step-8 MAN recollects the data or processes them by the method of data-changing, trend-exception, filtering and so on. And then, return to Step-1.

The characteristic points of this method are as follows. First, this method succeeded in shortening the modeling time, because of setting the self-learning organization in the computer, therefore, both a man and a computer become to be able to communicate with each other repeatedly and smoothly. Second, the tier-structural di-graph making a man possible to catch visually a bird's-eye view of the relations among the variables, and easy to find some unproper equations, he comes to understand steadily all the structure of system, discover something fresh and turn uncertainty into certainty. Thus through the process of IMDH, a man reflects his knowledge and experience enough in modeling, and on the other hand he also learns and develops himself. Therefore the more a man's knowledge and certainty of system increase, the more complete the model grows.

This method was successfully applied to build economic models of Kyoto Prefecture, Kyoto City and Shiga Prefecture from 1976 to 1985 (Ryobu and others, 1976-1985).

FAILURE DIAGNOSIS BY USE OF KNOWLEDGE ENGINEERING TECHNIQUES

The computer as a sequential computing machine is far superior to the human brain. There are, however, great many problems which cannot be solved by numerical processing only. Just suppose a specialist in some field try to solve a complex problem, he uses by all means his knowledge and experience as an expert as well as numerical computations. The computer so far has been incompetent in this kind of inferential processes in which it collects various information from its environment, understands it and then decides what to do. The research in the last 30 years to assist human intellectual activities by use of computers is called artificial intelligence. In 1977, Feigenbaum proposed the idea of knowledge engineering, which is a division of artificial intelligence and is more application oriented. The appearance and prosperity of knowledge engineering have been strongly supported by the very rapid progress of computers and the development of artificial intelligence oriented computer languages like LISP or Prolog. The knowledge engineering is, briefly to say, to make research on how human knowledge is represented and stored in a computer and how skillfully the stored human knowledge is retrieved and utilized.

An expert system is a subsidiary of artificial intelligence in which knowledge and experience of an expert in a specific field are stored in a computer and used as if the computer has almost the same ability of problem solving as the expert. The expert system is now the most prominent and prevailing system and thus is applied to various fields of engineering, medical, educational, economic, law and many other disciplines.

A distinctive feature of the knowledge engineering technique is that the knowledge which is necessary to solve a problem and the usage of the knowledge are completely separated. A collection of expert knowledge stored in a computer is called a knowledge base and the inferential process by use of the knowledge base is called an inferential engine. The independence of the knowledge base from the inferential engine makes the system very flexible, easy to manage and applicable to a variety of fields.

To check the possibility and validity of an expert system, the authors have constructed an expert system which can find causes of system failure in 1) a cooling system and 2) a start-up system of a marine diesel engine (Kumamoto and others, 1982, 1985). Novel and interesting points to be noted of the system are as follows:

(1) The system works in an interactive way. An operator sits up before a CRT terminal and answers a sequence of questions put by the computer, and eventually he obtains the cause of the trouble. This implies the system is not necessary for real time applications, rather it is quite fit for training non-experts.

(2) There are plenty of explanation functions. This means the computer can answer such questions asked by the operator as "Why did you conclude.....?", "How did you infer...?", "How can I fix it ?", etc. This helps training non-experts, too.

(3) It is very easy to add, to modify and to cut the knowledge base. This is due to the fact that the system is constructed on a general database management system, which contrasts to usual expert systems which are written in LISP or Prolog. This aspect well increases the

flexibility and adaptability of the system.

(4) The expert's knowledge is represented by a set of production rules or if-then rules. An if-then rule is expressed in terms of a prerequisite and a conclusion. Diagnosis of a plant failure is the search process of locating a cause in the form of a system component such as a valve or a pump through a successive classification of macro plant state into micro states. Successive classification of plant state is formulated by the following form of the if-then rule: If [plant state i] and (observable fact), then [plant state j].

(5) Two types of inference algorithms are implemented. The one is a forward or top-down algorithm and the other is a backward or bottom-up one. These algorithms are written in a database handling language, too.

(6) The system is implemented on a 16 bit personal computer. This implies that the system is portable, easy to tranfer and most of all not expensive.

(7) The cooling system failure diagnosis system has 22 rules and 15 basic causes of failure, and the start-up failure diagnosis system has 121 rules and 79 basic causes.

An example of the failure diagnosis process of the cooling system by operator-computer conversations is shown in Fig. 2. The underlined characters show operator's inputs to the computer. The input "?" denotes a question put by the operator to the computer: "Why do you ask me the fact?". The computer answers the followings. 1) the if-then rule currently being applied, 2) the proven states and facts in the prerequisite of the rule, and 3) a conclusion when observable facts turns out to be true.

MULTIPLE CRITERIA DECISION MAKING AND INTERACTIVE PROGRAMMING METHODS

In order to solve our decision problem by some systems-analytical methods, we usually require for the attainment-degree of objectives to be represented in some numerical terms, which may be of multiple kinds even for one objective. In order to exclude subjective value judgment at this stage, we restrict these numerical terms to physical measures, for example, money, weight, length, time and so on. As such a performance index or criterion for the objective P_i, an objective function $f_i:X\to R^1$ is introduced, where X and R^1 denote the set of alternatives and one dimensional Euclidean space, respectively. The value $f_i(x)$ indicates how much impact is given on the objective P_i by performing an alternative x. The impact modeling is performed to identify these objective functions from various viewpoints such as physical, chemical, biological, social, economical and so on. For convenience of mathematical treatment, we assume in this paper that the more of each objective function is preferred to the less. Now we can formulate our decision problems as a multiobjective optimization problem:

$$\text{Maximize} \quad f(x)=(f_1(x),\ f_2(x),\ldots,f_r(x)) \quad \text{over } x\in X.$$

Unlike traditional mathematical programming with a single objective function, there does not necessarily exist in multiobjective optimization problems an "optimal" solution in the sense that it maximizes all the objective functions simultaneously, and hence usually we are troubled with conflicts among objectives in decision problems with multiple objectives. The final decision should be made by taking the "total balance" of objectives into account. Therefore a new problem of value judgement, i.e., value trade-off arises. The evaluation modeling treats this problem peculiar to decision making with multiple objectives. Here we assume a decision maker who is responsable for the final decision.

In some cases there may be many decision makers, for which cases the decision problems are called group decision problems.

One approach to the evaluation modeling is to find a scalar valued function $u(f_1,\ldots,f_r)$ reperesenting the decision maker's preference, which is called a preference function in this paper: A preference function in decision making under risk is called a utility function, whereas the one in decision making under certainty is called a value function. Once we obtain such a preference function, our problem reduces to the traditional mathematical programming:

$$\text{Maximize} \quad u(f_1(x),\ldots,f_r(x)) \quad \text{over} \quad x\in X.$$

```
-- CAUSE ISOLATION OF "ENGINE OVERHEAT" --

Q1   **ENGINE ITSELF IS NOT THE CAUSE**
       Is this true? T
Rule 1 deduces that:
COOLING SYSTEM IS RESPONSIBLE FOR OVERHEAT

Q2   **PRESSURE GAUGE READING OF F/W IS LOW**
       Is this true? F

Q3   **PRESSURE GAUGE READING OF F/W IS NORMAL**
       Is this true? T
Rule 10 deduces that:
ENOUGH COOLANT CIRCULATES IN F/W LOOP

Q4   **WARM-UP STEAM VALVE IS OPEN**
       Is this true? F

Q5   **TEMPERATURE READING OF F/W AT EXIT OF HEAT EXCHANGER IS
       HIGH**
       Is this true? ?

/// I am trying to use /// Rule 12
/// I already know that: ///
ENOUGH COOLANT CIRCULATES IN F/W LOOP
/// If you inform me further that: ///
TEMPERATURE GAUGE READING OF F/W AT EXIT OF HEAT EXCHANGER IS
HIGH
/// Then I will conclude that: ///
HEAT EXCHANGER WORKS POORLY

Q5   **TEMPERATURE GAUGE READING OF F/W AT EXIT OF HEAT
       EXCHANGER IS HIGH**
       Is this true? T
Rule 12 deduces that:
HEAT EXCHANGER WORKS POORLY

Q6   **TEMPERATURE GAUGE READING OF S/W AT EXIT OF HEAT
       EXCHANGER IS LOW**
       Is this true? ?

/// I am trying to use /// Rule 13
/// I already know that: ///
HEAT EXCHANGER WORKS POORLY
/// If you inform me further that: ///
TEMPERATURE GAUGE READING OF S/W AT EXIT OF HEAT EXCHANGER IS
LOW
/// Then I will conclude that: ///
DIRTY HEAT EXCHANGER AND HENCE LOW HEAT EXCHANGE RATE IS
RESPONSIBLE FOR OVERHEAT

Q6   **TEMPERATURE GAUGE READING OF S/W AT EXIT OF HEAT
       EXCHANGER IS LOW**
       Is this true? F

Q7   **TEMPERATURE GAUGE READING OF S/W AT EXIT OF HEAT
       EXCHANGER IS HIGH**
       Is this true? ?

/// I am trying to use /// Rule 14
/// I already know that: ///
HEAT EXCHANGER WORKS POORLY
/// If you inform me further that: ///
TEMPERATURE GAUGE READING OF S/W AT EXIT OF HEAT EXCHANGER IS
HIGH
/// Then I will conclude that: ///
LOSS OF COOLANT OCCURS IN S/W LOOP

Q7   **TEMPERATURE GAUGE READING OF S/W AT EXIT OF HEAT
       EXCHANGER IS HIGH**
       Is this true? T
Rule 14 deduces that:
LOSS OF COOLANT OCCURS IN S/W LOOP

Q8   **S/W CONTROL VALVE IS CLOSED**
       Is this true? F

Q9   **S/W COOLANT PUMP IS NOT OPERATING**
       Is this true? F

Q10  **S/W COOLANT PUMP IS OPERATING**
       Is this true? T
Rule 17 deduces that:
LOSS OF COOLANT OCCURS IN S/W LOOP WITH PUMP OPERATION

Q11  **S/W DISCHARGE VALVE IS CLOSED**
       Is this true? F

Q12  **PRESSURE GAUGE READING AT SUCTION VALVE OF S/W COOLANT
       PUMP IS LOW**
       Is this true? T
Rule 19 deduces that:
BLOCKAGE OF STRAINER OR CLOSED SUCTION VALVE OF S/W COOLANT
PUMP IS RESPONSIBLE FOR OVERHEAT

**************************************************
This is the cause of "ENGINE OVERHEAT".
**************************************************
```

Fig. 2. Example of failure diagnosis process

Another popular approach is the interactive programming which performs simultaneously both the search of solution and the evaluation modeling:

In this approach, without identifying the preference function, the solution is searched by eliciting iteratively some local information on the decision maker's preference. We shall discuss interactive programming methods in more detail in the following.

As stated above, the aim of interactive programming methods is to support the decision maker to make their decision easily in a cooperative way with computers. Therefore, it goes without saying that it is very important in developing these interactive methods to make the best use of the strong points of man and computer. A computer is strong at iterative computation in routine and can treat large scale and complex computation with high speed. On the other hand, a man is good at global (but, possibly, rough) judgment, pattern recognition, flair and learning. With these points in mind, we impose the following properties on desirable interactive multiobjective programming methods:

(1) (easy) The way of trading-off is easy. In other words, decision makers can easily grasp the total balance among the objectives.
(2) (simple) The judgment and operation required to decision makers is as simple as possible.
(3) (understandable) The information shown to decision makers is as intuitive and understandable as possible.
(4) (quick response) The treatment by computers is as quick as possible.
(5) (rapid convergence) The convergence to the final solution is rapid.
(6) (explanatory) Decision makers can easily accept the obtained solution. In other words, they can understand why it is so and what it came from.
(7) (learning effect) Through the interaction process, decision makers can learn many things, for example, gaps between their desires and the real world, and mutual understanding of participants in group decisions.

Interactive programming methods seem promising in particular for design problems. However, in applying ordinary optimization techniques, we often encounter some difficulties: For example, in structural design problems such as bridges, function forms of some of criteria can not be obtained explicitly and their values are usually obtained by complex structural analysis. Similarly, values of criteria in design of camera lens are obtained by simulation of ray trace, and moreover the number of criteria is sometimes over one hundred. From such a practical viewpoint, many existing interactive optimization mehtods require too many auxiliary optimizations during the whole interaction process. Moreover, some of them require too high degree of judgment to decision makers such as the marginal rate of substitution, which seems to be beyond man's ability.

In many practical situations, decisions seem to be made on the basis of satisficing rather than optimization due to the limit of human ability and available information. However, rather than mere stisficing, it is more disirable to ensure that the obtained solution is satisfactory and in addition there is no other feasible solution superior to the obtained solution in terms of all criteria. For example, let us consider a case in which a decision maker ordered two designers to design some industrial product. Almost at the same time and almost at the same expense, these two designers completed their designs, which were all satisfactory to the decision maker. However, one of designs is superior to the other in terms of all criteria. In this case, it seems no doubt that the decision maker adopt the superior one. From this observation, we recently suggested "Satisficing Trade-off Method (Nakayama and Sawaragi, 1984a, 1984b). The outline of the satisficing trade-off method is as follows:

If the given aspiration level is feasible (i.e., there is a possibility to improve all criteria), then by solving an auxiliary Min-Max problem we show a Pareto solution which distributes the equal improvement to each criterion. On the other hand, if the aspiration level is not feasible, then by solving the same kind of Min-Max problem we show a Pareto solution for which each criterion shares an equal sacrifice. If the decision maker is not satisfied with the shown Pareto solution, then he answers his new aspiration level in view of some available trade-off information such as the Lagrangian multipliers in the auxiliary Min-Max problem. Repeating such a procedure, we finally obtain a satisfactory Pareto solution for the decision maker (Fig. 3.)

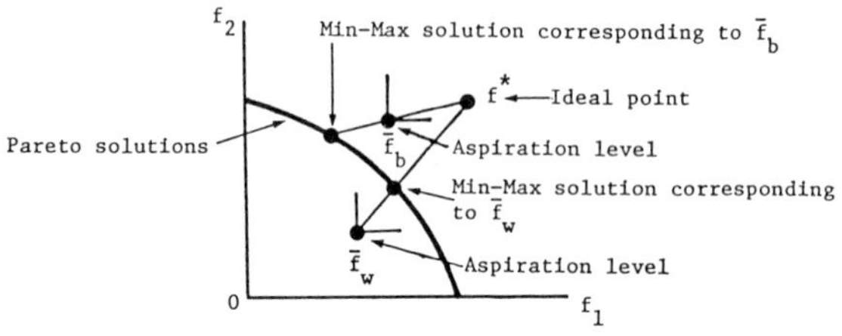

Fig. 3. Satisficing trade-off method

The satisficing trade-off method requires only the aspiration level of the decision maker which is very easy to answer. Therefore, it is very intuitive, easy and simple to carry out. Throughout many experiments, we have observed that the satisficing trade-off method holds almost of all desirable properties stated above. In addition, the method is also expected to be applicable to group decision problems as a tool of negotiation, because the aspiration level of each criterion can be considered the aspiration level of each person in the group. Many practical applications in the real world could encourage us to sophisticate the methodology in the future which is robustly flexible to multiplicity of human value scope.

GAMING SYSTEM FOR COMPLEX PROBLEMS

Gaming is a method that has been used with success for a variety of complex problems. Decision makers involved in a refined game could have a realistic situation in which they must collectively consider their strategies. Therefore, gaming appears to be a promising tool to deal with complex problems in which human decisions have far-reaching effects on others.

The authors believe that recent development of microcomputer will give a great impact on gaming techniques. We have been studying gaming approach by microcomputer in order to solve various complex environmental problems such as acid rain problem in Northwestern Europe, and we have made several microcomputer-based games (Baba and others, 1983, 1984, 1985).

Fig. 4 shows our microcomputer gaming system being used in our gaming experiments.

We can enumerate several advantages of this microcomputer gaming system.

(1) Since microcomputer calculates fast, players

can enjoy game playing in a dialogue mode.

(2) Players can grasp a vivid feature of the real situations through beautiful color graphic display. It helps players concertrate in the game playing.

(3) A floppy disk can be brought quite easily, and so, the game can be played in all the places where the computer can be available.

(4) The microcomputer can store various data in game playing. Therefore, the game director and/or players could utilize them successfully in the follow-up session.

(5) Line Printer provides game director all the necessary informations during the game playing.

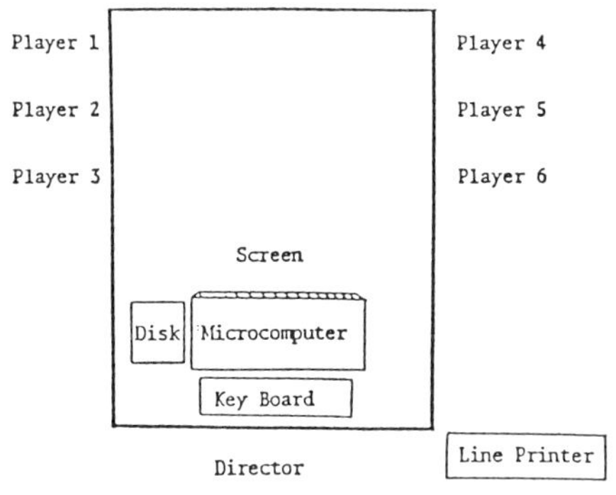

Fig. 4. Microcomputer gaming system

The availability of the microcomputer for the interactive use in game playing has been gradually recognized, and several microcomputer based-management games has been successfully constructed. The authors have the opinion that a large amount of efforts of the future's research should be directed toward the interactive use of microcomputer in order to let gaming be much more helpful for solving various complex problems around us.

CONCLUDING REMARKS

Some of the new trends in systems approach have been discussed from the view points that

(1) human judgement and experience must be incorporated in the system like an expert system, and

(2) microcomputers are more and more used interactively, or in a conversational manner.

These new trends will help people to resolve actual and complex problems which arise in the coming new society with revolutional information and communication netwerks. One of distinctive features of the new trends is that a positive or descriptive approach is combined to the usual normative approach. This direction of systems approach is, however, just on the horizon and much research is left to be done.

The authors express their sincere thanks to the contributions made by Mr. Ryobu, Researcher of JISR, Prof. Nakayama of Kohnan University and Prof. Baba of Tokushima University.

REFERENCES

Baba, N., M. Kaj, T. Hashimoto and Y. Sawaragi (1983). The greenhouse effect game by microcomputer. ISAGA Conference, Sofia.

Baba, N., H. Uchida and Y. Sawaragi (1984). A gaming approach to the acid rain problem. Simulation & Games, 15, 305-314.

Baba, N., K. Machida, E. Nakamura and Y. Sawaragi (1985). An environmental game using microcomputer. ISAGA Conference, Almata.

Kumamoto, H., K. Inoue, K. Ikenishi and Y. Sawaragi (1982). Plant failure diagnosis by if-then rules (in Japanese). Report of Ministry of Education Special Project Research [1], 45-53.

Kumamoto, H., T. Suzuki, K. Inoue and K. Ikenishi (1985). A plant failure diagnosis expert system written in relational database manipulation languages (in Japanese). Trans. The Society of Instrument and Control Engineers, 21, 842-843.

Nakayama, H. and Y. Sawaragi (1984a). Satisficing trade-off method for multiobjective programming. In M. Grauer and A.P. Wierzwicki (Ed.), Interactive Decision Analysis, Springer, pp. 113-122.

Nakayama, H. and Y. Sawaragi (1984b). Satisficing trade-off method for multiobjective programming. Preprints, 9th IFAC Congress, Budapest, 247-252.

Ryobu, M., Y. Nakamori, H. Fukawa, F. Shobayashi and Y. Sawaragi (1976-1985). Comunity and Economy (in Japanese).

Ryobu, M., Y. Yamato, K. Sibuya and Y. Sawaragi (1979). System analysis of a cement plant by the GMSM (in Japanese). Systems and Control, 23, 587-593.

Ryobu, M. and Y. Sawaragi (1982). The econometric model by group method of structural modeling (in Japanese). Trans. The Society of Instrument and Control Engineers, 17, 79-84.

ENGINEERING-ECONOMIC SYSTEMS: A PROBLEM-SOLVING DISCIPLINE

D. G. Luenberger

Department of Engineering-Economic Systems, Terman Engineering Center, Stanford University, Stanford, CA 94305, USA

Abstract. Engineering-Economic Systems represents a discipline that has an engineering, or problem-solving, orientation, but which encompasses application areas beyond those traditionally considered part of engineering. Applications include control engineering, energy policy analysis, urban service planning, medical decision making, investment strategy development, business analysis, algorithm design, as well as subtopics in all other branches of engineering. The discipline is based on a set of fundamental core concepts: dynamics, probability, optimization, economics, and decision analysis; these provide a framework for structuring and solving complex system problems in the same way that physical science forms the foundation for physical branches of engineering. The Department of Engineering-Economic Systems, Stanford University, was established to help create this discipline, through education, research, and involvement in applications. In the twenty years since its creation, the Department has pioneered the concept of broad analytical problem-solving as a discipline, and has evolved strong programs in systems engineering, public policy analysis, and business analysis -- a wide spectrum of areas cemented together by the foundation of core concepts and a commitment to apply these concepts wherever important issues arise.

Keywords. Economics; education, operations research; system analysis.

INTRODUCTION

Engineering is usually defined as the application of science for useful purposes. Originally, this definition emphasized the application of physical science, and, at least implicitly, referred to the design and construction of devices, machines, structures, and so forth. Yet, engineering and engineers have always been concerned with all of the technological aspects of an application; and many of these aspects, although not directly related to physical science, require the same level of rigorous analysis. In the 1950's it became widely recognized that for large complex systems the most challenging part of a design is often the overall structuring -- the creative connection of well-defined components -- rather than the development of new components. The field of "systems engineering," based mainly on mathematical science, soon emerged and was devoted to this important aspect of engineering problems. And, by the early 1960's, several systems engineering departments or programs were established at universities.

The central idea underlying the creation of systems engineering can actually be greatly extended. Rigorous analysis and an engineering, or problem-solving, viewpoint can be applied to a much wider assortment of problems than commonly considered by systems engineers. Not only can the idea be applied to the "nonphysical" aspects of physical systems, it can be applied to almost any complex system requiring organized problem-solving concepts, including operations planning, public policy analysis, business analysis, medical analysis, and so forth.

In 1964, the Institute of Engineering-Economic Systems (EES) was established at Stanford under the leadership of Prof. William K. Linvill. In many ways it closely paralleled other systems engineering departments, but it was based on the extended scope of analysis discussed above. The name of the Institute was itself unusual, containing the word "economic," and this signaled the broad charter that this new venture envisioned.

The Institute soon became a full graduate department and has since been following its self-defined charter, educating students in broad problem-solving concepts and helping to establish a new academic discipline.

In the intervening years EES has had the opportunity to explore many new problem areas, both for theory development and application. It addressed problems in industrial computer systems, the space program, and electric power in the early 1960's. It was a leader among engineering departments in bringing high-level analysis to the policy problems of pollution, housing, transportation, health, and energy in the 1970's. And it is developing new programs in business and decision systems in the 1980's. During this period EES also helped develop the systems discipline with major research contributions in control theory, decision analysis, optimization, and applied economics.

The Department continues to be devoted to developing this problem-solving discipline and to educating students in the discipline, encouraging them to apply their education to important problems wherever they occur. The Department has no special claim to the problem-solving principles it uses and teaches, but does have a unique charter and a unique educational structure that was purposely constructed to accomplish its objectives. This paper presents an overview of the Department and of the discipline that it helped create.

PROBLEM-SOLVING THEME

The EES philosophy treats problem solving as a discipline, and defines the discipline in terms of a set of broad fundamental principles. Specific application problems can be related to these principles. In other words, the EES philosophy states that deep understanding of a broad set of problem-solving concepts is ultimately more powerful than detailed knowledge of substance related to a narrower set of techniques. Once the broad concepts are understood, the narrower ones become special cases. The broad concepts provide a framework upon which a multitude of specific techniques for various applications can be assembled.

This philosophy is an argument for a "fundamental" viewpoint of problem solving -- it is a field based on fundamentals -- and this viewpoint has profound implications for the organization of an academic department in the field. Implementation of the philosophy demands that a broad set of concepts be taught, with emphasis on their interrelations, their implementation, and their application. This is opposed to an educational approach, for example, that stresses primarily an area of application, bringing in techniques and concepts secondarily as necessary; and to an approach that stresses a limited collection of techniques. It requires that excellence be achieved in a broad set of academic areas and that, simultaneously, meaningful involvement in practical problems be achieved.

In terms of substance rather than philosophy, it is difficult to define EES very specifically in a short space. It requires a more intense involvement, which I hope this paper can partially supply. We are often asked, "What is EES?," and the general reply that "we solve problems" doesn't seem to convey much meaning, for after all, every professional discipline solves problems. A more specific response to that question is that we analyze decisions, we optimize, we build computer models, we design algorithms, we perform cost-benefit analyses, and we develop strategy. These are perhaps more tangible responses, although they are narrower and do not fully capture the breadth and depth of our focus.

Alternatively, in response to that question, we sometimes describe EES in terms of specific applications. This helps paint a picture of the methods we employ as well as what we do.

The best explanation, I feel, relates the philosophical statement to the substance of what we do. After seeing the outline of our academic program alongside an assortment of specific applications, it is then usually clear that there is indeed a good deal of meaning to the phrase "we solve problems."

The structure of the EES academic program for Ph.D. students is illustrated in Fig. 1. A student first obtains, primarily as an undergraduate, a fundamental education in science and mathematics, the foundation disciplines. Then, beginning in graduate school, in an analogous fashion a student obtains a fundamental education in system or problem-solving principles. This is done principally through the core courses that were specifically designed for this purpose. The core courses are in many respects the focal point of the academic program, for they contain many of the fundamental concepts discussed before. More will be said about them later.

After mastering the core courses, a student takes courses related to particular technique areas, applications, or extensions of core concepts. Many of these courses are taken in other departments.

Next, a student is exposed to practical problems where the concepts of the core courses can be applied. This exposure is accomplished through classroom applications, special project courses,

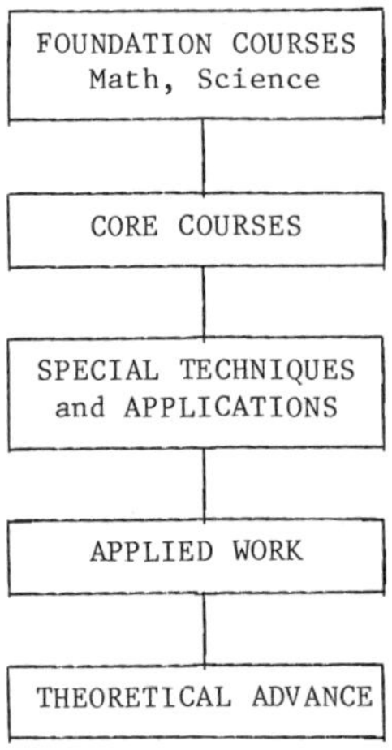

Fig. 1. Structure of the EES Ph.D. Program

individual tutorial work with a professor, and then, most intently, through an internship program.

Finally, the student conducts Ph.D.-level research to contribute to the field. This is often the most exciting phase. It in a sense closes the circle of the academic program, for results of this research nurture the core courses.

Overall, this approach to education, including the philosophy underlying it, is very similar to that of other engineering disciplines in strong universities. In those disciplines the equivalent of our core is found primarily in the sciences. Students take electromagnetic theory as well as many other basic sciences to provide general concepts before taking more specialized courses such as semiconductor circuit design. The EES approach thus mirrors this tried-and-true fundamental approach to education. EES sets the teaching of these fundamentals as its principal educational goal.

The remaining sections of this paper provide brief descriptions of the elements of the EES program more or less in the same order that a typical student progresses: core courses, course work in particular disciplines, exposure to significant applications, research, and finally entering a professional career.

THE CORE

The core is a central element of the unique EES program. The emphasis in the core courses is on general principles of systems analysis and problem-solving. In the Department we often refer to these concepts as portable concepts, since they can be carried from one problem area to another. Indeed "portability" is a term that seems to capture the spirit of the core concepts particularly well, since the the term implies simplicity as well as potency.

One might also refer to the core courses as stepladders. They provide individual viewing stations which, when climbed, provide heightened perspectives over the territory of technical detail below. The view from the top of each ladder somewhat overlaps that of the others, but each one exposes new territory and clarifies what was seen before. Once one has scaled the ladders and absorbed the view, it will be much easier to move around on the ground, through the tangle of detail.

Ideally, if one were designing a new core course structure from a fresh start, it would consist precisely of the collection of the most important fundamental (and portable) problem-solving concepts. Complete agreement on what exactly would constitute this collection seems to be impossible. So the present EES core is probably not perfect in this abstract sense, but it works well in the Stanford environment. The core consists of five major subjects spread over eight one-quarter courses. These subjects are described below.

Dynamic Systems (2 quarters). Almost all real system problems involve a dynamic (that is, time-evolutionary) aspect. Such problems include planning, control of moving objects, study of population growth, etc. The concepts associated with dynamic systems therefore form a fundamental component of the core. Some of the portable concepts related to dynamic systems are: stability, feedback, controllability, observability, linearity, and the concept of optimal control. The core courses in dynamic systems emphasize these concepts at three levels: (1) intuitive, so that the concepts are related to physical and mathematical insight; (2) mathematical theory, so that a student becomes facile with methodology; and (3) application, so that experience is gained in using the theory.

Probabilistic Analysis (1 quarter). Almost all system problems involve an element of uncertainty. Probability theory is the mathematical language of uncertainty and it is essential that students in our field have a strong background in this subject. But beyond the formalism of probability, it is essential that students learn to think correctly about uncertainty, and be able to represent uncertainty in probabilistic terms. This is a principal objective of this course.

Optimization (2 quarters). Optimization has in the past thirty years become a central and, in fact, almost dominating, theme for problem-solving. Even in loose conversation, one often hears comments like "we must optimize the utilization of our resources" in reference to an extremely large and complex system. The core course in optimization helps students develop an understanding of the pure theory, the ability to formulate optimization problems, and knowledge of the principal methods for solving optimization problems.

Economics (2 quarters). The core courses in economics are devoted to microeconomics, concerned with individual decisions made by firms and consumers. A good deal of mathematical structure is used in microeconomic theory to represent individual preferences, to describe technologies available to a firm, and to characterize markets; and this structure is presented in the course sequence. The course also contains, however, an important set of concepts relating these elements, and these form the basis for much of applied microeconomic analysis. These include the concepts of an economic equilibrium, Pareto efficiency, public goods, and externalities.

Decision Analysis (1 quarter). Decision analysis focuses primarily on the analysis of important one-of-a-kind decisions that must be made in the face of uncertainty. The rational approach of decision analysis is based on two fundamental concepts. The first is that uncertainty can be represented explicitly in probabilistic terms. The second is that of individual preference, especially with respect to money. The combination of these two ideas allows an ordered approach to complex decision problems.

The extremely broad collection of concepts, as evidenced by the above core course descriptions, can be divided into two basic types. This division is useful in understanding our approach to problem-solving. The first set are concepts related to problem formulation and the second set are related to problem solution. Concepts of problem formulation tend to be broad concepts that can be described, at least in general terms, with very little mathematics. They help focus the problem on definite issues and provide a framework for thinking about those issues. Such concepts include social efficiency, risk, incentives, etc. Concepts of problem solution tend to focus on method, and are usually somewhat mathematical, although again the general concept is often quite simple. Examples of solution concepts are iteration, convergence rate, decomposition, etc.

DISCIPLINES OF STUDY

Although all students in the Department take courses from the core, it is essential that they also develop an individual focus in some particular discipline. The details of such a focus vary widely from student to student, naturally, but one of three distinct disciplines of study is generally followed. The three disciplines are: systems engineering, public policy analysis, and business analysis.

These disciplines themselves represent broad areas -- areas that are usually thought of as defining separate departments within a university. Thus it is perhaps initially surprising to find that they coexist and flourish in our single small department. The reason that this is possible, of course, traces back to the underlying philosophy of a fundamental education and its realization in the core. With additional course work, on top of the core, a course program can be tailored in systems engineering, operations research, public policy analysis, or business analysis. Each of these appears distinct to the outside world, much like different branches of physical engineering appear to be quite different. But, like these other branches of engineering, the disciplines the Department supports are different only in secondary substantive knowledge; they overlap in that they are derived from a common base of fundamentals.

The coexistence of these disciplines within the Department is actually synergistic. Students with a special interest in one area see directly that the concepts upon which their discipline is built are also the basis for other areas as well. This reinforces the breadth of these concepts, and provides opportunity for cross-fertilization.

APPLICATIONS

One way that engineering departments are often characterized is by their position on an imagined applied-theoretical spectrum, as indicated in Fig. 2. A department, in this view, tends to develop a kind of departmental personality that tends to favor problems characterized by a unique blend of theory and application. Some pursue pure theory while others are much more applied.

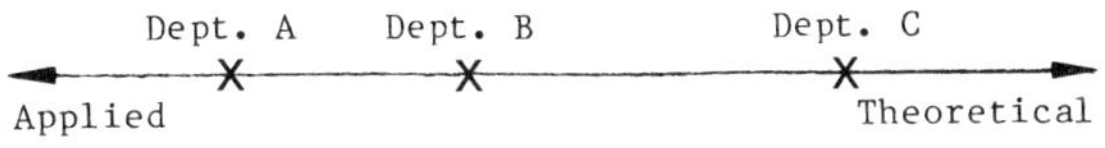

Fig. 2. The Applied-Theoretical Spectrum of Departments

When developing EES, a different view was taken. It was felt that rather than always selecting individual projects that represented a compromise between theory and practice, the Department should pursue activities from all parts of the spectrum and especially from both extremes. Thus, some activity is devoted to pure theory, without the distraction of application, and some is devoted to real application.

It is important, when developing theory, for example, to strive for the deepest and most conceptual aspects of that theory; that represents the right-hand extreme. Deep theory should not be confused with complication or obfuscation. Experience teaches us that the deepest and best theory, that which clarifies and unifies, is always fundamentally simple. But, the simplicity of good theory does not imply that it is trivial or easily attained. Quite the contrary, the path to simplicity is often difficult. It is somewhat analogous to the process of climbing a mountain in order to obtain the view from the top. The path up, especially the first time, may be steep and contain many switchbacks. But these hardships are not part of the magnificent view from the top.

At the other end of the spectrum are real applications. These are problems of industry, government, or business that involve the components of systems and/or economics. Examples range from the design of a control system to the design of decentralized computer systems, from the study of a city parking problem to evaluation of medical policy, or from determination of optimal financial portfolios to analysis of international hedging problems. The central focus, in such situations, is on the problem itself, not the method of solution. It is a real problem if someone or some group is responsible for it and wants a solution. This contrasts with abstract applications in the form of classroom examples designed to illustrate theory or case studies derived from previous experience. Such abstract applications certainly have a place in the university, but it is important to have some exposure to real applications as well.

THE INTERNSHIP

Since system problems are typically large and complex, it is not feasible to bring many of them as projects into the university without swamping the staffing capabilities. Such projects require a great deal of travel, reporting, and daily attention to detail, as well as a heavy level of technical work. A faculty devoted primarily to classroom instruction and research cannot also manage a full complement of such projects. It is possible to work on a few applied projects within the university, but in order to achieve a rich complement of real problems for students, it is most effective to work in conjunction with outside institutions that are better equipped to provide the total capability. This idea is the basis for our internship program.

In an internship a student spends an extended period, of about a year, in a company or government agency working on an applied project. It is important to recognize that an internship is not just a period away from the university to obtain practical experience. Rather, it is, ideally, an integral part of the educational program, with a continuing strong connection to the Department. Internships are usually tailored, or brokered, by a faculty member who knows both a student's abilities and the needs of a particular outside institution. The faculty member continues to stay involved with the student during the internship and in many cases has a continuing relation with the outside institution as an adviser or consultant.

Brief descriptions are given below of a few projects in which EES students and faculty have been involved either through internships or on-campus projects. They illustrate the range of applications and the scope of responsibility on particular projects.

Control of Processes. In the early 1960's Westinghouse was beginning to use computers for process control, particularly for power plants, steel mills, and paper mills. The company's original approach, based on traditional transform methods of control design, was found to be ineffective for many multivariable problems. A group of interns began to explore the applicability of modern state space methods. As a result, some fundamental results for modern control theory were developed there. Today these results are included in most textbooks on control design. This project thus played a significant role in launching the new era of digital process control.

Magnet Design Problem. In the early 1960's Westinghouse had been producing an excellent ceramic magnetic material but, surprisingly, had a very small share of the market. Top management concluded that the reason for the poor sales level was the long turn-around time in the design of specific devices using the material. Two interns then determined that a principal bottleneck of the design process was verification that specifications were satisfied by determining the magnetic flux pattern, which was accomplished by a tedious graphical technique. Effort thus focused on developing a computerized procedure to replace the graphical one. This amounted to solving a certain partial differential equation, an external Laplace equation. Computing in those days was archaic by today's standards, for it often required a full day to turn around a single trial run. Nevertheless, the program was developed in only a few weeks, and it, together with its extensions, was so successful that it accounted for a major share of computing time at Westinghouse Research Laboratories over the next few years.

The Decision to Seed Hurricanes. In 1970 a project was begun at Stanford Research Institute to investigate the decision problems inherent in the possibility of seeding hurricanes with silver iodide. There was evidence that such seeding could significantly reduce the destructive force of hurricanes. However, many other issues were involved in a decision. For example, the government might have some legal responsibility for the damage caused by a seeded hurricane. Or public outcry might result if a seeded hurricane caused an unusual amount of damage. The decision whether to seed a hurricane was thus a complex one involving uncertainty of meteorological consequences and its consequences to property damage and government responsibility. This problem was analyzed using decision analysis methods, which were at that time in an early stage of development. In fact this project, published in Science (1972), is regarded as an important illustration of decision analysis methods.

Agriculture Model. Working at Systems Control, Inc., several interns helped develop a model of the agricultural sector of the U.S. economy suitable for policy planning. The model was developed in the mid-1970's when issues of energy costs were of dominant concern, and this model was motivated by a desire to determine the likely impacts of higher energy costs on agriculture -- such as the types of crops grown, etc.

Screening for Cancer. Currently one of the most effective methods for reducing the incidence of cancer in the general population is through screening on a regular basis. Determining appropriate screening policies -- who should be screened, for what diseases, by what tests, at what frequencies -- is a complex problem with enormous ramifications. Beyond the purely medical consequences, screening policies typically lead to hundreds of screening centers, millions of lives, and billions of dollars. However, policies must be determined somehow. A major project in the Department addressed this issue rigorously. Some experimental data was already available regarding the effectiveness of various tests for different classes of individuals. However, screening is a complex dynamic and uncertain process, since disease progresses with time and test results are uncertain. Unscrambling the interrelation of these factors with experimental data was a challenging problem that was amenable to the concepts found in our core courses. The model developed at Stanford led to a new prepaid health benefit plan by the Blue Cross Association and to major new policy recommendations by the National Cancer Institute.

RESEARCH PROGRAMS

Research is an important and pervasive component of the Department (as suggested by the unusually high ratio of Ph.D. students to faculty). But in large measure research is closely related to other aspects of the program; it is the rejuvenating factor that provides focus for current students and new direction for future students. Research expands on the theme set by the Department. It is often focused on application, modification, and extension of the core concepts, and indeed an outstanding piece of research is often later incorporated into the core courses.

Aside from the general objectives of developing the core concepts and responding to applications, the overall direction of research in the Department is somewhat influenced by perceived national priorities and needs. Accordingly, in the 1960's much of EES research was related to the space program and the mainframe computer revolution. Much of the Department's work in those topics is considered pioneering today. In the 1970's EES built a strong public policy program addressing housing, energy, and other issues of high national priority. It is well recognized that EES was a leader in bringing public policy to schools of engineering. In the 1980's attention is turning to research related to the small-computer revolution, new engineering issues, and business strategy. We believe that we are again on a new frontier for engineering. In a sense we are somewhat opportunistic in our selection of areas, responding as we do to issues of the day. However, the one thing that has remained constant during these decades is our commitment to the development and application of the analytical concepts of problem solving.

PROFESSION

The students who come to EES tend to have what might best be called an adventurous spirit. This spirit may be related to their general high academic caliber, but I think it is also partially a reflection of the self-selection process inherent in a student's choosing to come to an unusual department such as EES. We on the faculty try to nurture that spirit, give it substance, and provide the opportunity for its exercise while they are here. It is our hope that when they leave us, they will maintain this spirit and will have attained confidence in their ability to contribute to the solution of important problems. With these two assets, an adventurous spirit and analytical confidence, they will find that adventure and worthwhile accomplishment can go hand-in-hand.

As pointed out earlier, there is no simple pattern of employment followed by EES graduates. Their careers are as diverse as our own program of teaching and research. Their special spirit and special education do seem to find suitable expression, however. Many of our students obtain leadership positions in industry or government very early in their careers. Some EES students have started companies. (In fact over twenty companies have already been founded by our students.) Many students have taken academic positions and a few have instituted programs similar to EES in other universities. EES is a special and rewarding discipline, with continuing challenges that respond to the challenges of society. Our graduates often find new areas of challenge, which we never foresaw and thereby expand our scope.

REFERENCE

Linvill, W.K. (1966). Engineering-economic systems: a new profession. IEEE Spectrum, 3 96.

SOME COMMON PROBLEMS IN CONTROL AND SIGNAL PROCESSING

M. Mansour

Institute for Automatic Control and Industrial Electronics, Swiss Federal Institute of Technology, CH-8092 Zürich, Switzerland

Abstract. In this paper, an overview of the problems in control systems and signal processing is given. The basic theory, namely signal and system theory, is the same for both areas. As examples, the stability, the estimation and realization problems are considered. It is shown that some control system results can be extended to deal with two-dimensional digital filtering. On the other hand, some control system problems can be solved using two-dimensional system results.

INTRODUCTION

Most of the problems dealt with in signal processing are digital filtering and multi-dimensional processing. We can also consider speech processing as a part of signal processing.

Signal and system theory as well as pattern recognition methods are the fundamental tools used to solve the above problems. On the other hand, in control we deal with problems like digital control, identification, and adaptive control. It is well known that the tools required to solve these problems are the same tools, namely signal and system theory as well as pattern recognition methods. The following setup, although not complete, shows these relations.

Mostly in signal processing we have a noisy signal and we try to retrieve the original signal through digital filtering. In other cases, we want to get some characteristics of the signal, e.g. spectrum, contour of an image or the speech behind the signal.

Fig. 1 shows some examples of signal processing and control systems. It is clear that in control systems we normally have a process which is given and which we control to achieve a certain goal. The input to the process is normally a function of the output, while in signal processing we normally receive a signal which we do not influence. However, the methods used to analyze the signal or to determine the filter are in general the same as the methods used to analyze the control system or to determine the controller.

In the following examples, stability, estimation and lattice realization are considered to demonstrate the close contact between signal processing and control.

Signal processing areas	Control system areas
digital filters	digital control
multidimensional filters	identification
speech processing	adaptive control
⋮	⋮

↓ ↓

Common methods and theory
- realization theory
- model reduction
- parameter and state estimation
- stochastic processes
- optimization theory
- adaptation
- stability theory
- oscillations
- sensitivity analysis
- controllability and observability
- pattern recognition
- ⋮

STABILITY

Here, we discuss two examples. The first one shows results of control systems transferred to 2-D signal processing, and the second one shows results of 2-D signal processing applied to control systems.

The stability of a digital control system can be investigated by considering the roots of a polynomial

$$f(z) = b_0 + b_z + b_2 z^2 + \dots + b_n z^n \qquad b_0 = 1 \tag{1}$$

to lie outside the unit circle. It is well known that the stability of such a polynomial can be investigated by the Jury test. Mansour (1965a) has shown that for a 1-D polynomial to be stable, it is necessary that the values of the coefficients should lie within certain ranges. In Table 1, the upper and lower limits for the coefficients are given for n = 1...6. This table was originally derived for control systems but was used afterwards for digital filters.

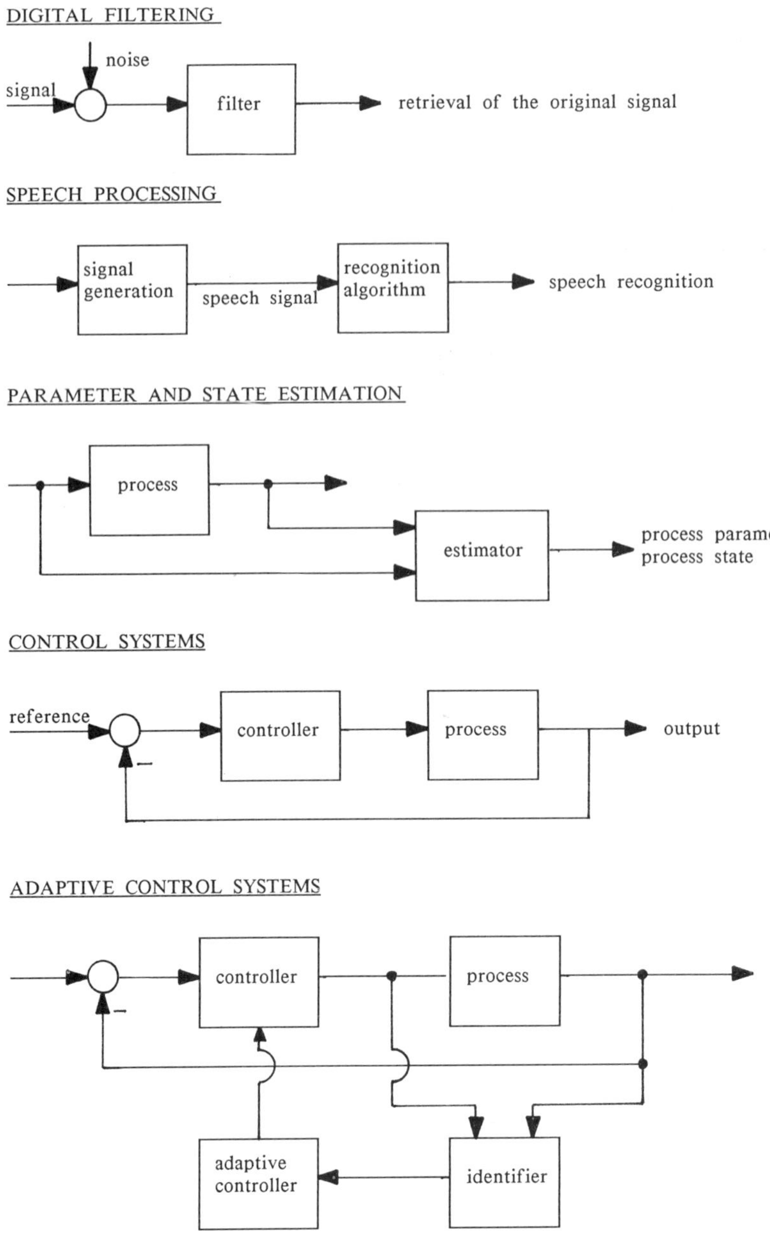

Fig. 1 Examples of signal processing and control systems

Table 1: Coefficient ranges for stability of a 1-D polynomial

	b_1	b_2	b_3	b_4	b_5	b_6
n=1	2 -1	1 -1				
n=3	3 -3	3 -1	1 -1			
n=4	4 -4	6 -2	4 -4	1 -1		
n=5	5 -4	10 -2	10 -10	5 -3	1 -1	
n=6	6 -6	15 -3	20 -20	15 -5	6 -6	1 -1

For the investigation of the stability of 2-D digital filters, this result was extended by Agathoklis & Mansour (1984).

Consider the 2-D polynomial

$$B(z_1,z_2) = \sum_{i=0}^{n}\sum_{j=0}^{n} b_{i,j} z_1^i z_2^i =$$

$$= \left[1, z_1, \ldots, z_1^n\right] \begin{bmatrix} b_{00} & \cdots & b_{0n} \\ \vdots & b_{ij} & \vdots \\ b_{n0} & \cdots & b_{nn} \end{bmatrix} \begin{bmatrix} 1 \\ z_2 \\ \vdots \\ z_2^n \end{bmatrix} \tag{2}$$

$$= \underline{z}_1^T \, B \, \underline{z}_2$$

where B is an (n+1)×(n+1) matrix with coefficients b_{ij} and $b_{00} = 1$.

The coefficient ranges necessary for the stability of a 2-D polynomial of the form (2) are given in Table 2.

The second example deals with the stability of systems with time delay. Kamen (1980) has shown that the stability of delay systems can be investigated using 2-D polynomials.

Let the delay system be given by

$$\frac{d^n}{dt^n} y(t) + \sum_{p=0}^{n-1} \sum_{q=0}^{m} q_{p,q} \frac{d^p}{dt^p} y(t-qh) = 0 \qquad (3)$$

The characteristic function of (3) is given by

$$P(s,\exp(-hs)) = s^n + \sum_{p=0}^{n-1} \sum_{q=0}^{m} q_{p,q} s^p \exp(-qhs) \qquad (4)$$

(3) is asymptotically stable if and only if

$$P(s,\exp(-hs)) \neq 0 \qquad \text{Re } s \geqslant 0 \qquad (5)$$

To the function $P(s,\exp(-hs))$ is associated a two-dimensional polynomial with the variable s and z given by

$$P(s,z) = s^n + \sum_{p=0}^{m} q_{q,p} s^p z^p \qquad (6)$$

Kamen (1980) has shown that asymptotic stability of (1) is equivalent to the stability condition of (6).

Jury and Mansour (1982) have shown that checking the stability of (6) can be reduced to checking positivity of a one-dimensional polynomial in the interval $x \in [-1,1]$.

This example shows how the results of signal processing can be used in investigating the stability of dynamic systems.

Table 2: Coefficient ranges for stability of 2-D polynomials of the form (2)

n=1	$B_{upper} = \begin{bmatrix} 1 & 1 \\ 1 & 1 \end{bmatrix}$	$B_{lower} = \begin{bmatrix} 1 & -1 \\ -1 & -1 \end{bmatrix}$
n=2	$B_{upper} = \begin{bmatrix} 1 & 2 & 1 \\ 2 & 4 & 2 \\ 1 & 2 & 1 \end{bmatrix}$	$B_{lower} = \begin{bmatrix} 1 & -2 & -1 \\ -2 & -4 & -2 \\ -1 & -2 & -1 \end{bmatrix}$
n=3	$B_{upper} = \begin{bmatrix} 1 & 3 & 3 & 1 \\ 3 & 9 & 9 & 3 \\ 3 & 9 & 9 & 3 \\ 1 & 3 & 3 & 1 \end{bmatrix}$	$B_{lower} = \begin{bmatrix} 1 & -3 & -1 & -1 \\ -3 & -9 & -9 & -3 \\ -1 & -9 & -9 & -3 \\ -1 & -3 & -3 & -1 \end{bmatrix}$
n=4	$B_{upper} = \begin{bmatrix} 1 & 4 & 5 & 4 & 1 \\ 4 & 16 & 24 & 16 & 4 \\ 6 & 24 & 36 & 24 & 6 \\ 4 & 16 & 24 & 16 & 4 \\ 1 & 4 & 6 & 4 & 1 \end{bmatrix}$	$B_{lower} = \begin{bmatrix} 1 & -4 & -2 & -4 & -1 \\ -4 & -16 & -24 & -16 & -4 \\ -2 & -24 & -36 & -24 & -6 \\ -4 & -16 & -24 & -16 & -4 \\ -1 & -4 & -6 & -4 & -1 \end{bmatrix}$

ESTIMATION

Signal processing is concerned in large part with the retrieval or the determination of the characteristics of a certain signal using noise currupted measurements. Control theory deals with the design of automated systems using noise corrupted measurements of a process and determining the control to be applied as input to it. In both cases, estimation of the signal or its parameters or the state of the process is essential. The elements of estimation theory which depend on probability theory and the theory of stochastic processes are the basis for dealing with these problems in signal processing, communication and control.

As an example of estimation theory, we discuss the application of the discrete Kalman filter. In control theory, Fig. 2 shows a control system with state variable feedback and a Kalman filter for the estimation of the state. This is obtained under certain statistical assumptions on the initial state, noise and measurement noise. The matrix H is obtained according to the minimum variance criterion.

The same structure of the Kalman filter can be used for extrapolation or interpolation. The Kalman filter has been applied in aerospace studies such as determination of the path of a spacecraft or of an airplane using the measurements at the earth station, in ship navigation systems and in communication systems. The Kalman filter is actually a generalization of the works of Kolmergoroff and Wiener which was done mainly for signal processing.

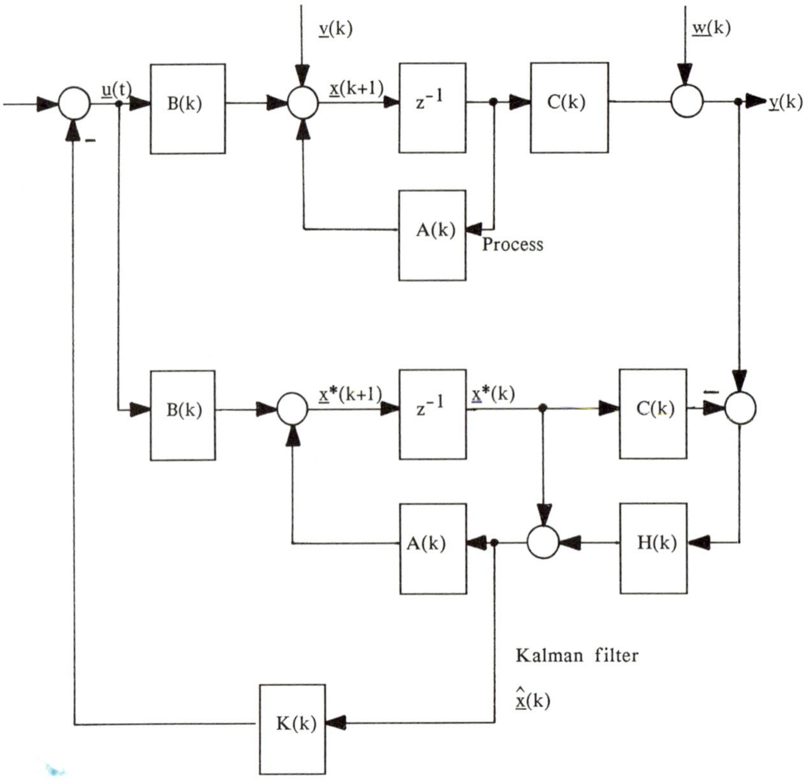

Fig. 2 State variable feedback using Kalman filter for state estimation

LATTICE REALIZATION

In the digital signal processing literature there are different filter structures that can be used to implement a linear filter of the transfer function

$$G(z) = \frac{K\left[1 + \sum_{j=1}^{m} b_j z^{-1}\right]}{1 - \sum_{i=1}^{n} a_i z^{-1}}$$

The lattice form is normally represented by the partial correlation coefficients k_i (Ahmad, 1984), or the reflection coefficients $r_i = -k_i$. $|k_i| < 1$ is necessary and sufficient for a stable filter. Lattice realization is also less sensitive to quantization effects.

Ahmad (1984) uses the lattice filter of Fig. 3 to model the vocal tract for speech processing.

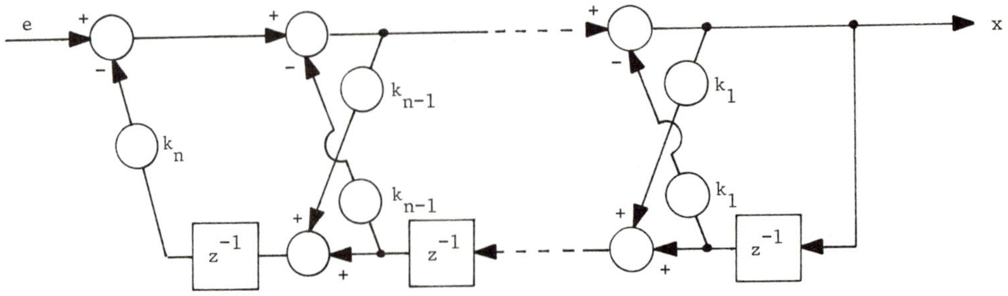

Fig. 3 Lattice realization of a linear filter

Mansour (1965b) developed a realization of discrete systems to prove the Schur-Cohn stability criterion using the Lyapunov direct method. This realization is actually a lattice realization as shown in Fig. 4 (Anderson, Jury & Mansour, subm.).

$$\begin{bmatrix} x_1(k+1) \\ x_2(k+1) \\ \vdots \\ x_{n-1}(k+1) \\ x_n(k+1) \end{bmatrix} = \begin{bmatrix} -\Delta_1 & (1-\Delta_1)^2 & 0 & \ldots & 0 \\ -\Delta_2 & -\Delta_1\Delta_2 & (1-\Delta_2)^2 & \ldots & \vdots \\ \vdots & \vdots & \vdots & & \vdots \\ -\Delta_{n-1} & -\Delta_1\Delta_{n-1} & -\Delta_2\Delta_{n-1} & \ldots & (1-\Delta_{n-1})^2 \\ -\Delta_n & -\Delta_1\Delta_n & -\Delta_2\Delta_n & \ldots & -\Delta_{n-1}\Delta_n \end{bmatrix} \begin{bmatrix} x_1(k) \\ x_2(k) \\ \vdots \\ x_{n-1}(k) \\ x_n(k) \end{bmatrix} + \begin{bmatrix} 1 \\ 0 \\ \vdots \\ \\ 0 \end{bmatrix} u(k)$$

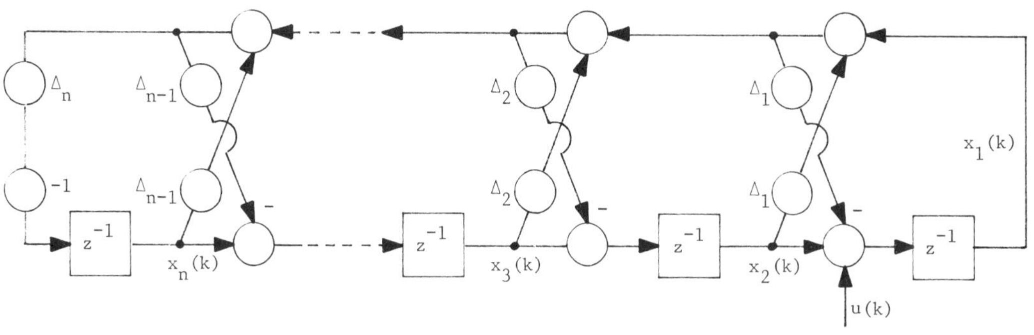

Fig. 4 Lattice realization of a discrete system equation (7)

Badreddin & Mansour (1980) used this realization for model reduction, and Anderson, Jury & Mansour (subm.) explained the method of model reduction very easily using the lattice realization.

CONCLUSIONS

It was shown with examples that different problems in signal processing and in control are the same or at least the methods of analysis and synthesis are the same. In my opinion, an effort should be made in order to unify the education in control and signal processing at least in the fundamental courses.

REFERENCES

Agathoklis, P., Mansour, M. (1984). Coefficient ranges necessary for stability of 2-D polynomials. Systems and Control Letters, 4, 169-173.

Ahmad, H.E. (1984). A study of efficient digital speech processing systems with applications to Arabic. Ph.D. Thesis, ETH Zürich.

Anderson, B.D.O., Jury, E.I., and Mansour, M. (submitted). On model reduction of discrete time systems.

Badreddin, E., Mansour, M. (1980). Model reduction of discrete-time systems using the Schwarz canonical form. Electronic Letters, 16, 782-783.

Jury, E.I., Mansour, M. (1982). Stability conditions for a class of delay differential systems. Int. J. Control, 35, 689-699.

Kamen, E.W. (1980). On the relationship between zero criteria for two-variable polynomials and asymptotic stability of delay differential systems. IEEE Trans. on Autom. Control, AC-25, 983-984.

Mansour, M. (1965a). Instability criteria of linear discrete systems. Automatica, 2, 167-178.

Mansour, M. (1965b). Die Stabilität linearer Abtastsysteme und die Zweite Methode von Ljapunov. Regelungstechnik, 13, 592-596.

NEW TRENDS IN IDENTIFICATION

P. Eykhoff

Department of Electrical Engineering, Eindhoven University of Technology, NL-5600 MB Eindhoven, The Netherlands

Abstract. This paper attempts to summarize some recent essential developments in the field of system identification and parameter estimation. The topics discussed include: the identification protocol, model representations, estimation methods, identification tools, identification applications, expectations on the further developments. It is recognized that identification in an engineering sense has both a 'science' and an 'arts' aspect.

Keywords. Identification; parameter estimation; modelling.

CONTENTS

1 INTRODUCTION

It is not easy to meet the challenge given by the symposium organizers: to present for a general, control engineering audience, a compact description of new developments in system identification and parameter estimation. These fields are developing at a high pace and over a very broad spectrum, ranging from very fundamental theory to down-to-earth applications. Probably many colleagues will share with me the feeling that it is almost impossible to keep track of even the most essential lines of progress. Yet, or just because of that feeling, the challenge given is worthwhile ...

The problem of keeping pace with the progress is partly due to the fact that three aspects of development have to be kept in mind, viz.:
- new theory / - new tools / - new applications, including new solutions to classic problems.

The interaction between these aspects can be indicated as in Fig. 1, where some of the feedforward (push from research) and feedback influences (pull from applications) can be recognized.

Another root of that problem is due to the fact that for practical applications of identification, a purely scientific approach is not sufficient. There is much dependence on the peculiar aspects of the particular application, on the engineering insight and know how. Often this cannot (yet) be formulated in precise scientific terms. Consequently, in applications there has to be a balance between the aspects 'science' and 'art'.
In section 2 the overall framework of identification will be considered, starting from a priori knowledge and decisions. Then in the Sections 3, 4 and 5 the model representations, theory- and tools (computer programs)-development will be considered. In Section 6 attention is given to applications. Expectations on the further development are discussed in Section 7.

2 IDENTIFICATION PROTOCOL

The ultimate motives behind the identification lie in its applications. Note that such applications can be divided into the following categories:

	effect:
diagnosis: the determination of such process quantities that cannot be measured in a direct way ('indirect measurements').	interpretation of the past
monitoring: the 'measurement' of e.g. quality, efficiency, wear, poisoning, in order to decide on overhaul, replacement, cleaning, etc. of (part of) the process.	interpretation of the present
prediction: the estimation of the development of process behaviour beforehand.	prediction of the future
control: collection of information for influencing the process behaviour	influencing the (immediate) future

The essential elements of the identification protocol are given in Fig. 2 (Eykhoff, 1984). Due to space limitations only a few remarks on this scheme have to suffice. The reader should particularly keep in mind the distiction between 'science'- and 'art'-aspects as mentioned before.

(Engineering) insight. At the outset it is essential that the 'model builder'/'identifier'/'experimenter' has an open mind for the real a priori knowledge as well as for the tacit assumptions that are embedded in his or her task. The first decision that requires careful consideration is the choice of the demarcations of the process under study:

- What is considered part of the process, what is non-process or 'environment'?
- What are (measurable) inputs, what has to be recognized as disturbances?

Clearly in real engineering situations such decisions require a knowledge/insight/intuition that, as yet, defies complete scientific argumentation.

(Physical) laws/modelling. The 'a priori' knowledge, associated with the modelling, depends on the 'artful' combination of techniques from many fields. Judicious simplifications such as linearizing, lumping/ reducing play an important role. An indication of the complexity of model-characterization is given by the following list of adjectives that, together with their obvious counterparts, do express some of the 'a priori' knowledge or assumptions:

static; time-continuous (non-sampled); time-invariant; linear dynamics; linear-in-the-parameters; single-input single-output (SISO); lumped parameters; deterministic; single layer; causal; one-dimensional; non-fuzzy; ...

The choice from these adjectives and their counterparts implies already a sizable (assumed) a priori knowledge.

Aims and circumstances. Of course the goals of the identification procedure as indicated before are of paramount importance. To a high degree they determine the answers to essential questions:

- For the particular type of application in mind, what kind of model would be adequate (explanatory model; representation model; prediction model)?
- Next to process dynamics, should the disturbances also be characterized?
- What complexity of the model (e.g. number of parameters) would be adequate?
- Are there closed loops to be considered, either directly recognizable or perhaps hidden, e.g. by human intervention in the process?
- Can test signals be applied to the process? If so, what would be an optimal choice?

Experiment design. Input signal design has been recognized as a useful tool for the improvement of the accuracy of parameter estimates. In the literature, a number of aspects of input signal design for system identification has been discussed for various classes of models. Overviews, also including some other aspects such as feasibility of the design methods, computation aspects, etc. can be found in survey papers of Mehra (1974, 1981), Gustavsson et al. (1972), Goodwin (1982), as well as in the books/papers by Goodwin and Payne (1977), Zarrop (1979), Söderström and Stoica (1983), Krolikowski and Eykhoff (1985).
An interesting challenge is the use of adaptive techniques for the generation of test signals; in this way the input signal can be created such that, within the constraints imposed on those signals, the information/estimation efficiency will be optimal.

The estimation method (Least Squares, Instrumental Variable, Markov, Maximum Likelihood, Bayes, ...) has to be chosen according to the a priori knowledge available.

The validation of the model found has several phases:

- auto validation/check of residuals, i.e. to test whether all information that is contained in the process input/output data has been 'explained' by the model;
- cross-validation, i.e. the test whether the model (prediction)performance is adequate for an independent set of measured input/output data;
- consistency check, i.e. the test whether the model corresponds to the engineering insight that is available on the process.

3 MODEL REPRESENTATIONS

In spite of advances in theory, capabilities and actual application of identification, one has to recognize that the major efforts and results have been for single-input single-output (SISO) processes. The case of multi-input multi-output (MIMO) processes has received much attention, but has not yet developed to the point where one may call it 'coming of age'. Among the reasons for this unfortunate situation are the following:

- the greatly increased complexity of MIMO-process-dynamics models compared to SISO situations;
- the additional complexity of model structure determination, viz. the estimation of structural parameters;
- the greatly increased complexity of MIMO-noise-dynamics models compared to SISO situations;
- the computer power/price ratio (speed, memory, capacity, word length) in the past;
- the limited availability of numerically reliable, well-tested software for such tasks.

Due to the pull from practice as well as the push from research and development, this situation is bound to change; Fig. 1.

In trying to establish a model for a particular process the most essential question is: what use will be made of that model? From this central question a number of other ones arise:

- can a priori knowledge be accommodated in the model in an acceptable way?
- what is of primary importance: -parameter accuracy, -output reconstruction; -output prediction?
- what is the minimal complexity of the process- and noise model for the application at hand, viz. what is the minimal number of parameters that may acceptably specify the model?

Not only the application, but also the estimation method chosen, has an impact on the choice of the type of model, e.g.:

- the properties of the error function (global and local minima?; uniqueness for the parameters to be estimated?);
- the unbiasedness of the estimator;
- the transformability of the model to other model(s);
- the error propagation under transformation of the model;
- computation time needed;
- numerical accuracy/stability.

The discussion will be restricted to linear, time-discrete, time-invariant models. The model representations that we will consider are the following:

Representation	I/O or structural characteristics
- Transfer function model	input-output
- Difference equation -	input-output
- Impulse response -	input-output
- State space -	structural

For each representation the SISO and MIMO case will be given in order to show their relations, as well as the increased complexity arising from the multivariable situation. In section 3.5 comparisons

will be made with respect to number of parameters, linearity-in-the-parameters, etc.
Also the transformations between these representations are indicated.

In the following:
q represents the shift operator in the time domain: $q\,y(k) = y(k+1)$

$$\underline{y}(k) = \begin{bmatrix} y_1(k) \\ y_2(k) \\ \vdots \\ y_q(k) \end{bmatrix} \qquad \underline{u}(k) = \begin{bmatrix} u_1(k) \\ u_2(k) \\ \vdots \\ u_p(k) \end{bmatrix} \tag{1}$$

So $\underline{y}(k)$ and $\underline{u}(k)$ are vectors with length q and p, the number of outputs and inputs, respectively.

3.1 Transfer function models

Notation:

SISO	MIMO	
time	domain:	
$y(k) = H(q)\,u(k)$	$\underline{y}(k) = K(q)\,\underline{u}(k)$	(2)
frequency	(z-)domain:	
$Y(z) = H(z)\,U(z)$	$\underline{Y}(z) = K(z)\,\underline{U}(z)$	(3)
$H(z) = \dfrac{b_0 z^n + b_1 z^{n-1} + \ldots + b_n}{z^n + a_1 z^{n-1} + \ldots + a_n}$	$;\ K(z) = [K_{ij}(z)]$ matrix q x p	(4)

Characteristics:
- simple physical interpretation;
- complete description of dynamic behaviour if the initial conditions are zero;
- SISO: order = degree of denominator polynomial of H(z);
- MIMO: order notion quite complex;
- large number of parameters;
- output not-linear-in-the-parameters;
- unique description of a process, provided there are no pole/zero cancellations.

3.2 Difference equation models

Notation:

SISO	MIMO	
time	domain:	
$A(q)\,y(k) = B(q)\,u(k)$	$A(q)\,\underline{y}(k) = B(q)\,\underline{u}(k)$	(5)
frequency	(z-)domain:	
$A(z)\,Y(z) = B(z)\,U(z)$	$A(z)\,\underline{Y}(z) = B(z)\,\underline{U}(z)$	(6)

A, B are polynomials	A,B are polynomial matrices
ARMA (auto-regressive	$K(z) = A^{-1}(z)\,B(z)$
moving average)	left matrix description
description	A(z) q x q matrix
	B(z) q x p matrix

Matrices can be written as a series:

$$A(z) = \sum_{i=0}^{g} A_i . z^{g-i} \tag{7}$$

g = 'order' of the auto-regressive part, i.e. the number of the last matrix in the series that does not contain zeros only.

$$B(z) = \sum_{i=0}^{\ell} B_i . z^{\ell-i} \tag{8}$$

ℓ = 'order' of the moving average part

Characteristics:
- representation not unique: $K(z) = \{M(z)\,A(z)\}^{-1}\,M(z)\,B(z)$; there is an infinite number of matrices M(z);
- degrees of determinants can be chosen minimal;
- if there are no common matrices like M(z) which lower the degree: non-reducable ARMA representation;
- not yet unique; still unimodular matrices M(z);
- linear-in-the-parameters; cf. Section 4.2.

3.3 Impulse response models

Notation:

SISO	MIMO	
time	domain:	
$y(k) = \sum_{j=0}^{\infty} h(j)u(k-j)$	$\underline{y}(k) = \sum_{j=0}^{\infty} M(j)\,\underline{u}(k-j)$	(9)
frequency	(z-)domain:	
$H(z) = \sum_{j=0}^{\infty} h(j)\,z^{-j}$	$K(z) = \sum_{j=0}^{\infty} M(j) z^{-j}$	(10)

M(j) Markov parameters
qxp sequences of impulse responses

Characteristics:
- unique representation;
- linear-in-the-parameters
- number of parameters is infinite.

However, if we consider a simple first order SISO process, then we know that one may write:

$$h(k) = a\,h(k-1) \qquad k \geqslant 2 \tag{11}$$

This implies that three parameters are sufficient to give the full characterization, viz. h(0), h(1) and a. In case of a strictly 'proper' process (h(0) = 0) even two parameters are sufficient. This notion can be extended to MIMO processes. In general it can be stated that for a finite dimensional process the following holds:

$$M(s+j) = \sum_{i=1}^{s} a(i)\,M(s+j-i) \qquad j \geqslant 1 \tag{12}$$

(Minimal polynomial). This means that the impulse response of such a finite dimensional process is completely determined by M(0), $\{a(i), M(i)\}_{i=1,s}$; a value of s exists such that the first s Markov parameters are independent and such that all following parameters can be derived from this initial series.

The three representations discussed so far provide a direct relation between input- and output variables. The next model gives this relation through an intermediate step, by way of the state space.

3.4 State space models

Notation:

$$\begin{aligned} \underline{x}(k+1) &= A\,\underline{x}(k) + B\,\underline{u}(k) \\ \underline{y}(k) &= C\,\underline{x}(k) + D\,\underline{u}(k) \end{aligned} \tag{13}$$

Now A,B,C,D are matrices with scalar elements of the sizes A[nxn], B[nxp], C[qxn], D[qxp]. n is the order of the process, i.e the dimension of the state space. Besides information on the input/output behaviour, this representation also provides insight into the structure of the process.
The set of matrices [A,B,C,D], is called a realization. This is not a unique representation. If [A,B,C,D] is the realization of a particular process, then for each non-singular matrix T of the proper dimensions the realization $[T^{-1}AT,\ T^{-1}B,\ CT,$

D] results in the same input/output behaviour. A realization describing the input/output behaviour with a state space of the smallest dimension is called a minimal realization. Such a realization is completely controllable and observable.

The relation between the state space- and the impulse response model can be shown in a simple way by combining both equations of the first representation:

$$\underline{y}(k) = \sum_{j=1}^{k} CA^{j-1}B\,\underline{u}(k-j) + D\,\underline{u}(k) + CA^{k}\underline{x}(0) \tag{14}$$

from which follows:

$$M(k) = \begin{cases} CA^{k-1}B & k \geq 1 \\ D & k = 0 \end{cases} \tag{15}$$

We note that the state space model also takes into account the initial conditions of the process, $\underline{x}(0)$, contrary to the previous models.

Also the transfer matrix can simply be found from the state space description through the relation:

$$K(z) = C(zI-A)^{-1}B + D \tag{16}$$

3.5 Model transformations; comparison of representations

Relations between the models discussed are schematically given in Fig. 3.
The relations indicated with a double arrow are, generally speaking, quite simple. The relations indicated with a single arrow are more complex and will not be explained.
For some time there has been a discussion on the question whether ARMA representations are suitable for MIMO processes. Recently in literature they are receiving more attention. Here the non-uniqueness plays a role.
The complex theory of polynomial matrices was an obstacle for the use of such models.
Considering the simpler relations, the state space and the impulse response model play a central role. A disadvantage of the impulse response model is the infinite number of parameters (this can be reduced, but then the linearity-in-the-parameters is lost). That (large) number of parameters is in contrast to the state space model.

Non-uniqueness of a model manifests itself through too large a number of parameters. The reverse need not be true; a model with a large number of parameters can be a unique representation (e.g. an impulse response model), in which then dependence exists between the various parameters.
In recent years the non-uniqueness of the state space model has been studied intensively, and such studies have led to the use of unique (canonical) forms. In all these forms special structures are imposed on the realizations [A,B,C,D] in such a way that, within a structure, each system has a unique representation.

We can compare the number of parameters of the various models (fully parametrized):

Table 1 Number of parameters in various models

Representation:	SISO model	example x) SISO	MIMO model	example x) MIMO
Transfer function/-matrix model	2n+1	7	pq(2s+1)	42
Difference equation/-matrix - (ARMA)	2n+1	7	q^2g+pq(ℓ+1)	36
Impulse response/Markov -	∞ *)	∞	∞ *)	∞
- minimal polynomial -	2n+1	7	(pq+1)(s+1)-1	27
State space -	n(n+2)+1	16	n(n+p+q)+pq	30
Canonical state space -	2n+1	7	n(p+q)+pq	21

*) In principle the number of parameters of the impulse response is infinite. Using the dependence as given by eq. (12) yet a finite number of parameters can be given.
X) Example: p = 3 (inputs); q = 2 (outputs); n = 2 = ℓ = g = 3.

In the SISO situation the number of parameters is slightly dependent on the model chosen. In the MIMO situation those differences are much bigger, as can be seen from the Table.

In the following Table three important properties of the models are compared: the number of parameters, the linearity-in-the-parameters of the output, and the uniqueness of the representation. The unbiasedness will be discussed in Section 4.1.

Representation:	Number of parameters	Output linear-in-the-parameters	Uniqueness	Unbiased if additive noise
Transfer function/-matrix model	-	-	+	+
Difference equation/-matrix - (ARMA)	-	+/-	-	-
Impulse response/Markov -	-	+	+	+
- minimal polynomial -	+	-	+	+
State space -	-	-	-	+
Canonical state space -	+	-	+	+

Table 2 Properties of the various models

As can be seen, there is no model that can be denoted as being the 'best'. The impulse response model has two attractive properties but has, for estimation purposes, an infinite number of parameters which, of course, is not practical. For identification purposes frequently ARMA models are being used; a transformation to a state model then provides a compact representation.

3.6 Noise models

So far only deterministic models have been discussed. In each of those models stochastic influences can be introduced. Coloured noise can always be assumed to have been formed from white noise by filtering. Such noise filters can be modelled by way of the models discussed in the previous paragraphs for process dynamics; see also Jakeman and Young (1981).

3.7 Nonlinear models

Unfortunately the huge class of nonlinear models still defies a coherent discussion. Progress is being made on particular approaches, e.g.:

- model with a simple nonlinearity and dynamics, e.g. Hammerstein model (see e.g. Stoica and Söderström, 1982), Wiener model;
- Volterra models;
- models with essential nonlinearities according to catastrophe theory; see e.g. Arnold, 1984;

- models built according to Ivakhnenko's Group Method of Data Handling (GMDH); see Farlow (1984).

4 ESTIMATION METHODS

4.1 Least squares (LS)

In the definition of identification a crucial aspect is the notion of equivalence between process and model. As mentioned before, depending on the type of application for which the model is needed, the emphasis may be on: - the values of physical parameters; - process/model output correspondence; - process output prediction.
Also the error function to be minimized can be chosen in various ways, e.g. squared error, absolute error, etc. The first one is the most popular due to its simplicity:

$$V = \sum_{k=1}^{N} \underline{e}^T(k) \underline{e}(k)$$

Denoting $\underline{u}(k)$ as the input vector, $\underline{y}(k)$ as the process output vector and $\underline{\hat{y}}(k)$ as the model output vector, the error vector can be chosen according to the various models in Table 3:

Representation	Error vector
Transfer matrix model	$\underline{e}(k) = \underline{y}(k) - K(q) \underline{u}(k)$ (17) with: $K(q) = A^{-1}(q) B(q)$
Difference equation model (ARMA)	$\underline{e}(k) = -A(q) \underline{y}(k) + B(q) \underline{u}(k)$ (18)
Impulse response model (Markov Parameters)	$\underline{e}(k) = \underline{y}(k) - \{M_0 q^{-1} + M_1 q^{-2} + ..\}\underline{u}(k)$ (19)
State space model	$\underline{e}(k) = \underline{y}(k) - \underline{\hat{y}}(k)$ (20) with: $\underline{\hat{y}}(k) = C\underline{x}(k) + D\underline{u}(k)$ $\underline{x}(k+1) = A\underline{x}(k) + B\underline{u}(k)$ or: $\underline{\hat{y}}(k) = Cq^{-1}\{A \underline{x}(k)+B\underline{u}(k)\}+D\underline{u}(k)$

From these expressions it is clear which cases can be considered to be linear-in-the-parameters, i.e. can be written as:

$$\underline{e}(k) = \underline{y}(k) - \Omega\underline{\theta} \qquad (21)$$

with Ω = matrix of observations and $\underline{\theta}$ = vector of parameters. For such linearity the explicit analytical expression for the estimator $\underline{\hat{\theta}}$ follows from the requirement that the first derivative with respect to $\underline{\theta}$ for the estimated values $\underline{\hat{\theta}}$ has to be zero, i.e.

$$\left.\frac{\partial V}{\partial \underline{\theta}}\right|_{\underline{\theta}=\underline{\hat{\theta}}} = -2 \sum_k \Omega^T \underline{e}(k) \bigg|_{\underline{\theta}=\underline{\hat{\theta}}}$$

$$= -2 \sum_k \Omega^T\{\underline{y}(k) - \Omega\underline{\hat{\theta}}\} = \underline{0} \qquad (22)$$

or

$$\underline{\hat{\theta}} = \left[\sum_k \Omega^T \Omega\right]^{-1} \left[\sum_k \Omega^T \underline{y}(k)\right] \qquad (23)$$

provided that the inverse matrix exists. Just in passing we note that this estimator can be easily brought into a recursive form, in which the new estimate is derived from the old one, updated/corrected on the basis of new observations.

For non-linearity-in-the-parameters the error has to be written as:

$$\underline{e}(k) = \underline{y}(k) - \underline{\hat{y}}(k,\Omega,\underline{\theta}) \qquad (24)$$

The situation is more complex now as $F = \partial \underline{e}/\partial \underline{\theta}$ is not simply Ω but has to be determined at each (intermediate) value of $\underline{\theta}$, and a hill-climbing (or rather valley-descending) algorithm has to be used; cf. fig. 4.

So far the influence of noise has not been discussed. If $\underline{y}$ contains additive noise which is independent of $\underline{u}$, then its input on the bias of the estimates also has to be considered.

Table 2 also indicates the (un)biasedness of the estimator in cases of additive noise.

Note that some authors also implement the transfer matrix and the state space model in terms of equation error, in order to attain linearity-in-the-parameters, and consequently come to different results with respect to biasedness. Again no simple 'best' choice can be given.

4.2 Pseudo-linear regression (PLR)

Due to the (mostly) small number of parameters, the ARMA models have become quite popular for the SISO-case. Their disadvantage with respect to biased estimators is being counteracted by either one, or a combination, of some of the following three methods (Van den Boom, 1982; Van den Boom and Eykhoff, 1984):
a. signal prefiltering
b. model extension
c. instrumental variable

In short, these methods are based on the following ideas.
- ARMA model for process dynamics

$$\left[1+A(z^{-1})\right]y(k) = \left[b_0+B(z^{-1})\right]u(k) + e(k) \qquad (25)$$

- ARMA model for additive noise

$$\left[1+D(z^{-1})\right]e(k) = \left[1+C(z^{-1})\right]\xi(k) \qquad (26)$$

$\xi(k)$ white

N.B. - note the change of interpretation of A,B compared to eq. (6).

ad a. Signal prefiltering: a combination of eq. (25) and (26) can be written as:

$$[1+A]\underbrace{y(k)}_{\tilde{y}(k)} = [b_0+B]\underbrace{u(k)}_{\tilde{u}(k)} + \underbrace{\frac{[1+C]}{[1+D]}\xi(k)}_{1 \; \xi(k) \; \text{white}} \times \frac{[1+D]}{[1+C]}$$

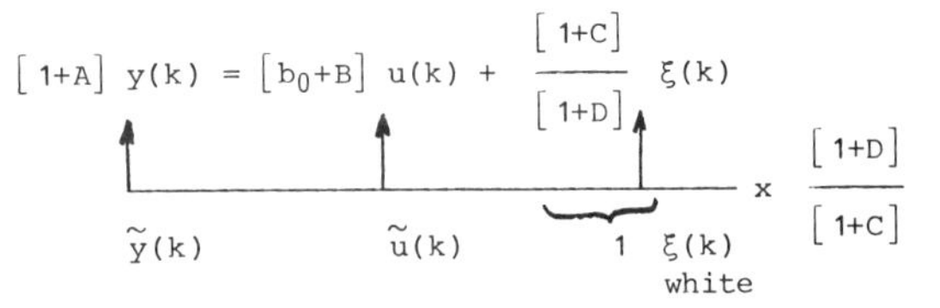

<u>signal prefilter</u>
(27)

By operating on the signals u and y with the noise-whitening filter as indicated, the residuals become white and the estimator unbiased. If the noise characteristics are not know, then in a bootstrap fashion the process parameter- and noise parameter estimates are improved alternately.

ad b. Model extension: a combination of eq. (25) and (26) can be written as:

$$y(k) = -A y(k) + [b_0+B]u(k) - \underbrace{[D]e(k) + [C]\xi(k)}_{\text{model extension}} + \xi(k) \qquad (28)$$

This can be noted as:

$$\underline{y} = \Omega(y,u,e,\xi)\underline{\theta} + \underline{\xi} \qquad (29)$$

Again the residual has become white and the estimator will be unbiased. As the variables e and ξ are not measurable, they have to be estimated, using estimates of the parameters. In a bootstrap fashion parameter- and e, ξ-estimates are improved alternately.

ad c. Instrumental variable: a combination of eq. (25) and (26) can be written as:

$$y(k) = -[A]y(k) + [b_0+B]u(k) + \frac{[1+C]}{[1+D]}\xi(k) \qquad (30)$$

x z(k)
instrumental variable

If z(k) fulfills the following conditions

$E[z(k)\ u(k)] \neq 0$ and $E[z(k)\ \xi(k)] = 0$

then the resulting parameter estimator

$$\hat{\underline{\theta}} = [Z^T\Omega]^{-1}[Z^T\underline{y}(k)] \qquad (31)$$

is unbiased.

In principle these techniques can be used for the MIMO-case as well. The feasibility and the quality of the estimates are strongly related to:

- the a priori knowledge available, e.g. the knowledge of the order of the process- and the noise dynamics;
- the complexity of the process dynamics;
- the complexity of the noise dynamics; are the noise contributions at the various outputs statistically related or not?
- the 'quality' of the input signals; are they 'persistently exciting' and sufficiently 'independent' of each other?

The problems that arise in the SISO-case in estimating the unknown order(structure) appear even more pronounced in the MIMO-case.

4.3 Maximum Likelihood (ML)

It is assumed that the principle of maximum likelihood estimation is well known: it is based on the probability density function of the observations, say Ω, given the parameter vector $\underline{\theta}$:

$$p(\Omega;\underline{\theta}) \qquad (32)$$

After measuring the observations, say Ω_t, then by substituting those actual samples in eq. (32) and by considering $\underline{\theta}$ as a variable, this expression is called a likelihood function:

$$L(\Omega_t;\underline{\theta}) \qquad (33)$$

Now this functional relation can be used to find those values $\underline{\theta} = \hat{\underline{\theta}}$ for which the likelihood of the occurrence of that particular Ω_t is a maximum.

Within this estimation scheme the model can be chosen in many different ways, and for each model a likelihood function can be found, provided the probability function of the additive stochastic (noise) influence is known. Mostly it is not known but assumed to be Gaussian.

In many cases the expression for the (log)likelihood function is such that the maximization cannot be done through an explicit algebraic expression and has to be done by way of a hill-climbing optimization algorithm. This implies that such estimation procedures can suffer from a number of associated problems such as: -the possible existence of local maxima; -computational complexity; -long computation time for the estimation of sizable parameter vectors; -numerical instability; etc. Also in this area much research is still going on and more experience is needed.

4.4 The element of choice

The problem of choosing an estimation method can be illustrated as follows for the IV and the ML method. Summarizing the main characteristics of these methods the following favourable (+) and comparatively unfavourable (-) properties can be mentioned.

IV:

+ rather simple estimation principle
+ rather simple implementation
+ no assumptions needed on the character of the disturbances
- optimality criterion is not clear

ML:

+ excellent statistical properties, certainly in asymptotical sense
+ well-defined optimality criterion
- complex and time-consuming minimization procedure
- very specific assumptions have to be made on the disturbances (mostly: white gaussian noise).

The weighing of these properties has to be done, keeping in mind the (almost) certainty that, for practical applications, the process will not be in the chosen model set.

5 IDENTIFICATION TOOLS

From the foregoing discussions a few provisional conclusions can be derived:

- In practical situations the identification is still a mixture of 'art' and 'science'. Consequently, it would be very useful to have means available that permit simple simulation and verification of assumptions.
- Experimental data have to be inspected and corrected (scaling, outliers, DC offset, trends) before the estimation step is done. Consequently, means for performing such inspection are needed.
- Many parameter estimation techniques have been presented and advocated. Consequently, a simple way of comparing the performance of these various techniques in particular situations would be very useful.
- The same holds for order estimation techniques.
- For many estimation schemes of practical interest, only the asymptotic properties can be derived theoretically. In practice only limited and even short data sequences are available. Consequently, simple means for studying the properties of finite sequences are of great importance.

These conclusions clearly point to the need for computer program packages that are user-friendly and simple to apply on simulated and experimental data. By interactive techniques, where the computer asks questions and the user answers them, the means (tools) provided really can be simple to exploit.

The presentation of (intermediate) results in a graphical display in an appropriate form provide the experimenter with a valuable aid for, e.g. deciding on the structure of the model.

A number of such program packages have been reported on. Examples are: IDPAC, developed at Lund University (see Åström, 1981); SATER, developed at the Eindhoven University of Technology (see van den Boom, 1982; van den Boom and Bollen, 1984); and CAPTAIN (see Young and Johnson, 1979). See also the survey of program packages for computer-aided design of control systems (IEEE, 1982).

For industrial applications of identification techniques robust and well-tested software is needed. This implies that the software is bound to be numerically reliable in the sense that, if no warnings are issued, the result satisfies strict reliability criteria. Accuracy, speed, flexibility and memory requirements have to be balanced.

New algorithmic developments present themselves, e.g. the Singular Value Decomposition, numerical lattice procedures for LS identification (see Graupe, 1984), etc. Computational efficiency and numerical stability can be enhanced by such judicious choices.

Another type of development is the software-adaptation to microprocessors/computers with limited word length for implementing dedicated identification systems. Again robustness and computational efficiency are of paramount importance, in spite of the complexity of multi-variable identification tasks that have to be tackled; see e.g. El-Sherief and Maud (1982).

6 IDENTIFICATION APPLICATIONS

With respect to the types of applications we wish to refer again to the categories mentioned in section 1:

diagnostics
monitoring, fault/failure detection
prediction: economy, weather, floods
control
- system design
- adaptive/self-tuning (Åström et al., 1977)

The nature and size of this paper does not permit an extensive elaboration on the multitude of (types of) applications. Yet it may be functional to mention a small selection:

- aerospace engineering
  - aircraft control (Hartmann and Brebs, 1980)
- biomedical engineering
  - overview (Eykhoff 1985)
  - respiratory and cardiovascular systems (Linkens, 1985)
  - neuromuscular systems (Bekey, 1985)
  - sensory and neuronal systems (O'Leary, 1985)
  - metabolic and endocrine systems (Cobelli 1985)
- chemical- and physical process engineering
  - heat exchanger (Bauer and Unbehauen, 1978)
  - steam superheater (De la Puente and Albertos, 1979)
- communication engineering
  - automatic channel equalization (Goodwin et al., 1980)
  - adaptive techniques in signal processing (Claasen and Mecklenbräuker, 1983)
- environmental control
  - stream flow and water quality (Whitehead et al., 1979)
- power engineering
  - turbogenerator (Sharaf and Hogg, 1981)
  - load forecasting (Mahalanabis et al., 1980; El-Sherief and Maud, 1982)
  - solar plant identification (Fortuna et al., 1980)
- speech processing (Morf et al., 1977)
- sysmic exploration
  - signal detection and parameter estimation (Ursin, 1979)
- transportation engineering
  - ship control (Van Amerongen, 1981; Åström, 1980)

This list can easily be extended with additional topics (e.g. human operator predictor; population control; economics) and with many more references. The size of the paper does not permit ....

7 EXPECTATIONS

A simple and straightforward way of expressing expectations is to note that the present trends are bound to continue:

- on theory
  - use of more sophisticated mathematical tools to describe the characteristics of models and to describe order- and parameter estimation methods (various abstract spaces; martingales);
  - a better understanding of convergence properties of estimation methods;
  - more emphasis on robustness of identification (Polyak and Tsypkin, 1980);
  - a closer interaction with information theory (e.g. Akaike, 1981).

- on tools
  - development of robust software with a flexible, modular structure, that will be well-tested and will provide 'signals' to the operator/experimenter in cases of transgressing preset reliability limits;
  - adaptation of algorithms and software to the limitations (word length; memory restrictions) of microcomputers.

- on applications
  - as is often the case in engineering, the really interesting industrial applications of identification are considered company-confidential, and are not or only partially published. Yet a sufficient number of applications has been and probably will be published to challenge and stimulate those who face the need for a particular application.

In contrast to such trends, fundamental new ideas are far more difficult, if not impossible, to predict. As an example: a new fundament for estimation has been published but has not yet been evaluated sufficiently by various colleagues (Kovanic, 1984).

8 CONCLUSIONS

Much work is devoted to extending the identification/estimation procedures from SISO- to MIMO-processes. In doing so, the problems grow considerably.
For model selection this is particularly due to the number of parameters needed for an adequate model, i.e. a model structure that leaves open all feasible (but probably non-existent) interconnections, interactions and noise influences. Consequently, this problem implies the desire to include as much a priori knowledge as possible in the selection of the model structure. This underlines the statement that, for the time being, engineering judgement is bound to remain an important ingredient in the identification procedure, i.e. an 'art' aspect besides (or: instead of) an approach purely based on 'science'.
Model structure estimation could not be discussed in this paper.

From the parameter estimation point of view, it is desirable to choose the model as being linear-in-the-parameters, in order to attain simple (recursive) estimation algorithms. Unfortunately, this desire is not compatible with the requirement formulated before with respect to the restriction of the number of parameters. For example, a Markov parameter model has that linearity property, but it has a large (infinite) number of parameters. Again human judgement plays an important role.

An appreciable amount of insight, theory, experience and know-how is already available. Yet the integration and extension of this insight, etc. towards a 'standard procedure' for MIMO-identification will still require a substantial effort by numerous researchers.

Acknowledgement
The author gratefully acknowledges the use of contributions by colleagues from the professional Group 'Measurement and Control' of the Eindhoven University of Technology.

REFERENCES

Akaike, H. (1981). Modern development of statistical methods. In: P. Eykhoff (edit.), Trends and Progress in System Identification. Pergamon, Oxford, p. 169-184.

Amerongen, J. van (1981). A model reference adaptive autopilot for ships - practical results. Proc. 8th IFAC World Congress, Kyoto. Pergamon, Oxford, p. 1007-1014.

Arnold, V.I. (1984). Catastrophe Theory. Springer Berlin. 79 pp.

Åström, K.J., U. Borisson, L. Ljung and B. Wittenmark (1977). Theory and applications of self-tuning regulators. Automatica, 13, 457-476.

Åström, K.J. (1980). Why use adaptive techniques for steering large tankers? Int. Journal of Control, 32, 689-708.

Åström, K.J. (1981). Maximum likelihood and prediction error methods. In: P. Eykhoff (edit.), Trends and Progress in System Identification. Pergamon, Oxford. p. 145-168.

Bauer, B. and H. Unbehauen (1978). On-line identification of a load-dependent heat exchanger in closed loop using a modified instrumental variable method. Proc. IFAC 7th World Congress, Helsinki. Pergamon Press, Oxford. p. 351-359.

Bekey, G.A. (1985). Identification of neuro-muscular systems. Preprints 7th IFAC/IFORS Symposium on Identification and System Param. Estim., York, July 1985, p.69-76.

Boom, A.J.W. van den (1982). System Identification; on the variety and coherence in parameter- and order estimation methods. Doctoral Dissertation, Eindhoven University of Technology, the Netherlands. 239 pp.

Boom, A.J.W. van den and P. Eykhoff (1984). A generalized diagram for various parameter estimation methods; the coherence in pseudo-linear regression schemes. Proc. 9th IFAC Congress, Budapest, July 1984, vol. 2, p. 615-620.

Boom, A.J.W. van den and R. Bollen (1984). The identification package SATER. Proc. 9th IFAC World Congress, Budapest, July 1984, vol. 1, p. 483-488.

Claasen, T.A.C.M. and W.F.G. Mecklenbräuker (1983). Overview of adaptive techniques in signal processing. In: H.W. Schüssler (edit.), Signal Processing II: theory and Applications, Elsevier, Amsterdam. p. 747-754.

Cobelli, C. (1985). Identification of metabolic and endocrine systems. Preprints 7th IFAC/IFORS Symposium on Identification and System Param. Estim., York, July 1985, 45-54.

El-Sherief, H. and M.A. Maud (1982). Dedicated microcomputers for real-time identification of multivariable systems. Proc. 6th IFAC Symp. on Identification and System Param. Estim., Washington, D.C., p. 1617-1622.

Eykhoff, P. (1974). System Identification, Parameter and State Estimation. Wiley, London, 555 pp. (also in Russian, Chinese, Polish, Rumanian).

Eykhoff, P. Editor (1981). Trends and Progress in System Identification. Pergamon Press, Oxford. 402 pp.

Eykhoff, P. (1984). Identification theory; practical implications and limitations. Proc. IMEKO Symposium on Measurement and Estimation, Bressanone (Brixen), It., May 1984. Also in: Measurement, 2, nr. 2, 75-85.

Eykhoff, P. (1985). Biomedical identification: overview, problems and prospects. Preprints 7th IFAC/IFORS Symposium on Identification and System Param. Estim., York, July 1985, 37-44.

Farlow, S.J. (1984). Self-Organizing Methods in Modeling. Dekker, New York. 368 pp.

Fortuna, L., A. Gallo, G. Cammarata and M. La Cava (1980). Micro-computer systems allow a real-time parameter estimation for the control of a solar plant. Proc. 3rd IFAC Symp. on System Approach for Development, Marocco. Pergamon, Oxford, p. 309-316.

Goodwin, G.C., H.B. Doan and A. Cantoni (1980). Application of ARMA models to automatic channel equalization. Information Sciences, 22, 107-129.

Goodwin, G.C. (1982). An overview of the system identification problem experiment design. Proc. 6th IFAC Symp. Ident. Syst. Parameter Estim. Washington. Pergamon Press, Oxford, p. 65-71.

Goodwin, G.C. and R.L. Payne (1977). Dynamic System Identification. Experiment Design and Data Analysis. Academic Press.

Graupe, D. (1984). Time Series Analysis, Identification and Adaptive Filtering. Krieger Publ. Comp., Malabar, Florida, 386 pp.

Gustavsson, I., L. Ljung and T. Söderström (1972). Identification of processes in closed loop - identifiability and accuracy aspects. Automatica, 13, 59-75.

Gustavsson, I., L. Ljung and T. Söderström (1981). Choice and effect of different feedback configurations. In: P. Eykhoff (edit)., Trends and Progress in System Identification. Pergamon, Oxford. p. 367-388.

Hartmann, U. and V. Krebs (1980). Command and stability system for aircraft; a new digital adaptive approach. Automatica, 16, 135-146.

IEEE (1982). Special issue on computer-aided design of control systems. Control Systems Magazine, 2, nr. 4, (Dec. 1982), p. 2-52.

Ivakhnenko, A.G. (1970). Heuristic self-organization in problems of engineering cybernetics. Automatica, 6, 206-219.

Jakeman, A.J. and P.C. Young (1981). On the decoupling of system and noise model parameter estimation in time-series analysis. Int. Journal of Control, 34, 423-431.

Kovanic, P. (1984). A new theoretical and algorithmical basis for estimation, identification and information. Proc. 9th IFAC Congress, Budapest, July 1984, vol. 3, p.1667-1676.

Królikowski, A. and P. Eykhoff (1985). Input signal design for system identification; a comparative analysis. Preprints 7th IFAC Symp. Ident. Syst. Parameter Estim., York, p. 915-920.

Linkens, D.A. (1985). Identification of respiratory and cardiovascular systems. Preprints 7th IFAC/IFORS Symp. on Identification and System Param. Estim., York, July 1985, p. 55-67.

Mahalanabis, A.K., M. Hanmandla and S. Rane (1980). Load modeling of an interconnected power system for short term prediction. Int. J. Systems Sci., 11, 445-453.

Mehra, R.K. (1974). Optimal inputs for linear system identification. IEEE Trans. Automat. Contr., AC-19, 192-200.

Mehra, R.K. (1981). Choice of input signals. In: P. Eykhoff (edit.), Trends and Progress in System Identification. Pergamon, Oxford. p. 305-366.

Morf, M., A. Vierira and D.T. Lee (1977). Ladder forms for identification and speech processing. Proc. 16th IEEE Conf. on Decision and Control, New Orleans.

O'Leary, D.P.O. (1985). Identification of sensory systems and neuronal systems. Preprints 7th IFAC/IFORS Symp. on Identification and System Param. Estim., York, July 1985, p. 77-84.

Peterka, V. (1981). Bayesian approach to system identification. In: P. Eykhoff (edit.), Trends and Progress in System Identification. Pergamon, Oxford. p. 239-304.

Polyak, B.T. and Ya. Z. Tsypkin (1980). Robust identification. Automatica, 16, 53-63.

Ponomarenko, M.F. (1981). Information measures and their application to identification; a bibliography. TH-Report, Eindhoven University of Technology, Nov. 1981.

Puente, J.A. de la and P. Albertos (1979). Closed loop identification of a steam superheater. Proc. 5th IFAC Symp. on Identification and System Param. Estim., Darmstadt. Pergamon, Oxford, p. 961-968.

Richalet, J. (1981). The model method. In: P. Eykhoff (edit.), Trends and Progress in System Identification. Pergamon, Oxford. p. 5-28.

Sharaf, M.M. and B.W. Hogg (1981). Evaluation of on-line identification methods for optimal control of a laboratory model turbogenerator. Proc. IEE, 128, 65-73.

Söderström, T. and P. Stoica (1981). Comparison of some instrumental variable methods - consistency and accuracy aspects. Automatica, 17, 101-115.

Söderström, T. and P. Stoica (1983). Instrumental Variable Methods for System Identification. Springer Verlag.

Stoica, P. and T. Söderström (1982). Instrumental variable methods for identification of Hammerstein systems. Int. Journal of Control, 35, 459-476.

Ursin, B. (1979). Seismic signal detection and parameter estimation. Geophysical Prospecting, 27, 1-15.

Van Amerongen, cf. Amerongen, J. van

Van den Boom, cf. Boom, A.J.W. van den

Vaněček, A. (1981). Models: equivalences, uses, extensions. In: P. Eykhoff (edit.), Trends and Progress in System Identification. Pergamon, Oxford. p. 29-66.

Whitehead, P., P.C. Young and G. Hornberger (1979). A systems model of stream flow and water quality in the Bedford-Ouse river-1. Stream Flow modelling. Water Research, 13, 1155-1169.

Young, P.C. and A.J. Jakeman (1979). The development of CAPTAIN: a computer aided program for time-series analysis and identification of noisy systems. Proc. IFAC Symp. on Computer Aided Design of Control Systems, Zürich. Pergamon, Oxford, p. 391-400.

Zarrop, M.B. (1979). Optimal Experiment Design for Dynamic System Identification. Springer Verlag.

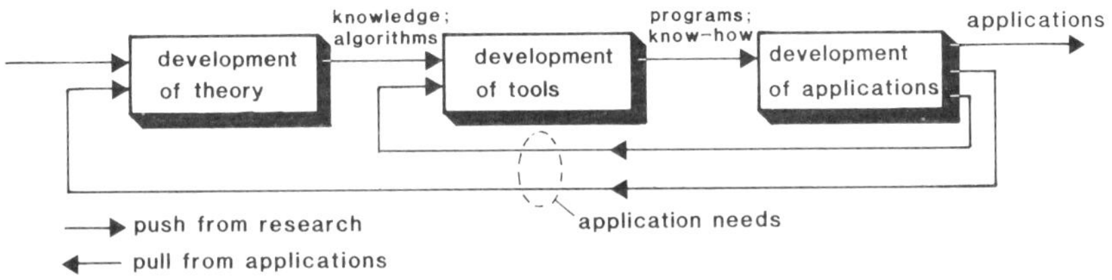

Fig. 1 Interaction of identification aspects

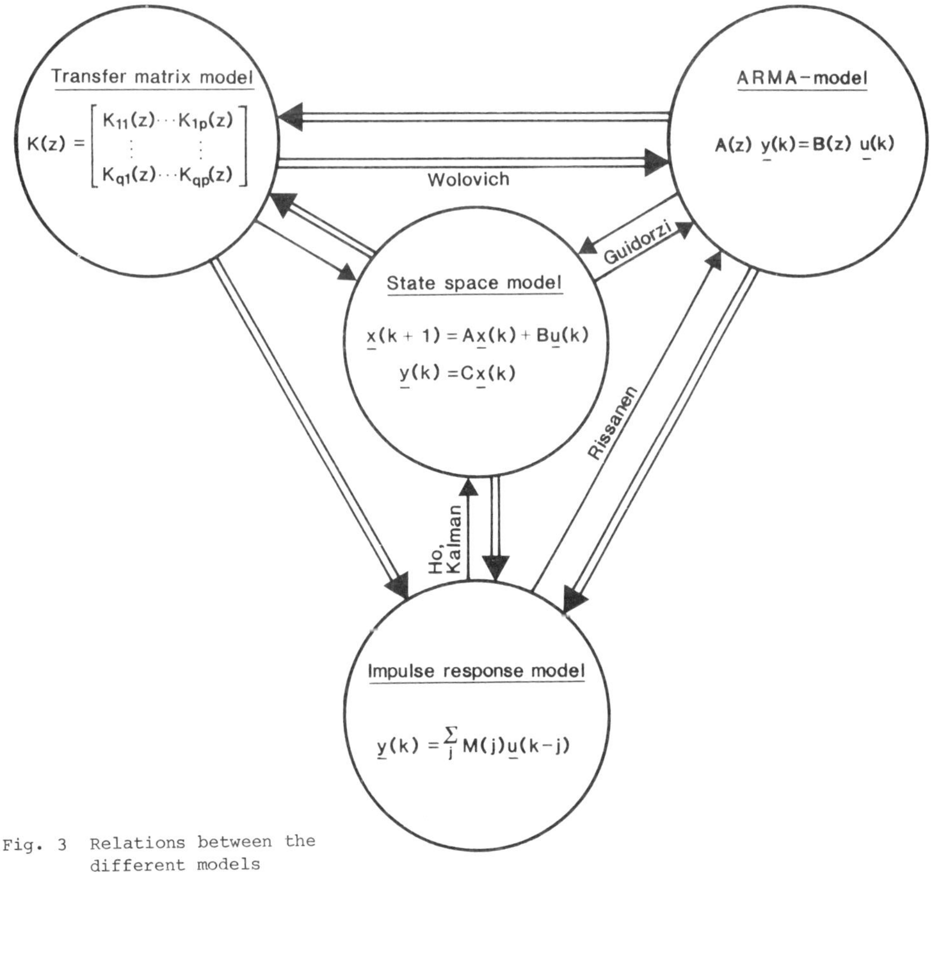

Fig. 3 Relations between the different models

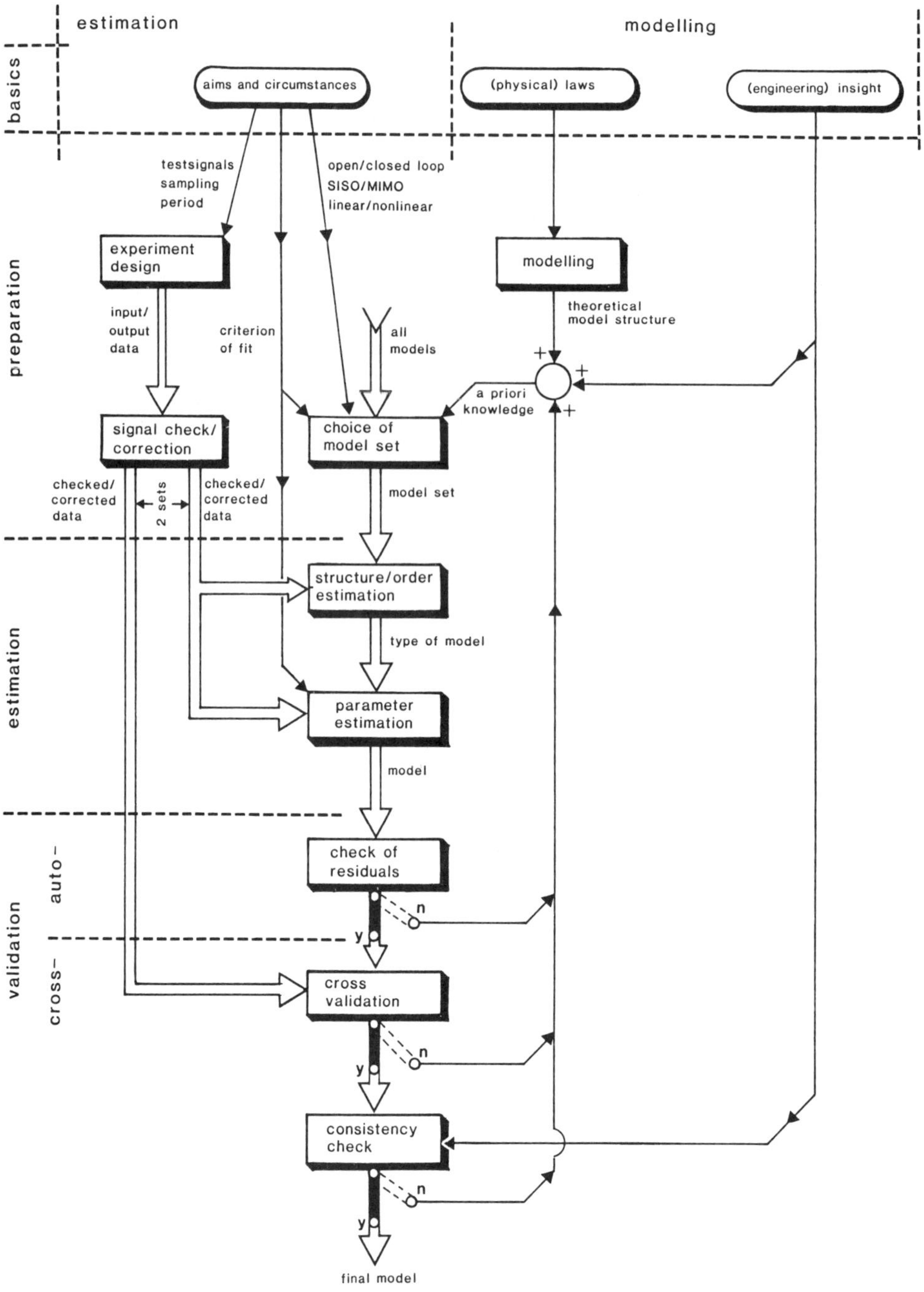

Fig. 2 The identification protocol

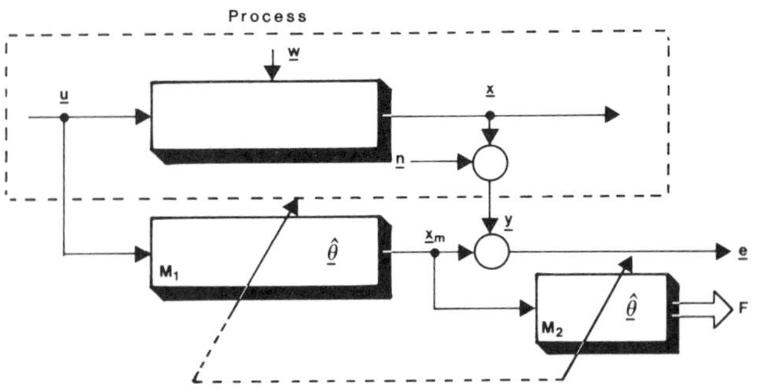

Fig. 4 Gradient determination

METHODOLOGIES FOR DESIGNING EXPERT SYSTEM FOR INDUSTRIAL MANAGEMENT

L. Pun

Laboratoire GRAI de l'Université de Bordeaux 1, Talence, France

Abstract. Industrial management can increase its efficiency by using the assistance of expert systems. The design of such system needs methodology. In this paper, we discuss : (a) Various levels of industrial management and the associated problems, (b) The constitution of an expert system, and (c) Some design methodology.

INTRODUCTION

The management of industrial production systems is a difficult task. Expert systems (ES), if they can be realized properly, are certainly useful as a management aid. The realizability of the ES depends upon the management levels. At each level, we have different problems to solve. We follow H.I. Ansoff (1), and distinguish three levels : (a) Strategical level ; (b) Organizational level, and (c) Executing level. To these three levels, corresponds roughly the industrial common terminology : long term, medium terme and short term.

In this paper, we shall discuss the following points : (a) The various levels of management and the associated problems, (b) The constitution of an expert system, where we shall notably clarify the understanding of various levels of knowledges, and various knowledge modelling tools, and (c) A methodology for designing an expert system.

2 - PRODUCTION MANAGEMENT AND THE ASSOCIATED EXPERT SYSTEMS

2.A - Production management problem.

The strategical level involves mainly three points : (a) product, (b) investment, and (c) production structure. These points involve a great number of correlated problems : prospective marketing, product choice, plant national or multinational implementation, firm merging, basic materials, staff, facilities, manufacturing technology... The implied universes are : economical, technological, financial, social, political and human. The problems are complicate. The problem elements are difficult to extract and to formalize. The expertise for solving these problems certainly exists, since these are many successful enterprises.

However, the successful and expert managers (receiving generally very high salaries) are certainly reluctant to yield their expertise. Besides, this type of expertise can be rarely generalized.

The executing level involves the executing activities of all the production functions. As "automation" and "computer" are concerned, there are two branches of improvements : (a) Office automation : document filing, text processing, language styling... (b) Technological automation : CAD, CAM, ROBOTS, NC, FMS, etc... In principle, executing activities are to be done under well-defined conditions (finality, frame, operating procedure). Therefor, there are very few managing tasks at this level, except some supervizing and local adjustings. In practice however, very often the human executor is also a decision-maker. At the organizational point of view, this is a mistake. When ever such a situation occurs, we have to clarify it out. In the technological branch, the problems are relatively easier to formulate, the problem elements, generally physical, can be extracted from the real world, and the expertise is a natural extension of the know-how of the various engineerings.

The organizatial level is also called tactical level. The managerial role of this level can then be clarified. From the strategical level, this level receives the general objective and framework (strategy). The managers have to find appropriate tactics, in order to organize sub-objectives, sub-frameworks for the various production functions. They have then accordingly to define the programs and to supervize their executing. In this paper, we are concerned by the ES of this level of management problem. To establish an efficient methodology for designing the ES, we have to understand clearly what are : the production management, the ES, and the connexions between them.

2.B - General configuration of expert-system assisted production management

The management system (MS) has to pilot the operations of the executing physical system (PS). The expert system (ES) is assumed to help the MS for solving its problems. The analyst and the designer (AD) are asked to design the ES. We must understand clearly the respective roles and problems of the MS, ES and AD.

The relations between various activity levels (strategical S, organizational or tactical O-T, executing E) are shown on Fig. 1. We have :

Oo : orders (objective, frame) from S to O-T.
IO : informations (reports, results) from O-T to S.
F1 .. Fn : various executing functions : (manufacturing, procurement, preparation, control, maintenance, staff, accounting, sales.. .).
O1 .. On : orders from O-T to E.
I1 .. In : informations from E to O-E.

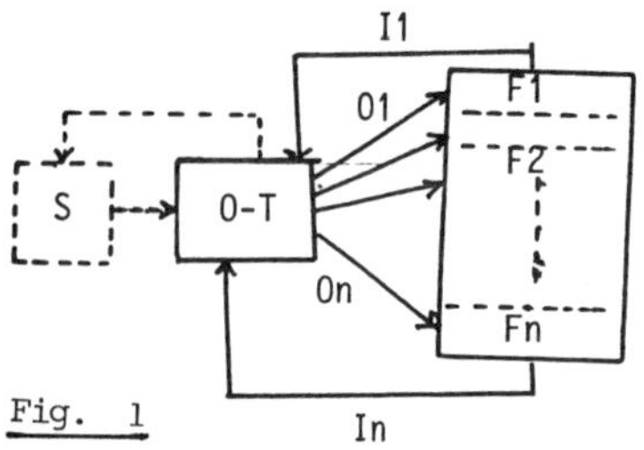

Fig. 1

We are only interested by O-T and E. We summarize thus Fig. 1 into Fig. 2a where we have :

Oa = (O1, O2, ... On)
Ia = (I1, I2, ... In).

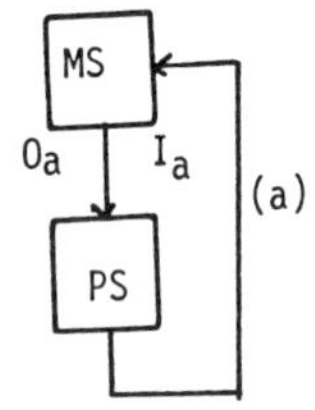

Fig. 2a : MS- PS

The O-T level MS has the roles and the problems of :

- planning (define the executing programs)
- supervizing
- adjusting (in case there are random disturbances).

For this, orders O1 to On are send to the executing managers F1 to Fn. Informations I1 to In are back to O-T. The operating criterion for the O-T level M-S is : appropriate multi-synchronization (of the activities) and multi-coordination (of the ressources) so that the general objective Oo is achieved. Clearly, there are many problems to solve to satisfy this criterion. An ES is supposed to help the MS with various degrees of effectiveness. We have thus :

- Ib : informations from PS to ES
- ab : aids from ES to MS

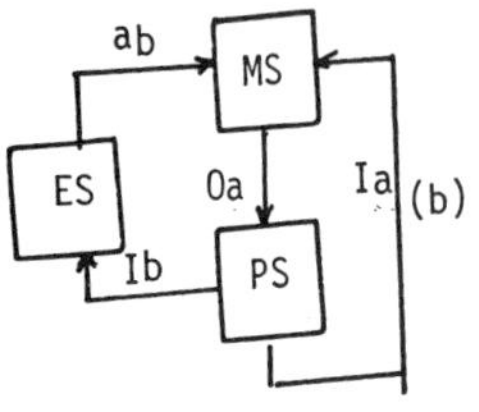

Fig. 2b : ES - MS - PS

The design of the ES requires many previous informations :

- Ic : the structure of PS and the executing problems
- Id : the structure of MS and the management problems, and eventually, some existing MS strategies
- Ie : strategies from some experts
- the structure of an ES

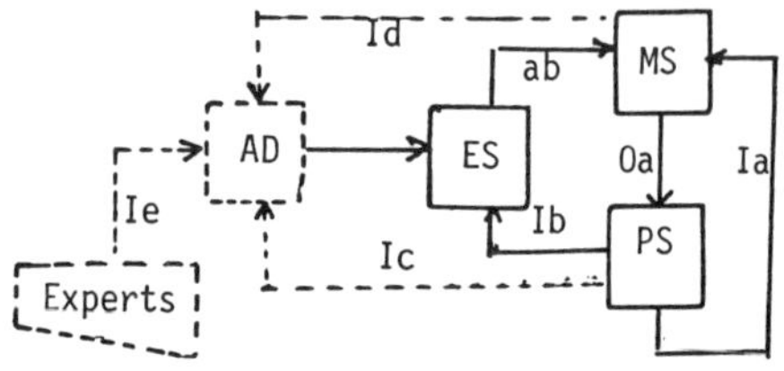

Fig. 2c : AD - ES - MS - PS

Then we can undertake the design activity Oc. In the field of automatic control, we have a known methodology. An ES assisted MS is a convivial control (2). The same methodology may be applied. However, we have to emphisize :

- the difficulties for understanding the complicated PS-MS structures (formulation problems),
- the difficulties for extracting the strategies (identification problems),
- the choice of a suitable programming language, easy to program and easy to interface (implementation problems).

3 - CONSTITUTION OF AN EXPERT SYSTEM

The design of an ES requires two clear understanding on two ES - constituting elements : contents and forms.

CONTENTS : informations or expressed knowledges upon the problem-elements, problem-solving methods, and utilization-prescriptions.

FORMS : structures of these informations, imposed by : (1) the mechanism of the computers, and (2) the adopted programming languages.
The two major design tasks are therefore :

- Modelling (expressing, formalizing) task, transforming the knowledges into informations, requiring thus suitable modelling tools.
- Programming (organizing) task, transforming the informations into ES structures, requiring thus suitable programming language.

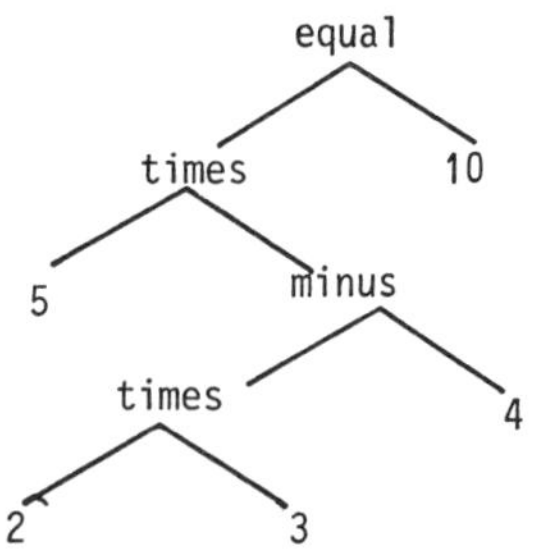

Fig. 3 : An example of semantic network
5 x (2 x 3 - 4) = 10

In the present-day litterature on AI and ES - (3, 4, 5, 6, 7, 8) - there are two important confusions.

- CONCEPTUAL CONFUSION (semantic problem), notably regarding the uses of the concepts : knowledge, information, data, rules, relations, metarules, etc...
- OPERATIVE CONFUSION (syntaxical problem) introduced by the interrelations between the languages (LISP, PROLOG, PASCAL, FORTRAN), the modelling tools (graphs, predicates, semantic trees, frames, scripts, prototypes, actors, scenarios, ...), and the problem-solving methods, (unifications, inference motors, etc..).

We attempt to clarify here these confusions.

3.A - Knowledges

For a ES - assisted production management, we need the following knowledges :

Level 1 - Object (KO), material or abstract : program, cycle, parts, delays, tolerance, suppliers, manpower,...

Level 2 - Attribute of the object (KA), characteristics which interest our problems : program feasibility, cycle length, part amount, supplier's fidelity...

Level 3 - Attribute evaluating scale (KAS) real number, binary or n-ary, fuzzy, non-ordered,... We need the scale to characterize the object behaviour or state, or to prepare decisional conditions.

Level 4 - Link between objects (KL) : (a machine and a tool) both for milling, (worker, board and electronic components) for a assembling, (technical processing, raw material) for machining.

Level 5 - Relation (KR) : transformation law between states, conditions.
- static, if the relation does not change with time : the manufacturing cycle is one month, a skilled worker assembles 20 boards per hour.
- dynamical, if the relation changes with time : the delivery delay of the supplier S1, the production efficiency of one workshop.

(The structure of a system can be defined by several classes of objects and relations, and the connexions between them).

Level 6 - Relations between relations (K(ROR)) : if (there is sub-systems out of tolerance) than (we need repairing).

Level 7 - Reasoning (KPS), problem-solving method implying the use of two types of (K(ROR)) :
(a) those which are known facts : (non-respect of delivery delay) implies (workshop-stop) ; (workshop WA has spare time) implies (we can give him more works).
(b) those which needs inference : (stock is missing) then (what shall we do ?) ; (we need more articles) then (which supplier shall we prospect ?).

Two types of reasoning are used actually in the IA and ES fields.

- Procedurial : All the required K(ROR) are well defined, and well organized (procedure, algorithm, meta-rule). The problem solving is deterministic.
- Paradigmatic : (example : pattern, complete conjugation or declension) : the required K(ROR) are unsufficient, not well organized. The problem solving uses heuristics, pragmatics, and common sense

Level 8 - Context (KCT) : a text explaining how the problem elements, relations, problem-solving methods are connected together (example : the factory, the MT & ST planning, how supply-delays occur, how the solution is found).

Level 9 - Interfaces (KIO) : the manners showing how the user introduces the parameters to the ES, and how he wishes the output of the ES be presented.

Remarks. There are many confusions between the actual uses of the expressions : state, production rule, rule, meta-rule, algorithm, information, data, and knowledge. We shall clarify them by adopting the following definitions.

Definition 1. State : a way of being. Generally, the value of an attribute of some object. Ex. : the milling machine is new, the delivery time is one week, (knowledge level 3 : KAS).

Definition 2. Knowledge : to be conscious of something. This applies to all the levels. Generally, a knowledge is assumed to exist in our mind, in our Universe.

Definition 3. Information: expression of a knowledge using either our daily language or in a coded form (modelling). Examples : (a) express a delivery time as: x = 7 ; (b) a relation as an equation or a predicate ; (c) a procedure as an algorithm.

Definition 4. Data : an expressed and formalized knowledge stored in a computer. In conventional information processing, data (in data files) represents knowledges on states. In AI, data may represent relations, programs, structures, etc...

Definition 5. Production rule (in AI) : a defined grammatical relation in an AI language showing how to generate states from states.

Definition 6. Rules, meta-rules (in ES) : expressed forms of K(ROR), and a reasoning

Comments. At this stage, we have to understand clearly the difference between knowledges and their expression. The knowledges are those elements that we must possess to fill the inside of the ES. They pose identification problems. (We have to identify not only the system elements, the problems, but also the expertise for solving the problems). Formalize the expressions of the knowledges is dictated by two objectives : (a) design efficiently an ES, and (b) design an efficient ES. This poses modelling problems. We have to model not only the lower-level knowledges, but also the reasoning, context and interfaces.

3.B - Knowledges modelling tools

An information is a knowledge expressed for communicating. A model is a form borne by an information, so that the latter is communicated appropriately. In this meaning, mathematics, graphics, the daily languages, the popular programming languages (FORTRAN, BASIC, COBOL, PASCAL, ADA), the AI languages (LISP, PROLOG...) are all modelling tools. The appropriateness of a modelling tool depends on the purpose of the communication. For designing an ES, we have many purposes : problem understanding, knowledge extraction, problem solving or reasoning, software implementation, ES interfacing. Two main questions arise :

Question 1 : Does it exist one unique modelling tool appropriate for all these purposes ?

Question 2 : If not, what are the characteristics of this ideal modelling tool, or of a group of interrelated modelling tools ?

We shall make two analysis to establishing the elements for the answers.

3.Ba - Reasoning appropriatness

We give here after the knowledge modelling tool classification, established by J.L. LAURIERE, and to which we have added pretentiously our GRAI-tools (10). All these tools model the knowledges level (KO) to level 6 (KROR). The arrow direction indicates the variation of the appropriateness with respect to reasoning. The first ones suit better procedural reasonings, the last ones suit better declarative reasonings.

The "finite automata" gives a mathematical form to a system, such as :

$$a = (qo, q, x, \delta, ql)$$

finite automata = (initial phase, phases, imputs, transformations, end phase).

The "program" gives a linguistic description of a reasoning procedure.
The "script" gives a linguistic description of a background.

Table 1 - Laurrière's classification of knowledge modelling tools

Behaviour	Tools	Author number	References (year ref.)
Procedural (closed, rigid)	1. Finite automata	Ginzburg 68	(11)
	2. Program	Meyer 78	(12)
	3. Script (scheme)	Schank 77	(13)
	4. Semantic networks	Minsky 75	(14)
↓	5. Frame (prototype)	Mc Carthy	(15)
	6. Graphs (PERT, PETRI GRAI)	Petri 62 Pun 82 (10)	(16)
	7. Formal specification	Germain 81	(17)
	8. Predicate calculus	Kowalski 79	(18)
	9. Theorems, writing rules	Huet 78	(19)
	10. Production rules	Shorliffe 76	(20)
Declarative (open, non ordered)	11. Phases in the natural languages	Pitrat 81	(21)

The "semantic networks" gives a graphical form of a knowledge Level 6, KROR. For instance, the knowledge "5 times the difference between 2 by 3 and 4 is equal to 10" (5 x (2 x 3 - 4) = 10) is represented by the graph of Fig. 3.

The "frame" gives a linguistic form of a well defined context, Level 8 (KCT). It is an enumerative process, using "slots" to contain state, parameters, links, criteria, constraints, etc... It does not contain, however, explicitly, a reasoning process for problem solving.

The "Graphs and networks" gives a visual tool for representing a structure of interrelated activities. Level 6 KROR.

The "Formal specification" is a tool similar to "frame", but oriented to problems.

The "Predicate calculus" is a tool giving both a linguistic and formal model of the knowledge Level 1 to 6 : KROR. Example : "IF (The horse can fly) and (the cheese comes from Holland) THEN (The moon is blue)" is transformed into :

CAN (HORSE, FLY) $\wedge$ COME (CHEESE, HOLLAND) $\rightarrow$ IS (MOON, BLUE)

The "Theorems" give a mathematical form to the knowledge Level 7 : KRS, where the problem and problem-solving reasoning process are well defined. Theorems are stated often with the daily language. This point is very much misleading. Because, the used alphabet and grammar belong to a special mathematical language. They remain obscur to non initiated persons.

The "Production rules" model Level 5 : KR, they can be symbolic, mathematical, graphical, linguistic. They are created especially in an AI language for a specific problem solving. Example : A language for constructing words. Alphabet = (A, B, X, O). Production rules : (R1 : A $\rightarrow$ AA) ; (R2 : BA $\rightarrow$ BAB) : (R3 : B $\rightarrow$ BA, or BB or BX). One application : initial state B, one possible state BABABBABX... From this simple inspection, it is remarkable to notice that none of these tools supply the reasoning process. This seems trivial. They are knowledge modelling tools. They are not reasoning tools.

Some of them may facilitate the building of a reasoning process. Notably those close to our daily language, such as program, script, frame, formal specification, predicates, production rules. We notice also that most of these tools only model the knowledge up to Level 6. The elements to answering Question 1 must be found in more general tools : languages.

3.Bb - Multi-appropriateness and languages

We consider, for the ES-design, the needs of 5 appropriatenesses :

- APPR1 : System structuring and understanding.
- APPR2 : Knowledge extraction and understanding specification.
- APPR3 : Problem formulation and solving.
- APPR4 : Programming and software implementing.
- APPR5 : Interfacing.

We consider 6 categories of languages, each one having their specific alphabet and grammar.

LANG 1 : Conventional mathematical languages (arithmetics, algebra, integro differential calculus) : variables, parameters, functions, operators, equations, integration, differenciation, inequalities).

LANG 2 : Modern-algebra type mathematical-languages (set theory, group theory, fuzzy theory (22, 23, 24, 25), binary logics, multi-valent logics, category theory (26, 27)) : elements, attributes, links, application, relations, connectives, ...

LANG 3 : Graphical languages (graphs, hypergraphs, PERT-CPM (28), GERT (29), GAN (30), Block-diagram, Flow-chart, GRAFCET (31)) : point, line, circle, rectangle, specific graphic symbols, paths, trees, lattices, networks.

LANG 4 : Common programming languages (BASIC, FORTRAN, COBOL, PASCAL, ADA, ...) : data, files, instructions, routines, subprograms.

LANG 5 : AI + ES type programming languages (LISP (32, 33), PROLOG (34, 35)...) : state spaces, production rules, control procedures, knowledges, premisses, rules, meta-rules, inference motors.

LANG 6 : Daily languages : alphabet, words, phrases, ponctuations, grammatical laws.

For each language, we shall analyze APPR1 to APPR5. Each appropriateness is evaluated with three possible values :

(+) : directly appropriate
(0) : do not appropriate or indirectly appropriate
(-) : opposinly appropriate

We summarize the results of our analyses in a table form (Table 2).

Table 2 : Appropriates of the languages as modelling tools

	APPR 1	APPR2	APPR3	APPR4	APPR5
LANG 1	–	0	+	0	–
LANG 2	–	0	+	0	–
LANG 3	+	+	–	0	0
LANG 4	0	0	0	+	0
LANG 5	+	0	0	+	+
LANG 6	+	0	0	0	+

Conclusions

Answer to Question 1 - There does not exist one unique language possessing all the appropriateness for an ES-design.

Answer to Question 2 - The languages are appropriate for some of the aspects of the ES-design. We must adopt a multi-modelling strategy.

Semiotics (36) - Some languages model the reasoning process. No language supplies the reasoning process. We go towards semiotical problems, i.e. : find a super-language, based on signs, and combining all the three aspects : semantics, syntactics and pragmatics (see a sister paper of the author) (37).

4 - EXPERT-SYSTEM DESIGN METHODOLOGY

4.A - Objective, background and procedure

We suggest here a design methodology of expert systems in production management.

Objective - Establish a systematic procedure to assist the analyst (control specialists, information-processing specialists, home engineers of the enterprises) in their design work.

Background Production - We assume the existence of a system consisting of its management system MS, physical system PS, from which various knowledges can be extracted, and to which the expert system will be used as a decision-aid.

Procedure - A sequence of organized and correlated activities (Fig. 4). Each activity is an operation (δ) transforming some initial state (go) into an end-state (q1) with the use of intelligent, human and material supports (X1, X2, X3).

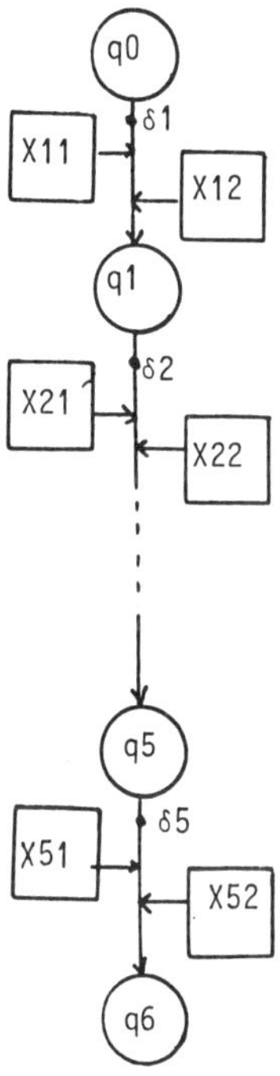

Fig.4 Design procedure of expert systems.

- Activity 1 - Structuration (δ1) : understand what are the (PS + MS) - elements, and how the problems can be stated. qo = existing (PS + MS), X11 = GRAI nets (modelling tools) ; X12 = operators = (analysts, managers, executors), q1 = an abstract structure.

- Activity 2 - Identification (δ2): extract the knowledges on the practical elements of the enterprise according to the frame q1.X21 = GRAI method (modelling tool) ; X21 = operators = X11, q2 = a specific structure corresponding to the reality of the enterprise, including the statement of the problems.

- Activity 3 - <u>ES specifying</u> ($\delta 3$) : specify the level and the content of the ES, according to its predicted feasibility, X31 = Finite Automata Terminology (modelling tools) ; X32 = analyst, q3 = specification.

- Activity 4 - <u>Problem solving</u> ($\delta 4$) : establish the reasoning procedure, either procedural or paradigmical, X41 = (Existing mathematical languages, New languages) (modelling tool) ; X42 = analyst. q5 = reasoning process.

- Activity 5 - <u>Programming preparation</u> ($\delta 5$) : formalize all the informations corresponding to various levels of knowledges, X51 = (HBDS, Predicates) (modelling tools), X52 = analyst, q6 = formalized informations ready for a portable programming activity.

4.B - Structuring

To structure a system is to analyze it in order to getting a formal representation of the underlying activities, and of the relations between them. In the ES - design for production management, we have to achieve several stages of structuring works, each one has its own aims. Thus :

- PS - structuring aims to understand the problem-elements of the executing activities.
- (PS + MS) - structuring aims to understand how the managers operate, what are their problems and specify consequently the desired ES.
- (PS + MS + IS) - structuring aims to understand what are the knowledges on the operating states of PS and MS, and design consequently the Information - processing system.
- (PS + MS + IS + ES) - structuring aims to understand what are the knowledges on the states, the PS operating relations, and the MS decisional relations needed for the ES-design.

During the period 1965-1980, many computerized MIS (Management Information Systems) have been developped without a previous careful structuring analysis. Many mismatches and inadequacies have resulted. The wake-up has occured around 1978. Several structuring - analysis methods have been developped : SADT, IDEF0, SSAD and GRAI.

<u>SADT</u> (Structured Analysis and Design Techniques). This method has been developped by the firm SOFTECH, and more specifically by DT Ross (39, 40). The initial aim of Ross was a method and a language permitting the communication of ideas. The method combines expressions of both technical and daily languages. The analysis yields a structured hierarchical model, where the elements are presented progressively from global to particular. At the present time, SADT is applied in USA and in Europe.

<u>IDEF0</u> (ICAM DEFinition). This method is inspired from SADT. It is developped in the framework of ICAM (Integrated Computer Aided Manufacturing) - project under the auspices of the US Air Force. It is used to developping advanced production techniques in american industries.

<u>SSAD</u> (Structured System Analysis and Design). Initially, this method has been developped by the Firm IST (Ch. Gam and Trish Sarson), and later by Mac Donnal Douglas Aviation. The method is inspired from "Data Flow Diagram" of Tom De Marco (40). At the present time, the method is applied to production system in the framework of "Factory-Management" project under the auspicies of CAM-I (International club "Computer Aided Manufacturing-International").

<u>GRAI</u> (Graphs with Results and Activities Interrelated). This is a combination of graphical tools (GRAI-tools), and of an identification method (GRAI-method) (41, 42,43). It has been developped by the authors from 1977 in the GRAI Laboratory (Groupe de Recherche en Automatisation Intégrée) of University Bordeaux 1. Since then, it has been applied to more than 20 french factories. At the present time, it is applied in PRC, Spain, Brazil, and is adopted by the European ESPRIT project to develop Computer Integrated Manufacturing (CIM) systems.

All these modelling tools possess in principle the necessary elements to represent the standardized structure of the production management system. The big difference is the existence of methods to utilizing this tool to help the other steps of the ES-design. In this sense, and using the multi-modelling concept, tue GRAI tool is associated with : (a) the GRAI method for identification ; (b) First-order predicates for calculus ; (c) Categorical algebra for analyzing structural properties : (d) a Situation theory for reasoning, and (e) Abstract types for establishing a modularized programming language (44,47).

4.C - The GRAI tools (42,46,48)

<u>An activity</u> : $a = (qo, x, \delta, ql)$; $\delta = (qo \wedge x) \to ql$ (qo = initial state, x = supports, δ = operating law, ql : final state or result).

<u>A program</u> : $A = (a_i) = (qo, X, Q, \Delta, qn)$;
$i = i..n$
$= q_o x\ X\ x\ \Delta \to qn$
(qo = initial state, $x \in X$ = set of supports, $q \in Q$ = set of states, $\delta \in \Delta$ = set of laws, qn = final state or desired results).

<u>A structure</u> : $STR = (A, B, C, \ldots, \Gamma IJ)$;
(A, $\propto$B, C..) = programs,
ΓIJ = connexions between programs. The connexions can be analogue or logical, multiplicative, additive and iteratif.

<u>A multi-structure</u> : M-STR = (STR1, STR2, .. ΨMN) ; (STR1, STR2, ..) = structures, ΨMN = connexions between structures.

<u>Structural representation</u>. The same elements are used to represent the activities of PS, MS, IS and ES and the connexions between them.
Example : (Fig. 5) PS : physical system, A = manufacturing program, B = procurement program ; MS = management systems ; IS = information system ; ES = expert system ; γAB = connexions between A and B ; ΨMP = criteria, framework of supports ; ψPI = executing informations ; ΨIE = decisional informations ; ΨME = interactions between MS and ES.

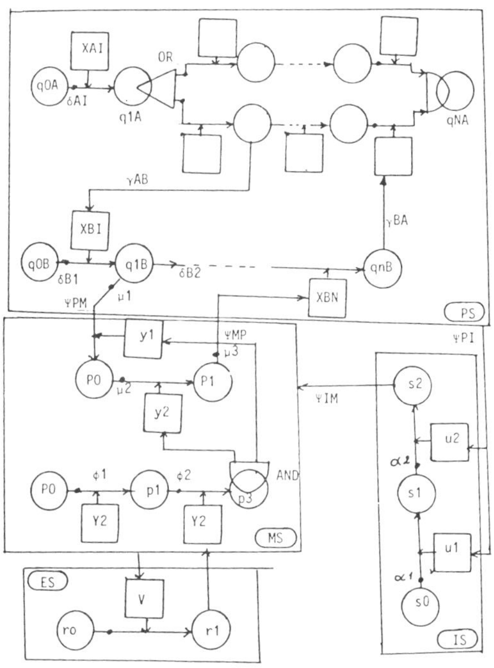

Fig. 5 Multi-structure representation with GRAI-tools.

Knowledge extraction. All the required knowledges for MS and ES are explicitly specified on the diagram. Knowledges on the states and on the supports (parameters) : Q, P, R, S, .. X, Z, Y, U, V, ... Knowledges on the relations : δ, μ, Φ, α, ... All these knowledges are required for the ES-design.

Problem formulation and statement. The structural understanding permits to formulating the problem of all the levels from PS to ES. For example : scheduling, ressource allocating, information logging, planning, adjusting, reasoning. The associate symbols permits to state clearly these problems. Example : scheduling, branch A, givern X1 (machines), X2 (manpower), technical processing X3, finality QN, organize (X1, X2, X3) into a program of operations : $\delta a \rightarrow \delta b \rightarrow .. \ \delta n, \ \delta \in \Delta$, so that qN is reached.

4.D - GRAI method

The GRAI method is a methodology which can be viewed as an identification language as : GRAI method = (objectives, tools, investigation teams, procedure).

Objectives. Represent the multi-structure (PS + MS + IS) of a production system, so that their problems, and the elements for their solving can be stated and specified adequately.

Tools. (a) The GRAI net used as an identification framework for the representation of the micro-structures ; and (b) The GRAI grid used as an identification framework for the representation of the macro structures. The GRAI grid is a matrix. The columns are the production functions : planning, manufacturing, procurement, maintenance, quality-control. The rows are the operating horizons and adjusting periods.

Investigation teams. (a) The Analysis team, for executing the investigation tasks ; (b) The Validating team, consisting of high level managers for validating the results of the investigations.

Procedure : Step 1 : Identify the existence of the decisional centers on a GRAI grid. Step 2 : Identify the decisional-liaisons and the informational liaisons between the centers. Step 3 : Identify, with GRAI-tools the structure of the executing activities, their planning problems, dynamical problems, and currently used solutions. Step 4 : Analyze the coherence of the MS. Step 5 : Specify the required IS. Step 6 : Specify the E.S.

Remark : We encounter here a control problem very much different from conventional automatic control problems. We have to identify here human behaviours with the help of human partners.

4.E - Expert-System Specification

Specify an ES is to determine : (a) its objective ; (b) the required elements to satisfy this objective. Among the latters, we must supply : the model of the underlying structure and the reasoning process. The higher is the level of the objective, the more difficult is the possibility for obtaining these elements, depending on available competence, investment and opportunity. The following levels can be distinguished.

Level 1 - Consulting, the solutions are predetermined.
Level 2 - Suggestion, the solutions are calculated according to the instantaneous situation.
Level 3 - Evolutive, the parameter of the models can be modified.
Level 4 - Learning, the strategy and the tactics of the reasoning process can be improved.

We summarize in Table 3 the requirements on the elements to be supplied.

Table 3 - Various levels of expert systems

Level	Required Model	Required Reasoning process
1. Consulting	no	predetermined
2. Suggestion	PS model	available optimization methods
3. Evolutive	- PS model - (MS + IS) models	- available methods - restructurating of the methods in a predetermined way.
4. Learning	- PS model - (MS + IS) models - Reasoning basic model	possibility for changing the tactics and the strategies.

Remark : To ensure feasability, it seems sensible to adopt a progressive approach. Start from a low level. Both the analyst-designer and the user can acquire skill by experience. Then progress to higher levels

4.F - Formal calculus and reasoning

A reasoning process, be it an algorithm, a procedure, or a paradigm, contains necessarily three types of calculus.

Calculus 1 - Cause to effect. In terms of the elements of an activity, we have :

Given (qo, X, δ) $\rightarrow$ What is (ql)

(calculate the result of an operation, calculate the value of a criterion, calculate the margin of a contraint.

Calculus 2 - Deterministic decisional calculus :

Given (qo, δ , ql) $\rightarrow$ What is (X)

(ressource allocating, scheduling).

Calculus 3 - Inference calculus : Given (X1 non successful) $\rightarrow$ What is (X2 successful).
An algorithm is a reasoning process where all the inference calculs are well defined, ie, we know exactly which X2 to choose. In the procedures and paradigms, we have to test the possible X2 in a set X (X2 $\subset$ X), leading to trials and errors. The latters are again calculus of the types 1 and 2. In continuous calculus, formal calculus and reasoning are made possible with algebraic and integro-differential calculus. The inference calculus is made possible mainly because of the gradient representations of the state variations. In discrete activities, formal calculus and reasoning are made possible with the First-order predicate logics (45). However, inference methods are absent because of the absence of gradients. In the field of Artificial Intelligence. "Unification method", "Forward tracking", "Back tracking" are frameworks for implementing inference techniques, but not inference techniques themselves.

4.G - Programming languages

Today, there is not an universally accepted programming language for the ES-design. We view such a programming language as a generalized modelling tool capable of representing all levels of knowledges (Section 3A). Two categories of problems thus arize.

Internal problems. How to model the knowledges in individualy or in a combined way so that efficient reasoning processes can be established.

External problems. How to formalize the knowledges so that the various steps of the ES-design, and the complete design itself, can be made easy and efficient.

The numbers of combinations, taking into account criteria of both categories, is so great that this is certainly a long-term research.

5 - CONCLUSIONS

In this paper, we have firstly analyzed the various levels of industrial management problems, and the contents of the desired expert systems. Secondly, we have to clarify what are the knowledges and the knowledge-modelling tools. Thirdly, we have presented a methodology for designing the expert systems. Three main research directions can be derived : (a) Knowledge (and expertise) identification in a human background ; (b) Reasoning-process establishment for solving complicate decisional problems, and (c) Multi-modelling language determination, enabling the building of "program generators", capable of generating a well-defined class of expert systems.

REFERENCES

1. Ansoff H.I., "Corporate strategy", Mc Graw Hill 1965, Penguin 1968.
2. Pun L., "Computer assisted static and dynamical, plannings for production activities", Automatica, v17, n2, march 1981.
3. Nilsson N.J., "Principles of Artificial Intelligence", Spring Verlag, New York, 1982, 476 p.
4. Hunt E.B., "Artificial Intelligence", Academic Press, New York, 1975.
5. Winston P.H., "Artificial Intelligence" Addison Wesley, 2nd Ed., 1979.
6. Hays-Roth F. et al Ed, "Building Expert Systems", Addison Wesley, 1983, 444 p.
7. Buchanan B.G. & E.H. Shortliffe Ed., "Rule-based Expert Systems", Addison Wesley, 1984, 748 p.
8. Fieschi M., "Intelligence Artificielle en médecine, des systèmes experts", Masson, Paris, 1984, 205 p.
9. Laurrière J.L. "Représentation et utilisation des connaissances", Techniques et Sciences Informatiques, AFCET, Paris, n° 1, mars 1982, pp. 25-42 ; n° 2, avril 1982, pp. 109-133.
10. Pun L., "GRAI net and applications", Workshop IFAC on CAD/PDP, june 22-24, 1982, Ankara.
11. Ginzburg A., "Algebraic theory of automata", Academic Press, 1968.
12. Meyer B., Baudoin C., "Méthodes de programmation, Eyrolles, n° 34, 1978.
13. Schank R., Albelson R., "Scripts, plans, goals and understandings", Laurence Erlbaum associates Inc., Milsdale N.J., 1977.
14. Mc Carthy I., "Epistemological problems in AI.", Proc. I.J.A.I., MIT, Cambridge 1977, 1038-1044.
15. Minsky M., "A framework for representing knowledge", in Winston Ed, "The psychology of computer vision", Mc Graw Hill, 1975.
16. Petri C.A., "Communication with automata", Ph. D. Bonn, 1962.
17. Germain M., Journées sur les spécifications, Congrès AFCET TTI, Nancy, 1981.
18. Kowalski R., "Logic for problem solving", North Holland, New York, 1979.
19. Huet G., "Unification dans les logiques d'ordre 1, 2, 3, ...52". Thèse d'état, Paris VI, 1978.
20. Shortliffe E., "Computer based medical consultation" MYCIN, American Elsevier 1976.
21. Pitrat J. "Une interprétation de connaissances déclaratives", Colloque IA, Toulouse 81, Publication GR, 22, n° 25.
22. Zadeh L.A. et al. Ed, "Fuzzy sets and their applications to cognitive and decision processes". Academic Press, 1975, 496 p.

23. Dubois D. & Prade H., "Fuzzy sets and systems, theory and applications", Academic Press, New York, 1980.
24. Sugeno M., "Fuzzy controls", Academic Press, to appear.
25. Gupta MM. et al Ed., "Fuzzy automata and decision processes", North Holland, 1977, 496 p.
26. Mac Lane S., "Categories for the working mathematician, Springer Verlag, 1971, 262 p.
27. Arbib M.A. & E.G. Mannes, "Arrows, structures and functions. The categorical imperative", Academic Press, 1975.
28. Moder J.J. & Phillips C.R., "Project management with PERT & CPM", Van Nostrand, 1970.
29. Pritsker A. & Alan B., "Modelling and analysis using Q-GERT", John Wiley, 1977.
30. Elmaghraby S.E., "Activity networks. Project planning and control by network models", John Wiley, 1977.
31. Blanchard M., "GRAFCET et applications", CEPADUES, Toulouse, 1981.
32. Siklossy L. "Let's talk LISP", Prentice Hall, 1976, 237 p.
33. Le Conte des Floris D. et Jouvelot P., "Intelligence Artificielle et systèmes experts : le langage LISP", Mini et Micros, n° 208 pp. 45-48 ; n° 209 pp. 75-78 ; n° 211 pp. 76-79.
34. Colmerauer A. et al, "Prolog, Bases théoriques et développements actuels", TSI, AFCET, v2, n° 4, 1983, pp. 211-271.
35. Colmerauer A. et C., "Prolog en 10 figures", Colloque Int. IA, octobre 1984, Marseille, Tome 1.
36. Pospelov D.A., "Semiotic Models in Control Systems", in "Cybernetics today", MIR Publishers, Moscow, 1984.
37. Pun L., "Situation recognition in Production Management", Invited paper, APMS COMPCONTROL 85, August 27-30, 1985, Budapest.
38. Ross D.T., "Structured Analysis (SA) : a language for communicating ideas, IEEE Transactions on Software Eng., VSE3, n° 1, janvier 1977.
39. Ross D.T. & Schoman R.E., "Structured Analysis for Requirements Definition", IEEE Tr. on software Eng., VSE3, n° 1, janvier 1977.
40. De Marco T., "Structured Analysis and System Specification", Prentice Hall, New Jersey, 1979.
41. Doumeingts G., Breuil D. & Pun L., "Gestion de Production Assistée par Ordinateur (GPAO)", Hermès, Paris, 1983.
42. Doumeingts G., "Méthode GRAI : Méthode de conception des systèmes en productique", Thèse d'Etat, Univ. Bordeaux 1 novembre 1984.
43. Breuil D., "Outils de conception et de décision dans les organisations de gestion de production", Thèse d'Etat, Univ. Bordeaux 1, novembre 1984.
44. Pun L., "Discrete activity multimodelling", IFROS Congress, July 20-24 1981 Hambourg.
45. Pun L. & Breuil D., "Calculus in the control of discrete activities involve in the Flexible Manufacturing Systems" IFAC Congress IX, July 2-6 1984, Budapest.
46. Pun L., "Systèmes Industriels d'Intelligence Artificielle - outils de Productique", Edi. Tests, Paris, 1984 (English version, Plenum, New York to appear ; Chineese version, Friendship Publ. Co. Beijing, to appear).

ADAPTIVE CONTROL IN CHEMICAL INDUSTRY

K. Najim*, M. Najim, B. Dahhou**, H. Youlal** H. Unbehauen*****

**Ecole Nationale Supérieure d'Ingénieurs de Génie Chimique, Toulouse, France*
***Leesa, Faculté des Sciences, BP 1014, Rabat, Morocco*
****Department of Electrical Engineering, Ruhr-University Bochum, FRG*

Abstract. Many chemical plants have to be controlled such that their responses are within given tolerances under various operating conditions including variation in plant parameters and disturbances. It is therefore desirable to have adaptive controllers which can compensate for changes in process parameters due to varying working conditions or initial indeterminateness. The high effort in development and implementation as well as operation of such control systems is justified economically if significant improvements in operation can be achieved.

The examples discussed in this paper are aimed to showing the different steps in the development and application of single variable and multivariable adaptive control for different types of processes.

The applications show that, at the present time, adaptive control has reached a high degree of maturity, making it a very powerful tool for the experienced control engineer. However, for making adaptive control accessible to the broad field of process engineers, the degrees of freedom have to be reduced by simplifying the elements of the adaptive controller. To clarify this, further research has to be undertaken.

Keywords. Adaptive control ; chemical industry ; moisture control ; identification.

INTRODUCTION

Adaptive control has been an active area of research and development in the last two decades. Much work has been directed to the establishment of basic techniques and theoretical tools for designing adaptive control systems. Also many feasibility studies carried out on laboratory scale pilot plants as well as a number of successful industrial applications have been reported (Aström, 1983 ; Unbehauen, 1980 ; Narendra and Monopoli, 1980). However, it is generally admited that a gapstill exist between theory and applications of adaptive control, although some special purpose adaptive regulators have found their way to the market among conventional systems.

The present paper is aimed to contribute to the effort for bridging this gap through the description of successful operation of both single variable and multivariable adaptive techniques on different laboratory scale pilot plants as well as industrial applications.

The paper is not intended to discuss the theory of the design and derivation of the control and adaptive laws which are employed in the reported to the description of actual applications and discussion of the results and improvements achieved with adaptive control.

The paper is organized as follows : first is considered the application of the single variable model reference adaptive control to a catalytic fluidized bed reactor. Then an industrial application of the single variable based minimum variance self-tuner and a model reference adaptive control scheme for moisture control in phosphate drying process is described. Finally multivariable adaptive control based on either generalised minimum variance and model reference ideas of distillation columns is presented.

MRAC OF A CATALYTIC FLUIDIZED BED REACTOR

1. The plant and control problem

The unsteady state operation of a catalytic fluidized bed reactor for ammoxidation of propylene to produce acrylonitrile was studied in a steel fluidized bed reactor. The reaction is rather complex because three reactants are needed : propylene, ammonia and oxygen. It gives several by products such as acroleine, acetonitrile and carbon oxides. Besides it is very exothermic (H=-515 KJ per mole acrylonitrile produced) and it involves the presence of water vapour in the reactant stream.

The reactor comprises a preheater and reaction chamber (Fig. 1). The preheater consists of a bed of sand fluidized by air and heated by two electric coils immersed in the sand. Water fed into the preheater is instantaneously vaporised on contacting the hot fluidized sand. Propylene and ammonia are heated separately in two coils immersed in the bed sand. Reactants are mixed in the zone just above the perforated plate distributor. Two rows of deflectors are incorporated to enhance mixing.

The reaction compartment contains a catalyst composed of tin, iron, antimony and copper. It is very fragile at high temperature and does not resist to large and sudden temperature variations. It has been found that the highest yield of acrylonitrile with this type of catalyst occurs in temperature range of 480-510 C.

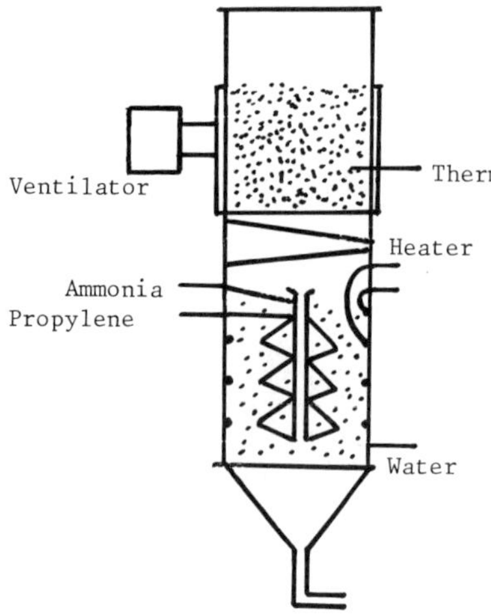

Fig. 1. The reactor

Maintaining a constant temperature in the reactor is very important because it allows to yield a good selectivity in the desired product, i.e. acrylonitrile. Furthermore, in our study temperature must be kept constant in order to discriminate between chemical and thermal effects under unsteady operation. The response of the system to change in reactant concentrations can indeed be hidden by the temperature evolution due to the variation of the gas flow rate through the preheater and the heat release in the reactor. Control of temperature is difficult to achieve due to the characteristics of the reactor.

Two possible techniques can be envisaged for the removal of the heat evolved during the reaction :

- variation of the rotational speed of the ventilator in the case of air cooling of the reactor
- cooling by increasing the rate of water vaporised.

It was observed that approximately twice the quantity of steam in the feed would be necessary to remove the heat evolved, but this would disturb considerably the bed hydrodynamics. We then have chosen the first technique for the heat removal.

2. Experiments and results

Among different adaptive techniques, the model reference scheme has been considered for temperature control in the reactor (Koutchoukali and co-workers 1983, 1984). The velocity of the ventilator was used as the manipulative variable, (this control device is aimed to carry away part of the heat produced by the reaction).

The reactor was interfaced with an Apple II microcomputer associated to digital-analog and analog-digital converters devices. The sampling period was chosen to be equal to 35 seconds. The time delay and the system order have been determined off-line from the transient response of the reactor.

The reactor was operated at the following steady state conditions :

Ammonia	5.79%
Propylene	3.59%
Oxygen	14.87% (air 70.82%)
Water vapour	19.80%

The total gas flow rate passing through the reactor is 5.0 N.m/h.

Figures 2 and 3 show the evolution of the reactor temperature and the rotation speed of the ventilator respectively when the model reference scheme is applied.

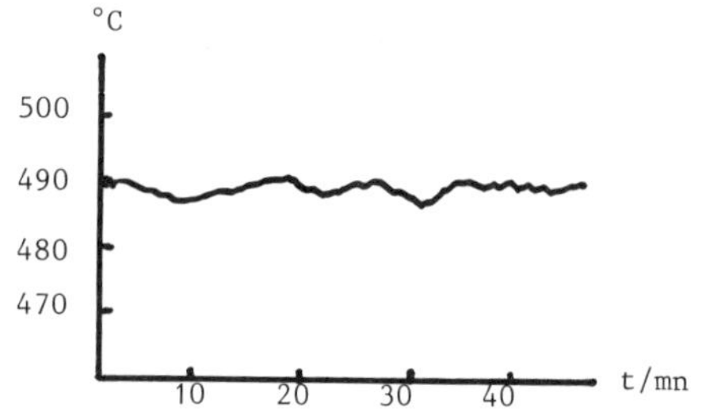

Fig. 2. Response using MRAS

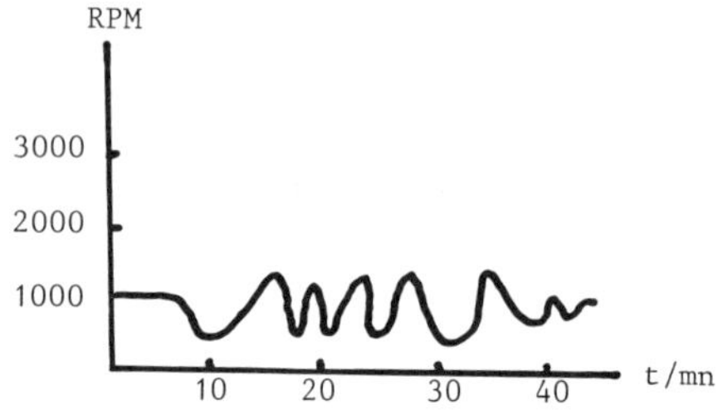

Fig. 3. Speed ventilator vs time

Figure 4 depicts the evolution of the reactor temperature when the analog PID regulator is used to regulate the fluidized bed reactor temperature. The PID regulator constants were chosen based on Ziegler-Nichols (Ziegler and Nichols, 1942) tuning procedure. The performance of this regulator is very poor because the process parameters are time varying. The total flow gas was maintained constant, then a step change in the feed flow defined by the following operation conditions :

Ammonia	7.18%
Propylene	2.50%
Oxygen	14.81% (air 70.52%)
Water vapour	19.80%

at time was introduced.

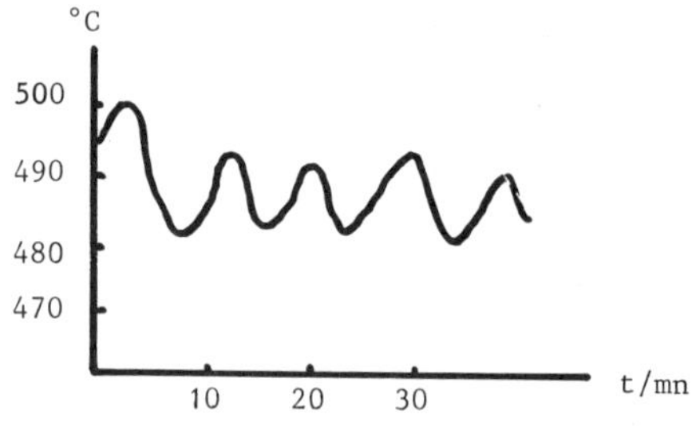

Fig. 4. Response using PID regulator

Figure 5 shows the isothermal operation of the reactor under model reference adaptive control algorithm. We can see that better results are obtained with this scheme, which yields in a better steady state operation of the reactor.

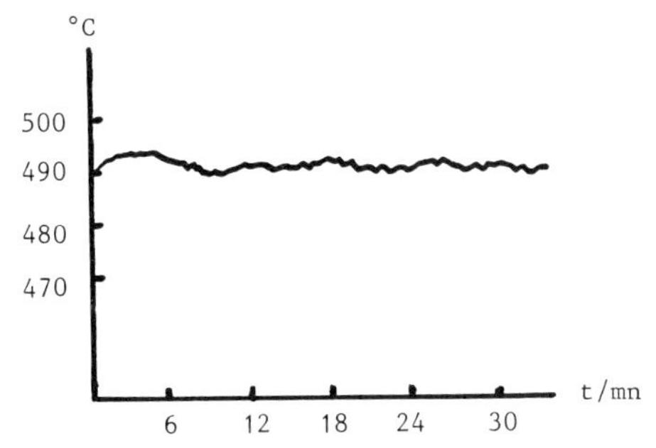

Fig. 5. Response using MRAS with step change in feed flow rate

ADAPTIVE CONTROL IN PHOSPHATE INDUSTRY

The process involved herein is an industrial phosphate dryer (fig. 6). The main control objective is to keep the moisture content of the dried phosphate close to a constant desired value, and at the same time to minimize the energy consumption, despite feed flow rate variations (100 to 240 t/h) and variable moisture content of the damp phosphate (7 to 20 per cent).

1. Motivation for adaptive control
Invariably there are some incertainties in the characteristics of the processed phosphate that can be attributed to the variable moisture content and the nature of the damp mineral. Therefore drying process is non-linear and non-stationnary in its nature. The change in dynamic characteristics with operating conditions is such that fixed parameter controller could not achieve satisfactory performances in the entire range over which the characteristics of the process may vary. Therefore, any solution of the control problem might be improved by tuning the parameters. On the order hand it was also apparent that the controller structure must be simple enough for implementation consideration. The adaptive control formulation fits most of these requirements and seems to be an appropriate alternative to the existing conventional control equipments. This motivated us to try the self-tuning and model reference approaches for moisture control in phosphate industry with the goal to achieve better performance and consequently energy saving in the drying process.

2. Implementation of adaptive control algorithms
The feasibility study of the application of adaptive control in phosphate drying industry has been carried out using two different approaches described already elsewhere (Dahhou and co-workers, 1982 ; Najim and co-workers, 1982 ; Youlal and co-workers, 1983), i.e. the self-tuning approach based on minimum variance strategy and least squares parameters estimation and the model reference approach. A simplified model of the complex dynamics of the dryer plant was assumed. In this model the key parameters to achieve the control objectives were manipulated, i.e. the fuel flow as the control variable and the moisture content of dried phosphate as the output variable, while the moisture content of dump mineral was used as the measurable process disturbances included in feedforward manner in control strategies.

The application software was developed during early part of the work. It was coded in a modulare form which makes updating and modification painless. It should be mentioned that only a small part of the software is actually involved with the adaptive control algorithms. A large proportion is included for operator interaction and industrial plant management. The fullest form of the application program includes the dialogue for a full initialisation procedure, data logging and full plant management including a careful presentation of the states of both the dryer plant and the adaptive controller. On-line dialogue is also included for setting up or altering crucial values.

3. Experiments and results
Evaluation of the adaptive control schemes for phosphate drying process control was focused on fuel flow control and the dried product quality. The existing conventional control equipment had been improved so that the performances make an adequate reference against which the on-line computer control can be judged. The operating conditions of the dryer were the most common ones. A product feed rate of about 220 tons per hour, subject to random variations in moisture content of damp phosphate which has found to belong in the interval 10 to 15 per cent. The conventional PID control was employed having fixed parameters and known to provide acceptable performances.

In table 1, are shown in the first column some statistics for 24 hours recording that were achieved with the conventional PID control. By comparison, in the last two columns are reported some typical statistics that were obtained by the use of the implemented adaptive controllers over 24 hours of actual plant operation. As it can be seen a much improved control performances have been achieved. This is illustrated by both the decreasing variance of the output product moisture content, and the increasing average product quality. On the other hand as can be seen from table 2, the comparison between conventional and adaptive control shows a clear decreasing in fuel consumption, under comparable operating conditions of the plant. Furthermore, we have to point out that the quality of regulation yields the diminution of the equipment deterioration for long term plant operation, which is an other source of economy.

The outcome of the application of adaptive control to moisture control in phosphate industry was successful in that over several weeks of actual plant operation a gain of about 4.3% in fuel consumption was obtained with comparison to conventional control. Due to large quantity of fuel consumption and to the increasing energy cost, this economy results in considerable profit which justifies the high efforts to introduce computer control in Moroccan phosphate industry.

ADAPTIVE CONTROL OF DISTILLATION COLUMNS

The application of adaptive control to distillation columns has been the topic of several papers (Dahlqvist, 1981 ; Morris and co-workers, 1982 ; Martin-Sanchez and Shah, 1984 ; Kiovo, 1980). In a typical distillation column (fig. 7) the feed flow enters near the center of the column and flows down. Vapours that are released by heating rise to the top, are condensed and can be removed as overhead product or distillate. Any liquid that is returned to the column is called reflux. The reflux flows down the column and joints the feed steam. Classical distillation columns often have large reflux rates and many trays for separating components with similar boiling points. The reflux rate has a major influence on the separation process. Too much reflux flow makes the product over pure, but wastes energy because more reflux liquid has to be vaporized, too little reflux flow causes an impure product.

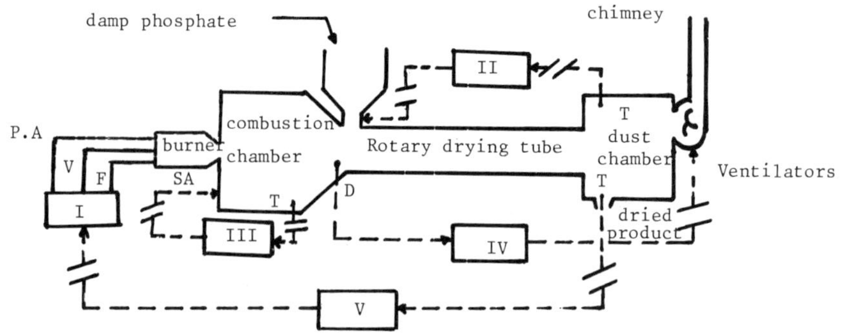

-----//---- : Measurements fed to process computer
I : Flow recorder/controller of fuel, vapour and air includes a servo drive of fuel valve
II : Product rate controller
III : Secondary air controller
IV : Ventilators speed controller
V : Fuel flow controller

Fig. 6. Schematic diagram of the phosphate dryer and the conventional control loops

		Conventional PID control	Self-tuning control	Model reference Adapt. control
Moisture content of damp phosphate	Average	12.3%	12.7%	12.7%
	Variance	0.18	0.47	0.71
Moisture content of dried phosphate	Average	1.78%	1.64%	1.42%
	Variance	0.57	0.09	0.15

Table 1 Comparison of 24 hours statistics of a phosphate dryer control

	Conventional PID control	Adaptive control
Average moisture content of damp phosphate	12.98	12.96
Average moisture content of dried phosphate	1.83	1.80
Fuel consumption per ton of dried phosphate in Kg Fuel/ton	11.38	10.89

Table 2 Results from several weeks of phosphate dryer plant operation

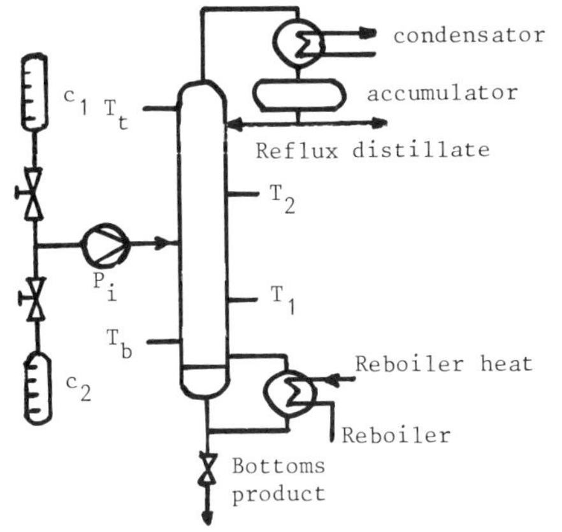

Fig. 7. Schematic diagram of a distillation column

Figure 8 shows how product purity is affected by a change in the reflux to feed ratio for a particular column. Changes in the feed flow cause only minor changes in product purity at the high reflux rates where as the sensitivity of the column near to the specification point, where the energy consumption is minimal, may be increased by factor of mine. To avoid off-specification products, distillation columns are often operated with higher than needed reflux rates and for that are consuming more energy than necessary for garanteeing product specification.

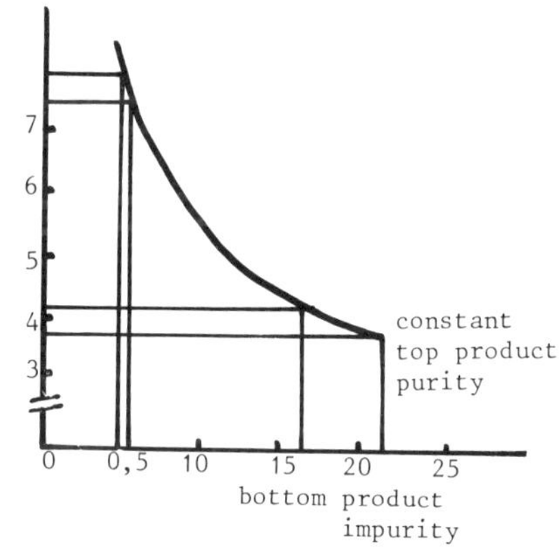

Fig. 8. Product impurity as a function of the reflux rate

From these points of view it becomes clear that energy saving can only be achieved when operating the column near to the specification point. As the sensitivity of the column with respect to changes in the feed flow or composition is highly increased at this operating point, the control problem is severe, and is still complicated by the fact that measurements on product specification can only be done by one-line chromatographs, which will add delays and maintenance problems. Conventional distillation column control mostly uses temperatures as controlled variables using a fixed reflux rate. It is generally only one temperature near to the feed, which is controlled via the reboiler heat. Additional control of a temperature near the top of the column by the reflux rate hardly produces an improvement since both controllers work independently and do not take into account the interactions of the two loops. This can result in both controllers working against each other producing oscillations or even distabilizing the control.

An improvement in distillation control, therefore, can only be achieved when using multivariable control, modelling the column as a two input/two output system which the inputs, reboiler heat and reflux rate, and two temperature correlated with the appropriate product concentrations forming the outputs. The main control objective is to maintain the top and bottom product compositions constant under varying feed flow and composition in the feed stream. Setpoint changes are of minor importance. As in binary distillation under isobaric conditions the concentrations are directly correlated to the temperature, the temperature near to the top T_t and the bottom T_b of the column are selected as controlled variables.

1. Multivariable adaptive control of distillation columns

The basic investigations for the application of different conventional and multivariable adaptive control schemes had been performed on a binary pilot distillation column, described already elsewhere (Wienner and co-workers, 1983 ; Hahn, 1983). The dynamic behaviour of this plant may be expressed by a 2x2 transfer matrix. Three elements of this matrix are estimated as transfer functions of 2nd order and one transfer function is assumed to be only of first order. For setpoint changes the two measured temperatures T_t and T_p have rise-time from 45 to 70 minutes. However, according to the dynamics of some fast disturbances a sampling time between 10 sec (for the self-tuning controller) and 30 sec (for the model reference adaptive controller) is chosen. A (fast) reference model of the form

$$Y_{mi}(z) = z^{-1} W_i(z) \qquad i=1,2$$

has been applied.

The experimental results are shown in figures 9 and 10. The ordinate-axes are scaled in relative values related to a certain setpoint. Because of the plotter used, the time bases of the different plots are slightly shifted to each other.

Figure 9 shows the optimally adapted cases for :

a) conventional PID-controllers
b) linear state feed back controller with PI-action
c) generalized multivariable self-tuning minimum variance controllers (Koivo, 1980)
c) multivariable model reference adaptive controller (Wienner and co-workers, 1983 ; Hahn, 1983).

When disturbances in the composition and feed flow at the time are applied. The influence of the disturbances is well rejected by both adaptive schemes as well as by the linear state feed-back controller with PI-action. The linear state feedback controller, however is sensitive to changes in the operation point so that its parameters have to be retuned from time to time. The results obtained by the conventional PID-controller are in all cases improved by the modern controller principles. It should, however, be mentioned that the full least squares estimation combined with self-tuning control leads to larger storage requirements and a computation time which is about five times greater than that for the model reference scheme.

The results of the comparison of these 4 different controllers are listed in Table 3.

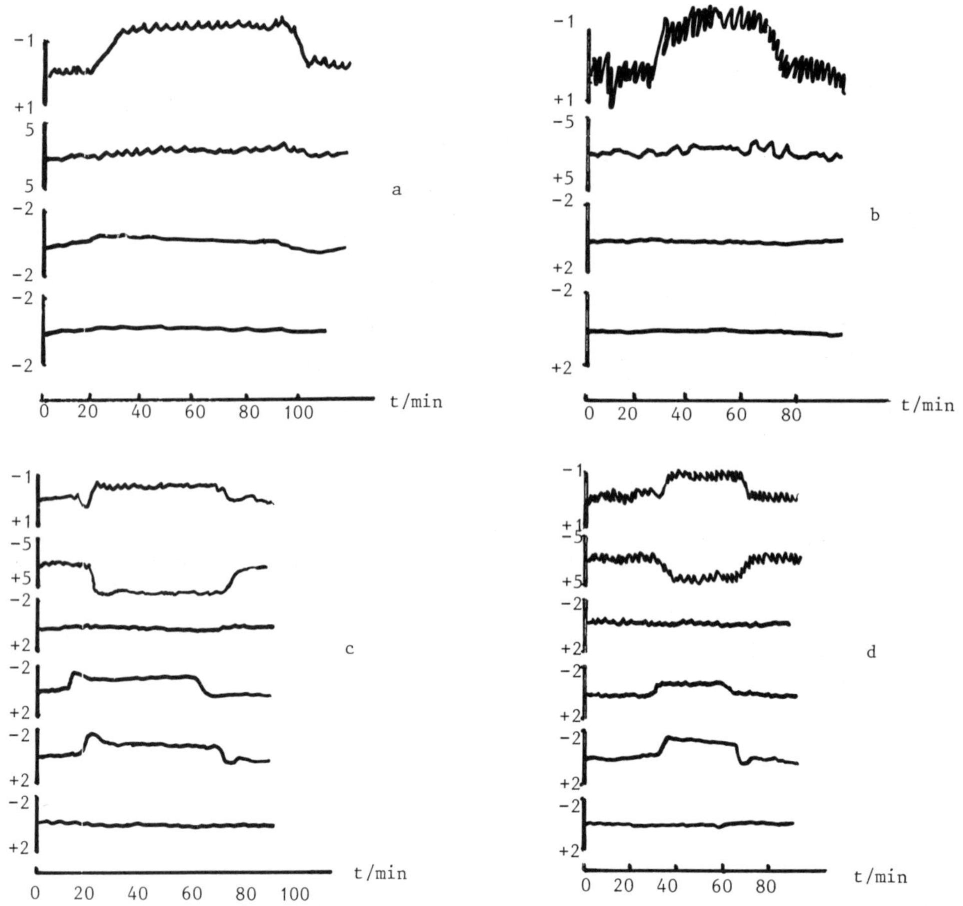

a) classical PID controllers
b) linear state feedback controller with PI-action
c) generalised multivariable self-tuning minimum variance controller
d) multivariable model reference adaptive controller

Fig. 9. Performance with well adjusted controller parameters

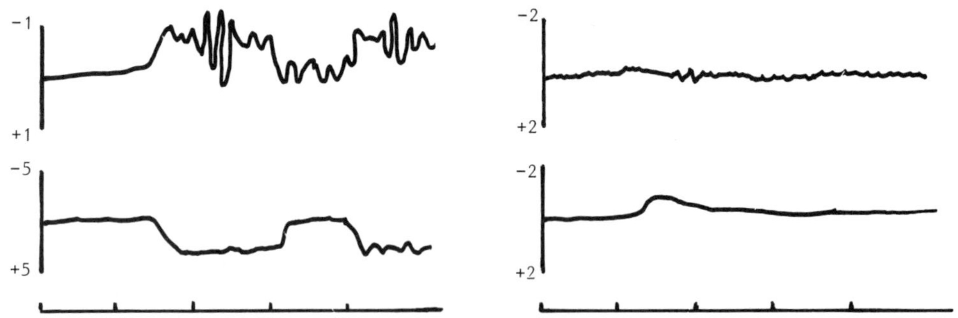

Fig. 10. Performance of adaptive transients for the multivariable model reference adaptive controller with zero initial parameters

Type	Settling time (min)		Overshoot (c)	
	Top	Bottom	Top	Bottom
a	25	10	0.5	0.2
b	0	0	0	0
c	0	0	0	0
d	10	5	0.2	0

Table 3 Settling time and overshoot for all 4 types of controllers (a=PID-controller b=state feed-back controller, c=model reference adaptive controller, d=self-tuning controller)

From figure 10 the adaptation procedure can be seen for the multivariable model reference adaptive controller. At time t=0, all initial values of the controller are started with null values. After some disturbances in the concentration and feed flow the parameters are well adapted so that no overshoot and settling time are observed.

The adaptive controllers discussed here have been successfully applied since 1981 to large technical recycling columns in Germany. As a result energy has been saved in the range of 10% compared to conventional control. At the same time the product continuity has been increased and the bottom's product purity has been improved from 200 ppm with conventional PID control to 20 ppm.

CONCLUSION

We tentatively tried to present herein an overview of some successful applications of adaptive control in chemical industry. Theoretical aspects of the adaptive control schemes investigated are purposely omited, as they are well covered in the quoted references.

Undoubtdely there are a number of practical details that might be drawn from this concise paper. However, implementation considerations and actual operation of different adaptive control schemes on different laboratory scale chemical plants as well as industrial plants are covered.

In phosphate industry, significant improvements in operation of the drying plants and energy saving have been achieved with simple adaptive control schemes. On the other hand, different multivariable adaptive control schemes have been successfully applied to binary distillation columns. Though model reference schemes have originally been designed for servo problems, with slight modifications they can also be used for regulation. The comparison with a self-tuning controller shows nearly identical results. However, the realization of the model reference scheme is found to be simpler. In contrast to linear controllers, the implementation of adaptive controllers is simplified because only little a priori knowledge about the plant dynamics is needed. A retuning of the controller parameters, which has to be done with linear controllers from time to time, is not necessary.

REFERENCES

Aström, K.J. (1983). Automatica, Vol. 19, 5, 471-486

Dahhou, B., K. Najim, M. M'saad (1982). RAIRO Aut. Syst. Analy. and Control, Vol. 16, 3, 221-231

Dahlqvist, S.A. (1981). Can. J. Chem. Eng., Vol. 59 118-127

Hahn, V. (1983). Ph. D. Dissertation, Ruhr-University, Bochum

Koivo, K. (1980). Automatica, Vol. 16, 351-365

Koutchoukali, M.S., C. Laguérie and K. Najim, (1983). IFAS Workshop on Adaptive Systems in Control and Signal Processing, San Francisco

Koutchoukali, M.S., C. Laguérie and K. Najim, (1984). 9th IFAC World Congress, Budapest

Martin-Sanchez, J.M., S.L. Shah, (1984). Automatica Vol. 20, 5, 607-620

Morris, A.J., Y. Nazer and R.K. Wood, (1982). Optimal Control Applications & Methods, Vol. 3 363-387

Najim, K., M. Najim and H. Youlal, (1982). Optimal Control Applications & Methods, Vol. 3, 435-442

Narendra, K.S., R.V. Monopoli, (Eds.), (1980). Applications of Adaptive Control, Academic Press, New York

Unbehauen, H. (Ed.), (1980). Methods and Applications in Adaptive Control, Springer, Berlin

Wienner, P., V. Hahn, Ch. Schmid and H. Unbehauen, (1983). Proceedings IFA Workshop Adaptive Systems in Control and Signal Processing, San Francisco

Youlal, H., K. Najim, (1983). Moroccan J. of Control Computer Science and Signal Processing Vol. 1, 1, 35-59

Ziegler, J.G., N.B. Nichols, (1942). Trans. ASME, Vol. 64, 759

FEEDBACK LINEAR ESTIMATION OF ARMA PROCESSES

Xu Yanhua

Department of Electrical Engineering, Tongji University, Shanghai, China

Abstract. A four-step algorithm of FLE (Feedback Linear Estimation) based on the feedback control principle for ARMA processes is presented in this paper. The proposed algorithm uses three linear least square estimators and a linear filter. Linear estimations are utilized as a tool throughout the algorithm. The physical meaning of FLE is discussed and some simulation examples are shown.

Keywords. Feedback; ARMA model; least square estimator; recursive form; convergence.

INTRODUCTION

There are two shortcomings in estimating the parameters of an ARMA model by nonlinear algorithms:
1. The initial values of parameters to be estimated should be chosen properly.
2. The nonlinear algorithm requires much calculation and much memory space in the computer.

For this reason, some authors proposed approximate algorithms with linear estimations which need no iterative calculation. One of them is the five-step linear estimation algorithm presented by Mayne and Firoozan (1982). This algorithm has been extended into a controlled recursive form by Åström and Mayne (1983). The convergence of the algorithm has been proved, but a prerequisite for using this algorithm (that is, for the parameters to approach the true values well) is that the order of the AR(ρ) model in the first step should be large enough. Therefore, the choice of ρ, in fact, is restricted by the following:

A. The value of ρ should not exceed half the length of the samples (N/2).
B. When the value of ρ increases, the noise influence on the parameters becomes serious.
C. The value of ρ is required to tend to infinity when the polynomial of the moving average part of a system possesses some zeros near the unit circle.

Thus the generality of this algorithm is limited. By use of this approach, it is known that the accuracy of the fitted ARMA model is chiefly determined by that of the initially estimated residues. In order to obtain an accurate ARMA model with initial residues having deviations to a certain extent, the algorithm itself should possess an ability to correct the residues automatically. In this paper the FLE algorithm is developed. The idea of feedback control is introduced to the modelling of ARMA processes in FLE. At first, an AR(ρ) model is fitted to the stationary data samples from a system (the value of ρ does not affect the modelling result). Then the residues before the time t are considered as known inputs and are put into an ARMA model to be fitted. Thus, the rough values of model parameters can be estimated on the basis of the least-square sum. The residues of the ARMA process are linearly filtered, and the errors of initial residues are calculated which are used to correct the initial residues.

Finally, the model parameters are estimated again by linear least-squares. Therefore, the three shortcomings in the choice of ρ in the five-step algorithm are overcome and the accuracy of the model can be improved.

FLE FOUR STEP ALGORITHM

The FLE algorithm consists of three linear least square estimators and a linear filter.

Step 1: Fit an AR(ρ) model to observed data x_t

$$x_t + \psi(B)x_t = a_t \tag{1}$$

where $\psi(B) = \psi_1 B + \psi_2 B^2 + \ldots\ldots + \psi_\rho B^\rho$

B is a back operator

$$Bx_t = x_{t-1}\ , \quad Ba_t = a_{t-1}$$

Estimate the parameters $\hat{\psi}_i$ (i = 1,2,......,) by Yule-Walker equation such that the performance index

$$J_1 = \sum_{i=1}^{N-\rho} a_{t-i+1}^2 = \min \tag{2}$$

Meanwhile calculate the corresponding residues

$$\hat{a}_t = x_t + \hat{\psi}(B)x_t \tag{3}$$

Step 2: Suppose the practical process can be described by an ARMA (n,m) model

$$x_t + \phi(B)x_t = a_t + \theta(B)a_t \tag{4}$$

where

$$\phi(B) = \phi_1 B + \phi_2 B^2 + \ldots\ldots + \phi_n B^n$$

$$\theta(B) = \theta_1 B + \theta_2 B^2 + \ldots\ldots + \theta_m B^m$$

put $\hat{a}_t$ into (4)

$$x_t + \phi(B)x_t = \theta(B)\hat{a}_t + R_t \tag{5a}$$

$$R_t = \theta(B)\Delta a_t + a_t \tag{5b}$$

$$a_t = \hat{a}_t + \Delta a_t \tag{5c}$$

Estimate the parameters $\hat{\phi}_i$ (i = 1,2,... n) and $\hat{\theta}_j$ (j = 1,2, ... m) such that

$$J_2 = \sum_{i=1}^{N-n} R^2_{t-i+1} = \min \tag{6}$$

Similarly calculate the corresponding residues

$$\hat{R}_t = x_t + \hat{\phi}(B)\, x_t - \hat{\theta}(B)\, \hat{a}_t \tag{7}$$

step 3: Substitute $\tilde{R}_t$, $\hat{\theta}(B)$ for R_t, $\theta(B)$ in (5b) separately and consider (5c)

$$R_t = \hat{\theta}(B)\, \Delta a_t + \Delta a_t + \hat{a}_t \tag{8}$$

Then calculate Δa_t , In practice the following recursive formulae can be used:

$$\left.\begin{aligned}
&\Delta a_{t-N'} = R_{t-N'} - a_{t-N'} \\
&\Delta \hat{a}_{t-i} = \hat{R}_{t-i} - \hat{a}_{t-i} - \sum_{j=1}^{N'-i} \theta_j\, \Delta a_{t-i-j} \\
&\qquad i = N'-1, \ldots\ldots, N'-m \\
&\Delta \hat{a}_{t-i} = \hat{R}_{t-i} - \hat{a}_{t-i} - \sum_{j=1}^{m} \hat{\theta}_j\, \Delta \hat{a}_{t-i-j} \\
&\qquad i = N'-m-1, \ldots\ldots, 0
\end{aligned}\right] \tag{9}$$

where $N' = N - n + 1$ and supposing the values $\hat{a}_t$ before he time $t - N' + 1$ take their mean value of zero.

step 4: Substitute $\Delta\hat{a}_t$, $\hat{\phi}(B)$ and $\hat{\theta}(B)$ for the corresponding terms separately in (5a) and (5b). The result becomes

$$x_t + \hat{\phi}(B)\, x_t = \hat{\theta}(B)\, (\hat{a}_t + \Delta a_t) + a_t \tag{10}$$

Estimate the parameters $\overline{\phi}{}^1_i$ $(i = 1, 2, \ldots n)$, $\overline{\theta}{}^1_j$ $(j = 1, 2, \ldots m)$ again by linear least-squares such that

$$J_3 = \sum_{i=1}^{N-n} a^2_{t-i+1} = \min \tag{11}$$

Thus $\overline{\phi}{}^1_i$, $\overline{\theta}{}^1_j$ can be taken as an estimation on the true parameters. Practically in order to improve the model accuracy, the step 3 and step 4 should be repeated after estimating $\overline{\phi}{}^1_j$ and $\overline{\theta}{}^1_j$. Estimate step by step $\overline{\phi}_i$, $\overline{\theta}{}^2_j$; $\overline{\phi}{}^3_i$, $\overline{\theta}{}^3_j$ ..... and so on. The simulations in the latter part of this paper show a considerably higher modelling accuracy can generally be obtained by $\overline{\phi}{}^3_i$, $\overline{\theta}{}^3_j$. The following two results of FLE can be proved.

1. $\hat{a}_t + \Delta\hat{a}_t$ converges to a_t in mean squares.
2. $\overline{\phi}_i$, $\overline{\theta}_t$ are asymptotic unbiased estimations of ϕ_i , θ_j .

The physical meanings of FLE can be described by Fig.1. It is known that a forward and a feedback linear filters, in fact, are introduced into FLE.

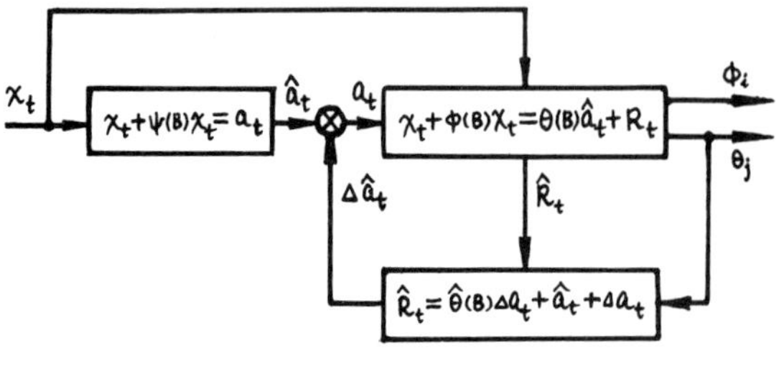

Fig. 1 The principle of FLE

The observed sequence x_t, after the filtering of $\psi(B)$, yields a rough estimation of the system residues ($\hat{a}_t$), by which the system is excited. Then the system parameters $\hat{\phi}_i$, $\hat{\theta}_j$ and the residues $\hat{R}_t$ are roughly obtained. Afterwards the feedback signal is taken from $\hat{\theta}_j$ and $\hat{R}_t$ which pass through the feedback filter and produce a correcting sequence $\Delta\hat{a}_t$. Although $\hat{a}_t + \Delta\hat{a}_t$ asymptotic converges to a_t, the estimated parameters $\overline{\phi}_i$, $\overline{\theta}_j$ converge to the true values.

Now consider the feedback filter

$$\hat{R}_t = \hat{\theta}(B)\, \Delta a_t + \Delta a_t + \hat{a}_t \tag{12}$$

Take matrix form

$$R = \Theta\, \Delta A \tag{13}$$

where

$$R = (\hat{R}_t - \hat{a}_t, \ldots\ldots, \hat{R}_{t-N+n-1} - \hat{a}_{t-N+n-1})^T$$

$$\Delta A = (\Delta a_t,\ \Delta a_{t-1}, \ldots\ldots \Delta a_{t-N+n-1})^T$$

$$\Theta = \begin{bmatrix} 1 & \hat{\theta}_1 \cdots\cdots & \hat{\theta}_m & & 0 \\ & 1 & \ddots & \ddots & \hat{\theta}_m \\ & & \ddots & \ddots & \vdots \\ & & & \ddots & \hat{\theta}_1 \\ 0 & & & & 1 \end{bmatrix}$$

Both sides of (13) are postmultiplied by the transposed matrix of themselves and then take mathematical expectation

$$E\{R R^T\} = \Theta E\{\Delta A\ \Delta A^T\}\Theta^T$$

or

$$\gamma_R = \Theta \gamma_{\Delta A} \Theta^T \tag{14}$$

where γ_R, $\gamma_{\Delta A}$ are the variance matrices of $\hat{R}_t - \hat{a}_t$ and Δa_t respectively. Pay attention to Θ, it is a constant matrix and its determinant is 1. Both sides of (14) take determinant

$$|\gamma_R| = |\gamma_{\Delta A}| \cdot |\Theta|^2$$

$$= |\gamma_{\Delta A}|$$

It is known by this fact that after the parameters $\hat{\theta}_j$ $(j = 1, 2, \ldots m)$ have been estimated, the determinant of the variance matrix of the residue error sequence of the system is equal to that of the correcting sequence. As $\hat{a}_t$ is known, when the rough estimated parameters are more accurate and the $|\gamma_R|$ is less, the $|\gamma_{\Delta A}|$ is less, too. Otherwise it turns to the opposite. So FLE does not demand the order of the forward filter to be larger. When the errors between $\hat{a}_t$ and a_t are larger, then there are Δa_t with larger variance to compensate the $\hat{a}_t$ automatically.

SIMULATIONS AND CONCLUSIONS

Example 1. FLE was used to model the sun spots yearly data from 1719 to 1924. We chose the model structure ARMA (2.1). The purposes of this experiment were to investigate the effects caused by different orders of the forward filter to the modelling results, and to compare the modelling accuracy by using FLE with that by using the nonlinear least-squares (NLS). In this experiment, the order of the forward filter (p) was taken as 1, 2, 3, 5, 8 and 10 respectively. The results are shown in Table 1.

Example 2. We made a comparison between FLE and

the five steps algorithm. The second example of reference 3 was taken. It was an ARMA (1.1) model.

$$x_t - 0.5x_{t-1} = a_t + 0.5a_{t-1} \qquad (15)$$

There are two purposes of this examination. The first is to compare with the modelling accuracy between FLE and the five steps algorithm. The second is to verify the effectiveness of FLE when the polynomial of the moving average part of a system possessed some zeros near the unit circle. In this simulation, we let the value of θ_1 move towards the boundary of the inversable range and investigate the situation when θ_1 was taken as 0.9, 0.95 and 0.99 besides 0.5 . The simulating results are shown in table 2 and table 3.

From the results of above two simulations the conclusions are following:

A. The different orders of the forward filter only affect the repeating times and do not hamper the convergence of parameters towards the true values.

B. When the characteristic roots of the moving average part of a system are far away from the boundary of the inversable range, only a few repeating times are demanded to make the parameters approach to the true values. Otherwise on the condition of the same modelling accuracy, the repeating times should correspondingly be increased.

C. The modelling accuracy is related to the repeating times. The repeating times should be properly increased as the modelling accuracy is required higher. If the fast modelling speed is expected instead of higher modelling accuracy the repeating times should properly be decreased.

D. In the example 1, as the repeating times is about 3 the residue square sum of the fitted model is only about 0.71% larger with FLE than those with NLS.

E. In the example 2, as $\phi_1 = -0.5$ and $\theta_1 = 0.5$, the absolute errors of the parameters are about 90% less with FLE than with the five steps algorithm.

In brief, the three shortcomings in chosing the forward filter order in the five steps linear estimating algorithm are overcomed by introducing a feedback linear filter into FLE. So the parameters can eventually converge to the true values well on different initial residues. Repeating the step 3 and step 4 only a few times, a higher modelling accuracy can be obtained. FLE can also be used for the case completely that the characteristic roots of the moving average part of a system lie near the boundary of the inversable range. The program of FLE is simple. It needs not much calculation and takes only a less memory space. It can be realized on a personal microcomputer.

Table 1 The simulating results to the sun spots data (1719 ∿ 1924, N = 176) Type ARMA (2,1)

Algorithm						
FLE						
p	$\bar{\phi}_1$	$\bar{\phi}_2$	$\bar{\theta}_1$	RSS	RSS error ratio	**
1	-1.401± 0.17	0.703±0.15	-0.115±0.22	41077.930	0.71%	3
2	-1.401±0.17	0.703±0.15	-0.116±0.22	41078.035	0.71%	2
3	-1.401±0.17	0.703±0.15	-0.116±0.22	41078.047	0.71%	4
5	-1.401±0.17	0.703±0.15	-0.116±0.22	41078.051	0.71%	3
8	-1.401±0.17	0.703±0.15	-0.116±0.22	41078.031	0.71%	3
10	-1.401±0.17	0.703±0.15	-0.116±0.22	41078.047	0.71%	3
NLS						
	-1.41±0.16	0.71±0.14	-0.14±0.23	40788.00	0	

** repeating times

Table 2 The simulating results of the five steps algorithm

filtering times	1	2	3	4	5
$\bar{\phi}_1$	-0.546	-0.578	-0.573	-0.574	-0.574
$\bar{\theta}_1$	0.458	0.393	0.404	0.402	0.402
$\|\phi_1\ \ \bar{\phi}_1\|$	0.046	0.078	0.073	0.074	0.074
$\|\theta_1\ \ \bar{\theta}_1\|$	0.042	0.107	0.096	0.098	0.098

$\phi_1 = -0.5 \quad \theta_1 = 0.5 \quad N = 500 \quad p = 5$

Table 3 The simulating results of FLE (N = 500)

true parameters	$\phi_1 = -0.5$	$\theta_1 = 0.5$		$\phi_1 = -0.5$	$\theta_1 = 0.9$	
p	1	2	3	1	2	3
$\bar{\phi}_1$	-0.497	-0.497	-0.497	-0.497	-0.496	-0.497
$\bar{\theta}_1$	0.503	0.503	0.503	0.903	0.905	0.903
$\|\phi_1 - \bar{\phi}_1\|$	0.003	0.003	0.003	0.003	0.004	0.003
$\|\theta_1 - \bar{\theta}_1\|$	0.003	0.003	0.003	0.003	0.005	0.003
repeating times	2	4	3	5	6	6
true parameters	= - 0.5	=0.95		= - 0.5	= 0.99	
p	1	2	3	1	2	3
$\bar{\phi}_1$	-0.496	-0.495	-0.496	-0.457	-0.497	-0.457
$\bar{\theta}_1$	0.954	0.955	0.954	0.981	0.992	0.981
$\|\phi_1 - \bar{\phi}_1\|$	0.004	0.005	0.004	0.043	0.003	0.043
$\|\theta_1 - \bar{\theta}_1\|$	0.004	0.005	0.004	0.009	0.002	0.009
repeating times	7	6	8	4	6	5

REFERENCES

An, H.Z., Chen, S.G., Du, J.G. and Pan, Y.M. (1983). Time Serise Analysis and Applications. Scientific Press, Beijing.

Åström, J.K. and Mayne, D.Q. (1983). 6th IFAC Symposium on Identification and System Parameter Estimation.

Eykhoff, P. (1977). System Identification Parameter and State Estimation. John Wiley & Sons, New York.

Goodwin, G.C. and Payne, R.L. (1977). Dynamic System Identification, Experiment Design and Data Analysis. Academic Press, New York.

Hannen, E.J. and Kvalieris, L. (1983). Refined Instrument variable Methods in ARMA Processes. Automatica, 19. 447-449.

Mayne, D.Q. and Firoozan, F. (1982). Linear Identification of ARMA Processes. Automatica, 18, 461-464.

APPENDIX THE CONVERGENCE PROPERTY OF FLE

Following two results can be obtained by the FLE four steps algorithm:

1. $\hat{a}_t + \Delta\hat{a}_t$ converges to a_t in mean squares.
2. $\overline{\phi}_i$, $\overline{\theta}_j$ are asymptotic unbiased estimations for ϕ_i, θ_j $(i = 1,2,...n;\ j = 1,2,...\ m)$.

We give the heuristic prooves.

Proof 1.:
Substitute $\hat{R}_t$, $\hat{\phi}_i$, $\hat{\theta}_j$ for the corresponding terms separately in (5a) and subtract the result from (4) by both sides

$$\tilde{\phi}(B)\,x_t = \theta(B)\,a_t + \tilde{a}_t - \hat{\theta}(B)\,(\hat{a}_t + \Delta\hat{a}_t)$$

$$= \tilde{\theta}(B)\,a_t + (1 + \hat{\theta}(B))\,\tilde{a}_t \qquad (A.1)$$

where

$$\tilde{\phi}(B) = \phi(B) - \hat{\phi}(B), \quad \tilde{\theta}(B) = \theta(B) - \hat{\theta}(B),$$

$$\tilde{a}_t = a_t - (\hat{a}_t + \Delta\hat{a}_t).$$

$$\because\ E\{\tilde{a}_t x_{t-i}\} = E\{a_t x_{t-i}\} - (\hat{a}_t + \Delta\hat{a}_t)\,E\{x_{t-i}\}$$

$$= 0 \qquad i = 1,2,......$$

$$E\{a_t a_{t-i}\} = E\{a_t a_{t-i}\} - (\hat{a}_t + \Delta\hat{a}_t)\,E\{a_{t-i}\}$$

$$= 0 \qquad i = 1,2,......$$

Both sides of (A.1) are multiplied by a_t and then take the mathematical expectation

$$E\{\tilde{a}_t^2\} + \hat{\theta}(B)\,\tilde{a}_t^2\} = 0$$

that is

$$E\{\tilde{a}_t^2\} + E\{\sum_{j=1}^{m} \hat{\theta}_j\, \tilde{a}_{t-j}^2\}$$

$$= E\{\tilde{a}_t^2\} + \sum_{j=1}^{m} \theta_j\, E\{\tilde{a}_{t-j}^2\}$$

$$= E\{\tilde{a}_t^2\}\,(1 + \sum_{j=1}^{m} \hat{\theta}_j) = 0 \qquad (A.2)$$

$\because\ 1 + \sum_{j=1}^{m} \hat{\theta}_j$ is not always equal to zero,

$$\therefore\ E\{\tilde{a}_t^2\} = E\{(a_t - (\hat{a}_t + \Delta\hat{a}_t))^2\} = 0 \qquad (A.3)$$

Proof 2.:
Subtract (10) from (4) and substitute ϕ_i, θ_j for the corresponding terms

$$\overline{\phi}'(B)\,x_t = \theta(B)\,a_t - \overline{\theta}(B)\,(\hat{a}_t + \Delta\hat{a}_t)$$

$$= \tilde{\theta}'(B)\,a_t + \overline{\theta}(B)\,\tilde{a}_t \qquad (A.4)$$

Take the matrix form

$$X\,\Lambda = H\,¢ \qquad (A.5)$$

where

$$\overline{\phi}'(B) = \phi(B) - \overline{\phi}(B)$$

$$\tilde{\theta}'(B) = \theta(B) - \overline{\theta}(B)$$

$$\Lambda = (\tilde{\phi}_1', \tilde{\phi}_2', ..., \tilde{\phi}_n', \tilde{\theta}_1', \tilde{\theta}_2', ... \tilde{\theta}_m')^T$$

$$¢ = (\tilde{a}_{t-1}, \tilde{a}_{t-2}, ... \tilde{a}_{t-N+1})^T$$

$$\Lambda = \begin{bmatrix} -x_{t-1} & -x_{t-n+1} & a_{t-1} & a_{t-m+1} \\ \vdots & \vdots & \vdots & \\ \vdots & \vdots & \vdots & \\ \vdots & \vdots & \vdots & \\ -x_{t-N+n-1} & -x_{t-N+1} & a_{t-N+n-1} & a_{t-N+n-m+1} \end{bmatrix}$$

$$H = \begin{bmatrix} \overline{\theta}_1 \cdots\cdots \overline{\theta}_m & & \\ & \ddots & \ddots \\ & & \cdot\overline{\theta}_m \\ 0 & \ddots & \vdots \\ & & \cdot\overline{\theta}_1 \end{bmatrix}$$

$X^T X$ is full rank, by (A.5)

$$E\{\Lambda\} = E\{\lim_{N\to\infty} (X^T X)^{-1} H \lim_{N\to\infty} ¢\}$$

$$= 0$$

that is $\overline{\phi}_i \Rightarrow \phi_i$, $\overline{\theta}_j \Rightarrow \theta_j$ $\quad i = 1,2,...\ n$; $j = 1,2,...\ m$; $N \to \infty$

DIRECT IDENTIFICATION OF A CLASS OF NONLINEAR SYSTEMS/APPLICATION TO A NEUTRALIZATION PROCESS

B. Neyran*, D. Thomasset and J. Dufour*****

GRAP, "Groupe de Recherche Rhône Alpin en Automatique et ses Applications à la Production"

**Laboratoire d'Automatique, Bât. 721, Université de Lyon 1, 69622 Villeurbanne Cedex, France*

***Laboratoire d'Energétique et d'Automatique, Bât. 404, Institut National des Sciences Appliquées, 69621 Villeurbanne Cedex, France*

****Laboratoire d'Automatique et de Microinformatique Industrielle, Université de Savoie, BP 806, 74016 Annecy Cedex, France*

ABSTRACT

This paper describes the direct identification of a varylinear system that is linear with variable parameters. This approach leads to a least square estimation computed with the input and output sequences filtered through a non linear filter, in order to obtain an unbiased estimation. An application to a neutralization process illustrates the method.

KEYWORDS

Identification, Varylinear Systems, State-affine Systems.

INTRODUCTION

Modelling by linear identification is only valid around a working point and cannot represent dynamical behaviour for important disturbances, so adaptive methods are often chosen for design control law. Our approach is quite different, the next sections allow to find a global model that will lead to a global control law valid for all operating points. With Stone-Weierstrass theorem, it has been shown (Bourdon, 1982, Fliess and Normand-Cyrot, 1980) that discrete non linear systems can be approached by state affine models.

Identification of state-affine systems using linear identifications at different working points has given good results in several applications (Normand-Cyrot D., and Dang Van Mien H., 1980, Normand-Cyrot D., 1978).The section 2 develops this previous method. Section 3 presents a direct global approach in order to avoid privileging the working points where linear identification have been done : it uses a non linear filter in order to obtain an unbiased estimation. It differs from the previous works because it does not require a serie of linear identifications. In section 4, an application to a neutralization process allows to compare the two methods.

In this paper, we consider state-affine systems introduced by E.D. Sontag (1979) with a state space representation of the form :

$$\begin{cases} x(k+1) = \left[A_0 + \sum_i P_i(u_1(k),\dots,u_m(k))A_i\right]x(k) \\ y(k) = C\,x(k) \end{cases}$$

where :

$$\begin{cases} x(k) : \text{state vector, n dimensional ; } x(0) \text{ is given} \\ y(k) : \text{output vector, p dimensional} \\ u_1(k),\dots,u_m(k) : \text{the m inputs} \\ P_i : \text{monomials in the inputs } u_j (j=1,m) \\ A_0,\ A_i,\ C : \text{appropriate dimensional matrices} \end{cases}$$

CONSTRUCTION OF A STATE-AFFINE MODEL USING SEVERAL LINEAR IDENTIFICATIONS (Normand-Cyrot and Dang Van Mien, 1980, Normand-Cyrot, 1978)

General one-output linear model can be described by the state form :

$$\begin{cases} \nu(k+1) = \alpha\,\nu(k) + \beta\,U(k) \\ y(k) = \gamma\,\nu(k) \end{cases}$$

with $\beta = [\beta^1\ \beta^2\ \beta^m]$ nxm dimensional

$$U(k) = \begin{bmatrix} u_1(k) \\ \vdots \\ u_m(k) \end{bmatrix}$$

α : nxm dimensional $\quad\gamma$: 1xn dimensional

This system is indiscernible from a bilinear one of the form :

$$\begin{cases} X(k+1) = \begin{bmatrix} \nu(k+1) \\ 1 \end{bmatrix} = \left[\begin{bmatrix} \alpha & 0 \\ 0 & 1 \end{bmatrix} + u_1(k)\begin{bmatrix} 0 & \beta^1 \\ 0 & 0 \end{bmatrix} + u_2(k)\begin{bmatrix} 0 & \beta^2 \\ 0 & 0 \end{bmatrix} + \dots + u_m(k)\begin{bmatrix} 0 & \beta^m \\ 0 & 0 \end{bmatrix}\right] X(k) \\ y(k) = [\gamma\ \ 0]\,X(k) \end{cases} \quad (1)$$

Let us consider that the α and β matrices coefficients change with an inputs variation or with a modification of measurable parameters Q_i. This variation may be described by two sets P_α , P_β of monomials in the new inputs $u_1,\dots,u_m,Q_1,\dots,Q_1$.

$$\begin{cases} \alpha = A_0 + A_1 P_{A_1}(u_1,\dots,u_m,Q_1,\dots,Q_1) + \dots \\ \quad + A_q P_{A_q}(u_1,\dots,u_m,Q_1,\dots,Q_1) \\ \beta = B_0 + B_1 P_{B_1}(u_1,\dots,u_m,Q_1,\dots,Q_1) \\ \quad + B_q P_{B_q}(u_1,\dots,u_m,Q_1,\dots,Q_1) \end{cases} \quad (2)$$

The choice of the monomials P_{A_i} , P_{B_i} may be changed in order to improve the model precision.

Without loss of generality, the companion or Jordan canonical form is considered, leading to a γ constant matrix. With (1) and (2) we obtain :

$$X(k+1)=\left\{\begin{bmatrix}A_0 & 0\\ 0 & 1\end{bmatrix}+\ldots+P_{A_q}(u_1,\ldots,Q_1)\begin{bmatrix}A_q & 0\\ 0 & 0\end{bmatrix}\right.$$

$$+u_1(k)\begin{bmatrix}0 & B_0^1\\ 0 & 0\end{bmatrix}+\ldots+P_{B_q}(u_1,\ldots,Q_1)\begin{bmatrix}0 & B_q^1\\ 0 & 0\end{bmatrix}+\ldots$$

$$\left.+u_m(k)\begin{bmatrix}0 & B_0^m\\ 0 & 0\end{bmatrix}+\ldots+P_{B_q}(u_1,\ldots,Q_1)\begin{bmatrix}0 & B_q^m\\ 0 & 0\end{bmatrix}\right\}\begin{bmatrix}X(k)\end{bmatrix}$$

$$y(k) = \begin{bmatrix}\gamma & 0\end{bmatrix} X(k)$$

with : B_i^j $(i=1,q)$ $(j=1,m)$ the jth column of B_i matrix

This method gives a state-affine model from a serie of linear identifications. In order to avoid privileging the working points where linear identifications were obtained, the next section presents a global approach.

DIRECT IDENTIFICATION OF STATE-AFFINE MODEL

A one-input linear system may be described by a canonical state representation of the form :

$$\begin{bmatrix}x_1(k+1)\\ \cdot\\ \cdot\\ \cdot\\ \cdot\\ x_n(k+1)\end{bmatrix}=\begin{bmatrix}\alpha_1 & 1 & 0 & \ldots & 0\\ \vdots & 0 & 1 & \ldots & 0\\ \vdots & & \ddots & & \vdots\\ \vdots & & & \ddots & 1\\ \alpha & 0 & \ldots & \ldots & 0\end{bmatrix}\begin{bmatrix}x_1(k)\\ \cdot\\ \cdot\\ \cdot\\ x_n(k)\end{bmatrix}+\begin{bmatrix}\beta_1\\ \vdots\\ \vdots\\ \beta_n\end{bmatrix}u(k)$$

$$y(k) = \begin{bmatrix}1 & 0 .. 0\end{bmatrix}\begin{bmatrix}x_1(k)\\ \vdots\\ x_n(k)\end{bmatrix}$$

noted : $\begin{cases}x(k+1) = \alpha\, x(k) + \beta\, u(k)\\ y(k) = \gamma\, x(k)\end{cases}$

Consider that the coefficients of α and β matrices change with a measurable parameter Q, but for each value of Q, this system is a linear one. In order to obtain only one model valid for all operating points, we write a non linear state-affine system, like in section 2, using two sets of monomials in the inputs u and Q.

$$\begin{cases}\alpha = A_0 + A_1\, P_{A_1}(Q) + A_2\, P_{A_2}(Q) + \ldots + A_q\, P_{A_q}(Q)\\ \beta = B_0 + B_1\, P_{B_1}(Q) + B_2\, P_{B_2}(Q) + \ldots + B_q\, P_{B_q}(Q)\end{cases}$$

So :

$$\alpha=\begin{bmatrix}a_{10} & 1 & 0 . 0\\ . & 0 & 1 . 0\\ \vdots & \vdots & \ddots\\ \vdots & 0 & \cdot 1\\ a_{n0} & 0 & .. 0\end{bmatrix}+P_{A_1}(Q)\begin{bmatrix}a_{11} & 0 .. 0\\ \vdots & \vdots\\ a_{n1} & 0 .. 0\end{bmatrix}+\ldots+P_{A_q}(Q)\begin{bmatrix}a_{1q} & 0 .. 0\\ \vdots & \vdots\\ a_{nq} & 0 .. 0\end{bmatrix}$$

$$\beta = \begin{bmatrix}b_{10}\\ \vdots\\ b_{n0}\end{bmatrix}+P_{B_1}(Q)\begin{bmatrix}b_{11}\\ \vdots\\ b_{n1}\end{bmatrix}+\ldots+P_{B_q}(Q)\begin{bmatrix}b_{1q}\\ \vdots\\ b_{nq}\end{bmatrix}$$

which gives a model of the form :

$$X(k+1)=\begin{bmatrix}x_1(k+1)\\ \vdots\\ x_{n_1}(k+1)\end{bmatrix}=\left[\begin{bmatrix}A_0 & 0\\ 0 & 1\end{bmatrix}+P_{A_1}(Q(k))\begin{bmatrix}A_1 & 0\\ 0 & 0\end{bmatrix}+\ldots+P_{A_q}(Q(k))\begin{bmatrix}A_q & 0\\ 0 & 0\end{bmatrix}\right.$$

$$\left.+u(k)\begin{bmatrix}0 & B_0\\ 0 & 0\end{bmatrix}+u(k)P_{B_1}(Q(k))\begin{bmatrix}0 & B_1\\ 0 & 0\end{bmatrix}+\ldots+u(k)P_{B_q}(Q(k))\begin{bmatrix}0 & B_q\\ 0 & 0\end{bmatrix}\right]\begin{bmatrix}X(k)\end{bmatrix}$$

$$y(k) = \begin{bmatrix}\gamma & 0\end{bmatrix} X(k)$$

So, we obtain a relationship between inputs u and Q and output y :

$$\begin{aligned}y(k) = {} & a_{10}y(k-1)+a_{11}y(k-1)P_{A_1}(Q(k-1))+\ldots\\ & + a_{1q}y(k-1)P_{A_q}(Q(k-1))\\ & + b_{10}u(k-1)+b_{11}u(k-1)P_{B_1}(Q(k-1))+\ldots\\ & + b_{1q}u(k-1)P_{B_q}(Q(k-1))\\ & + a_{20}y(k-2)+a_{21}y(k-2)P_{A_1}(Q(k-2))+\ldots\\ & + a_{2q}y(k-2)P_{A_q}(Q(k-2))\\ & + \ldots\\ & + a_{n0}y(k-n)+a_{n1}y(k-n)P_{A_1}(Q(k-n))+\ldots\\ & + a_{nq}y(k-n)P_{A_q}(Q(k-n))\\ & + b_{n0}u(k-n)+b_{n1}u(k-n)P_{B_1}(Q(k-n))+\ldots\\ & + b_{nq}u(k-n)P_{B_q}(Q(k-n))\end{aligned}$$

Note :

$$R_0(z^{-1}) = 1 - a_{10}z^{-1} - a_{20}z^{-2} - \ldots - a_{n0}z^{-n}$$

$$R_i(z^{-1}) = a_{1i}z^{-1} + a_{2i}z^{-2} + \ldots + a_{ni}z^{-n}$$

for i = 1 to q

$$S_j(z^{-1}) = b_{1i}z^{-1} + b_{2i}z^{-1} + \ldots + b_{ni}z^{-n}$$

for j = 0 to q

it gives :

$$\begin{aligned}R_0(y(k)) = {} & R_1(y(k)P_{A_1}(Q(k)))+\ldots+R_q(y(k)P_{A_q}(Q(k)))\\ & + S_0(u(k))+S_1(u(k)P_{B_1}(Q(k)))+\ldots\\ & + S_q(u(k)P_{B_q}(Q(k)))\end{aligned}$$

Let us consider an additive output white noise :

$$y_b(k) = y(k) + b(k)$$

$$\begin{aligned}R_0(y_b(k)) = {} & R_1(y_b(k)P_{A_1}(Q(k)))+\ldots R_q(y_b(k)P_{A_q}(Q(k)))\\ & + S_0(u(k))+S_1(u(k)P_{B_1}(Q(k)))+\ldots\\ & + S_q(u(k)P_{B_q}(Q(k)))+R_0(b(k))\\ & - R_1(b(k)P_{A_1}(Q(k)))-\ldots-R_q(b(k)P_{A_q}(Q(k)))\end{aligned}$$

The residue is noted E(k)

$$E(k) = R_0(b(k))-R_1(b(k)P_{A_1}(Q(k)))-\ldots-R_q(b(k)P_{A_q}(Q(k))) \quad (3)$$

To identify the parameters of this model, the previous equations may be rewritten for N different sampling times : $(N>2n(q+1))$

$$\begin{bmatrix} y_b(k) \\ \vdots \\ y_b(k+N) \end{bmatrix} + \begin{bmatrix} y_b(k-1)..y_b(k-n)P_{A_q}(Q(k-n))u(k-1)..u(k-n)P_{B_q}(Q(k-n)) \\ \vdots \qquad\qquad\qquad \vdots \\ y_b(k+N-1).......................u(k+N-n)P_{B_q}(Q(k+N-n)) \end{bmatrix} \begin{bmatrix} a_{10} \\ \vdots \\ a_{nq} \\ b_{10} \\ \vdots \\ b_{nq} \end{bmatrix} \begin{bmatrix} E(k) \\ \vdots \\ E(k+N) \end{bmatrix}$$

And a matricial form gives $Y = X\theta + E$

The least square estimator of θ is $\hat{\theta}=(X^T X)^{-1} X^T Y$ (4)

In order to obtain an unbiased estimator, the X and E matrices must not be correlated and a zero mean value residue is necessary (that's the case if we suppose a zero mean value noise). But the first condition is not true : the parameters are correlated with the X matrix (3)(4). The residue must be filtered to obtain an unbiased estimator.

From (3) the residue may be rewritten :

$$\begin{aligned} E(k) = {} & b(k)-a_{10}b(k-1)-a_{20}b(k-2)-\ldots-a_{n0}b(k-n) \\ & - a_{11}b(k-1)P_{A_1}(Q(k-1))-\ldots-a_{n1}b(k-n)P_{A_q}(Q(k-n)) \\ & \ldots\ldots\ldots\ldots \\ & - a_{1q}b(k-1)P_{A_1}(Q(k-1)-\ldots-a_{nq}b(k-n)P_{A_q}(Q(k-n)) \end{aligned}$$

With such notations :

$$\begin{cases} g_0 = 1 \\ g_1 = -(a_{11}P_{A_1}(Q(k-1))+\ldots+a_{1q}P_{A_q}(Q(k-1))+a_{10}) \\ \ldots\ldots\ldots\ldots \\ g_n = -(a_{n1}P_{A_1}(Q(k-n))+\ldots+a_{nq}P_{A_q}(Q(k-n))+a_{n0}) \end{cases}$$

we have :

$$E(k) = b(k)(g_0+g_1z^{-1}+\ldots+g_nz^{-n}) = G(z^{-1})b(k)$$

$G(z^{-1})$ is a nonlinear filter, but a linear one for each value of the parameter $Q.G(z^{-1})$ known, we search a filter $F(z^{-1})$, such as :

$b(k) = F(z^{-1})E(k)$ with $b(k)$ the white output noise.

For each k we have :

$$b(k) = E(k)-g_1b(k-1)-\ldots-g_nb(k-n)$$

We start with : $\begin{cases} b(0) = E(0) \text{ with } k=0 \\ \ldots\ldots\ldots \\ b(n-1) = E(n-1) \end{cases}$

then : $\begin{cases} b(n) = E(n)-g_1b(n-1)-\ldots-g_nb(0) \\ b(n+1) = E(n+1)-g_1b(n)-\ldots-g_nb(1) \\ \ldots\ldots\ldots \\ b(N) = E(N)-g_1b(N-1)-\ldots-g_nb(N-n) \end{cases}$

In practice, we filter X and Y columns matrices and we build the following algorithm :

1. compute the first estimation of θ:$\hat{\theta}=(X^TX)^{-1}X^TY$,
2. find the residue : $E = Y - X\hat{\theta}$,
3. compute the nonlinear filter, and apply it to X and Y columns matrices,
4. compute the new estimation of θ with the filtered matrices,
5. return to step 2.

The method stops when the parameters variations is less than the suitable precision.

Like in the generalized least squares method (Clarke, 1967), the convergence has always been obtained but cannot be proved.

APPLICATION TO A NEUTRALIZATION PROCESS

Let us consider the continuous neutralization of a strong base (Na OH) and a strong acid (HCl) in a constant volume vessel (local level close loop). The configuration of the plant is described Fig. 1.

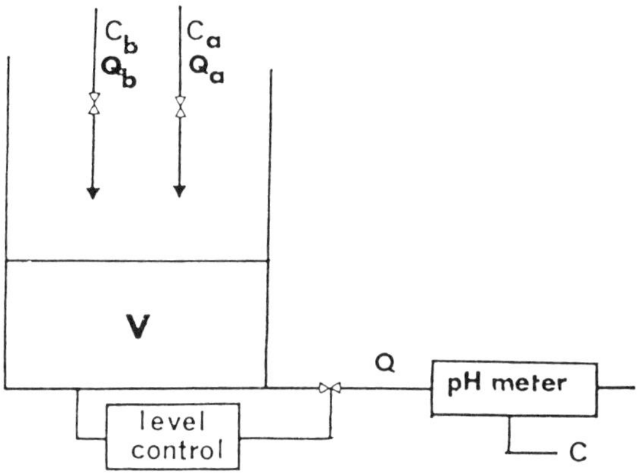

Fig. 1. Chemical pilot plant

$$\begin{cases} Q_a, C_a & : \text{acid flow-rate and concentration} \\ Q_b, C_b & : \text{base flow-rate and concentration} \\ Q & : \text{output flow-rate} : Q = Q_a + Q_b \\ & \quad \text{(local level close loop)} \\ C & : \text{output concentration} : C = C_{sa}-C_{sb} \end{cases}$$

with...: C_{sa} : output acid concentration

C_{sb} : output base concentration

The neutralization is considered instantaneous with a sufficient agitation in the vessel.

The conservation equations show that the evolution around a working point Q_{a0}, Q_{b0}, C_0 of the concentration $c(t)$ with respect to a base flow-rate variation $q(t)$ is given by the differential equation :

$$\frac{V}{Q_0}\frac{d\,c(t)}{dt} = \frac{-(C_b+C_0)}{Q_0}q(t)-c(t)-\frac{c(t)q(t)}{Q_0} \qquad (5)$$

with : $\begin{cases} C_0 = \dfrac{C_aQ_{a0}-C_bQ_{b0}}{Q_0} \\ Q_0 = Q_{a0}+Q_{b0} \end{cases}$

$c(t)q(t)$ is a bilinear term, negligible if we consider small values of $q(t)$.

In order to have a normalized model, we consider :

$$c_n(t) = \frac{c(t)}{-(C_b+C_0)}$$

So, the equation (5) may be rewritten :

$$\frac{V}{Q_0}\frac{dc_n(t)}{dt} = -\frac{1}{Q_0}q(t) - c_n(t)$$

With Q_0 constant, this equation represents a first order linear system with variable parameter V.

We have used the two previous methods on this neutralization process :

- we make five linear identifications for V = 4 l., 6 l., 8 l., 10 l., 12 l. in the vessel and build a state-affine model of the form :

$$\begin{cases} x(k+1) = (A_0+V(k)A_1+V(k)^2A_2+V(k)^3A_3+q(k)B_0 \\ \qquad + q(k)V(k)B_1+q(k)V(k)^2B_2+q(k)V(k)^3B_3)x(k) \\ c_n(k) = x(k) \end{cases}$$

This model is valid for all V belonging to [4 l., 12 l.] (Neyran and others, 1984). For the linear identification, we have used the model method on impulse responses obtained by intercorrelation.

- we make the direct identification of the state-affine model with a linear variation of V(t) from 3 l. to 13 l., and a pseudo random binary sequence on q(t).

The Fig. 2 to 6 show the step response of the two models for different values of V : 4 l., 6 l., 8 l., 10 l., 12 l.

For one value of V, we can consider the time constant and the gain of the tangent linear system of each state-affine model. Theoritically, the gain is constant and equal to 0.018 and the time constant has a linear variation with V. The Fig. 7 and 8 show the variation of this two parameters with V for each state-affine system. Note that for the small values of V, the direct method gives a better model but rather gets away from the theoritical parameters for the high values of V : the amplitude and the period of the pseudo random sequence are constant for all the identification, but as the time constant becomes higher, the signal noise ratio gets worse with the increase of V.

CONCLUSION

This global method, with a more complex formalism, needs a little more computation time, but does not require several linear identifications and does not privilege any working points because the state-affine model is obtained in passing on all the working points. The application to a neutralization process illustrates the validity of the method.

REFERENCES

Bourdon, P., M. Fliess, H. Dang Van Mien and D. Normand-Cyrot (1981), Méthodes d'identification et de réalisation non linéaires par espace d'état appliquées à des centrales nucléaires, Congrès AFCET, Nantes, pp. 517-529.

Bourdon, P., (1982), Techniques non linéaires en temps discret d'identification et de réalisation minimale pour modèle à état affine, Thesis, Paris.

Clarke, D.W., (1967), Generalized least squares estimation of the parameters of dynamic model, IFAC Symposium, Prague, paper 3.17.

Fliess, M., and D. Normand-Cyrot, (1980), Vers une approche algébrique des systèmes en temps discret. Analysis and Optimisation of Systems (A. Bensoussan and J.L. Lions Ed.), Lectures Notes in Cont. and Info. Sciences, 28, Springer Verlag, Berlin, pp. 594-603.

Normand-Cyrot, D., (1978), Utilisation de certaines familles algébriques de systèmes non-linéaires à quelques problèmes de filtrage et d'identification, Thesis, Paris.

Normand-Cyrot, D., and H. Dang Van Mien, (1980), Non-linear state-affine identification methods, applications to electrical power plants, IFAC Symposium on Aut. Cont. in Power generation, Distribution and Protection, Pretoria, pp. 449-462.

Neyran, B., J. Saade-Castro, D. Thomasset, R. Reynaud, (1984), Modélisation dynamique non-linéaire d'un procédé chimique de neutralisation, Colloque Automatique Appliquée S.E.E., Nice, pp. 255-259.

Sontag, E.D., (1979), Realization theory of discrete time non-linear systems, Part I. The bounded case, IEEE Trans. on Circuits and Systems, 26.

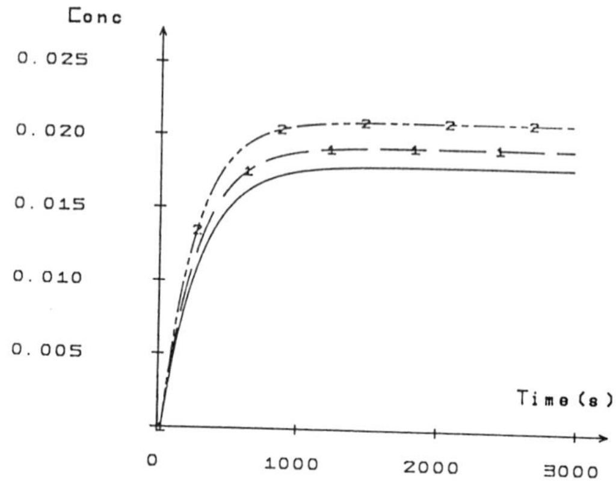

Figure 2 . V = 4 l

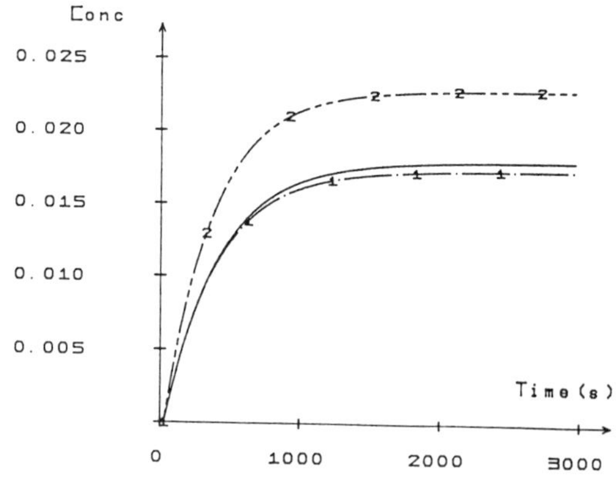

Figure 3 . V = 6 l

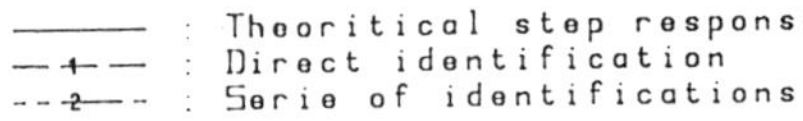

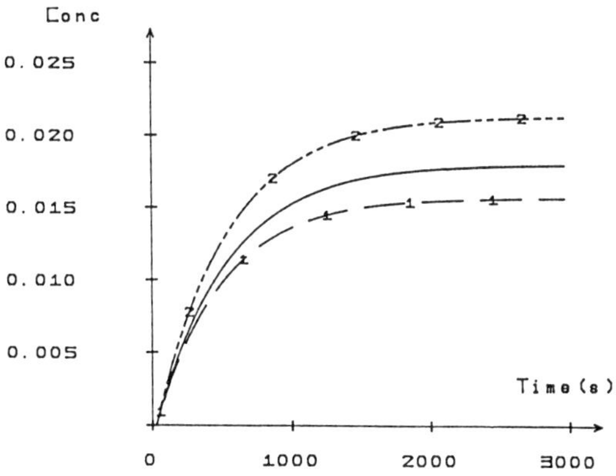

Figure 4 . V = 8 1

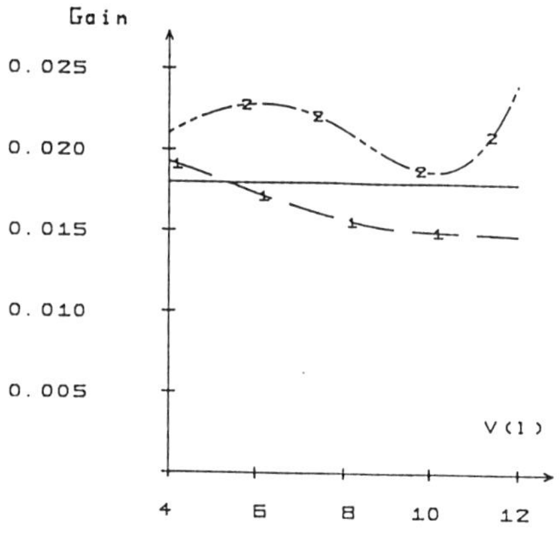

Figure 7

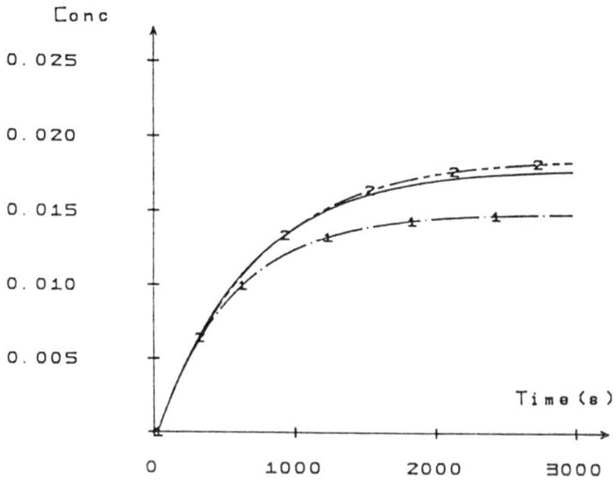

Figure 5 . V = 10 1

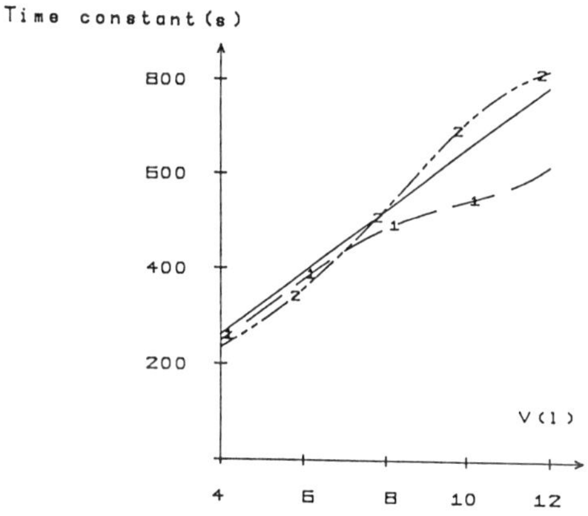

Figure 8

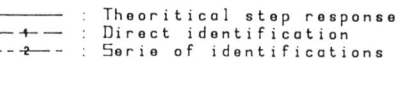

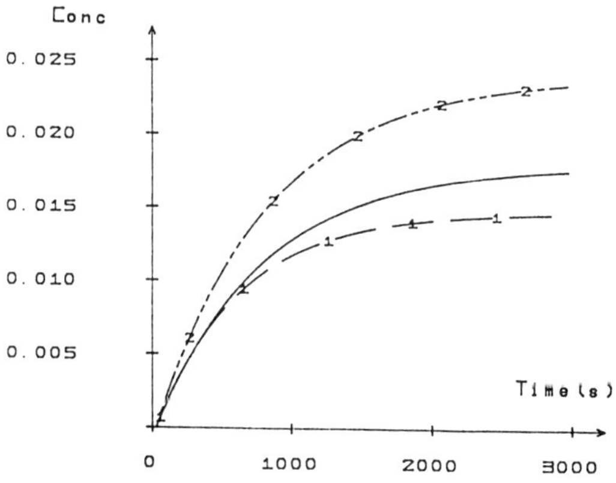

Figure 6 . V = 12 1

——— : Theoritical step response
—1— : Direct identification
- -2- - : Serie of identifications

BILINEAR DYNAMICAL MODEL OF A TUBULAR FIXED-BED REACTOR

Hua Xiangming and Jiang Weisun

Department of Automatic Control and Electronic Engineering, East China Institute of Chemical Technology, Shanghai, China

Abstract. The aim of the present work is to investigate the bilinear modeling of a tubular fixed-bed reactor, with an adiabatic ammonia reactor as an illustrative example. Through orthogonal collocation technique and bilinear approximation, a bilinear model is obtained from nonlinear partial differential equation describing the dynamics of the reactor. Furthermore a low order bilinear model is also developed. The validity of the bilinear model over the linear one is established via simulation. It is shown that this bilinear model is more accurate in matching the dynamics of the reactor at much wider range around the operating point. The methods suggested here are helpful for the modelization of similar complex processes and other tubular fixed-bed reactors such as methanol reactors.

Keywords. Nonlinear equations; distributed parameter systems; modelling; bilinear systems; chemical industry; ammonia reactor.

INTRODUCTION

In process industry there exist many nonlinear distributed parameter processes, among which the tubular fixed-bed reactors are typical. Because of the difficulty in mathematic treatment for these processes, the standard practice is first to linearize them about the operating steady state, then to lump them and finally to deal with the resulting linear system (Georgakis, Aris, and Amundson, 1977; Michelsen, Vakil, and Foss, 1973; Patnaik, Viswanadham, and Sarma, 1980). For a tubular ammonia reactor, Patnaik, Viswanadham and Sarma (1980) obtained a linear state space formulation using linearization and difference approximations, and studied its control schemes. However, the linearized lumped parameter model is sometimes unsuitable, especially for this case. Kopple (1965) affirmed that the range of the linearized model that can be used may be narrow, because of the nonlinearity introduced by a nonisothermal reaction with an Arrhenius type of temperature dependence. Some workers (Shah, 1967) have also pointed out this limitation. Therefore, this linearized model will be invalid when there exist larger disturbances or a transition of working point from one to another, thus a better local approximate model is expected.

Especially in the control problems of some reactors such as the multi-bed quench-type reactor, the flowrate of feedgas is manipulated so as to maintain the temperature along the length of the reactor around its optimum profile in the face of various disturbances. Therefore, the forcing functions are not inlet boundary conditions but are variations in the coefficients of the nonlinear partial differential equations describing the dynamics of the reactor, and thus the modelization of this kind of tubular fixed-bed reactors becomes more complex.

Recently the attention on bilinear system has been more and more focused. It is considered that this kind of system is a good compromise between linear models and nonlinear models. This is not only because the bilinear systems occur naturally in most industrial processes, such as heat exchangers (Jiang and Jiang, 1982), distillation columns (Espana and Landau, 1978), but also because for a general nonlinear dynamical system with linear inputs, bilinear approximation in the neighbourhood of a steady state can make a considerable improvement over the linearization (Svoronos, Stephanopoulos, and Aris, 1980). In addition, since the control function in bilinear systems acts as a multiplicative control, one of the most significant benefits is improvement of controllability. Some workers have pointed out that bilinear control is a useful tool for designing robust control systems. So far, the control theory for this kind of systems has been developed to a considerable extent, and an important theoretical support exists already (Mohler and Kolodziej, 1980).

Furthermore it may be noted that the orthogonal collocation method (OCM) has been proved to be a very efficient numerical technique in solving the mathematical models describing the combined effects of diffusion and reaction occuring in chemical engineering systems (Michelsen, Vakil and Foss, 1973; Finlayson, 1972). One of the most important advantages is that lumping using OCM for a given degree of accuracy can considerably reduce the number of collocation points.

In this paper, the nonlinear partial differential equations describing the dynamics of a adiabatic ammonia reactor are first given. Based on these equations, a bilinear model is obtained through lumping by OCM and using bilinear approximation. Then a low order bilinear model and linearized model are developed. Through simulation of the lumped nonlinear model, the bilinear model and the linear model given above, it is shown that the bilinear model can more accurately approach the real systems than linear model in spite of large variations of the input . Therefore this is a better approximate model of the original nonlinear distributed parameter systems.

DYNAMIC MODEL OF THE REACTOR

The process under consideration is chosen from an adiabatic section of some kind of ammonia reactor (Zhu, 1981). Ammonia is formed under high pressure following the exothermic reaction, $N_2+3H_2 \rightleftharpoons 2NH_3$. The dynamic model is derived under the following assumptions: (i) unimportant radial gradients and axial diffusion; (ii) no temperature difference between the catalyst particles and the gas phase; (iii) uniformly constant pressure throughout the reactor; (iv) negligible accumulation of mass because of the small residence time of the gas in the reactor; (v) no heat loss. The fixed-bed reactor is modeled by two partial differential euqations, namely, the mass and energy balance

$$\frac{\partial y}{\partial l} = \frac{COR\ A}{22.4\ N_{in}(1+y_{in})} R(y,\ T_b) \quad (1)$$

$$\frac{\partial T_b}{\partial \theta} + \frac{N_{in}\ C_{pb}\ (1+y_{in})}{\rho A\ C_{pc}\ (1+y)} \frac{\partial T_b}{\partial l} = \frac{COR\ (-\Delta H_R)}{22.4\ \rho\ C_{pc}(1+y)^2} R(y,\ T_b) \quad (2)$$

Boundary conditions

$$y(o,\theta) = y_{in};\qquad T_b(o,\theta) = T_{bo} + T_{b,\ in}. \quad (3)$$

Equations (1)-(2) can be transformed into the following dimensionless form

$$(1+u)\ (1+y_{in})\ \frac{\partial y}{\partial z} = \alpha\ r(y,\ T) \quad (4)$$

$$\frac{\partial T}{\partial t} + (1+u)(1+y_{in})\ f(y,T)\ \frac{\partial T}{\partial z} = \beta\, g(y,T)\ r(y,T) \quad (5)$$

The boundary conditions are now

$$y(o,t) = y_{in};\quad T(o,t) = 1+T_{in} \quad (6)$$

where

$$f(y,T) = \frac{C_{pb}(y,\ T_b)}{C_{po}(1+y)};$$

$$g(y,T) = \frac{(-\Delta H_R)}{(1+y)^2}.$$

In the above equations, the reaction rate, r(y,T), the mole heat capacity of the feed-gas, C_{pb}, and the enthalpy; ΔH_R, etc. are calculated according to the expressions given by Zhu (1981). The catalyst activity factor COR is assumed to be constant. The definitions of the symbols are listed at the end of this paper. It can be seen that this is a distributed parameter, nonlinear, parameterically forced system. Because of its complexity, it becomes imperative to simplify these equations.

LUMPING USING OCM

Since the control theory of finite dimensional systems is more complete than that of infinite dimensional ones, there is a great incentive in lumping the distributed parameter processes. A large number of lumping methods are available. One of the most efficient lumping methods for chemical reactor model may be viewed as the OCM.

Lumping using OCM is to force a set of approximate solutions, written as polynomials in z with time-dependent coefficients, to satisfy the boundary conditions and the partial differential equations at n interior collocation points (Finlayson, 1972). In present case, the collocation points are chosen as the roots of shifted Legendre polynomials. Consequently, the original set of partial differential differential equations (4)-(6) is converted to the following set of nonlinear ordinary differential equations

$$o = -(1+u)(1+y_{in})(y_{in}A_{jo}+\sum_{i=1}^{n} A_{ji}y_i) + \alpha\, r(y_j,\ T_j) \quad (7)$$

$$\frac{dT_j}{dt} = -(1+u)(1+y_{in})\ f(y_j,T_j)((1+T_{in})A_{jo}+\sum_{i=1}^{n} A_{ji}T_i)+\beta\, g(y_j,T_j)r(y_j,T_j) \quad (8)$$

where, $T_j=T(z_j,t)$, $y_j=y(z_j,t)$, $z_j(j=1,2,\ldots n)$ are n interior collocation points. The matrix A with $A_{ij}(i,j=0,1,\ \ldots\ n)$ is calculated according to the methods given by Finlayson (1972).

The number of collocation points n has to be chosen to keep model order as low as possible, while keeping the model accuracy high. By OCM, this goal has been satisfactorily realized. Through investigations of the steady and dynamics characteristics of the lumped model in different numbers of collocation points, say 2, 3,4,5,6, with the step responses at z=0.5 shown in Fig. 1, we find that accuracy of the model with n=4 is enough.

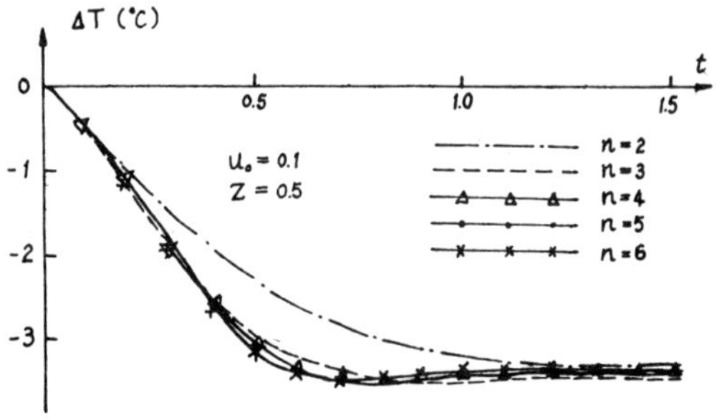

Fig. 1 Step responses in different number of collocation points.

It is worthy to point out that although equations (4)-(6) are collocated only at n interior collocation points, to collocate them including endpoint is possible.

Besides, the state values with the exception of collocation points can be obtained by Lagrange extrapolation.

BILINEAR MODEL

The lumped model equations (7)-(8) are still too complicated for control purpose, and must be further simplified. The ordinary method is linearization, but as has been early pointed out, the linearized model may be unsuitable in this case.

For simplicity, only the flowrate of feed u is assumed to be input as control function. Expanding the lumped dynamic equations (7)-(8) into Tayler series around the steady state profile, and keeping only the zero and first order terms, we obtain

$$o=-y_a\Phi_j(A,\Delta y)+\alpha\overline{DRy}_j\Delta y_j+\alpha\overline{DRT}_j\Delta T_j -(a_j+y_a\,\Phi_j(A,y_s))u-u\,y_a\Phi_j(A,\Delta y) \quad (9)$$

$$\frac{d\Delta T_j}{dt} = (-b_j\overline{Dfy}_j-y_a\overline{DFOy}_j+\beta\overline{DF1y}_j)\Delta y_j +(-b_j\overline{DfT}_j+\beta\overline{DF1T}_j)\Delta T_j-y_a\Phi_j(\overline{DFOT},\Delta T)-(b_jf_{js}+y_af_{js}\Phi_j(A,T_s))u+u(-b_j\overline{Dfy}_j-y_a\overline{DFOy}_j)\Delta y_j-u\,b_j\overline{DfT}_j\Delta T_j-u\,y_a\Phi_j(\overline{DFOT},\Delta T) \quad (10)$$

where, $\Delta T_j=T_j-T_{js}$, $\Delta y_j=y_j-y_{js}$, $r_j=r(y_j,T_j)$, $f_j=f(y_j,T_j)$, $FO_j=f_j\sum_{i=1}^{n}A_{ji}T_i$, $F1=g_jr_j$, $y_a=1+y_{in}$, $b_j=y_aA_{jo}$, $a_j=b_jy_{in}$, $\Phi_j(P,Q)=\sum_{i=1}^{n}P_{ij}Q_i$, (j=1,2,... n). The variables with the form DVS_j stand for the first partial derivatives of V_j with respect to S_j, and their values depend on the steady state values of temperatures and concentrations, namely T_{js}, y_{js} (j=1, 2, ... n).

Let $$Y \triangleq (\Delta y_1\Delta y_2 \,\ldots\ldots\, \Delta y_n)^T$$

$$T \triangleq (\Delta T_1\Delta T_2 \,\ldots\ldots\, \Delta T_n)^T.$$

The above equations can be compactly written as

$$\dot{T} = A_1T+A_2Y+B_1\,u+u\,N_1\,Y+u\,N_2T \quad (11)$$

$$0 = A_3T+A_4Y+B_2\,u+u\,N_3\,Y \quad (12)$$

where, A_1-A_4, N_1-N_3 are nxn matrices, B_1, B_2 are nx1 vectors.

So far, the bilinear model of an ammonia reactor has been derived. From equations (11)-(12), it can be noted that products of state and control variables are a characteristic feature of this model. For the original systems, it will therefore be a closer approximation to reality, and thus it will be expected to have a greater area of validity than a linear one.

Equations (11)-(12) can be also expressed in the full bilinear form. Since the material transport in the reaction can be assumed to be at quasi steady state, it is possible to write equation (12) as

$$\varepsilon\dot{Y}=A_3T+B_2u+u\,N_3T \quad (13)$$

where ε is small positive scalar, and in this case, about 0.1, which is given from some manipulations to the fundamental equations related to reaction mechanism. Hence, according to equations (11) and (13) we get

$$\dot{X}=\overline{A}\,X+\overline{N}\,X\,u+\overline{B}\,u \quad (14)$$

where, $X^T=(T^T, Y^T)$, the matrices $\overline{A}$, $\overline{N}$, and $\overline{B}$ are formed of the submatrices A_1-A_4, N_1-N_3 and B_1, B_2, respectively. It can be shown that the solutions of the equations (11)-(12) approach ones of the equation (14).

Furthermore, the order reduction of equations (11)-(12) is possible, too, which is later discussed.

ORDER REDUCTION AND LINEAR APPROXIMATION

The model equations (11)-(12) contain 2n state variables, the concentration variables of which may be not easy to measure in practice. However, it can be noted that due to the quasi steady state of concentration filed, as soon as a step input is exerted, immediate changes of the field will occur. Assume that the concentration variables before and after their changes are represanted by $y_j(t=0^-)$ and $y_j(t=0^+)$, (j=1, ... n), respectively. Because the difference between $y_j(t=0^+)$ and $y_j(t=0^-)$ for some input disturbance is constant, Y can be divided into two parts

$$Y = Y^* + Y' \quad (15)$$

where $Y^*=(\Delta y_1^*\Delta y_2^* \,\ldots\ldots\, \Delta y_n^*)^T$, $Y'=(\Delta y_1'\Delta y_2' \,\ldots\ldots\, \Delta y_n')^T$, $\Delta y_j'=y_j(t)-y_j(t=0^+)$, $\Delta y_j^*=y_j(t=0^+)-y_{js}$, $y_j(t=0^-)=y_{js}$ (j=1, 2, ... n), and the vector Y^* can be calculated by means of the following equation, which is from equation (12)

$$Y^* = -(A_4+u_oN_3)^{-1}B_2u_o \quad (16)$$

Thus, equations (11)-(12) may be rewritten as

$$\dot{T} = \breve{A}\,T+\breve{B}\,u+u\,\breve{M}\,Y+u\,N_2\,T \quad (17)$$

where, $N_1=0$ is negelected, $\breve{A}=A_1-A_2A_4^{-1}A_3$, $\breve{B}=B_1-A_2A_4^{-1}B_2$, and $\breve{M}=-A_2^{-1}A_4N_3$. Then, substituting equation (15) in (17) and negelecting those terms relating to Y', because y_j' (j=1,2, ... n) is here very small, we get

$$\dot{T} = A^*\,T+B^*\,u+u\,N^*\,T \quad (18)$$

where, $A^*=\breve{A}$, $B^*=\breve{B}+\breve{M}\,Y^*$, $N^*=N_2$.

Since $y_j(t=0^+)$ is related to input u, the equation (18) can be known as a variable-parameter bilinear model, in which only temperature vector T occurs.

The linear model is also required in order to investigate the characteristics of bilinear models develped in this paper. However it is convenient in this case to obtain the linearized model. From the equation (17), negelecting the products term of state and control variables, directly yields

$$\dot{T} = \breve{A}\,T+\breve{B}\,u\,. \quad (19)$$

SIMULATION RESULTS AND VERIFICATION OF MODELS

In order to verify the bilinear model (11)-(12) (BLS), including the low order variable-parameter bilinear model (BLS*), the dynamic simulations on these models, together with nonlinear lumping model (7)-(8) (OCE), and linearized model (19) (LS) are carried out on IBM 5150 computer. The Newton-Raphson method and the fourth Runge-Kutta method are used to solve nonlinear algebra equations and ordinary differential equations, respectively. The simulations are performed under different numbers of collocation points and various amplitudes u_o and directions of step disturbance in the feedrate. Some results for n=4 are shown in Figs. 2-4. A simple discussion is given as follows.

Dynamic Characteristics of the Reactor

The step response curves of OCE in Figs. 2 - 4 closely represent dynamic characteristics of the reactor. Based on the various OCE curves, the following main conclusions may be drawn.

1. The step responses at the two locations (Z=0.67, 0.93) show that the dynamic characteristics for the same n and u_o but at different points along the reactor length are not identical, and the dynamic responses near the exit of the reactor are larger than at other points.
2. For the same n but at different points, the positive and negative responses to the step inputs with the same amplitude u_o are seriously non-symmetrical, moreover the non-symmetry strengthens with the increase in u_o, and the variation of the negative responses is larger than that of the positive ones. It can be concluded that the ammonia reactor system is seriously nonlinear.
3. When a positive step disturbance is exerted, the temperature and the ammonia mole fraction will decrease, while when a negative step disturbance is exerted they will increase.

Efficiency of BLS (BLS*)

Figs. 2-4 give the temperature responses

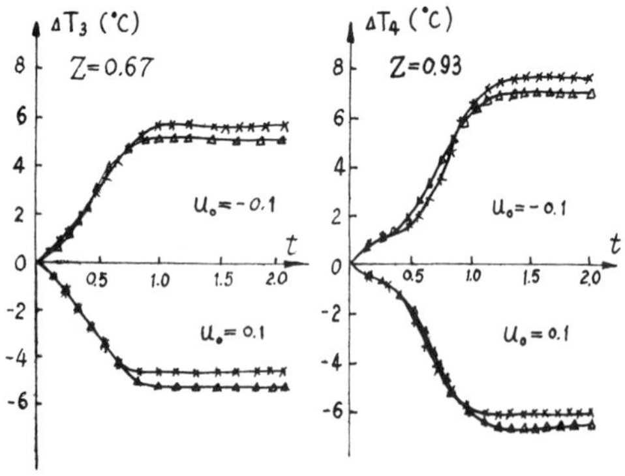

Fig. 2 Temperature responses to a 10% step change in feedrate.

Response	Model
—•—	OCE
—×—	BLS
—△—	LS
----	BLS*

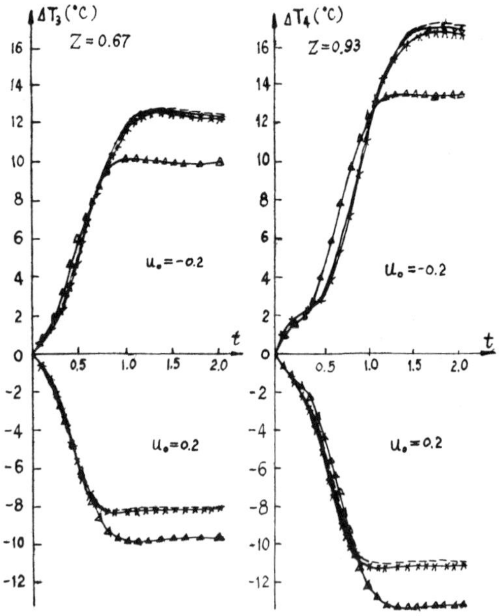

Fig. 3 Temperature responses to a 20% step change in feedrate.

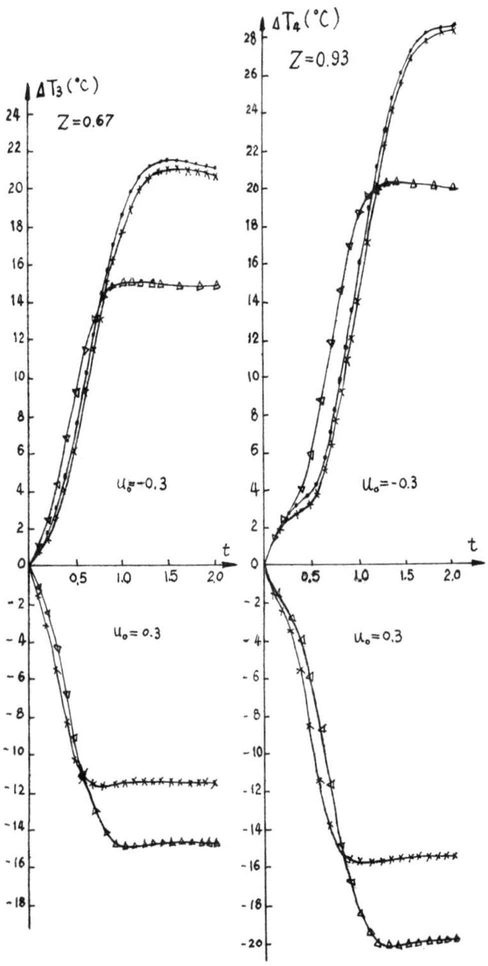

Fig. 4 Temperature responese to a 30% step change in feedrate.

for n=4, to the step input with an amplitude of 10%, 20% and 30%, respectively. From the curves shown in Fig. 2, it can be seen that the BLS response curves at two different locations are quite close to the corresponding OCE curves but the LS curves deviate from OCE ones, when the system is subjected to a step disturbance of 10% in feedrate. Fig. 3 indicates that when a 20% step disturbances is exerted, the accuracy of the positive BLS response is still very high, although that of the negative BLS response becomes a little bad. However, the difference between LS and OCE responses becomes so large that LS can not be used in this case. Fig. 4 shows that even with a 30% step disturbance, BLS approximation is still satisfactory. The positive BLS temperature responses are very close to the corresponding OCE responses, although the negative BLS responses are a little apart from the OCE responses.

It follows that (i) the range of linearization that can be used is narrow in this case. In fact, the simulation results show that the LS responses to even a 5% step disturbance do not agree with the corresponding OCE responses distinctly, and (ii) BLS approximation is superior to linearization. Especially for large step disturbances, BLS can still more accurately approach the real system.

From the BLS approximations made at two different locations, we notice that the positive approximation is better than the negative one. Even though the amplitude of step disturbance becomes rather big, say 50%, the positive approximation is still fairly good, The reason why the negative approximation becomes bad when a large disturbance is exerted, is mainly due to the serious nonlinearity of the reactor. Because the variation of negative OCE responses is much larger than that of positive ones, the negative approximation becomes even more difficult.

Finally, the validity of a low order variable-parameter bilinear model has also been established via simulation. From Fig. 3 (see BLS*), comparing these curves with the corresponding BLS curves, it can be seen that the loss of accuracy of this approximation model is less.

CONCLUSION

In this paper, for an adiabatic ammonia reactor where the feedrate is manipulated, the modelization using the orthogonal collocation technique and bilinear approximations is presented. It has been shown via simulation that because of the complexity of this nonlinear distributed parameter system, the linearization is valid only for small range around operating point. Whereas the accuracy of the bilinear model is much higher than that of linear model. Moreover, due to some advantages of the bilinear model itself, it is more suitable for the ammonia reactor.

The bilinear model given here may be very helpful in presenting more adequate control algorithms and improving the control performance of the reactor, thus increasing the yield of the reactor. As for the bilinear modeling method suggested in this paper, it can easily be extended to the modelization of similar or more complex tubular fixed-bed reactors, such as methanol reactors.

NOTATION

A	area of reactor (m^2)
COR	catalyst activity factor
C_{pb}	mole heat capacity of feed gas (Kcal/Kmol ^oK)
C_{pc}	specific heat of catalyst (Kcal/Kg ^oK)
C_{po}	average heat capacity of feedgas (Kcal/Kmol ^oK)
ΔH_R	enthalpy of ammonia reaction (Kcal/Kmol)
l	length of reactor (m)
N_{in}	mole flowrate of feed (Kmol/hr)
P	operating pressure (atm)
r	reaction rate of ammonia
t	normalized time, $\theta \cdot \theta_L$
T	normalized catalyst bed temperature, T_b/T_{bo}
T_b	catalyst bed temperature (^oC)
u	normalized flowrate, $(N_{in}-N_{ino})/N_{ino}$
u_o	step amplitude
V_R	volume of the catalyst bed (m^3)
y	NH_3 mole fraction
z	normalized distance, l/L $(0 \leq z \leq 1)$

Greek Symbols

α	COR $V_R/22.4/N_{ino}$
β	$\alpha/T_{bo}/C_{po}$
ρ	density of catalyst (kg/m^3)
θ	time (hr)
θ_L	$N_{ino}\, C_{pbo}/C_{pc}/\rho/V_R$ (1/hr)

Subsript

o, in	inlet state
s	steady state

REFERENCES

Espana, M., and I. D. Landau (1978). Reduced order bilinear models for distillation columns. Automatica, 14, 345-355.

Finlayson, B. A. (1972). The Method of Weighted Residuals and Variational Principles. Academic Press, New York and London.

Georgakis, C., R. Aris, and N. R. Amundson (1977). Studies in the control of tubular reactors - I, II. Chem. Engng. Sci., 32, 1359-1369, 1371-1379.

Jiang, Z. H., and W. S. Jiang (1982). A new approach to bilinear systems identification and its application. Journal of East China Institute of Chemical Technology, 8 , 359-367(in Chinese).

Koppel. L. (1965). Dynamics of a class of nonlinear, distributed-parameter, chemical reactors. Ind. Engng Chem. Fundamentals, 4, 269-275.

Michelsen, M. L., H. B. Vakil, and A. S. Foss (1973). State-space formulation of fixed-bed reactor dynamics. Ind. Engng Chem. Fundamentals, 12, 323-328.

Mohler, R. R., and W. J. Kolodziej (1980). An overview of bilinear system theory and applications. IEEE Trans. Syst., Man, Cybern., SMC-8, 683-388.

Patnaik, L. M., N. Viswannadham, and I. G. Sarma (1980). State space formulation of ammonia reactor dymanics. Computers & Chem. Engng, 4, 215-222.

Shah, M. J. (1967). Control simulation in ammonia production. Ind. Engng Chem., 59, 72-83.

Svoronos, S., G. Stephanopoulos, and H. Aris (1980). Bilinear approximation of general non-linear dynamic systems

with linear inputs. Int. J. Control, 31, 109-126.
Zhu, B. C. (1981). Inorganic Chemical Reaction Engineering. Chem. Engng Publishing House (in Chinese).

IDENTIFICATION OF A PRESSURIZED WATER REACTOR STEAM GENERATOR BY STATE-SPACE MULTI-VARIABLE MODELS

S. Bittanti*, R. Cori, F. Pretolani**, L. Rassu*** and D. Roncaglioni*****

**Dipartimento di Elettronica, Politecnico di Milano, Piazza L. da Vinci 32, 20133 Milano, Italy*
***ENEL, Italian Electricity Board, Research and Development Department, Automation and Computing Research Center, Via A. Volta 1, Cologno Monzese, Milano, Italy*
****Politecnico di Milano, Milano, Italy*

Abstract. The identification of a pressurized water reactor steam generator (Westinghouse F-type) and its feedwater system is dealt with. The identification experiments are performed via simulation using an accurate validated nonlinear model of the overall plant. Three state-space multivariable models are worked out for the 10%, 50% and 100% loads respectively. The identified models are validated taking into account their intended use for the steam generator regulation design.

Keywords. Process control, nuclear plants, steam generators, parameter identification.

INTRODUCTION

The Italian Electricity Board (ENEL) is the general supervisor for the construction of its pressurized water reactors (PWR-PUN project - Westinghouse licency). As is well known, the steam generator is a critical component of any PWR plant in that its complex dynamics is a source of many operating and availability problems. The aim of this paper is to contribute to the understanding of the dynamics of such a component, mainly for control purposes. As such, it represents a first report on the present state of the Steam Generator Advanced Control (SGAG) project which is jointly carried out by the the ENEL and Politecnico di Milano. The process constituted by the steam generator (SG) and its feedwater system is considered. THe influence of the various systems using the steam produced by the SG is also taken into account.

By means of a suitable deterministic identification procedure, three state-space linear multivariable models are worked out in order to describe the process dynamics about the steady-state operating points corresponding to the 10%, 50% and 100% loads. The data used in the identification procedure makes reference to a number of step responses of an accurate nonlinear model of the overall plant, acting here as the "true" system.

PROCESS DESCRIPTION

The functional scheme of the steam generators and their feedwater system is represented in Fig. 1. For each SG, the reactor coolant water flows through INCONEL U tubes. The feedwater is injected through a toroidal tube at the top of an anular space (downcomer) between the external SG shell and an internal shroud. Inside the shroud, which contains the U tubes system, the heat exchange between the primary and secondary circuits takes place. This is the so-called riser area. The water of the downcomer-riser circulation loop is constituted by that provided by the turbopumps (TPA) as well as that coming from the centrifugal riser moisture separators. The steam-water mixture pro-

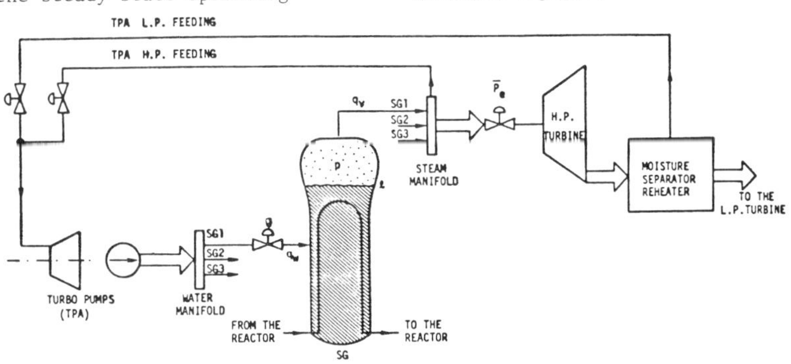

Fig. 1. Functional scheme of the Steam Generator and its feedwater system

duced by the water boiling in the riser region is lighter than the water contained in the downcomer. This gives rise to a natural circulation effect which causes the water motion into the circulation loop balancing the friction losses.
The steam-water mixture at the riser outlet enters the centrifugal separators system which separates most of the water from the mixture. After a further dehumidification, the steam is conveyed to the turbine through the steam manifold. The steam flow-rate is denoted by q_v. The separated water is collected into the so-called drum region surrounding the separators and standing above the downcomer. The water level l is determined by the total mass of water contained in the drum and downcomer. In the normal operating condition, the toroidal tube is covered by the drum water.

For each SG, the feedwater flow rate q_w is modulated by the opening θ of a regulating valve. The feedwater comes from a water manifold fed by two turbopumps. In turn the turbopumps turbines feeding is provided by the steam coming from the High Pressure (H.P.) main turbine outlet. Whenever the pressure is not sufficiently high (lower plant loads), the steam manifold supplies the necessary feeding.

SG DYNAMICS - SOME CRITICAL ISSUES

The main problems arising from the SG dynamics are connected with the water level variation during the plant transients.
Keeping the drum water at the proper level is essential for the following reasons:

(i) Turbine protection against humidity excess in the inlet steam caused by a too high level.
(ii) Reactor protection against insufficient primary water cooling caused by a too low level.
(iii) Protection against water hammers in the toroidal tube whenever the feedwater flow-rate increases while the water level is below the toroidal tube.
(iv) U tubes thermal stresses protection against a too low level.

As is well known, the level regulation is a nontrivial task for the following basic reasons:

(a) The intrinsic process nonlinearity due to the dependence of the dynamic behaviour upon the operating condition.
(b) The nonminimum phase characteristic at low plant loads in the relationship between the water level and the feedwater mass flow-rate, which is its regulating variable.
(c) The water "shrink and swell" effect due to variations in the demanded electric power. For instance, a decrease in the demanded power results in a reduction of the SG demanded steam. The consequent increase in the SG pressure produces the well known water "shrink" effect. Such a level decrease is in contrast with the long term level ramp behaviour resulting from the unbalance between the SG steam and feedwater flow-rates.

Another problem connected with the possible coupling of three SG's acting in parallel is due to variations in the pressure difference Δp between the steam and water manifolds. To avoid instability effects producing masses interchanges amongst the SG's, a Δp regulation is needed. Such a regulation is also useful to keep the feedwater valves in a suitable operating range in spite of the plant load variations.

SYSTEM CHARACTERIZATION AND INPUT/OUTPUT VARIABLES SPECIFICATIONS

The scheme of the considered system is given in Fig. 2.
As in any level regulation problem, it is advisable to feed back the inlet and outlet flow-rates, besides the level itself. Consequently q_v and q_w are dealt with as output variables. As a matter of fact, the chosen output variables are q_w and $q_w - q_v$ (Fig. 2). The motivation is that the time variation of variable $q_w - q_v$ turns out to be much smoother than the one of q_v.

For the reasons explained in the previous section, the pressure difference Δp between the steam and water manifolds is a further output variable.
Finally, for a possible improvement of the feedback control system, the steam generator pressure Δp has been considered as system output too.

Obviously, the plant main disturbance is the demanded variation in the electrical power i.e. SG thermal power. Such a variation is determined by the turbine throttle valve through its regulation system. Thus, the demanded electrical power $\bar{P}_e$ is an obvious input variable.

It is also apparent that the feedwater valve opening θ plays the role of an input variable.

As for the turbopumps speed ω, a preliminary observation is required concerning the influences of θ and $\bar{P}_e$ on ω itself. Basically, the turbopumps act as an actuator. However, the dynamics of the process controlled by such an actuator has an influence on the effective speed of the turbopumps themselves. Indeed, a feedwater valve opening variation results in a variation of the TPA load. Consequently other things being equal, variations in θ results in perturbations in ω. Moreover, variations in $\bar{P}_e$ produce variations in the moisture separator-reheater pressure, acting in turn on the TPA feeding. Hence, $\bar{P}_e$ influences ω too. These perturbation effects on ω are schematically represented in Fig. 2 (Part A).

The TPA speed ω is determined by its set point $\bar{\omega}$ through a regulation system. Notice that such a regulation system is designed independently of the overall SG regulation system since the TPA response is much quicker that the SG one.

In conclusion, as indicated in Fig. 2, the system under consideration is constituted by two parts. Part A is the TPA together with its feeding and regulation systems. Part B is the SG with all its surrounding components. Obviously, even though this is not pointed out in Fig. 2, the behaviour of part B is influenced by the reactor operating condition.

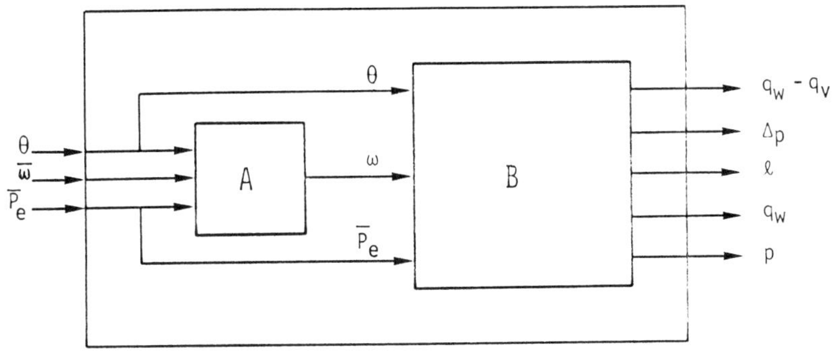

Fig. 2. Input-Output characterization of the considered system

IDENTIFICATION EXPERIMENT DESIGN

In the process identification, a black-box deterministic procedure has been adopted. A digital nonlinear model worked out by the Italian Electricity Board on the basis of the SICLE package (EdF - French Electricity Board), named STRIP (Busi, Cori, Pretolani, 1983), has been used as "true" model.

As suggested by Fig. 2, two alternative approaches can be considered. The first consists in working out a model for the overall process with θ, ω and $\bar{P}_e$ as input variables. The second one amounts in deriving a model for each of the two parts (A and B) constituting the process. THe overall model will then be derived by a suitable composition of the two partial models.
In this respect, note that the PUN project is not fully consolidated. This especially applies to the TPA. Moreover, the possibility of changing the TPA regulation system should also be taken into consideration for a better SG regulation. On the apposite, it is unlikely that Part B will be subjected to any change in the project development. Consequently, two models for Part A and B have been separately identified.

Part B Identification-Experiment Design

The digital nonlinear model STRIP used as "true" system includes an accurate description of all the plant components, from the nuclear reactor to the condenser, together with all their regulation loops. In the identification, the level regulation system (acting on the feedwater valve opening θ) and the Δp regulation system (acting on the turbopumps speed set point $\bar{\omega}$) have been excluded. Moreover, the TPA speed regulation has also been excluded. This means that θ, ω, $\bar{P}_e$ act as independent input variables for Part B. All the remainder feedback loops are normally operating, but the one acting on the reactor control rods. As is well known, the rods speed is determined by a nonlinear characteristic exhibiting a dead band. Notice that the variation of any part B input causes a temperature variation of the primary water SG outlet, which calls for the rods intervention. Hence, the output responses will depend on the input perturbation amplitude insofar it produces or not an effective rod position variation (dead band effect). To avoid such a strong nonlinearity, STRIP has been modified for identification purposes by eliminating the dead band. As input functions, step changes in θ, ω and $\bar{P}_e$ have been considered one at a time. Their amplitude have been suitably chosen according to the particular operating conditions (10%, 50%, 100% nominal load). See (Rassu and Roncaglioni 1984) for details.

Part A Identification-Experiment Design

As "true" system, the digital model STRIP has been again used. As for the various feedback loops in operation, the same experimental conditions as for Part B identification have been considered. Step changes in θ, $\bar{\omega}$ and $\bar{P}_e$, acting one at a time, have been taken as input functions. For the chosen amplitudes, (see Rassu and Roncaglioni, 1984).

IDENTIFICATION METHODOLOGY

The same identification methodology has been adopted both for Part A and Part B. The methodology can be concisely described as follows (see Ragni and Righetti, 1981, for more details). As previously seen, the typical experimental condition consists of keeping all input variables at their nominal values but one, say u_i, which is subject to a step variation. Let y_j be the correspondent jth output variable variation. The transfer function $G_{ij}(s)$ relating y_j to u_i is assumed to be a rational one, with p_{ij} and q_{ij} as numerator and denominator degrees respectively. Letting α be the parameters of $G_{ij}(s)$, the identification is carried out for a fixed pair (p_{ij}, q_{ij}) by minimizing the quadratic criterion:

$$J(\alpha) = \int_0^T [\tilde{y}_j(t) - \bar{y}_j(t,\alpha)]^2 \, dt.$$

Here

$\tilde{y}_j(t)$ is the j-th output variable step response of the "true" system

$\bar{y}_j(t,\alpha)$ is the output fo the assumed model, with $G_{ij}(s)$ as transfer function, relative to an input step variation with the same magnitude of the given step change in u_i

T is the output observation interval.

The optimal parameter estimate has been achieved by minimizing J via a suitable mathematical programming technique. Degrees p_{ij} and q_{ij} have been selected by a trial-and-error procedure.

The above identification procedure leads to 3 and 15 transfer functions at each operating point for Part A and B respectively. From these transfer function models, a global state-space model has been derived, at each operating point, for the overall system of Fig. 2.
As is well known, the passage from a transfer function model to a state-space one requires to be carefully performed in order to avoid uncontrollable and/or unobservable realizations (Kalman, Falb, Arbib, 1969). Such a care is important in the present application since any transfer fucntion relating the water level l to θ, ω or $\bar{P}_e$ will obviously contain an integrator. A minimal realization calls for a unique integrator representative of the three previous ones. This corresponds to the fact that a variation in either θ, ω or $\bar{P}_e$ gives rise to a variation in the balance between the water and steam mass flow-rates. The water level integration effect is due to such an unbalance.

MODEL VALIDATION

The state-space linear models derived by means of the described methodology turn out to be of orders 50 - 65 at 10%, 50% and 100% operating points for the overall system (the Part A model is of order 9).

Account taken of the intended use of these models, a closed-loop validation procedure has been adopted. Precisely, the responses of the "true" process with the standard regulation (STRIP) has been compared with the corresponding ones of the identified models equipped with the same regulations, at the three operating points. The comparison has been carried out about the operating points by using the appropriated identified model.

The closed-loop systems present as inputs the set points of the water level and the pressure Δp together with disturbance $\bar{P}_e$. In Fig. 3-5, the responses of l and θ to stepchanges in $\bar{P}_e$ are reported for the "true" closed loop system and that relative to the identified model (—•—). The reponses correspond to the following variations in $\bar{P}_e$: -0.2% at the 10% operating point (Fig. 5) and -1% at the 50% and 100% operating points (Fig. 4 and 3).

ACKNOWLEDGEMENTS

This paper has been supported by ENEL, Centro di Teoria dei Sistemi (CNR-Milano), M.P.I. and A.P.A.S.C. .

REFERENCES

Busi,T., R. Cori,and F. Preolani (1983).Analysis of Anticipated Plant Transients as an Aid for PWR Plant Verification, Anticipated and Abnormal Plant Transients in Light Water Reactors (ANS, Jackston - Wyo).

Kalman, R.E., P.L. Falb, and M.A. Arbib (1969). Topics in mathematical system theory, Mc Graw-Hill, New York.

Procaccia, H., J. David, L. De Penguern, P. Hamon, A.R. Wazzan (1982). Tests of Types 51A and 51M Steam Generators at Bugey 4 and Tricastin 1 Nuclear Power Plants (EPRI-NP-2869, Project S154-1).

Ragni, M. and M. Righetti (1984). Identification and order reduction methods of linear time-invariant dynamic system. Doct. Thesis, Politecnico di Milano (in Italian).

Rassu, L. and D. Roncaglioni (1984). Identification and control of the steam generator of a nuclear PWR power plant. Doct. thesis, Politecnico di Milano (in Italian).

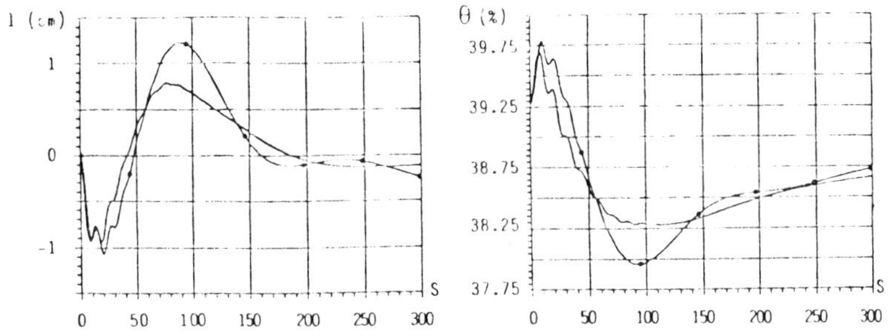

Fig. 3 - Responses to a stepchange in $\overline{P}_e$ at the 100% operating point (—•— identified model).

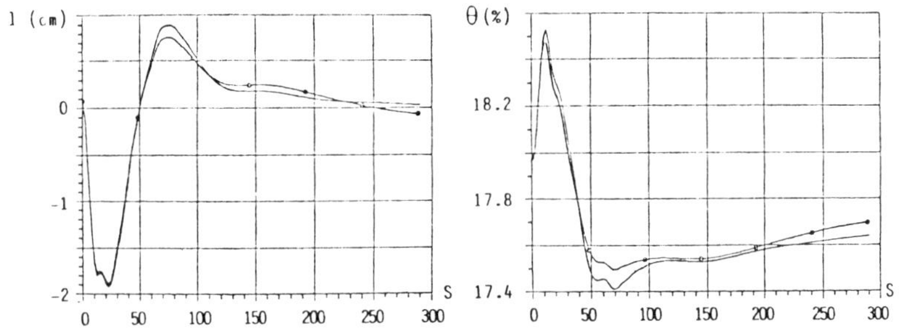

Fig. 4 - Responses to a stepchange in $\overline{P}_e$ at the 50% operating point (—•— identified model)

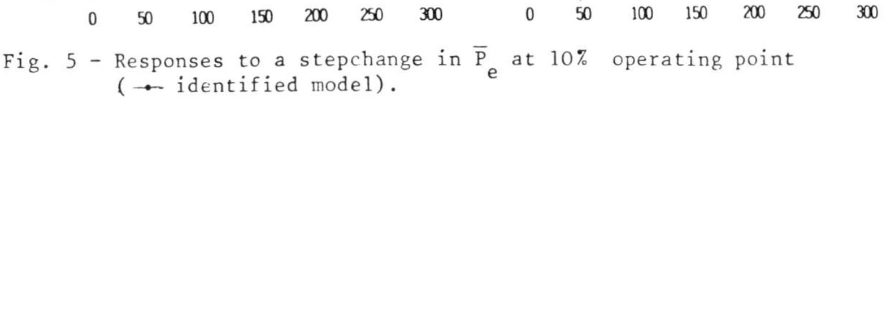

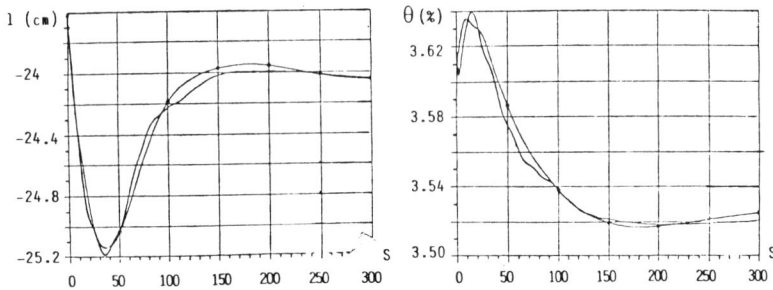

Fig. 5 - Responses to a stepchange in $\overline{P}_e$ at 10% operating point (—•— identified model).

MODELING AND CONTROL OF AUTOMATIC VEHICLE STEERING SYSTEM USING MICROPROCESSORS

R. Shoureshi and D. Carey

School of Mechanical Engineering, Purdue University, West Lafayette, IN 47907, USA

Abstract. This paper presents the initial results of a research on development of an automatic steering system for vehicles. The objective is to reduce the driver's role as being the only commander and controller of a vehicle for traffic surveillance and safety, especially for cases like drunk drivers. A nonlinear model for a rotary steering system is developed and reduced to a linear model. The reduced model is used to design a regulator for the steering system. Since position is the only available measurement, an optimum observer based on the dual system is designed. The results are implemented on a microprocessor. The required software and hardware are presented. The results of the model combined with the regulator are compared with the experimental data.

Keywords. Modeling; Nonlinear Equation; Microprocessors, Signal Processing, Kalman Filter Vehicles.

INTRODUCTION

Transportation, which is one of the major aspects of today's mobile type of living, has been studied in great detail in terms of automotive vehicle design, higher productivity and fuel economy. The primary objectives of transportation research are the passenger's comfort, safety, and speed. Nevertheless, until recently, not enough effort has been devoted to investigation of all possible means for passenger safety, especially for cases like drunk drivers.

Quite a few interesting concepts pertaining to safety have been considered and brought to different levels of development. Some of these have been initiated by proposed government regulation. However, if accidents and collisions can be pre-monitored and stopped in advance, then more passenger safety can be gained.

This paper presents the results of the first phase of a research on development of an automatic steering system for vehicles. The objective is to reduce the driver's role as being the only commander and controller of a vehicle for traffic surveillance and safety.

Application of control theories to the development of strategies for commanding vehicle (cars, trains, etc.) was introduced in the 1970's. Most efforts have concentrated on automatically-driven trains on specified rails. Jet Propulsion Laboratory (JPL) studied cars controlled by an autopilot and driven automatically by following a signal from a wire buried just beneath the road surface. Recently, research on automatic guidance of farm machinery has increased substantially in the search for guidance systems which would help prevent operator fatigue caused by wearisome driving. This is mostly due to the advent of microprocessors used for system monitoring and control.

Modelling and dynamics of a rotary valve steering system is presented in this paper and the results are used for design of a regulator and an optimal observer required for an automatic steering system. Simulation results are compared with the experimental data and the required hardware and software are presented.

MODELLING OF STEERING SYSTEM

Development of advanced power steering systems started in the early 1950's mainly at GM Saginaw Steering Gear. An accurate model of this system is a necessity for the design of the controller. Derivation of the model requires a good understanding of operation of the steering system. Fig. 1 shows a rotary steering gear. The rotary valve is the essential mechanism of the gear. The valve changes flow restrictions according to the difference in angular positions between the input shaft and the worm gear which directly controls the output shaft. The input shaft and worm gear are connected by a torque shaft. Driver effort twists the torque shaft and changes the valve position. This angular position depends on the load on the output shaft. The valve position is changed such that hydraulic pressure is directed to the cylinder volume that will restore the torque shaft to zero deflection.

Fig. 2 shows the steering gear in the complete hydraulic circuit. The pump contains a flow control valve so that the supply flow rate through the gear is essentially constant at about ten litters per minute. The supply pressure will vary depending on the load at the output shaft of the gear. This pressure is limited to 1850 psig by a pressure relief valve located either in the pump or in the gear itself. The fluid that is commonly used in the gear is automotive transmission fluid with a density of about 861.1 kg/m^3.

The analysis described in this paper is concerned only with the operation of the steering gear. The system dynamics associated with the pump and hoses are not considered.

NONLINEAR MODEL

Steering gear models are based on flow through the valve and into the cylinder and on forces on the rack-piston. Summing forces on the rack-piston yields

$$\Sigma F = dp^{\cdot}A + K_1 K_T(\theta - K_1 y) = \text{load} + f\dot{y} + M_e \ddot{y} \quad (1)$$

Solving for dp, the differential cylinder pressure:

$$dp = \frac{1}{A}\,[\text{load} + f\dot{y} + M_e\ddot{y} - K_1K_T(\theta - K_1y)] \qquad (2)$$

Flow into the cylinder was defined as the sum of flow resulting from piston movement, leakage flow, and displaced flow due to compressibility of the flow, namely

$$q_{cyl} = \rho A\dot{y} + L\ dp + \frac{\rho As}{K}\,(\dot{dp}) \qquad (3)$$

The model equation is obtained by substituting equation (2) into equation (3). The nonlinearities arise in defining the cylinder flow, q_{cyl}. This flow was determined by considering the flow passages in the valve. Fig. 3 shows the flow passages in the valve labeled with variables that are explained here. The supply flow rate was defined as:

$$q_s = q_1 + q_e = C_1A_1\sqrt{\rho(P_s - P_2)} + C_3A_3\sqrt{\rho(P_s - P_1)} \qquad (4)$$

Since the supply flow is constant, the following must be satisfied:

$$q_3 - q_4 = q_2 - q_1$$

or

$$DA\sqrt{\rho(P_s - P_1)} - C_4A_4\sqrt{\rho P_1}$$

$$= C_2A_2\sqrt{\rho P_2} - C_1A_1\sqrt{\rho(P_sP_2)} \qquad (5)$$

Based on the experimental results of the steering gear used in this study, it is found that

$$C_1 = C_3 = .79$$

$$C_2 = C_4 = .60$$

Equations (4) and (5) are two nonlinear equations with two unknowns, P_s and P_2 provided that $P_1 = dp + P_2$. The flow areas are functions of valve displacement. The approximating functions for the areas are derived based on data obtained from the Ross Gear Division of TRW.

Equations (4) and (5) are solved numerically for given dp and valve displacements using a Newton-based iterative method. Fig. 4 shows the resulting cylinder flow as a function of dp and valve displacement, where displacement is shown in terms of degrees of rotation.

LINEAR MODEL

In order to use linear control techniques for derivation of a commanding signal for the steering system, a linear model based on the results from nonlinear analysis is derived. Nonlinearity of the system is due to the equations used for cylinder flow. Fig. 4 shows variations of cylinder flow as a function of pressure difference and valve displacement. These results are based on the nonlinear analysis stated before. However, for practical operating range of the steering gear, these curves can be linearized and represented by

$$q_{cyl} = K_n(X) + K_p(dp) \qquad (6)$$

The least square technique is used to find the best values for K_n and K_p such that equation (6) would follow the contours of Fig. 4 as close as possible. Combination of this equation with the earler equations results in the following transfer function.

$$\frac{Y(S)}{\theta(S)} = \frac{N_1S + N_2}{D_1S^3 + D_2S^2 + D_3S + D_4} \qquad (7)$$

where:

$$N_1 = -\frac{\rho s}{K}\,K_1K_T$$

$$N_2 = -K_n\left(\frac{180}{\pi}\right) + \frac{K_1K_T}{A}\,(L - K_p)$$

$$D_1 = \frac{\rho s}{K}\,M_e$$

$$D_2 = \frac{\rho s}{K}\,f + \frac{L}{A}\,M_e - \frac{K_p}{A}\,M_e$$

$$D_3 = \frac{\rho s}{K}\,K_1^2\,K_T + \rho A - \frac{K_p}{A}\,f + \frac{L}{A}\,f$$

$$D_4 = K_1\left(\frac{LK_1K_T}{A} - \frac{K_pK_1K_T}{A} - K_n\,\frac{180}{\pi}\right)$$

Y(S) = Laplace transform of the piston movement

Fig. 5 shows responses of linear and nonlinear models to a 60 degree step input of the steering wheel. As shown, the linear model follows the nonlinear model very accurately. The nonlinear system has 93 miliseconds settling time whereas for the linear model it is 100 miliseconds. Since the results of the linear model are so close to the nonlinear model, then further analysis of this system is performed based on this linear model.

CONTROLLER DESIGN

The piston movement is related to the turning of the vehicle wheel by kinematic relationship resulting from the linkages between the piston and the wheels. Since the mass of the linkage mechanism and friction between the joints are negligible compared with those in the steering gear, then the piston motion can be assumed linearly related to the turning of the wheels. Therefore, the only dynamics taken into account for design of a controller for the automatic steering system of the vehicle are those given by the transfer function of equation (7). Furthermore, those disturbances coming from the road (including road surface friction torque) are assumed to be noises affecting the system. Finally, for this stage of the research, those disturbances caused by the lateral motion of the vehicle are neglected.

In order to represent this system in a state space form, equation (7) is transformed into a phase variable form as

$$\dot{\underline{X}} = \begin{bmatrix} 0 & 1 & 0 \\ 0 & 0 & 1 \\ -D_4/D_1 & -D_3/D_1 & -D_2/D_1 \end{bmatrix}\underline{x} + \begin{bmatrix} 0 \\ 0 \\ K \end{bmatrix}U \qquad (8)$$

$$y = [N_2\ ,\quad N_2\ ,\quad 0]\underline{X} \qquad (9)$$

Controller design can be interpreted as a tracking problem by identifying a desired path. Assuming the road edge (center lane) is the reference point at every instant and the objective is to keep the vehicle about 0.25 meters from that reference point, then a new variable can be defined as:

$$y^* = (\Delta t)(V_c)(K_c)(y) - 0.25 \qquad (10)$$

where Δt is the maximum allowed time to correct the vehicle direction (assumed 0.01 seconds), V_c is the vehicle speed (assumed 80 km/hr), and K_c is the conversion factor from stroke to degrees which is 2400 for the steering gear under study.

The above discussion results in the following equation for K given in equation (8)

$$K = \frac{(\Delta t)(V_c)(K_c)}{D_1} \qquad (11)$$

and a quadratic performance index written as

$$P.I. = \int_0^\infty [y^* Q y^* + UpU]dt \qquad (12)$$

This performance index calls for minimization of y^* (tracking error) and optimum input energy. p and Q are weighting factors, and it can be shown that the optimal controller depends only on the ratio of p and Q. Therefore, Q is assumed unity and p is selected based on the resulting closed loop eigenvalues and the feedback gains. The optimal feedback gains are obtained by solving the Riccati equation written as

$$A^T R + RA - RBP^{-1}B^T R + C^T QC = 0 \qquad (13)$$

for the system equations (8) and (9). It was found that p = .01 results in a suitable closed loop system with the following feedback gains and eigenvalues

$$U = G\underline{X} \qquad (14)$$

$$G = [-18.14, \quad -8.7\times10^{-3}, \quad -2.2\times10^{-6}] \qquad (15)$$

Closed Loop Eigenvalues

$$\lambda = -0.126 \times 10^4$$

$$\lambda_{2,3} = -(0.304 \times 10^4) \pm j(0.621 \times 10^4)$$

OBSERVER DESIGN

In real case of an automatic steering system for a vehicle there will be one measurement available for the controller and that would be the vehicle position with respect to the road edge and/or center lane. The second phase of this research deals with design of a sensor for vehicle position measurement. However, an observer has to be designed to provide the controller with good estimates of all three states of the system. A Kalman filter approach (optimum observer design) is used.

The observer equations are presented as

$$\dot{\hat{\underline{x}}} = A\hat{\underline{x}} + B[u + v] \qquad (16)$$

$$\hat{\underline{Y}} = C\hat{\underline{x}} + W \qquad (17)$$

where u represents the optimum input, v is the road disturbance, and W is the measurement noise involved in the vehicle position measurement. v and W are stochastic components and are assumed to e stationary white noises with Gaussian distribution and zero mean values. Therefore, their autocorrelation functions can e written as

$$E[v(t)v^T(t + \tau)] = \bar{v}\delta(\tau) \qquad (18)$$

$$E[W(t)W^T(t + \tau)] = \bar{W}\delta(\tau) \qquad (19)$$

where $\delta(\tau)$ is the dirac delta function, E denotes the expected value, and $\bar{v}$ and $\bar{W}$ are noise powers.

The optimal observer gain matrix is obtained by solving the Riccati equation for the dual system, written as

$$AR + RA^T - RC^T\bar{W}^{-1}CR + B\bar{v}B^T = 0 \qquad (20)$$

The relative magnitudes of the noise intensities $\bar{v}$ and $\bar{W}$ determine the optimum balance between the speed of the states reconstruction and the immunity to measurement noise. A method similar to the one used for the selection of p and Q in the design of the controller is used. The resulting observer gains are

$$K = [-9.2 \times 10^{-2}, \quad 4.10 \times 10^2, \quad 6.6 \times 10^5] \qquad (21)$$

and the observer eigenvalues are

$$\lambda_1 = -0.163 \times 10^3$$

$$\lambda_{2,3} = -(0.24 \times 10^3) \pm j(0.534 \times 10^4)$$

It is worth noting that the observer eigenvalues are about 10 times faster than the closed loop system eigenvalues. Fig. 6 shows a block diagram of the combined steering system with the optimum observer.

EXPERIMENTAL MODEL

Fig. 7 shows the experimental setup and hardware used for verification of the model and the controller. It consists of a TRW HBF52 power steering gear and a hydraulic steering pump, an electric drive motor, a SLO-SYN stepping motor and gear train connected to the HBF52 input shaft. Two potentiometers are used as sensors on the input and output shafts of the steering system. An 8085 microprocessor along with an A/D-D/A board are used to develop the control function and the optimum input for the input shaft.

The stepping motor is used to precisely steer the system based on the commanded value. Fig. 8 shows how the input from this motor is transmitted to the steering gear. The stepping motor is pulsed by digital signals coming from the microprocessor in proper sequence. Fig. 9 shows the interfacing board between the microprocessor output port and the stepping motor coils.

A series of tests based on step and ramp inputs were performed and the results were very well predicted by the linear model. The only exception was the small errors due to backlash in the gears and dry friction which were not accounted for in the system model. However, the results indicated a high accuracy for the linear model.

EXPERIMENTAL RESULTS

A Z-151 microcomputer is used to load the controller and the optimum observer. Since the results of the second phase of this study dealing with the road sensing was not available, then the road function was preprogrammed into the memory. This is similar to assuming the road sensor has a faster dynamic than the steering system. This is usually the case for optical sensors, also the noises which may be involved in the measurement are already included in the observer design.

As an input was received from the road sensor indicating that the vehicle was no longer travelling parallel to the road centerline a corrective action was undertaken by the controller. The optimum controller starts and ends the corrective moves such that the vehicle ride seems to remain gentle. Fig. 10 show the experimental results.

The corrective action continues until a certain distance that needed to be corrected for the given vehicle speed and travelling direction. The output is then driven to zero to correct the remaining part of the required distance. If the road sensor indicates that the vehicle required additional correction before a previous action was completed, the magnitude of the new distance from the present distance is sensed and an appropriate correction is commanded. Figs. 10 and 11 show the steering gear output in response to various road inputs.

CONCLUSION

A nonlinear model describing dynamics of a vehicle steering system was derived. By a least-square technique, an accurate linear model was developed from the nonlinear model. A linear quadratic controller for road tracking by the automatic steered vehicle was designed. Due to availability of only vehicle position measurement with respect to the road and center lane, an optimum observer was designed to estimate the other states accurately and as fast as possible. The resulting controller and observer were implemented on a microcomputer. The required interfacing hardware was presented. Satisfactory experimental results of automatic steering based on the road input was produced. These results indicated a successful completion of the first phase of the research for development of an automatic vehicle steering system to improve traffic surveillance and safety.

NOMENCLATURE

A = piston area (.005772 m^2)

A_1 = area associated with q_1 (m^2)

A_3 = area associated with q_3 (m^2)

C_1 = discharge coefficient associated with q_1

C_3 = discharge coefficient associated with q_3

K_1 = worm gear pitch (459.348 rad/m)

K_T = torque shaft spring constant (84.33 N m/rad)

L = leakage coefficient (0.10 m s)

M_e = equivalent mass (6.21 kg)

P_s = supply pressure (N/m^2)

P_1, P_2 = cylinder pressures (N/m^2)

dp = differential cylinder pressure (N/m^2)

f = equivalent viscous friction constant (877.5 N·s/m)

k = bulk modulus (N/m^2)

q_{cyl} = mass flow rate into cylinder (kg/sec)

q_s = constant supply flow rate (kg/sec)

s = piston stroke (m)

y = position of rack-piston (m)

ρ = density (kg/m^3)

θ = position of input shaft (rad)

REFERENCES

1. A. Pne, Johns, "A State Constraint Approach to Vehicle Follower Control for Short Headway Automated Transit Systems," Proceedings of 1977 Joint Automatic Control Conference.

2. S. Schladover, D. Wormley, H. Richardson, and R. Fish, "Steering Controller Design for Automated Guideway Transit Vehicles," Proceedings of 1977 Joint Automatic Control Conference.

3. Y.K. Kwak and C.C. Smith, "An Active and Passive Steering Controller Study of Rubber-Tired Automated Guideway Transit Vehicles," ASME Journal of Dynamic Systems Measurement, and Control, Sept. 1980.

4. R. Fenton, G. Melocik, and K. Olson, "On the Steering of Automated Vehicles: Theory and Experiment," IEEE Transactions on Automatic Control, Vol. AC-21, June 1976.

5. B. Upchuch, B. Tennes, and T. Surbrook, "Development of a Microprocessor-Based Steering Controller for an Over-The-Row Apple Harvester," SAE Paper No. 810940.

6. P. Nine and H. Griffin, "Microcomputer Control of Steering on a Small Tractor (Mower)," SAE Paper No. 821052.

7. B. Larson, "Load Sensitive Steering for Energy Savings," SAE Paper No. 800995.

8. M. Kenny, "Applying Microprocessors to Off-Highway Vehicles," SAE Paper No. 810939.

9. D. Marquis, "How Fundamentals are Applied to the Design of Safe, Efficient, Automotive Steering Systems," Saginaw Steering Gear Division, General Motors Engineering Journal, 1957.

10. R. Shoureshi, "Theory and Design of Control Systems," Course Notes, ME 575, Fall, 1983, School of Mechanical Engineering, Purdue University.

11. H. Kwakernaak and R. Sivan, "Linear Optimal Control Systems," John wiley and Sons, 1972.

12. Elbert Owens, Ross Gear Division of TRW Inc., Lafayette, IN, personal communications, 1984.

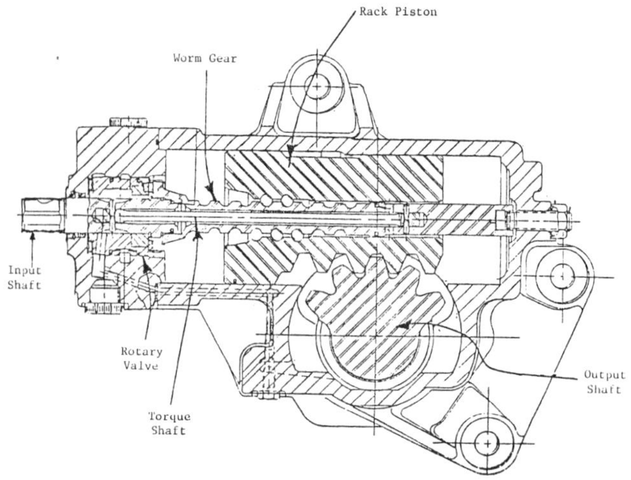

Fig. 1: Steering Gear Mechanism

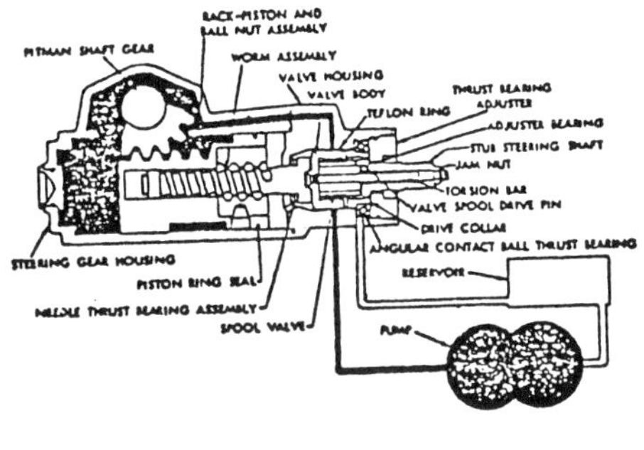

Fig. 2: Hydraulic Circuit of Steering System

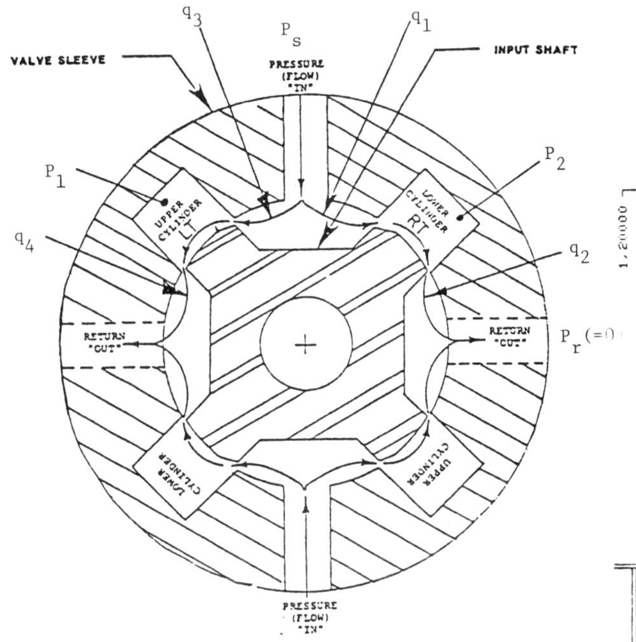

Fig. 3: Cross-Section of Rotary Valve

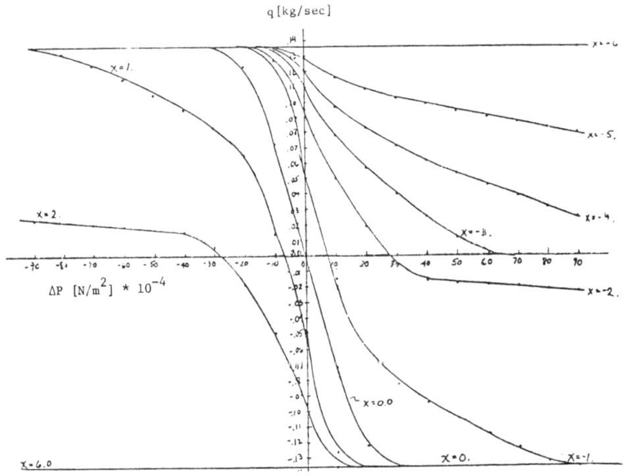

Fig. 4: Variation of Flow with Gear Pressure Difference Based on Nonlinear Model

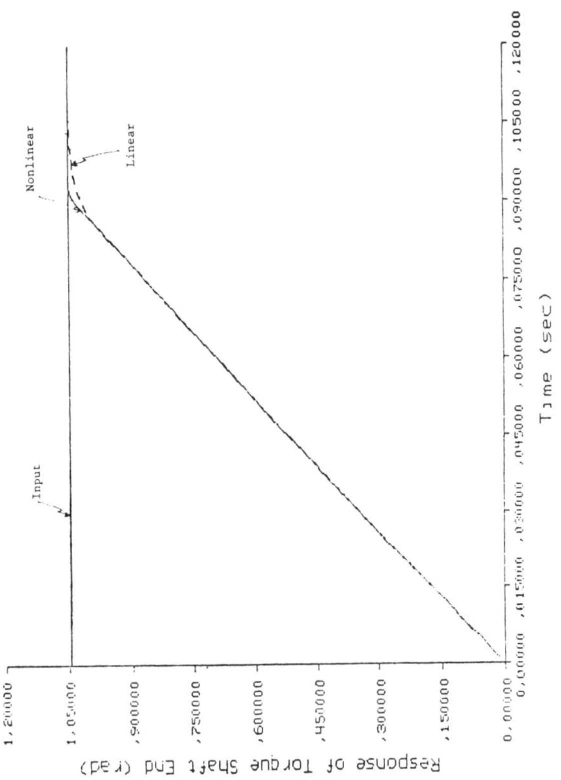

Fig. 5: Comparison of Linear and Nonlinear Model Responses to a Step Input.

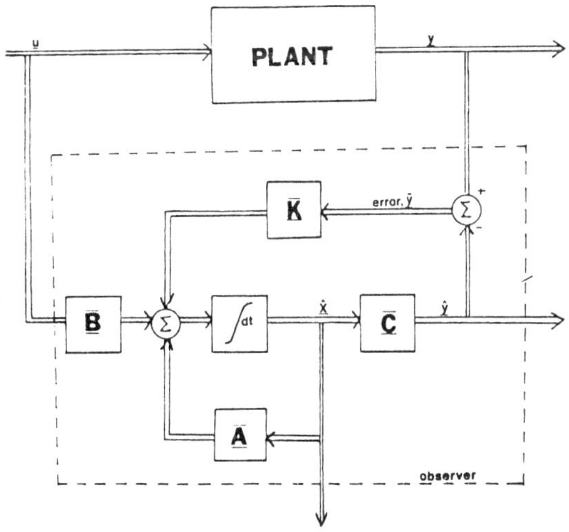

Fig. 6: Block Diagram of Steering System with Observer

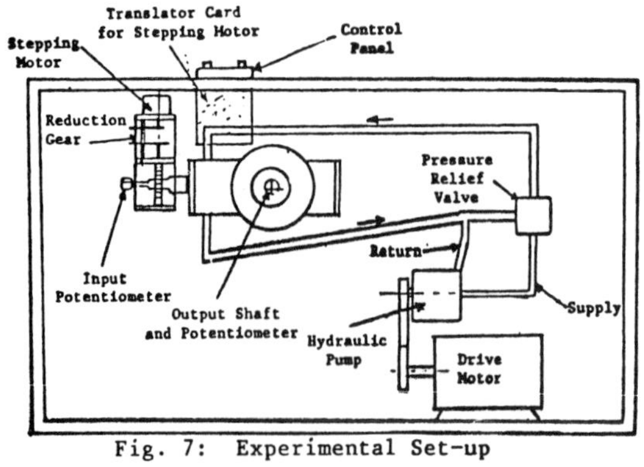

Fig. 7: Experimental Set-up

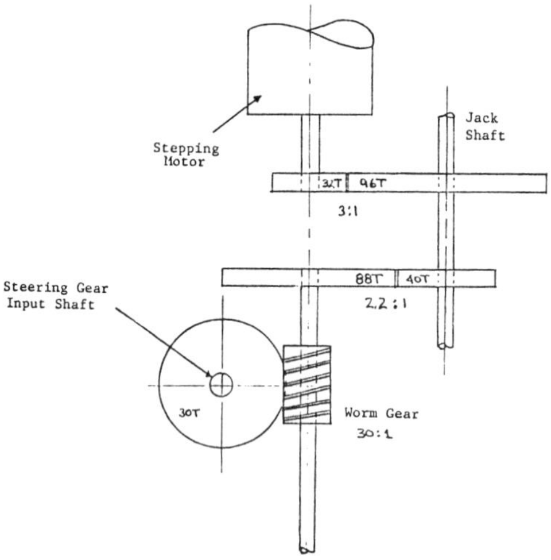

Fig. 8: Hardware Between Steering Gear and Stepping Motor

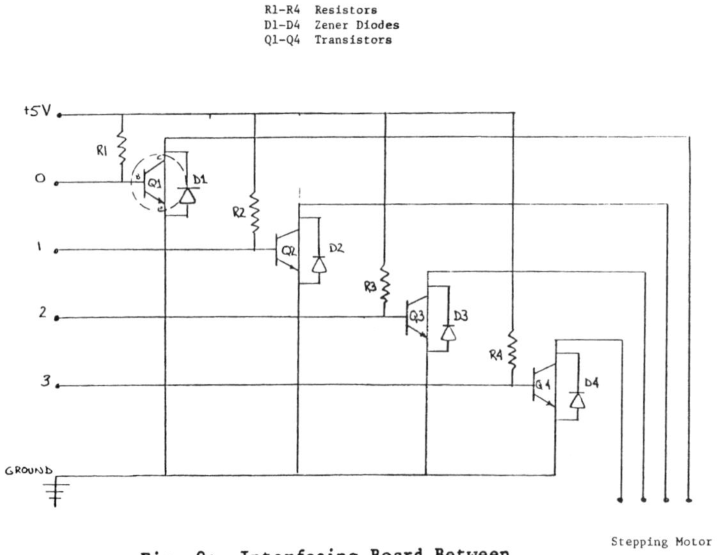

Fig. 9: Interfacing Board Between Steering System and Microprocessor

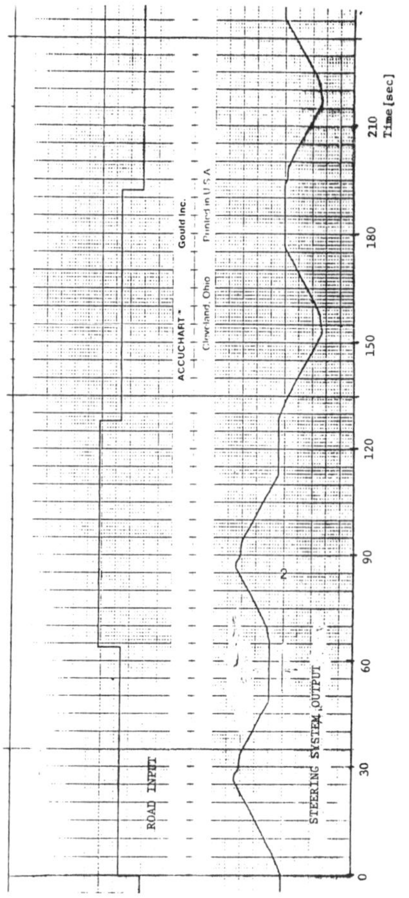

Fig 10: **Steering System Response to Successive Road Variations**

SELF-ADAPTIVE FORECAST AND CONTROL OF THE VOLUME OF PURCHASE TO FARM PRODUCTS AND SIDELINE PRODUCTS OF DEVELOPING COUNTRIES

Tang Bingyong

Institute of Applied Mathematics, Heilongjiang University, Harbin, China

Abstract. In this paper, the forecast and control problems of the volume of purchase to farm products and sideline products of developing countries have been discussed, a forecast and control method of dynamic system which has greater self-adaptive ability is introduced, and its application is explained concretely through living example.

Keywords. Self-adaptive control method of parameter forecasting; a forecasting and control model of the volume of purchase; example.

INTRODUCTION

At present, the vast developing countries are promoting enthusiastical a great advance in national economy, in the light of their reality, self-reliance, and through cooperating each other within themselves and with the developed countries. But, the most developing countries are still on the footing of agricultural countries and their industries yet are not very flourishing. So far as now we say it is a very important and urgent mission to develop energetically national agricultural economy which can suit their own characteristics and their own propel to rapidly develop their national industry step by step. Thus, the forecast and control problems of the volume of purchase to farm produce and sideline products on the developing countries are raised. To solve the problems with the advanced scientific forecast and control method, it takes very important effect notional economic prosperity of every country, for the making of the developing plan of industry and agriculture, and brings their superiority of natural resources into full play. Since these years, it has brought great importance to the problems for each country.

In this paper, a new method[1] which suits to solve the forecast and control problems of the volume of purchase to farm produce and sideline products is introduced. This new method has clearer self-adaptive feature, its foundation is forecasting the states (or outputs) of the systems and seeking optimal control inputs on the base of forecasting time-varying parameters of the systems. The forecasting precision is improved substantially, the control medium is simplified and it is advantageous to spread in every aspect, because the method is in enough consideration of time-varying feature of the systems in the whole course and overcomes the bigger error defect resulted from the general method by taking fixed parameter model to handle the problems. Giving an example about the forecast and control of the volume of purchase to the sheep's wool, the application of the above-mentioned method is explained concretely in this paper, it can be referred to the experts and persons of the same trade or occupation in each country.

THE SELF-ADAPTIVE PREDICTION AND CONTROL METHOD OF FORECASTING PARAMETERS

Now we summarize the prediction method of the dynamic system, which is proposed in reference [1], and the self-adaptive control method of parameter forecasting spreaded therefrom as follows.

Considering the following system model:

$$y(k)+a_1(k)y(k-1)+\cdots+a_n(k)y(k-n) = b_0(k)u(k)+b_1(k)u(k-1)+\cdots+ +b_m(k)u(k-m)+e(k) \qquad (1)$$

where k is discrcte time, $y(k)$ is output of the system, $u(k)$ is input of the system, $a_1(k),\cdots,a_n(k),b_0(k),b_1(k),\cdots,b_m(k)$ are unknown time-varying parameters, $e(k)$ is white noise, n and m are suitable constants.

Let $\quad \phi(k)^T=\{-y(k-1),\cdots,-y(k-n),u(k), u(k-1),\cdots,u(k-m)\}$

$$\theta(k)^T=\{a_1(k),\cdots,a_n(k),b_0(k), b_1(k),\cdots,b_m(k)\}$$

then (1) can be rewritten as follows:

$$y(k)=\phi(k)^T\theta(k)+e(k) \ . \qquad (2)$$

1. Estimating unknown parameters

For system (2), in order to seek the estimate value $\hat{\theta}(k)$ of the time-varying parameter vector $\theta(k)$ by the observational data, and consider fully its time-varying instantaneous feature, we can apply the recursive extended gradient method (see [2]):

$$\hat{\theta}(k)=\hat{\theta}(k-1)+\frac{1}{\|\phi(k)\|^2}\cdot\phi(k)\cdot \left\{y(k)-\phi(k)^T\hat{\theta}(k-1)\right\}. \qquad (3)$$

2. Forecasting time-varying parameters

Supposing a series of estimate values of the time-varying parameter $\theta(k)$ has been obtained by applying formula (3) to pursue the parameter of the system (2) :

$$\hat{\theta}(1)^T=(\hat{a}_1(1),\cdots,\hat{a}_n(1),\hat{b}_0(1),\hat{b}_1(1),\cdots,\hat{b}_m(1))$$

$$\hat{\theta}(2)^T=(\hat{a}_1(2),\cdots,\hat{a}_n(2),\hat{b}_0(2),\hat{b}_1(2),\cdots,\hat{b}_m(2))$$

$$\cdots \quad \cdots \quad \cdots \quad \cdots$$

$$\hat{\theta}(N)^T=(\hat{a}_1(N),\cdots,\hat{a}_n(N),\hat{b}_0(N),\hat{b}_1(N),\cdots,\hat{b}_m(N))$$

here N is the number of the observational data.

Through analysing $\{\hat{\theta}(k)\}$ $(k=1,2,\cdots,N)$, looking for its rule, setting up its mathematical model or forecasting formula by proper mathematical approach and method, we can obtain a series of forecasting values

$$\hat{\theta}^*(N+1),\hat{\theta}^*(N+2),\cdots,\hat{\theta}^*(N+h)$$

of the time-varying parameter $\theta(k)$, where h is forecasting step length (for the detailed discussion, see [1] [3])

3. Forecasting the output of the system

On the base of the forecasted parameters, we can further obtain the forecasting formula of the output $y(k)$ of the system as follows :

$$\begin{aligned}\hat{y}(N+1)= &-\hat{a}_1^*(N+1)y(N)-\cdots-\hat{a}_n^*(N+1)y(N+1-n)+\\ &+\hat{b}_0^*(N+1)u(N+1)+\cdots+\hat{b}_m^*(N+1)u(N+1-m)\ ;\end{aligned}$$

$$\begin{aligned}\hat{y}(N+2)= &-\hat{a}_1^*(N+2)\hat{y}(N+1)-\cdots-\hat{a}_n^*(N+2)y(N+2-n)+\\ &+\hat{b}_0^*(N+2)\hat{u}(N+2)+\cdots+\hat{b}_m^*(N+2)u(N+2-m)\ ;\end{aligned}$$

$$\cdots \qquad \cdots \qquad \cdots$$

$$\begin{aligned}\hat{y}(N+h)= &-\hat{a}_1^*(N+h)y^*(N+h-1)-\cdots-\hat{a}_n^*(N+h)y^*(N+h-n)+\\ &+\hat{b}_0^*(N+h)\hat{u}(N+h)+\cdots+\hat{b}_m^*(N+h)u^*(N+h-m)\ . \qquad (4)\end{aligned}$$

where $\hat{y}(k)$ denotes the forecasting value of $y(k)$, $\hat{u}(k)$ denotes the forecasting value or scheduled value of $u(k)$,

$$y^*(N+h-i)=\begin{cases}\hat{y}(N+h-i) & \text{if } h>i,\\ y(N+h-i) & \text{if } h\le i, i=1,\cdots,n;\end{cases}$$

$$u^*(N+h-j)=\begin{cases}\hat{u}(N+h-j) & \text{if } h>j,\\ u(N+h-j) & \text{if } h\le j, j=1,\cdots,m.\end{cases}$$

4. Seeking the optimal control

Assume $y(N+h)=r$ be the expected output on the time N+h of the system, thus, under the least square variance sense the self-adaptive control tactics $u(N+h)$ satisfies as below:

$$\begin{aligned}r= &-\hat{a}_1^*(N+h)y^*(N+h-1)-\cdots-\hat{a}_n^*(N+h)y^*(N+h-n)+\hat{b}_0^*(N+h)u(N+h)+\\ &+\hat{b}_1^*(N+h)u^*(N+h-1)+\cdots+\hat{b}_m^*(N+h)u^*(N+h-m)\end{aligned}$$

or

$$\begin{aligned}u(N+h)= \frac{1}{\hat{b}_0^*(N+h)}\Big[&r+\hat{a}_1^*(N+h)y^*(N+h-1)+\\ &+\cdots+\hat{a}_n^*(N+h)y^*(N+h-n)-\hat{b}_1^*(N+h)u^*(N+h-1)-\cdots-\hat{b}_m^*(N+h)u^*(N+h-m)\Big]\ . \end{aligned} \qquad (5)$$

In order that formula (5) has sence, we can choose $\hat{b}_0^*(N+h)\neq 0$ in the course of identification.

THE MATHEMATICAL MODEL OF THE VOLUME OF PURCHASE TO FARM PRODUCTS AND SIDELINE PRODUCTS

There are larger differences between the volumes of purchase to farm products and sideline products in the distinct countries or areas, and there are many factors which affact them, these factors are very complicated. Through analysis, we think the main influential factors are the purchasing price, the stockpiled volume of the raw materials, the volume of purchase to other farm products and sideline products relating to them, the weather elements relating to them, the sales volume (or including the export volume) last year, the stock in the end of last year, and so on. And in fact, themselves are the results which are resulted synthetically from every influential factor.

Therefore, in order to express their actual situation as objectively as possible and seize the most essential things, when we set up their mathematical model, we not only consider the development law of themselves, but also choose the above-mentioned main influential factors which is related to them. Thus, on grounds of historical data of these volumes we have obtained, we can determine the mathematical model (input-output) of the volume of purchase to farm products and sideline products as follows :

$$y(k)= a_1(k)y(k-1)+a_2(k)y(k-2)+ +b_1(k)w_1(k-1)+b_2(k)w_2(k-1)+ +c_1(k)u_1(k-1)+c_2(k)u_2(k)+ +c_3(k)u_3(k-1)+c_4(k)u_4(k-1)+ +c_5(k)u_5(k)+e(k) \qquad (6)$$

where k is discrcte time, here denotes year; $y(k)$ denotes the volume of purchase to farm products and sideline products at year k; $w_1(k-1)$ and $w_2(k-1)$ denote the main weather elements at year k-1 which is related to $y(k)$ (example: precipitation, temperature and so on); $u_1(k-1)$ denotes a certain main influential factor (example: the stockpiled volume of the raw materials) at year k-1; $u_2(k)$ denotes a certain main influential factor (example: the volume of purchase to other farm products and sideline products which relates to $y(k)$) at year k; $u_3(k-1)$ denotes the sale volume (or including the export volume) at year k-1; $u_4(k-1)$ denotes the stock at the end of year k-1; $u_5(k)$ denotes the purchasing price at year k; $a_1(k),a_2(k),b_1(k),b_2(k),$ $c_1(k),c_2(k),c_3(k),c_4(k),c_5(k)$ are all unknown time-varying parameters; $e(k)$ is white noise.

If let

$$\varphi(k)^T= \{ y(k-1),y(k-2),w_1(k-1), w_2(k-1),u_1(k-1),u_2(k), u_3(k-1),u_4(k-1),u_5(k)\}$$

$$\theta(k)^T= \{ a_1(k),a_2(k),b_1(k),b_2(k), c_1(k),c_2(k),c_3(k),c_4(k), c_5(k) \}$$

then (6) can be rewritten as:

$$y(k)=\varphi(k)^T\theta(k)+e(k) \ . \qquad (7)$$

This is the basic mathematical model which is used by us to forecast and control.

EXAMPLE: THE FORECAST AND CONTROL OF THE VOLUME OF PURCHASE TO SHEEP'S WOOL

On grounds of relevant historical data which we have obtained in the course of the purchasing sheep's wool somewhere, and through synthetical analysis, we can determine the mathematical model of the volume of purchase to sheep's wool as formula (6) (or (7)).

Where $y(k)$ denotes the volume of purchase to sheep's wool at year k (unit: dan= 50 kilograms), $w_1(k-1)$ denotes the average winter temperature at year k-1 (unit: C^o), $w_2(k-1)$ denotes the winter precipitation at year k-1 (unit: millimetre), $u_1(k-1)$ denotes the amount of sheep at the end of year k-1 (unit: hundred), $u_2(k)$ denotes the volume of purchase to sheepskin at year k (unit: a whole), $u_3(k-1)$ denotes the sale volume to sheep's wool at year k-1 (unit:dan), $u_4(k-1)$ denotes the stock at the end of year k-1 (unit: dan), $u_5(k)$ denotes the purchasing price to sheep's wool at year k (unit: yuan, the monetary unit of China).

For (7), applying the recursive formula (3) to parameter $\theta(k)$ by the historical data, we can obtain a series of estimate values as in the following table :

TABLE 1. The Estimate Values of the Time-varying Parameters

k	year	$\hat{a}_1(k)$	$\hat{a}_2(k)$	$\hat{b}_1(k)$	$\hat{b}_2(k)$
3	1958	0.6875	0.6311	0.0992	-0.9975
4	1959	0.7036	0.6456	0.0992	-0.9975
5	1960	0.7032	0.6453	0.0992	-0.9975
...		...	...	...	...
10	1965	0.6509	0.6097	0.0992	-0.9975
11	1966	0.6090	0.5708	0.0992	-0.9975
...		...	...	...	...
21	1976	0.4902	0.5451	0.0992	-0.9976
22	1977	0.5021	0.5574	0.0992	-0.9976
23	1978	0.4968	0.5528	0.0992	-0.9976

k	$\hat{c}_1(k)$	$\hat{c}_2(k)$	$\hat{c}_3(k)$	$\hat{c}_4(k)$	$\hat{c}_5(k)$
3	0.2600	-0.1259	0.5590	-0.3836	200
4	0.2645	-0.0444	0.5723	-0.3807	200
5	0.2644	-0.0461	0.5720	-0.3808	200
...	...	...	...	...	...
10	0.2577	-0.1167	0.5308	-0.3847	200
11	0.2475	-0.2203	0.4941	-0.3946	200
	...	...	...	...	...
21	0.1994	-0.1520	0.4264	-0.4230	200
22	0.2016	-0.1195	0.4387	-0.4200	200
23	0.2007	-0.1337	0.4335	-0.4215	200

To analyse $\{\hat{a}_i(k)\}$ $\{b_j(k)\}$ and $\{\hat{c}_l(k)\}$ (i=1, 2; j=1,2; l=1,2,3,4,5) by [1][3], we can apply the " Approximate Method of Average Value " for $\{\hat{b}_1(k)\}\{\hat{b}_2(k)\}$ and $\{\hat{c}_5(k)\}$, and obtain:

$$\hat{b}_1^*(N+h)= \overline{\hat{b}_1(k)}= \frac{1}{N-3}\sum_{k=3}^{N} \hat{b}_1(k)= 0.0992 \ ,$$

$$\hat{b}_2^*(N+h)=\overline{\hat{b}_2(k)}=\frac{1}{N-3}\sum_{k=3}^{N}\hat{b}_2(k)=-0.9976,$$

$$\hat{c}_5^*(N+h)=\overline{\hat{c}_5(k)}=\frac{1}{N-3}\sum_{k=3}^{N}\hat{c}_5(k)=200\ ,$$

here N=23 (year 1978); h=1,2,···.

Applying the " Multi-stratum Recursive Method of AR Model " to $\{\hat{a}_1(k)\}$, after regulating fitly, we can obtain :

$$\hat{a}_1^*(N+h)=0.2880\hat{a}_1^*(N+h-1)+0.1977\,\hat{a}_1^*(N+h-2)+0.2017\hat{a}_1^*(N+h-3)+0.1521\hat{a}_1^*(N+h-4)+0.1531\,\hat{a}_1^*(N+h-5)\ .$$

Calculating the forecasting value of $a_1(k)$ by the above-mentioned formula, we have:

$$\hat{a}_1^*(24)_{1979}=0.4949\ ,$$
$$\hat{a}_1^*(25)_{1980}=0.4972\ ,$$
$$\hat{a}_1^*(26)_{1981}=0.4986\ ,$$
$$\hat{a}_1^*(27)_{1982}=0.4980\ ,$$
$$\hat{a}_1^*(28)_{1983}=0.4933\ ,$$
$$\cdots\qquad\cdots\ .$$

For $\{\hat{c}_2(k)\}$, we can apply the "Increment Method of Separable Stage", in other words, studying the varying law of the increment:

$$\Delta\hat{c}_2(k)=\hat{c}_2(k)-\hat{c}_2(k-1)\ ,$$

here been found that it has distinct cycticity (the cycle: T=5) by its increasing and decreasing feature. Thus, the estimate value of each stage in a cyclic varying limits is determined separablely and the incremental forecasting value: $\Delta\hat{c}_2^*(k)$ will be choosed after regulating suitablely, then the forecasting formula of $c_2(k)$ will be obtained:

$$\hat{c}_2^*(k)=\hat{c}_2^*(k-1)+\Delta\hat{c}_2^*(k)$$

where

$$\Delta\hat{c}_2^*(k)=\begin{cases}-0.0161 & k=24,29,34,\cdots;\\ +0.0576 & k=25,30,35,\cdots;\\ +0.0234 & k=26,31,36,\cdots;\\ -0.0144 & k=27,32,37,\cdots;\\ -0.0508 & k=28,33,38,\cdots.\end{cases}$$

Thus we get:

$$\hat{c}_2^*(24)_{1979}=-0.1498\ ,$$
$$\hat{c}_2^*(25)_{1980}=-0.0922\ ,$$
$$\hat{c}_2^*(26)_{1981}=-0.0688\ ,$$
$$\hat{c}_2^*(27)_{1982}=-0.0832\ ,$$
$$\hat{c}_2^*(28)_{1983}=-0.1340\ ,$$
$$\cdots\qquad\cdots\ .$$

Then, for other series of estimate values, all applying the "Cyclic Variable Method of Separable Stage", it is easy to find these series are varied by the cycle (T=2) on the whole, so through proper arrangement, we can obtain approximatively their forecasting estimate as following in table 2 :

TABLE 2. The Forecasting Estimate of Some Time-varying Parameters

k	$\hat{a}_2^*(k)$	$\hat{c}_1^*(k)$	$\hat{c}_3^*(k)$	$\hat{c}_4^*(k)$
24,26,28,	0.5838	0.1969	0.4384	-0.4233
25,27,29,	0.5857	0.2030	0.4459	-0.4227

On the base of the forecasting parameters, applying further the forecasting formula to the output which is similar to formula (4) :

$$\hat{y}(N+h/N)=\hat{a}_1^*(N+h)y^*(N+h-1)+\hat{a}_2^*(N+h)\,y^*(N+h-2)+\hat{b}_1^*(N+h)w_1^*(N+h-1)+\hat{b}_2^*(N+h)w_2^*(N+h-1)+\hat{c}_1^*(N+h)\,u_1^*(N+h-1)+\hat{c}_2^*(N+h)u_2^*(N+h)+\hat{c}_3^*(N+h)u_3^*(N+h-1)+\hat{c}_4^*(N+h)\,u_4^*(N+h-1)+\hat{c}_5^*(N+h)u_5^*(N+h)\quad(8)$$

where the meaning of $w_j^*(k-1)$ and $u_l^*(k)$ (j=1,2; l=1,2,3,4,5) are the same as $y^*(k)$ and $u^*(k)$ in the formula (4), we can obtain the forecasting values of the volume of purchase to sheep's wool (1979-1983) and their errors with the real values as the following table :

TABLE 3.

k	year	$\hat{y}(k)$	$y(k)$	$\delta(k)$	$\nabla(k)$
24	1979	134067	130038	4039	3.1%
25	1980	161030	160017	1013	0.6%
26	1981	181708	185771	-4063	2.2%
27	1982	201463	205090	-3627	1.8%
28	1983	213254	210682	2572	1.2%

where $\hat{y}(k)$ is the forecasting value, $y(k)$ is the real value, $\delta(k)=\hat{y}(k)-y(k)$ is the error, $\nabla(k)=\delta(k)/y(k)$ is relative error. By the above-mentioned table, it can be seen the forecasting mean relative error (1979-1983) is 1.8%. This result is satisfactory.

Then proceeding to repeat the similar method, we can continue to forecast the volume of purchase to sheep's wool after 1984. At this moment, the unknown input values must be determined, where $w_1(k)$, $w_2(k)$ and $u_1(k)$ all choose the forecasting values which are obtained by their relevant AR models, other inputs can be chosen their scheduled values or calculated values. For example, price $u_5(k)$ can be chosen the price of 1983, through summarization for $u_2(k)$ $u_3(k)$ $u_4(k)$, we have the approximative recurrence formula by their practical meaning as follows :

$$\hat{u}_2(k)\doteq 14.6017\hat{u}_1(k-1)+0.1954\hat{w}_2(k-1)$$

$$\hat{u}_3(k) \doteq 0.9113\hat{y}(k) - 0.0013\hat{w}_1(k)$$

$$\hat{u}_4(k) \doteq 0.9432\hat{u}_4(k-1) + 0.9680\hat{y}(k) - 0.9940\hat{u}_3(k) \quad .$$

Thus, we can obtain:

$$\hat{y}(29)_{1984} = 212504 \;,$$
$$\hat{y}(30)_{1985} = 242257 \;,$$
$$\hat{y}(31)_{1986} = 278529 \;,$$
$$\hat{y}(32)_{1987} = 311889 \;,$$
$$\hat{y}(33)_{1988} = 329923 \;,$$
$$\cdots \qquad \cdots \qquad .$$

Now we consider seeking the optimal control of the volume of purchase to sheep's wool (in other words, seeking the optimal measures which can achieve the regulargoal). For example, we hope the volume of purchase will achieve 280000 dan in 1986, but it only achieves 278529 dan by the forecasting value. What optimal measures must be taken? Here the weather elements $w_1(k)$ $w_2(k)$ can not be controled and the price can be raised 15% compared with 1983, thus by the above-mentioned method, we have :

$$\begin{cases} 0.4906\hat{y}(30) + 0.5857\hat{y}(29) + 0.0992 \\ \quad \hat{w}_1(30) - 0.9976\hat{w}_2(30) + 0.2030\hat{u}_1(30) - \\ -0.0691\hat{u}_2(31) + 0.4459\hat{u}_3(30) - \\ -0.4227\hat{u}_4(30) + 200\hat{u}_5(31) = 280000 \\ \hat{u}_2(31) = 14.6017\hat{u}_1(30) + 0.1954\hat{w}_2(30) \\ \hat{u}_3(30) = 0.9113\hat{y}(30) - 0.0013\hat{w}_1(30) \\ \hat{u}_4(30) = 0.9432\hat{u}_4(29) + 0.9680\hat{y}(30) - \\ \qquad -0.9940\hat{u}_3(30) \\ \hat{u}_5(31) = (1+15\%)u_5(28) \end{cases}$$

where $\hat{y}(30)_{1985}$, $\hat{y}(29)_{1984}$, $\hat{w}_1(30)_{1985}$, $\hat{w}_2(30)_{1985}$, $\hat{u}_4(29)_{1984}$, $u_5(28)_{1983}$ are all known. So the optimal control solation can be found :

$$\hat{U}(31)_{1986} = \begin{bmatrix} \hat{u}_1(30) \\ \hat{u}_2(31) \\ \hat{u}_3(30) \\ \hat{u}_4(30) \\ \hat{u}_5(31) \end{bmatrix} = \begin{pmatrix} 34972 \\ 510659 \\ 220769 \\ 80683 \\ 2.79 \end{pmatrix}$$

In other words, if the volume of purchase is wished to achieve 280000 dan, then the amount of sheep must be 34972 at the end of 1985,the volume of purchase to sheepskin must be 510659 in 1986, the sales volume to sheep's wool must be 220769 in 1985, the stock must be 80683 in 1985, the purchasing price to sheep's wool must be raised to 2.79 . Similarly, we can also seek the optimal control for the other years.

REFERENCES

[1] Han. Z. G, (1983) A New Method on Dynamic System Prediction, ACTA Automatics Sinica,Vol.9, No.3, pp.161-168.

[2] Han. Z. G, (1984) Identification of Time-varying Parameters on Dynamic System, ACTA Automatics Sinica, Vol.10, No.4, pp.330-337.

[3] Tang. B. Y, (1983) The Prediction Computational Method of Time-varying Parameters on Dynamic System, J.Natural Science, Heilongjiang University, No.1, pp.18-26.

SPECTRAL ESTIMATION OF SPEECH CORRUPTED BY COLORED NOISE

H. Morikawa and H. Fujisaki

Department of Electronic Engineering, Faculty of Engineering, University of Tokyo, Tokyo, Japan

Abstract. A modified SEARMA method is proposed for estimating the speech spectrum in the presence of colored background noise. The following assumptions are used in developing the analysis. The speech production process is represented by an autoregressive moving-average (ARMA) model. The background noise is represented by an AR process, and is always present regardless of presence and absence of speech activity. The noise process is locally stationary during speech activity. Following these assumptions, the process during speech activity can be represented by an extended ARMA model. In this formulation, unique estimation of AR parameters of the vocal tract transfer function is always possible if the AR parameters of the noise process can be estimated separately, but the estimation of MA parameters requires further assumption of a high SNR. The validity of the proposed method is demonstrated by spectral estimation of synthetic speech sounds in the presence of additive colored noise, and by comparing the results with those obtained by a method using an AR model.

Keywords. Speech analysis; Identification; Parameter estimation; Spectral analysers; Iterative methods; ARMA model; Modified SEARMA method.

INTRODUCTION

In practical speech transmission systems, the speech signal is always accompanied by some noise. Depending on the type and the amount of noise, the quality of the reconstructed speech can range from being slightly degraded to being annoying to listen to, and finally to being totally unintelligible (Sambur and Jayant, 1976). The speech signal can be reconstructed by the vocal tract information and the source information. The performance from a voice system is dependent mainly on the estimation accuracy of the speech spectrum. Therefore, accurate estimation of speech spectrum in the presence of background noise is an important problem in speech communication.

The autoregressive power spectral density estimate, which is known as the linear prediction (LPC) spectral estimate in speech research (e.g., Atal and Hanauer, 1971; Makhoul, 1975), has been shown to possess excellent performance for noise-free input speech. The addition of white noise to an AR process, however, results in an incorrect power spectral density estimate (e.g., Kay, 1979; Tierney, 1980). The reason for the degradation of LPC analysis in the presence of noise is that the diagonal elements of the Toeplitz matrix are modified by the variance of the additive noise (Kay, 1980; Morikawa and Fujisaki, 1982b, 1984). Therefore, the lower the signal-to-noise ratio (SNR), the poorer the spectral estimate is obtained from LPC analysis. One previous approach to enhance speech spectrum in the presence of white noise is based on an AR model of speech (Kay, 1980) while another approach is based on an ARMA model (Morikawa and Fujisaki, 1982b, 1984). Since the overall observation process does not include spectral information of the background noise, however, neither of these approaches are applicable to the estimation of the speech spectrum in the presence of colored noise.

In this paper we propose a method for the accurate estimation of speech spectrum in the presence of colored background noise. The estimation is accomplished in three stages. The first stage is the detection of the speech activity interval based on stochastic properties of the autocorrelation function of speech. The second stage is the spectral estimation of the additive noise during the absence of speech activity. The third stage is the estimation of AR parameters of the vocal tract transfer function in the extended ARMA model. The validity of the proposed method is demonstrated by spectral estimation of synthetic speech sounds in the presence of additive colored noise, and by comparing the results with those obtained by a method using an AR model.

STATEMENT OF THE PROBLEM

Before going into details, we will give a rough idea of the procedure as shown by the block diagram of Fig. 1. At the first step, a decision is made on the presence/absence of speech activity within the current frame of about 30 msec duration, which is the standard length of one analysis frame in speech analysis. In the case of a decision for "silence" where speech is absent but the background noise is present, we estimate noise spectrum. In the case of a decision for "speech" where speech activity and background

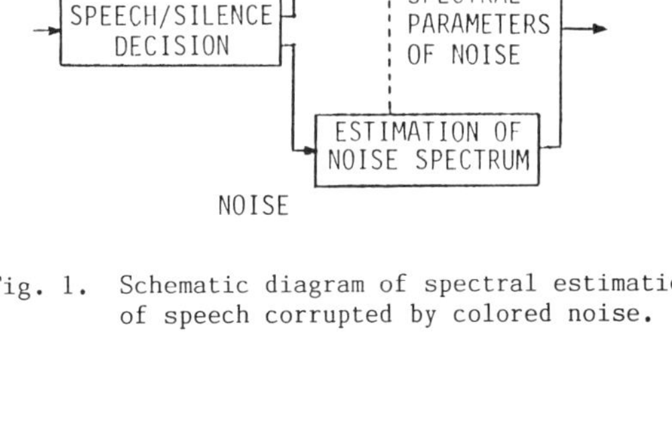

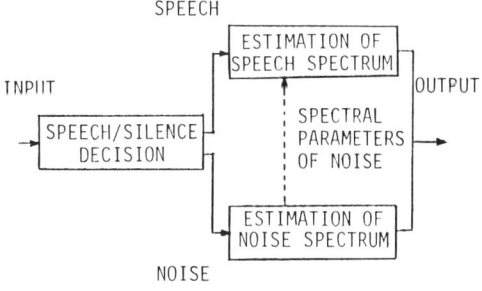

Fig. 1. Schematic diagram of spectral estimation of speech corrupted by colored noise.

noise are both present, we estimate the speech spectrum using the noise spectrum estimated from the "silence" intervals.

Model of Additive Noise

The following assumptions are used with regard to the background noise. The noise process v_i is represented by an AR process

$$v_i = w_i + \gamma_1 w_{i-1} + \cdots + \gamma_\ell w_{i-\ell} \qquad (1)$$

and is always present regardless of presence or absence of the speech activity. The noise process is locally stationary during speech activity. Hence, the values of γ_i's are constant and w_i's are represented by a zero-mean white Gaussian noise with constant variance R during speech activity. Even if the background noise process changes to a new state, there exists enough time (more than 300 msec) to estimate new noise spectral parameters. The background noise is added to the speech signal s_i, and the observed process can be expressed as:

$$y_i = s_i + v_i \,. \qquad (2)$$

In all practical situations, conversational speech is accompanied by silent intervals. Some of background noise such as room noise, traffic noise and speech-like noise may be approximated by an AR process. Therefore, above assumptions with regard to the background noise have a generality for the practical noise environments.

Model of the Speech Production Process

A model of the vocal tract can be constructed by representing it as a discrete time-varying linear system. The entire process of conversion from the source to the speech signal can be somewhat idealized and represented by the response of a linear system whose input is either a train of volume velocity impulses or random pressure fluctuations, if we include all other characteristics into the system characteristics. The process of speech production is represented by an ARMA process that converts the source signal u_i into the speech signal s_i(Morikawa and Fujisaki, 1982a):

$$s_i + \alpha_1 s_{i-1} + \cdots + \alpha_n s_{i-n} = u_i + \beta_1 u_{i-1} + \cdots + \beta_m u_{i-m} \,. \qquad (3)$$

The order n of the AR scheme and the order m of the MA scheme vary with the vocal tract configuration and the source location, and m is smaller than n. It is assumed that the source signal u_i is either a zero-mean white Gaussian noise or a train of randomly spaced impulses with variance Q. The transfer function of (3) can be represented by a z-transformation:

$$G(z) = \frac{1 + \beta_1 z^{-1} + \cdots + \beta_m z^{-m}}{1 + \alpha_1 z^{-1} + \cdots + \alpha_n z^{-n}} \,. \qquad (4)$$

Corresponding to the transfer function G(z) of the vocal tract, the power spectrum $f(\omega)$ at the angular frequency ω is given by

$$f(\omega) = \frac{Q}{2\pi} \cdot \frac{B_0 + 2B_1\cos\omega + \cdots + 2B_m\cos m\omega}{A_0 + 2A_1\cos\omega + \cdots + 2A_n\cos n\omega} \,, \qquad (5)$$

where

$$A_i = \sum_{j=0}^{n-i} \alpha_j\alpha_{i+j} \,, \quad \alpha_0 = 1 \,,$$

$$B_i = \sum_{j=0}^{m-i} \beta_j\beta_{i+j} \,, \quad \beta_0 = 1 \,.$$

Therefore, the power spectrum of the speech signal can be obtained from the AR and MA parameters. These parameters are then used to obtain the poles and zeros of the transfer function of the speech production process.

ALGORITHMS FOR NOISE PROCESSING

Accurate detection of presence/absence of speech activity is rather difficult at high noise levels, while detection of voiced speech intervals presents less problems. In the present study, therefore, we adopt the following two-step procedure for speech activity detection. In the first step (Algorithm 1), voiced/voiceless detection is made on the basis of the autocorrelation function of the observed signal within each frames of 30 msec duration. If more than nine consecutive frames are judged as "voiceless" by this algorithm, there is a quite high possibility that at least a few frames in the middle should be really "silence" (i.e. devoid of speech activity), and thus could be used for modeling the background noise itself and for estimating its spectral parameters, although there may be a few frames each of voiceless speech sounds at both ends of such an interval. We thus disregard the first and last few frames of such an interval but use the remaining frames for estimating noise parameters.

In the second step (Algorithm 2), we utilize the obtained knowledge on the background noise to exactly detect the speech activity. This is accomplished by examining, for each of the remaining frames that have been judged as "voiceless" by Algorithm 1, whether or not the same noise model applies. If the estimation error for a given frame exceeds a certain threshold, then the frame is judged as containing speech activity (i.e., the presence of voiceless speech sounds but not silence). Finally, the running spectra of speech are estimated from all the frames that have been judged as containing speech activities, by using an extended ARMA model and by using the knowledge on the spectral properties of the background noise. The algorithm for speech/silence detection will be described in the following, while the spectral estimation algorithms will be described in the next section.

An Algorithm for Voiced/Voiceless Decision (Algorithm 1)

This algorithm is based on the properties of the short-time autocorrelation function of both speech and noise. If the observed signal is a white Gaussian noise, the values of the normalized autocorrelation function at nonzero delay values will have a Normal distribution with a mean value of zero and a standard deviation of $1/\sqrt{N}$ (Morikawa, 1976), where N denotes the number of samples within one frame (e.g., 300 samples). Therefore, the following rule can be applied to the voiced/voiceless decision.

$$\left.\begin{array}{ll} \text{If } \Pr(|\hat{\lambda}_k| \geq \theta/\sqrt{N}) \leq \varepsilon_1, & \{y_i\}_{i=1}^{N} \text{ is voiceless.} \\ \text{If } \Pr(|\hat{\lambda}_k| \geq \theta/\sqrt{N}) > \varepsilon_1, & \{y_i\}_{i=1}^{N} \text{ is voiced.} \end{array}\right\} \qquad (6)$$

where $\hat{\lambda}_k$ denotes the normalized autocorrelation function at delay k, and $Pr(|\hat{\lambda}_k| \geq \theta/\sqrt{N})$ denotes the percentage of the numbers of times $|\hat{\lambda}_k|$ lies outside the bound $\theta/\sqrt{N}$ to N-1. By counting the number of points at which $|\hat{\lambda}_k|$ exceeds a certain threshold (here denoted by $\theta/\sqrt{N}$), and by setting another threshold (here denoted by ε_1) on the percentage of the number of points at which the autocorrelation function exceeds the threshold, the algorithm can make decision as to whether the observed signal can be regarded as "voiced" or "voiceless". It is of course to be noted that a frame judged to be "voiceless" by this Algorithm 1 may contain voiceless speech sounds, or may correspond to a "silence" with only background noise.

There is always the possibility of two kinds of error in the statistical decision described by (6): namely a "miss" (or a failure in detecting voiced sounds) and a "false alarm". In the present scheme, however, we do not try to minimize the overall error rate but try to avoid mainly the "false alarms", since the "miss " can be rejected at the next step by Algorithm 2. The values of autocorrelation function of speech are considerably smaller for voiceless sounds than for voiced sounds. In noise-free cases, the threshold values for θ and ε_1 can be determined experimentally, and are equal to 8 and 5, respectively. In the noisy situations these thresholds brings to lowered but the threshold lowest bounds are 2 and 1, respectively.

Figure 2 shows an example of a voiced/voiceless decision for a natural utterance of /si/ in a noisy environment (with a SNR of 6 dB for the vowel). The uppermost waveform indicates the noise-free speech signal, and the middle waveform is the observed signal with additive noise. The lowermost figure shows the percentage of the autocorrelation function which exceeds a threshold of $2/\sqrt{300} = 0.12$. It can be seen that the vowel /i/ is judged as voiced sound by Algorithm 1 with a threshold of $\varepsilon_1 = 3$.

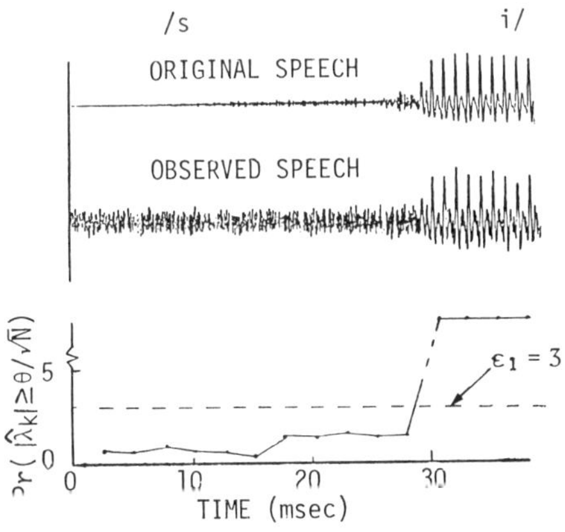

Fig. 2. $Pr(|\hat{\lambda}_k| \geq \theta/\sqrt{N})$ as function of time for an utterance of /si/.

Estimation of Noise Spectrum and Detection of Speech Onset

The AR parameters of the noise, which is regarded as an AR process with white noise excitation, can be estimated using the well known sequential identification technique (Lee, 1964). The algorithm is a subset of the modified SEARMA method to be described in the next section. These estimated noise parameters are then utilized in the second step (i.e., Algorithm 2) in order to exactly detect the speech activity. This is accomplished by examining, for each of the remaining frames that have been judged as "voiceless" by Algorithm 1, whether or not the same noise model applies.

Let us denote the estimate of γ_k at k th frame by $\hat{\gamma}_k$. Taking the norm of the difference between $\hat{\gamma}_k$ and $\hat{\gamma}_{k+1}$, we have

$$\|\hat{\gamma}_k - \hat{\gamma}_{k+1}\| \leq \|\hat{\gamma}_k - \gamma\| + \|\hat{\gamma}_{k+1} - \gamma\| = 2x . \quad (7)$$

where x denotes expectation of estimation error of γ. $\|\hat{\gamma}_k - \hat{\gamma}_{k+1}\|$ takes a larger value than 2x if the observed signal contains a speech signal. Hence, we can apply the following algorithm (Algorithm 2) detect the onset of speech activity:

$$\left.\begin{array}{l} \text{If } \|\hat{\gamma}_k - \hat{\gamma}_{k+1}\| \geq \varepsilon_2 \text{, then the k+1 th frame} \\ \text{includes the onset.} \\ \text{If } \|\hat{\gamma}_k - \hat{\gamma}_{k+1}\| < \varepsilon_2 \text{, then the k+1 th frame} \\ \text{does not include the onset.} \end{array}\right\} \quad (8)$$

The threshold ε_2 should be set to a larger value than 2x, which is the theoretical lowest bound.

ALGORITHM FOR SPEECH PROCESSING

After detecting the onset of speech activity, the algorithm estimates the speech spectrum by using the estimated AR parameters of the noise process. From (1) and (2), we have

$$y_i = s_i + w_i + \gamma_1 w_{i-1} + \cdots + \gamma_\ell w_{i-\ell} . \quad (9)$$

Substituting (9) into (3) and rearranging with respect to the regression of y_i, we have

$$\sum_{j=0}^{n} \alpha_j y_{i-j} = \sum_{j=0}^{m} \beta_j u_{i-j} + \sum_{j=0}^{n+\ell} \delta_j w_{i-j} , \quad (10)$$

where

$$\delta_j = \sum_{k=0}^{j} \alpha_k \gamma_{j-k} , \quad \alpha_0 = \beta_0 = \gamma_0 = 1 ,$$

$$\alpha_k = 0 \quad (k > n) , \quad \gamma_k = 0 \quad (k > \ell) .$$

The whole model of the observation process is represented as an extended ARMA model with two sources. We have proposed a method for simultaneous estimation of ARMA parameters (i.e., the SEARMA Method) which estimates the orders as well as the parameter values of the transfer function of the vocal tract (Morikawa and Fujisaki, 1982a). The SEARMA method is capable of controlling the location and the length of the analysis interval to adapt to speech sounds with rapid spectral changes and a high fundamental frequency (Morikawa and Fujisaki 1983). However, the method cannot be directly applied to the spectral estimation of speech corrupted by colored noise.

We have also shown that using zeros in the spectral model does not improve the spectral estimation at low SNR conditions (e.g., less than 5 dB in the case of white noise) (Morikawa and Fujisaki, 1982b, 1984). In this paper we therefore replace all the zeros of the ARMA formulation of (10) by a single compensation factor and then try to obtain accurate estimates only for the AR parameters.

Under low SNR conditions, we may assume that

$$\sum_{j=0}^{m} \beta_j u_{i-j} \ll \sum_{j=0}^{n+\ell} \delta_j w_{i-j} . \tag{11}$$

We then assume that a value of ξ exists such that it satisfies the following relation:

$$\sum_{j=0}^{m} \beta_j u_{i-j} \approx \xi w_{i-1} . \tag{12}$$

Substituting (12) into (10), (10) reduces to

$$y_i = Y_{i-1}^T \Phi + W_{i-1}^T \Omega + w_i , \tag{13}$$

where

$$Y_{i-1}^T = [y_{i-n}, \ldots, y_{i-1}, w_{i-1}] ,$$
$$\Phi^T = [-\alpha_n, \ldots, -\alpha_1, \xi] ,$$
$$W_{i-1}^T = [w_{i-n-\ell}, \ldots, w_{i-1}] ,$$
$$\Omega^T = [\delta_{n+\ell}, \ldots, \delta_1] .$$

The problem is to obtain an estimate $\hat{\Phi}$ for Φ which minimizes the following performance index:

$$I_k(\Phi) = \sum_{i=1}^{k} (y_i - Y_{i-1}^T \Phi - W_{i-1}^T \Omega)^2 + \| \Phi - \Phi_0 \|^2_{P_0^{-1}} , \tag{14}$$

where Φ_0 is an a priori estimate of Φ, P_0 is the initial value of the symmetric and positive weighting matrix P_k. For a given set of n, the parameter values can be estimated by setting $\partial I_k(\Phi)/\partial \Phi$ equal to zero and by solving for Φ, and then replacing Y_k by $\hat{Y}_k$.

$$\hat{\Phi}_{k+1} = \hat{\Phi}_k + K_k (y_{k+1} - \hat{Y}_k \hat{\Phi}_k) \tag{15}$$

$$K_k = P_k \hat{Y}_k (\hat{Y}_k^T P_k \hat{Y}_k + \hat{R})^{-1} \tag{16}$$

$$P_k = P_{k-1} - P_{k-1} \hat{Y}_{k-1} (\hat{Y}_{k-1}^T P_{k-1} \hat{Y}_{k-1} + \hat{R})^{-1} \cdot \hat{Y}_{k-1}^T P_{k-1} , \tag{17}$$

where

$$\hat{Y}_{k-1}^T = [y_{k-n}, \ldots, y_{k-1}, \hat{w}_{k-1}] ,$$
$$\hat{\Phi}^T = [-\hat{\alpha}_n, \ldots, -\hat{\alpha}_1, \hat{\xi}] ,$$
$$\hat{W}_{k-1}^T = [\hat{w}_{k-n-\ell}, \ldots, \hat{w}_{k-1}] ,$$
$$\hat{\Omega}^T = [\hat{\delta}_{n+\ell}, \ldots, \hat{\delta}_1] ,$$
$$\hat{w}_k = y_k - \hat{Y}_{k-1}^T \hat{\Phi}_k - \hat{W}_{k-1}^T \hat{\Omega}_k ,$$
$$\hat{\delta}_j = \sum_{k=0}^{j} \hat{\alpha}_k \hat{\gamma}_{j-k} .$$

The initial values Φ_0 and P_0 for the formulas above are arbitrary as long as P_0 is positive definite. $\hat{R}$ and $\hat{\gamma}$ are obtained from the analysis of background noise in the silence interval. We initiate the computation process after the first n+1 measurements by assigning the initial values to Φ_k and P_k(e.g., $\Phi_0 = 0$ or estimate at the immediately preceding frame, and $P_0 = 100I$), Starting from k = n+1, the above formulas can be applied recursively to renew the estimate $\hat{\Phi}_k$ at every sampling point. The convergence of estimation is detected and the estimated parameters are obtained at the end of an analysis time window. The orders of the noise process and of the speech production process are estimated by the previously proposed method which minimize the variance of residuals. We call this method as the Modified SEARMA method (the MSEARMA method).

Under a high SNR condition, we can write

$$\sum_{j=0}^{m} \beta_j u_{i-j} \gg \sum_{j=0}^{n+\ell} \delta_j w_{i-j} . \tag{18}$$

Then, we have the same formulas as described in (15) - (17), where

$$\hat{Y}_{k-1}^T = [y_{k-n}, \ldots, y_{k-1}, \hat{u}_{k-m}, \ldots, \hat{u}_{k-1}] ,$$
$$\hat{u}_k = y_k - \hat{Y}_{k-1}^T \hat{\Phi}_k ,$$
$$\hat{\Phi}^T = [-\hat{\alpha}_n, \ldots, -\hat{\alpha}_1, \hat{\beta}_m, \ldots, \hat{\beta}_1] .$$

The difference between this algorithm and the SEARMA method is only $\hat{R}$ in (16).

EXPERIMENTAL RESULTS

In order to assess the validity of the theory, the speech analysis system described above was tested by using synthetic speech sounds generated with a sampling frequency of 10 kHz and a quantization accuracy of 10 bits. The parameters of the noise process and the speech production process were then determined for each analysis frame according to the procedure described in above sections.

Estimation of Noise Spectrum

We show a result of the spectral estimation of noise. The parameters of the noise process are shown in Table 1. Figure 3 shows the spectral

TABLE 1 Parameters of Additive Noise

First Pole Frequency	Bandwidth of First Pole	Real Pole Frequency
500	300	2500

TABLE 2 Parameters of Synthetic Speech Sounds

	First Pole Frequency	Second Pole Frequency	Third Pole Frequency	Fourth Pole Frequency	First Zero Frequency	Second Zero Frequency
/i/	290	2130	2640	3450	——	——
/m/	220	1050	2380	4100	1600	3320

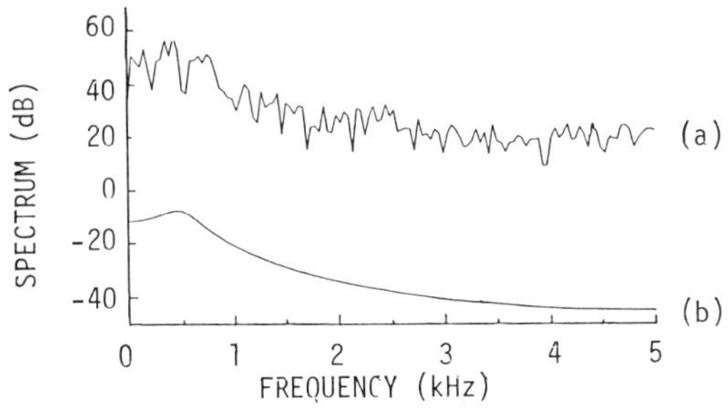

Fig. 3. Illustration of (a) short-time spectrum and (b) spectral envelope of additive noise.

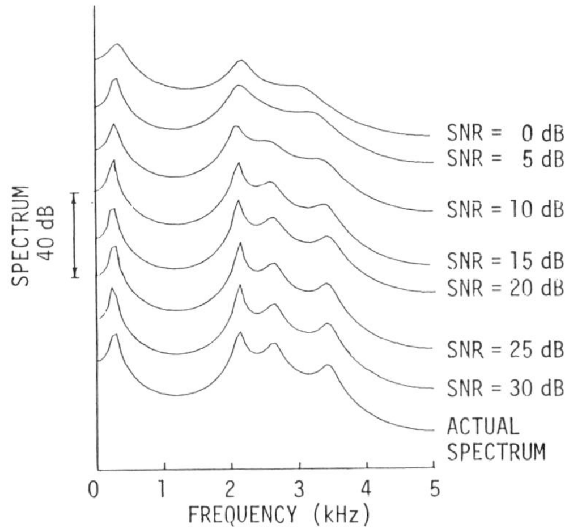

Fig. 4. Spectral estimation of synthetic vowel /i/ at various SNR conditions.

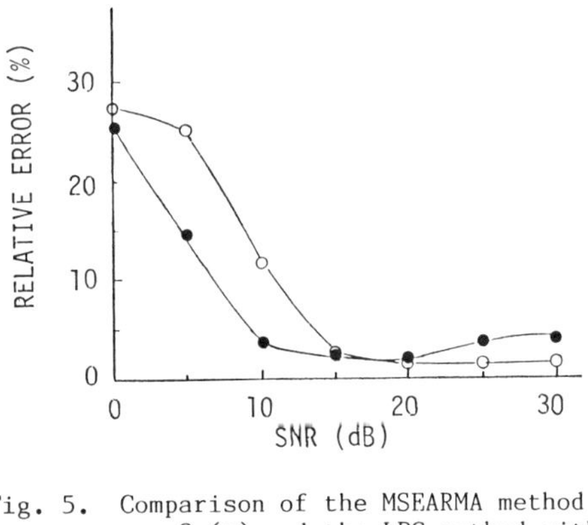

Fig. 5. Comparison of the MSEARMA method with n = 8 (●) and the LPC method with n = 11 (○) in the first pole frequency estimation, for the synthetic vowel /i/ at various SNR conditions.

envelope given by the AR model for the noise process. In this figure, (a) is the short-time spectrum obtained by the Fast Fourier Transform (FFT), (b) is the spectral envelope given by the AR model. The spectral envelope is obtained from the denominator of (5), replacing α_i by $\hat{\gamma}_i$. These estimated parameters are then used in the estimation of speech spectrum.

Estimation of Speech Spectrum

We show here the results of the spectral estimation for the synthetic speech sounds /i/ and /m/. The parameters of the synthetic speech sounds are shown in Table 2.

Figure 4 shows the estimated spectral envelope for the synthetic vowel /i/ with an SNR varying from 0 dB to 30 dB. The estimated speech spectrum is slightly degraded by the additive noise at every SNR condition as shown in Fig. 4. The noise spectrum has a resonance at 500 Hz which is close to the first pole frequency of the synthetic speech sound. Figure 5 shows the comparison of the MSEARMA method and the LPC method in the first pole frequency estimation. The estimation accuracy of the MSEARMA method is superior to that of the LPC method at all SNR values below 15 dB. The MSEARMA method with lower order than that of the LPC method yields more accurate estimated poles in the analysis of speech sound with a high noise level.

In the LPC analysis with the order 11, which is equal to the order 8 of the speech process plus the order 3 of the noise process, five complex poles are estimated. The additional pole, which appears when using larger order than the actual order in the speech process, approximates the noise component in the spectrum. The estimation accuracy of the first pole frequency obtained by the MSEARMA method is inferior to that obtained by the LPC method with order 11 at an SNR larger than 20 dB. The reason is that the estimate of the MSEARMA method has a bias caused by the estimation error of noise spectrum. However, this difference is no serious problem to apply the speech reconstruction, since the error is always less than 5 percent.

As example of speech sounds in which the role of zeros are more conspicuous, Fig. 6 shows the estimated spectral envelope for the synthetic nasal /m/. The spectral envelope is obtained by the AR scheme of the ARMA model. Figure 7 shows the comparison of the MSEARMA method and the LPC method in the first pole frequency. A large error of the first pole frequency obtained by the LPC method at high SNR conditions is caused by the zeros in the vocal tract transfer function. The estimate of AR parameters obtained by the MSEARMA method is scarcely affected by the zeros of the vocal tract transfer function and the noise.

CONCLUSIONS

We have proposed a new method for estimating the speech spectrum corrupted by colored noise. The speech production process is represented by an ARMA model. The background noise is represented by an AR model. The whole model of the observation process is then represented as an extended ARMA model with two sources. As preliminary processing of speech, we proposed a two-step procedure for speech/silence decision. The first step is based on the properties of the short-time autocorrelation function of both speech and noise. The second step is based on the properties of estimation error of AR parameters. After detecting the onset of speech activity and estimating the noise spectrum, the method can estimate AR parameters of the vocal tract transfer function.

The validity of the proposed method was demonstrated by spectral estimation of synthetic speech sounds. Since the method can estimate the parameters required for speech synthesis, it can be applied to noise-reduction of speech.

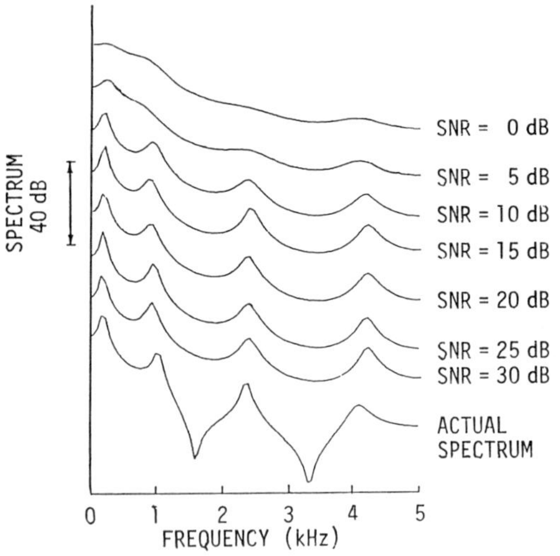

Fig. 6. Spectral estimation of synthetic nasal /m/ at various SNR conditions.

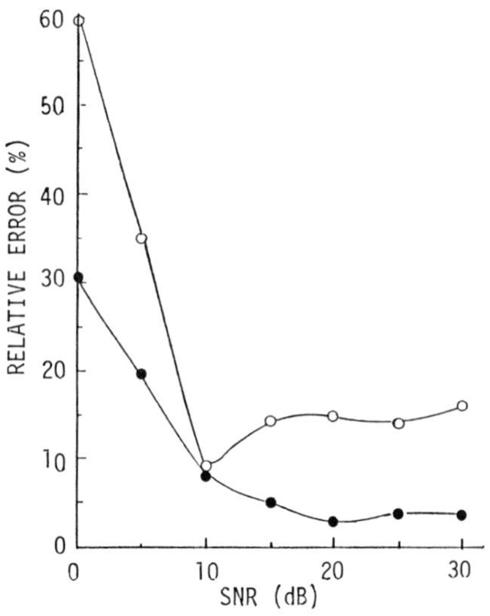

Fig. 7. Comparison of the MSEARMA method with n = 8 (●) and the LPC method with n = 11 (○) in the first pole frequency estimation, for the synthetic nasal /m/ at various SNR conditions.

REFERENCES

Atal, B. S., and S. L. Hanauer (1971). Speech analysis and synthesis by linear prediction of the speech wave. J. Acoust., Soc. Amr., 50, 637-655.

Kay, S. M. (1979). The effects of noise on the autoregressive spectral estimator. IEEE Trans. Acoust., Speech, Signal Processing, ASSP-27, 478-485.

Kay, S. M. (1980). Noise compensation for autoregressive spectral estimates. IEEE Trans. Acoust., Speech, Signal Processing, ASSP-28, 292-303.

Lee, R. C. K. (1964). Optimal Estimation, Identification, and Control. M. I. T. Press, pp. 79-126.

Makhoul, J. (1975). Linear prediction: A tutorial review. Proc. IEEE, 63, 561-580.

Morikawa, H. (1976). A multicategory classification machine with self-growing function. Trans. IECE Japan, J59-D, 9-16.

Morikawa, H., and H. Fujisaki (1982a). Adaptive analysis of speech based on a pole-zero representation. IEEE Trans. Acoust., Speech, Signal Processing, ASSP-30, 77-88.

Morikawa, H., and H. Fujisaki (1982b). Spectral estimation of speech in noisy environments. Trans. IECE Japan, J65-A, 987-994.

Morikawa, H., and H. Fujisaki (1983). Adaptive control of analysis interval in the SEARMA speech analysis methods. J.Acoust. Soc.,Japan, 39, 512-520.

Morikawa, H., and H. Fujisaki (1984). System identification of the speech production process based on a state-space representation. IEEE Trans. Acoust., Speech, Signal Processing, ASSP-32, 252-262.

Sambur, M. R., and N. S. Jayant (1976). LPC analysis/synthesis from speech inputs containing quantizing noise or additive white noise. IEEE Trans. Acoust., Speech, Signal Processing, ASSP-24, 488-494.

Tiereny, J. (1980). A study of LPC analysis of speech in additive noise. IEEE Trans. Acoust., Speech, Signal Processing, ASSP-28, 389-397.

INVERTIBILITY OF BILINEAR DISCRETE-TIME SYSTEMS

U. Kotta

Institute of Cybernetics, Tallinn 200108, USSR

Abstract. This paper gives necessary and sufficient conditions for the (left) invertibility of discrete time single-input single-output bilinear systems extending previous work on the continuous time systems. For invertible systems the (left) inverse system is constructed. The class of sequences which can appear as outputs of a given system is described and the inverse system is used to generate the required control.

Keywords. Discrete time systems; nonlinear control systems; invertibility; tracking systems.

INTRODUCTION

The control system is said to be invertible if the corresponding input-output map is injective, i.e. if whenever we have distinct control sequences then the corresponding output sequences are different. There is a considerable amount of literature dealing with invertibility and connected problems of construction inverses for continuous time and discrete time linear systems, see for example (Brockett, 1965; Sain and Massey, 1969; Silverman, 1968; Silverman, 1969; Willsky, 1974). Some results have been recently extended to continuous time bilinear (Hirschorn, 1977; Hsu and Mohler, 1981) and nonlinear (Hirschorn, 1979; Nijmeijer,1982) systems. The interest can be explained by the fact that invertibility is related to the problem of determination the input of control system from the knowledge of its output, the problem of functional reproducibility, the decoupling problem, the problem of immersion the nonlinear system by feedback into a linear one and so on.

Unfortunately, the question of invertibility of nonlinear discrete time systems has received only slight attention. To the author's knowledge, the only contributions are due to Kotta (1983), which considers the special class of dyadic bilinear systems and Monaco and Normand-Cyrot (1983), where invertibility definition requires only different controls at time instant t=0.

The purpose of this paper is to extend some previous work on continuous time systems to general discrete time single-input single-output systems. We will give necessary and sufficient conditions for the (left) invertibility and construct the (left) inverse for invertible systems. The class of sequences which can appear as outputs of a given system will be described and the inverse system will be used to generate the required control.

INVERTIBILITY CONDITIONS

In this section we derive necessary and sufficient conditions for (left) invertibility of discrete time bilinear systems. Consider a system described by equations

$$x(t+1) = Ax(t) + Bx(t)u(t) \qquad (1)$$

$$y(t) = cx(t) \qquad (2)$$
$$x(0) = x_0,$$

where the state $x(t) \in R^n$, $u(t)$ is the scalar control, $y(t)$ is the scalar output, A, B are the $n \times n$ real matrices and c is the $1 \times n$ real vector.

Definition 1. The index α of the bilinear system (1), (2) is the least positive integer such that $cA^{\alpha-1}B \neq 0$ or $\alpha = \infty$ if $cA^kB = 0$ for all $k \geqslant 0$.

It is easy to show that $t = \alpha$ is the first instant of time at which the output is affected by the input at time $t = 0$. Now, denote $H = \{x \mid cA^{\alpha-1}Bx = 0\}$, $X = R^n - H$. For any $x_0 \in X$ introduce a set U_{x_0}, containing the sequences $u = \{u(t), t=0,1,\ldots\}$ that yield the state of the system (1) $x(t) \in X$ for all $t > 0$. We denote by $y(x_0,u)$ (by $y(t,x_0,u(0),\ldots,u(t-1))$ the output sequence (the output at time instant t) of the system (1), (2) initialized at x_0 and driven by the control sequence u (the controls $u(0),\ldots,u(t-1)$).

Definition 2. The discrete time bilinear system (1), (2) is invertible at $x_0 \in X$ with respect to U_{x_0} if whenever $u \in U_{x_0}$ and $\tilde{u} \in U_{x_0}$ are distinct input sequences (i.e. $\exists\, t_k$ such that $u(t_k) \neq \tilde{u}(t_k)$) the output sequences $y(x_0,u)$ and $y(x_0,\tilde{u})$ are different.

We have proved the following theorem.

Theorem 1. The discrete time bilinear system (1), (2) is invertible at $x_0 \in X$ with respect to U_{x_0} if and only if the index of the system $\alpha < \infty$.

Proof. Sufficiency. Suppose $\alpha < \infty$. Let us assume that $u = \{u(t), t=0,1,\ldots\}$ and $\tilde{u} = \{u(t), t=0,1,\ldots\}$ are distinct input sequences i.e. $\exists\, t_k$ such that $u(t_k) \neq \tilde{u}(t_k)$. Let β be the least among them (β is arbitrary). We have to show that the output sequences which correspond to input sequences $u \in U_{x_0}$ and $\tilde{u} \in U_{x_0}$ are different, i.e. $\exists\, t_r$ such that $y(t_r, x_0, u(0),\ldots,u(t_r-1)) \neq y(t_r,x_0,\tilde{u}(0),\ldots,\tilde{u}(t_r-1))$. We will show that t_r can be chosen equal to $\alpha + \beta$. By the definition of α we have

$$y(t,x(t)) = cx(t)$$

$$\vdots$$

$$y(t+\alpha-I,x(t),u(t),\ldots,u(t+\alpha-2)) = c\,A^{\alpha-I}x(t)$$

$$y(t+\alpha,\ x(t),\ u(t),\ldots,\ u(t+\alpha-I)) =$$

$$= cA^{\alpha}x(t) + cA^{\alpha-I}\ x(t)\ u(t).$$

By the definition of β, $\alpha + \beta$ is the first instant of time at which the output sequences $y(x_o,u)$ and $y(x_o\tilde{u})$ differ (if only $x(\beta) \in X$ as it is by the choice of x_o, u and $\tilde{u}$):

$$y(\alpha+\beta,x_o,u(o),\ldots,u(\alpha+\beta-I)) -$$

$$- y(\alpha+\beta,x_o,\tilde{u}(o),\ldots,\tilde{u}(\alpha+\beta-I)) =$$

$$= cA^{\alpha-I}Bx(\beta)\ [u(\beta) - \tilde{u}(\beta)] \neq 0.$$

The sufficiency has been proved.

Necessity. Suppose $\alpha = \infty$. Then it follows from the definition of α that for all $k \geq 0$ $cA^kB = 0$. That implies $y(t,x_o,u(0),\ldots,u(t-I)) = cA^tx_o$ for all $t > 0$, which is independent of the controls. This completes the proof.

Remark I. The result somewhat differs from the corresponding result for continuous time bilinear systems, where it is required only that $x_o \in X$. In the case of discrete time bilinear systems invertibility depends besides the initial state on the applied input sequence - it must guarantee that $x(t) \in X$ for every $t > 0$.

Remark 2. In the case when $\alpha < \infty$, $x_o \in X$ and $u \in U_{x_o}$ are sufficient conditions for invertibility, but they are far from being necessary. Let us see the example I.

Example I. Consider the system (I), (2) and suppose $x(\beta) \notin X$, $u(\beta+I) = \tilde{u}(\beta+I)$, $cA^{\alpha-I}B^2x(\beta) \neq 0$. We then have

$$y(\alpha+\beta+I,\ x_ou(0),\ldots,u(\alpha+\beta)) =$$

$$= cA^{\alpha+I}x(\beta) + cA^{\alpha}BA\ x(\beta) + cA^{\alpha}Bx(\beta)\ u(\beta) +$$

$$+ cA^{\alpha-I}B^2x(\beta)\ u(\beta)\ u(\beta+I)$$

which implies

$$y(\alpha+\beta+I,\ x_o,\ u(0),\ldots,u(\alpha+\beta)) -$$

$$- y(\alpha+\beta+I,\ x_o,\ \tilde{u}(0),\ldots,\ \tilde{u}(\alpha+\beta)) =$$

$$= cA^{\alpha-I}B^2\ x(\beta)\ u(\beta+I)\ [\ u(\beta) - \tilde{u}(\beta)] \neq 0.$$

Example 2. Consider the system (I), (2), where

$$A = \begin{bmatrix} 0 & 0 & I \\ I & I & 0 \\ 0 & I & 0 \end{bmatrix},\quad B = \begin{bmatrix} 0 & 0 & 0 \\ 0 & 0 & 0 \\ 0 & I & 0 \end{bmatrix},\ c = [0\ \ 0\ \ I].$$

Since the vector $cB = [0\ \ I\ \ 0] \neq 0$, the index of the system $\alpha = I$. According to Theorem I this system is invertible at $x_o \in R^3 - \{\ x|x_2 = 0\ \}$ with respect to input sequences from U_{x_o}, where x_2 is the second component of the state vector. Let $x_o = [0\ \ I\ \ 0]$. In this case, the system is invertible, for example, with respect to input sequence $\{I, -2, 0, 2\}$. But nothing can be said by Theorem I if the system will be driven by the input sequence $\{I, -2, 0, -2\}$ which will not belong to $U_{[0\ I\ 0]}$, because $x_2(6) = 0$ in that case.

INVERSE SYSTEM

In the case when the system is invertible, the output sequence uniquely determines the input sequence. One then faces the practical problem of determining the input sequence given only the output sequence of the system. This problem can be solved in the following manner. A second system, called a left inverse, will be constructed. This inverse system, when driven by the output of the given system with appropriate shift forward, produces the input of the given system as an output. In this section we will construct the (left) inverse for discrete time bilinear system (I), (2).

We will look for a inverse in the class of nonlinear systems of the form

$$\begin{aligned} z(t+I) &= a(z(t)) + b(z(t))\ v(t) \\ w(t) &= d(z(t)) + e(z(t))\ v(t), z(0) = z_o, \\ &\qquad z \in Z, \end{aligned} \tag{3}$$

where a, b, d and e are real analytic functions on Z.

Definition 3. **The system (3) is called a left inverse for the given system (I), (2) if $v(t) = y(t+\alpha)$ implies that $w(t) = u(t)$. We have proved the following theorem.**

Theorem 2. **The (left) inverse of the invertible** bilinear system (I), (2) with index α is defined by (3), where $Z = X$, $z_o = x_o$, $z(t) = x(t)$ and

$$\begin{aligned} a(z(t)) &= Az(t) - [cA^{\alpha}z(t)/cA^{\alpha-I}Bz(t)]\ Bz(t), \\ b(z(t)) &= [I/cA^{\alpha-I}Bz(t)]\ Bz(t), \\ d(z(t)) &= -\ [cA^{\alpha}z(t)/cA^{\alpha-I}Bz(t)], \\ e(z(t)) &= [I/cA^{\alpha-I}Bz(t)]. \end{aligned} \tag{4}$$

Proof. Applying the shift operator on the output of the system (I), (2) $y(t)$ until the control $u(t)$ appears, we obtain by the definition of α

$$y(t+\alpha) = cA^{\alpha}x(t) + cA^{\alpha-I}Bx(t)\ u(t). \tag{5}$$

As we consider only invertible bilinear systems and invertibility is defined at $x_o \in X$ with respect to U_{x_o}, we have $x(t) \in X$. Therefore, the solution of equation (5) is

$$\begin{aligned} u(t) &= -\ [cA^{\alpha}x(t)\ /\ cA^{\alpha-I}Bx(t)] + \\ &+ [I/cA^{\alpha-I}Bx(t)]\ \ y(t+\alpha) = \\ &= d(x(t)) + e(x(t))\ y(t+\alpha) \end{aligned} \tag{6}$$

Substituting the right hand side of equation (6) into equation (I) we get

$$x(t+I) = Ax(t) - [cA^{\alpha}x(t)/cA^{\alpha-I}Bx(t)]\ Bx(t) +$$

$$+ [Bx(t)/cA^{\alpha-I}Bx(t)]\ y(t+\alpha) =$$

$$= a(x(t)) + b(x(t))\ y(t+\alpha).$$

We can see from here that the output of the system (7), (6) is equal to u(t), if $x(0) = x_o$ and its input is equal to $y(t+\alpha)$. Consequently, the system (7), (6) (which is identical to system (3), (4)) is the (left) inverse of the system (I), (2).

Example 3. Consider the system, described in Example 2. According to Theorem 2 the inverse of this system is the system (3), where

$Z = R^3 - \{x | x_2 = 0\}$, $z_o = x_o$, $z(t) = x(t)$ and

$$a(z(t)) = [z_3(t),\ z_I(t) + z_2(t),\ 0]^T$$

$$b(z(t)) = [o\ \ o\ \ I]^T$$

$$d(z(t)) = -I$$

$$e(z(t)) = I / z_2(t).$$

This is in accordance with the known result of Kotta (I983), where it has been proved that if the rank of matrix B is equal to unity, then the inverse is the linear system with nonlinear output.

FUNCTIONAL REPRODUCIBILITY

The problem of constructing an input which will yield some desired output is referred as the problem of functional reproducibility. In the case of continuous time systems this problem can be solved by the second system, called a right inverse. This inverse system, when driven by the certain desired output of the given system with appropriate shift forward computes the required input of the given system as an output. We shall show that the same situation occurs in the case of discrete time systems.

Definition 4. The system (3) is called a right inverse for the given system (I), (2) if $v(t)=f(t+\alpha)$ implies $w(t)=u(t)$, such that $y(t, x_o, u(0),\ldots, u(t-I)) = f(t)$.
Actually we shall prove that the left inverse (3), (4) acts as a right inverse for discrete time bilinear system (I), (2).

Theorem 3. If the desired output sequence $\{f(t), t=0,I,\ldots\}$ satisfies the following conditions:

I) $f(t) = cA^t x_o$, $t=0,I,\ldots,\alpha-I$,

2) the state of the system (3), (4) $z \in X$, if its input sequence $v(t) = f(t+\alpha)$, $t=0,I,\ldots$,
then there exists the input sequence $u = \{u(t), t=0,I,\ldots\}$ which will guarantee $y(t) = f(t)$ for the system (I), (2). This input sequence can be generated as the output of the system (3), (4) driven by $f(t+\alpha)$.

Proof. Let $f(t+\alpha)$ be the input of the inverse system: $v(t) = f(t+\alpha)$. If f(t) satisfies the condition 2) of Theorem, then $cA^{\alpha-I}Bz(t) \neq 0$ and the output of the inverse system will be

$$w(t) = -[cA^{\alpha}z(t) / cA^{\alpha-I}Bz(t)] + $$
$$+ [I / cA^{\alpha-I}Bz(t)]\ \ f(t+\alpha).$$

We must show that $y(t, x_o, w(0),\ldots,w(t-I)) = f(t)$.

When $u(t) = w(t)$, x(t) satisfies the difference equation $x(t+I) = Ax(t) + Bx(t)\,w(t)$, $x(0)=x_o$.

We will show that z(t) satisfies the same equation:

$$z(t+I) = a(z(t)) + b(z(t))\ f(t+\alpha) =$$
$$= Az(t) - [cA^{\alpha}z(t) / cA^{\alpha-I}Bz(t)]\ Bz(t) +$$
$$+ [I / cA^{\alpha-I}Bz(t)]\ Bz(t)\ f(t+\alpha) =$$
$$= Az(t) + Bz(t)\ w(t),$$
$$z(0) = x_o.$$

By the definition of α and the assumption I) of Theorem we have

$$y(0, x_o) = cx_o = f(0),$$
$$\vdots$$
$$y(\alpha-I, x_o, w(0),\ldots,w(\alpha-2)) = cA^{\alpha-I}x_o = f(\alpha-I).$$

Now,

$$y(t+\alpha, x_o, w(t),\ldots,w(t+\alpha-I)) =$$
$$= cA^{\alpha}x(t) + cA^{\alpha-I}Bx(t)\ w(t),$$

$t=0,I,\ldots$.

Since $z(t) = x(t)$, we get

$$y(t+\alpha, x_o, w(t),\ldots,w(t+\alpha-I)) =$$
$$= cA^{\alpha}x(t) - [cA^{\alpha}x(t)/cA^{\alpha-I}Bx(t)]\ cA^{\alpha-I}Bx(t) +$$
$$+ [I/cA^{\alpha-I}Bx(t)]\ cA^{\alpha-I}Bx(t)\ f(t+\alpha) = f(t+\alpha),$$

$t=0,I,\ldots$.

This completes the proof.

Remark 3. In continuous time case (Hirschorn, I979) the condition analogical to condition 2) of Theorem 3 has been omitted.

CONCLUSIONS

For a class of bilinear discrete-time systems a necessary and sufficient conditions for (left) invertibility were obtained. A (left) inverse system was constructed to determine the input sequence given only the output sequence of the system. The class of sequences which can appear as output of a given system is described and the inverse system is used to generate the control such that the output tracks a given reference sequence..

REFERENCES

Brockett, R.W. (I965). Poles, zeros and feedback: state-space interpretation. IEEE Trans. Autom. Contr., I0, I29-I35.

Hirschorn, R.M. (I977). Invertibility of control systems on Lie groups. SIAM J. Contr. and Optimiz., I5, I034-I049.

Hirschorn, R.M. (I979). Invertibility of nonlinear control systems. SIAM J. Contr. and Optimiz., I7, 289-297.

Hsu, C.S. and R.R.Mohler (I98I). On the inverse of a special class of bilinear systems. Trans.of ASME. J. of Dynamic Systems, Measurement and Control, I02, I03-I05.

Kotta, U. (I983). On the inverse of a special class of MIMO bilinear systems. Proc. of the Academy of Sciences of the Estonian SSR. Physics, Mathematics, 32, 323-326.

Monaco, S. and D. Normand-Cyrot (I983). Some remarks on the invertibility of nonlinear discrete-time systems. In Prepr. I983 American Control Conference, vol. I, IEEE, New York, pp.324-328.
Nijmeijer, H. (I982). Invertibility of affine nonlinear control systems: a geometric approach. Systems and Control Letters, 2, I63-I68.
Sain, M.K. and J.L.Massey (I969). Invertibility of linear time-invariant dynamical systems. IEEE Trans. Autom. Contr., I4, I4I-I49.
Silverman, L.M. (I968). Properties and application of inverse systems. IEEE Trans. Autom. Contr., I3,436-437.
Silverman, L.M. (I969). Inversion of multivariable linear systems. IEEE Trans. Autom. Contr., I4, 270-276.
Willsky, A.S. (I974). On the invertibility of linear systems. IEEE Trans. Autom. Contr., I9, 272-274.

A MULTIVARIABLE DECOUPLING POLE-ZERO PLACEMENT SELF-TUNING CONTROLLER

Lang Shi Jun, Gu Xing Yuan and Chai Tian You

Department of Automatic Control, Northeast University of Technology, Shenyang, Liaoning, China

Abstract. This paper describes a new multivariable self-tuning controller which is able to handle the system with different time delays between each of the input-output pairs, and deal with the system with unknown or varying time delays. The proposed controller not only moves the closed-loop-system poles to the prespecified locations but also decouples the multivariable system. It can track time varying reference signals and eliminate tracking errors. In addition an implicit algorithm is proposed under certain conditions. The controller is demonstrated using a concentration-flow process, a paper machine head box and an unstable and nonminimum phase system having different time delays.

Keywords. Self-tuning control; multivariable control system; pole-zero placement; decoupling; identification.

INTRODUCTION

For the multivariable case tuning of controllers has also been considered. The control of multiple-input/multiple-output processes is one of the most challenging and important control problems today. The application of conventional single-loop controllers to complex, multivariable processes has been only partially successful because of complexities such as time delays and control loop interactions. The use of multivariable controllers can improve the situation greatly; however, an accurate model of the process is invariably required. Because they are easy to implement and no a priori process model is required the multivariable self-tuning controllers are particularly attractive. The multivariable self-tuning regulators (MVSTR) based on the single-input/single-output minimum variance regulator of Astrom and Wittenmark (1973) have been proposed by Borisson (1979). However,this mltivariable STR suffers from same problems that detract from the single-input/single-output STR. It is not applicable to nonminimum-phase systems; it is not able to track time varying reference signals and reduce fluctuations and peaking in control signals. The multivariable self-tuning controller (MVSTC) , based on the SISO generalized minimum variance controller of Clarke and Gawthrop (1975) has been proposed by Koivo (1980) to avoid these problems. However, these controllers have restrictions on the time delays,i.e., they require that all of the time delays to be equal, to be known and constant. The MIMO pole placement regulator of Prager and Wellstead (1980) is not sensitive to unknown or varying time delays. Their approach allows for different time delays between the various inputs and outpurs,and variable time delays. However, this regulator is not able to track time varying reference signals. Its main disadvantage lies in the complexity of the computation required which restricts maximum allowable sampling frequency because it does not have a corresponding implicit algorithm. A common disadvantage of above self-tuning controllers is that the control loops are not decoupled.

The constribution of this paper is to show how these difficulties can be avoided by a multivariable decoupling pole-zero placement self-tuning controller. To this end, the paper begins by reviewing the offline design for multivariable decoupling pole-zero placement controller of a stochastically disturbed system. The self-tuning version of the algorithms consisting of explicit and implicit ones is then developed. Simulated examples illustrate advantages of the derived controller compared with that of Koivo.

OFFLINE CONTROLLER DESIGN

Before discussing the self-tuning controller, the offline design of the decoupling pole-zero placement controller to be employed is presented. It is assumed that the plant, which is both controllable and observable, may be modelled by the difference equation.

$$A(z^{-1})y(t)=z^{-k}B(z^{-1})u(t)+C(z^{-1})e(t) \quad (1)$$

where $u(t)$ and $y(t)$ are n-vectors defining the system input and output respectively, and $e(t)$ is a n-vector representing a zero-mean white-noise process with covariance $E(e(t)e^T(t))=r_e$. $A(z^{-1})$, $B(z^{-1})$ and $C(z^{-1})$ are polynomial matrices in the backward shift operator z^{-1}, and are of the form

$$A(z^{-1}) := I+A_1z^{-1}+...+A_{n_a}z^{-n_a}$$

$$B(z^{-1}) := B_0 + B_1 z^{-1} + \ldots + B_{nb} z^{-nb}$$

$$C(z^{-1}) := I + C_1 z^{-1} + \ldots + C_{nc} z^{-nc}.$$

The component of the samllest system pure time delay that is an integer multiple of the sampling time is modelled by the term z^{-k}. Further time delays (integer or non-integer of the sampling time) are absorbed in the $B(z^{-1})$ polynomial.

As shown in Fig. 1. Introduce a control law of the form

$$u(t) = F^{-1}(z^{-1})(H(z^{-1})w(t) - G(z^{-1})y(t)) \quad (2)$$

where

$$F(z^{-1}) := F_0 + F_1 z^{-1} + \ldots + F_{nf} z^{-nf}$$

$$H(z^{-1}) := H_0 + H_1 z^{-1} + \ldots + H_{nh} z^{-nh}$$

$$G(z^{-1}) := G_0 + G_1 z^{-1} + \ldots + G_{ng} z^{-ng}$$

and $w(t)$ is n-vector defining time varying reference signals. Substituting eqn. 2 into eqn. 1 the closed-loop system becomes

$$y(t) = (F(z^{-1})B^{-1}(z^{-1})A(z^{-1}) + z^{-k}G(z^{-1}))^{-1} \times$$

$$(H(z^{-1})w(t-k) + F(z^{-1})B^{-1}(z^{-1})C(z^{-1}) \times$$

$$e(t)). \quad (3)$$

Introduce $\tilde{A}(z^{-1})$ and $\tilde{B}(z^{-1})$ given by

$$A(z^{-1})\tilde{B}(z^{-1}) = B(a^{-1})\tilde{A}(z^{-1}) \quad (4)$$

where $\det \tilde{B}(z^{-1}) = \det B(z^{-1})$ and $\tilde{B}(0) = B(0)$. The polynomial matrices $B(z^{-1})$ and $A(z^{-1})$ always exist but they are not unique (Wolovich, 1974). The orders of the polynomial matrices $\tilde{A}(z^{-1})$ and $\tilde{B}(z^{-1})$ are equal to n_a and n_b. The coefficients of polynomial matrices $F(z^{-1})$ and $G(z^{-1})$ are chosen so that

$$F(z^{-1})\tilde{A}(z^{-1}) + z^{-k}G(z^{-1})\tilde{B}(z^{-1}) = T(z^{-1}) \quad (5)$$

$$H(z^{-1})\tilde{B}(z^{-1}) = T(z^{-1}) \quad (6)$$

where the orders of polynomial matrices $F(z^{-1})$, $G(z^{-1})$, and $T(z^{-1})$ are governed by

$$n_f = k-1+n_b, \quad n_g = n_a - 1, \quad n_t \leqslant n_a + n_b + k - 1 \quad (7)$$

and $F(z^{-1})$ and $T(z^{-1})$ are given by

$$F(z^{-1}) = I + F_1 z^{-1} + \ldots + F_{nf} z^{-nf}$$

$$T(z^{-1}) = I + T_1 z^{-1} + \ldots + T_{nt} z^{-nt}$$

where T_i is a diagonal matrix which is chosen by the designer. The solution to eqn. 5 requires the solution of the set of simultaneous linear equations

$$[F_1, \ldots, F_{nf}, G_0, \ldots, G_{ng}] \times$$

$$\left[\begin{array}{ccccccc} I & \tilde{A}_1 & \ldots & \tilde{A}_{na} & & & \\ & & \ddots & & & & \\ & & & I & \tilde{A}_1 & \ldots & \tilde{A}_{na} \\ \underbrace{0 \ldots 0}_{(k-1)n} & \tilde{B}_0 & \ldots & \tilde{B}_{nb} & & & \\ & & \ddots & & & & \\ & & & \tilde{B}_0 & \ldots & \tilde{B}_{nb} & \end{array}\right] \begin{array}{l} \left.\vphantom{\begin{array}{c}1\\1\\1\end{array}}\right\} (n_b+k-1)n \\ \left.\vphantom{\begin{array}{c}1\\1\\1\end{array}}\right\} n_a n \end{array}$$

$$\underbrace{\hphantom{xxxxxxxxxxxxxxxxxxxx}}_{(n_a+n_b+k-1)n}$$

$$= [T_1, \ldots, T_{nt}] - [\tilde{A}_1, \ldots, \tilde{A}_{na}, 0, \ldots, 0]. \quad (8)$$

For the solution to exist, the matrix on the left-hand side of the equation must be nonsingular. This condition is met if, for example the system is generic. Thus the closed-loop system can be written as

$$y(t) = w(t-k) + \tilde{B}(z^{-1})T^{-1}(z^{-1})F(z^{-1}) \times$$

$$B^{-1}(z^{-1})C(z^{-1})e(t). \quad (9)$$

It is obvious that this controller decouples the loops and eliminates the tracking errors. In order to reduce the amount of computation the value of matrix H is chosen so that

$$H = T(1)B^{-1}(1) \quad (10)$$

In this case the steady-state errors are eliminated.

When the system is minimum-phase and its zeros have good damping all process zeros can cancelled. The coefficients of polynomial $F(z^{-1})$ are chosen so that

$$F(z^{-1}) = R(z^{-1})B(z^{-1}) \quad (11)$$

where B_0 is nonsingular and $R(z^{-1})$ is a polynomial matrix.

In order to decouple the loops and eliminate the tracking errors the coefficients of polynomials R, G and H are selected as

$$R(z^{-1})A(z^{-1}) + z^{-1}G(z^{-1}) = T(z^{-1}) \quad (12)$$

$$H(z^{-1}) = T(z^{-1}) \quad (13)$$

where the orders of polynomials R, G and T are governed by

$$n_r = k-1, \quad n_g = n_a - 1, \quad n_t = n_a + k - 1$$

(where this condition is required for a solution of eqn. 12 to exist) and T is a prespecified polynomial matrix which has the same form metioned above. The solution to eqn. 12 requries the solution of the set of simulataneous linear equations

$$[R_1, \ldots, R_{nr}, G_0, \ldots, G_{ng}] \times$$

$$\underbrace{\begin{bmatrix} I\;A_1\cdots A_{n_a} & & \\ \ddots & & \\ & IA_1\cdots A_{n_a} & \\ 0 & \ddots & \\ & & I \end{bmatrix}}_{(n_a+k-1)n} \begin{matrix} \}\,(k-1)n \\ \\ \}\,n_a n \end{matrix}$$

$$=[T_1,\dots,T_{n_t}\,I-IA_1,\dots,A_{n_a},0,\dots,0]. \quad (14)$$

The solution to eqn. 14 always exists since the matrix on the left-hand side of the equation is nonsingular. Now the closed-loop system becomes

$$y(t)=w(t-k)+T^{-1}(z^{-1})F(z^{-1})B^{-1}(z^{-1})\times C(z^{-1})e(t).$$

If for simplification eliminating the steady-state error is only pursued the H can be chosen as

$$H=T(1). \quad (15)$$

SELF-TUNING CONTROLLER

Consider now the control of the system (1), when the system parameters are unknown or varying slowly. The basic idea when using the separation principle to design self-tuning controller can be expressed as follows: start with a design procedure for systems with known parameters; when the parameters are not known they are estimated recursively and the controller is redesigned in each step, using the estimated parameters instead of the true ones. This means that the certainty equivalence hypothesis is used to determine the controller.

For minimum-phase systems multiplying eqn. 1 by $R(z^{-1})$ from the left and using eqns. 11 and 12 results in

$$T(z^{-1})y(t)=G(z^{-1})y(t-k)+F(z^{-1})u(t-k)+N(z^{-1})e(t) \quad (16)$$

where

$$N(z^{-1}):=R(z^{-1})C(z^{-1}).$$

Defining the data vector X(t) by

$$X(t):=[y^T(t),y^T(t-1),\dots;u^T(t),u^T(t-1)\; e^T(t+k-1),\dots,]$$

and the parameter matrix θ by

$$\theta=[e_1,\dots,e_n]:=[G_0,G_1,\dots;F_0,F_1,\dots;N_1,N_2,\dots]^T$$

where the column vector e_i is

$$e_i=[g_{i1}^0,\dots,g_{in}^0,g_{i1}^1,\dots,g_{in}^1;f_{i1}^0,\dots,f_{in}^0,f_{i1}^1,\dots,f_{in}^1,\dots;n_{i1}^1,\dots,n_{in}^1,\dots]^T,\; i=1,\dots,n.$$

Eqn. 16 can be written componentwise as

$$(T(z^{-1})y(t))_i=X(t-k)e_i+e_i(t)$$

where the components of X(t-k) are uncorrelated with $e_i(t)$. The parameters θ are estimated by a recursive extended least-squares algorithm (RELS) in the manner, given by

$$\hat{\theta}_i(t)=\hat{\theta}_i(t-1)+K(t)((Ty)_i(t)-\hat{X}(t-k)\hat{\theta}_i(t-k)),i=1,\dots,n \quad (17)$$

$$K(t)=P(t-1)\hat{X}^T(t-k)(\beta+\hat{X}(t-k)P(t-1)\hat{X}^T(t-k))^{-1}, \quad (18)$$

$$P(t)=(P(t-1)-K(t)\hat{X}(t-k)P(t-1))/\beta, \quad (19)$$

$$e_i(t)=(Ty)_i(t)-\hat{X}(t-k)\hat{\theta}_i(t-1),\quad i=1,\dots,n \quad (20)$$

where $\hat{X}(t-k):=[y^T(t-k),y^T(t-k-1),\dots;u^T(t-k),u^T(t-k-1),\dots;\hat{e}^T(t-1),\dots]$

and is the exponential forgetting factor, the values of which should be between $0.9<\beta\leq 1$.

The following self-tuning algorithm is now obtained.

Algorithm 1: (implicit algorithm with all process zeros cancelled)

1. Read new output y(t), set point w(t).
2. Compute Ty(t).
3. Update the controller parameters by RELS algorithm (17) - (20).
4. Generate new control from eqn. 2.
5. Set t=t+1 and go to 1.

Since all process zeros are cancelled in the above algorithm, the controller will not be satisfactory for nonminimum-phase system or for system with zeros having poor damping. Such systems can, however, be handled using the explicit algorithm. The corresponding self-tuning control algorithm is given by

Algorithm 2: (explicit algorithm with no process zeros cancelled)

1. Read new output y(t), set point w(t)
2. Estimate the parameters matrices A, B, an C of the model (1) using RELS algorithm.
3. Computer the polynomial matrices $\tilde{A}$ and $\tilde{B}$ from eqn. 4.
4. Computer the controller parameters F, G, and H from eqns. 5 and 6 or 10.
5. Generate new control from eqn. 2.
6. Set t=t+1 and go to 1.

SIMULATION RESULTS

Example 1. A model for a laboratory-scaled concentration-flow process was used by

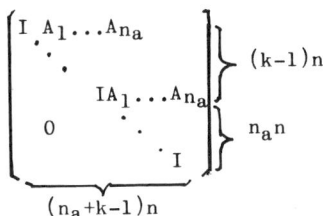

Koivo (1981) to illustrate the performance of his self-tuning controller. Here the same example is treated. The system model is

$$y(t)+A_1y(t-1)=B_0u(t-1)+Ce(t-1)+e(t)$$

where

$$A_1=\begin{bmatrix}-0.98 & 0\\ 0.01 & -0.98\end{bmatrix}, B_0=\begin{bmatrix}0.012 & 0.012\\ -0.023 & 0.023\end{bmatrix},$$

$$C=\begin{bmatrix}0.02 & 0.7\\ -0.9 & -0.2\end{bmatrix}.$$

Note that B_0 has quite small elements which can cause numerical difficulties.The noise covariance r=diag [0.01, 0.01] was used. The forgetting factor was β=1. In Fig. 2 system output, control signals, and the reference signals are shown, when the proposed self-tuning controller (implicit algorithm) and the poles of the closed-loop system are placed at $z_1=z_2=0.5$. One can readily see that the steady-state errors eliminaled. Besides, the proposed algorithm minimizes the interactions between control loops. In this example the output responce was quite oscillatory and the settling time extremly long when Koivo's controller with an integrator was used to reduce the steady-state error to zero, because the integrator moves the closed-loop-system poles to the unit circle.

Example 2. A model for a head-box of a paper machine was used (Koivo, 1980; Borisson, 1979) to illustrate the performance of their self-tuning controllers. Here the same example is treated. The system model is

$$y(t)+A_1y(t-1)=B_0u(t-1)+e(t)$$

where

$$A_1=\begin{bmatrix}-0.99191 & 8.80512X10^{-3}\\ -0.80610 & -0.77089\end{bmatrix},$$

$$B_0=\begin{bmatrix}0.89889 & -4.59329X10^{-3}\\ 19.390 & 0.88052\end{bmatrix}$$

$$E[e(t)e^T(t)]=diag[0.1, 0.1].$$

In Fig. 3 system output and reference signals are shown for the proposed self-tuning controller and the poles of the closed-loop system are placed at $z_1=z_2=0.5$. Obviously the interactions between control loops is greatly reduced using the suggested controller.

Example 3. An unstable and nonminimum-phase system is studied next:

$$y(t)=A_1y(t-1)+B_0u(t-1)+B_1u(t-2)+e(t)$$

where

$$A_1=\begin{bmatrix}0.9 & -0.5\\ -0.5 & 0.2\end{bmatrix}, \quad B_0=\begin{bmatrix}0.2 & 0\\ 0.25 & 0\end{bmatrix},$$

$$B_1=\begin{bmatrix}1 & 0\\ 0 & 1\end{bmatrix},$$

$$E[e(t)e^T(t)]=diag[0.1, 0.1].$$

The system has different time delays between each of the input-output pairs. Figure 4 shows system outputs, control signals, and computed parameter estimates when the derived explicit algorithm is used. The closed-loop poles are to be placed at $z_1=z_2=0.5$. The suggested controller works quite satisfactorily and the controller parameters converge to their required values very fast when either the implicit algorithm or the explicit algorithm is used.

CONCLUSIONS

The proposed self-tuning controller that achieves pole-zero assignment is able to control systems with different, unknown and/or slowly varying time delays. It can not only track time-varying reference signals without using an integrator but also decouble control loops without a specified decoupling unit. It is more robust than self tuners based on optimal control stratety. Using the implicit self-tuning algorithm it is possible to decrease the amount of the computation required. Simulation examples illustrate the effectiveness of the self-tuning controller presented.

REFERENCES

Astrom, K.J., and B. Wittenmark (1973). Self-tuning regulators. Automatica, 9, 185-199.

Borisson, U. (1979). Self-tuning regulators for a class of multivariable systems. Automatica, 15, 209-215.

Clarke, D.W. and P.J. Gawthrop (1975). Self-tuning controller. Proc. IEE, 122, 929-934.

Koivo, H.N. (1980). A multivariable self-tuning controller. Automatica, 16, 351-366.

Koivo, H.N. (1981). Experimental comparison of self-tuning controller methods in multivariable case. Proceedings IFAC World Congress, Kyoto, Japan.

Prager, D. and P.E. Wellstead (1980). Multivariable pole-assignment self-tuning regulators. Proc. IEE, 128, 9-18.

Wolovich, W.A. (1974). Linear Multivariable Systems. Springer-Varlag, New York, p. 159.

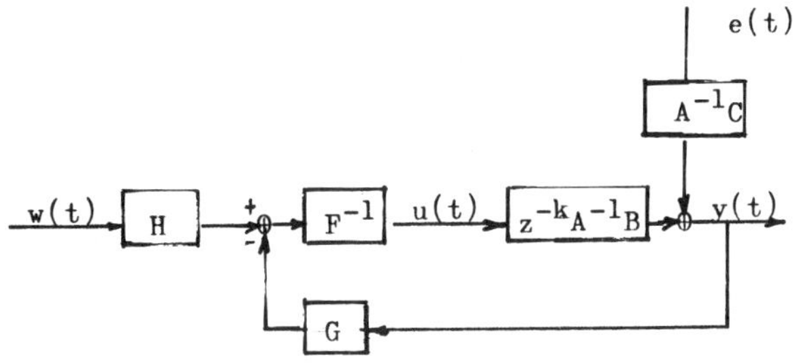

Fig. 1. Block diagram of controller.

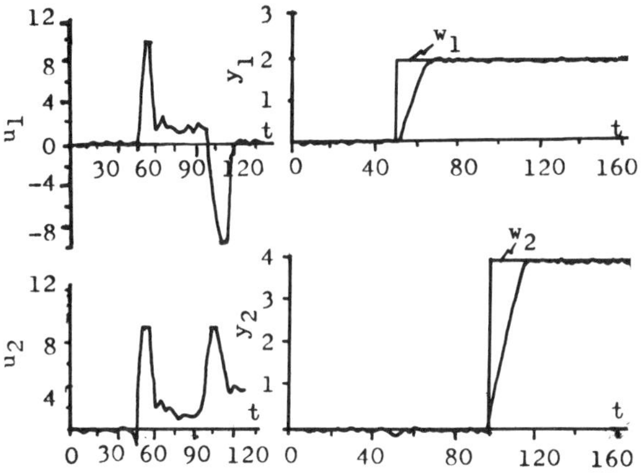

Fig. 2. Example 1 with the proposed control.(a) Output y_1,reference signal w_1, and control, u_1. (b) Ouptut y_2, reference signal w_2,and control u_2.

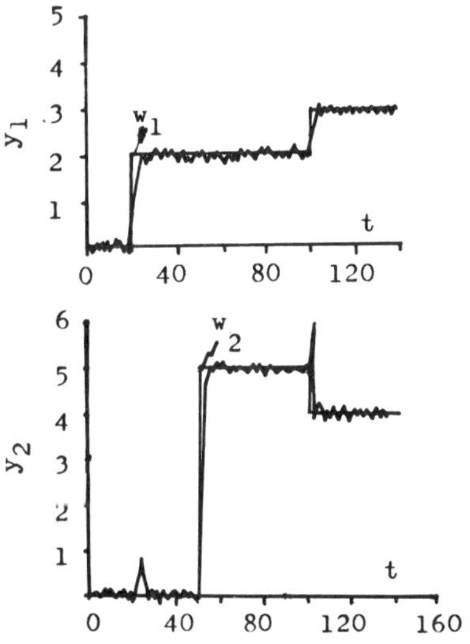

Fig. 3. Example 2 with the proposed control. (a) Output y and reference signal w_1 (b) Output y_2 and reference signal w_2.

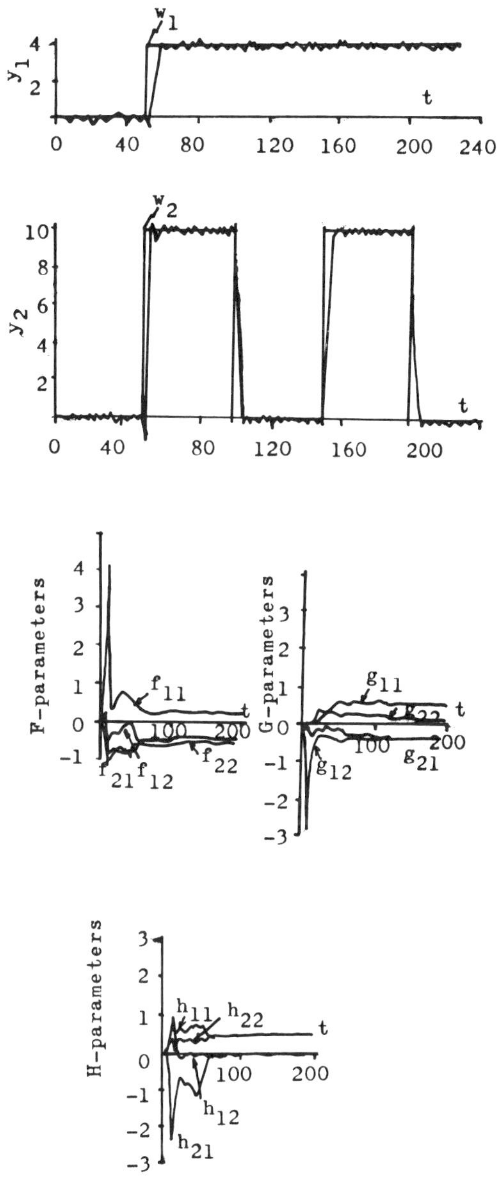

Fig. 4. Control of the unstable and nonminimum-phase system having different time delays of example 3 with the proposed control. (a) Output y_1, and reference signal w_1 (b) output y_2, and reference signal w_2 (c) F-, G-, and H- parameters.

ADMISSIBLE DIFFERENCE CONTROL

Gu Guangen

Wuhan Iron and Steel University, Wuhan, Hubei, China

Abstract. This pamphlet deals with the concepts and methods used in determining the major admissible difference control and the additive admissible difference control in the dynamic discrete time system. The working principle is based on the applications of nonlinear techniques. The property indices are taken to be key determinatives of the former type of contpol, which determine the make-up of the control vector u(k), and the economic indices, the key determinatives of the latter type of control, which determine the precision of the control vector. This pamphlet provides practical steps in designing the optimal level of the major control; institution of orthogonal layouts (both primaries and extensions), data analysis, variance analysis and increment calculation. The merits of the institution of major control are stressed.
A description of the methods in designing major admissible difference control with closed loop, and the methods in designing additive control is also presented.
The last part of the pamphlet is on the prospects of the application of the design of major control.

Keywords. Control theory; Optimal Control; Nonlinear systems; Major admissible difference control; Additive admissible difference control; Primary orthogonal layouts; Extention orthogonal layouts.

FUNDAMENTAL CONCEPTS

The task in determining the optimal control in dynamic discrete time system is to establish a control vector sequence $\{u(0),\ u(1),\ \ldots u(l-1)\}$ so as to obtain an objective function with the smallest value. Ordin arily, the sum of the quadric homogeneous function of the state vector sequence and of the control vector sequence is taken to be the objective function. In practice, the state equation of the system is established on the basis of controllability, observability, stability, and the result is that the precision of the system, and the optimal control of the system are often limited to the consideration of obtaining the objective function with the smallest value, while the admissibility of difference and the freedom from jamming, and the economic factors are being least considered.

The concepts and methods used in determining the major and additive admissible difference control are discussed in the following. These methods are basically applications of nonlinear techniques.

Definition 1. The product space $\mathop{\times}\limits_{i=0}^{l-1} u(i)$ produced by the stretch resulted from the control vector sequence $\{u(0),\ u(1),\ \ldots u(l-1)\}$--which assures the operation of the system to be within the admissible region--is called the functionally bounded space of system control (Or the admissible control region) and designated as Ω, and the Lebesgue measurement of Ω is called the functional boundary measurement, and designated as $|\Omega|$.

Definition 2. Let $\omega^* = \mathop{\times}\limits_{i=0}^{l-1} u^*(i) \in \Omega$, and obtain an objective function J with the smallest value, in which ω^* is to be considered as one of the key conditions of the optimal control, briefly, the ideal control.

The loss function L caused by the deviation of from ω^* is the function of $\Delta\omega = \omega - \omega^*$ and ω^*. Since ω^* depends upon the objective function J, the loss function L must be equal to $L(\Delta\omega, J)$, and when J is chosen (ω^* being determined simultaneously), the loss due to the $\omega\bar{\in}\Omega$ is Q, thereby the loss function $L(\Delta\omega)$ that came from the deviations of systematic control from ideal control ω^* can be determined.

Definition 3. The deviation $\Delta u(i)$ caused by the precision of the control vector u(i) itself is called the innate admissible difference. $\Delta\sigma = \prod_{i=0}^{l-1} u(i)$ is the innate admissible difference space of the control, the Lebesgue measurement of which is called the innate admissible difference measurement, designated as $|\Delta\sigma|$.

Definition 4. $$\min_{\substack{\omega\in\Omega \\ \Delta\omega\in\Delta\sigma}} L(\Delta\omega, J) \qquad (1)$$

which is determined by $(\omega^*, \Delta\omega)$, is called the region of the major admissible difference control in the dynamic discrete system, and ω^* is called the major admissible difference control of the system.

Apparently, major admissible difference control is a method which takes into account the objective function, loss function, and the innate admissible difference space, and may be regarded as the optimal combination of the control vector sequence $\{u(0), u(1), \ldots u(l-1)\}$ that is required for the purpose of promoting stability and reliability.

Definition 5. The ratio of the Lebesgue measurement $|\Delta\omega|$ of the region of major admissible difference control to the functional boundary measurement $|\Omega|$ is called the rate of admissible difference, designated as ε, thus $\varepsilon = |\Delta\omega| / |\Omega|$.

Let ε be the maximun major admissible difference control, that means that it has a greater stability range and possibility of arriving at a definition in the following.

Definition 6. The control in region $(\omega^*, \Delta\omega)$ that satisfies

$$\min_{\substack{\omega\in\Omega \\ \Delta\omega\in\Delta\sigma \\ \max\varepsilon}} L(\Delta\omega, J) \qquad (2)$$

is called maximum-stability-region major admissible difference control.

Definition 7. Let $P(\Delta(\Delta\sigma))$ be the designation for the investment used for the reduction of the innate admissible difference space, and $L(\Delta(\Delta\sigma))$ be the designation for the amount of loss represented by the corresponding loss function, and $(\omega^*, \Delta\sigma)$ be the region of major admissible difference. The region $(\omega^*, \Delta\sigma+\Delta(\Delta\sigma))$ derived from

$$\max_{\Delta(\Delta\sigma)} \{ L(\Delta(\Delta\sigma)) - P(\Delta(\Delta\sigma)) \geq 0 \} \qquad (3)$$

is called the region of the additive admissible difference control in the dynamic system.

Hence, it is quite clear that the economic indices are the key determinatives of the additive admissible difference control, which determine the precision $\Delta u^*(i)$ of the controller vector $u^*(i)$.

MAJOR ADMISSIBLE DIFFERENCE DESIGN FOR THE OPTIMAL CONTROL

Let

$$X(k+1) = f(x(k), u(k), k) \qquad (4)$$

where $k = 0, 1, \ldots l-1$,

$$X(0) = X_0 = \text{const}$$

be a controlable dynamic discrete time system, where X(k) be a n-dimensional vector, representing the state of kth step, u(k) be a r-dimensional vector, representing the control added to the system between the kth step and the (k+1)th step.

The task in choosing an optimal control is to determine the control vector sequence $\{u(0), u(1), \ldots u(l-1)\}$, assume $\prod_{K=0}^{l-1} u(k) \in \Omega$, where Ω is the admissible control region (functionally bounded space), so as to arrive at a smallest value in the objective function,

$$J = \phi[x(l)] + \sum_{K=0}^{l-1} \psi[x(k), u(k), k] \qquad (5)$$

We assume that the terminal state be zero. The first term on the right in (5), $\phi[x(l)]$ is the terminal index function, which may be the loss function, or risk function, ect., when $x(l)$ deviates from zero state. The second term on the right is the dynamic index function, when a steady control is the only factor considered it comes to zero.

Under the constraint of formula (4), the methods, which may be used to arrive at a u(k) whose J will be the smallest, include the Lagrange multiplication method, calculus variation, nonlinear programming, dynamic programming, ect. The calculation is very difficult when they are applied to nonlinear systems. One may take the approximate calculation, but then greater errors will appear in the results arrived at, and it's difficult to estimate the effects produced on the objective function by the control vector $u(k)$, (where $K = 0, 1, \ldots l-1$), if a slight change $\Delta u(k)$ exists in the control vector.

In the following we are going to propose a method to chose the optimal admissible difference control on the basis of orthogonal design.

Assume that the ideal control obtained by satisfying (4) with (5) at the smallest value is $u^*(k)$, where $k=0, 1, \ldots l-1$, that is $f(x(l-1), u^*(l-1), l-1)=0$. actually, $u^*(k)$ may stand for any approximate ideal control.

1. Primary Orthogonal Layouts and Extention Orthogonal Layouts

The purpose in designing a major admissible difference control is to establish a highly stable control vector sequence, which may be obtained by an analysis of the changes in the composite deviations due to the interferences from the inside and outside of the system, and by searching for an objective function at the smallest value which is least affected by these changes, when the level of control factors are of a large amptitude of change. The method to realize this to construct orthogonallayouts of the leveles of rl factors (where rl stands for the number of factors) by the designations $u^*(k)$, where $k=0, 1, \ldots l-1$. Let $u^*(k)$ be the value of level 2, $\hat{u}^*(k)$ be the lower boundary of the corresponding vector in Ω, the value of level 1; $\hat{\hat{u}}^*(k)$ be the upper boundary of the corresponding vector in Ω, the value of level 3. With rl factors (three levels for each factor) we can construct the primary orthogonal layout L_N (where N can be determinated by rl factors), The layout thus made is called the first table of the first round. We can construct five tables for the first round on the basis of the following formulas,

$$\sqrt{\frac{6-t}{5}}(u_s^*(k)-\hat{u}_s^*(k)),\ \sqrt{\frac{6-t}{5}}(\hat{\hat{u}}_s^*(k)-u_s^*(k)),$$

where $s= 1, 2, \ldots, r$; $t= 1, 2, 3, 4, 5$. Similar extention orthogonal layouts can be constructed. This can be done by taking into account the deviations for each factor (due to the degree of precision of the controllers), from the results obtained in the ith test of the tth table of primary orthogonal layout. This new layout is called the extention orthogonal layout of the ith test of the tth table.

2. Data Analysis

With the extention layout table, the data in the j row of which are taken from the i row of the tth primary orthogonal layout table, and with the use of formulas of (4) and (5) we can obtain the objective function J_{ij}, where $j=1, 2, \ldots, N$.

The sum of squares due to the noises of the factors

$$S_e=\sum_{j=1}^{N}(J_{ij}-J_{i\cdot})^2, \quad \text{where } J_{i\cdot}=\frac{1}{N}\sum_{j=1}^{N}J_{ij}.$$

The estimate of the mean square of the signals of the factors is designated as

$$V_m=\frac{1}{N}(S_m-V_e), \quad \text{where } S_m=\frac{1}{N}\Big(\sum_{j=1}^{N}J_{ij}\Big)^2, \quad V_e=S_e/(N-1).$$

The signal-noise ratio (SN Ratio) is designated as η, and $\eta=V_m/V_e$, then the calculation of $10\log\eta$ is represented by SN value (the unit is called decibel), hence the SN value can be obtained for the ith test on the tth primary orthogonal layout table. Byrepeating this process, we can obtain all the SN values of the N tests of the tth primary orthogonal layout in the first round.

3. Variance Analysis and Increment Calculation

The variance of each of the factors can be obtained by calculating the SN values of the tth primary orthogonal layout tables of the first round, and then the significance of each factor can be tested. The optimal control level of each controlling factor can be obtained by applying the following formula,

$$SN^*=\max_{t,i}(SN)_{t,i} \quad \text{where } i=1, 2, \ldots, N;\ t=1, 2, 3, 4, 5;$$

When these factors to be tested are significant with regard to the SN values. This is to be regarded as the result of the first round. The average signal-noise ratio of i^* level (designated as $(\overline{SN})_{i^*}$) is obtained by the calculation on the significant factors with the chosen t, and $(\overline{SN})_0$ is the corresponding designation of the average signal-noise retio of the 2nd level in the layouts. With these data we can obtain the increment of the design of the major admissible difference control of the first round layouts, represented by the formula,

$$P = \sum_{j=1}^{r\ell} \left[(\overline{SN})_i^* - (\overline{SN})_o\right] \pm R,$$

where R is the estimated radius,

and

$$R = \sqrt{F_\alpha(1, n_e) \cdot \frac{S_e}{n_e \cdot d_e}} \quad ;$$

n_e being the degree of freedom on the errors of the factors.
$F_\alpha(1, n_e)$ being F statistical value at α level.
d_e being the number of the effective repetition.

This increment comes from a good choice of controlling factors in combination.

4. The Calculation of the Second and the Third Round Layouts

Take the value of the factors with the best control level of the first round as the key condition (as the value of the second level) for the second round, and repeat the same processes of analysis and calculation, we can obtain the results of the second and third round. The end results obtained in the third round are the optimal controlling conditions contributed by the design of major admissible difference control.

The merits of this design are as follows:

A. The stabilitv, and the freedom from jamming of the control system are pretty good, because the reliability of each of the controller u_i has been considered, the stability being indicated by the signal-noise ratio. The method used is basically an application of non-linear techniques.

B. The optimal control process $u^*(k)$ may find various solutions in the admissible control region, when the methods of choosing the objective functions are varied. These solutions may be easily obtained by the design of the major admissible difference control.

C. The amount of calculation when using the design of orthogonal layouts with $u_i(k)$ at different times is much less than that of the corresponding dynamic programming method, and the method is simpler and with a possibility of obtaining greater amount of information, as compared to the other methods.

Note 1: The optimal control (or the ideal control) obtained by other methods, such as dynamic programming, are not necessarily the optimal control in the practical system, because these methods are characterized by the appoximation in the state equation of the system itself and by the errors appearing in the process of solution, nor can they be the major admissible difference control. Owing to the complexities of the problem in obtaining the ideal control, the result is that the ideal control can not be obtained in many situations. The major admissible difference control is a design that makes it possible to approach the major admissible difference control step by step with the application of non-linear techniques.

MAJOR ADMISSIBLE DIFFERENCE DESIGN FOR THE OPTIMAL CONTROL WITH CLOSED LOOP

In a controllable discrete time system, the optimal control with closed loop is to determine a control vector sequence $\{u(k)\}$, where $k=0, 1, \ldots l-1$. It is the function of the state vector $\{x(k)\}$.

Assume $\bigtimes_{k=0}^{\ell-1} u(k) \in \Omega$, where Ω is the region of admissible control. The purpose of the determination of $u(k)$ is to make the objective function

$$J = \phi(x(l)) + \sum_{k=0}^{\ell-1} \psi(x(k), u(k), k) \text{ the smallest.}$$

The solution $x^*(k)$ as the optimal state for (4) may be obtained by the application of Hamilton-Belman Equation (on the basis of Pontryagin's principle of the smallest value), and its corresponding optimal control is $u^*(k)$. Since the system is controllable, we will formalate the following definition.

Definition 8. The product space produced by the stretch resulted from the state vector sequence $\{x(0), x(1), \ldots, x(l-1)\}$--the operation of system (4) has been proved to be within the admissible region --is called the functionally bounded space of system state (or the admissible region of state), designated as $\mathcal{X} = \bigtimes_{i=1}^{\ell-1} x(i)$. From $\mathcal{X}$ we can obtain the functional boundary region Ω of the system control.

Definition 9. The estimate of optimal state $x^*(k)$ ($k=0, 1, \ldots, l-1$) of (4) is called the estimation of the ideal state of (4). That is to say when the estimated ideal state is $x^*(k)$, the control to be used is the ideal control $u^*(k)$.

The region of the estimated state $x^*(k)$ determined by the innate admissible difference space of the ideal control $u^*(k)$ is called the innate admissible difference space of the ideal state.

Definition 10. $(x^*, \Delta x)$ obtained from the $(\omega^*, \Delta\omega)$ determined by the corresponding $\min\limits_{\substack{\omega\in\Omega \\ \Delta\omega\in\Delta\sigma}} L(\Delta\omega, J)$ is called the estimated region of the major admissible difference of the state, and x^* is called the major admissible difference estimation of the state.

Based on the definitions formulated above, the major admissible difference estimation x^* of the state can be obtained by repeation the processes in obtaining the optimal (the objective function is formula (5)) by the orthogonal design of the state as what is being done in Part 2 (The Design for an Optimal Admissible Difference Control) with the ideal state estimated and innate admissible difference of the state in $\mathfrak{X}$. The corresponding control u^* thus arrived is the function of the state, and can be used in the closed loop control.

Note 2: The estimation of the ideal state, similar to that of the ideal control, can be obtained by solving the Hamilton-Belman Equation, or by other methods, such as application of state recurrence estimation in a linear system, application of Kalman's filtering estimation in a random linear dynamic system, etc. But similar to what is being siad in Note 1, the ideal estimate is , not necessarily the optimal one, and rather hard to obtain in most casas. The major admissible difference control is a design that provides a method to approach step by step the estimation of major admissible difference of the state.

ADDITIVE ADMISSIBLE DIFFERENCE CONTROL

As that has been stated above, when ω^* is given a certain value, the loss function L is the function of $\Delta\omega(=\omega-\omega^*)$, and when $\omega\bar{\in}\Omega$, the constant in $L(\Delta\omega)$ is determinable. Let

$$L(\Delta\omega)=\begin{cases}\frac{1}{rl}(\omega-\omega^*)^{\tau}K(\omega-\omega^*), \quad K=kI=k\begin{pmatrix}1 & & 0\\ & \ddots & \\ 0 & & 1\end{pmatrix}_{rl\times rl} & \text{when } \omega\in\Omega;\\ \max\limits_{i} K_i(\omega_i-\omega_i^*), & \text{when a certain } \omega_i \text{ is not in } \Omega\end{cases}$$

$i=1, 2, \ldots rl.$

When $\omega_i=\tilde{\omega}_i$, where $i=1, 2, \ldots, rl$; ($\tilde{\omega}_i$ representing the corresponding component of u^* lies on the boundary), the loss produced from the control system is Q, and matrix K and the value of K_i as the only solutions can be determined by the above formula.

The optimal innate admissible difference space $\Delta\sigma_i^*$ of the controller u_i can be obtained by the application of formula (3), with definite loss function and the promotion of the precision of the controller u_i (corresponding to the change in the innate admissible difference space $\Delta\sigma_i$).

OTHER USES OF THE DESIGN OF MAJOR ADMISSIBLE DIFFERENCE

The design of major admissible difference can be used for various other purposes apart from determining the optimal control and the optimal state as described above. These uses are:

1. This design may be used to replace the second method propose by Lyapounov in obtaining the information of the stability of the system. In essence, the design of major admissible difference is the orthogonal decomposition of the stability, from which the signal-noise ratio between the value of the state and that of interference may be determined, and the upper boundary of oscillation of state, too. This is a better method than that of Lyapounov's in determining quantitive index.

2. The major admissible difference design is a statistical method that can be used in establishing the optimal state regulator.

3. In case a dynamic system suffers from several interferences, set each of the interfering factors in their corresponding column on the orthogonal layout table and obtain their sums of square. The relative importance of the influence exerted on the objective function by the interfering factors can be readily discriminated by comparing their sums of square.

4. Dynamic programming is the chief methods in optimalization of the dynamic system. The design of major admissible difference is the orthogonal design of the dynamic programming, so as to provide the best condition that is with the highest stability and freedom from jamming.

5. The design of major admissible difference provides the main directions for the adjustment of the control system and the correction of the dynamic quality (i.e. the main state components or control components.)

6. In the establishment of an outer equation (i.e. the equation of input-output), the design of major admissible difference contributes a profitable character to the orthogonal design of the regression.

REFERENCES

Genichi Taguchi, and yu-in Wu (1979). Introduction to off-line Quality Control, Central Quality Control Assoiation.
Katsuhiko Ogata (1970). Modern Control Engineering, Prentice-Hall.
Tu Xuyen (1982). The problem on structural synthesis of the most economical control systems. Acta Automatica Sinica, 8(2), 103.

A MICROCOMPUTER-BASED WAVE GENERATION SYSTEM

Zhang Quan, Zhao Jing, Pei Run and Lan Pusen

Department of Control Engineering, Harbin Institute of Technology, Harbin, China

Abstract. A microcomputer-controlled wave generation system has been designed at the Harbin Institute of Technology and is being tested at the Chinese Ship Science Research Center (CSSRC) to meet the ever-increasing requirements of coastal engineering and ship building engineering.

This system, which is used to generate single frequency periodic, irregular and "random" waves in a 46mx70m towing tank, consists of an electro-hydraulic servosystem and a microcomputer-based controller.

Two methods--the recursive approach in the frequency domain and the shaping filter approach--are applied to simulate ocean waves. In the frequency domain, the desired wave is specified by a spectral distribution and the frequency response of the water tank is identified recursively. In the shaping filter approach, a PRBS generator is used to generate a "random" signal which is then filtered by a software-implemented digital shaping filter with a recursively modified mathematical model. Although the system is specially designed for ocean wave generation, the control scheme can also be directly applied to generating "random" signals in many fields of test engineering.

Keywords. Microcomputer, ocean wave simulation, random signals, identification, towing tank, coastal and test engineering.

INTRODUCTION

With the rapid development of ship building and coastal engineering, the traditional wave test approach in towing tanks for model tests is not satisfactory anymore. A new approach of generating "random" waves with specified spectral content has been recently proposed. A microcomputer controlled wave generation system has been designed and is being tested to solve the problem mentioned above.

The system, as shown in Fig. 1, consists of an electro-hydraulic servosystem, eight air-operated wavemakers, waveheight meters (WHM) and a microcomputer system.

The microcomputer system consists of a Z-80-based microcomputer PS-85 with 256 RAM, dual disk drive, 12-bit A/D and D/A converters, a printer and six-pen plotter.

The electro-hydraulic servosystem drives eight wavemakers to allow the pumped air to get into or get out from the air-room which generates the simulated ocean waves in the water tank.

The computer samples wave-height data from the wave-height meter by an A/D converter and stores it in the memory or the disk. The driving signal g(t) generated by the computer is first filtered by a Butterworth filter to prevent high frequency noise and then is fed to the servosystem to drive the wavemakers.

Two approaches of generating simulated waves are proposed:

1. Recursive approach in frequency domain
2. Shaping filter approach

Each of these approaches will be discussed in detail as follows.

RECURSIVE APPROACH IN FREQUENCY DOMAIN

The recursive approach in frequency domain is shown in Fig. 2.

Because of rapid response performance of the electro-hydraulic servosystem and the low-pass characteristic of the Butterworth filter, the driving signal g(t) is close to X(t); thus g(t) = X(t).

If nonlinearity of the wavemakers is ignored, then

$$S_y(\omega) = |G(j\omega)|^2 \, S_x(\omega) \tag{1}$$

where $S_x(\omega)$ -- power spectrum of X(t)
$S_y(\omega)$ -- power spectrum of Y(t)
$G(j\omega)$ -- transfer function of wavemakers

and

$$G(j\omega) = Y(j\omega)/X(j\omega) \tag{2}$$

Actually, $G(j\omega)$ is unknown before tests. So a selected $G_o(j\omega)$ should be given as an origin. When the desired wave spectrum in the tank is given by $S_d(\omega)$, then Eq. (1) becomes

$$S_x(\omega) = S_d(\omega)/|G_o(j\omega)|^2 \tag{3}$$

According to Danis-Pierson's formula

$$X(t) = \int_0^{\infty} \cos(\omega t + \varepsilon(\omega)) \cdot \sqrt{S_x(\omega)}\, d\omega \tag{4}$$

where $\varepsilon(\omega)$ is a random number between 0 and 2π.

Based on Eq. (4), the driving signal X(t) can be obtained either by FFT or by numerical computation.

Since the transfer function $G_o(j\omega)$ is originally selected, the spectrum of Y(t) cannot be close to the desired. Therefore, modification of the transfer function must be made. A new transfer function is calculated by using the formula:

$$|G_N(j\omega)| = \sqrt{S_y(\omega)/S_x(\omega)} \quad (5)$$

Eq. (5) gives a modified transfer function which can be used to correct the spectrum of the driving signal. It may continue until the desired spectrum is duplicated with an acceptable accuracy in the operator's judgement. After that, the time history of the driving signal g(t) can be stored in a disk data file, which may be recalled by the operator.

The program flow chart is shown in Fig. 3 and the simulation result is given in Fig. 4.

The result indicates that after several times of modification the actual wave spectrum in the water tank can be close to the desired.

Using the recursive approach in frequency domain this system can generate periodic irregular waves. When the total number of harmonic components is very large, the simulated wave can be considered as "random". But this would take long time to compute the time history and require large memory space in the computer to store the time history file. Since the memory capacity of the microcomputer is limited, the experiment duration of model test can not be very long. In order to improve that mentioned above, the shaping filter approach is proposed.

SHAPING FILTER APPROACH

The shaping filter approach is shown in Fig. 5. This wave generating system suitable for producing "random" ocean waves consists of a pseudo-random noise generator (PRBS generator), a shaping filter (both are software implemented), and other equipments which are the same as shown in Fig. 2.

Pseudo-random Noise Generator

There are many methods of producing noise. But the analog noise generators must be subject to amplitude drift. Eventhough hardware implemented digital noise generator is not subject to drift, it must require extra hardware cost. Thus software implemented pseudo-random noise generator is reasonable to be used, because it is not subject to drift, it can produce repeated noise patterns, which is not possible with analog noise generator and is necessary to ship model tests, it can be implemented without any extra hardware cost, and if the duration of the sequence is long enough compared with experiments using this generator, the sequence can be considered as a "true random" signal.

The noise generator is based on a software implememted shift register with feedback from correctly selected stages via an exclusive OR gate. Thus a binary maximum length sequence (PRBS) will be produced.

If the signal (PRBS) is filtered by a properly designed software implemented digital shaping filter, then a more accurately required spectrum of the driving signal can be produced, which drives the electro-hydraulic servosystem to produce a specified wave spectrum in the tank.

Software Implemented Shaping Filter

When the frequency response of the wavemakers is given by $G(j\omega)$, then the power spectrum of the driving signal can be calculated by Eq. (2). In order to compute the amplitude spectrum, the cutoff frequency of S is divided by N. Thus the resolution is $\Delta f = f_{max}/N$ and the amplitude spectrum of S_x is

$$S_{xm}(\omega k) = \sqrt{2 \cdot \Delta f \cdot S_x(\omega k)}\ , \quad k=0,1..,N \quad (6)$$

The phase spectrum may be selected in different ways, for example it may be 0 or $\pm\pi/2$.

The frequency response of the shaping filter is denoted by $H(j\omega)$

$$H(j\omega) = \alpha(\omega)+j\,\beta(\omega) = H(j\omega)\, e^{j\theta(\omega)} \quad (7)$$

if $H(j\omega)=H(-j\omega)=S(j\omega)$ and $\theta = \pi/2$, then $\alpha(\omega)=0$. Thus the impulse response of the shaping filter is

$$h(n) = \sum_{n=0}^{M-1} H(\omega k)e^{j\omega kn}, \quad n=0,1,\ldots, M-1 \quad (8)$$

where $\omega_k = 2\pi k/M \qquad M=2N+1$

The output of the PRBS generator is C_i (i=0,1,..), and then the output of the shaping filter is

$$x_n = \sum_{k=0}^{M-1} h_k \cdot C_{n-k} \qquad (n=0,1,..) \quad (9)$$

Because C_i can only be either "0" or "1", only logical additions are required for Eq. (9). This can save a lot of time in real time control and this may be another reason why to select the PRBS generator.

Spectrum Modification and Program Design

Since the frequency response of the wavemakers is selected before tests, it can not be the same as theactual response. So there must be some difference between the expected and the actual wave spectrum in the tank. Thus the spectrum of the driving signal should be modified. The modified spectrum is S_{xk}

$$S'_{xk} = S_{xk} + \overline{W}_k(S_{dk}-S^*_{ok}) \quad (10)$$

where S_{xk}--the spectrum of actual driving signal

S_{dk}--the desired spectrum

S^*_{ok}--the measurement of the spectrum

$\overline{W}_k$ --weighting factor

After several times of modification the actual wave spectrum in the tank can be close to the desired. The program flow chart is shown in Fig. 6 and the simulation results are given in Fig. 7 and Fig. 8.

These plots indicate that the approach discussed above is viable for duplicating wave spectrum in the tank equiped with a microcomputer and an electro-hydraulic servosystem.

CONCLUSION

To automatically control the wavemakers to generate various kinds of wave patterns, a microcomputer based control system has been proposed and the system is being tested.

Two approaches, the recursive approach in frequency domain and the shaping filter approach, are used to duplicate defined wave spectrum with acceptable accuracy. The shaping filter approach seems to be better than the recursive approach in frequency domain, because it doesn't require large memory space in the computer and it can generate repeated wave patterns, which are nearly "random".

To realize the control system both assembly and FORTRAN languages are used. Thus consideration can be given to both powerful arithmetic capability and real time functions, such as data acquisition, sampling control, analog signal output and interrupt service, etc.

It is important to point out that the non-linearity of wave makers would make it difficult to control spectral shapes in the tank very accurately. But this problem can be solved by using a non-linear compensator. This is one of problems on which we will focus attention in our further work.

ACKNOWLEDGEMENTS

The authors would like to express their sincere glatitude to Mr. Zhao Zundao, Mr. Zhao Yuzhuo, Mr. Li Hongren, Mr. Zhou Lianshan and Mr. Qiu Huazhou who designed the electro-hydraulic servosystem for this project.

REFERENCES

Anderson, C.H. and B. Jonson (1977). Proceedings of 18th American towing tank conf.
Funke, R. (1974). Proceedings of 14th conf. on coastal eng., 340-351
Kimura, A. and Y. Iwagaki (1976). Proceedings of 15th conf. on coastal eng., 368-387

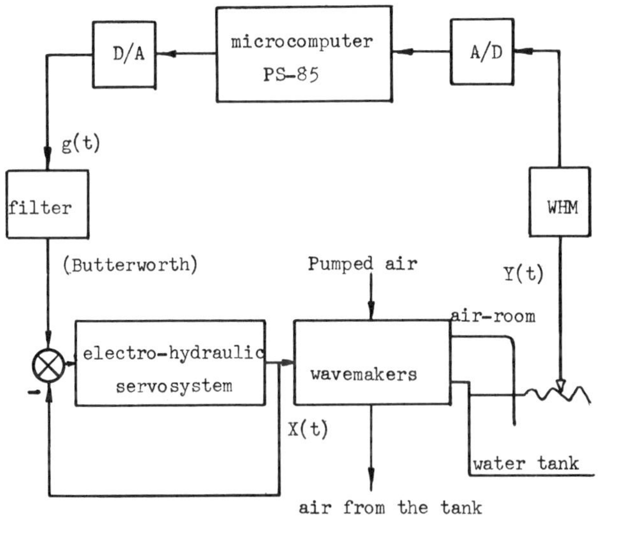

Fig. 1 Ocean wave generation system

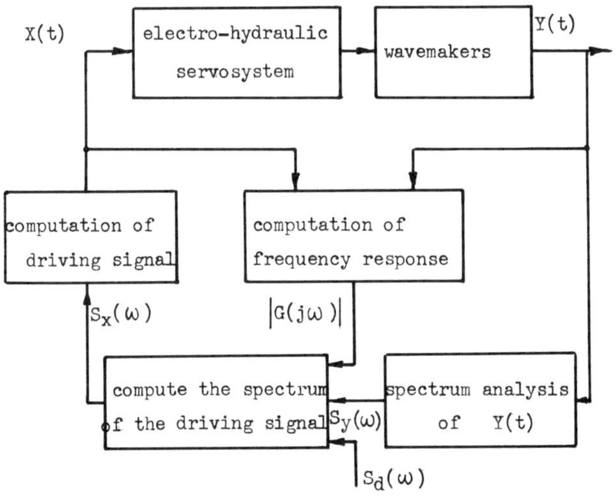

Fig. 2 Recursive approach in frequency domain

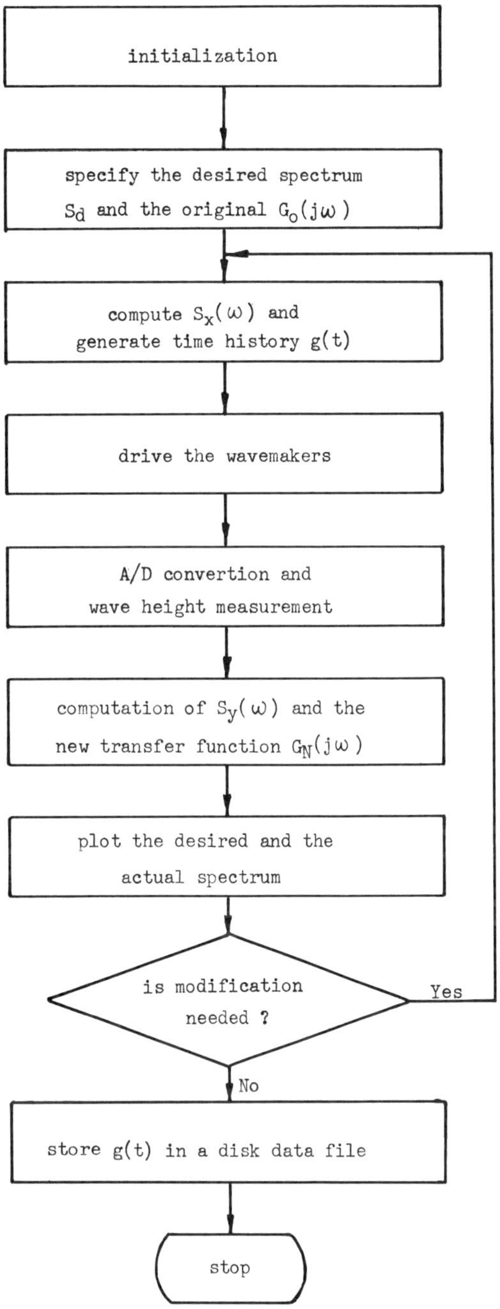

Fig. 3 Program flow chart

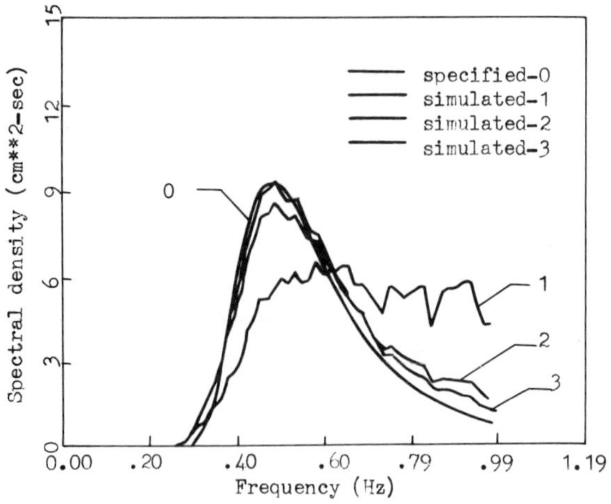

Fig. 4 Simulation result (spectrum)
(Reduced scale in the tank: 1/30)

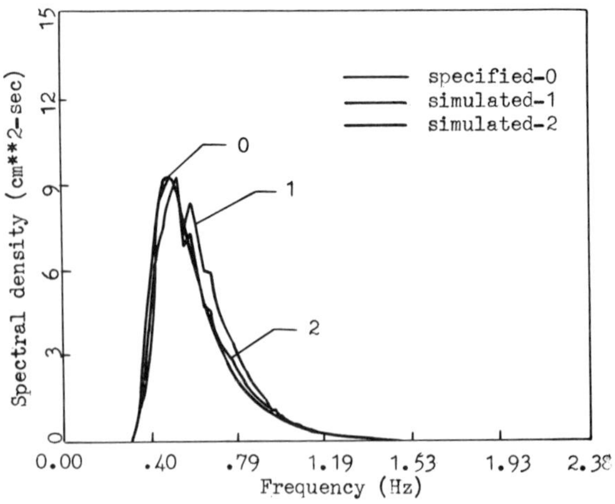

Fig. 7 Simulation result (spectrum)
(Reduced scale in the tank: 1/30)

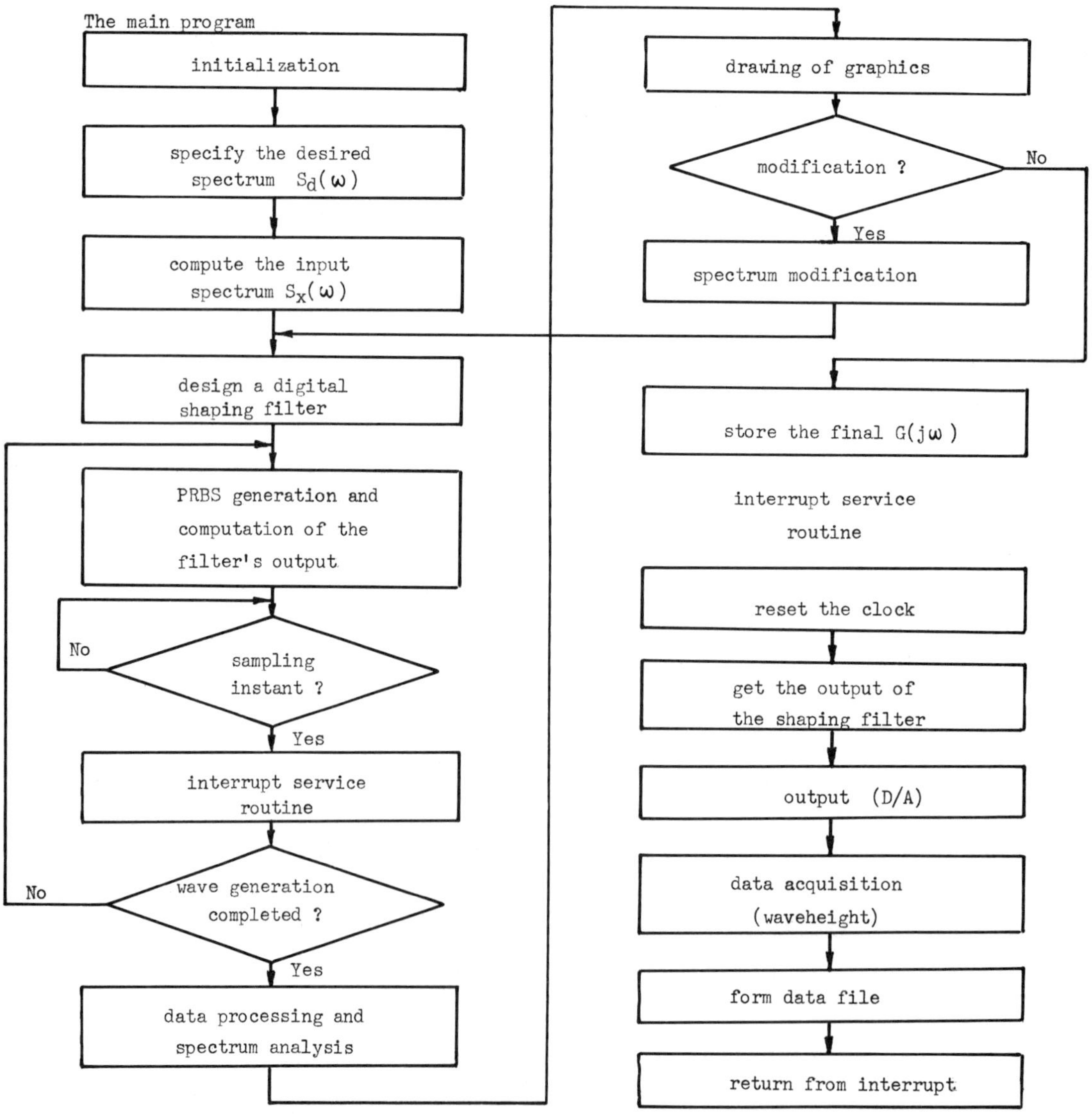

Fig. 6 Program flow chart

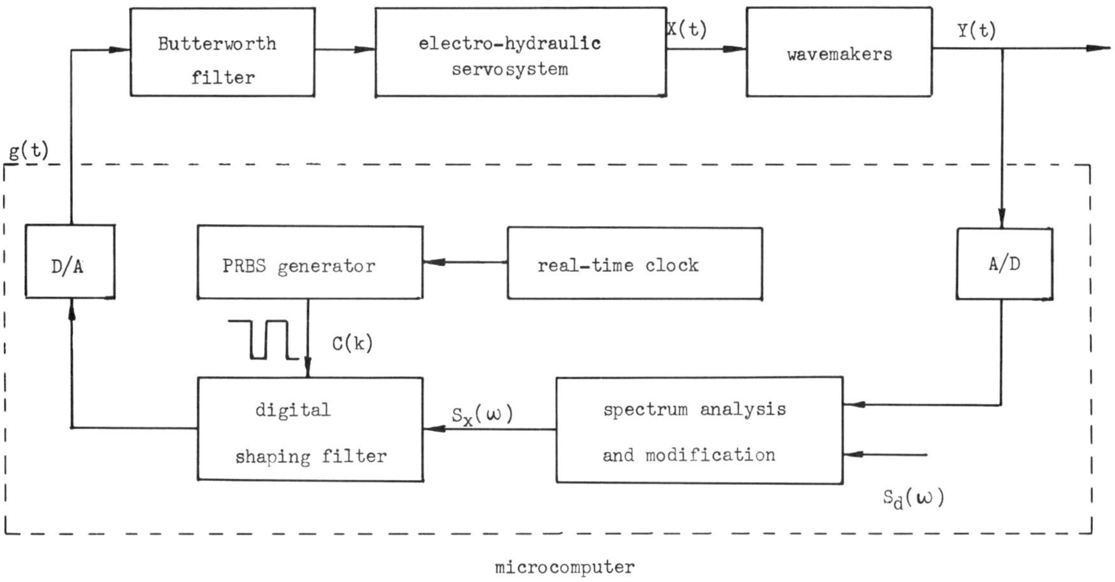

Fig. 5 Shaping filter approach

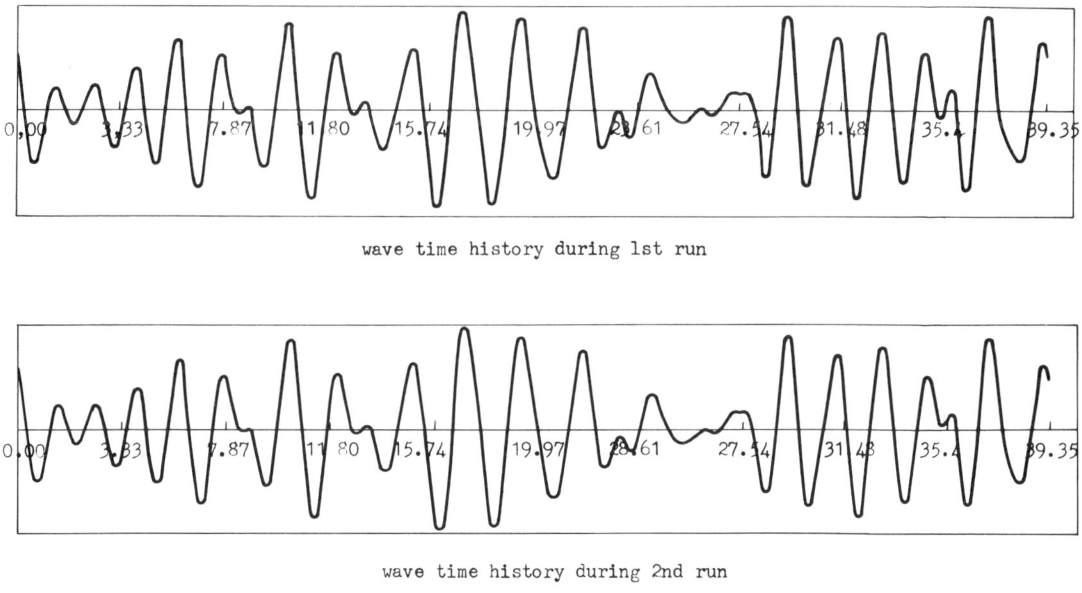

Fig. 8 Simulation result (wave pattern)

DYNAMICS AND CONTROL OF PAPER MACHINE

You-Xian Sun, Qing-Guo Wang, Yi-Qun Ying and Chun-Hui Zhou

Laboratory for Industrial Process Modelling and Control, Zhejiang University, Hangzhou, China

Abstract. A dynamic model of a paper machine is built by the mechanismic analysis and some mathematical treatments, and a new frequency response approach to multivariable control system design is developpped by optimization technique and applied to the design of a basis weight and moisture content control system. This system offers satisfactory technical performances and significant economical profits.

Keywords. Paper industry; modelling; multivariable control system; PID control; frequency response; computer-aided design.

INTRODUCTION

Many authors (Astrom, 1977; Field, 1978; Gentil, 1975) have studied modelling and control for paper machines and obtained some useful results with the methods of identification and time-domain design. Different from that, we will place our emphysis on the physical analysis of papermaking parocesses and frequency domain designs. In next section the model of a paper machine is built. In the following sections the concept of objective transfer function matrix is proposed, and a new frequency response approach to multivariable system designs is developped. Some results and conclusions from the practice of control for a paper machine are presented.

MODELLING

In order to develop basis weight and moisture content control for a paper machine, whose dynamic model must be built first. So far, several kinds of paper machine models have been obtained by identification, but no physical model have been seen. It can not only used in control, but also used in analysis, design and management of paper machine. The crutial problems in building physical model of a paper machine are how to calculate mass transfer rates for the whole papermaking process from pulp to paper and how to simplify the original model of nonlinear and distributed parameter properties. The reasonings, methods and results of solving these problems are briefly introduced in this section.

Process Description Fig.1 shows the flow diagram of a certain paper machine. The thick pulp is mixed with white water in a mixing tank, and mixture flows through a pipe into headbox. The consistency and liquid level of the headbox are very important because they almost determine surface properties and basis weight of paper. Pulp from the headbox is gradually dewatered and taken a shape of web while passing through table rolls, suction boxes and couches. When web is in drying section, it is heated up and water in it is evapourated by a series of drums, into which steam is supp-

supplied from a common steam source with pressure P_s. After drying the desired basis weight and moisture content will be reached.

Dynamic Equations Take an elementary length Δl of web and study longitudinal change pattern of its basis weight z (total weight of web per unit area) and moisture content x (ratio of water weight to total weight) starting from pulp onto wire to the formation of paper at reel. By mass balance, we get

$$dz/dt=-g \quad (1)$$

$$d(xz)/dt=-(1-c)g \quad (2)$$

where g is the amount of dewatered white water from unit area of web per unit time, and c is the consistency of the white water (ratio of bone-dry weight to total weight).

Because z and x are function of position l (away from the headbox) and time t, that is z=z(t,l) and x=x(t,l), and dl/dt=V (machine velocity), then the dynamic equations of web take the partial differential form as

$$\partial z(t,l)/\partial t+V\partial z(t,l)/\partial l=-g(t,l) \quad (3)$$

$$\partial(x(t,l)z(t,l))/\partial t+V\partial(x(t,l)Z(t,l))/\partial l =-g(t,l)(l-c(t,l)) \quad (4)$$

whose initial conditions are determined by the pulp consistency and level in the headbox.

Discretization Because of the difficulty in calculation and application, equations (3) and (4) must be simplified. They are discretizised about both time and position. Using differences to approximate differentials, we have

$$\partial z(t,l)/\partial l= z(t,l)-z(t,l-L_s))/L_s =(z_j(k)-z_{j-1}(k))/L_s$$

$$\partial z(t,l)/\partial t=(z(t+T_s,l)-z(t,l))/T_s =(z_j(k+1)-z_j(k))/T_s$$

where k=1,2,....,n,n is the number of the discrete interval of position. Then equation (3) becomes

$$(z_j(k+1)-z_j(k)/T_s+V(z_j(k)-Z_{j-1}(k)/L_s = -g_j(k) \quad (5)$$

in order to reduce the error introduced by using differences in place of differentials and ensure a certain accurancy of equation (5), the discrete interval T_s and L_s should not be taken too large. But the smaller L_s is, the higher the order of the system would be. To overcome this conflict and to get a model of low order but with high accuracy, we select the position and time interval satisfying the following equation:

$$T_s=L_s/V$$

Then equation(5) are greatly simplified to

$$z_j(k+1)=z_{j-1}(k)-g_j(k)L_s/V \quad (6)$$

It should be pointed out that equalion (6) not only is very simple but also have a clear physical significance. The basis weight of web at the position $(j-1)L_s$ and time kT_s is $z_{j-1}(k)$. After one period $T_s=L_s/V$, the web is rightly travelled to the posi tion jL_s and its basis weight is changed into $z_j(k+1)$. This means that the second term on the right hand of equation (6) should be equal to the amount of water lost by web per unit area during the travel. Therefor,we can represent the equation (6) in a more accurate form as

$$z_j(k+1)=z_{j-1}(k)-W_j(k) \quad (7)$$

where $W_j(k)=\int_{(j-1)L_s}^{jL_s} g(kT_s,l)dl/V$. The similar result holds for equation(4)

$$x_j(k+1)z_j(k+1)=x_{j-1}(k)z_{j-1}(k)-S_j(k) \quad (8)$$

where $S_j(k)=\int_{(j-1)L_s}^{jL_s}(l-c(kT_s,l))g(kT_s,l)dl/V$.

Mass Transfer Calculations It is very important to calculate the amount W of dewatered white water and its consistency c in equations (7) and (8). Due to the limited space, only partial rather than all calculations are presented.

A. Web between drums When web passes through between two drums, the water in it is evapourated. The evapourating rate may be expressed as

$$g=mk(c_w-c_a)$$

where m is the molecular weight of water, c_a and c_w are water concentrations of air and web surface respectively. They are related to partial pressures P_a and P_w according to the ideal gas law. And P_w is a function of moisture content of web and saturated vapour pressure $P_s(t)$ at web temperature t

$$P_w=P_s(t_w)e(-k(1/x-1/x_c))$$

where x_c is the critical moisture content. Mass Transfer coifficient K is evaluated from the relationships amoung dimenssionless numbers Sh.,Sc. and Re.

B. Web in coating liquid Coating is to improve the surface properties of web. The water absorbed by web from coating liquid obeys the Fick's law 2:

$$\partial x(y,t)/\partial t=D\partial^2 x(y,t)/\partial y^2$$

where D is the diffusional coefficient of water in web. The initial condition is the moisture content x_o of web just before enterring coating liquid. Because both surfaces of web is in contact with coating liquid, boundary conditions are

$$x(0,t)=x(Y,t)=x_h$$

where x_h is water content of coating liquid and Y is the thickness of web. Solving the above finite problem, we get

$$x_1=(x_o z_o+4z_o(x_h-z_o)/Y)\cdot(D\ V/\Pi L)^{\frac{1}{2}}$$

where x_1 is the moisture content of web when it just comes out of coating liquid and L is the liquid level of coating liquid.

Linearization Unsally, equations (7) and (8) are linearized for each j, and the iterative calculations are carried out to elimi nate internediate variables z_1, x_1,..., z_{n-1}, x_{n-1}. Final result is

$$\begin{bmatrix}\Delta z_n(k+1)\\ \\ \Delta x_n(k+1)\end{bmatrix}=A\begin{bmatrix}\Delta z_o(k+1-n)\\ \\ \Delta x_o(k+1-n)\end{bmatrix}+B\Delta U \qquad (9)$$

where A and B are constant matrices.

Unfortunately, the results obtained in this way are quite different from reality. We find out that the error comes from the seperate linearization of each processing unit. The linearization error of the former unit is amplified by the latter. Through such a seqential linearization the accumulated errors make the integrated model behaves entirely different from the way it ought to be. Therefore a new method of linearization must be developped. In equation (9), if only one increment is nonzero (for example Δz_o), and all the rest are zeros, then

$$a_{11}=\Delta z_n/\Delta z_o, a_{21}=\Delta x_n/\Delta z_o \qquad (10)$$

Following this reasoning a new method is developped as follows. First maintaining z_o^o, x_o^o and U_o at values of steady operating point, evaluate outputs z_n^o and x_n^o by nonlinear iterative equations (7) and (8). Next giving only z_o an increment Δz_o, so that $z_o^1=z_o^o+\Delta z_o$. The corresponding output z_n^1 and x_n^1 are again evaluated in the same way. Hence $\Delta z_n=z_n^1-z_n^o$, $\Delta x_n=x_n^1-x_n^o$ and a_{11} and a_{21} can be calculated from equation (10). Other elements are obtained in the same way.

Since the linearization of the integrated function relating z_n and x_n to z_o and x_o and u is carried out with nonlinear calculations, the model obtained in the above method has a much higher accuracy than that from usual unitwise linearization.

Based on above discusions, together with other mathematical treatments, hear and mass transfer calculations and production data, we establish the model of a paper machine as follows:

$$X(k+1)=AX(k)+BU(k) \qquad (11)$$

$$Y(k)=CX(k) \qquad (12)$$

In the model input vector U has 14 elements including thick pulp flow rate, white water flow rate, machine velocity, steam pressure and air temperature etc.. State vector X has 15 elements. Output vector has 4 elements, which are the consistency and liquid level of the pulp in headbox, basis weight and moisture content of paper at reel.

For verification of the model, many experiments have been carried out in a laboratory centre (with 1/1000 measurement accuracy). Actual measured outputs of paper machine and those calculated from the model are given in Fig.2 for comparison, which shows that this model has high accaracy.

CONTROLLER DESIGN

In the last two decades, there have been considerable advances in the theory and design of linear multivariable control systems such as pole assignment, linear guadratic optimal control and inverse Nyquist arrays method. They all achieved some new objectives of the control problem. But most of them have not been found suitable for pactical multivariable control system applications where the requirements are chiefly satisfactory transcient and steady-state performances and where the feedback variables should be meassurable outputs.

It is essential to seek for a unified mathematical representation of performance specifications which can reflect the above requirements directly. If this mathematical representation existed the theoretical design and analysis of a control system could be made easily.

We suggest to use the objective transfer funcion matrix $H_R(s)$ (shortened as OTFM) as performance specifications which is defined by:

$$Y(s)=H_R(s)V(s)$$

where Y is mx1 output vector, and V mx1 command vector. In fact, OTFM is also the the desired closed-loop transfer function matrix. $H_R(s)$ may also be expressed as $H_R(s)=N(s)/d(s)$, where d(s) is the monic least common demominator of all elements of $H_R(s)$ and N(s) is a polynomial matrix. It should be pointed out that almost all performance specifications can be represented by OTFM. For the case of pole assignment, d(s) should be specified and N(s) arbitrary. For steady-state accuracy, $H_R(0)$ should be specified. For decoupling, the off-diagonal elements of $H_R(s)$ are made zero and diagonal ones nonzero. For certain required transcient behaviour, both d(s) and N(s) should be specified accordingly.

Now consider a multivariable unit feedback control system, whose closed-loop transfer function matrix relating outputs Y(s) to commands V(s) is

$$H(s)=G(s)K(s)(I+G(s)K(s))^{-1} \qquad (13)$$

The design task is to determine controller K(s) so that H(s) processes the desired closed-loop performance specifications represented by the OTFM $H_R(s)$ of this system.

Now the closed-loop system is expected to meet the following four specifications:(a) the closed-loop system is stable; (b) output vector returns to zero vector at the steady-state when it is responded to constant disturbances; (c) for any constant command vector V(t)=V the output vector becomes equal to V at the steady-state;(d) the system exhibits small overshoots, fast response speed when responded to step command. These four general performance specifications are often required in industrial systems.

For the construction of OFTM, consider a familiar second-order system $h_{ij}(s)=(a_{ij}s+b_{ij})/(s^2+2\omega_{ij}\xi_{ij}s+\omega_{ij}^2)$,whose parameters are directly related to performance specifications. For instance specifiation (a) requires all ω_{ij} and ξ_{ij} strictly larger than zero. Specification (b) and (c) need parameter b_{ij} in each off-diagonal position being zero. In the case of $a_{ii}=0$ each diagonal element $h_{ii}(s)$ becomes a standard second-order system where b_{ii} should equal to ω_{ii}^2 in order to meet specifications (b) and (c). Finally for specification (d), parameters ω_{ii} and ξ_{ii} in $h_{ii}(s)$ should be some simple functions of the desired overshoot and response time of the ith closed-loop. Grouping all above considerations, OTFM can taken as $H_R(s)=\{h_{ij}(s)\}$.

As to the structure of controller K(s), specifications (b) and (c) need every off-diagonal element of K(s) being nonzero in order to satisfy different requirements of strengthening or weadening the possible interactions exsisted in the controlled system. Specification (b) and (c) also demand a integral term in every diagonal element of K(s). To improve the system stability reduced due to integral action, and also to meet specifications(a) and (d) proportional and derivative terms are added to every element. All these suggest the choice of multivariable PID controller for K(s) which can meet requirements of lead and /or lag conpensations also.

$$K(s)=K_p+K_I/s+sK_D = [\,I \;\; sI \;\; s^2 I\,]\; K/s \qquad (14)$$

where $K=[K_I^T \; K_p^T \; K_D^T]^T$, a constant matrix. Under the ideal situation, $H(s)=H_R(s)$. Hence combining eqution (13) and (14), we have

$$G_p(s)K=Q_R(s) \qquad (15)$$

where $G_p(s)=G(s)\,[\,I \;\; sI \;\; s^2 I\,]$ and $Q_R(s)=[H_R^{-1}(s)-I]^{-1}$. In ordinary instances, however, $H(s)\neq H_R(s)$ and $G_p(s)K\neq Q_R(s)$. But equation (15) can improved as

$$G_p(s)K=Q_R(s)+\Delta Q(s) \qquad (16)$$

in which $\Delta Q(s)$ depends on the selection of controller parameter matrix K. This relation shows that the smaller $\Delta Q(s)$ is, the closer $G_p(s)K$ to the desired characteristics $Q_R(s)$ would be. Clearly, K would be optimal if it is evaluated by minimizing $\Delta Q(s)$.

Let $s=j\omega_i$ i=1,2,..., n, after the approaprate rearrangement equation (16) becomes

$$AK=B+E$$

where A,B and E depend on the frequency characteristics of $G_p(s)$, $Q_R(s)$ and $\Delta Q(s)$ respectively. The evaluation of K by minimizing the norm of E is a least square problem. Its solution is

$$K=(A^TA)^{-1}\,A\;B$$

The method described above is applied to the design of a basis weight and moisture content control system of a paper machine. The open-loop transfer function matrix is obtained form the last section as follows.

$$G(s)=\begin{bmatrix} \dfrac{8.79(s+5)\exp(-s)}{(s^2+3s+15)} & \dfrac{-.081\exp(-.1s)}{(s+.25)} \\[2ex] \dfrac{-.075(s+5)\exp(-s)}{(s^2+3s+15)} & \dfrac{-.213\exp(-.1s)}{(s+.25)} \end{bmatrix}$$

In the design of basis weight and moisture content control system, we specify from the production practice that for unit step change in command v_1 (or v_2) the corresponding output y_1 (or y_2) should have a fast transcient response lasting no more than 6 (or 8) minites with no offsets. Furthermore, interaction between y_1 and y_2 and their over-shoots should be small. To satisfy all these requirements, OTFM of the following form is proposed:

$$H_R(s)=\text{diag}\{\,\omega_{ii}^2/(s^2+2\xi_{ii}\omega_{ii}s+\omega_{ii}^2),\quad i=1,2\,\}$$

where $\omega_{11}=1.5;\xi_{11}=0.8;\omega_{22}=2;\xi_{22}=0.95$. Controller parameters are calculated at ten selected frequencies $\{\omega_i:k=1,2,...10\}$ = {0.1,0.3,0.6,0.9,1.2,1.4,1.8,2.3,2.8,4.0}. They are

Proporional term

$$K_p=\begin{bmatrix} .0250 & -.0075 \\ -.8524 & -.3.6755 \end{bmatrix}$$

Integral term

$$K_I=\begin{bmatrix} .2034 & -.0704 \\ -.1292 & -1.213 \end{bmatrix}$$

Derivative term

$$K_D=\begin{bmatrix} .0223 & -.0096 \\ -.1044 & -.4812 \end{bmatrix}$$

Fig.3 gives simulation results of the closed-loop system. The responses are fast with small overshoots and zero offsets. Compared with the objective system, there are some differences at the starting portion. The reason is the delays existed in the actual system causing its outputs to respond slower at beginning. But considering all aspects the designed system is still satisfactory and meets all design specifications. It has been operating continuously in production line for more than a year. After this system is put in operation paper breaking rate is reduced, and production rate is increased by 2.3%. The quality of paper is also significantly improved. Thick pulp consumption is decreased by 1.3%. The heating steam saving is 27.6%. For production of 60g/M^2 writing paper the variation of basis weight is less than ±0.5g/M^2 and that of moisture content less than ±0.25%.

CONCLUSION

A dynamic mathematical model of a paper machine has been established based mainly on the physical analysis of the papermaking process. The test data from a paper mill agree with the model closely so that it is accurate enough to be used in control system design. Furthermore, because the physical analysis emphasize common characteristics of papermaking processes, models of different paper machines would have the same structure as this model. The only difference is that they take different parameters. But model parameters of any paper machine can be determined from its production data at the steady-state, and no dynamical experiment is necessary.

Objective transfer matrices can be used to represent general performance specifications of multivariable control systems. On the other hand, based on the given specifications a suitable objective transfer matrix can be constructed. With the help of objective transfer matrix, the theoretical treatment in design will be easier and design objective will appear clearer and more practicable. A simple and fast computing frequency response method for the optimal design of multivariable PID controllers has been proposed. This controller will assure both desired transcient and steady-state performances. The steady-state characteristics are inherent properties of a control system. They are not affected by system or controller parameter changes provided that the closed-loop system remains stable. Because of the generality of OTFM and only use of meassurable outputs as feedbacks without rquirement of any state variable, this design method can widely applied to practical multivariable control systems. Furthermore, PID units are commercially available, thus the designed controllers can readily be implemented without the need for elaborate realization and implementation. The design procedures are straight forward with no iteration. Therefore the computer time consumed for computation of parameter matrix K is much shorter than that needed for solving Riccati equation of the same order system. From the successful application to the control of a paper machine, it is believed that the proposed design method would offer a system which is simpler in construction and easier in implementation, yet with good control qualities as sophisticated control schemes do.

REFERENCES

Astrom,K.J.(1977). Theory and applications of self-tuning regulators. Automatica, 13, 457.

Fjeld,M.(1978). Application of modern control concepts on a kraft paper machine. Automatica, 14, 107.

Gentil,S.(1975). Different methods for dynamics identification of a experimental paper machine. Proc. 3th IFAC sym.on identification and system parameter estimation, Part I, 478.

Rosenbrock,H.H.(1974). Computer-aided control system design. Academic Press, London.

Zhou,C.H.(1980). Principles of chemical process control. (in Chinese). Chemical Industry Press, Beijing.

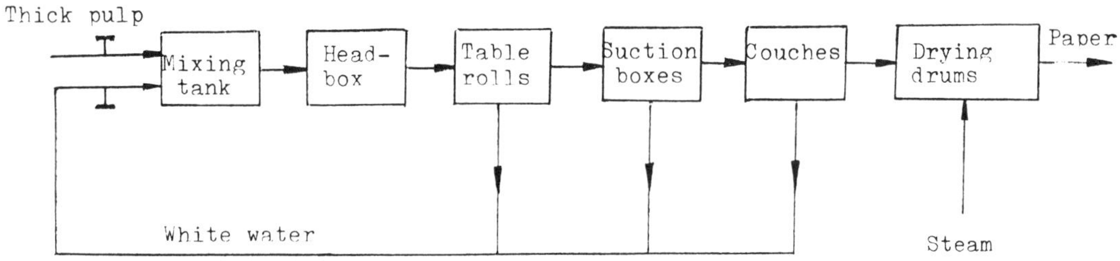

Fig.1 Flow diagram of a paper machine

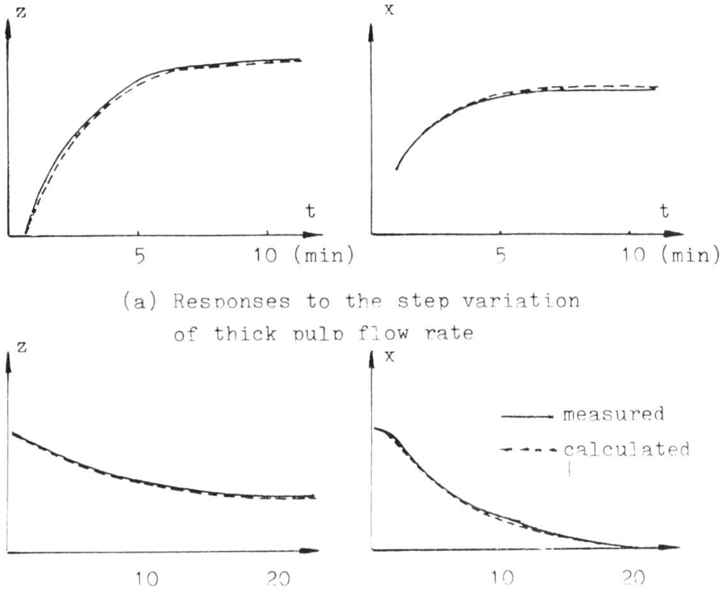

(a) Responses to the step variation of thick pulp flow rate

(b) Respones to the step variation of steam flow rate

Fig.2 Model simulation

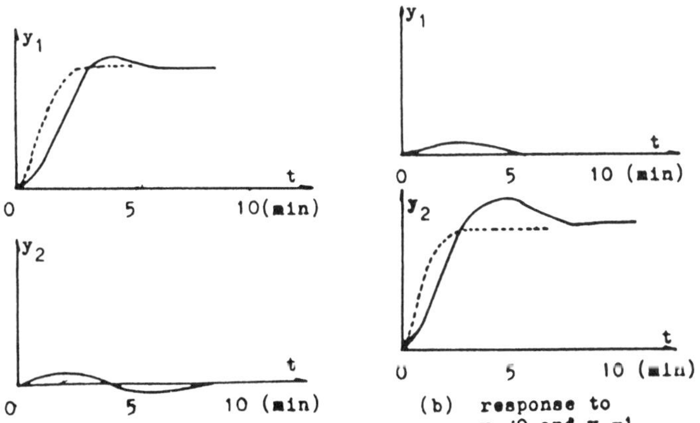

(a) Response to $v_1=1$ and $v_2=0$

(b) response to $v_1=0$ and $v_2=1$

Fig. 3 System simulation results

Note: —— for the designed system

···· for the objective system

COMPUTER CONTROLLING FOR THE ESR PROCESS – AN APPLICATION OF MODERN CONTROL THEORY

Ruan Rong-Yao, Hu Qi-Di, Tang Xian-Xiang and Ma Guo-Xuan

Department of Mathematics, East China Normal University, Shanghai, China

Abstract. This paper is to introduce an example of applying modern control theory to industrial producing process -- the control of the electroslag remelting (ESR) process. Acording to the relations between the performance of solidified ingot and the smelting current, the smelting voltage and the melt rate, we have suitably determined the structure of the control system. And a matematical model with which the system is controlled with optimal feedback is established after identifying the system. With this model, the ESR process is controlled by a microcomputer and the quality of the solidified ingot is greatly raised.

Keywords. Computer control; control theory; identification; parameter estimation; optimal filtering; optimal stochastic control; adaptive tracking.

DETERMINING THE STRUCTURE OF THE CONTROL SYSTEM

Electroslag remelting (ESR) is a smelting method to produce alloy steel with high quality. Using this method, the material (electrode) to be refined is heated and remelted by passing a current through it into the molten slag, and melted metal forms an ingot in the water-cooling crucible. This process is shown in Fig. 1.

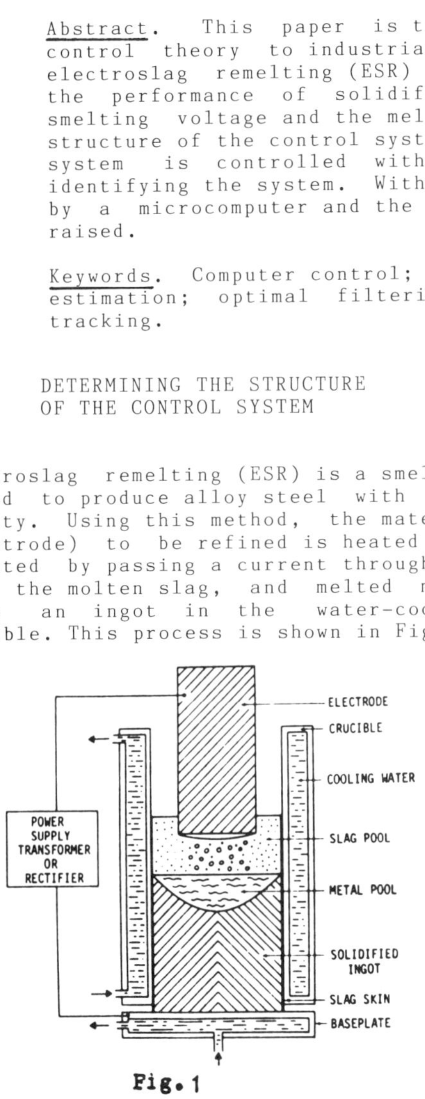

Fig. 1

The goal for ESR is to improve the structure status of metal and ingot quality. In the ESR process, the main factors to affect the performance of electroslag steel are the depth and shape of the metal pool and the quality of the ingot skin and they can not be measured directly in smelting. There is an approximately linear relation between a melt rate and a depth of the metal pool, while a resistance of the slag pool is related to a shape of the metal pool to a certain extent. If the melt rate and the resistance of the slag pool keep constant, basically the metal pool will not change its depth and shape. In this case, the crystallized status of the solidified ingot will be uniform and its quality will be higher. For this reason, we take it as a objective to be controlled to keep the melt rate and the resistance of slag pool constant basically.

R_s, resistance of the slag pool, can not be measured directly also, but it can be obtained approximately by measuring the smelting current I and voltage V (to omit the voltage-drop on the electrode and the solidified ingot, so the smelting voltage is the voltage-drop on the resistance of the slag pool), that is,

$$R_S = P/I^2 = VI\cos\varphi /I^2 \doteq V/I$$

where P is the power to be used in smelting (smelting power).

The melt rate can be given by measuring the weight of the electrode. Usually, it is adopted to regulate the smelting current in order to make the melt rate keep about a setting value(The smelting voltage is also regulated by an other control loop,so that the resistance of the slag pool will keep its value to be a constant). And the smelting current and melt rate are taken as the input and output of the system. See Chen et al.(1979) and Li et al.(1982). In this case, the realation between the input and output of this system is nonlinear (a change in melt rate depends on a change in I^2, but does not depend linearly on a change in I), so that the control accuracy for a model with PID control may not be so high. Now, we take smelting power as the system input, and still melt rate as the system output. Here, the system input is a dummy input. In fact, it is the smelting current and voltage. Thus, the control system for ESR has been reduced into the linear one with single input (the smelting power) and single output (the melt rate).

The smelting power is regulated by regulating the smelting current and voltage simultaneously in order to keep the resistance of slag pool constant. If the smelting power is to be changed in ΔP, then the smelting current must be changed in

$$\Delta I = \text{sign}(\Delta P)\sqrt{|\Delta P| / R_s}$$

and the smelting voltage in

$$\Delta V = R_s \Delta I$$

respectively.

A change in smelting current is given by regulating the controlled current of a saturation reactor. And a change in the smelting voltage is obtained by regulating the velocity with which the electrode comes down (That is, changing the depth by which the electrode inserts into the slag pool).

Besides current and voltage, there are composition of the slag, temperature of cooling water, quantity and temperature of the slag and filling coefficient to affect the quality of solidified ingot. We have not considered how to do optimal control to these technique parameters yet.

MODELLING OF THE CONTROL SYSTEM

The test shows that the relation between the input and the output of this system can be described approximately by a second order diffrence eqation with stochastic disturbance as follows:

$$y_k = a_1 y_{k-1} + a_2 y_{k-2} + b_1 u_{k-1} + b_2 u_{k-2} + \xi_k \quad (1)$$

where y_k is the diffrence between the melt rate M_k at time t_k and the setting melt rate M_0, u_k is the difference between the input power P_k at time t_k and the setting input value P_0. And ξ_k is the linear combination of dynamic noise W and measuring noise V, that is,

$$\xi_k = V_k - a_1 V_{k-1} - a_2 V_{k-2} + b_1 W_{k-1} + b_2 W_{k-2} \quad (2)$$

We obtained the required data for system identification by doing experiment of step response. After slag had been melted, the ESR shifted from manual control to automatic control with a microcomputer. First, let the smelting current and voltage be steady at certain fixed values I_0 and V_0 respectively. The corresponding

melt rate, on the whole, is also steady at a certain value, after a period of time. At the same time, step signs ΔI and ΔV are added to the system. Here, ΔI and ΔV satisfy a relation as below:

$$\Delta V / \Delta I = V_0 / I_0$$

One experiment will be over when the melt rate has a steady value again after a period of time (not shorter than setting time T_s of the system). Such experiment we

have done several times. Under normal condition, the experiment result has a good reproducibility. So, for system identification, only one experiment result to be chosen is enough. Of course, the sampling data must be filtered appropriately so as to eliminate the non-regular values caused by measuring noise.

Estimating the parameters, we have reduced the data of step response into the ones of pulse response. Assume the sampling values of step response are M_0, M_1, ... , M_{n-1}

$(n\Delta t > T_s)$ then

$$y_k = M_k - M_{k-1}, \quad k=0, 1, \ldots, n-1,$$

where $M_{-1} = 0$. These are the data of pulse response of this system.

Let

$$\theta = \begin{bmatrix} a_1 \\ a_2 \\ b_1 \\ b_2 \end{bmatrix}, \quad X = \begin{bmatrix} y_0 & y_{-1} & 1 & 0 \\ y_1 & y_0 & 0 & 1 \\ y_2 & y_1 & 0 & 0 \\ \vdots & \vdots & \vdots & \vdots \\ y_{n-1} & y_{n-2} & 0 & 0 \end{bmatrix}, \quad Y = \begin{bmatrix} y_1 \\ y_2 \\ y_3 \\ \vdots \\ y_n \end{bmatrix}$$

then the least square estimate of parameter θ is

$$\hat{\theta} = (X^T X)^{-1} X^T Y ,$$

where T represents transpose of matrix or vector. Besides we have adopted also instrument variable method (Wong and Polak, 1967) to estimate θ , the result is, on the whole , as same as that one obtained by the least square method.

The variance of the dynamic noise of the system, $EW_k^2 = Q$, and the variance of measuring noise, $EV_k^2 = R$, are estimated by the formula given (See Ruan et al. 1982).

The relation between the input and output of this system also can be described approximately by a third order diffrence equation, that is,

$$y_k = a_1 y_{k-1} + a_2 y_{k-2} + a_3 y_{k-3} + b_1 u_{k-1} + b_2 u_{k-2} + b_3 u_{k-3} + \xi_k$$

$$\xi_k = V_k - a_1 V_{k-1} - a_2 V_{k-2} - a_3 V_{k-3} + b_1 W_{k-1} + b_2 W_{k-2} + b_3 W_{k-3} \,.$$

We have got the estimate values a_1, a_2 , a_3, b_1, b_2, b_3, R and Q by the same method used above. Also, we have compared the third order model with the second one by Akaike Information Criterion (AIC) (See Akaike, 1974) The computation (The software of Ruan (1982) is employed) shows that in the sense of AIC, the third one is better, but the difference between them is not large. In practice, either will do. Now we are using the second order one. And we are going to introduce the control law designed to the second order.

DESIGNING OF THE OPTIMAL FEEDBACK CONTROL LAW

If we take the state variable as

$$x_k=(y_k-V_k, a_2 y_{k-1}-a_2 V_{k-1}+b_2 u_{k-1}+b_2 W_{k-1})^T$$

difference equations (1) and (2) are equivalent to following state and measuring equations (that is, they have the same Z — transfer function and the same noise statistics) :

$$x_{k+1} = Ax_k + Bu_k + BW_k \qquad (3)$$

$$y_k = Hx_k + V_k \qquad (4)$$

where

$$A = \begin{bmatrix} a_1 & 1 \\ a_2 & 0 \end{bmatrix}, \quad B = \begin{bmatrix} b_1 \\ b_2 \end{bmatrix}, \quad H = [\ 1,\ 0\ ]$$

We take the quadratic performance index as a criteria for choosing optimal control strategy. That is to say, we want to find the admissible control $(u_0, u_1, u_2, \ldots, u_{n-1})$ such that objective function

$$J(u)=E\sum_{k=1}^{n}(u_{k-1}^2 + x_k^T Q_0 x_k) \qquad (5)$$

is minimized, where

$$Q_0 = \begin{bmatrix} q & 0 \\ 0 & 0 \end{bmatrix}.$$

Its actual significance is to minimize the sum of the variances of fluctuations occuring in the heating power and the deviations of the melt rate. The choice of weight q is related to the proportionality factor betweeen u and x, and is also determined appropriately according to reducing

$E\sum_{k=1}^{n} x_k^T x_k$ and $E\sum_{k=1}^{n-1} u_k^2$ as small as possible.

If the admisible control u_k is restricted to the linear functions of $y_0, y_1, \ldots, y_k$, according to the separation theorem (See e.g., Astrom, 1970), the control strategy which minimizes J(u) is

$$u_k = -L_k \hat{x}_{k/k} \qquad (6)$$

where $\hat{x}_{k/k}$ is the optimal linear estimate of state x_k based on $y_0, y_1, \ldots, y_k$, and

$$L_k = (1 + B^T Q(k) B)^{-1} B^T Q(k) A \qquad (7)$$

$$Q(k) = Q_0 + A^T R(k+1) \qquad (8)$$

$$R(k) = Q(k)(A-BL_k) \qquad (9)$$

where k=n, n-1, ..., 1, and $Q(n)=Q_0$.

When n is sufficiently large,

$$L_k \longrightarrow L \ (k \longrightarrow 0).$$

It can be obtained by computing formulae (7)-(9) recursively. Also, it can be obtained by letting $n=\infty$ in (5) and solving Riccati equation. L , so obtained , is a coefficient (a row vector) for optimal feedback control at steady state. In this case, the computing formulae (6)-(9) for feedback control on-line can be reduced to

$$u_k = -L\hat{x}_{k/k} \qquad (10)$$

where L does not depend on k, but on q. After digit simulating, analizing and comparing, we choose the optimal feedback coefficient L corresponding to an appropriate value of q.

For simulating off-line (or controlling on-line), the recursive filter is computed according to the following formulae:

$$\hat{x}_{k/k} = \hat{x}_{k/k-1} + K\Delta y_k \qquad (11)$$

$$\Delta y_k = y_k - H\hat{x}_{k/k-1} \qquad (12)$$

$$\hat{x}_{k+1/k} = A\hat{x}_{k/k} + Bu_k \qquad (13)$$

where Δy_k is an innovation obtained from random simulating value y_k (or the measuring y_k), and $\hat{x}_{k+1/k}$ denotes the one-step optimal predication, that is, it is the optimal linear estimate of x_{k+1} based on $y_0, y_1, \ldots, y_k$. The initial filtering value $\hat{x}_0$ can take any possible value of state x_0.

K in (11) is a gain coefficient (a column vector) of filtering at steady state. It can be computed recursively by the following formulae (See , e. g., Anderson et al. 1979):

$$P_{k/k-1} = AP_{k-1}A^T + BQB^T \qquad (14)$$

$$K_k = P_{k/k-1}H^T(HP_{k/k-1}H^T+R)^{-1} \qquad (15)$$

$$P_k = P_{k/k-1} - K_k HP_{k/k-1} \qquad (16)$$

where $P_{k/k-1}=E[(x_k - \hat{x}_{k/k-1})(x_k - \hat{x}_{k/k-1})^T]$ and $P_k = E[(x_k - \hat{x}_{k/k})(x_k - \hat{x}_{k/k})^T]$ denote the covariance matrices of predicting error and filtering error respectively, and K_k is filtering gain (a column vector) at time t_k. Any positive matrix can be taken as the iterating initial value P_0. Usually, iterating formulae (14)-(16) for certain times cyclically and recursively, the first five significant digits of corresponding components in vectors K_k and K_{k-1} will be the same. This means that the matrix sequence $\{K_k\}$ is convergent. And the limit of K_k is K in (11) as $k \rightarrow \infty$.

Because the feedback control for the melt rate is a control problem with setting value so it is required to consider optimal filter and feedback for the steady state only. In this case, the formulae used in on-line computation of Kalman filtering and optimal feedback control are very simple, that is, to compute according to formulae (11)-(13) and (10).

ADAPTIVE TRACKING OF THE SETTING VALUE OF INPUT POWER

The optimal feedback control signal u_k calculated from (10) is superposed on the setting value P_0 of input power, that is, the input power of the system at t_k should be

$$P_k^* = P_0 + u_k \tag{17}$$

When the process was controlled manually in the past, the technique adopted was to fix the smelting current I_0 and voltage V_0 (so was $P_0 \doteq I_0 V_0$) such that the melt rate rose gradually, see Fig. 2 (The period of time for melting slag is not included). Now to require the melt rate being constant, the setting value of input power must be corrected timely according to the fluctuation of the melt rate.

Let the power setting value of input power at t_k is P_k. We use the self-adaptive tracking method of dynamic parameter (See, Ruan, Xia and Wang, 1982) to estimate P_k on-line. The approximately computing formulae are as follows:

$$\begin{aligned} \hat{m}_k &= \alpha \hat{m}_{k-1} + (1-\alpha)\Delta y_k \\ \Delta\hat{P}_k &= c_1 \hat{m}_k - c_2 \Delta\hat{P}_{k-1} \\ \hat{P}_k &= \hat{P}_{k-1} - \Delta\hat{P}_k \end{aligned} \tag{18}$$

where $0 < \alpha < 1$. $c_1 = \beta/(b_1 + b_2)$, $0 < \beta < 1$ and $c_2 = c_1 b_2$, Δy_k is given by (12), then $\hat{m}_k$ is a value of recursive least square estimate with a exponential weight of the slow time-varying parameter $m = E\Delta y_k$. The values for α and β are chosen appropriately according to the condition of real time control.

Replacing P_0 in (17) by $\hat{P}_k$, we can obtain the actual input power P_k^* for the system in time interval $[t_k, t_{k+1}]$. For real time control, the values of u_k, $\hat{P}_k$ and P_k^* will all be restricted to appropriate ranges. For example, $a \leqslant \hat{P}_k \leqslant b$. If, computed from (18), there is that $\hat{P}_k < a$ (or $\hat{P}_k > b$), then it should be done that $\hat{P}_k = a$ (or $\hat{P}_k = b$) a and b are determined from the possible and admissible fluctuating range of P_k.

THE CONTROL EFFECT

Using a microcomputer to control the closed loop system automatically according to the mathematical model introduced above, the requirements that the melt rate is constant and resistance of the slag pool is on the whole constant are all reached (See Fig. 3).

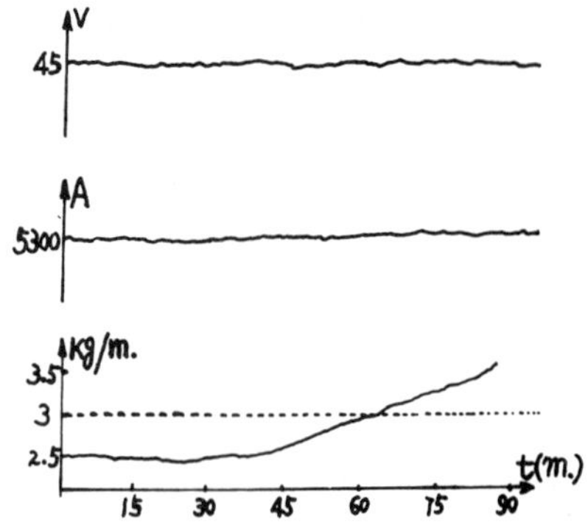

Fig. 2. Original technique (melting slag is not included)

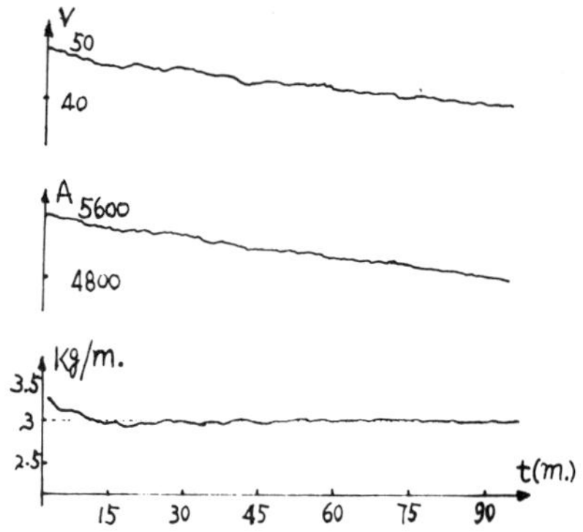

Fig. 3. Computer control (melting slag is not included)

Both of structure status and quality of the ingot skin are better than before. Not only the performance of every solidified ingot is of a uniformity, but also the melt rate, time used for smelting and power consumption are the same when the same sort of steel is smelted at different time in the same furnace. The stableness, reproducibility and quality of solidified ingot of the ESR are raised. When the smelting process was controlled manually in the past, there was difference with different person. It was difficult to show consistency. Hence there is no possibility for all to reach the optimal state of technique.

We did experiments for the ESR by this mathematical model with different sorts of steel and different sizes of ingot, setting values of the melt rate and technique parameters. Because of the setting value of input power is corrected by adaptive tracking method so that the model for closed loop control has a strong ability of self-adaptation. Both of the melt rate and inserting depth of the electrode are well-controlled without correcting any model parameters manually. The deviations of the melt rate, except a very little part, are within ±0.2 kg/min. (measuring error caused by the electronic-weighing is not included).

During probation period, we got 1537 statistical data of melt rate for 5 sorts of steel, 4 sorts of ingot and 24 furnace-times. The average value of absolute deviation is 0.08 kg/min. There are 1498 of the data which have a deviation within ±0.2 kg/min., 97.46% of the total, and 1329 data of it have a deviation not exceeding ±0.1 kg/min., 86.47% of the total. There are only 2.54% of the total whose absolute deviation is greater than 0.3 kg/min.. The computer has been put in use for more than three year already. The control effect is better than that time of probation period. Almost no deviation of the melt rate which is greater than 0.2 kg/min. occurs.

CONCLUTIONS

After taking the smelting power as a dummy input and the melt rate as a output of ESR process, the optimal feedback control can be implemented by using a microcomputer according to the mathematical model as mentioned above. The effectiveness demonstrates that the high control accuracy is obtained and the quality of the solidified ingot is greatly raised. This is a successful example of applying modern control theory to the solution of process control in industry.

REFERENCES

Akaike,H.(1974). A new look at stochastical model identification. IEEE Trans., Autom.Control, 19.

Anderson, B. D., and J. B. Moore (1979). Optimal Filtering, Prentice-Hall, Inc. Englewood Chiffs, New Jersey. pp. 36-128.

Åström, K. J. (1970). Introduction to Stochastic Control Theory, Academic Press, pp. 256-290.

Chen,J.H., R.C.Myers, and D.R.Engel(1979). Computer control of the ESR process. Proceedings of the 6th International Vacuum Metallurgy Conference, San Diego, California. pp. 831-847.

Li, Y.G., Y.S. Wu, and B.K. Xiao (1981). An application of the minicomputer to ESR. A paper presented at the National Scientific Conference for Automatic Technique and Its Application.

Ruan,R.Y.(1982). Software of identification and modeling for linear multivariable stochastic systems. 6th IFAC Symposium on Identification and System Parameter Estimation, Pergamon Press.

Ruan, R.Y., T.C. Xia, and X.W. Wang(1982). The Mathematical Model of Temperature Control for Creep-testing Furnaces. Recent Developments in Contol Theory and Its Application -- Proceedings of Bilateral Meeting on Control Systems, Science Press, Beijing, China, Gordon and Breach, Science Publishers, Inc., New York. pp. 277-293.

Wong, K.Y., and E. Polak (1967). Identification of linear discrete time systems using the intrumental variable method. IEEE Trans. Autom. Control, 12, 643-652.

MICROPROCESSOR BASED AUTOMATIC START UP SYSTEM FOR FLUIDIZED BED COMBUSTION PROCESS

E. Tulunay* and H. Midoğlu**

**Electrical and Electronics Engineering Department, Middle East Technical University, Ankara, Turkey*
***MEKO, Microcomputer and Electronics Control Corporation, Ankara, Turkey*

Abstract. A manual start up procedure has been developed for a 25x25 cm^2 fluidized bed lignite combustor which heats an 80 000 kcal/h conventional fire tube hot water boiler. Logical steps are formulated in a way suitable for the implementation by using fuzzy logic in an expert system in future. A 6502 microprocessor based automatic start up controller and the necessary interface and instrumentation have been designed and constructed.

Keywords. Digital Control, Industrial Control, Microprocessors, On Line Operation, Boilers.

INTRODUCTION

Fluidization is a term used for describing the fluid like behavior of solid particles when flow of a fluid is passed through them. In this application, inert ash particles of 2-3 mm diameter are placed in a container with 25x25 cm^2 cross section to form a bulk with a height of about 15 cm. This bed material resides on a distributor plate with certain amount of holes of 1.0 mm diameter perforated on it. Air is passed through the holes of the distributor and the bed material in the upwards direction. When the velocity of the fluidizing air exceeds the minimum fluidization velocity, the particulate mass behaves like a fluid. At the higher air velocities bubbles are formed and the sight is similar to a boiling liquid.

In this combustion application lignite particles of about 3 mm diameter are used as fuel. Inert ash particles form the combustion medium and the oxygen of the air is used for combustion. Fluidized bed is a process when gas and solid interaction is at a very high level. High combustion efficiencies can be achieved and sulfur dioxide emission can be controlled by adding limestone to the bed. Such favorable properties make fluidized bed a unique process in modern combustion applications. Detailed information concerning the fluidization and fluidized combustion can be found in the literature. (Kunii and Levespiel, 1969; Howard (ed),1983).

In this part of the work an algorithm is proposed for the automatic start up of the fluidized bed lignite combustor. A 6500 family microprocessor based card rack control system is designed and constructed.

EXPERIMENTAL

Fluidized bed lignite combustor boiler system together with the associated instrumentation and control devices is shown in Fig. 1. A fluidized bed lignite combustor has been designed and constructed to drive a conventional fire tube hot water boiler with a capacity of 80 000 kcal/h (Tulunay and Midoğlu, 1983).

Turkish Seyitömer lignite of 2-5 mm average diameter is used. It is fed to the bed by a screw feeder coupled to a dc compound motor. Coal feed rate is controlled by varying the dc voltage across the motor terminals.

The fluidizing air is supplied by an air fan. Air flow rate is controlled manually by a valve or automatically by a butterfly valve driven by an induction motor. Air flow rate is measured by an orifice, two manometers and water level to dc voltage converters.

Liquid petroleum gas (LPG) is used for initial heating of the bed which is turned on or off manually by a valve or automatically by a solenoid valve. The gas is ignited by a spark.

Temperatures at 5 different points in the bed are measured by using 5 NiCr/Ni thermocouples. The inlet and outlet water temperatures are measured by using Platinium 100 thermoresistive elements.

START UP PROCEDURE

Fluidized bed combustion process has a complicated structure involving a lot of interacting variables. The nonlinear behavior and the fast variations in the transient conditions of the fluidized bed combustion further increase the complexity of the system. This is the main difficulty in deriving an easy to implement start up procedure. The basic idea underlying this procedure is to simplify the required operations as much as possible. The number of manipulated variables is kept at a minimum. Continuous adjustments are replaced by on off controls whenever possible. The details of the start up procedure are given elsewhere (Mıdoğlu, 1985). Here only a short summary is given. A simple flow chart illustrating the manual start up procedure is given in Fig. 2. Start up is achieved by passing through the following four basic steps.

Step 1 : Initializing and checking all the devices.
Step 2 : Igniting sparks and starting to feed the LPG.
Step 3 : Approaching the steady state conditions.
Step 4 : Starting to feed the lignite.
Special safety interlocks are also provided.

Figure 2 shows that the start up procedure is based on the facts, rules of thumb and the necessary knowledge concerning the fluidized bed combustion system that are collected and judged by an expert. The procedure is basically composed of "if then" rules. Thus the controller designed for implementing the start up procedure may be viewed as a "rule based expert system" (Gevarler, 1983; Zadeh, 1984). It is also noted that the rules of the start up procedure contains "fuzzy sets" such as "low", "high", "small enough" etc. In this study the start up procedure is aimed to be implemented by a rule based expert system by using bivalent logic. Later on, the fluidized bed combustion system may be the subject of a fuzzy logic application.

HARDWARE OF THE SYSTEM

Detailed information concerning the hardware details of the instrumentation of the system and the start up controller are given elsewhere (Mıdoğlu, 1985). Here only the block diagram of the 6502 microprocessor based controller is given in Fig. 3. A single card microcomputer is designed and constructed which contains the central processing unit, address decoding circuitry, memory and input/output (I/O) ports. Additional printed circuit cards containing the analog input/output ports, keyboard and display interface circuitry, indicator and relay driving circuitry as well as other necessary circuits such as power supplies are also designed and constructed. These printed circuit cards use a common bus. They are rack mounted on a single back pannel.

RESULTS AND DISCUSSIONS

A manual start up procedure has been developed for the fluidized bed lignite combustor. This procedure has been simplified such that only the air flow rate is manipulated continuously, all of the other process variables are on off controlled.

The choice of 6502 system is heavily affected by the availability of softwave development tools at the time.

Automatic start up system is one of the key factors in the small scale applications of fluidized bed lignite combustors which are very important in the applications where a large scale central heating is not applicable. The automatic start up procedure developed can also be applied to large scale systems in modular form with some modifications.

REFERENCES

Gevarler, W.B. (1983). Expert Systems: Limited but powerful. IEEE Spectrum, 20.

Howard, J.R. (ed)(1983). Fluidized Beds, Combustion and Applications. Applied Science Publishers, London.

Kunii,D., and D.Levenspiel (1969). Fluidization Engineering, John Wiley and Sons, Inc., New York.

Mıdoğlu, H. (1985). Investigation of Automatic Start Up Possibility for a Fluidized Bed Lignite Combustor. Master's Thesis. Electrical and Electronics Engineering Department , Middle East Technical University Ankara, Turkey.

Tulunay, E. and Mıdoğlu, H. (1983). Preliminary Experiments on a Fluidized Bed Combustion System. Heat Science and Technology, 6, 47-50. (In Turkish).

Zadeh, L.A. (1984). Making Computers Think Like People, IEEE Spectrum, 21, 26-32.

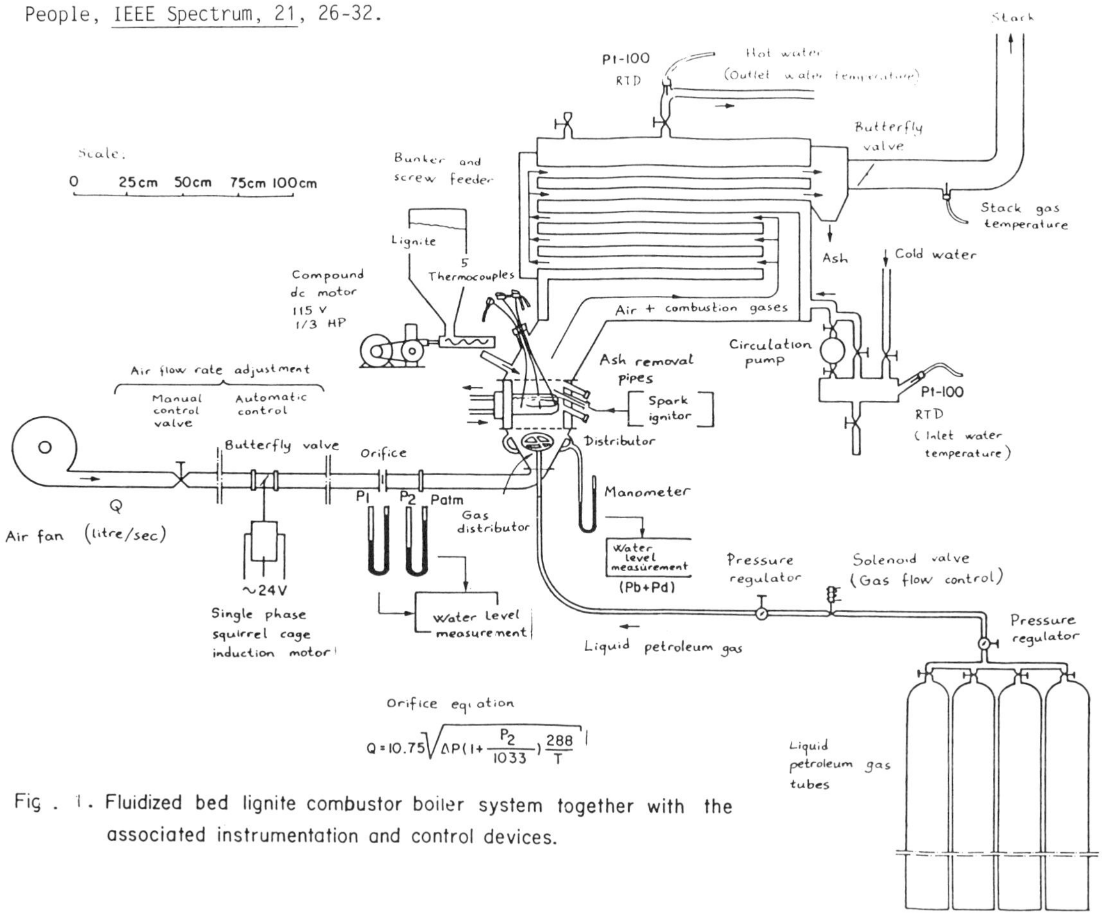

Fig. 1. Fluidized bed lignite combustor boiler system together with the associated instrumentation and control devices.

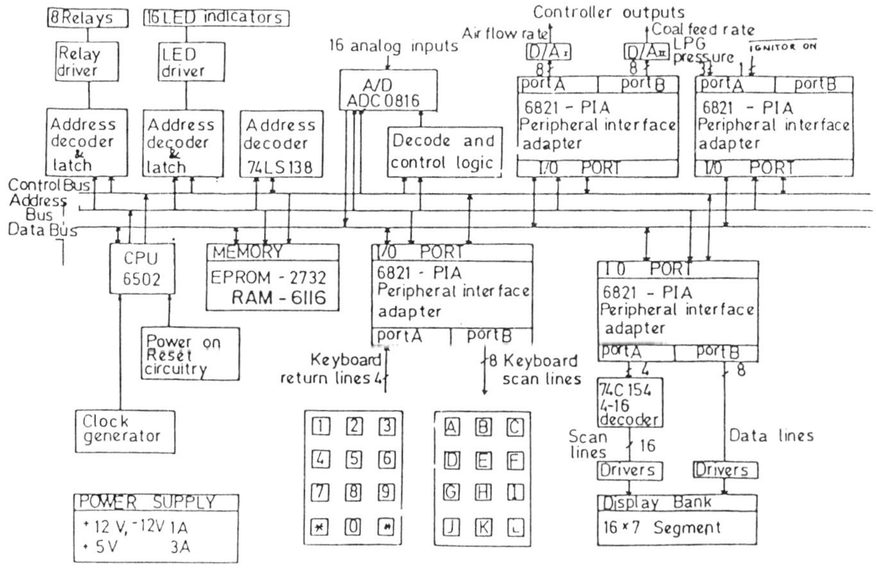

Fig. 3. Block diagram of the controller

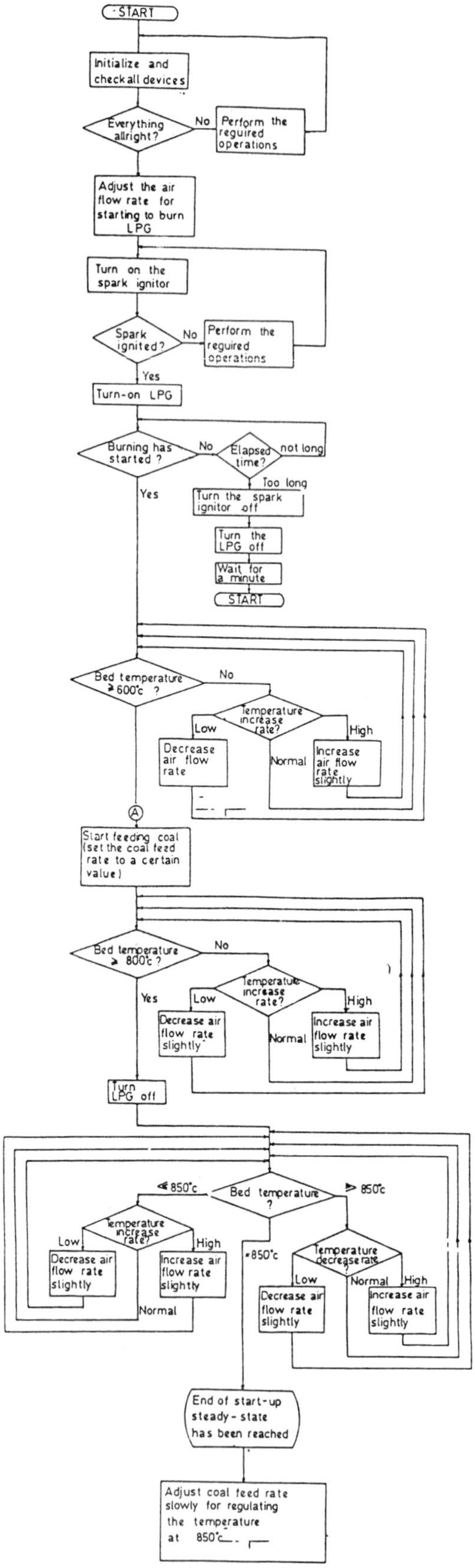

Fig. 2. Flowchart for the start up procedure

INTRINSIC SAFETY RULES OK

I. C. Hutcheon

Measurement Technology Ltd., Luton, UK

Abstract
The technique of intrinsic safety was originated in England for mine signalling during 1914-16, and is now the predominant solution for process measurements and control in hazardous locations.[1] This paper reviews its basic principles and advantages.

Introduction
In the oil, gas and petrochemical industries, onshore and offshore, many wires carry signals between the control room, which is always a 'safe area', and apparatus in 'hazardous areas' where flammable liquids and gases are present. In order to prevent explosions due to a combination of gas leaks and electrical faults, the system can be made 'intrinsically safe' as shown in Fig. 1.

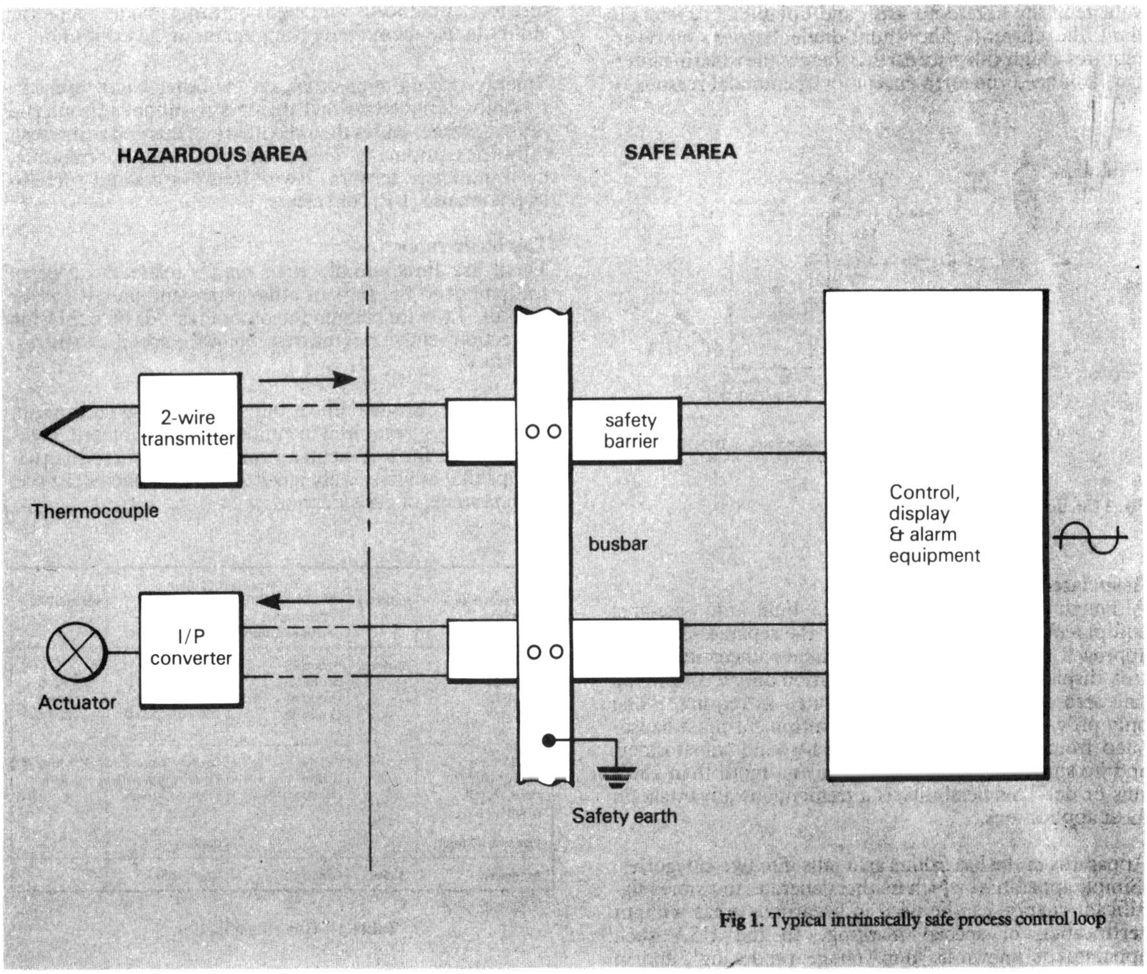

Fig 1. Typical intrinsically safe process control loop

The 'safety barriers' pass the small measurement and control signals without shunting them, but restrict any fault energy to a level which cannot ignite an explosive atmosphere. They may be simple 'shunt-diode' barriers as shown in Figs 2 and 3, or 'interface units' providing galvanic isolation between the safe and hazardous-area circuits, together with simple amplifying, signal conditioning, switching or logic functions, Fig. 4.

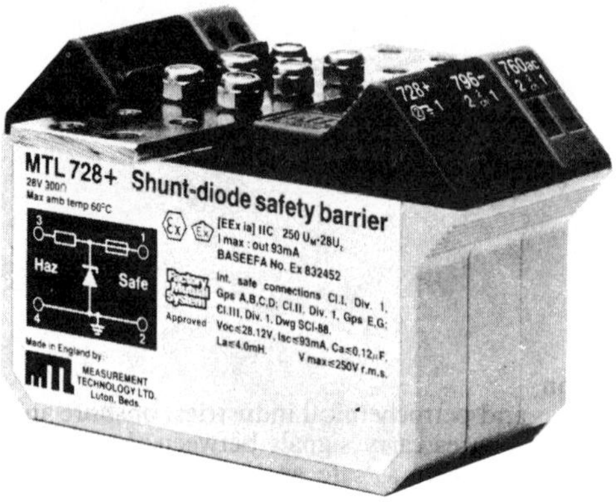

Fig 2. Typical shunt-diode safety barriers

Both devices rely on a fuse and Zener diodes to limit the voltage in the hazardous area, and 'infallible' resistors to limit the current. The shunt-diode barrier, however, requires a high-integrity earth, whereas the isolating interface unit needs no earth except for operational reasons.

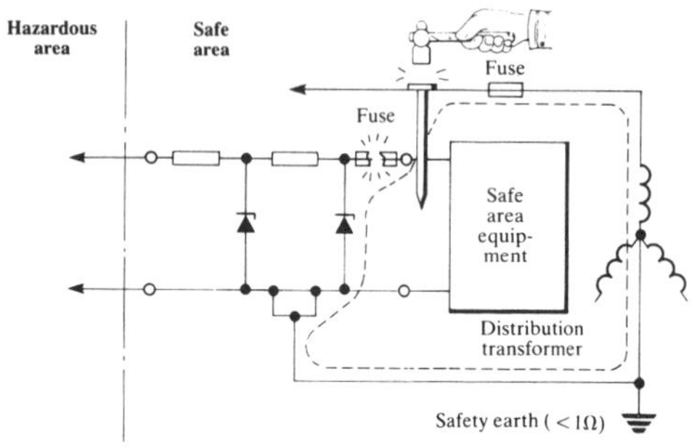

Fig. 3 Circuit and operation of safety barrier

Associated equipment

Although safety interfaces can be built into safe-area equipment, the great advantage of the separate-interface approach shown above is that ordinary uncertified control, display and alarm instrumentation can be used in the safe area, and be changed or updated as required. The only proviso is that the safe-area equipment must be isolated from the supply by a double-wound transformer, and be supplied from (and contain) no more than 250V rms or dc. This flexibility is a tremendous advantage in most applications.

Apparatus in the hazardous area falls into two categories. 'Simple apparatus', which neither generates nor stores significant energy, can be used in hazardous areas without certification or special marking. In the USA such apparatus is known as 'non-voltage producing', and in

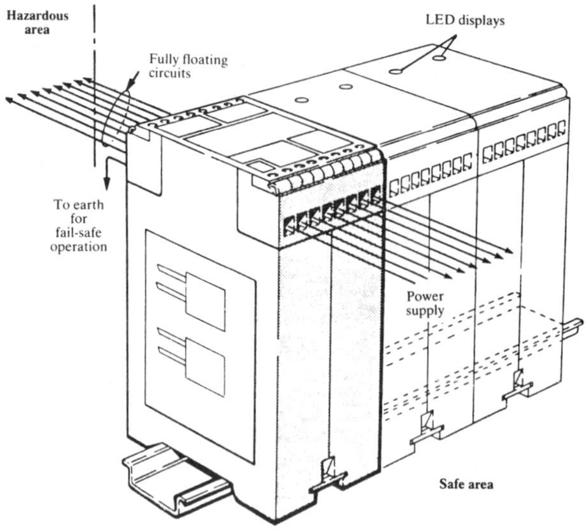

Fig 4. Typical 'interface unit'

Europe it is defined by CENELEC standard EN50 014 as 'devices in which . . . . none of the values 1.2V, 0.1A, 20μJ or 25mW is exceeded'. These definitions embrace a wide range of devices such as thermocouples, switches, resistors, photocells and light-emitting diodes; and the ability to use them freely brings great joy to everyone.

'Energy-storing apparatus', on the other hand, must be specially designed so that the energy cannot get out and cause damage, and is then certified and marked as intrinsically safe equipment. This category includes, for example, most inductive sensors, 2-wire transmitters and current-to-pneumatic (I/P) converters.

The basic rules

There are three simple rules which relate the type of equipment to the type of atmosphere and area it can be used in. Two are common to other methods of 'explosion protection': only one (the first, below) is specific to intrinsic safety.[2]

1. Gases are classified into groups according to the amount of spark energy required to ignite them. Equipment is certified as safe for a particular group (and of course all other groups that are less easily ignited). Table 1 shows the two main systems of classification:

Representative (test) gas	Industry	Gas classification		Ignitability
		IEC countries (includes Europe)	USA & Canada	
Acetylene Hydrogen Ethylene Propane	Surface	Group IIC Group IIC Group IIB Group IIA	Class I, Group A Class I, Group B Class I, Group C Class I, Group D	↑ More easily ignited
Metal dust, carbon dust, flour, starch, grain		(Dusts under consideration)	Class II, Group E Class II, Group F* Class II, Group G	↑ More easily ignited
Fibres & flyings			Class III	
Methane	Mining	Group I	Unclassified	

* Canada only

Table 1—Gas classification

2. Equipment is classified according to the maximum surface temperature that can be produced under fault conditions at an ambient temperature of 40°C, Table 2. The user must ensure that the temperature class of the equipment is below the ignition temperature of any gas-air mixture that may be present.

Temperature classification	Maximum surface temperature	Desirability
T1	450°C	Better equipment ↓
T2	300°C	
T3	200°C	
T4	135°C	
T5	100°C	
T6	85°C	

Table 2—Temperature classification

3. Hazardous areas are classified according to the probability that an explosive atmosphere will be present: Table 3 shows the two main systems. Equipment must be designed and certified to an appropriate standard of reliability. In Europe there are two standards: Ex ia for Zone 0 and Ex ib for Zone 1. In the USA there is just a single standard, for Division 1 locations.

IEC countries	USA & Canada
Zone 0: explosive gas-air mixture continuously present, or present for long periods Zone 1: explosive gas-air mixture is likely to occur in normal operation	Division 1: hazardous concentrations of flammable gases or vapours continuously, intermittently or periodically present under normal operating conditions
Zone 2: explosive gas-air mixture not likely to occur and, if it occurs, it will exist only for a short time	Division 2: volatile flammable liquids or flammable gases present, but normally confined within closed containers or systems, from which they can escape only under abnormal operating or fault conditions

Table 3—Area classification

Advantages

Unlike all other methods of explosion protection, intrinsic safety is a low-energy technique, which does not rely on physical barriers between the electrical circuits and the explosive atmosphere. It is therefore much safer since ignition cannot occur if cables are cut, or seals fail, or the covers of enclosures are replaced improperly. It is safer for personnel as well, since the low voltages employed (typically less than 30V) cannot harm them. Intrinsically safe systems are also much lower in cost, since lightweight enclosures can be used instead of the heavy flameproof variety, 'simple' field apparatus such as thermocouples and switches can be to ordinary (weatherproof) specifications, and ordinary wiring can replace expensive armoured cable. They have the additional advantage that they can be maintained and calibrated 'live' without first closing down a section of the plant and switching off the control loop. Last, intrinsic safety is the only explosion protection technique that can be used in Zone 0 (high risk) hazardous areas. It is this combination of advantages that has led to the widespread use of intrinsic safety for process control thoughout the world in recent years.

Certification

In most countries, responsibility for the safety of a plant falls on the plant owner, who must convince the local factory inspector that his plant is safe: this is much easier if the equipment is formally certified. Many countries now have their own national certifying authorities, and there is a gradual harmonisation of the rules they work to. Nearly all countries follow the International Electrotechnical Commission Standard IEC 79–11: 1976 and 1984. Most countries in Europe are members of the European Committee for Electrotechnical Standardisation (CENELEC) and now certify equipment to CENELEC standards. Even the USA, which for a long time went its own way, is increasingly following European methods. Table 4 lists the main national affiliations.

Country	Voted for IEC 79-11 1976	IEC79-11 1984	CENELEC member country	National CENELEC standards published	EEC member country	EEC approved test house	National certifying authority
Australia	√	√					SAA
Austria	√		√				ETVA
Belgium	√	√	√	√	√	√	INIEX
Brazil		√					
Canada	√	√					CSA
China		√					
Denmark	√	√	√	√	√	√	DEMKO
Egypt		√					
Finland	√	√	√	√			
France	√	√	√	√	√	√	LCIE & CERCHAR
Germany, W	√	√	√	√	√	√	PTB & BVS
Greece			√		√		
Hungary	√	√					
Ireland			√		√		
Israel	√	√					
Italy	√	√	√	√	√	√	CESI
Japan	√	√					RIIS
Korea	√						
Luxembourg			√		√		
Netherlands	√	√	√	√	√		
Norway	√	√	√	√			NEMKO
Poland	√						
Portugal	√		√		Due 85		
Rumania	√						
S. Africa	√						SABS
Spain		√	√	Allocated	Due 85		LOM
Sweden	√	√	√	EN 50 020			SP
Switzerland	√	√	√	√			SEV
Turkey	√						
USA	Against	√					FM & UL
USSR	√						
UK	√	√	√	√	√	√	BASEEFA
Yugoslavia	√	√					S. Comm.

Table 4—National affiliations

References

1. Hutcheon, IC, Wisitex Symposium, Bombay (1981). MTL Reprint TP1051
2. 'A user's guide to intrinsic safety'. Application Note AN9003, MTL, Luton, England.
3. 'List of all MTL Technical Papers and Application Notes'. MTL, Luton, England.

ON AN ADAPTIVE MODEL FOLLOWING CONTROL SYSTEM FOR A PLANT WITH DEAD TIME

K. Inoue and S. Tamura

Department of Electrical Engineering, Ritsumeikan University, Kyoto, Japan

Abstract. This paper deals with an adaptive model following control system. In model reference adaptive control systems, a system which has adaptive action by signal composition is called an adaptive model following control system. Also, such a control system is called a passive adaptive control system. In these systems, an adaptive compensator having inverse characteristics of the model is approximately realized using an adaptive observer. In this paper, an adaptive model following control system using a combination of a Smith predictor and a passive adaptive control loop is proposed and discussed for systems with dead time and varying plant parameters. Stability and adaptation of these systems are discussed. Moreover, simulation results obtained for some examples are shown.

Keywords. Adaptive control; predictive control; self-adjusting systems; time lag systems; time-varying systems.

INTRODUCTION

Model reference adaptive control systems have attracted considerable interest recently and have been discussed in many publications since the paper by Landau (1974). Within the general framework of model reference adaptive control, a system that achieves adaptive effects by the composition of the plant input signal is called model following control or passive adaptive control (Tyler, 1964; Morse, 1973; Nishimura 1966). These systems have a simple structure and fast adaptivity compared to the more usual adaptive parameter adjusting systems. But these systems are not expected to provide adaptive or compensated effects for plants with large lags or dead time because of a lag in the modificatory signal. This paper describes the principle of an adaptive model following control system which uses the inverse function of the model as a compensator. As a method applicable to plants containing dead time, a combined system with a Smith predictor and a passive adaptive control loop is proposed. Stability of this system and adaptation to parameter variations of the plant are discussed. This paper deals with single-input single-output systems, but it seems that this study is extendible to multivariable systems.

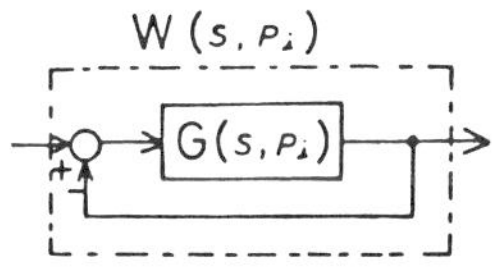

Fig.1. A feedback control system.

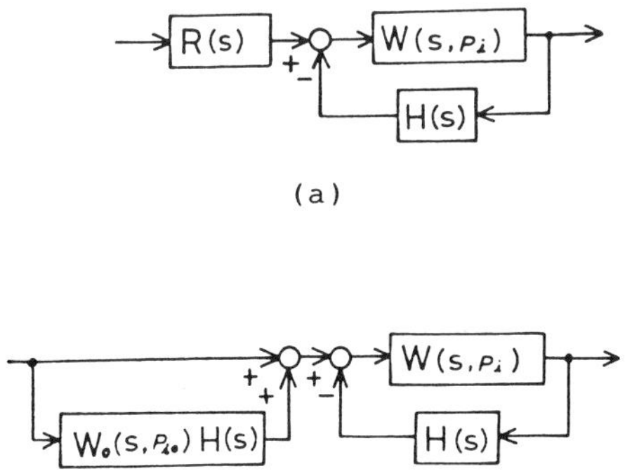

(b)

Fig.2. An insensitive control system.

AN ADAPTIVE MODEL FOLLOWING CONTROL SYSTEM USING THE INVERSE FUNCTION OF A MODEL AS A COMPENSATOR

Constitution of the System

Fundamentally, a feedback control system is expected to exhibit small parameter sensitivity. In order to achieve small parameter sensitivity, we consider a system shown in Fig. 2(a) which has more design flexibility than the system shown in Fig. 1.

The overall transfer function $W^*(s)$ of the system shown in Fig.2.(a) is written as

$$W^*(s)= \frac{W(s)R(s)}{1+W(s)H(s)} \tag{1}$$

When the system shown in Fig.1. is equivalent to the system of Fig.2.(a), $R(s)$ must be as follows.

$$R(s)=1+W(s)H(s) \tag{2}$$

In Fig.2.(b), $W_0(s,p_{i0})$ is $W(s,p_i)$ before parameter varying. When $W_0(s)$ is $W_m(s)$, Fig.2.(b) is equivalent to Fig.3. . The transfer function of the system shown in Fig.3. is written as follows.

$$W^*(s)= \frac{1+W_m(s)H(s)}{1+ W(s)H(s)} W(s) \tag{3}$$

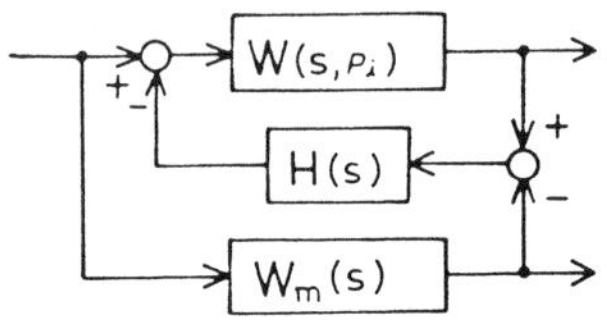

Fig.3. A model following control system.

The parameter sensitivity of $W^*(s)$ for p_i is as follows.

$$S_{pi}^{W*}= S_W^{W*}\cdot S_{pi}^{W} = \frac{\partial \ln W^*(s)}{\partial \ln W(s)} S_{pi}^{W}$$

$$= \frac{1}{1+W(s)H(s)} S_{pi}^{W} \tag{4}$$

Namely the parameter sensitivity of the system shown in Fig.3. for p_i becomes to $1/(1+W(s)H(s))$ compared with the system shown in Fig.1. . If a compensator $H(s)$ is selected as Eq.(5), parameter sensitivity becomes as Eq.(6) in $p_i= p_{i0}$ ($W(s)=W_m(s)$).

$$H(s)= \beta W_m(s)^{-1} \tag{5}$$

$$S_W^{W*} = \frac{1}{1+W(s)H(s)}$$

$$= \frac{1}{1+W_m(s)\ \beta/W_m(s)} = \frac{1}{1+\beta} \tag{6}$$

That is, the sensitivity reduces to $1/(1+\beta)$. In the system shown in Fig.3., it is able to consider that $W(s)$ is a controlled system and $W_m(s)$ is a kind of a model to follow. In Eq.(3), if $H(s)$ is settled to satisfied Eq.(7), characteristics of the system coincidence to one of the model.

$$\| W_m(s)H(s)\| \gg 1$$
$$\| W(s)H(s) \| \gg 1 \tag{7}$$

If $H(s)$ is an inverse function of the model as Eq.(8) and gain β is increased, characteristics of the system approximate to these of the model as Eq.(9).

$$H(s)= \beta W_m(s)^{-1}, \qquad \beta \gg 1 \tag{8}$$

$$W^*(s)= \frac{1 + \beta}{W_m(s)/W(s) + \beta} W_m(s) \tag{9}$$

Consideration by Root Locus

In the system shown in Fig.3. , assume that for both of $W(s)$ and $W_m(s)$, orders of denominator and numerator in s are n and m respectively, and orders of denominator and numerator of $H(s)$ are k and ℓ respectively. That is

$$W(s)= K \prod_{j=1}^{m} (s-z_j)/ \prod_{i=1}^{n} (s-p_i) \tag{10}$$

$$W_m(s)= K_1 \prod_{j=1}^{m} (s-z_j')/ \prod_{i=1}^{n} (s-p_i') \tag{11}$$

$$H(s)= \beta \prod_{h=1}^{\ell} (s-z_h)/ \prod_{h=1}^{k} (s-p_h) \tag{12}$$

Where $n \geq m$ and $n+k \geq m+\ell$. Then we consider how poles and zero points of the control system vary by constitution of Fig.3. .

In case that $H(s)$ is expressed by Eq.(12). $W^*(s)$ is written as follows using Eq.(3), (10), (11), and (12).

$$W^*(s)$$
$$= \frac{\Pi(s-p_i')\ \Pi(s-p_h)+\beta K_1\Pi(s-z_j')\ \Pi(s-z_h)}{\Pi(s-p_i)\ \Pi(s-p_h)\ +\beta K\ \Pi(s-z_j)\ \ \Pi(s-z_h)}$$
$$\cdot K \frac{\Pi(s-z_j)}{\Pi(s-p_i')} \tag{13}$$

Then poles and zero points of $W^*(s)$ are expressed as follows.

(a) Poles of $W^*(s)$ are roots satisfying $1+W(s)H(s)=0$ and poles of $W_m(s)$. ((2n+k)th order in s)
(b) Zero points of $W^*(s)$ are roots satisfying $1 + W_m(s)H(s)=0$ and zero points of $W(s)$. ((n+m+k)th order in s)

Fig.4. shows the location of poles and zero points of $W^*(s)$ when β of $H(s)$ increases in case of $W(s)$, $W_m(s)$ and $H(s)$ having Eq.(14), (15) and (16) respectively.

$$W(s)= \frac{p_1\ p_2}{(s-p_1)(s-p_2)} \tag{14}$$

$$W_m(s)= \frac{p_1'\ p_2'}{(s-p_1')(s-p_2')} \tag{15}$$

$$H(s)= \frac{\beta(s-z_h)}{(s-p_h)} \tag{16}$$

In Fig.4.(c), poles starting from p_1 (p_2) and zero points starting from p_1' (p_2') become dipoles, similarly a pole starting from p_h becomes dipole with a zero point starting from p_h. Therefore when β increases, the characteristics of the system are dominated by poles (p_1', p_2') of the model.

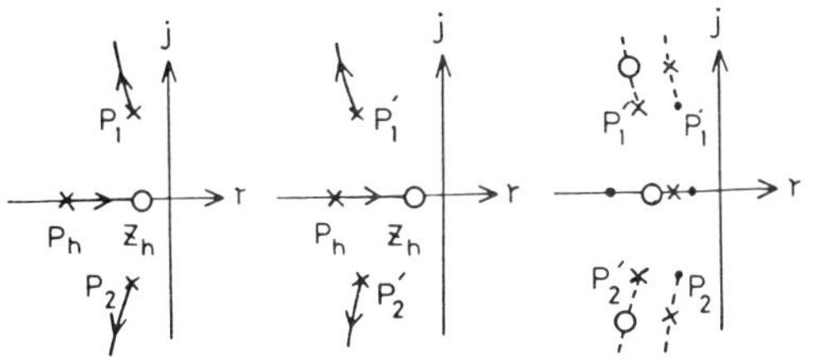

(a) Movement of poles (b) Movement of zero points (c) Movement of poles and zero points

Fig.4. Root loci for the system with Eq.(14), (15), and (16).

<u>In case that H(s) is inverse function of $W_m(s)$.</u>

In case that W(s), $W_m(s)$ and H(s) are given as Eq.(10), (11) and (8) respectively. Then W*(s) is written as Eq.(17) from Eq.(9).

$$W^*(s) = \frac{(1+\beta)KK_1 \Pi(s-z_j)\ \Pi(s-z_j')}{K_1\ \Pi(s-p_i)\ \Pi(s-z_j')+\beta K\ \Pi(s-p_i')\ \Pi(s-z_j)} \qquad (17)$$

Namely, poles and zero points of W*(s) are expressed as follows.

(a) Poles of W*(s) are roots satisfying $1+\beta W(s)/W_m(s)=0$ ((n+m)th order in s).

(b) Zero points of W*(s) are zero points of W(s) and $W_m(s)$ (2m th order in s).

Fig.5. shows the location of poles and zero points of W*(s) for increasing β in case of Eq.(14), (15) and (8). That is, poles of W*(s) approach to poles of $W_m(s)$ starting from poles of W(s) for increasing β. Where β becomes infinitely large, poles of W*(s) coincidence to poles of the model.

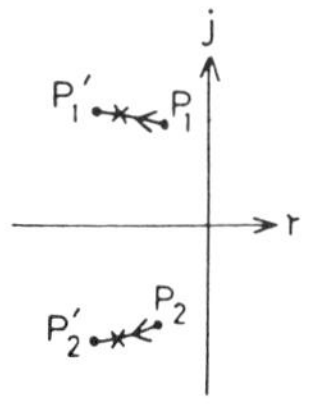

Fig.5. Root loci for the system with inverse function of the model.

REALIZATION OF A COMPENSATOR WITH INVERSE TRANSFER FUNCTION BY ADAPTIVE OBSERVER

In former session, it was shown that when an inverse transfer function of the model is used as a compensator and loop gain is increased, characteristics of the system coincide to the model. Then we consider realization of an inverse function by use of an adaptive observer. Consider that a system is expressed by Eq.(18).

$$\dot{x}=Ax+Bu \qquad y=Cx \qquad (18)$$

And the system has state feedback as follows.

$$u=Kx+\xi v \qquad (19)$$

From Eq.(18) and (19)

$$\dot{x}=(A+KB)x + \xi v \qquad (20)$$

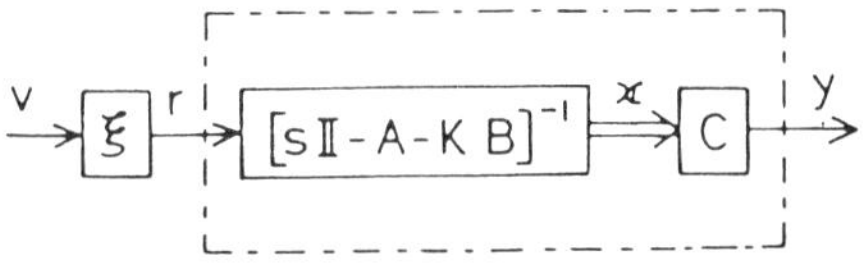

Fig.6. A control system having state feedback

Consider that the relation of y and r in Fig.6. is given as Eq.(21).

$$y^{(n)}+a_{n-1}y^{(n-1)}+ \cdots +a_1y^{(1)}+a_0y = b_m r^{(m)}+b_{m-1}r^{(m-1)}+ \cdots +b_1r^{(1)}+b_0r \qquad (21)$$

where, a_i(i=0,1,2,---,n-1) and b_j(j=0,1,2,---,m) are expressed as follows using p_i and z_j in Eq.(10).

$$\begin{cases} a_0=(-1)^n \prod_{i=1}^{n} p_i \\ a_1= -a_0 \sum_{i=1}^{n} \frac{1}{p_i} \\ \vdots \\ a_{n-1}= -\sum_{i=1}^{n} p_i \end{cases} \qquad \begin{cases} b_0=(-1)^m K \prod_{j=1}^{m} z_j \\ b_1= -Kb_0 \sum_{j=1}^{m} \frac{1}{z_j} \\ \vdots \\ b_m= K \end{cases}$$

Expressing Eq.(21) by use of state variables,

$$\begin{bmatrix} \dot{x}_1 \\ \dot{x}_2 \\ \vdots \\ \dot{x}_n \end{bmatrix} = \begin{bmatrix} 0 & 1 & 0 & & 0 \\ 0 & 0 & 1 & & 0 \\ & & & & \\ -a_0 & -a_1 & & & -a_{n-1} \end{bmatrix} \begin{bmatrix} x_1 \\ x_2 \\ \vdots \\ x_n \end{bmatrix} + \begin{bmatrix} 0 \\ 0 \\ \\ \alpha_{n-m} \\ \alpha_n \end{bmatrix} r$$

$$y = \begin{bmatrix} 1 & 0 & 0 & \cdots & 0 \end{bmatrix} \begin{bmatrix} x_1 & x_2 & \cdots & x_n \end{bmatrix}^T \qquad (22)$$

Where

$$\begin{cases} \alpha_{n-m} = b_m \\ \alpha_{n-m+1} = b_{m-1} - a_{n-1}\,\alpha_{n-m} \\ \vdots \\ \alpha_{n-1} = b_1 - a_{n-m+1}\,\alpha_{n-m} - \cdots \\ \qquad - a_{n-1}\,\alpha_{n-2} \\ \alpha_n = b_0 - a_{n-m}\,\alpha_{n-m} - a_{n-m+1}\,\alpha_{n-m+1} - \cdots \\ \qquad - a_{n-1}\,\alpha_{n-1} \end{cases}$$

Now, assume that the plant and the model have no zeros, and that the order of the model is same that of a plant in s, and that state equation of the model is expressed same as Eq.(22) whose parameters are $a_i'(i=0,1,2,---,n-1)$ and α_j' $(j=n-m,n-m+1,---,n)$. From the state equation of the model using Eq.(22), differential of x_n of the model is written as follows.

$$\dot{x}_n = -\sum_{i=1}^{n} a_{i-1}'\, x_i + \alpha_n' r \tag{23}$$

Therefore

$$r = (\dot{x}_n + \sum_{i=1}^{n} a_{i-1}'\, x_i) \,/\, \alpha_n' \tag{24}$$

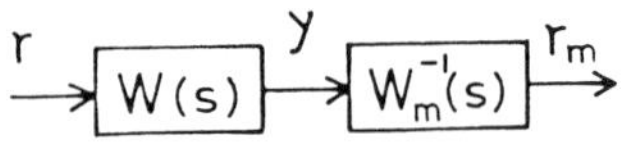

Fig.7. Expression of inverse function

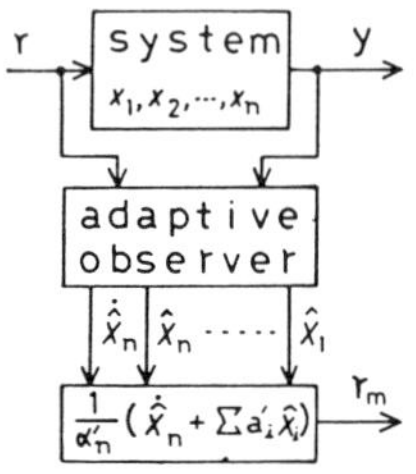

Fig.8. Realization of inverse function using an adaptive observer

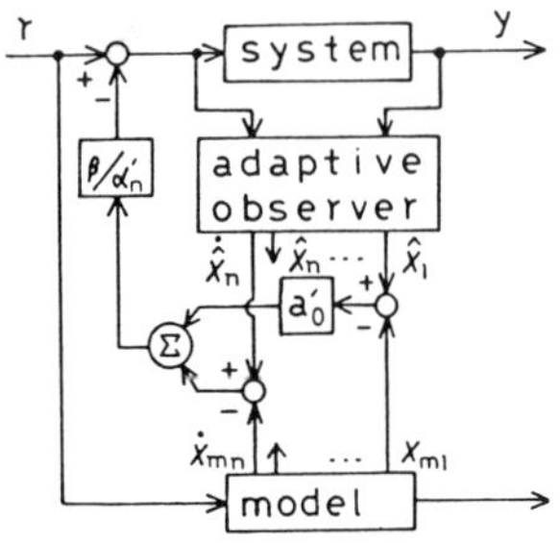

Fig.9. Constitution of an adaptive model following control system with inverse function of the model

Then the system shown in Fig.8. is equivalent to Fig.7. . Therefore an adaptive model following control system with inverse function as a dynamic compensator is constituted as shown in Fig.9. . Where, assume that the input signal of the plant and the adaptive observer is uniformly bounded and sufficiently rich.

In Fig.8., $\dot{\hat{x}}_n$ is given to differentiate $\hat{x}_n$ with respect to t. When $\dot{\hat{x}}_n$ is taken by Eq.(23), compensating effects of the system is expected for parameter variations, but is not gotten for disturbance.

APPLICATION FOR SYSTEM WITH DEAD TIME

Combined System with Smith Predictor Method

In order to synthesize a feedback control system with a plant having pure dead time, Smith predictor method was proposed (Smith, 1958). The method is used a controller incorporating a model of the plant as shown in Fig.10. .

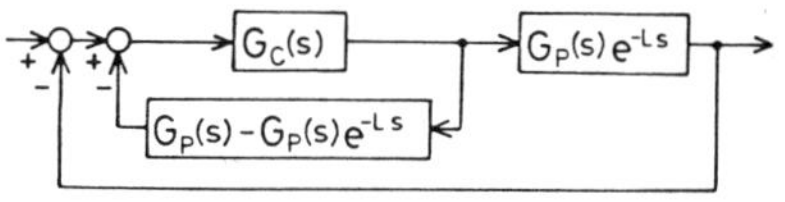

Fig.10. A control system using Smith predictor method.

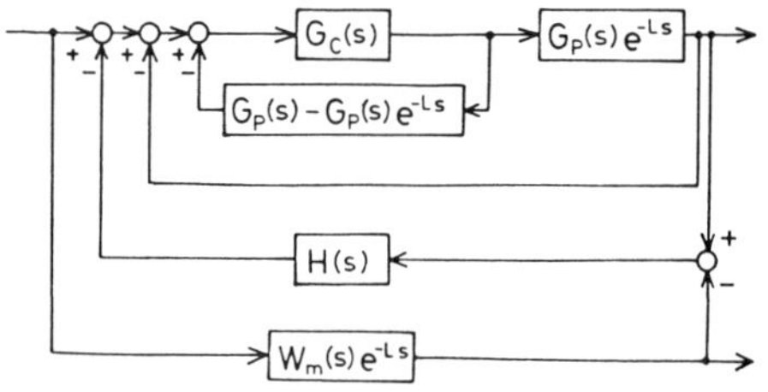

Fig.11. Application of a model following control system for Smith predictor method

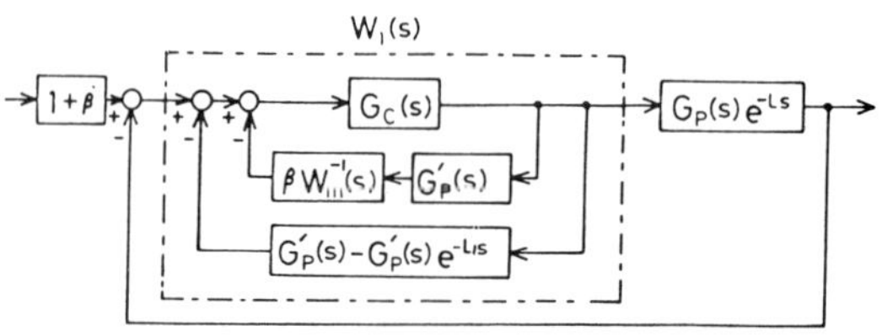

Fig.12. A model following control system using Smith predictor method

If transfer function of the plant is identified and used a exact model of the plant, the overall transfer function of the system shown in Fig.10. is written as follows.

$$W(s)=\frac{G_c(s)G_p(s)}{1+G_c(s)G_p(s)}\,e^{-Ls} \qquad (25)$$

Namely, the design of the system is easy because dead time of the plant is pushed out of feedback loop seemingly. Then we apply a passive adaptive control to the system which is used Smith predictor method as shown in Fig.11. .

When the compensator H(s) with the inverse transfer function of reference model is used, H(s) is as follows.

$$H(s)=W_m^{-1}(s)\,e^{Ls} \qquad (26)$$

Then the system shown in Fig.11. is equivalent to the system shown in Fig.12. (where $G_p'(s)=G_p(s)$). Where, it is assumed that $G_p(s)$ is stable regardless the parameter variation. The overall transfer function of the system shown in Fig.11. is as follows.

$$W^*(s)=\frac{1+\beta}{1+\beta W_m(s)^{-1}G_c(s)G_p(s)/(1+G_c(s)G_p(s))}\cdot\frac{G_c(s)G_p(s)}{1+G_c(s)G_p(s)}\,e^{-Ls}$$

$$=\frac{1+\beta}{1+\beta W_m(s)^{-1}W(s)}\,W(s)e^{-Ls} \qquad (27)$$

From Eq.(27), transfer function of the system approaches to that of the model for increasing β.

Discussion of Stability

Respect to stability of a system with Smith method, some papers have been reported (Sawano, 1962; Ioanides, 1979; Hang, 1980). Then we discuss stability of the adaptive model following control system applying Smith method as shown in Fig.12. . When transfer function of the plant with dead time is known exactly, as mentioned at former session, the stability of the system is same as not having dead time. But generally, it is difficult that the transfer function of the plant measured accurately, therefore $G_p(s)e^{-Ls}$ and $G_p'(s)e^{-L s}$ in Fig.12. are different. $W_1(s)$ in Fig.12. is written as follows.

$$W_1(s)=\frac{G_c(s)}{1+\beta W_m(s)^{-1}G_c(s)G_p'(s)+G_c(s)G_p'(s)(1-e^{-L_1 s})} \qquad (28)$$

Therefore overall transfer function is as follows.

$$W^*(s)=\frac{(1+\beta)W_1(s)G_p(s)e^{-Ls}}{1+W_1(s)G_p(s)e^{-Ls}} \qquad (29)$$

From Eq.(29), poles of the system are roots of equation whose denominator of Eq.(29) equals zero. Involving dead time in denominator, generally the system becomes unstable for increasing loop gain.

As an example, consider the stable region of the system which has transfer function as follows.

$$\begin{cases} G_p(s)e^{-Ls}=\dfrac{K}{s}e^{-Ls} \\ G_p'(s)e^{-L_1 s}=\dfrac{K_1}{s}e^{-L_1 s}, \quad W_m(s)=\dfrac{K_m}{s+K_m} \\ G_c(s)=1, \quad K_m=K_1 \end{cases} \qquad (30)$$

Fig.13.(a) and (b) show stable regions in case of $L=L_1$ and $K=K_1$ in Eq.(30) respectively.

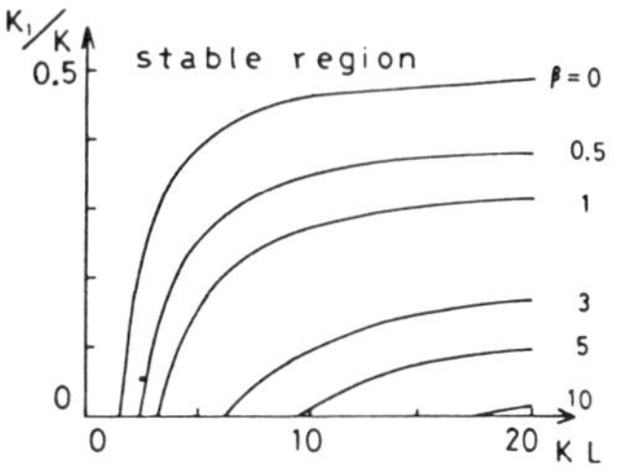

(a) In case of Eq.(30) and $L=L_1$

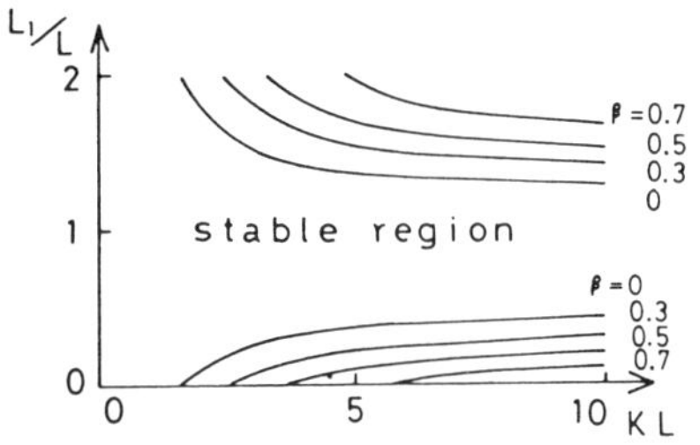

(b) In case of Eq.(30) and $K=K_1$

Fig.13. Stability region of the model following control system.

From Fig.13.(a), it is found that the system using only Smith method ($\beta=0$) is stable for about $K_1/K>0.5$ when dead time is not vary, but using model following control system, the stable region is spread for increasing β.

From Fig.13.(b), in case that dead time varys, the system using Smith method is stable for about $0.5<L_1/L<1.3$, but in this system, stability region is expanded for increasing β.

As examples of these system, experimental results by computer simulation are shown as follows. Fig.14. and Fig.15.(a),(b) show indicial responses of the system shown in Fig.12. which has transfer function of Eq.(31) and Eq.(32) respectively.

$$\begin{cases} G_p(s)e^{-Ls} = G_p'(s)e^{-L_1 s} \\ = \dfrac{1}{s(s+0.5)} e^{-0.14s} \\ W_m(s) = \dfrac{10}{(s+2)(s+5)}, \\ G_c(s) = 1 , \quad H(s) = \beta W_m(s)^{-1} \end{cases} \tag{31}$$

$$\begin{cases} G_p(s)e^{-Ls} = \dfrac{1}{s} e^{-s} \\ G_p'(s)e^{-L_1 s} = \dfrac{1}{s} e^{-L_1 s} \\ W_m(s) = \dfrac{1}{s+1} , \quad G_c(s)=1 , \quad H(s)=\beta W_m(s)^{-1} \end{cases} \tag{32}$$

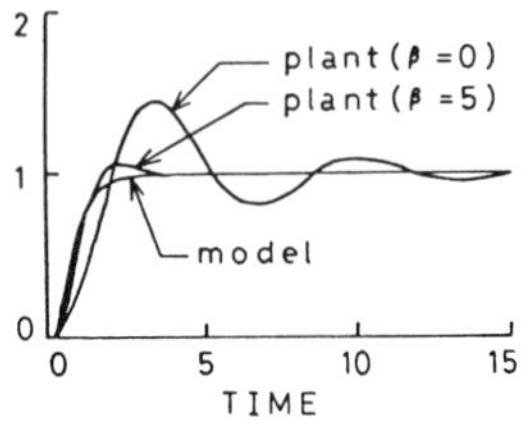

Fig.14. Indicial response of the system having Eq.(31) in Fig.12. .

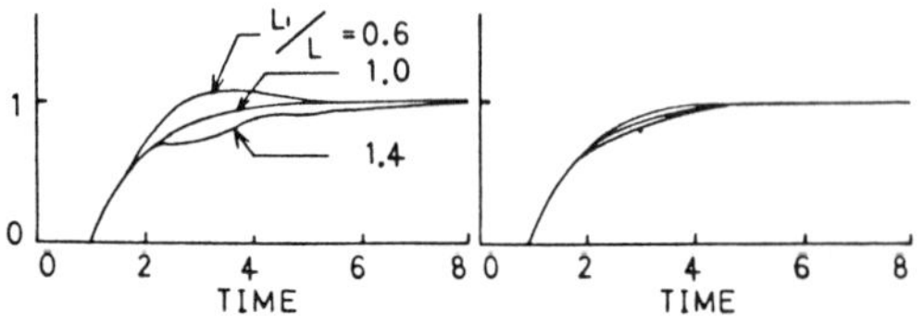

(a) In case of β=0 (b) In case of β=5

Fig.15. Indicial response of the system having Eq.(32) in Fig.12. .

CONCLUSION

With respect to an adaptive model following control system using inverse function of the model, constitution of the system, aspect of model matching and realization of inverse function were considered. And for plants containing dead time, a combined system with Smith predictor method and model following system was proposed. This combined control system is expected compensating effects even case of plant parameter varying. And it is found that stability region for parameter variation is expanded than Smith method. Such a control system is a kind of high gain feedback systems having model following function. The problem of this system is that state variable of the plant must be observed using adaptive observer which has restriction for input signal. And model following effects for disturbance are the subject of a current investigation by the authors.

REFERENCES

Landau, I.D. (1974). A Survey of Model Reference Adaptive Techniques. Automatica, 10, 353-379.

Tyler, J.S. (1964). The Characteristics of Model Following Systems as Synthesized by Optimal Control. IEEE Trans., AC-9, 485-498.

Morse, A.S. (1973). Structure and Design of Linear Model Following Systems. IEEE Trans., AC-18, 346-353.

Nishimura, M., K. Fujii, and K. Inoue (1966). Synthesis of Passive Adaptive Control Systems. J. of the Japan Assoc. of Automatic Control Engineers. 10, 304-311.

Smith, O.J.M. (1958). Feedback Control Systems. McGraw-Hill.

Sawano, S. (1962). Analog Study of Process Model Control Systems. J.Soc. of Instrument and Control Engineers of Japan. 1, 198-203.

Ioanides, A.C., G.J. Rogers, and V.Latham (1979). Stability Limits of a Smith Controller in Simple Systems containing a Time Delay. Int. J. Control. 29, 557-563.

Hang, C.C., C.H. Tan, and W.P. Chan (1980). A Performance Study of Control Systems with Dead Time. IEEE. Trans., IECI-27, 234-241.

HIGH-PRECISION, FAST-RESPONSE MICROCOMPUTER-BASED SPEED REGULATOR MSR-1

Zhong Ye Zhu, Gao Feng and Hu Jiang

Beijing Research Institute of Automation for Machine-Building Industry, Ministry of Machine-Building Industry, Beijing, China

Abstract. The principle and construction of a high-precision fast-response microcomputer based speed regulator M1, recently developed in our Institute, is described in this paper. Though MSR-1 has been developed for a hydraulic speed-regulating system, it can also be used in other speed-regulating systems. The MSR-1 regulator is of modular construction and is easy to use with a 12-bit or 16-bit digital speed-detecting unit, according to the precision required. With a 12-bit digital speed-detecting unit, the precision is about 0.5%.

Keywords. Hydraulic system; digital speed-detecting unit; speed control; computer control; modular construction.

INTRODUCTION

With the progress of four modernizations in China there is an increasing demand for high quality automatic control equipment. For example, the generally adopted analog speed regulators often can not meet the requirements of new equipment; it may be necessary to use digital speed regulators, particularly those based on microcomputers.

Hydraulic drives have been used for a long time, yet their rapid development took place only after the Second World War, especially in the last two to three decades. Hydraulic systems, due to their fast response, good speed-load characteristics, small size and light weight, have found an increasing use in military applications; they have also wide applications in industry, e.g., in the speed-regulating system of steam turbines and hydraulic turbines.

The microcomputer-based speed regulator MSR-1, described in this paper, has been developed for high-precision hydraulic drive systems. It can also be used in other speed-regulating systems.

SYSTEM ANALYSIS

The speed regulator MSR-1 is of modular construction and consists of a digital speed detecting unit, microcomputer, interface modules and an analog control part. Therefore, it is actually an analog-digital hybrid system. System analysis is performed on the Z-plane after sampling and discretization of the analog part.

The regulating signal of the system, through D/A conversion and signal and power amplification, is sent to the hydraulic servovalve and effects the control of speed of the hydraulic drive. Consider the electro-hydraulic system to be linear, and the time constant of the hydraulic drive to be much larger than that of the other inertial loops. Then the transfer function in the S-plane of this drive system may be expressed approximately as

$$W(S) = \frac{K}{S(S + 1/T_S)} \tag{1}$$

where K ——— the amplification of the hydraulic drive.

T_S ——— the equivalent time constant of this drive system, neglecting the small time constant of inertia loops.

The overall transfer function of the analog part in the S-plane will be

$$W_c(S) = K_c \frac{1 - e^{-T_c S}}{S} \cdot \frac{K}{S(S + 1/T_S)}$$

$$= K_S \frac{1 - e^{-T_c S}}{S^2(S + 1/T_S)} \tag{2}$$

where $K_c \dfrac{1 - e^{-T_c S}}{S}$ is the transfer function of D/A converter, and

K_c —— the amplification of D/A converter,

T_c —— the sampling period for microcomputer used.

$K_S = K_c \cdot K$

The transfer function of the analog part in the Z-plane will be

$$W_c(Z) = K_S \frac{T_c Z(Z - e^{-T_c/T_S}) + Z(Z-1)(e^{-T_c/T_S} - 1)}{(Z-1)\ (Z - e^{-T_c/T_S})} \tag{3}$$

If the sampling period for the microcomputer is very short, i.e., $T_c < T_S$, the transfer function of the analog part of the speed regulating system can be simplified as

$$W_c(Z) = K_S \frac{T_c Z}{Z - 1} \tag{4}$$

Suppose the speed detecting unit to be a first order inertia loop, and its transfer function F(S) to be Z-transformed as F(Z). It can be shown

$$F(Z) = \frac{C}{Z + K_F - 1} \quad (5)$$

where C — the preset constant

$$K_F = \frac{\Delta t}{\tau}$$

Δt and τ — the sampling period and the time constant of the speed detecting unit respectively.

If in this system, PI regulation is adopted (if necessary, PID regulation may also be adopted), according to PI algorithm, one obtains the transfer function in Z-plane of the correction loop

$$G(Z) = K_I \frac{Z - P/K_I}{Z - 1} \quad (6)$$

where P and K_I — the proportional factor and the integral factor of the correction loop respectively. Thus the block diagram of closed loop speed regulating system can be shown as Fig. 1.

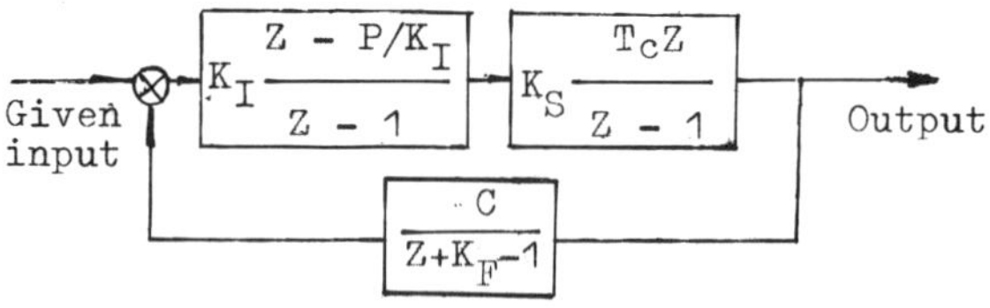

Fig. 1 Block diagram of speed regulator MSR-1

From Fig. 1, it is seen that the above closed loop system is approximately a typical second-order integrating loop plus a small inertia one. Such a system can have high precision at steady-speed. By adjusting parameters and proportionality factors, one can easily control the system transient, thus achieving satisfactory dynamic characteristics.

The above analysis shows that the key to achieve a high precision regulating system is to ensure high precision and small time constant of the speed detecting unit.

HIGH-PRECISION FAST-RESPONSE DIGITAL SPEED DETECTING UNIT

The method of speed detection generally used now is to count the number of pulses per unit time. To get high-precision and fast-response in detection, the pulse frequency of pulse transmitter must be very high, while the pulse frequency is limited by the dimension of photoelectric coded disk. Furthermore, the photoelectric coded disk can not work satisfactorily or even fails to work in polluted environment such as in dusty, vaporous atmosphere, or under seriously bad weather conditions. Therefore, we developed a new speed detecting unit using magnetic speed sensor.

In above analysis, transform first the equation (5) into recurrent form.

Let $\Delta t = 1/f_n$, where f_n is the signal frequency of the magnetic speed sensor. The following recurrent equation can be obtained:

$$F(K+1) = F(K) - \frac{F(K)}{\tau} \cdot \frac{1}{f_n} + C \quad (7)$$

where F(K) and F(K+1) — detected speed value at sampling time K and (K+1) respectively.

Take $\tau = \frac{M}{f_s}$, (8)

where f_s — the standard frequency selected,
M — the capacity of speed detecting unit.

Equation (7) then becomes

$$F(K+1) = F(K) - \frac{f_s}{M} \cdot F(K) \cdot \frac{1}{f_n} + C \quad (9)$$

This equation expresses the basic working principle of the speed detecting unit.

When the speed detecting unit enters into steady state operation, let F(K+1) = F(K) in equation (9); one obtains

$$F(K) = F = \frac{M}{f_s} \cdot C \cdot f_n = \frac{M}{f_s} \cdot C \cdot \frac{n}{60} \cdot N \quad (10)$$

where $f_n = \frac{n}{60} \cdot N$,
n — the shaft speed (rpm) to be detected,
N — the number of teeth of speed detecting gear wheel.

It can be seen from equation (10) that the reading of speed detecting unit F is directly proportional to the shaft speed n to be detected, and the precision is determined by the capacity M of the speed detecting unit. For example, if $M = 2^{12}$, the precision calculated theoretically is about 0.025%, while the actual precision measured is about 0.05%.

Equation (8) shows that the time constant of speed detecting unit is directly proportional to its capacity M and inversely proportional to the selected standard frequency. If we use 12-bit speed detecting unit, i.e. $M = 2^{12} = 4096$, and select standard frequency equal to 250 KC, then from equation (8), the time constant of the speed detecting unit will be

$$\tau = \frac{4096}{250000} \doteq 16 \text{ ms}$$

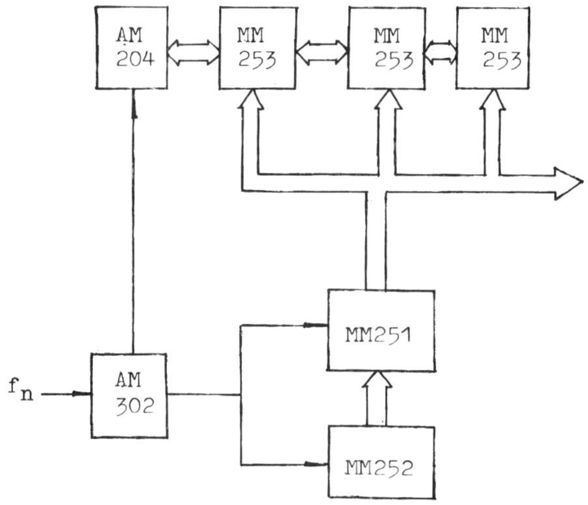

Fig. 2 Block diagram of the construction of speed detecting unit

Figure 2 is the block diagram to realize above principle of the recently developed speed detecting unit. In this diagram, each block is a module. AM 204 is the standard frequency generator module of 1000 KC, 500 KC and 250 KC, using quartz crystal oscillator ZWB-1 with temperature compensation as the source of standard frequency generation. The stability of frequency is $\pm 5 \times 10^{-6}$ in the temperature range from -40^oC to $+70^oC$. AM 302 is the shaping module to shape and transform the signal of the speed detecting unit and to produce the control signal together with the AM 204 module. MM 253 is the 4-bit frequency multiplier module, and three MM 253 modules construct jointly a 12-bit frequency multiplier for performing the operation of $f_s/M \cdot F(K)$ term in equation (9). MM 253 is the reversible counter module that completes the operation of $f_s/M \cdot F(K) \cdot 1/f_n$ term in equation (9). MM 251 is the adder module that completes the arithmetic adding operation of equation (9), its output being F(K + 1), or F as defined in steady state.

The test results of response characteristics to step input of the above-mentioned speed detecting unit are shown in Fig. 3.

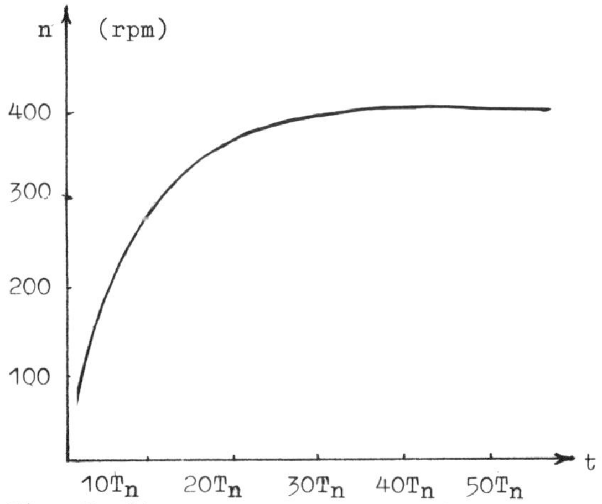

Fig. 3 Step response curve of speed detecting unit, $T_n = 1/f_n = 1.928$ ms

It is seen from Fig. 3 that the speed detecting unit can be regarded as a first order inertia loop with time constant 16.4 ms, conforming basically with the theoretically calculating value.

THE CONSTRUCTION AND OPERATION OF REGULATOR MSR-1

The regulator MSR-1 is made in modular form and consists of various functional modules. Each module is a plug-in circuit board. The main frame uses a single-board microcomputer TP-801. The data, address and control buses of the microcomputer and other relevant leads are led to each sub-rack of plug-in module board through two switching modules AM 502. All interfaces are made into standard plug-in modules. Settings of speed, speed raising or lowering time and overspeed limit are connected to above three buses through each own interface module MM 211. The detected speed is sent to three buses also by interface module MM 211. Enter the start command signal into microcomputer through dejittering module MM-401. The regulated value output given by the microcomputer is converted into analog value by 12-bit D/A conversion module MM 222. The analog control part consists generally of operational amplifier module AM 401 for adding operation and power amplifier module AM 101. The microcomputer has a self-checking program to ensure that it can be allowed to start only when its operation is normal. If each setting value is "reasonable", the microcomputer, on accepting the start (stop) command signal, will effect to increase (or decrease) the speed of hydraulic drive system according to the setting speed and the time requirement of raising (or lowering) the speed uniformly. As the setting speed is closely reached, the microcomputer will turn to work in PI mode (or PID mode) and give out the regulated value, thus performing the PI (or PID) closed loop regulation.

THE SYSTEM SOFTWARE DESIGN FOR REGULATOR MSR-1

The main factors considered in designing the system software are reliability, simplicity, economy and real time operation.

1. Program structure
The system software includes: Self-checking programs for CPU and RAM, PIO interrupt processing program, raising (or lowering) speed course CTC interrupt service program, steady-speed operation program and stop course CTC interrupt service program. In each program, there are some relevant subprograms. The whole set of programs occupies about 3K internal memory cell and is written in EPROM of 2 pieces of 2K bytes.

2. Program specification
In programming, two control modes of operation, i.e., the control of speed raising (or lowering) course and the control at steady-speed operation, are designed to meet the ordinary requirements of speed regulation. During the speed raising (or

lowering) course, the incremental type of regulation is adopted. The amount of increment is calculated from the setting speed and the time of speed raising (or lowering). At each CTC interrupt, CPU will make corresponding regulation to increase or decrease speed according to the polarity of speed deviation. When the absolute value of speed deviation is less than 1 rpm, the microcomputer will turn to work in steady-speed control mode. In this mode, CTC will interrupt at every 20 ms; CPU then responses, carries out sampling and performs the PI (or PID) operation for closed loop regulation in order to meet the requirement on steady-speed precision of the system.

The transfer of the above control modes of operation is brought about by modifying the interrupt vector of CTC.

CONCLUSIONS

1. The regulator MSR-1 uses microcomputer and its mode of operation is determined by programming. It can thus meet the requirements of different controlled objects.

2. The newly developed high-precision, fast-response digital speed detecting unit has been successfully used in this regulator. With a 12-bit speed detecting unit, the precision of the regulator is about 0.05% and the time constant is about 16 ms. If necessary, the precision of the regulator can be fuether increased by using a 16-bit speed detecting unit.

3. The newly developed interface modules for microcomputer are completely compatible with analog modules (modules of EMS system) developed by our Institute, hence they are applicable to different regulating systems, thus extending the field of application of regulator MSR-1.

4. Speed raising is smooth, the amount of overshoot is very small (even no overshoot at all). The mode of raising or lowering speed and the acceleration or deacceleration value may be selected at will.

5. Speed setting, the time of raising or lowering speed, overspeed limit and speed display are all digitized, thus it is easy to operate and monitor.

6. Fault self-diagnosing program is provided, thereby further improving the reliability of operation.

7. The regulator, made of modules and using standard hinged bay, is compact in construction, reliable and easy to maintain.

ACKNOWLEDGEMENT

This work had been done under the direction of Dr. Chi-yung Lin, Professor of our Research Institute. We are very grateful to him for his valuable suggestions, discussions and his help in preparing this paper.

REFERENCES

Takemura, Miura, Oyama (1972)"High Precision Speed Control and Position Control" YASUGAWA DENKI Vol. 36, No. 138

LINEARIZING CONTROL OF A CLASS OF NON-LINEAR CONTINUOUS PROCESSES

G. Gilles and N. Laggoune

Laboratoire d'Automatique, Université Lyon 1, 43 boulevard du 11 novembre 1918, F 69622 Villeurbanne Cedex, France

Abstract. Most of the continuous industrial processes working inside a large operating range can be modelled by non-linear state equations that are linear in control. A way for controlling those processes can be a closed loop linearization. It is shown that, when the solution exists, we get an explicit non-linear state feedback control law. The linearizability condition is demonstrated and it is shown that the control law may involve some singularities. Two practical applications on chemical and hydraulic pilot plants are presented.

Keywords. Nonlinear systems ; nonlinear control systems ; optimal control ; bilinear control ; hydraulic systems ; pH control.

NON-LINEAR MODELLING OF CONTINUOUS INDUSTRIAL PROCESSES

Most of the continuous processes (electrical, mechanical, hydraulic, thermal, chemical, biochemical, etc...) present in general a non-linear global behaviour. Only in some special working conditions, it is possible to approach their behaviour by linear models, mathematical abstraction then allowing to get simple properties and analytical developments. Since the variation magnitude of the descriptive variables is large, the use of linear models becomes totally unsuitable. In general, a knowledge mathematical model of a continuous process is got by writing infinitesimal balances of quantities like mass, energy, etc... :

$$\begin{Bmatrix}\text{storage inside}\\ \text{the volume of the system}\end{Bmatrix} = \begin{Bmatrix}\text{what enters}\\ \text{through the surface}\end{Bmatrix} - \begin{Bmatrix}\text{what goes out}\\ \text{the surface}\end{Bmatrix} + \begin{Bmatrix}\text{what is generated}\\ \text{inside the volume}\end{Bmatrix} - \begin{Bmatrix}\text{what is consumed}\\ \text{inside the volume}\end{Bmatrix}$$

that is to say :

$$d\underline{x} = \underline{f}_i(\underline{x},\underline{u})dt - \underline{f}_o(\underline{x})dt + \underline{g}_i(\underline{x})dt - \underline{g}_o(\underline{x})dt,$$

$\underline{x}$ denoting the state vector of the system (dimension n) and $\underline{u}$ being the control vector (dimension m).

The input vector modulates some terms, of the vectorial differential equation, mainly the input flux $\underline{f}_i$. Mostly, this modulation is expressed as a product of a "flow-rate" by the ratio of the element which is carried, that can be written under the following type of state equation :

$$\dot{\underline{x}} = \underline{f}(x) + \sum_{j=1}^{m} g_j(x).u_j \text{ , or}$$

$$\dot{\underline{x}} = \underline{f}(x) + G(\underline{x}).\underline{u} \qquad (1)$$

$G(\underline{x})$ being a nxm matrix non-linearly depending upon the state-vector $\underline{x}$ as well as the vector $\underline{f}$.

Such a non-linear state equation has the property to be linear with respect to the control vector. This quasi-general property is very interesting for designing control systems. In particular, it will be shown that, under certain conditions, an explicit non-linear control law can linearize those non-linear dynamic processes.

LINEARIZING CONTROL LAW

Introduction

Many recent works have been devoted to systems linear in control, also called "linear-analytic systems". Among the works related to their control, most of them are based on differential geometry and Lie algebra. In this field, let us specially mention the immersion theory of a non-linear system into another system by Claude, Fliess and Isidori (1983), and the model matching of non-linear systems by Di Benedetto and Isidori (1984). Recently, the idea of pseudo-linearization, local concept extended to a global area, has been developed by Mouyon, Champetier and Reboullet (1984). Extremely rare the papers dealing with concrete applications. Apart from this last work applied to an asynchronous motor control, let us mention the work of Alvarez Gallegos brothers to digital computer optimal control of fermentation processes (Alvarez Gallegos, 1982). Much more works are devoted to a special subclass of this type of systems : bilinear systems.

Finally, we can recall the work of Gilles (1984) which is extended here and which concerns the idea of linearization by considering only state equations. In the case where $\underline{f}(\underline{x}) = A\underline{x}$, Gilles (1983) showed that the linearizing control law minimizes a quadratic criterion. For the special subclass of bilinear processes, the authors (Gilles and Laggoune, 1985) derived a digital linearization by using a discrete model. After having recalled the linearization principles, the authors will emphasize on some pratical aspects of the use of the linearization control law. Applications to 2 pilot plants will be presented.

Principles

The control problem of systems linear with respect to the control vector (Eq. (1)) can be solved in order to get, by means of an appropriate feedback, a linear system whose dynamic properties are more well-known. Moreover, if such a control can be managed, the fact that the closed loop system belongs to the linear class allows to set dynamic performances which remain constant inside the whole operating range.

The closed loop system is linear if its error $\underline{\varepsilon}(t)$

satisfies a linear dynamic equation :

$$\dot{\underline{\varepsilon}}(t) = D\,\varepsilon(t) \quad (2)$$

D is a desired closed loop matrix whose eigenvalues guarantee the system stability and prescribe an arbitrary dynamics for the system.
$\underline{\varepsilon}(t)$ is an error vector between the state $\underline{x}$ and a reference on the state

$$\underline{\varepsilon}(t) = \underline{x}_c - \underline{x}(t) \quad (3)$$

or can be an error vector between the output $\underline{y}(p\times 1)$ and a reference input :

$$\underline{\varepsilon}(t) = \underline{y}_c - \underline{y}(t) \quad (4)$$

In the first case, the control system is said to be a strong linearization (between input and state). In the second one, it is called a weak linearization (between input and output). In order to simplify the presentation, this paragraph will be only devoted to strong linearization. In the case of regulation problems or step changing in the reference, the input $\underline{x}_c$ is constant, so :

$$\dot{\underline{\varepsilon}} = -\dot{\underline{x}} = -\underline{f}(\underline{x}) - G(\underline{x}).\underline{u} = D(\underline{x}_c - \underline{x})$$

or $$G(\underline{x}).\underline{u} = D(\underline{x} - \underline{x}_c) - \underline{f}(\underline{x}) \quad (5)$$

The algebraic set of equations (5), non-linearly depending on the state $\underline{x}$, is structurally linear with respect to the unknown $\underline{u}$. Then, it can be solved explicitly in order to get the expression of the control law. Nevertheless, as the solution does not always exist, its existence must be discussed.

Linearizability condition

Let us call r ={rank $G(\underline{x})$}.
If r = n = m, the Cramer system leads to the unique solution :

$$\underline{u} = G^{-1}(\underline{x})\left[D(\underline{x} - \underline{x}_c) - \underline{f}(\underline{x})\right]$$

If r = n < m, the system is undetermined of order m − r = m − n. Then, (m − n) control variables can be arbitrarily chosen, the n other control variables expressing themselves by means of an inversion of a submatrix (n × n) of $G(\underline{x})$.
If r < n, the system is undetermined of order (m − r) iff the (n − r) characteristic determinants are equal to zero. If this condition is not satisfied, the solution is impossible.

This leads to the system under the following form :

$$\begin{bmatrix} G'(\underline{x}) \\ G''(\underline{x}) \end{bmatrix} .\underline{u}' = D(\underline{x}-\underline{x}_c) - \underline{f}(\underline{x}) - \sum_{i=r+1}^{m} u_i \underline{g}_i(\underline{x})$$

$G'(\underline{x})$ being the square (r × r) submatrix of G, $G''(\underline{x})$ being another (n − r) × r submatrix and $\underline{u}'$ being a(r × 1) subvector of $\underline{u}$. By applying a linear transformation T on the rows, we can cancel the matrix $G''(\underline{x})$ and get the system :

$$\begin{bmatrix} G'(\underline{x}) \\ 0 \end{bmatrix} . \underline{u}' = \begin{bmatrix} \underline{\varphi}'(\underline{x}) \\ \underline{\varphi}''(\underline{x}) \end{bmatrix} \begin{matrix})r \\ )n-r \end{matrix} \quad (6)$$

The system being solvable iff $\underline{\varphi}''(\underline{x}) = 0$, we deduce:
Theorem : A system linear in control defined by Eq. (1) is closed loop linearizable in the sense of Eq. (5) iff, after the application of a linear transformation T leading to Eq.(6), the second member vector $\underline{\varphi}(\underline{x})$ is such that :

$$\underline{\varphi}''(\underline{x}) = 0 \quad (7)$$

This linearizability condition leads to the following remark. If $\underline{\varphi}''(\underline{x})$ does not involve non-linear elements, $\underline{\varphi}''(\underline{x})$ can be cancelled by setting some elements of the D matrix. But a sufficient non-linearizability condition is that, after applying the linear transformation T, $\underline{\varphi}(\underline{x})$ involves more than r non-linear elements.

When the linearizability condition is satisfied, we get the explicit control law :

$$\underset{(r\times 1)}{\underline{u}'} = \underset{(r\times r)}{\left[G'(\underline{x})\right]^{-1}} . \underset{(r\times 1)}{\underline{\varphi}'(\underline{x})} \quad (8)$$

with $$\underline{\varphi}(\underline{x}) = D'(\underline{x} - \underline{x}_c) - \underline{f}'(\underline{x}) - \sum_{i=r+1}^{m} u_i \underline{g}_i'(\underline{x}) \quad (9)$$

Some elements of D'' are, in general, set in order to cancel $\underline{\varphi}''(\underline{x})$. The remaining elements of D are chosen in order to set the eigenvalues, which is close to modal control.

Existence of singularities

The control law involving a matrix inversion, the control vector norm may become infinite for the values of $\underline{x}$ such that

$$\det\{G'(\underline{x})\} = 0 \quad (10)$$

So, a system which is linear in control and which satisfies the linearizability condition is linearizable almost everywhere except in a finite number of singular points.

In order to get a robust control, it is necessary to detect tendencies towards a singularity and apply a temporary auxiliary control allowing to avoid the singularity.

Nevertheless, in fact, we must say that for some industrial applications, the singularities are not accessible. Moreover, the non-linear processes such that $G(\underline{x})$ = G = Cte present no singularities.

Optimality properties

Let us recall that, for the subclass of systems such that $\underline{f}(\underline{x}) = A\underline{x}$, it has been shown that the linearization control law minimizes a quadratic criterion :

$$J = \int_0^{+\infty} \{\underline{x}^T P\,\underline{x} + \left[G(\underline{x}).\underline{u}\right]^T R\left[G(\underline{x}).\underline{u}\right]\}\,dt \quad (11)$$

which is an extension of the well-known result for linear systems and where the 2 terms respectively represent a precision criterion and a control energy. This optimal control law is especially very useful for bilinear systems (Gilles, 1983).

Examples

Compartmental system (bilinear system followed by an integrator).
If $\underline{x}_c = 0$

$$\dot{\underline{x}} = \begin{pmatrix} 0 & 1 \\ 0 & a \end{pmatrix}\underline{x} + \begin{pmatrix} 0 \\ b+cx_2 \end{pmatrix} u = \begin{pmatrix} d_{11} & d_{12} \\ d_{21} & d_{22} \end{pmatrix} \underline{x} \quad \text{leads to :}$$

$$\begin{cases} ax_2 + (b + cx_2)u = d_{21}\,x_1 + d_{22}\,x_2 \\ x_2 = d_{11}\,x_1 + d_{12}\,x_2 \end{cases}$$

Obviously, $d_{11} = 0$, $d_{12} = 1$ and

$$u = \frac{d_{21}\,x_1 + (d_{22} - a)x_2}{b + cx_2} .$$

d_{21} and d_{22} are adjusted in order to set eigenvalues which are solutions of the equation $\lambda^2 - d_{22}\lambda - d_{21} = 0$.

Compartmental system (linear system followed by a bilinear system).
If $\underline{x}_c = 0$, x_2 being the output of the first system controlling the second one, the state equations :

$$\dot{x}_1 = a_1x_1 + b_1x_2 + c_1x_1x_2 = d_{11}x_1 + d_{12}x_2$$

$$\dot{x}_2 = a_2x_2 + b_2u = d_{21}x_1 + d_{22}x_2$$

show that

$$\underline{\varphi}''(\underline{x}) = (d_{11} - a_1)x_1 + (d_{12} - b_1)x_2 - c_1x_1x_2 \neq 0$$

$\forall\ \underline{x}$, and this system is not linearizable.

WEAK CLOSED LOOP LINEARIZATION

In the most important practical cases, a weak linearization from input to output is sufficient. In this case, we must consider Eq. (1) followed by the output equation that we will consider to be linear :

$$\underline{y} = C \cdot \underline{x} \quad (12)$$

Then Eq. (2) and (4) lead to :

$$\dot{\underline{\varepsilon}} = - \dot{\underline{y}} = - C\dot{\underline{x}} = - C.\underline{f}(\underline{x}) - C.G(\underline{x})u$$
$$= D(\underline{y}_c - \underline{y})$$

$$\text{or } C.G(\underline{x})\underline{u} = - C\ \underline{f}(\underline{x}) - D(\underline{y}_c - \underline{y}) \quad (13)$$

As in the previous paragraph devoted to strong linearization, we have to solve a set of linear equations of the same structure in order to find the control law. So, the same results can be applied : if the second (n × r) part of the second member of Eq.(13) transformed by T is zero, the linearization control law exists and explicitly expresses as :

$$\underline{u} = -\left[(C\ G(\underline{x}))'\right]^{-1}\left[(C\ \underline{f}(\underline{x}))' + D'(\underline{y}_c - \underline{y})\right] \quad (14)$$

which corresponds to the control structure shown on Fig. 1 which points out the 3 important parts : the non-linear pre-compensator $[(C\ G(\underline{x}))']^{-1}$, the non-linear state feedback and the closed loop gain matrix D' which sets the system performances.
Let us notice that the cases where r = m = p are very common. Then, D' = D and (C.G)' = C.G and a decoupling control can be achieved by taking a diagonal D matrix. Moreover, this kind of linearization also presents singularities and optimaly properties.

APPLICATION TO A HYDRAULIC PILOT PLANT

Presentation

Figure (2) shows a part of the hydraulic pilot plant used for level and temperature control. A tank (area S) is water supplied by the input flow-rate u ; the output flow-rate, denoted q(x), is a square root function of the level x in the tank. Thus, a mass balance leads to the following model without any disturbance :

$$\frac{dx}{dt} = \frac{1}{S}(u - q(x))$$
$$\text{and} \quad q(x) = k\sqrt{x} \quad (15)$$

By applying the linearization technique, we get the control law :

$$u = S.d(x - x_c) + q(x) \quad (16)$$

representing a combination of state (level) and flux (output flow-rate) feedbacks as shown in Fig. 3.

An analog realization of the control law becomes very simple because it only needs the use of gain factors and adders-substracters. This realization takes into account the instrumentation technology involving sensors and actuators in the 4-20 mA industrial standard.

Experimental results

Large operating range linearization. Fig. 4 shows the closed loop step responses of the level related to input magnitudes of 30 % and 55 % of the tank height, the desired dynamics being $d = -1/40\ sec^{-1}$ (Eq. 2). Those reference magnitudes make the plant working outside its linearity domain. From the tangent to the origin of the step response, we get the time constant $\tau = 1/d$ for which $x(\tau)$ is roughly 64 % of the final value in both cases ; so we can consider the system to be well linearized. Static errors can be observed ; they can be explained by modelling errors (instrumentation conversion factors). Other arbitrary dynamics can be set and similar results are got. Nevertheless, we cannot go over $d = -1/20\ sec^{-1}$ because then the initial control value reaches the maximum flow-rate delivered by the wide open valve.

Disturbances effect. Let us assume the existence of a leak acting as a disturbance $q_p \neq 0$. The state equation now becomes :

$$\dot{x} = \frac{1}{S}\left[u - q(x) - q_p\right] \quad (17)$$

Replacing u by its expression (Eq. 16), we get the new static level :

$$x_\infty = x_c + q_p/Sd \quad (18)$$

So, $x_\infty < x_c$ in the case of a disturbance. Fig. 5 shows this effect when $d = -1/20\ sec^{-1}$.
In order to reduce the influence of modelling errors and to cancel static errors, the use of a reference model and an integrator must be prescribed.

APPLICATION TO A CHEMICAL PILOT PLANT

Presentation

The pilot plant (Gilles and Laggoune, 1985 ; Neyran and co workers, 1984) (Fig. 6), is a continuous neutralization process of a strong base (NaOH, concentration C_B) by a strong acid (HCl, concentration C_A). The reactor volume (V = 6ℓ) is maintained constant by local regulation, as well as for the acid flow-rate (Q_A = 35 ℓ/h). The description variables are C (output concentration) and Q_b (base flow-rate). The monovariable state model is of bilinear type :

$$\dot{x} = a \cdot x + b\ u + n\ u \cdot x \quad (19)$$

where $u = Q_b(t) - Q_b(0)$ is the input variable deviation, $x = -c(t)/(C_B + C_o)$ is the state representing the normalized concentration in the reactor (C_o is the initial concentration) and c(t) is the output concentration deviation.

By applying Eq. (8), the linearizing control law is:

$$u = \frac{d(x - x_c) - ax}{b + n\ x} \quad (20)$$

This control law involves the singularity x = -b/n but we can easily show on the knowledge model that this corresponds to $c(t) = C_B$ (or the symmetrical $C(t) = C_A$) which is unaccessible.

As shown in Fig. 6, the state (normalized concentration) must be deduced by converting the Ph value into concentration what can be done by using an antilog analog chip. The desired non-linear control law (Eq. (20)) can be generated by adders/substracters, gain elements and a divider. We prefered using a digital computer, the control variable being holden during each small sampling period (2 sec.).

Experimental results

In the following, the constant parameters in equations (19) and (20) are : $a = -9.27\ h^{-1}$, $b = 0.17\ \ell^{-1}$ and $n = -0.185\ \ell^{-1}$. All results are given for an initial base flow-rate $Q_b(0)$ = 20 ℓ/h and the Ph excursion is limited to the acid field. The average open loop time constant is greater than 300 sec.

Reference changing. Figures 7 and 8 respectively show the evolution of the state x in case of an input reference $x_c = 0.1$ and $x_c = 0.15$. The dynamics is $d = -25h^{-1}$ which represents a closed-loop system approximately 2 times faster than the open-loop one. The results denote that the linearized system is slightly delayed (20 to 30 sec.) and the final value is also slightly different from the reference. The static error is due to the fact that the model is not exact (Gilles and Laggoune, 1985), and the use of integrators must improve the result.

The delay is simply due to the Ph measurement location (Fig. 6). The desired time constant is 144 sec. ($d = - 25\ h^{-1}$) and the experimental dynamics represents approximately a time constant of 128 sec. in Fig. 7 and 176 sec. in Fig. 8.

Thus, with regard to the modelling errors and the Ph meter precision, we can say that the results are very satisfactory : the relative errors are 13 % and 9 % for the final value, 10 % and 22 % for the dynamics respectively for $x_c = 0.1$ (Fig. 7) and $x_c = 0.15$ (Fig. 8).

Fig. 9 shows the linearizing control variable evolution in the both former cases.

Arbitrary dynamics. Let us assume that the desired time constant for the closed-loop linearized system is 80 sec. (4 times faster than the open-loop). When choosing $d = - 45\ h^{-1}$, Fig. 10 shows the state evolution. The experimental time constant is approximately 118 sec. which cannot be considered as satisfactory (48 % relative error between the desired and the obtained values).

This illustrates the limits of the method in the presence of an important noise and of non negligeable modelling errors. When a time delay is not small enough with respect to the desired time constant, it has to be taken into account in the model in order to build a new control law.

CONCLUSION AND PROSPECTS

It has been shown that, under certain conditions, continuous processes that are linear with respect to the control vector can be linearized, inside its whole operating range, by means of a non-linear state feedback control law. Two applications on different monovariable pilot plants have shown the interesting possibilities presented by the linearizing method. Assigning a linear behaviour to the closed loop system simplifies further studies on the system among a more complex plant, guarantees its stability and leads to a same dynamic behaviour inside the whole operating range. Simple for low order processes, this type of control avoids designing an adaptive control with respect to the operating point variation. Moreover, as the linearizing control law is still valid for non-linear processes involving measurable variable parameters, it would be very interesting to set up experiments on this point of view ; for example, on the neutralization pilot plant, it would be possible to take into account the measurable variations of the volume. The differences observed between the theoretical and practical results are mostly due to modelling errors. In order to reduce their influence, it would be useful to introduce a reference model into the linearizing control structure. For higher order systems, the non-linear state feedback control law needs an observation of the whole state vector : then it is necessary to extend the results got on observers for bilinear systems (Laggoune, 1984) over the class of systems linear in control. On multivariable processes, the most illustrative application would be the possibility of decoupling control ; this could be shown on the hydraulic pilot plant by using two tanks. By applying this type of control on different multivariable industrial processes, a deep analysis of the singularities could be managed in order to known their degree of concrete importance. Finally, on the realization point of view, we showed that we can use analog or digital technologies in the case of simple processes. For low order monovariable processes, the design of analog non-linear controllers may be preferable. In the case of high order multivariable processes, the pseudo-analog non-linear controllers have to be approached with a digital technology if the sampling period can be small enough, that is to say if the real-time computations are rapidly processed in a parallel way by means of multimicroprocessor structures.

ACKNOWLEDGMENTS

This research is supported by grant from Centre National de la Recherche Scientifique for a cooperation project between three teams on non-linear process control. The authors are grateful to this institution and to their colleagues J. Biston, B. Neyran and D. Thomasset for their fruitful help on the use of the pilot plants.

REFERENCES

Alvarez Gallegos, J. (1982). Optimal Control of a class of discrete multivariable non-linear systems. Application to a fermentation process, Journal of Dynamic Systems, Measurement and Control, vol 104, pp. 212-217.

Claude, D., M. Fliess and A. Isidori, (1983). Immersion directe et par bouclage d'un système non-linéaire dans un linéaire. Compte-rendu à l'Académie des Sciences, série 1, t. 296, pp. 237-240.

Di Benedetto, M. and A. Isidori, (1984). The matching of non-linear models via dynamic state feedback, research report, universita di Roma

Gilles, G., (1983). Une loi de commande optimale d'une classe de systèmes bilinéaires multivariables, Optimization days, Montreal, Canada.

Gilles, G., (1984). Commande optimale linéarisante d'une classe de systèmes non-linéaires continus, workshop on non-linear system theory, Grenoble, France.

Gilles, G. and N. Laggoune, (1985). Digital control of bilinear continuous processes. Application to a chemical pilot plant. 7th IFAC Symp. on Digital computer Applications to Process Control, Sept. 1985, Vienna.

Laggoune, N., (1984). Sur la discrétisation de l'observateur de Hara et Furuta. Internal note, Nov. 1984, Laboratoire d'Automatique de Lyon 1

Mouyon, P., C. Champetier, and C. Reboullet, (1984). Application d'une nouvelle méthode de commande des systèmes non-linéaires - la pseudo-linéarisation - à un exemple industriel, Sixth INRIA international conference on Control and Optimization, Nice, France.

Neyran, B., J. Saade-Castro, D. Thomasset and R. Reynaud, (1984). Modélisation dynamique non-linéaire d'un procédé chimique de neutralisation, Congrès S.E.E., Nice, France.

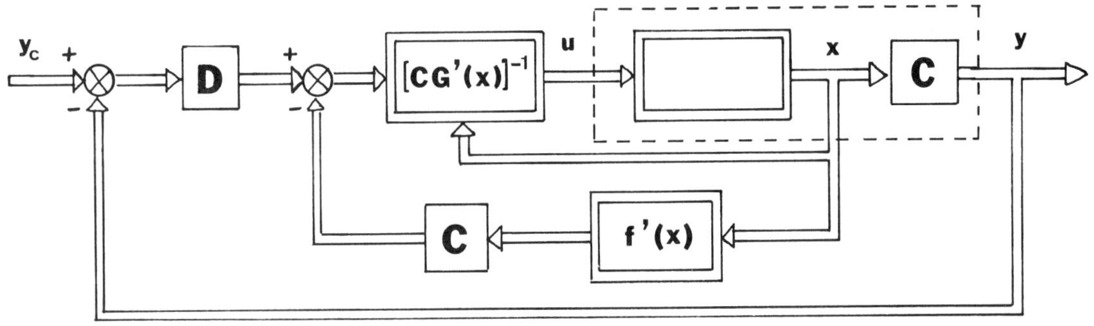

Fig. 1 Weak closed-loop linearization block-diagram

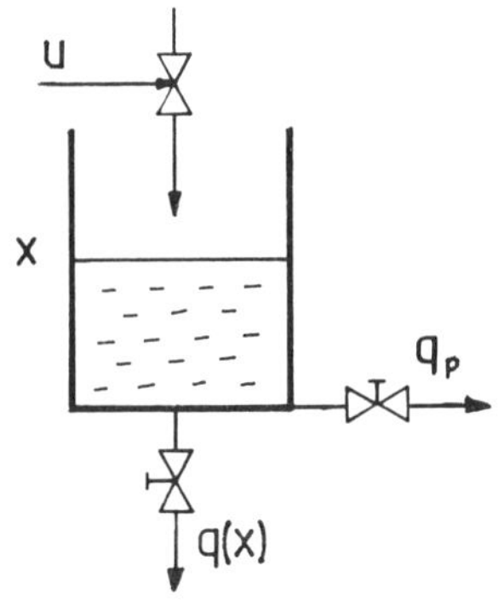

Fig. 2 Descriptive scheme of the tank

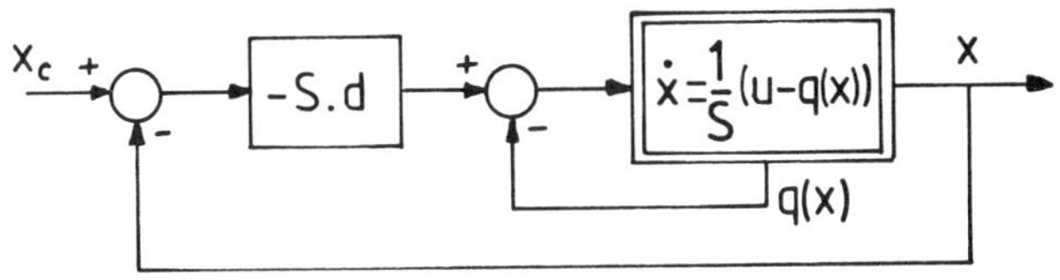

Fig. 3 Linearizing level control structure

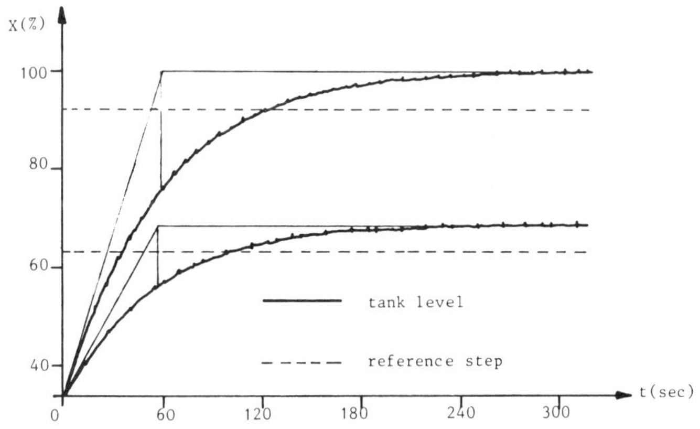

Fig. 4 Closed loop level step responses

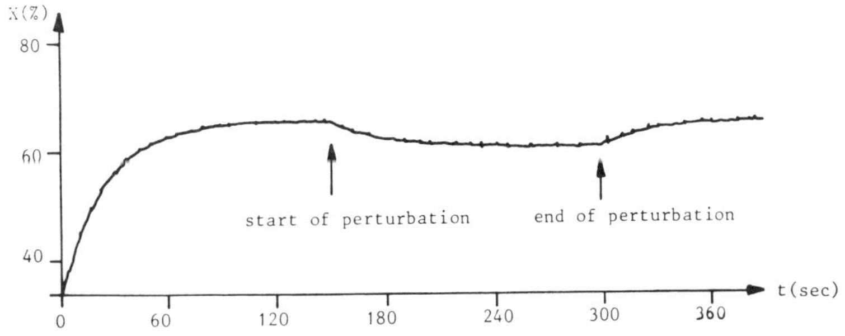

Fig. 5 Level regulation-perturbation effect

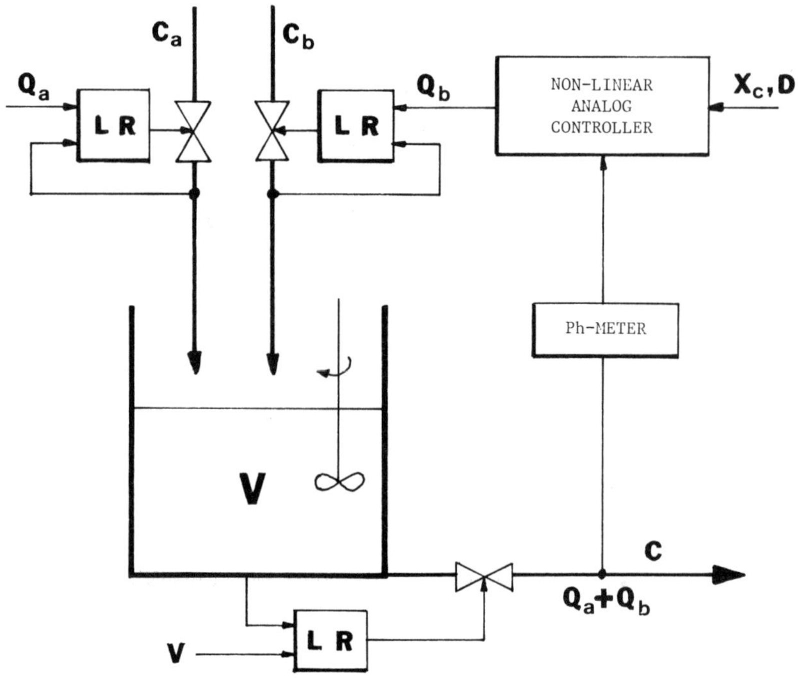

LR : Local Regulation

Fig. 6 Continuous neutralization pilot plant and control structure

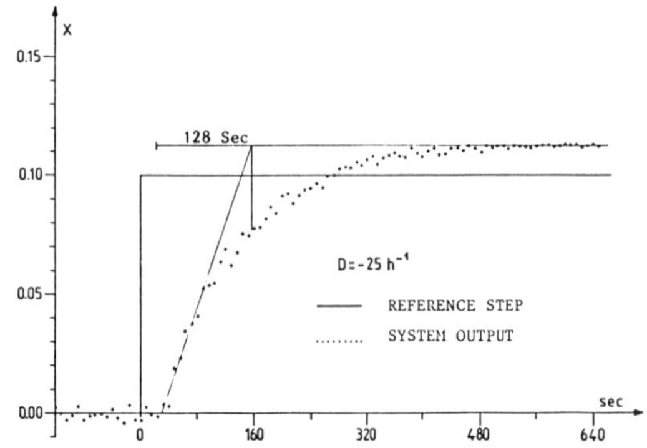

Fig. 7 Closed loop step response (reference : $x_c = 0.1$)

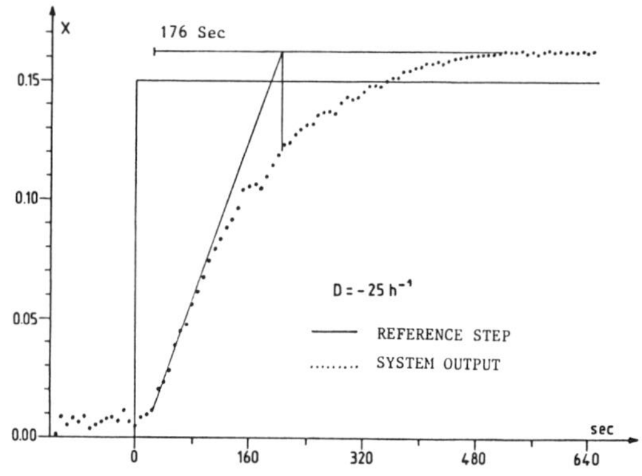

Fig. 8 Closed loop step response (reference : $x_c = 0.15$)

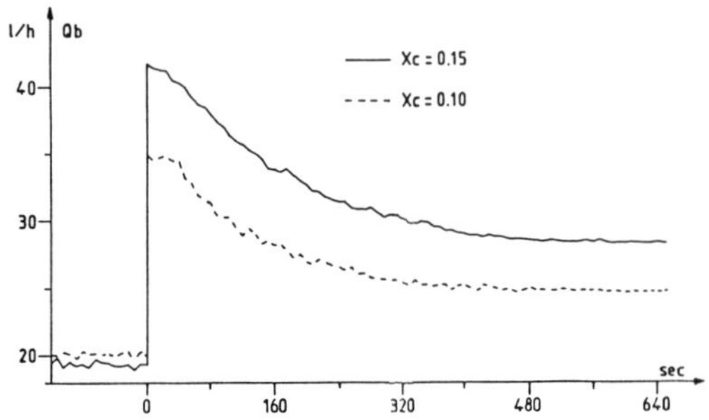

Fig. 9 Input variable in case of two reference steps ($D = -25h^{-1}$)

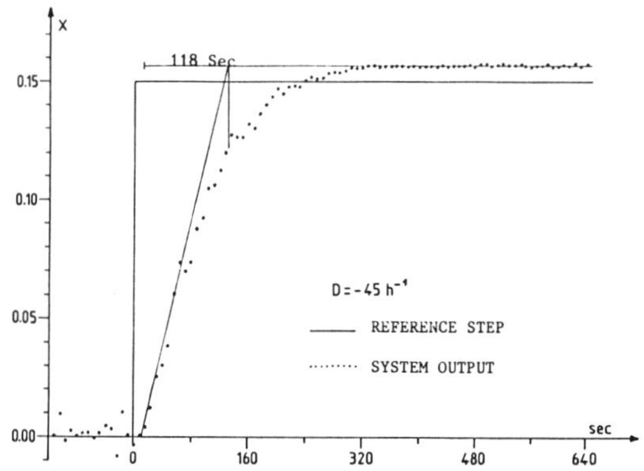

Fig. 10 Closed loop step response with a faster dynamics

MICROPROCESSOR-BASED FAILURE DETECTION OF HEAT PUMPS

R. Shoureshi* and K. McLaughlin**

*School of Mechanical Engineering, Purdue University, West Lafayette, IN 47907, USA
**TRW Space Technology Group, One Space Park, Redondo Beach, CA 90278, USA

Abstract. Thermofluid processes and system undergo spatial and temporal variations in both fluid and thermal properties. Therefore, they are highly nonlinear and coupled. This paper presents an approach to failure detection of a heat pump. The technique involves modeling of the system, design of an observer and statistical analysis for failure identification. This paper focuses on the first two major steps. A bond graph approach is used to model the heat pump and an extended linear observer technique is used to derive an optimal state estimator for the nonlinear thermofluid system. A Z-151 microcomputer is used to implement the observer and other required software. The results of the model and the observer are compared with experimental data and very good agreement have been obtained.

Keywords. Graph Theory; Refrigeration state estimation; observability Kalman Filter; state-space methods.

INTRODUCTION

Thermofluid processes and system undergo spatial and temporal variations in both fluid and thermal properties. The necessary coupling and mutual dependency of these two sets of equilibrium and transport properties make analysis of such systems far more complex. As processes have become more complex and performance more demanding, there has been an inexorable growth in more rational, quantitative approaches to system design and operation. Furthermore, thermofluid processes are mostly found in energy systems. With recent energy crisis more effort has been concentrated on dynamics and control of thermofluid processes and systems. In general, HVAC equipment are good examples of energy systems used in all sectors of society. Previous studies show the effects of the dynamic response of environmental control processes in industrial and residential buildings on energy savings. Recently, interest in using control strategies for failure detection and accommodation has significantly increased. One objective is to apply microprocessor technology to develop a time-history of failures that a system undergoes when a component fails. Such time-history records will be of great help to detect and locate the failed component(s) and ease the process of preventing future failures of a similar type.

As explained in the next section, the failure detection scheme used in this study composed of four main steps. This paper presents the results of the first two steps namely modeling and observer design for a heat pump.

FAILURE DETECTION

There are several failure detection techniques presented in the literature. They are mostly dealing with the state space representation of systems and assume the system can be linearly represented. The failure detection process used in the present study consists of the following steps:

1. A priori knowledge of the normal state of the system.

2. On-line process observation, identification, and estimation.

3. Statistical likelyhood analysis (Baysian approach) for fault detection based on a sequential test on the parameters variances.

4. Fault localization using the properties of the identified parameter variations.

Figure 1 shows a block diagram for implementation of the above scheme on a heat pump.

The first step dealing with the normal state of the system depends on a particular system or process. Usually manufacturing data, or a simple experiment on a non-fault system provides the normal state information.

A major step in failure detection is on-line process identification. Although the control literature is very rich in the design of an observer for various linear systems; however, there is no analytical method for handling nonlinear systems. This paper presents an approach to design of an observer for a nonlinear heat pump by using temperature-entropy bond graphs.

Figure 2 shows a schematic of a heat pump. A heat pump consists of two primary heat exchangers, evaporator and condenser, an expander, and a compressor. Estimation and on-line identification of the heat pump can be achieved in real time by derivation of an accurate model and a fast observer. The objective is to apply bond graph and derive a model for the system of Fig. 2. Authors in references [2,7,8,11 published in 1984 and 1985] have shown a true variable, temperature-entropy, bond graph can be applied to thermofluid processes and systems. In summary, the analysis of every thermofluid system begins with selection of a set of control volumes. In general each control volume has a fluid flow, with incompressible or variable density fluid, it has heat transfer and work transfer with its environment, and it has the inherent dependency between thermal and fluid energy domains. Figure 3 shows a control volume and a bond graph constructed for this control volume. Details about derivation of such bond graph is given by the authors in references.

The components included in the model are compressor, condenser, evaporator, and capillary tube. Four control volumes are considered for these four components. The bond graph of Fig. 3 is applied and simplified for each of these control volumes. Thereby a complete bond graph of the heat pump is developed and it is shown in Fig. 4. The inner loop of the bond graph represents the fluid energy domain, and the other loop represents the thermal domain. More complex and detailed models can be created and inserted for different components of this model (if necessary) as long as the causalities are preserved. As indicated by the integral causalities of the bond graph, the model is 6th order. Therefore, six nonlinear differential equations that describe dynamics of the heat pump can be obtained from the bond graph of Fig. 4 as listed below.

$$\dot{\lambda} = e_2 - e_1 - e_R \tag{1}$$

$$\dot{m}_{ev} = \dot{m}_2 - \dot{m}_1 \tag{2}$$

$$\dot{S}_1 = \frac{\dot{m}_2 h_2 - \dot{m}_1 h_1}{T_1} - \frac{\eta_1 \mu_1}{T_1} - \frac{T_1 - T_{p_1}}{R_{i_1} T_1} \tag{3}$$

$$\dot{S}_2 = \frac{\dot{m}_1 h_1 - \dot{m}_2 h_2}{T_2} - \frac{\dot{\eta}_2 \mu_2}{T_2} - \frac{T_2 - T_{p_2}}{R_{i_2} T_2} \tag{4}$$

$$\dot{S}_{p_1} = - \frac{T_1 - T_{p_1}}{R_{i_1} T_2} - \frac{T_{p_1} - T_{evap}}{R_{o_1} T_2} \tag{5}$$

$$\dot{S}_{p_2} = \frac{T_2 - T_{p_2}}{R_{i_2} T_2} - \frac{T_{p_2} - T_{cond}}{R_{o_2} T_2} \tag{6}$$

where h is enthalpy, λ is fluid momentum, $\dot{m}$ is mass flow rate, μ is chemical potential, R is the thermal resistance, and S is entropy.

One of the main difficulties in dealing with thermofluid systems is that there are no closed form constitutive laws between properties of the working fluid. In case of a temperature-entropy bond graph the capacitive field with integral causalities requires temperature and pressure as functions of specific volume and specific entropy. This means, in order to find temperature and pressure of the heat pump from equations (1) through (6) one requires the following functions:

$$T = T(v,s) \quad , \quad P = P(v,s)$$

A new technique for derivation of such function is described in the next section.

PIECEWISE CLOSED FORM METHOD

This method provides an estimate of temperature and pressure of a two phase flow given the specific entropy and specific volume. This method is very useful for real time failure detection applications because of the simplicity of the resulting equations. The complete details of the method is given in reference [14, McLaughlin 1984].

In an entropy-volume plane points with constant temperature can be shown to lie in a straight line. The slope of that line has a one to one correspondence to the temperature. It was found that for most refrigerants these constant temperature lines all intersect at almost a pivot point on the s-v plane. Figure 5 shows this plane. Then by derivation of that point (s_o, v_o), for any given point in the two phase region (s,v) the slope of the line connecting (s,v) point and the pivot point (s_o, v_o) can be calculated by

$$m = \frac{s - s_o}{v - v_o} \tag{7}$$

The temperature is defined uniquely by the slope m, so it is merely a matter of performing a least square fit of the temperature in terms of the slope to arrive at the temperature that corresponds to this slope.

In reality, the lines do not all intersect at the same point, but rather there is a small neighborhood within which all of the lines intersect. However, once the temperature region is identified it can be divided to smaller subregions and a pivot point for each subregion can be identified. The accuracy of this technique is so high that even in the extreme cases the error is less than one degree celcius.

MODEL VERIFICATION

Combination of the modeling results shown by equation (1) through (6) and the piecewise technique for on-line calculations of thermofluid properties of the working fluid in the heat pump forms the basis for a model of the heat pump. This model was implemented on a Z-151 microcomputer and the software required for integration of the nonlinear equations was developed. The software is in hybrid form, namely combines assembly instructions and BASIC, high level language programming format.

To check the accuracy of the model its time response was compared with actual data. Figure 6 shows the time response of the condenser and evaporator during two transient operations of the compressor, namely, during off-cycle and on-cycle. As shown, the agreement is very good. The model has a smaller rise time which is due to the use of one lump model for every component.

Similar agreements have been obtained between model predictions and experimental results for pressure variations and mass flow rate through each component. The comparison with experimental results verifies the accuracy of the model. Figure 7 shows the model predication after a compressor failure. It represents the temperature-entropy diagram of the heat pump. The area underneath of this diagram corresponds to the performance of the heat pump. This figure shows how the performance changes during the transient operation of working fluid caused by the compressor failure.

STATE RECONSTRUCTION

The task of an observer is to reconstruct the state variables of the system, as accurately as possible, based on a minimum number of measurements. The chosen measurements must satisfy the observability requirements. In case of thermofluid systems those measurements are preferred to be quantities that can be measured without having to break into the system. This is especially important in case of closed loop cycles such as a refrigerator or other HVAC equipment where breaking into the system results in a loss of working fluid and many leakage problems.

A schematic diagram for the general operation of an observer is shown in Fig. 8. The observer is located within the dashed lines. The state equations of the plant are given by equations (1) through (6). In general they can be written as:

$$\dot{\underline{X}} = \underline{f}(\underline{X},\underline{U}) \tag{8}$$

$$\underline{Y} = \underline{g}(\underline{X}) \tag{9}$$

where $\underline{X}$ is the state vector with its components being fluid momentum in the capillary tube, mass flow rate in the condenser, entropies in the evaporator and condenser, and entropies of wall materials in the evaporator and in the condenser.

State equations for the observer can be written as

$$\dot{\hat{\underline{X}}} = \underline{f}(\hat{\underline{X}},\underline{U}) + K(\underline{Y} - \hat{\underline{Y}}) \tag{10}$$

where K is the gain matrix, to be determined, and $\hat{X}$ and $\hat{Y}$ are estimate state and output vectors. The output vector (available measurements) are assumed to be exit temperatures of the condenser and evaporator. These have nonlinear relationships with the state vector.

In order to determine the entries of the gain matrix K, which is a 2 by 6 matrix, while using linear observer theory, the nonlinear state equations (8) must be linearized. In the linearized form the A matrix is simply the Jacobian matrix of $\underline{f}(\underline{X},\underline{U})$.

To calculate the feedback gains, the optimal observer problem was solved [4,5,6,14]. The following is assumed for the system

$$\underline{U} = \underline{U}' + \underline{V} \tag{11}$$

$$\underline{Y} = C\underline{X} + \underline{W} \tag{12}$$

This means the input vector ($\underline{U}$) has a deterministic component ($\underline{U}'$) and a stochastic component ($\underline{V}$). The output noise in measurements is shown by vector $\underline{W}$. Furthermore, it is assumed that $\underline{V}$ and $\underline{W}$ are stationary white noises with Gaussian distribution and zero mean values. Therefore, the correlation functions can be written as

$$E[\underline{V}(t)\underline{V}^T(t + \tau)] = V\delta(\tau) \tag{13}$$

$$E[\underline{W}(t)\underline{W}^T(t + \tau)] = W\delta(\tau) \tag{14}$$

where $\delta(\tau)$ is the dirac delta function, and V and W are noise powers.

The optimal feedback gains are obtained by solving the Riccati equation for the dual system, written as

$$AR + RA^T - RC^T WCR + BVB^T = 0 \tag{15}$$

and

$$K = RC^T W^{-1} \tag{16}$$

The relative magnitudes of the white noise intensities V and W determine the optimum balance between the speed of the state reconstruction and the immunity to measurement noise. Reference [14. McLaughlin 1984] describes a method for selection of V and W along the trajectory of operation of the heat pump. There will be a series of gain matrices obtained along the trajectory. Out of those the one which best guarantees the observer requirements during the whole transient operation should be selected. The criteria is to have stability from both dynamic sense and computational sense.

It was found that $\rho = 1000$ produces a desired closed loop solution. Furthermore, it was observed that although the optimal feedback gains vary along the trajectory, but the magnitudes of the entries in each of the gain matrices are similar and the signs do not change.

Based on this method the optimum gain matrix was determined and the observer was constructed. The results of the observer are compared with the actual measurements in Figure 9. As shown, the observer performs very well even though the observer starts with zero initial conditions while the heat pump has non-zero initial state.

CONCLUSION

A failure detection scheme for thermofluid processes was described from the four main steps in the detection process, the results of the first two steps, namely, modeling and states estimation were described. The results of these two steps will be used to compare the instantaneous state of the heat pump with the no-fault state and calculate the error. Based on the resulting error and statistical analysis such as Baysian likelyhood ratio a decision will be made for detection and localization of a failure. These steps are part of the current study and the results will be reported in future publications.

REFERENCES

1. D.R. Tree and M.F. McBride, "The Dynamic Response of Environmental Control Processes in Buildings," Purdue University, March 13-15, 1979.

2. R. Shoureshi, "Dynamic Analysis and Failure Detection of HVAC Systems Using Temperature-Entropy Bond Graphs," ASHRAE Workshop on HVAC Control Modeling and Simulation, Feb. 2-3, 1984, Atlanta.

3. D.F. Farris, "Energy Conservation by Adaptive Control for a Solar Heated Building," Proceedings of the International Conference on Cybernetics and Society, Sept. 1977.

4. M. Athans, P.L. Falb, "Optimal Control," McGraw-Hill, New York, 1966.

5. H. Kwakernaak, R. Sivan, "Linear Optimal Control Systems," John Wiley, New York 1972.

6. R. Shoureshi, "Theory and Design of Control Systems," Lecture Notes, ME 575, School of Mechanical Engineering, Purdue University, Fall 1983.

7. R. Shoureshi, K. McLaughlin, "Analytical and Experimental Investigation of Flow-Reversible Heat Exchangers Using Temperature-Entropy Bond Graphs," Journal of Dynamic Systems, Measurement, and Control, Vol. 106, June 1984.

8. R. Shoureshi, K. McLaughlin, "Modeling and Dynamics of Two-Phase Flow Heat Exchangers Using Temperature-Entropy Bond Graphs," Proceedings of 1984 American Control Conference, San Diego, California.

9. R.C. Rosenberg, D.C. Karnopp, "Introduction to Physical System Dynamics," McGraw-Hill, New York, 1983.

10. F.R. Hildebrand, "Advanced Calculus for Applications," 2nd Edition, 1976, Prentice-Hall publications.

11. R. Shoureshi, K. McLaughlin, "Application of Bond Graph to Thermofluid Processes and Systems," Proceedings of 1985 American Control Conference, Boston, MA, June 1985.

12. U. Tsach, "Failure Detection and Location Method Applied to a Simulated Fossil Power Plant," Proceedings of 1983 ACC, pp. 946-950.

13. A. N. Madiwale, B. Freidland, "Comparison of Innovations-Based Failure Detection and Correction Methods," Proceedings of 1983 ACC, pp. 940-945.

14. K. McLaughlin, "MSME Thesis, School of Mechanical Engineering, Purdue University, December 1984.

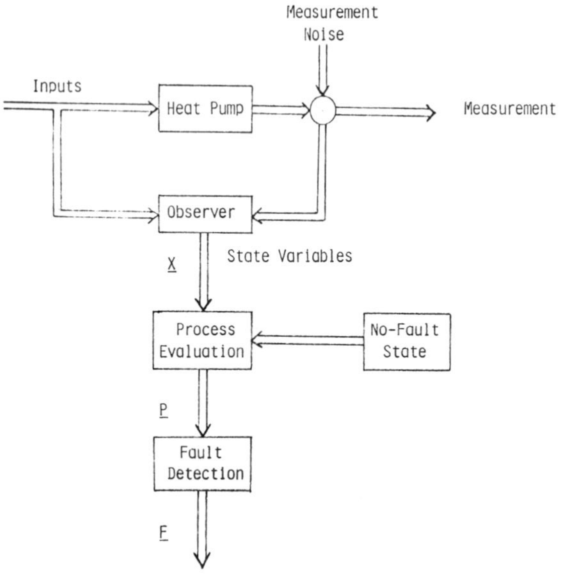

Fig. 1: Fault Detection Process

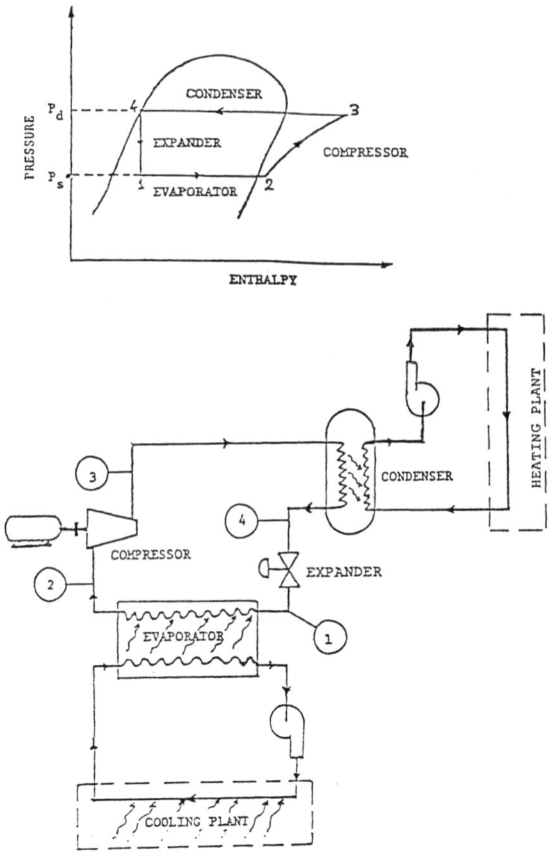

Fig. 2: Schematic and Pressure-Enthalpy Diagrams of a Heat Pump

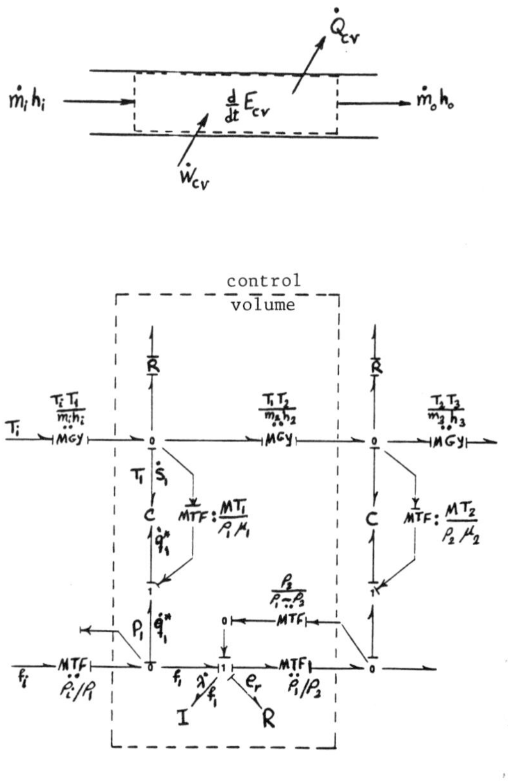

Fig. 3: Schematic of a General Control Volume and Corresponding Bond Graph.

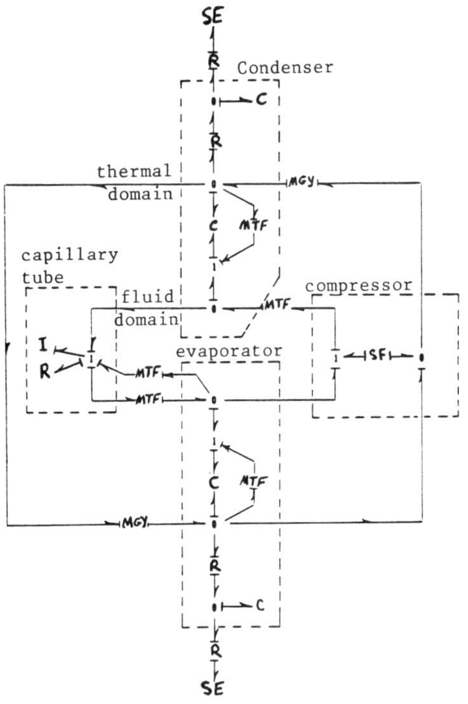

Fig. 4: Bond Graph of a Heat Pump

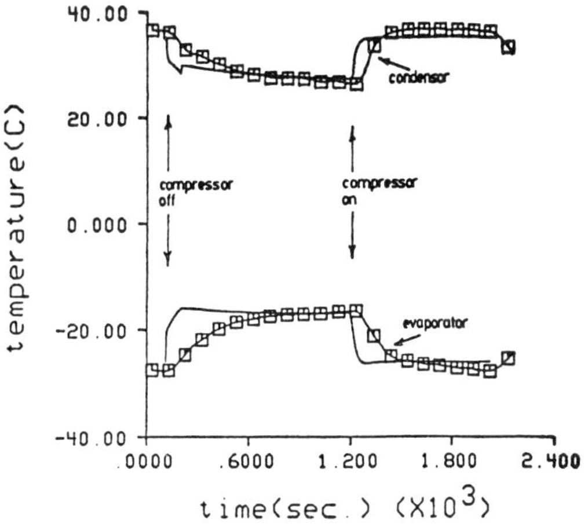

Fig. 6: Comparsion of Heat Pump Model Response and Experimental Data

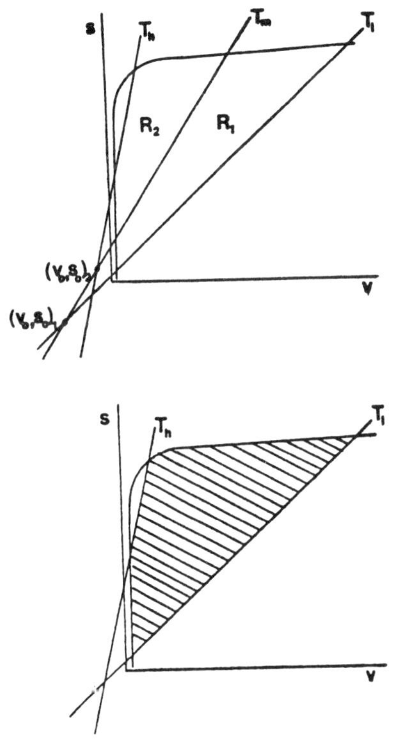

Fig. 5: Piecewise Closed Form Method for Evaluation of Temperature and Pressure

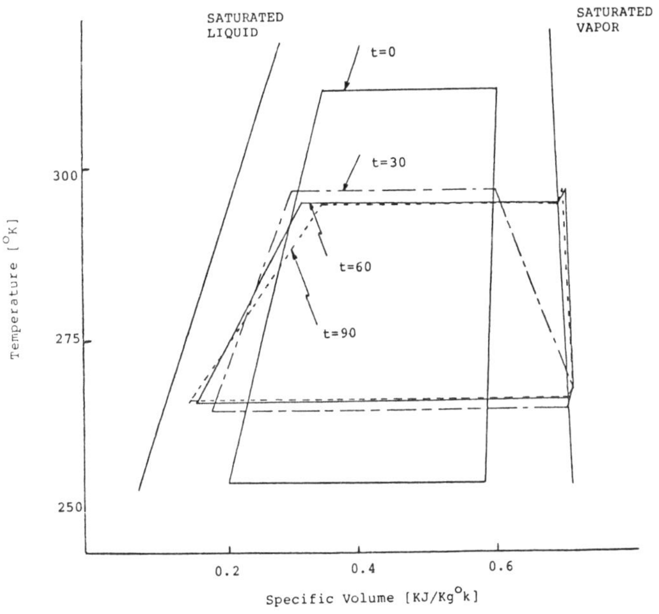

Fig. 7: Dynamics of T-S Diagram after Compressor Failure

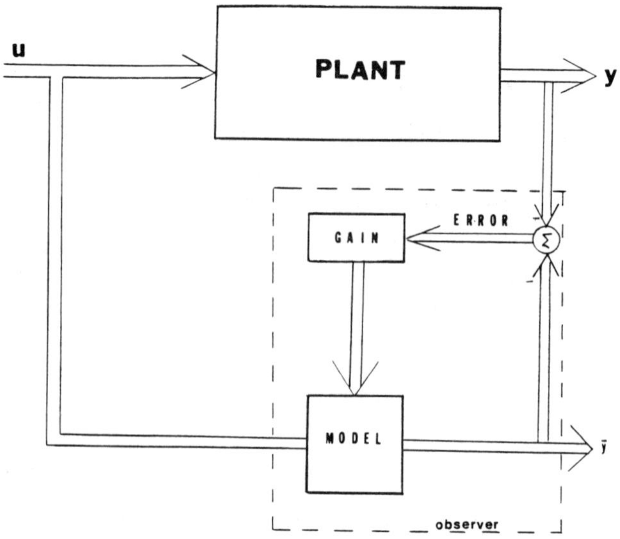

Fig. 8: Block Diagram of an Observer for a Nonlinear System

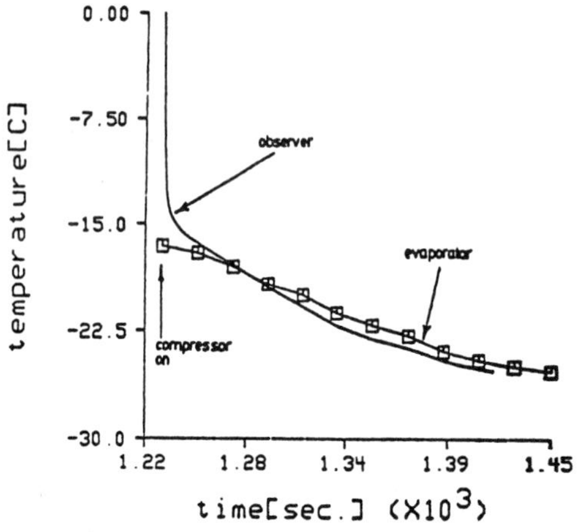

Fig. 9: Comparison of Observer Estimated Evaporator Temperature with Experimental Data.

TEMPERATURE SELF-TUNING CONTROL FOR A HEAT-VACUUM CABIN

Zheng Qin, Sun Demin, Peng Lixing, Yang Xueshan, Zhang Desong and Zhang Shanjian

Department of System and Management Science, University of Science and Technology of China, Hefei, Anhui, China

Abstract. This paper introduces a successful example of temperature self-tuning control for a heat-vacuum cabin. To obtain the desired process when the temperature increases and decreases and to fulfil the steady accuracy requirement of the temperature control, a new self-tuning control algorithm is proposed which combines the minimum variance control strategy for reducing noises and the pole assignment strategy for following reference input. The self-tuning controller has been implemented on a single board microprocessor. The experiment results showed the controller has excellent properties.

Keywords. Adaptive control; computer control; discrete-time systems; Heat systems; identification; optimal control.

1. INTRODUCTION

Self-tuning control has been greatly developed in the last more than ten years. In theory, many new algorithms were proposed and some of them were proved to have the convergence property. In applications, a number of successful applications were also reported. These applications showed the self-tuning controllers to have many excellent properties over the widely used PID controllers. This paper introduces a new self-tuning control algorithm first and then applies the algorithm to a heat-vacuum cabin temperature control system. The experiment results showed that the control accuracy of the temperature was over ± 0.1 °c which was one order superior than ± 1 °c required and the controller was adaptable.

The paper is organized as follows: A new self-tuning control algorithm is proposed in section 2. In section 3, the structure of the heat-vacuum cabin is described and the mathematical model is derived. Section 4 determines the control algorithm and simulates it on a digital computer. Section 5 presents the experiment results and analyzes the data obtained.

2. A NEW SELF-TUNING CONTROL ALGORITHM

In this section, a new self-tuning control algorithm is proposed which has the following features: (a) The algorithm combines the adantages of the pole-assignment self-tuning and the optimal self-tuning algorithms, i.e., it possesses both the good tracking property and noise reduction property. (b) Comparing with other pole-assignment self-tuning controllers, this controller needs less computation. (c) The difficulty of choosing the weighting factor λ in the optimal self-tuning controllers is overcome to some extent. (d) The stability and convergence results can be obtained in theory.

Suppose that the system to be controlled can be described by the following equation

$$AY(t) = q^{-k}BU(t) + Ce(t) \qquad (2.1)$$

where

$$A = 1 + a_1q^{-1} + \dots + a_{na}q^{-na}$$

$$B = b_0 + b_1q^{-1} + \dots + b_{nb}q^{-nb}$$

$$C = 1 + c_1q^{-1} + \dots + c_{nc}q^{-nc}$$

q^{-1} is the backward shift operator. A, B, C are stable polynomials. $b_0 \neq 0$. Y(t), U(t) are the output and control, respectively. e(t) is a zero-mean white noise process and is uncorrelative with the control up to time t-1.

If the following control strategy is applied to system (2.1):

$$U(t) = -\frac{G}{F}Y(t) + \frac{G_1}{F_1}Yr(t) \qquad (2.2)$$

where $Y_r(t)$ is the reference input. G, F, G_1, F_1 are polynomials of q^{-1} with proper orders.

Then the following closed-loop equations can be obtained:

$$Y(t) = \frac{q^{-k}BG_1F}{F_1(AF + q^{-k}BG)} Y_r(t) + \frac{CF}{AF + q^{-k}BG} e(t) \qquad (2.3)$$

$$U(t) = \frac{AFG_1}{F_1(AF + q^{-k}BG)} Y_r(t) - \frac{GC}{AF + q^{-k}BG} e(t) \qquad (2.4)$$

We separate Y(t) into two parts:

$$Y(t) = Y_1(t) + Y_2(t) \qquad (2.5)$$

where

$$Y_1(t) = \frac{q^{-k}BG_1F}{F_1(AF + q^{-k}BG)} Y_r(t) \qquad (2.6)$$

is the part tracking the reference input and

$$Y_2(t) = \frac{CF}{AF + q^{-k}BG} e(t) \qquad (2.7)$$

is the part caused by the noise.

Similarly, we separate U(t) into two parts

$$U(t) = U_1(t) + U_2(t) \qquad (2.8)$$

where

$$U_1(t) = \frac{AFG_1}{F_1(AF + q^{-k}BG)} Y_r(t) \qquad (2.9)$$

is the control needed to track the reference input and

$$U_2(t) = -\frac{GC}{AF + q^{-k}BG} e(t) \qquad (2.10)$$

is the control needed to reduce the disturbance of the noise.

In order to reduce the disturbance of the noise, $Y_2(t)$ should be as small as possible. So we choose the following cost function:

$$J = E\{(Y_2(t+k))^2 + (\bar{\lambda}U_2(t))^2\} \qquad (2.11)$$

where $\bar{\lambda}$ is a constant.

Now we choose F, G, F_1, G_1 to both minimize the cost function J and set the transfer function from $Y_r(t)$ to $Y_1(t)$ to be the desired one:

$\frac{q^{-k}BZ}{T}$, i.e., to make

$$Y_1(t) = \frac{q^{-k}BZ}{T} Y_r(t) \qquad (2.12)$$

where T, Z are polynomials specified by the designer, which determine the desired poles and zeroes, respectively.

Substituting (2.5), (2.8) into (2.1), get:

$$AY_2(t) = q^{-k}BU_2(t) + Ce(t) \qquad (2.13)$$

From the generalized minimum variance control theorem, it is known that for system (2.13) the control strategy which minimizes the cost function J is:

$$U_2^*(t) = -\frac{G}{\lambda A + B} e(t) \qquad (2.14)$$

where G is determined from the equation $A\bar{F} + q^{-k}G = C$ with $\deg\bar{F}=k-1$, $\deg G=\max(n_c-k,\ n_a-1)$. $\lambda = \bar{\lambda}^2/b_0$.

Comparing with (2.10), the following G, F can be chosen to make $U_2(t)=U_2^*(t)$:

$$G = G \tag{2.15}$$

$$F = B\bar{F} + \lambda C \tag{2.16}$$

Comparing with (2.6), the following F_1, G_1 can be chosen to make (2.12) satisfied:

$$F_1 = TF \tag{2.17}$$

$$G_1 = ZC(\lambda A+B) \tag{2.18}$$

From the above discussion, the following self-tuning control algorithm is obtained:

1. At time t, use the recursive extended least squares to estimate system parameters and get the estimators A_t, B_t, C_t.

2. Solve the polynomial equation $A_t\bar{F}_t + q^{-k}G_t = C_t$ with respect to $\bar{F}_t$, G_t with $\deg\bar{F}_t=k-1$, $\deg G_t=\max(n_a-1,\ n_c-k)$. Then compute:

$$F_t = \bar{F}_t B_t + \lambda C_t \tag{2.19}$$

$$F_{1t} = TF_t \tag{2.20}$$

$$G_{1t} = ZC_t(\lambda A_t+B_t) \tag{2.21}$$

3. Let M_1 represent the control U(t) determined by the equation:

$$U(t) = -\frac{G_t}{F_t}Y(t) + \frac{G_{1t}}{F_{1t}}Y_r(t) \tag{2.22}$$

then take the control at time t as follows:

$$U(t) = \begin{cases} M_1 & \text{if } M_1 \le M \\ \mathrm{sgn}(M_1).M & \text{if } M_1 > M \end{cases}$$

where M is an engineering restriction on the control.

The criterion of choosing λ in the algorithm is such that the polynomial $\lambda A+B$ is made to be a stable polynomial and $\lambda b_0 \ge 0$. Under the assumption that A is a stable polynomial, $\lambda A+B$ will be always stable if we choose λ large enough. But from eq. (2.11) we see that an increase in λ will result an increase in the disturbance of the noise. So we shouldn't choose λ too large in practice. In other optimal self-tuning control algorithms, the determination of λ is rather difficult. But for the algorithm proposed in this paper, λ can be much easier determined since we have the estimators of A and B. λ can be modified on-line. For example, if $A=1+a_1q^{-1}+a_2q^{-2}$, $B=b_0+b_1q^{-1}+b_2q^{-2}$, $b_0>0$, then λ can be chosen such that the following inequalities are satisfied:

$$(1+a_1+a_2)\lambda + (b_0+b_1+b_2) > 0 \tag{2.23}$$

$$(1-a_1+a_2)\lambda + (b_0-b_1+b_2) > 0 \tag{2.24}$$

$$(1-a_2)\lambda + (b_0-b_2) > 0 \tag{2.25}$$

If A is stable, λ can be simply chosen as:

$$\lambda = \begin{cases} k_0 \cdot \lambda_0 & \text{if } \lambda_0 > 0 \\ 0 & \text{if } \lambda_0 \le 0 \end{cases} \tag{2.26}$$

where $\lambda_0 = \max(-(b_0+b_1+b_2)/(1+a_1+a_2),\ -(b_0-b_1+b_2)/(1-a_1+a_2),\ -(b_0-b_2)/(1-a_2))$, and $k_0 > 1$.

The stability and convergence theorems and other details of the algorithm were presented in the papers of Zheng Qin (1985)

3. MATHEMATICAL MODEL OF THE HEAT-VACUUM CABIN

The heat-vacuum cabin is an experimental set-up which is used to do heat-vacuum simulation experiments for various objects to be tested. The structure of the cabin is shown in the following graph.

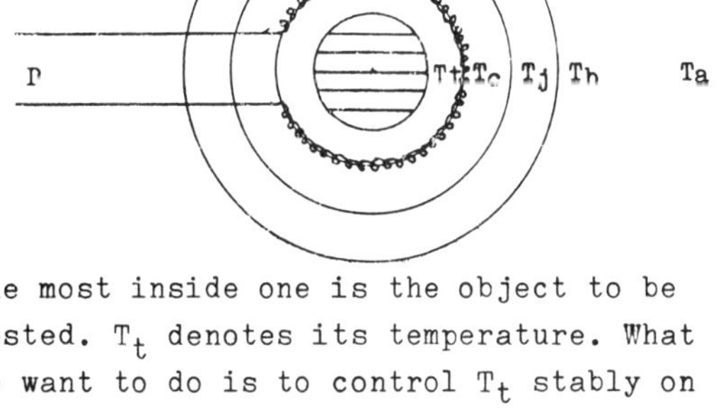

The most inside one is the object to be tested. T_t denotes its temperature. What we want to do is to control T_t stably on some value specified. The second layer is the temperature compensator on which the heater is winded. T_c is its temperature and P is the electric power applied on the heater. The third layer is a jacket. The cooled alcohol or liquid nitrogen is pumped in it when a low temperature expriment is going on. T_j denotes the jacket's temperature. The fourth layer is the outer body of the cabin inside which a high vacuum of 10^{-6} mmHg is maintained. T_b denotes its temperature and T_a denotes the temperature of the surrounding air.

According to the principles of thermophysics and the structure parameters of the cabin, the nonlinear mathematical model can be obtained:

$$\begin{bmatrix} T_c(t+1) \\ T_t(t+1) \\ T_j(t+1) \\ T_b(t+1) \end{bmatrix} = \begin{bmatrix} T_c(t) \\ T_t(t) \\ T_j(t) \\ T_b(t) \end{bmatrix} + \begin{bmatrix} 0 \\ 0 \\ 0 \\ 8.2E-4 \end{bmatrix} (T_a(t)-T_b(t)) +$$

$$+ \begin{bmatrix} -9.0E-11 & 6.9E-11 & 2.1E-11 & 0 \\ 1.8E-10 & -3.1E-10 & 1.3E-10 & 0 \\ 0.1E-11 & 0.4E-11 & -7.4E-11 & 6.9E-11 \\ 0 & 0 & 3.7E-11 & -3.7E-11 \end{bmatrix} \cdot$$

$$\cdot \begin{bmatrix} T_c^4(t) \\ T_t^4(t) \\ T_j^4(t) \\ T_b^4(t) \end{bmatrix} + \begin{bmatrix} 5.8E-3 \\ 0 \\ 0 \\ 0 \end{bmatrix} P(t) \qquad (3.1)$$

From this equation, we see that one of the stable points of the system is $T_t(t)=T_c(t)=T_j(t)=T_b(t)=T_a(t)=T_o$. Denote $\bar{T}_t(t)=T_t(t)-T_o$, $\bar{T}_j(t)=T_j(t)-T_o$, then the linearized I/O model can be derived:

1. $$(1-1.946q^{-1}+0.946q^{-2})\bar{T}_t(t) = =1.423E-3q^{-2}P(t)+q^{-1}(1.723E-2 - -1.695E-2q^{-1})\bar{T}_j(t) \qquad (3.2)$$

when T_o = 50 ℃ = 323 K.

2. $$(1-1.980q^{-1}+0.980q^{-2})\bar{T}_t(t) = = 5.359E-4q^{-2}P(t) + q^{-1}(8.613E-3 - - 8.544E-3q^{-2})\bar{T}_j(t) \qquad (3.3)$$

when T_o = -40 ℃ = 233 K

Detail derivation of the mathematical models was presented by Peng Lixin(1984).

4. DETERMINATION OF THE CONTROL ALGORITHM AND COMPUTER SIMULATIONS

The design requirements are as follows: $T_t(t)$ can be controlled on 50 ℃ or -40 ℃ within ±1 ℃ for more than ten hours and the time needed to convert $T_t(t)$ from one set point to another should be as short as possible.

Since the system to be controlled is time-varying, the self-tuning control strategy must be chosen. There are two main reasons cause the variation of the system characteristics: (a) The nonlinear property of of the system. From section 3 we see that the value of b_o when T_o=50 ℃ is 3.27 times greater than that when T_o=-40 ℃. (b) The differences of the objects tested. As the heat-vacuum cabin is an experimental set-up, the objects tested often change. Obviously, the characteristics of the system will change as the sizes, shapes and materials of the objects change.

From the discussions in section 3, the mathematical model of the system can be taken as:

$$A\bar{T}_t(t) = b_o q^{-2}P(t) + q^{-1}B_1\bar{T}_j(t) + +Ce(t) \qquad (4.1)$$

where $A = 1 + a_1q^{-1} + a_2q^{-2}$
$B_1 = b_1 + b_2q^{-1}$

Ce(t) is the term of the noise. For simplicity, here we choose $C = 1 + c_1q^{-1}$.

Let $\bar{P}(t) = P(t) + B_1\bar{T}_j(t)/b_o$, then eq.(4.1) can be rewritten as:

$$A\bar{T}_t(t) = q^{-2}b_o\bar{P}(t) + q^{-1}B_1(\bar{T}_j(t)- -\bar{T}_j(t-1)) + Ce(t) \qquad (4.2)$$

Here $\bar{T}_j(t)$ is the temperature of the jacket. Its variation is slow. So $\bar{T}_j(t)-\bar{T}_j(t-1)$ approximately equals zero. Then we get the system equation of the standard form:

$$A\bar{T}_t(t) = q^{-2}b_o\bar{P}(t) + Ce(t) \qquad (4.3)$$

Applying the self-tuning control algorithm proposed in section 2 to this system, we get the control algorithm as follows:

1. At time t, use the recursive extended least squares to estimate system parameters and get the estimators of a_1, a_2, b_o, b_1, b_2, c_1.

2. Get the control power P(t) from equation:

$$P(t) = (a_1-c_1)P(t-1)+(-(a_1^2-a_1c_1-a_2)\cdot \bar{T}_t(t)+a_2(c_1-a_1)\bar{T}_t(t-1)+T_{rt}(t)+ +c_1T_{rt}(t-1)-b_1\bar{T}_j(t)-(b_2+(b_2+(c_1-a_1))\bar{T}_j(t-1)-b_2(c_1-a_1)\bar{T}_j(t-2))/b_o$$

where $T_{rt}(t)=t_1T_{rt}(t-1)+(1-t_1)(T_r-T_o)/b_o$, $T_{rt}(0)=\bar{T}(0)$, t_1 is the closed-loop pole assigned.

We see that the computational efforts needed are small.

Simulations were carried out on a digital computer to verify the control algorithm obtained. The block diagram of the simulation program is shown as follows:

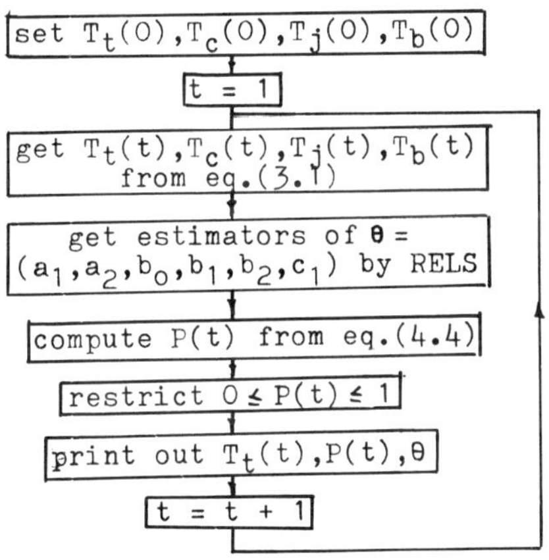

Simulation results showed the algorithm is effective.

5. EXPERIMENT RESULTS

The general structure of the temperature self-tuning control system for the heat-vacuum cabin is shown in the following graph.

The computer used is the TP-801B single board microprocessor. The temperature of the object tested ($T_t(t)$) and the jacket ($T_j(t)$) are detected by thermocouples and

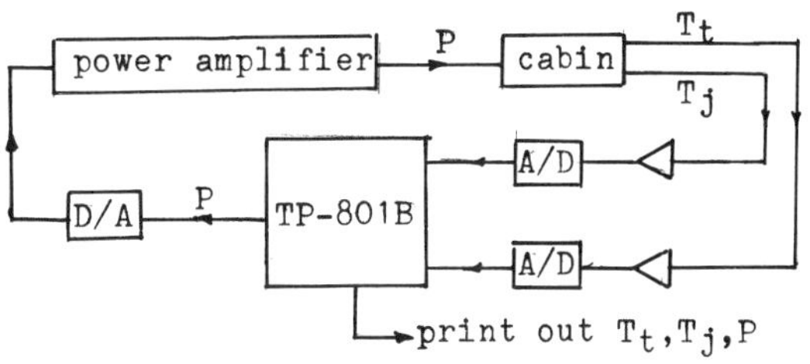

the output electric signals of the thermocouples are amplified by the amplifiers and converted to digital signals by A/D converters. Then the microprocessor computes the control power P(t) according to the self-tuning control algorithm obtained in section 4. The digital signal P(t) from the output of the microprocessor is converted to the analog signal by a D/A converter and then amplified by a power amplifier. The output of the power amplifier is directly applied to the heater in the heat-vacuum cabin.

The control program and the management program used about 4K EPROM. The running time of the control program is less than 0.3 second which can be ignored since the sample period is taken to be 60 seconds. The printer prints out all the data needed every one minite. The control system works safely and is easy to operate.

Now we present experiment results.

1. The first experiment

In the first experiment the jacket is cooled by alcohol. $T_j(t)$ reached -70 °c. First a low temperature test was carried out in which T_r was set to be -25 °c. The values of $T_t(t)$ are shown in Fig. 1.1 and the values of P(t) are shown in Fig. 1.3. $T_t(t)$ is maintained on -25 °c for about an hour and then the high temperature test begun in which T_r was set to be 50 °c. The value of $T_t(t)$ are shown in Fig. 1.2 and P(t) in Fig. 1.3.

2. The second experiment

In this experiment the jacket was cooled by the liquid nitrogen. T_j reached about -150°c The values of $T_t(t)$ obtained are shown in Fig. 2.1. For the convenience of analysis, the values of parameters estimated are also

shown in Fig. 2.2 - Fig. 2.7. We see that the parameters varied a lot with the set point's changing. So the necessity of choosing a self-tuning strategy is obvious.

6. CONCLUSIONS

This paper presents a successful example of self-tuning control for a heat-vacuum cabin. The control algorithm, modelling, simulations and experiment results are discussed in details and the necessity of applying self-tuning control strategy for the cabin is verified.

The authors would like to acknowledge Mr. Liu Guangjun and Li Yaozhong for their works in building assembly programes and also acknowledge Mr. Liu Yuezhi, Li Peiyuan, Cai Donghai, Zhang Enchao and Miss Zhou Li for their cooperations.

REFERENCES

Peng Lixin (1984). Modelling of a heat-vacuum cabin. USTC Report 84006.

Zheng Qin (1985). A generalized minimum variance self-tuning controller with pole-assignment. 7th IFAC Symposium on Identification and System Parameter Estimation.

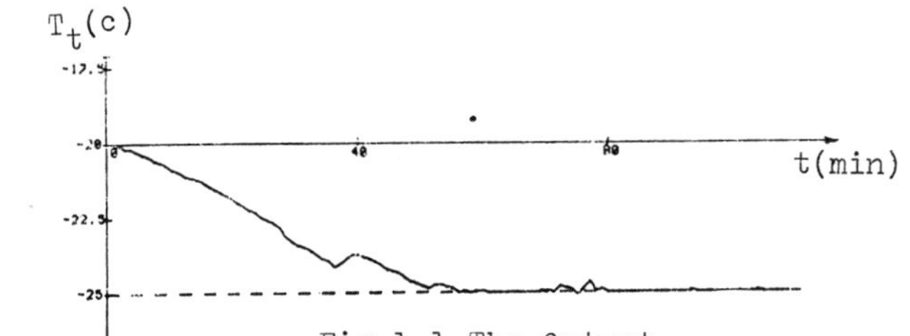

Fig.1.1 The Output

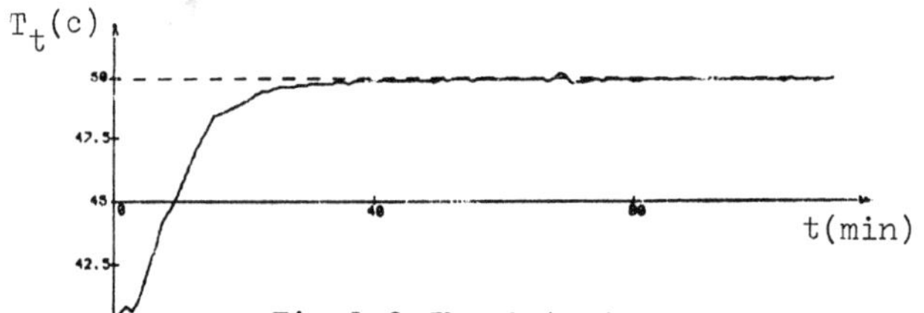

Fig.1.2 The Output

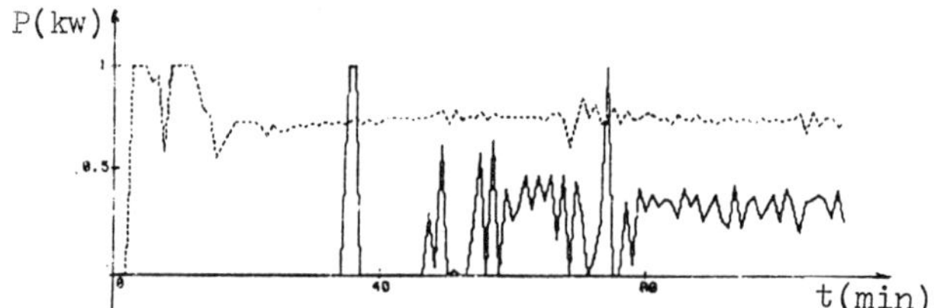

Fig.1.3 The Control

——— value of power when Tr= -40 c
------ value of power when Tr= 50 c

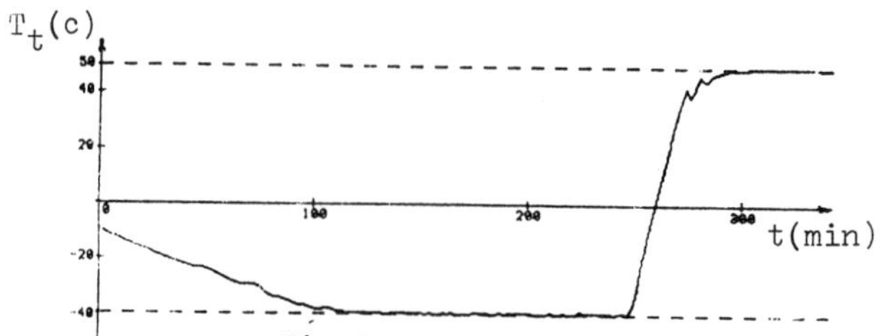

Fig.2.1 The Output

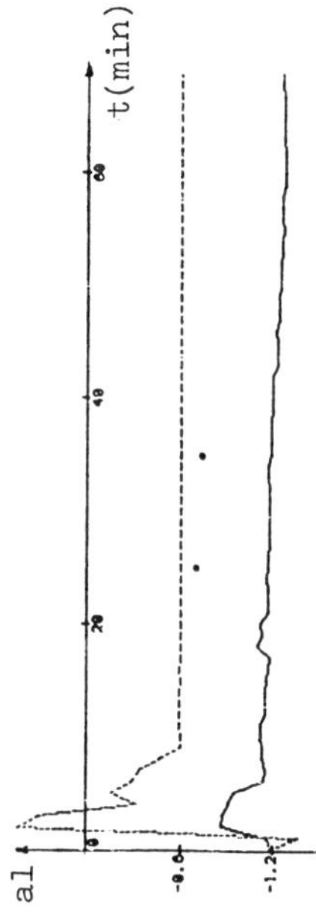

Fig.2.2 The Convergence of a1

—— value of a1 when Tr = -40 c
---- value of a1 when Tr = 50 c

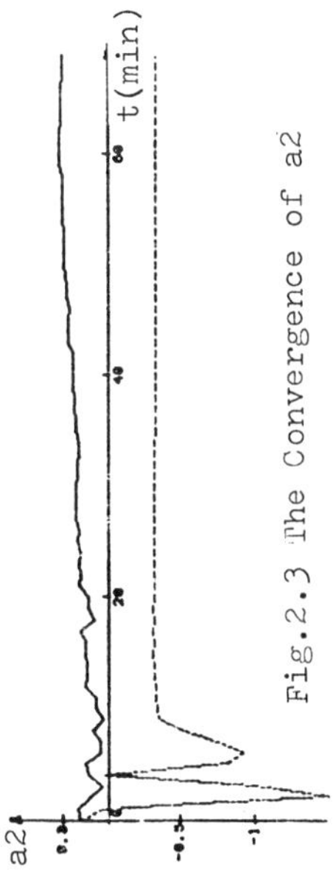

Fig.2.3 The Convergence of a2

—— value of a2 when Tr = -40 c
---- value of a2 when Tr = 50 c

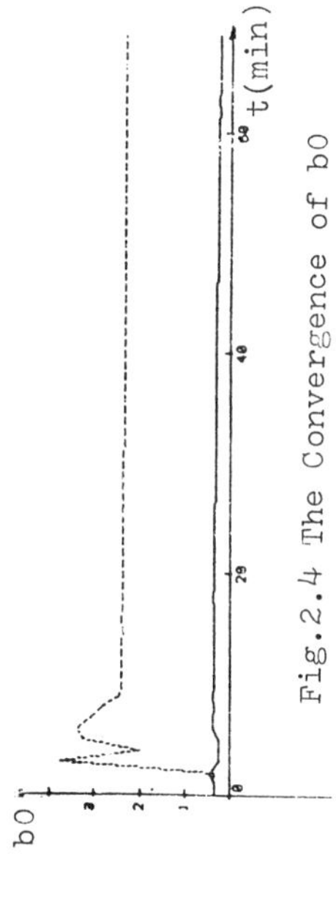

Fig.2.4 The Convergence of b0

—— value of b0 when Tr = -40 c
---- value of b0 when Tr = 50 c

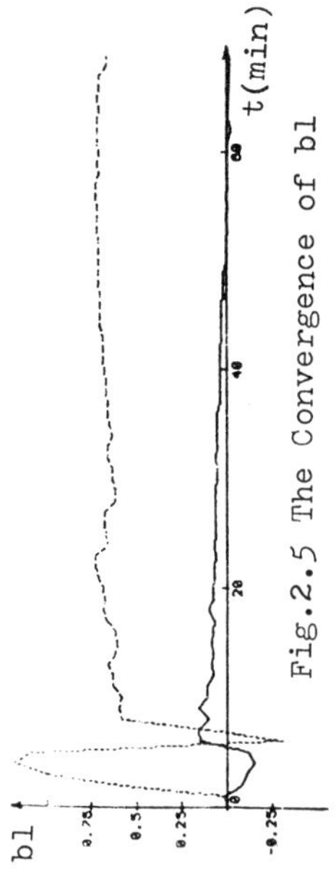

Fig.2.5 The Convergence of b1

—— value of b1 when Tr = -40 c
---- value of b1 when Tr = 50 c

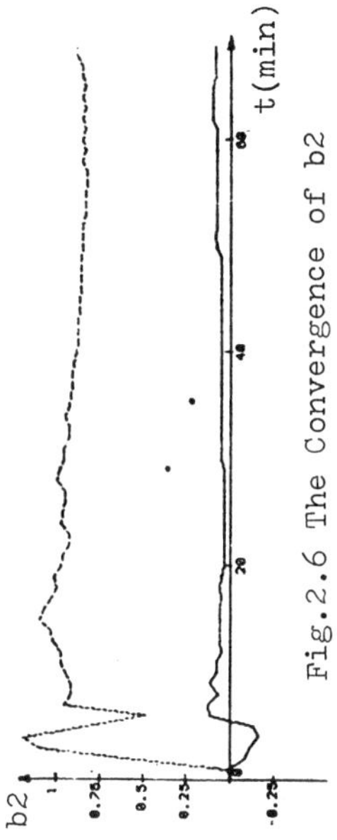

Fig.2.6 The Convergence of b2

—— value of b2 when Tr = -40 c
---- value of b2 when Tr = 50 c

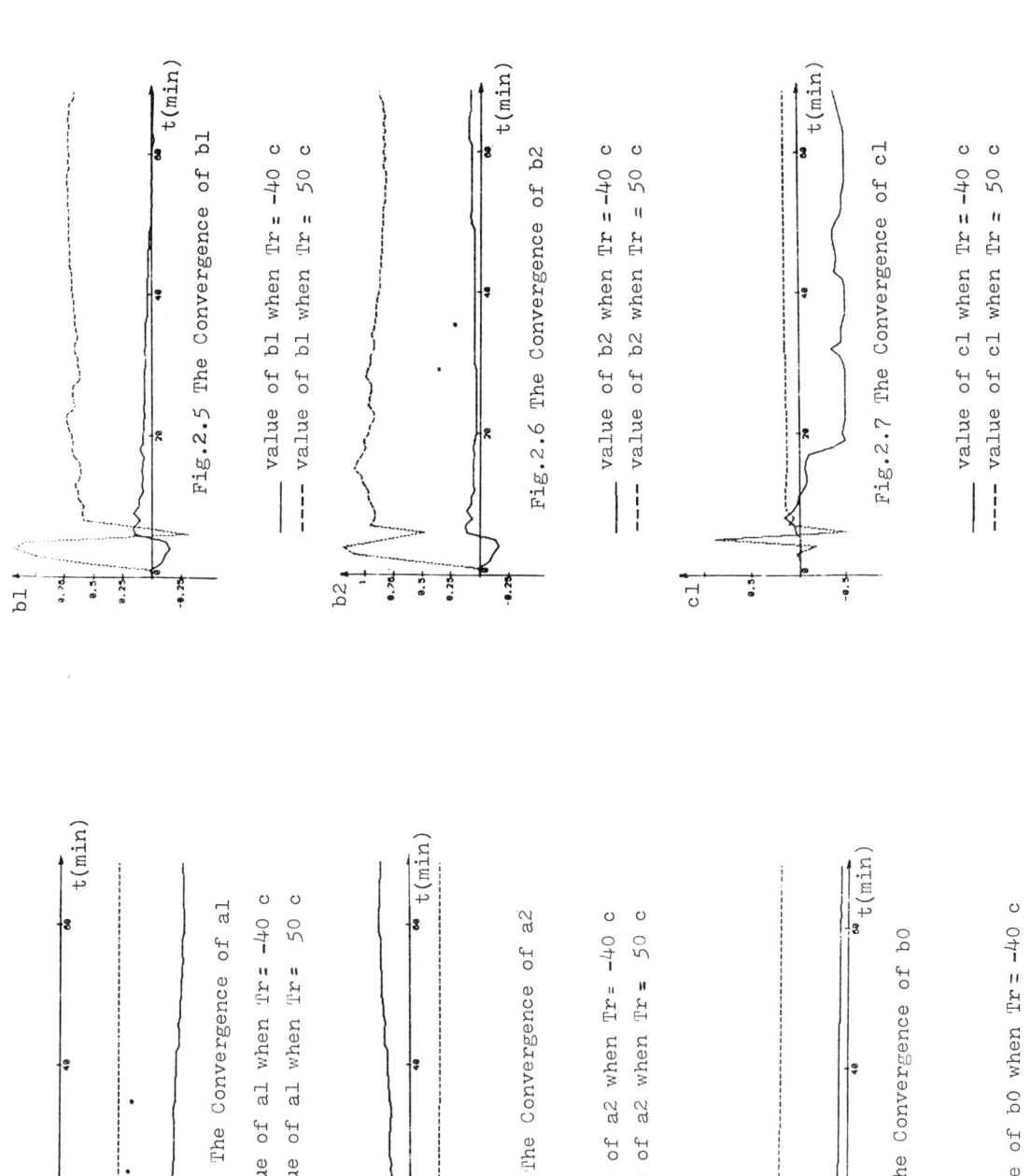

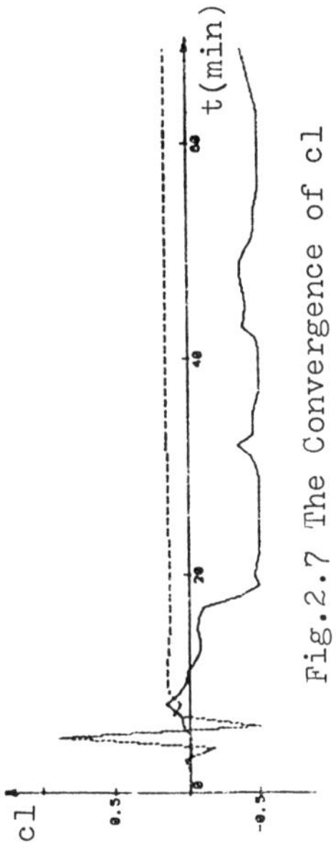

Fig.2.7 The Convergence of c1

—— value of c1 when Tr = -40 c
---- value of c1 when Tr = 50 c

WHAT FACTORS AFFECT THE DYNAMICS OF HEAT EXCHANGERS?

S. Kawata, H. Kanoh and M. Masubuchi

Osaka University, 2–1 Yamadaoka, Suita, Osaka 565, Japan

Abstract. It is clarified why one heat exchanger shows various types of dynamics and what factors affect the dynamics of heat exchanger by introducing the spatial weighting function. Especially, in one particular heat exchanger, it is found that the ratio of N.T.U. which is known to show the static behavior governs the dynamic behavior.

Keywords. Heat exchangers; parameter estimation; dynamic response; steady state; N.T.U.

1. INTRODUCTION

Heat exchangers have been widely used in many places such as chemical, steel-making and air-conditioning processes where heat transfer is the main factor. Particularly in these days, attention have been paid to the energy saving problems and heat exchanger control is supposed to be one of the important topics.

However, it seems that pertinent heat exchanger control system design has not been done owing to the lack of the knowledge about dynamic performance in the design engineers. This tendency is sometimes found in general process design too. The main reason is that although the parameters governing the steady state and dynamic responses are separately known, the common parameters which relates steady state and dynamic performances have rarely been discussed.

Though numbers of heat exchangers such as parallel- and counterflow, crossflow, plate and spiral types have widely been studied [1-4], parallel- and counterflow heat exchange processes are considered below. Now, a particular dynamic response of some heat exchanger may vary extensively according to such change of operating point as flow rate. This can be considered to be based on the structual factor that the basic equations governing the temperature variation is nonlinear with respect to the flow rate and also based on the physical factor that such constant as the heat transfer coefficient varies according to the flow rate and temperature changes. These factors are combined with the nature of a distributed parameter system and constitute the variation of dynamic response.

The existence of this variation means that readjustment of controller parameters may often become necessary when operating point changes during operation. For this purpose, it is important how to quickly determine or identify the variation of dynamics of heat exchanger.

This paper presents a new method which checks the parameters showing the corelation between the steady state and dynamic responses and which shows the minimum number of necessary parameters governing the dynamic performance or its variation. This approach is based on the concept of the spatial weighting function of the manipulated variable profile defined in a simple wall-fluid heat exchange process [3].

2. NOTATION

x : temperature [°C], V : volume of flowing pass [m^3], A : heat transfer surface [m^2], k : over all heat transfer coefficient [$kJ/m^2h°C$], F : flow rate [m^3/h] or [l/min], c_p : specific heat [kJ/kg°C], ρ : density [kg/m^3], t : time [h], subscripts : 1 = the 1st fluid, 2 = the 2nd fluid, i = inlet, o = outlet, s = steady state
m is shown in Sec.5.

$$a_1 = \frac{k_s A}{c_{p1}\rho_1 F_{1s}}, \quad a_2 = \frac{k_s A}{c_{p2}\rho_2 F_{2s}}, \quad r = \frac{V_2 F_{1s}}{V_1 F_{2s}}$$

$$b_1 = \frac{mAF_{2s}}{c_{p1}\rho_1 F_{1s}}, \quad b_2 = \frac{mA}{c_{p2}\rho_2}, \quad F_1 = F_{1s},$$

$$F_2 = F_{2s}(1+u), \quad \tau = \frac{F_{1s}}{V_1}t, \quad z = l/L,$$

$$\theta_1 = \frac{x_1 - x_{1s}}{x_{2is} - x_{1is}}, \quad \theta_2 = \frac{x_2 - x_{2s}}{x_{2is} - x_{1is}},$$

$$\mu = \begin{cases} \dfrac{a_1 - a_2}{a_1 - a_2 e^{(a_2 - a_1)}} & \text{(for counterflow)} \\ 1 & \text{(for parallelflow)} \end{cases}$$

3. A SIMPLE WALL-FLUID HEAT EXCHANGE PROCESS WITH DISTRIBUTED WALL TEMPERATURE AS THE MANIPULATED VARIABLE

Fig.1 shows a simple heat exchange process which outlet fluid temperature θ(τ,1) is the controlled variable and the fluid temperature in the process is manipulated by u(τ) having distributed gain g(z) along the flowing pass. The fundamental equations are

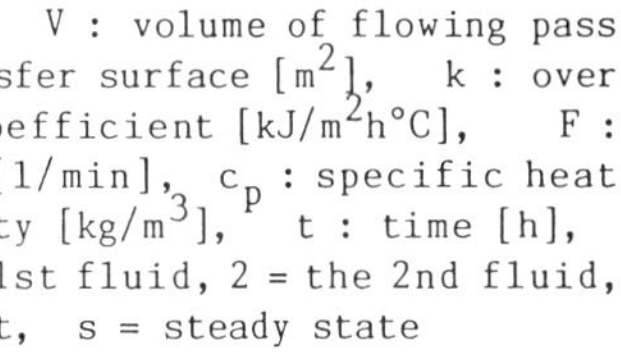

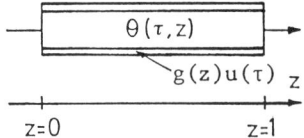

Fig.1 Wall-fluid heat exchange process

$$\frac{\partial\theta(\tau,z)}{\partial\tau} + \frac{\partial\theta(\tau,z)}{\partial\tau} = -a\theta(\tau,z) + ag(z)u(\tau) \quad (1)$$

$$\theta(\tau,0) = 0, \quad 0 \leq z \leq 1, \quad \tau \geq 0 \quad (2)$$

where g(z) is defined as the spatial weighting function. The manipulated variable $u(\tau)$ is distributed along the whole length of the tube pass and gives influence on the fluid temperature. The influence is divided into two categories ; the effect of the wall temperature manipulation at place(a) in Fig. 2 has mainly heat transfer lag and has no appreciable lag. This effect is called here as quick effect. The effect at place(b) far from the point of controlled variable has heat transfer lag plus lag due to fluid transport lag and has appreciable lag effect on the controlled variable. This is called slow effect.

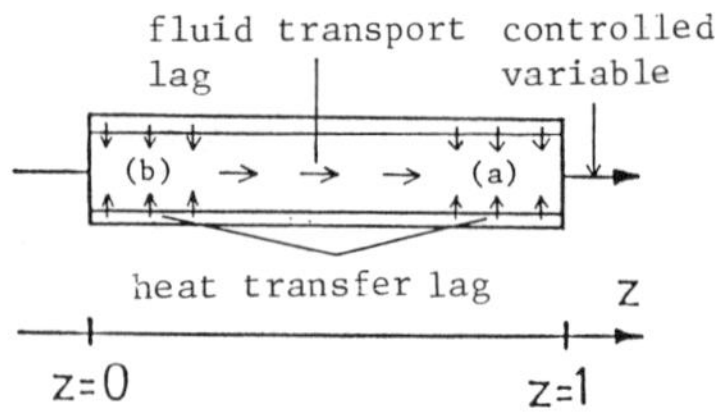

Fig.2 Quick and slow effects

Now, consider how the quick and slow effects give influence on the process dynamics when the spatial weighting function g(z) is varied. g(z) is defined by the following form.

$$g(z) = d\{\exp(\alpha z)\} \quad (3)$$

$$d = \begin{cases} \alpha/(e^{\alpha}-1) & (\alpha \neq 0) \\ 1 & (\alpha = 0) \end{cases} \quad (4)$$

where d is a constant defined as $\int_0^1 g(z)dz=1$ is hold.

Consider the following three cases.

1) Case 1 $\alpha = -5$ (slow)
2) Case 2 $\alpha = 5$ (quick)
3) Case 3 $\alpha = 0$ (mixed)

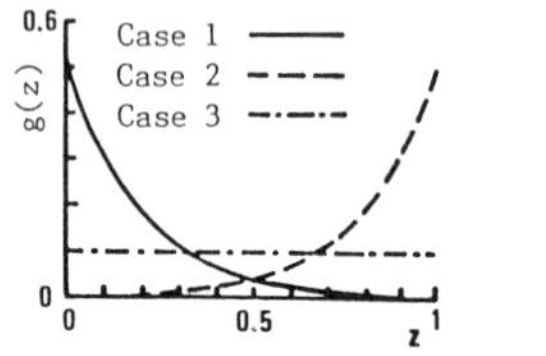

Fig.3 Spatial weighting functions

It is seen from Fig.3 that in case 1 slow effect predominates, in case 2 quick effect does and that in case 3 mixed effect of slow and quick ones exists. Fig.4 shows corresponding frequency response for each case. In the case 1, phase lag is appreciable and is over -360°for high frequency range, which fact shows the property of dead time is evident. Meanwhile, in case 2 where quick effect predominates phase curve approaches -90° and the property of 1st order lag is noticed. In case 3 where quick and slow effects are mixed equally, both phase and gain curves show oscillatory behavior with respect to frequency.

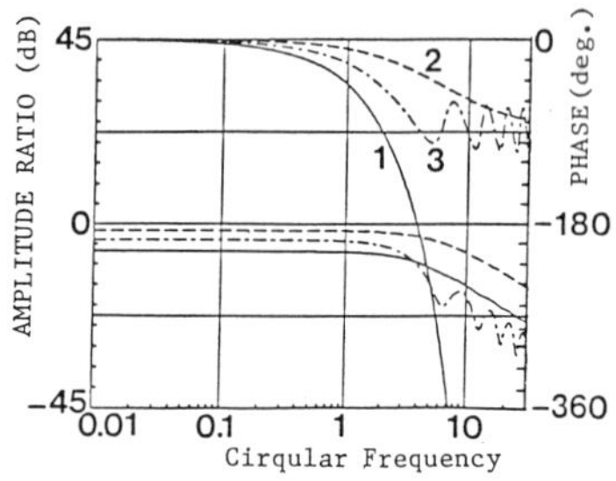

Fig.4 Frequency responses (Case 1, 2, 3)

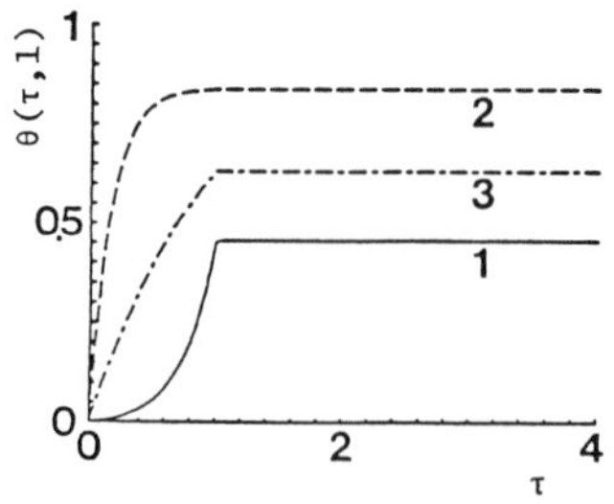

Fig.5 Step responses (Case 1, 2, 3)

Fig. 5 shows corresponding step responses for these cases. It is seen in all cases that the responses settle well after residence time. However, it is noted that the rise time becomes shorter in the order of cases 1, 3 and 2 and that transient responses have varied behaviors corresponding to the variation of the spatial weighting function g(z).

4. BASIC EQUATIONS AND DEFINITION OF THE SPATIAL WEIGHTING FUNCTION OF PARALLEL- AND COUNTERFLOW HEAT EXCHANGERS WITH FLOW RATE MANIPULATION

Fig.6 shows parallel- and counterflow heat exchangers in which two fluids exchange heat bilaterally. Outlet temperature of the 1st fluid is the controlled variable and the fluid flow rate of the 2nd fluid is the manipulated variable. It is interesting to note that the feature of the dynamic performances of these heat exchangers can be understood by using the similar concept of quick and slow effects. For this object, the spatial weighting function of these two heat exchangers (with flow rate manipulation) is discussed in this section.

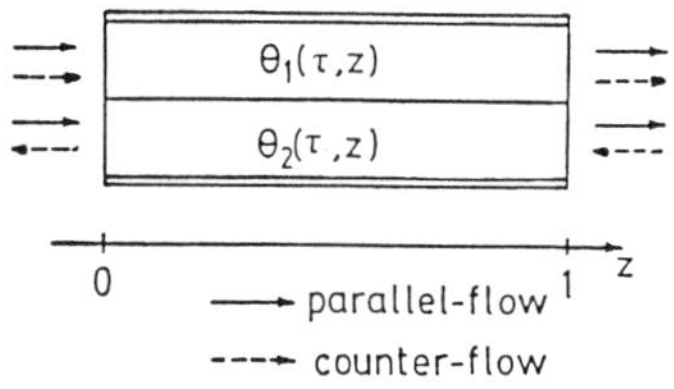

Fig.6 Parallel- and counterflow heat exchangers

The linearized basic equations are

$$\frac{\partial\theta_1}{\partial\tau} + \frac{\partial\theta_1}{\partial z} = -a_1\theta_1 + a_1\theta_2 + b_1\mu e^{-(a_1 \pm a_2)z}u \quad (5)$$

$$r\frac{\partial\theta_2}{\partial\tau} \pm \frac{\partial\theta_2}{\partial z} = -a_2\theta_2 + a_2\theta_1 + a_2\mu e^{-(a_1 \pm a_2)z}u - b_2\mu e^{-(a_1 \pm a_2)z}u \quad (6)$$

$$\tau \geq 0 \quad , \quad 0 \leq z \leq 1 \quad ,$$

$$\theta_1(\tau,0) = 0 \quad , \quad \theta_2(\tau,\{\begin{smallmatrix}0\\1\end{smallmatrix}) = 0 \quad (7)$$

where + in the double sign indicates parallelflow, - counterflow, the outlet temperature $\theta(\tau,1)$ of the 1st fluid is the controlled variable, and the flow rate change $u(\tau)$ of the 2nd fluid is the manipulated variable. $u(\tau)$ is nondimensionalized as in sec.2. The heat capacity of the heat transfer surface is assumed to be neglected.

Flow rate change $u(\tau)$ of the 2nd fluid gives mainly the variation of the overall coefficient of heat transfer (3rd term of the right hand side of eq.(5) and 4th term of the right hand side of eq.(6)) and the variation of the heat supplied by the 2nd fluid.

Through the comparison of these heat exchanger equations with the former simple process eq.(1), it is found that the 1st fluid corresponds to the fluid in the wall-fluid heat exchange process, since the controlled variable is assumed to be the outlet fluid temperature of the 1st fluid.

By considering the manipulated variable $ag(z)u(\tau)$ in eq.(1), we get a corresponding governing term of heat to the 1st fluid as

$$a_1\theta_2(\tau,z) + b_1\mu e^{-(a_1\pm a_2)z}u(\tau) \tag{8}$$

Laplace transforming eq.(8) with respect to time τ yields

$$a_1\Theta_2(s,z) + b_1\mu e^{-(a_1\pm a_2)z}U(s) \tag{9}$$

$\Theta_2(s,z)$ can be expressed by the product of transfer function $G_f(s,z)$ and $U(s)$ as (see App.A)

$$\Theta_2(s,z) = G_f(s,z)U(s) \tag{10}$$

By substituting eq.(10) into eq.(9), we get

$$a_1G_f(s,z)U(s) + b_1\mu e^{-(a_1\pm a_2)z}U(s)$$
$$= [\ a_1G_f(s,z) + b_1\mu e^{-(a_1\pm a_2)z}\ ]U(s) \tag{11}$$

As eq.(11) is separated into the product of manipulated variable $U(s)$ and function of place z in the s domain, we can consider the term

$$a_1G_f(s,z) + b_1\mu e^{-(a_1\pm a_2)z}$$

in eq.(11) as the corresponding term $g(z)$ in eq.(1).

This spatial weighting function is a function of s, and $g(z)$ shows a distributed gain. Therefore, by substituting $s=j\omega$ and by considering absolute value of this term, we can define a new spatial weighting function as

$$g_f(\omega,z) \triangleq |a_1G_f(j\omega,z)+ b_1\mu e^{-(a_1\pm a_2)z}| \tag{12}$$

In the following sections, it is discussed how the spatial weighting function affects the process dynamics by considering several ω from some important frequency band.

5. INDEPENDENCY OF NON-DIMENSIONALIZED PARAMETERS

Parameters a_1, a_2, b_1, b_2 and r are found in the basic dynamic equations (5), (6), and (7). If these equations are considered simply as partial differential eauations, each of these five non-dimensionalized parameters can be varied independently with one degree of freedom. However, if the object is considered as a real heat exchanger, it is found that any of these five parameters are composed of flow rate, coefficient of heat transfer and so on as defined in Sec.2 . So that, some intimate relations can be found in each parameter.

From the definition of a_1, a_2, b_1, and b_2, we get

$$\frac{a_2}{a_1} = \frac{b_2}{b_1} \ (\triangleq \xi) \tag{13}$$

where ξ is the ratio of N.T.U. which is well-known to show temperature effectiveness.

By using this parameter ratio ξ, we get simply

$$a_2=\xi a_1 \ , \quad b_2=\xi b_1 \tag{14}$$

and independent parameters are reduced to ξ, a_1, b_1 and r. Can the number of these parameters be reduced further ?

First, consider b_1. Since $b_1=mAF_{2s}/c_{p1}\rho_1F_{1s}$ from sec.2, a new parameter m which is not found in a_1 and r is contained in b_1. m is the ratio of the change of the coefficient of heat transmission to the flow rate of the 2nd fluid and changes according to the property of the fluids and geometrical construction (for example, sectional profile) of heat exchanger.

Then, consider parameter r.

$$r = \frac{V_2F_{1s}}{V_1F_{2s}} = (\frac{V_2}{V_1})(\frac{C_{p2}\rho_2}{C_{p1}\rho_1})\xi \tag{15}$$

r is a function of the ratio of the fluid volume V_2/V_1 of the flow pass which is not contained in a_1, r and b_1.

Therefore, considering in general, we get four independent parameters ξ, a_1, b_1 and r which govern the characteristics of parallel- and counterflow heat exchangers with flow rate as the manipulated variable.

Next, a control problem of one particular heat exchanger is considered. The mechanical construction and operating fluids in this case are given beforehand and when the dynamic performance varies according to the operating conditions, it is necessary to readjust the controller parameters for maintaining the optimal control.

For this control problem, the variation of dynamic performance of some particular heat exchanger should be considered. In this case, geometrical construction is fixed (if heat exchanger does not change) and we get

$$A = \text{const.},\ V_1 = \text{const.}\ ,\ V_2 = \text{const.} \tag{16}$$

When the working fluid is selected and the operating range is given to some extent, the properties c_p and ρ are supposed not to vary largely in a case when the working fluid is liquid. Therefore, as can be seen in eq.(15) in the limited small range of the variation of c_p and ρ, r can be determined by ξ and independent parameters are reduced to three, ξ, a_1 and b_1.

Then, the variation of b_1 for the variation of ξ and a_1 is discussed by using the two examples of heat exchangers. (see. App.B)

1) Laboratory scale heat exchanger (working fluids are water)
   $A = 0.147$, $V_1 = 1.766 \times 10^{-3}$,
   $V_2 = 1.766 \times 10^{-3}$, $L = 12.25$
2) Double-pipe type heat exchanger (working fluids are benzen and toluene)
   $A = 2.919$, $V_1 = 2.780 \times 10^{-3}$,
   $V_2 = 2.162 \times 10^{-3}$, $L = 6.096$

In the former case, a_1, a_2 and b_1 are obtained for the sets of various flow rates of the working fluids from 1 [l/m] to 9 [l/ m], and shown in

Fig.7 where b_1 is on the ordinate and on the abscissa. It is noted that the variation of b_1 is small. In order to check the effect of this variation of b_1 on the dynamic performance, frequency responses for b_1 = 0.52(max.) ∿ 0.34(min.) and ξ = 1/3 are shown on Fig. 8. No appreciable differences are noticed.

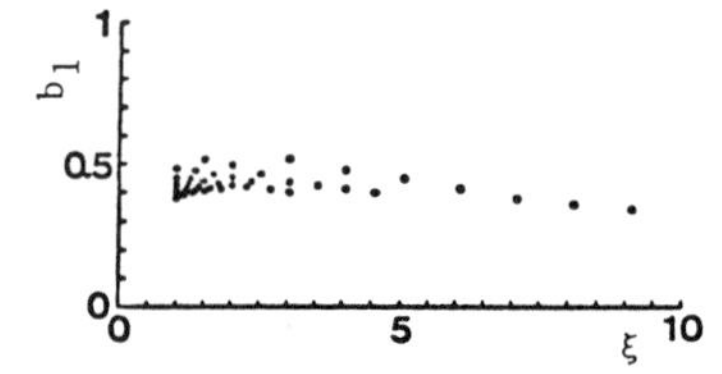

Fig.7 The value of b_1 for various ξ (laboratory scale)

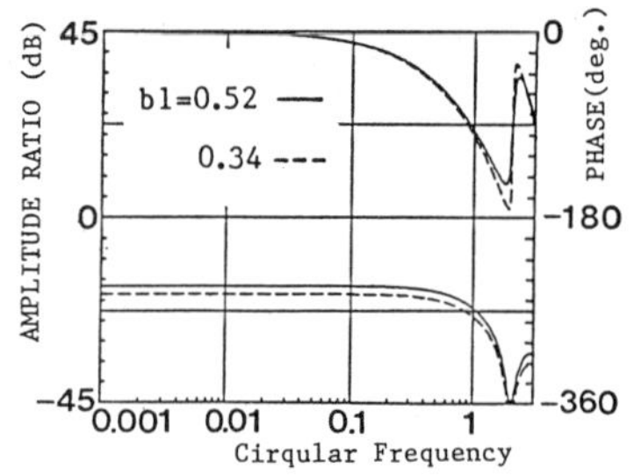

Fig.8 Frequency responses of counterflow heat exchanger

In the latter case , Fig.9 shows similar result and the variation of b_1 vs ξ is small.

In these two cases , the variation of parameter a_1 is also considered for each ξ. Therefore, it is concluded that the variation of b_1 is small for the variation of a_1 and ξ.

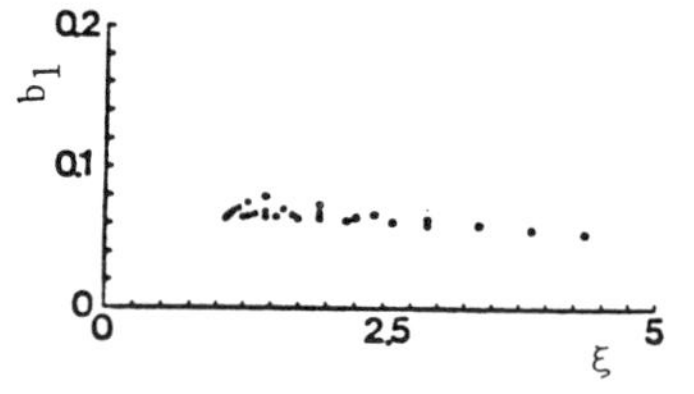

Fig.9 The value of b_1 for various ξ (double-pipe type)

6. WHAT FACTORS GOVERN THE DYNAMIC PERFORMANCE OF COUNTERFLOW HEAT EXCHANGER ?

Spatial weighting function of this heat exchanger is defined by eq.(12). This profile can not be adopted at will like g(z) of the simple wall-fluid process, but only can be varied by the physical parameters such as a_1, a_2, b_1, b_2 and r. In the following , by using the independent parameters ξ and a_1 discussed in the former section, the relation between the spatial weighting function and the dynamic performance is considered.

The 2nd term of the right hand side of eq.(12) is an exponential function of z and $-(a_1-a_2)$ governs the profile. From the relation

$$- (a_1 - a_2) = - a_1(1 - \xi) \qquad (17)$$

and from $a_1 > 0$, the exponential function monotonically decreases for $\xi < 1$ and increases for $\xi > 1$. Therefore, it is found that the 2nd term of eq.(12) is mainly governed by ξ. Then, the following three cases of ξ are considered.

Case 1	ξ = 0.1976	a_1 = 2.108
Case 2	ξ = 4.936	a_1 = 0.443
Case 3	ξ = 0.9875	a_1 = 1.098

From case 1 to 3, non-dimensionalized parameters are shown in Table 1. Three kinds of circular frequencies $\omega = 10^{-3}$ [rad/s], 10^{-1} [rad/s] and 10 [rad/s] are selected from important frequency region and the spatial weighting functions of in equ. (12) are shown in Figs. 10 to 12.

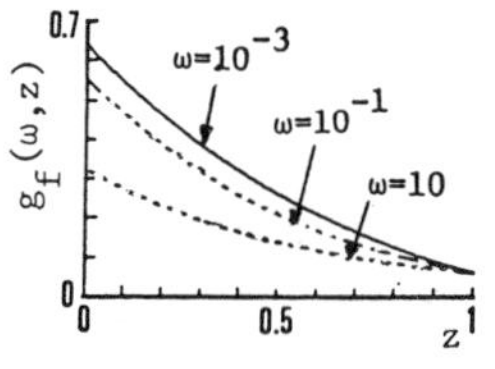

Fig.10 Profiles of the spatial weighting function for case 1

By comparing Figs. 10 to 12 with Fig. 3, it is seen that above three cases correspond to the former three cases respectively. To check the correspondency in the frequency domain too, three kinds of frequency responses for case 1 to 3 are shown in Fig.13.

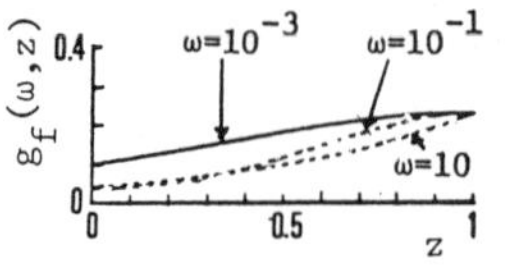

Fig.11 Profiles of the spatial weighting function for case 2

By comparing Fig.13 and Fig. 4 it is seen that vivid correspondency exists. Namely, even in counterflow heat exchanger, when slow effect dominates for flow rate manipulation (Fig.10) as in case 1, phase curve exceeds -360° for higher frequency. And when quick effect dominates (Fig. 11) as in case 2, the curve shows approximate 1st order lag though some oscillatory behavior for frequency is noticed. In case 3 where slow and quick effects are mixed (Fig.12), oscillatory behavior for frequency is shown.

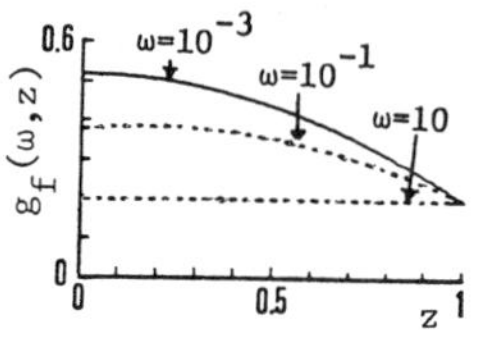

Fig.12 Profiles of the spatial weighting function for case 3

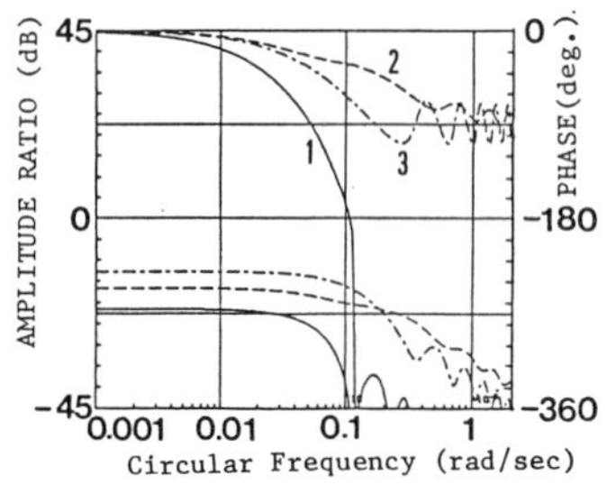

Fig.13 Frequency responses for various ξ

Table 1 Values of the parameters

	a_1	a_2	b_1	b_2	r
Case 1	2.108	0.4164	0.3889	0.0768	5
Case 2	0.443	2.186	0.2775	1.3698	0.2
Case 3	1.098	1.084	0.4143	0.4091	1

As can be observed from these figures, the dynamic response of conterflow heat exchanger changes widely and this change can be explained by using the difference of quick effect and slow effect and by considering the variation of the spatial weighting function. It is also found that the parameter which governs the types of distinctive dynamics is ξ (= a_2/a_1).

In the following, effect of one more independent parameter a_1 is discussed. Fig. 14 shows frequency responses of three cases with different a_1 and the same ξ = 1.013. In this figure, it is to be noted that the unit of abscissa is rad/s. It is found from this figure that although the frequency characteristics belong to the same type owing to the equal value of ξ, the phase and gain curves are shifted in the direction of ω axis because of the difference of residence time in each fluid.

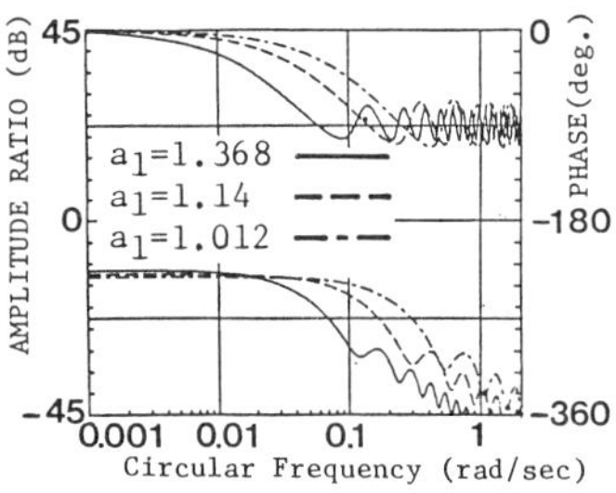

Fig.14 Frequency responses for various ξ

Non-dimensionalizing the frequency response by using the residence time of the 1st fluid, we get Fig.15 from Fig. 14 and it is seen that the patterns of frequency responses for equal value of ξ are the same irrespective the value of a_1.

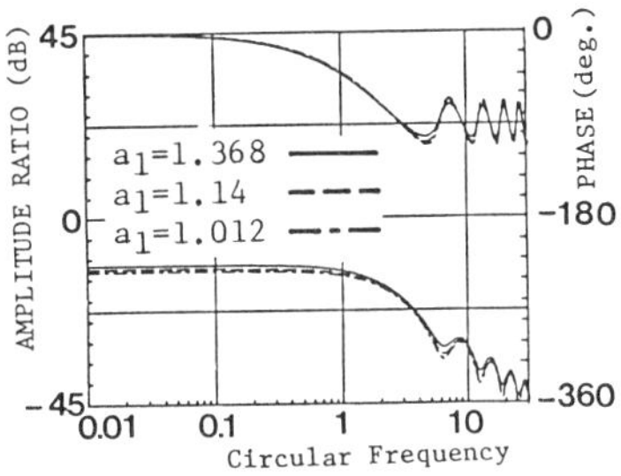

Fig.15 Frequency responses for non-dimensionalized abscissa

7. WHAT FACTORS GOVERN THE DYNAMIC PERFORMANCE OF PARALLELFLOW HEAT EXCHANGER ?

The parameter which governs the exponential function of the spatial weighting function, that is, the 2nd term of the right hand side of eq.(12) is $-(a_1+a_2)$. Since $a_1+a_2>0$, above spatial weighting function always decreases monotonically. In order to see the whole tendency of eq.(12), the spatial weighting functions are obtained for the following four cases having different values of a_1 and for $\xi = 0.988$ and are shown on Figs. 16, 17, 18, 19.

Non-dimensionalized parameters and flow rates for these four cases are shown on Table 2.

From case 1 to 4, the flow rate changes by eight times, but the rate of change of a_1+a_2 is about

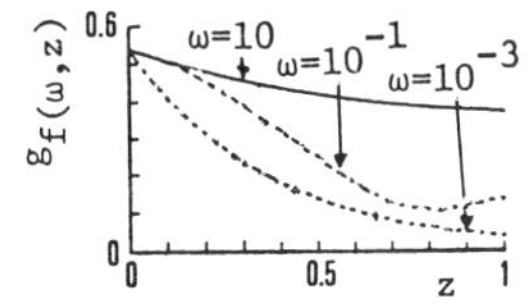

Fig.16 Profiles of the spatial weighting function for case 1

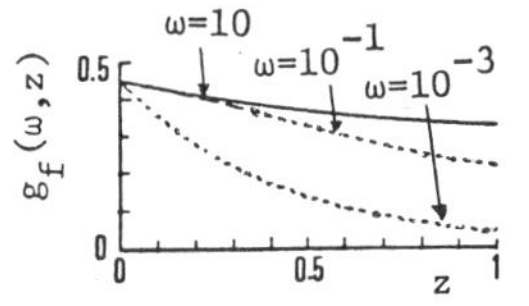

Fig.17 Profiles of the spatial weighting function for case 2

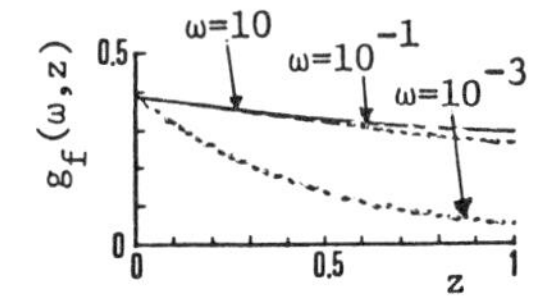

Fig.18 Profiles of the spatial weighting function for case 3

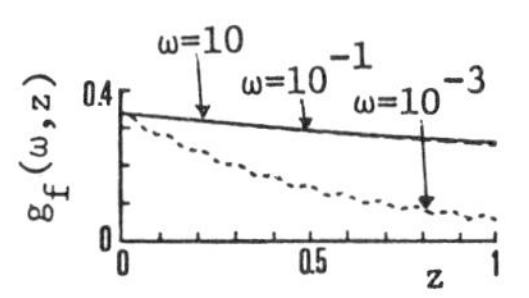

Fig.19 Profiles of the spatial weighting function for case 4

Table 2 Values of the parameters

	Case 1	Case 2	Case 3	Case 4
a_1	1.368	1.191	1.036	0.9020
a_2	1.351	1.176	1.024	0.8910
b_1	0.5408	0.4518	0.3903	0.3392
b_2	0.5341	0.4462	0.3855	0.3350
r	1	1	1	1
F_1	2	4	8	16
F_2	2	4	8	16

1) Case 1 $\xi=0.988$ $a_1=1.368$ $a_1+a_2=2.718$
2) Case 2 $\xi=0.988$ $a_1=1.191$ $a_1+a_2=2.366$
3) Case 3 $\xi=0.988$ $a_1=1.036$ $a_1+a_2=2.060$
4) Case 4 $\xi=0.988$ $a_1=0.902$ $a_1+a_2=1.790$

34%. This is because when the flow rates which are the denominators of a_1 and a_2 increase the coefficient of heat transfer in the numerator also increase.

By noting Fig.16 to Fig.19, it is seen that all the spatial weighting functions of this heat exchanger decrease from tube inlet (z=0) to tube outlet (z=1) of the 1st fluid and that these spatial weighting functions are slow effect type.

Fig. 20 shows frequency responses for case 1 to 4. It is found from Fig. 20 that lag of the phase curves increases over -360° for high frequencies and that the same features are shown as the frequency response of case 1 of Sec.3 having slow effect. This tendency corresponds to the fact that the spatial weighting functions for Figs. 16

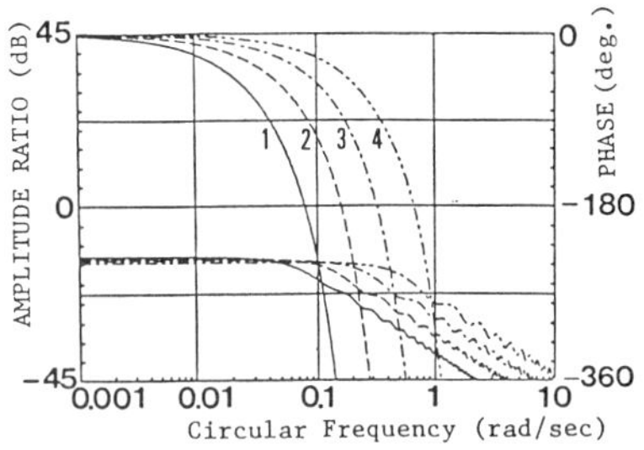

Fig.20 Frequency responses

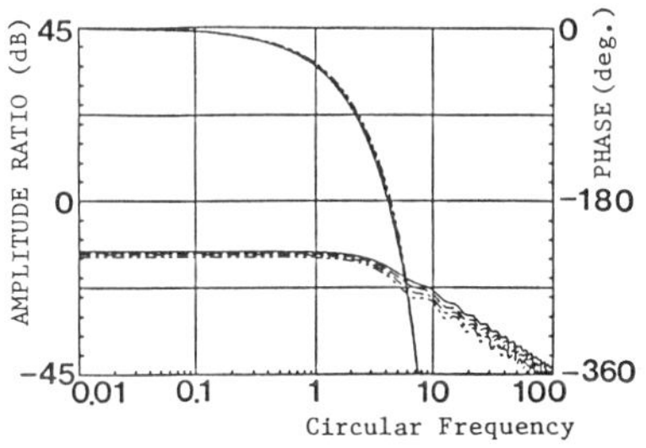

Fig.21 Frequency responses for non-dimensionalized abscissa

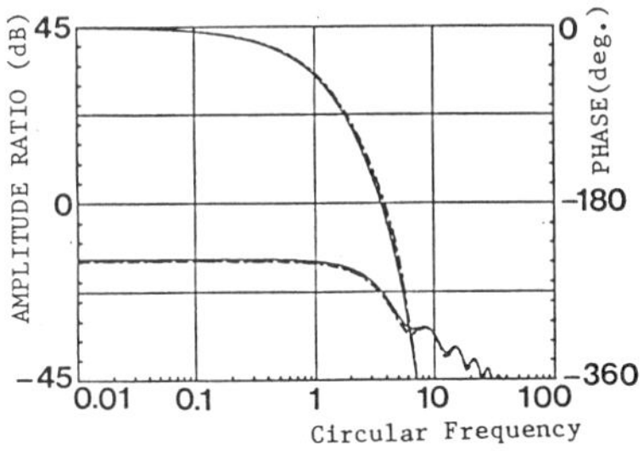

Fig.22 Frequency responses for ξ=1.974

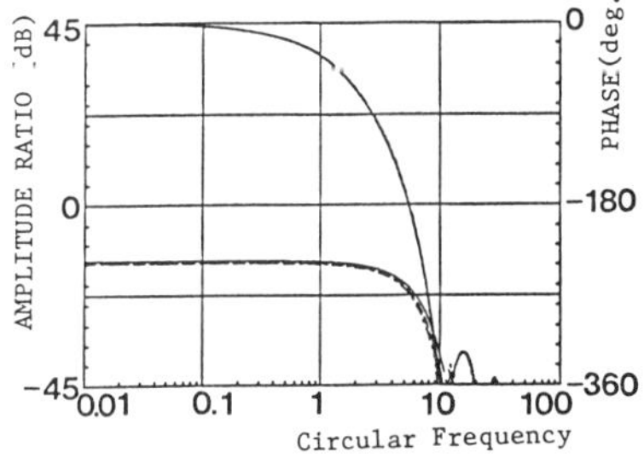

Fig.23 Frequency responses for ξ=0.494

to 19 are governed by slow effect. Four kinds of frequency responses shown on Fig. 20 seem to coincide when the abscissa is shifted. By redrawing axis as in the case of counterflow heat exchanger, we get Fig. 21 in which four responses coincide as has been suggested.

Discussions continue to the two cases having different ξ values and each having three different values of a_1.

1) $\xi = 1.974$	$a_1 = 0.8838$	$(F_1=4,\ F_2=2)$	———
	$a_1 = 0.7694$	$(F_1=8,\ F_2=4)$	– – – –
	$a_1 = 0.7095$	$(F_1=12,\ F_2=6)$	—·—
2) $\xi = 0.494$	$a_1 = 1.714$	$(F_1=2,\ F_2=4)$	———
	$a_1 = 1.492$	$(F_1=4,\ F_2=8)$	– – – –
	$a_1 = 1.376$	$(F_1=6,\ F_2=12)$	—·—

For these values, we get Figs. 22 and 23.

It is found from these Figs. 21, 22 and 23 that both phase and gain curves almost coincide when the values ξ are the same irrespective of the different values of a_1. Then, it is concluded that even in this parallelflow heat exchanger with flow rate manipulation ξ is the parameter which governs the dynamic response.

8. CONCLUSIONS

The effects of the spatial weighting function of the prallel-and counterflow heat exchangers on the dynamic performance have been discussed from the view points of quick effect and slow effect and the following results have been obtained.

(1) In the case where dynamic response has the character of 1st order lag, quick effect predominates.

(2) In the case where dynamic response has the character of dead time, slow effect predominates.

(3) And in the case where quick and slow effects are mixed, oscillatory characteristics are observed on the frequency response curves.

The consideration on the governing parameters of the spatial weighting function has been further developed to the following result that non-dimensionalized dynamic performance can be determined by using the value ξ $(=a_2/a_1)$ of this paper when a particular heat exchanger is considered.

It has been thus verified that steady state and dynamic performances are closely corelated each other by using the value of ξ which is shown to be equal to the ratio of N.T.U. [4], that is, the steady state temperature effectiveness.

REFERENCES

[1] Masubuchi M. (1977). Heat Exchangers - Dynamics and Control, J. SICE., 16(2) (in Japanese).

[2] Kanoh H. (1982). Distributed Parameter Heat Exchangers Modeling, Dynamics and Control, in : Distributed Parameter Control System. Edited by Tzafestas, S.G., Pergamon-press, pp. 438.

[3] Kawata S., Kanoh H. and Masubuchi M. (1982). On the Relation between the Spatial Weighting Function of Manipulated Variable and the Positions of Zeros in Some System of First Order Partial Differential Equation, Trans. SICE., 18(10), 974. (in Japanese).

[4] Kawata S., Kanoh H. and Masubuchi M. (1985). On the Relation between the Spatial Weighting Function of Manipulated Variable and the Positions of Zeros for the Distributed Heat Exchangers, Trans. SICE., 21(1), 35. (in Japanese).

[5] McAdams W. H. (1954). Heat Transmission 3rd ed., McGraw-Hill.

APPENDIX A TRANSFER FUNCTIONS

(1) parallel-flow

$$\frac{\theta_1(s,z)}{u(s)} = J_{p11}e^{p_1 z} + J_{p12}e^{p_2 z} + J_{p13}e^{p_3 z}$$

$$G_f(s,z) = \frac{\theta_2(s,z)}{u(s)} = J_{p21}e^{p_1 z} + J_{p22}e^{p_2 z} + J_{p23}e^{p_3 z}$$

where

$$J_{p11} = \frac{b_1p_1 + h_1}{(p_1-p_2)(p_1-p_3)} \qquad J_{p21} = \frac{(a_2-b_2)p_1 + h_2}{(p_1-p_2)(p_1-p_3)}$$

$$J_{p12} = \frac{b_1p_2 + h_1}{(p_2-p_1)(p_2-p_3)} \qquad J_{p22} = \frac{(a_2-b_2)p_2 + h_2}{(p_2-p_1)(p_2-p_3)}$$

$$J_{p13} = \frac{b_1p_3 + h_1}{(p_3-p_1)(p_3-p_2)} \qquad J_{p23} = \frac{(a_2-b_2)p_3 + h_2}{(p_3-p_1)(p_3-p_2)}$$

$$f_1 = s + a_1 \ , \quad f_2 = rs + a_2 \ , \quad p_3 = -(a_1 + a_2)$$

$$h_1 = b_1f_2 + a_1(a_2 - b_2) \qquad h_2 = (a_2 - b_2)f_1 + a_2b_1$$

and p_1, p_2 are the roots of

$$(p + f_1)(p + f_2) - a_1a_2 = 0$$

(2) counter-flow

$$\frac{\theta_1(s,z)}{U(s)} = \mu\{J_{c11}e^{p_1 z} + J_{c12}e^{p_2 z} + J_{c13}e^{p_3 z} + \frac{J_{c21}e^{p_1} + J_{c22}e^{p_2} + J_{c23}e^{p_3}}{J_{c24}e^{p_1} + J_{c25}e^{p_2}}(J_{c14}e^{p_1 z} + J_{c15}e^{p_2 z})\}$$

$$G_f(s,z) = \frac{\theta_2(s,z)}{U(s)} = -\mu\{J_{c21}e^{p_1 z} + J_{c22}e^{p_2 z} + J_{c23}e^{p_3 z} - \frac{J_{c21}e^{p_1} + J_{c22}e^{p_2} + J_{c23}e^{p_3}}{J_{c24}e^{p_1} + J_{c25}e^{p_2}}(J_{c24}e^{p_1 z} + J_{c25}e^{p_2 z})\}$$

where

$$J_{c11} = \frac{b_1p_1 + h_1}{(p_1-p_2)(p_1-p_3)} \qquad J_{c21} = \frac{(a_2-b_2)p_1 + h_2}{(p_1-p_2)(p_1-p_3)}$$

$$J_{c12} = \frac{b_1p_2 + h_2}{(p_2-p_1)(p_2-p_3)} \qquad J_{c22} = \frac{(a_2-b_2)p_2 + h_2}{(p_2-p_1)(p_2-p_3)}$$

$$J_{c13} = \frac{b_1p_3 + h_1}{(p_3-p_1)(p_3-p_2)} \qquad J_{c23} = \frac{(a_2-b_2)p_3 + h_2}{(p_3-p_1)(p_3-p_2)}$$

$$J_{c14} = \frac{a_1}{p_1 - p_2} \qquad J_{c15} = \frac{a_1}{p_2 - p_1} \qquad J_{c24} = \frac{f_1 + p_1}{p_1 - p_2}$$

$$f_1 = s + a_1 \ , \quad f_2 = -rs - a_2 \ , \quad p_3 = -(a_1 - a_2)$$

$$h_1 = b_1f_2 - a_1(a_2 - b_2) \ , \quad h_2 = a_2b_1 + (a_2 - b_2)f_1$$

and p_1, p_2 are the roots of

$$(p + f_1)(p + f_2) + a_1a_2 = 0$$

APPENDIX B CALCULATION OF OVERALL HEAT TRANSFER COEFFICIENT

$$x_{1is} = 80\ [^\circ C] \qquad x_{2is} = 10\ [^\circ C]$$

$$k_s = \frac{1}{\frac{1}{h_1} + \frac{1}{h_2}} \qquad h = \frac{N_u\lambda}{D}$$

(1) Laboratory scale heat exchanger

$$N_u = 0.037\,R_e^{0.8}\,P_r^{1/3}$$

(2) Double-pipe type heat exchanger

$$N_u = 0.023\,R_e^{0.8}\,P_r^{0.4}$$

STABILIZING CONTROL OF HEAT EXCHANGERS

H. Kanoh and M. Yoshida

Osaka University, 2–1 Yamadaoka, Suita, Osaka 565, Japan

Abstract. This paper investigates stabilizing control of heat exchangers which are apt to be unstable for a simple feedback control law. The instability is caused by the time delay effect due to the pure delay time element and the right-hand side zeros in their transfer functions. It is shown that parallel type heat exchangers such as a parallel heat exchanger or one-shell two-pass heat exchangers may have these difficulties. A Smith type predictor which can be realized on a small micro-computer is derived to eliminate the time delay effect. A control system using it shows considerable improvement of stability and control performance. Because a heat exchanger has an irrational transfer function which does not express a pure time delay element explicitly, the dynamics of such a heat exchanger is approximated with a transfer function which consists of a pure time delay element and a rational transfer function to apply the predictor. By using an asymptotic approximation method, such approximate transfer functions are derived to extract a pure delay element.

INTRODUCTION

Many investigations of the dynamics of heat exchangers have been done for a long time. It has been clarified that types of dynamics are determined depending on the structures and operating conditions of heat exchangers. Kanoh (1980) studied types of dynamics of heat exchangers on a point of view of frequency response and defined two types of dynamics; parallel flow type for which the phase lag exceeds -180 line,and counter flow type for which the phase lag does not exceed -180 line. It can be said equivalently that when the parallel type heat exchangers are controlled by a proportional feedback control law, the control systems are apt to be unstable at large feedback gain and good control performance is not expected. Two causes are considered for the instability from a study of types of the dynamics on a point of view of the configuration of root loci of a closed loop system of the heat exchangers controlled by a proportional feedback (Kanoh, 1982); an existence of time delay effect and right-hand side zeros in their transfer functions. The time delay effect makes root loci running horizontally from left hand-side infinity to the right hand-side infinity. Another kind of root loci run horizontally from finite or infinite poles and terminates at right hand side zeros. Both root loci cause instability to parallel type heat exchangers.

This paper investigates stabilizing control of such parallel type heat exchangers. One effective method is to use a Smith predictor which was developed to control a process having a rational transfer function with a pure time delay element (Smith, 1959). Because the heat exchangers have irrational transfer functions which do not express the time delay element explicitly and are not suitable form for digital control, we have to seek approximate models whose transfer function consists of a pure time delay element and a rational transfer function. Such a model suit for designing Smith predictor which is easy to realize on a small micro-computer. By using an asymptotic approximation method (Schone, 1966), we obtain such approximate transfer functions with coefficients which are functions of the original system parameters. The control system including a Smith predictor eliminates the root loci relating to the time delay effect and enlarges stability region. When the original transfer function has right hand side zeros, the standard Smith predictor cannot remove the time delay effect perfectly and there remains an extra time delay due to the right hand side zero effect. For this case, a Smith type predictor eliminating a high frequency part of transfer function is shown to be effective for improvement of stability and control performance.

MODELS OF PARALLEL FLOW TYPE HEAT EXCHANGERS

1. Basic Equations and Exact Transfer Functions

A parallel flow heat exchanger and a two-pass and one-shell C-P type heat exchanger are shown in **Fig.1.**

A parallel flow heat exchanger and a counter flow heat exchanger are described by the following equation:

$$r\frac{\partial\theta_1}{\partial t} \pm \frac{\partial\theta_1}{\partial x} = a_1(\theta_2 - \theta_1) + \mu b_1 e^{\mp(a_1 \pm a_2)x} u(t) \tag{1}$$

$$\frac{\partial\theta_2}{\partial t} + \frac{\partial\theta_2}{\partial x} = a_2(\theta_1 - \theta_2) + \mu(a_2 - b_2) e^{\mp(a_1 \pm a_2)} u(t)$$

$$\theta_1(t, \{{0 \atop 1}) = 0 , \quad \theta_2(t, 0) = \theta_{2i}(t) \tag{2}$$

where the upper term corresponds parallel flow and the lower term to the counter flow, u(t)

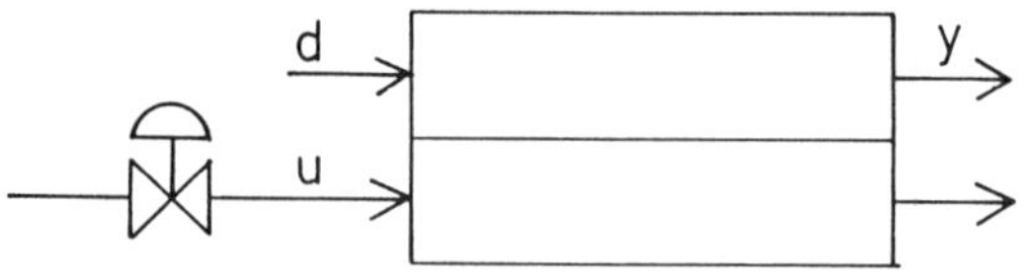

Fig.1 (a) A parallel flow heat exchanger.

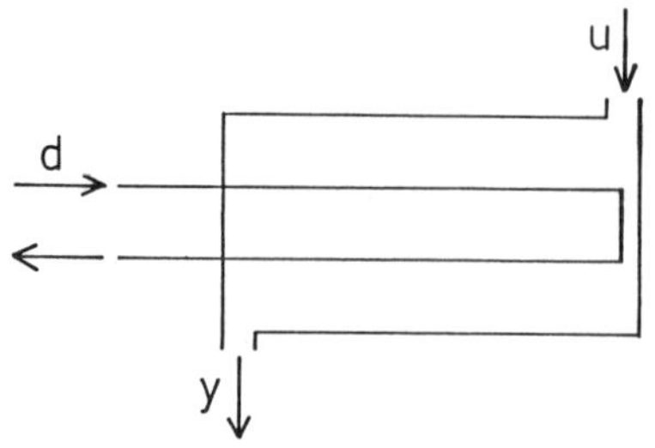

Fig.1 (b) A C-P type 2-pass 1-shell heat exchanger.

represents the flow rate input and $\theta_i(t)$ inlet temperature input, $\theta_1(t)$ and $\theta_2(t)$ are temperatures of fluids. Taking two dimensional Laplace transformation and inverse Laplace transformation with respect to spatial variable yield the following transfer functions:

$$G_p(s) = \frac{a_1(e^{p_1} - e^{p_2})}{p_1 - p_2} \tag{3}$$

$$G_{pf}(s) = \frac{b_1(e^{p_1} - e^{p_2}) + g\left[\frac{e^{p_1} - e^{p_3}}{p_1 - p_3} - \frac{e^{p_2} - e^{p_3}}{p_2 - p_3}\right]}{p_1 - p_2} \tag{4}$$

$$G_{cf}(s) = \mu \frac{b_1(e^{p_1} - e^{p_2}) + g\left[\frac{e^{p_1} - e^{p_3}}{p_1 - p_2} - \frac{e^{p_2} - e^{p_3}}{p_2 - p_3}\right]}{(f_2 + p_1)e^{p_1} - (f_2 + p_2)e^{p_2}} \tag{5}$$

where p_1, p_2 are the roots of the following equation:

$$p^2 \pm (f_1 \pm f_2)p \pm (f_1 f_2 - a_1 a_2) = 0 \tag{6}$$

$$\mu = \frac{a_2 - a_1}{a_2 - a_1 e^{a_1 - a_2}} \text{ (counter)}, \quad \mu = 1 \text{ (parallel)}$$

$$f_1 = rs + a_1, \quad f_2 = s + a_2, \quad p_3 = (\mp a_1 - a_2),$$

$$g = b_1 s + a_1(a_2 \mp b_1 - b_2)$$

A two-pass and one-shell C-P type heat exchanger is described by the following equation:

$$\frac{\partial\theta_1}{\partial t} + \frac{\partial\theta_1}{\partial x} = a_1(\theta_3 - \theta_1)$$

$$\frac{\partial\theta_2}{\partial t} - \frac{\partial\theta_2}{\partial x} = a_1(\theta_3 - \theta_2) \tag{7}$$

$$r\frac{\partial\theta_3}{\partial t} - \frac{\partial\theta_3}{\partial x} = a_2(\theta_1 + \theta_2 - 2\theta_3)$$

$$\theta_1(t, 0) = 0, \quad \theta_3(t, 1) = \theta_i(t),$$
$$\theta_2(t, 1) = \theta_1(t, 1) \tag{8}$$

We obtain a transfer function as follows:

$$G_{cp}(s) = \frac{\det\begin{bmatrix} \alpha_{11} & \alpha_{12} & \alpha_{13} \\ (\alpha_{11}-\alpha_{21})e^{p_1} & (\alpha_{12}-\alpha_{22})e^{p_2} & (\alpha_{13}-\alpha_{23})e^{p_3} \\ \alpha_{21} & \alpha_{22} & \alpha_{23} \end{bmatrix}}{\det\begin{bmatrix} \alpha_{11} & \alpha_{12} & \alpha_{13} \\ (\alpha_{11}-\alpha_{21})e^{p_1} & (\alpha_{12}-\alpha_{22})e^{p_2} & (\alpha_{13}-\alpha_{23})e^{p_3} \\ e^{p_1} & e^{p_2} & e^{p_3} \end{bmatrix}} \tag{9}$$

where p_1, p_2 and p_3 are the solutions of the following equation:

$$p^3 - f_2 p^2 - f_1^2 p + f_1^2 f_2 - 2a_1 a_2 f_1 = 0 \tag{10}$$

$$f_1 = s + a_1, \quad f_2 = rs + 2a_2,$$

$$\alpha_{1j} = \frac{a_1}{(p_j + f_1)}, \quad \alpha_{2j} = \frac{-a_1}{(p_j - f_1)}, \quad j = 1, 2, 3$$

2. Approximate Models by Asymptotic Approximation Method

An asymptotic approximation method makes an approximate model using the high frequency asymptote and the low frequency asymptote of the exact transfer function. The degree of approximation is fairly good. (Friedly,1972) The approximate model obtained has a rational transfer function with pure delay elements and its coefficients are expressed by the original system parameters. Therefore, the model is useful to design a control system and the resulting control system is easy to install by a small process computer.

For the parallel flow heat exchanger, as s tends to infinity, we obtain

$$p_1 \to -f_1, \quad p_2 \to -f_2$$

thus we have the following approximate transfer function:

$$G_{pa}(s) = a_1 \frac{m + Ts}{1 + Ts} \cdot \frac{(e^{-(rs + a_1)} - e^{-(s + a_2)})}{(1 - r)s + (a_2 - a_1)}$$

$$= G_{pa1}(s) \cdot G_{pa2}(s) \tag{11}$$

where the first part of the transfer function $G_{pa1}(s)$ is added to express the low frequency asymptote and $m = G(0)/[a_1 G_{pa2}(0)]$, T is taken an appropriate value to combine both asymptotes. Similarly we have approximate transfer functions for flow forced parallel flow and counter flow heat exchangers and for C-P type heat exchanger as follows:

$$G_{pfa}(s) = \frac{m + Ts}{1 + Ts} \cdot \left\{ \frac{b_1(e^{-(rs+a_1)} - e^{-(s+a_2)})}{(1 - r)s + (a_2 - a_1)} \right. \tag{12}$$

$$\left. + \frac{g\left[\frac{e^{-(rs+a_1)} - e^{-(a_1+a_2)}}{a_2 - rs} - \frac{e^{-(s+a_2)} - e^{-(a_1+a_2)}}{a_1 - s}\right]}{(1 - r)s + (a_2 - a_1)} \right\}$$

$$G_{cfa}(s) = \mu \frac{m + Ts}{1 + Ts} \cdot \left\{ \frac{b_1(1 - e^{-\{(r+1)s + a_1 + a_2\}})}{(r + 1)s + a_1 + a_2} \right. \tag{13}$$

$$\left. \frac{g\left[\frac{1 - e^{-(rs+a_2)}}{rs + a_2} + \frac{e^{-\{(r+1)s+a_1+a_2\}} - e^{-(rs+a_2)}}{s + a_1}\right]}{(r + 1)s + a_1 + a_2} \right\}$$

$$G_{cpa}(s) = -a_1\frac{m+Ts}{1+Ts}\cdot\left\{\frac{\{(r+1)s+a_1+2a_2\}e^{-(rs+2a_2)}}{(rs+2a_2)^2-(s+a_1)^2} - \frac{2(rs+2a_2)e^{-(s+a_1)}}{(rs+2a_2)^2-(s+a_2)^2} + \frac{\{(r-1)s+2a_2-a_1\}e^{-\{(r+2)s+2a_1+2a_2\}}}{(rs+2a_2)^2-(s+a_1)^2}\right\} \quad (14)$$

DESIGN OF PREDICTORS

Let a transfer function G(s) be approximated with $G_1(s)G_2(s)$ where $G_2(s)$ is a high frequency asymptote of G(s) and $G_1(s)$ represents a low frequency asymptote, i.e.

$$\lim_{s\to\infty} G(s) = G_2(s)\ ,\ \lim_{s\to 0} G(s) = \lim_{s\to 0} G_1(s)G_2(s) \quad (15)$$

We assume that $G_2(s)$ represents the time delaly effects in G(s). Then a Smith type predictor is developed to eliminate the effect of $G_2(s)$. If $G_2(s)$ is a pure time delay, the Smith type predictor coincides with the standard Smith predictor. The implementation of it is shown in **Fig.2**. **Fig.3** shows a standard Smith predictor.

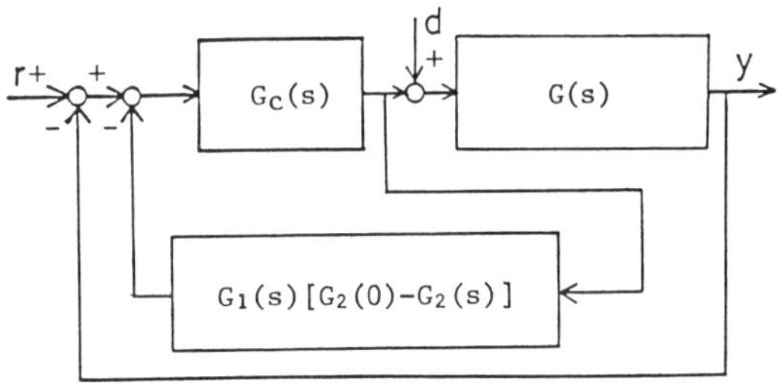

Fig.2 Implementation of a Smith type predictor.

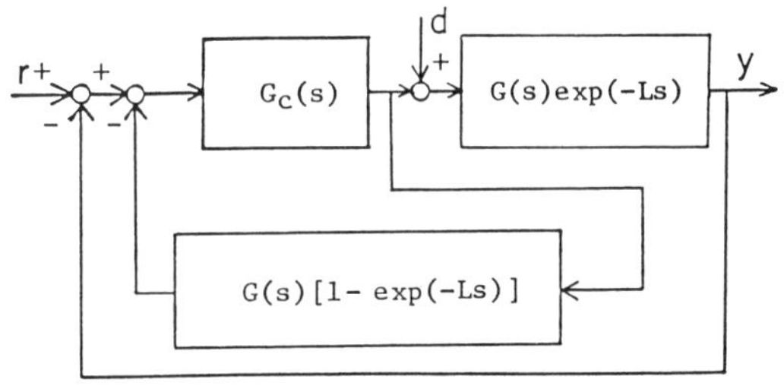

Fig.3 Implementation of a Smith predictor.

The output y(s) of the control system may be written

$$y(s) = \frac{G_cGr(s) + G\{1+G_c(G_1G_{20}-G)+G_c(G_1G_2-G)\}d(s)}{1+G_cG_1G_{20}+G_c(G_1G_2-G)}$$

$$G_{20} = G_2(0) \quad (16)$$

Thus, if C_c becomes a large constant, the output for step change of disturbance converges to zero, and the output for input becomes;

$$y(s) \to [G_2(s)/G_{20}]r(s)$$

1. Parallel Flow Heat Exchanger

The approximate transfer function (11) shows that its high frequency part $G_{pa2}(s)$ consists of two time delays. Then a control law is made to eliminate both time delays by using Smith type predictor. Another control law is made to eliminate the smaller time delay of the two time delays using the standard Smith predictor. In the former, as $G_{pa2}(s)$ represents almost all part of $G_{pa}(s)$ except the low frequency part, the control object is only its low frequency part, then improvement of control performance is not expected for intermediate frequency. In the latter the the predictor eliminates only the pure time delay effect but a problem is that the right hand-side zeros cannot be eliminated.

a) A Smith type predictor

We remove all high frequency asymptote in this case; i.e.

$$G_1(s) = G_{pa1}(s) = a_1\frac{m+Ts}{1+Ts}\ ,\quad G_2 = G_{pa2}(s) \quad (17)$$

b) A Smith predictor to eliminate a single time delay

The approximate transfer function (11) contains two delay elements e^{-rs} and e^{-s}. It can be expressed as following:

$$G_{pa}(s) = \begin{cases} a_1e^{-a_2}\dfrac{m+Ts}{1+Ts}\cdot\dfrac{(e^{-\{(r-1)s+a_1-a_2\}}-1)}{(1-r)s+(a_2-a_1)}e^{-s}, & r>1 \\[2ex] a_1e^{-a_1}\dfrac{m+Ts}{1+Ts}\cdot\dfrac{(1-e^{-\{(1-r)s+a_2-a_1\}})}{(1-r)s+(a_2-a_1)}e^{-rs}, & r<1 \end{cases} \quad (18)$$

We write this as

$$G_{pa}(s) = G_{pa3}(s)e^{-Ls}\ ,\quad L = \min(1, r) \quad (19)$$

From this expression, we can construct a standard Smith predictor putting,

$$G_1(s) = G_{pa3}(s)\ ,\quad G_2(s) = e^{-Ls} \quad (20)$$

c) Control results

There is no finite pole in $G_p(s)$. The zeros can be found from $e^{p_1} - e^{p_2} = 0$ ($p_1 \neq p_2$) as follows:

$$s = \frac{a_2-a_1}{r-1} \pm \frac{2\sqrt{a_1a_2+\pi^2n^2}}{r-1}j\ ,\quad n=1,2,\cdots \quad (21)$$

Hence, the location of zero is possible to be in the left hand side or in the right hand side of s-plane depending on the parameters. **Fig.4** shows

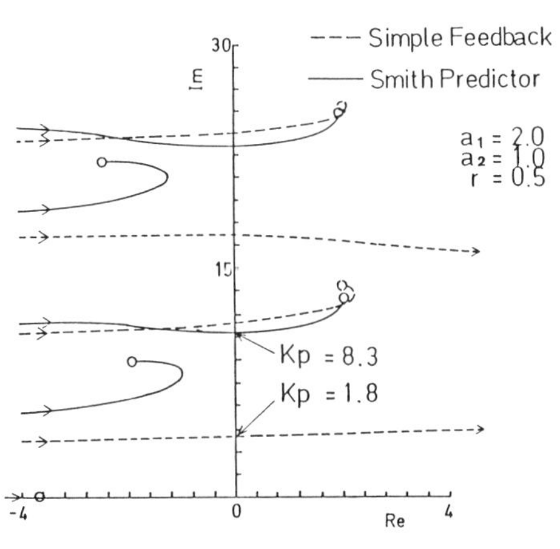

Fig.4 Root loci of a parallel flow heat exchanger controlled by a simple feedback control and a Smith predictor.

root loci (dotted lines) of a parallel flowheat exchanger controlled by a simple feedback control law.

There are two kind of root locus, half lines running from the left-side infinity and terminating at the zeros, and lines running from left hand side infinity to the right hand side infinity which are consistent with the pattern of the root locus of a system of a pure time delay element and make the system unstable for a large feedback gain. The solid lines in Fig.4 represent root

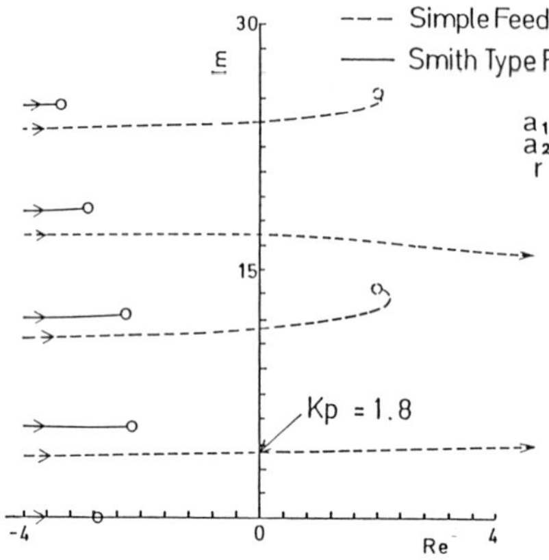

Fig.5 Root loci of a parallel flow heat exchanger controlled by a Smith type predictor.

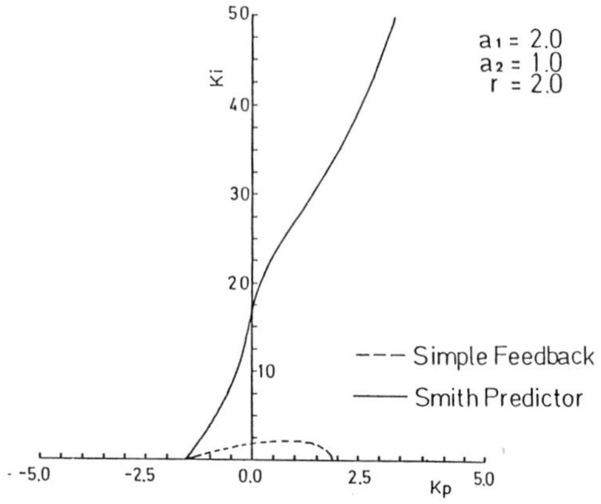

(a) Left hand zero type

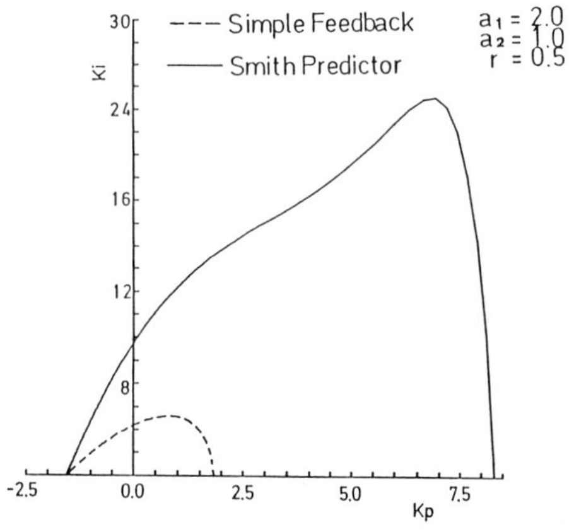

(b) Right hand zero type

Fig.6 Stability regions of parallel flow heat exchanger controlled by a PI feedback controller and the Smith predictor.

loci of the control system with the Smith predictor. Although all root loci running form left infinity to right infinity are eliminated, root loci terminating at zeros are unchanged. This means that Smith predictor eliminates a pure time delay element but there remains extra delay due to the right hand side zeros. **Fig.5** shows root loci of the control system using the Smith type predictor. All root loci running from left infinity to right infinity are eliminated and the right hand side zeros are moved to the left hand side of s-plane. Hence a parallel flow heat exchanger can be always stabilized by use of the Smith type predictor. When the standard Smith predictor is used, the heat exchanger having left hand side zeros can be always stabilized. The heat exchanger having right hand side zeros is still unstable for large feedback gain, but the stable region enlarges. **(Fig.6)**

Fig.7 shows transient responses to changes of set-point and disturbance. Since the Smith type predictor control law cannot control intermediate frequency range, the transient curve after the dead time cannot be improved. While,the response of the system using the Smith predictor has quick rise time but there remains a hunting.

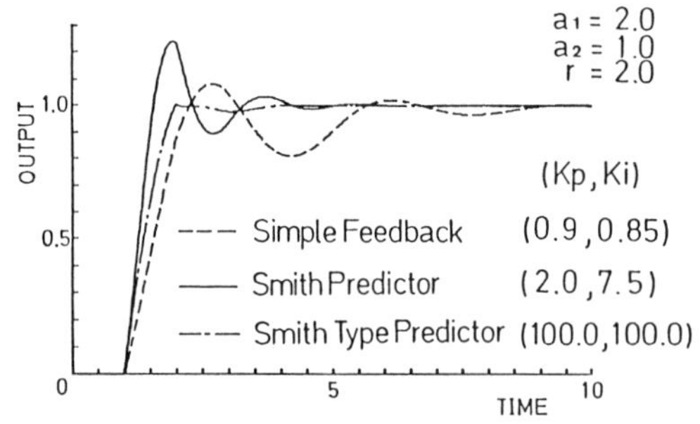

(a) Step input

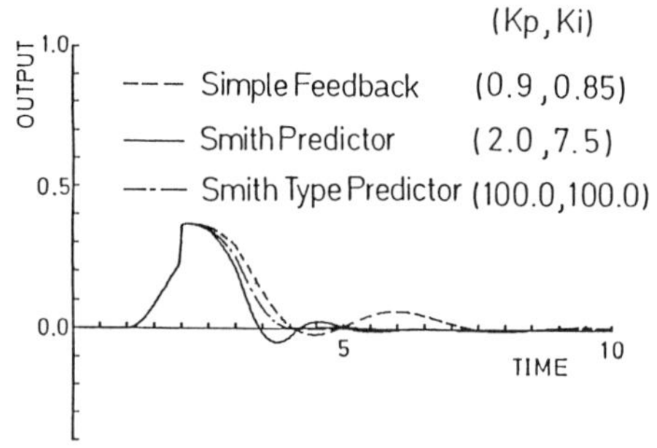

(b) Step disturbance.

Fig.7 Transient responses of the control systems.

2. Control of Flow-Forced Heat Exchangers

When a parallel flow or a counter flow heat exchanger is controlled by manipulation of a flow rate, there is no root locus due to a pure time delay element but all root loci terminate at zeros which are in the right half plane of s-plane for a parallel flow heat exchanger for almost all operating conditions and may be in the right half plane for a counter heat exchanger depending on operating conditions. As shown previously, the time delay effect due to the right hand side zero can be eliminated by the Smith type predictor, it is expected that the Smith type predictor will work well for the case of control of flow-forced heat exchangers. **Fig.8** and **Fig.9** show step responses for set-point change of a parallel flow and a counter flow heat exchangers controlled by the manipulation of a flow rate respectively.

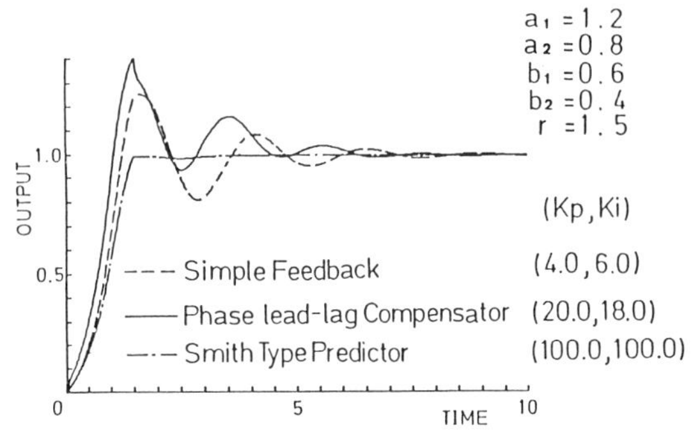

(a) Step input

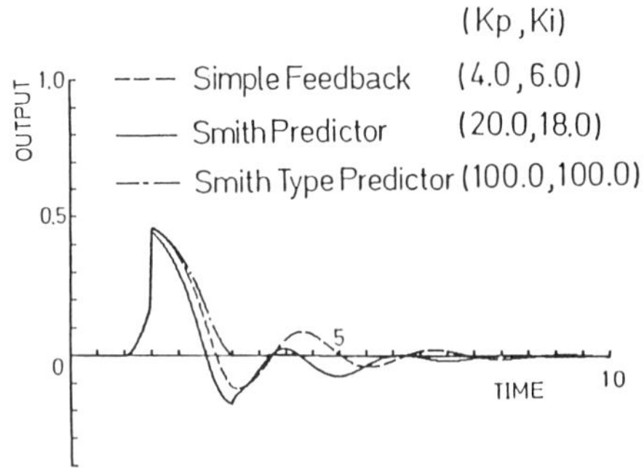

(b) Step disturbance.

Fig.8 Transient responses of a flow forced parallel flow heat exchangers controlled by a Smith type predictor.

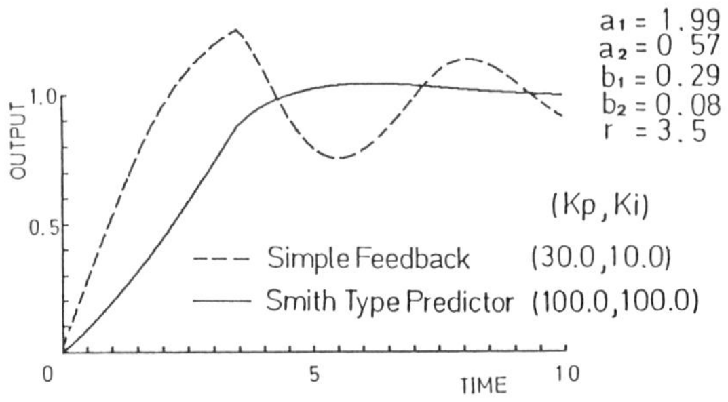

(a) Step input

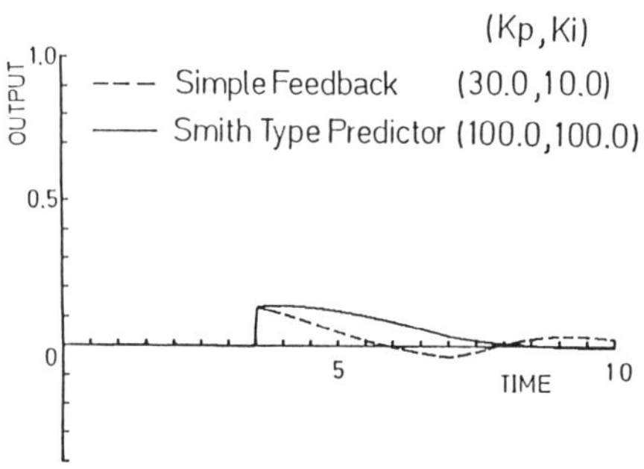

(b) Step disturbance.

Fig.9 Transient responses of a flow forced counter flow heat exchangers controlled by a Smith type predictor.

In Fig.8, as examples of the conventional control law, PI-feedback control and phase lag-lead compensator control are considered. The latter can shift phase lag and is effective to increase a stability region. The convensional control systems improve rise times of the responses but there remain considerable hunting, which contrast well with those of the predictor controllers.

Fig.10 shows comparison among root loci of a parallel flow and a counter flow heat exchangers controlled by a Smith type and the Smith predictor. The root loci do not go to the right hand side zeros and stop at points located in the left hand side half plane. Then the systems never be unstable for any feedback gains.

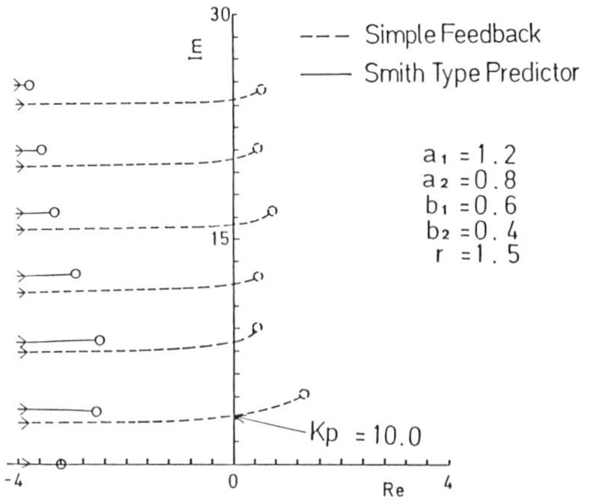

(a) Root loci of a parallel flow heat exchanger

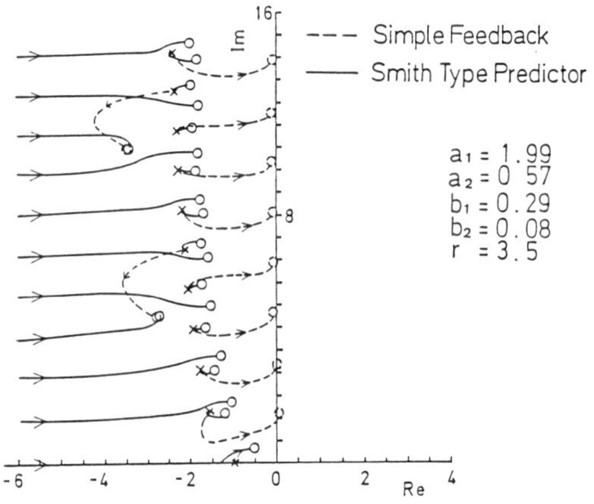

(b) Root loci of a counter flow heat exchanger

Fig.10 Root loci of flow forced heat exchangers.

3. A Two-Pass and One-Shell C-P Type Heat Exchanger

C-P type heat exchanger has root loci similar to those of a parallel flow heat exchanger; there are two kind of root locus, half lines running from the left-side infinity and terminating at zeros and lines running from left hand side infinity to the right hand side infinity. Therefore, the Smith type predictor like one shown in Fig.2 will give a good performance. However,as seen from equation (14), the transfer function is more complicated to install the predictor than that of

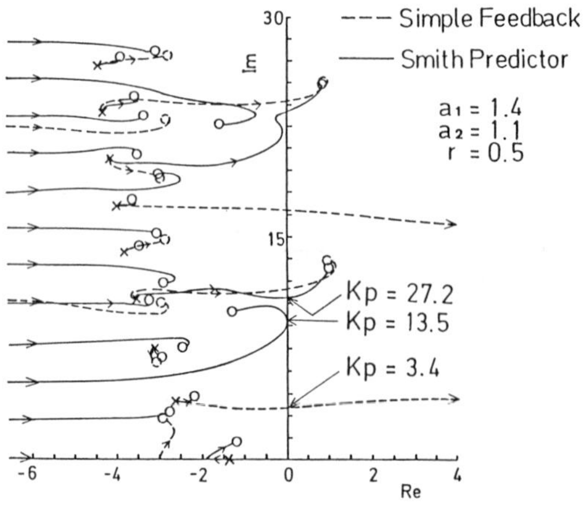

Fig.11 Root loci for Smith predictor.

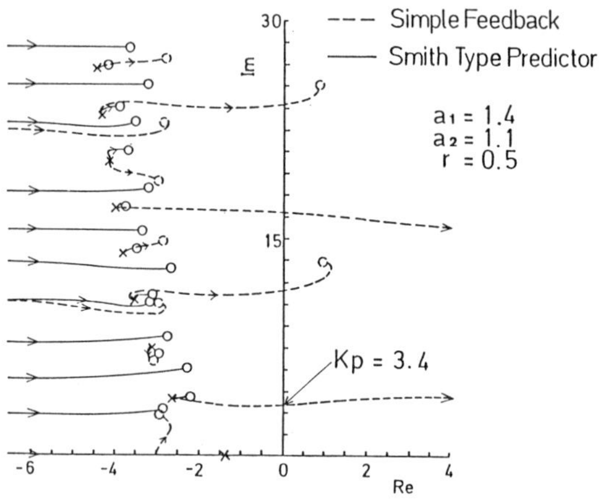

Fig.12 Root loci for Smith type predictor.

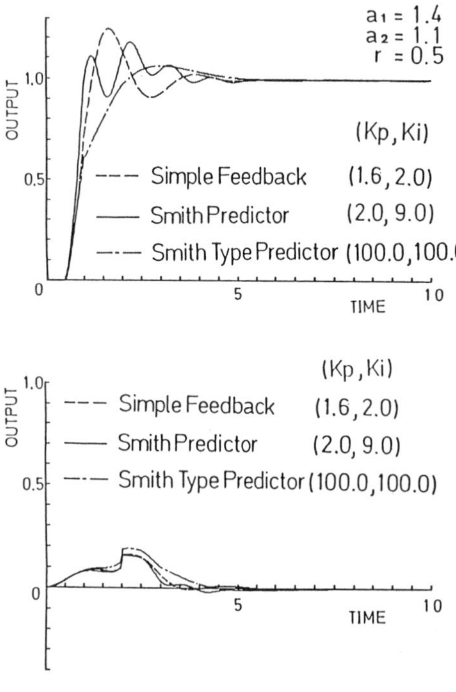

Fig.13 Transient response of C-P type heat exchanger controlled by a Smith type, a Smith predictor and PI feedback.

a parallel flow heat exchanger. At first, we use the standard Smith predictor to remove only the pure delay time delay effect. Extracting the pure delay time element e^{-Ls} ($L=\min(r,1)$) from the approximate transfer function (14), we obtain

$$G_{cpa}(s) = G_{cpa1}(s)\cdot e^{-Ls} \tag{22}$$

from which a Smith predictor is constructed.

Fig.11 shows root loci of resulting control system, where the right hand-side zeros are not removed. Secondly, we use a Smith type predictor to remove all root loci entering into the right half plane. (**Fig.12**)

Fig.13 shows transient responses to step changes of set-point and disturbance. The Smith type predictor control law gives very good control performance. The Smith predictor gives quick rise but there is hunting because of the right hand side zeros and the response is similar to that of PI-feedback control.

CONCLUSION

This paper investigates stabilizing control of the parallel flow type heat exchangers which are apt to be unstable for a conventional feedback control law. To overcome this difficulty, the following control laws are proposed.

1) A Smith type predictor is the most effective to eliminate the influence of the time delay effect due to either the pure time delay element or the right-hand side zeros. This method improves the control performance in the high and low frequency range but cannot improve that in intermediate frequency.

2) If the transfer function has not the right hand side zero, we can extract a pure delay element from the approximate transfer function and the standard Smith predictor which eliminates the pure time delay improves control performance and guarantees its stability. But this method cannot eliminates the delay effect due to the right hand side zeros.

3) The asymptotic approximation method is useful to construct the Smith type predictor realized by a small micro-computer, because it makes such approximate model that its transfer function consists of pure time delay elements and rational transfer functions and its coefficients are the functions of the original system parameters.

REFERENCES

1. H.Kanoh, Distributed Parameter Heat Exchangers Modeling, Dynamics and Control, in Distributed Parameter Control System, Edited S.G.Tzafestas, Pergamon-press (1982), 438.
2. H.Kanoh, IFAC 3rd Symposium Control of Distributed Parameter Systems, (1982), 427.
3. J.M.Smith, IAS J., 6(2) (1959), 28.
4. A.Schone, Proc 3rd IFAC Congress (London), (1966), 10B.
5. J.C.Friedly, Dynamic Behaviour of Process, Prentice-Hall,inc.,(1972), 444.

TOOL BREAKAGE DETECTION IN TURNING USING A MULTI-SENSOR STRATEGY

A. Galip Ulsoy and Erliang Han

Department of Mechanical Engineering and Applied Mechanics, The University of Michigan, Ann Arbor, MI 48109–2125, USA

Abstract. To meet the industrial goal of autonomous manufacturing, a multi-sensor strategy for tool breakage detection is proposed and a microcomputer based system which monitors (i)cutting force (ii)cutting force rate, and (iii) acoustic emission count is implemented. Tool breakage tests for impact, thermal stress at tool exit or entry, and tool wear are conducted and observations of cutting forces, acoustic emission count, and temperature are made. Relevant cutting process failure phenomena and the current practice for detection techniques are also discussed. Computer evaluation tests were successful and the methods are promising. It is observed that the force rate signal is particularly useful but the acoustic emission count is not. The proposed system is also flexible enough to incorporate other signal detection techniques.

Keywords. Tool breakage detection; autonomous manufacturing; machine tool diagnosis; cutting processes.

NOMENCLATURE

F_f: Feed force component;
f_f: Feed force average over a short time;
$\underline{F}_f$: Feed force average over a long time;
F_c: Cutting force component;
AE: Acoustic Emission;
S: Slope of the acoustic emission count signal;
s : Short time average of S ;
$\underline{S}$: Long time average of S ;
D_f: Force rate or derivative absolute value;
d_f: Short time average of force rate;
$\underline{D}_f$: Long time average of force rate;
α_1: Upper criteria for f_f;
α_2: Lower criteria for f_f;
β: Upper criteria for d_f;
γ_1: Upper criteria for s ;
γ_2: Lower criteria for s ;
P_1: Upper process limit for F_f;
P_2: Lower process limit for F_f;
P_d: Upper process limit for D_f;
P_3: Upper process limit for S;
P_4: Lower process limit for S;
L_1: Lower absolute limit for F_f;
L_2: Upper absolute limit for F_f;
L_3: Lower absolute limit for S;
T_d: Time constant for differentiator;
K_d: Differentiator gain.

INTRODUCTION

Tool breakage detection is receiving increasing attention in industry due to the desire for autonomous manufacturing systems. Successful detection of tool breakage events can prevent damage to the workpiece and machine tool and reduce down-time. Breakage detection systems can also be incorporated into process control systems for maximizing production rates.

Various cutting process malfunctions(e.g., chatter, chip entanglement, etc.) will, under autonomous circumstances, eventually lead to breakage events. Immediate on-line detection of a tool breakage event can reduce the potential damage. To prevent breakage events, however, it is essential to forecast such events. Breakage events due to developing cutting process malfunctions can be predicted, and prevented using on-line sensing systems.

Many systems and methods have been studied to meet the challenge of breakage detection and prevention(Tlusty,1983; Kegg,1984; Jetly,1984), but none of them has proven satisfactory(See,1982). A common deficiency of these methods is that they are all based upon the monitoring of a single cutting process state, therefore, the method is not reliable. By monitoring multiple signals on-line during cutting, we propose to develop an intelligent system robust enough to detect as well as forecast various catastrophic damage events. So far, our study is limited to a continuous cylindrical turning operation, but can be adapted to more complicated operations.

Three series of tests were performed to observe tool failure phenomena under impact, thermal stress, and wear situations by monitoring cutting forces, temperature, and acoustic emission(AE) signals. A microcomputer based system with a strategy of monitoring feed force and acoustic emission signals was developed and tested. The computer evaluation tests are promising, but require further development.

PHENOMENA AND DETECTION

Cutting tool breakage is due to fracture or large plastic deformation of the cutting tool. A breakage event is likely to occur when large cutting forces and high temperatures are present for a worn tool. Excessive wear will weaken the cutting tool or increase the cutting force(Micheletti). The conditions under which breakage occurs depend upon cutting force, tool geometry, material properties and cutting temperature. Mechanical stresses induced by impact at tool entry, and thermal stresses due to sudden unloading at tool exit can cause tool fracture(Lee,1984). Crater wear, due to a sharpening effect, decreases cutting force but greatly weakens the cutting tool. Flank wear, on the other hand, increases the cutting force(Micheletti). Events such as chatter and chip entanglement can also lead to tool breakage, especially under untended conditions.

A fool-proof detection system should be able to detect all kinds of failure events related to the cutting tool. The ultimate goal in tool breakage detection is to completely prevent such events by prediction, or at least to minimize the damage by immediate detection and response. The latter function is essential because events like breakage could occur with no warning signals that can be detected. But a detection system cannot be satisfactory without the first function because failure events like chatter and gradually developing wear do provide information for tool breakage prediction. The challenge is to develop a fool-proof system which can minimize false alarms.

Current practice includes monitoring based upon cutting force(Tlusty, 1983; Mouri,1981), acoustic emission(Tlusty, 1983; Kakino,1981), motor current(Matsushima,1982; Uhlman,1981) and other methods(Kegg,1984). Most of these methods have incorporated special signal processing techniques like low and high pass filters, RMS averaging, burst-type signal counters, and spectrum analysis.

Cutting force signals the best developed sensing technique, but fail to give a complete picture of the cutting process or tool wear. Cutting force may either increase or decrease as the tool wears depending upon the types of wear developed (Uhlman,1981). The force may not be sensitive to certain types of breakage, such as that due to thermal effects.

Temperature sensing methods include natural or semi-natural thermocouples (Xu,1978), remote sensing, and infrared ray sensors. None have been successfully applied to tool breakage detection due to various practical problems.

Acoustic emission is gaining increasing attention, particularly its relationship with the cutting process(Dornfeld,1982;Emel,1983). Acoustic emissions result from plastic deformation and crack propagation. This technique could be useful for all types of breakage and its application in tool breakage detection appears promising(Kakino,1981). Because the useful signal frequency range is about 100-300 KHz, dedicated high frequency processing equipment is required.

Motor current or power are easy to sense, and some commercial devices have been developed based upon this technique(Tlusty, 1983). Motor current or power is proportional to the spindle torque or cutting force, and therefore, contains the same information as the force signal. But it has a time lag and a reduction in sensitivity due to the machine tool dynamics(Tlusty, 1983).

Optical methods are difficult for on-line sensing, but are also receiving increasing attention due to development of computer vision systems. Machine tool vibration signals are relatively easy to pick up, but often the signal is coupled with machine tool dynamic response. These and other techniques were discussed in (Jetly,1984).

The fact that there are so many detection techniques available indicates that there is not yet a single measurement which can be used as the basis for a satisfactory detection system.

EXPERIMENTAL SETUP AND PROCEDURES

Three series of turning tests were performed and several signals were recorded using an instrumentation recorder for later playback and computer analysis. The cutting conditions are summarized in Table 1. The three types of tests were performed are: (i) failure due to the impact between the tool and a hardened insert in the workpiece; (ii) failure due to excessive wear, and (iii) failure due to thermal stresses induced by rapid cooling when a hot tool is rapidly retracted from the cut. These different types of tests are denoted in Table 1 by I,W, and T respectively.

The series A tests were performed on a LeBlond NC lathe. The test sequence is a series of interrupted "parts" until failure, where one part is defined by one cubic inch of metal removal. Two tests were run, and in both cases the tool developed significant crater wear and finally failed due to excessive wear.

Table 1 Cutting Conditions

Series	Test #	Cutting Velocity (m/min)	Feed (mm/r.)	Depth of cut (mm)	Tool Type	Material
A	1&2(W)	260	0.51	2.03	TNMP434	C370
	1 (I)	122	0.325	1.52		
	2 (I)	152	0.325	1.78		
B	3 (T)	198	0.325	1.52	TNMG434	4140
	4 (I)	152	0.325	1.78	E48	
	5 (W)	244	0.325	1.52		
	6 (T)	213	0.325	1.91		
	1 (W)	244	1.78	1.27		
C	2 (W)	262	3.43	1.27	TNMG434	4145

The series B and C tests were performed on a conventional 20 HP American lathe using a continuous cut, and several passes were required to achieve failure.

A strain gauge dynamometer was used to measure the feed and cutting components of the force. Signals recorded in the series C were later used for computer analysis. Force signals were directly recorded from the dynamometer using FM recording. The acoustic emission signal frequency is, however, too high to record, so an acoustic emission counter processor was employed and the acoustic emission count signal was recorded.

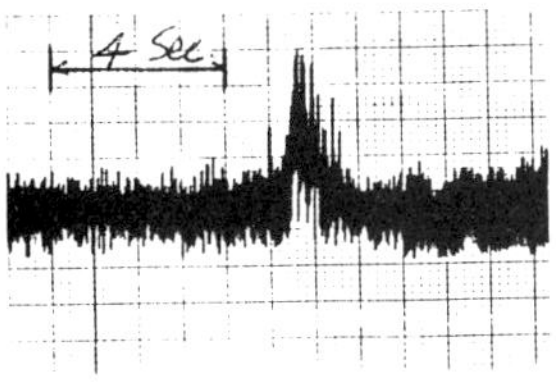

Fig. 1. High frequency component and burst-type force signal

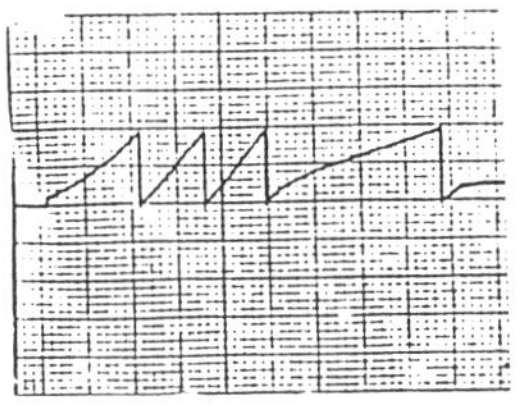

Fig. 2. reset rate of AE count change in test 4 of series B

Those signals observed are: the feed and cutting components of the force, and the acoustic emission count. As shown in Fig. 1, the force signal often contains high frequency noise and must be filtered to avoid aliasing, so a low pass filter was employed. Fig. 2 illustrates the acoustic emission count output. The slope of the saw tooth-like signal is proportional to the number of burst-type events in the acoustic emission signal (Emel,1983).

<u>Observations From the Force Signals</u> The following observations are based on the feed force and cutting force signals during the series A, B, and C tests:

1) During the tool wear process, the force average showed apparent changes in several tests, as in Fig. 3 from series C. The high frequency component usually increases with the wear, as in Fig. 4 from series B, test 6. The force signal shows low frequency components and burst type signals when the cutting tool is especially worn, as in Fig.1 and Fig.5 from series C.

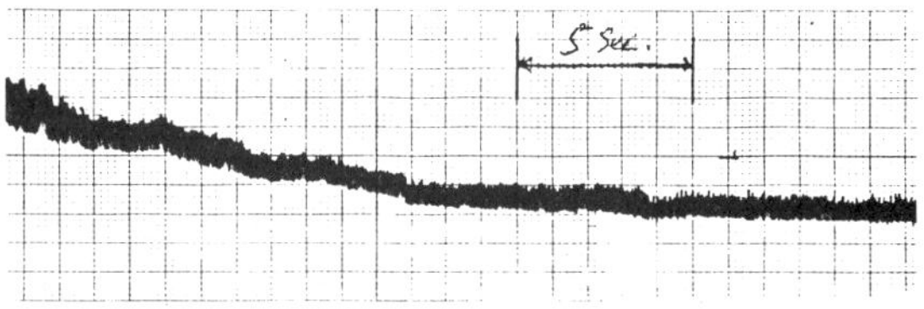

Fig. 3. Feed force changes during test 2 of series C

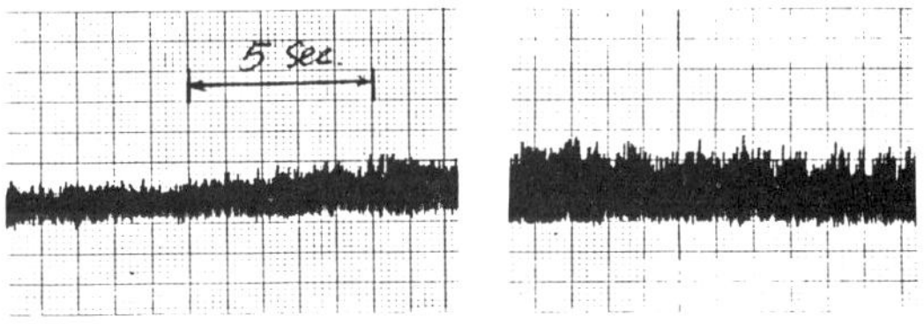

Fig. 4. Alternating component of the feed force at the beginning and end of a pass in test 2 of series C

2) Fig. 6 shows that when impact occurs, the force signal shows a sudden large increase.

3) In series B test 3, a thermal stress type test, there was no readily observable effect.

4) The cutting force component is similar to the feed force, but not as sensitive as the feed force to tool wear.

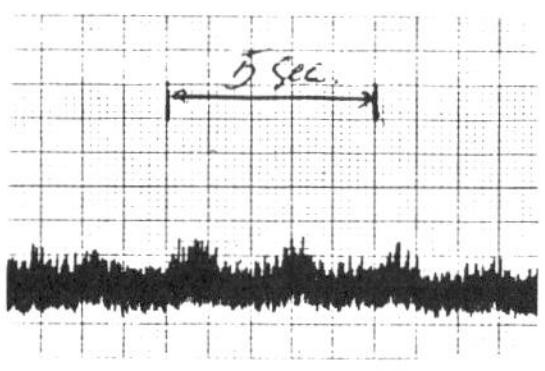

Fig. 5. Low frequency fluctuation in the force signal

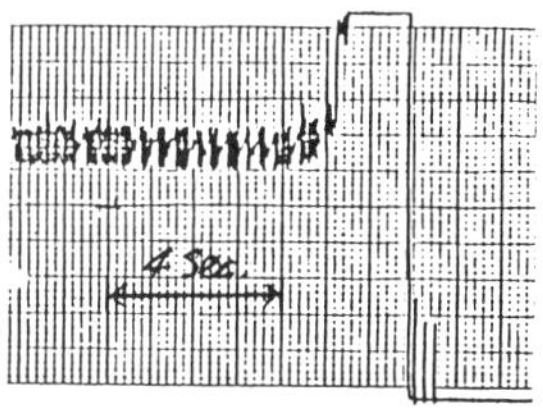

Fig. 6. Sudden increase of force at impact in 1 of series B

<u>Observations from the Acoustic Emission Signal</u> The following observations are based on the acoustic emission count signals during the series A, B, and C tests:

1) In the "wear to failure" type test, the burst-type events showed a reasonable increase from 2 to 4 resets per part as we went from

part 1 to part 16.
2) When impact occurred, the AE count signal reset rate first increased then decreased, as in Fig. 2 from series B test 4.
3) When thermal stress was introduced, the AE count signal again first increased then decreased as shown in Fig. 7 from series B test 6.

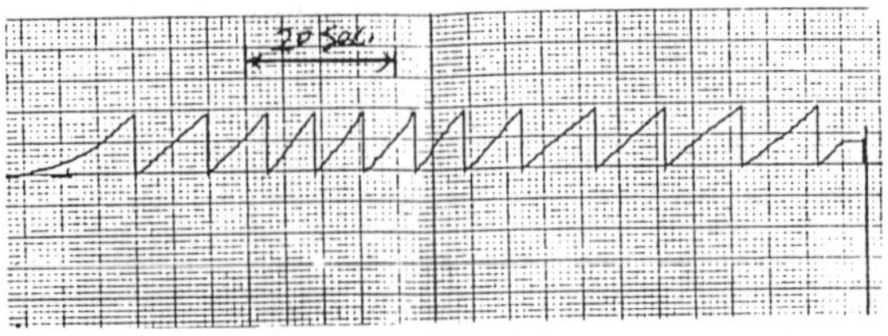

Fig. 7. Reset rate of AE count first increased then decreased in test 6 of series B

The force and acoustic emission count signals show certain features as discussed above when the tool is worn, impact occurs, or thermal stress is introduced. It was noticed, however, that these features are not always consistent with the particular event, and one can not be convinced that a certain event has occurred by observing only one feature. On the other hand, one can be more confident when several confirming features are observed.

Computer Evaluation Test In the computer evaluation test, the feed force and AE count signals recorded during series C tests were played back to simulate a real cutting process. The signals input to the computer via an analog-to-digital converter are the AE count signal, the feed force average, and feed force rate signal which was obtained by processing the feed force signal through a low pass filter and approximate differentiator respectively.

The relationship between the differentiated signal D_f and feed force signal F_f is(Ulsoy,1983):

$$D_f(s)/F_f(s)=K_d*s/(T_d*s+1) \quad (1)$$

where $T_d \ll K_d$, T_d is the differentiator time constant, and K_d is the differentiator gain. The force rate, D_f, can be positive or negative. For convenience, D_f will exclusively be used to designate its absolute value through out the following text. The relation between feed force, F_f, and feed force rate, D_f, is such that any increase in the alternating component of the feed force will be reflected as an increase in D_f, and the burst-type event or sudden change in F_f will lead to a "peak" in D_f.

While the recorded signals are for series C tests in which a breakage event did not occur, it is anticipated that the force average and acoustic emission count may go to zero immediately, the signal D_f will show a "peak" because of the sudden change in the force signal. By suddenly unplugging the force signal or stopping the recorder, effects similar to breakage were simulated in the signals. A 4-beam storage oscilloscope was used to observe simultaneously the signals F_f, D_f, and AE. Signal signatures were observed at the moment when a failure event was indicated by the computer.

The computer evaluation test also involves tedious tuning of many program parameters, which will be discussed later. For the two groups of data signals of series C, the tuned system indicates failure at a reasonable moment when the cutting tool is seriously worn, and the force rate, D_f, signature is very large. When either the force, or acoustic emission signal was interrupted, a failure event was immediately indicated.

The force rate signal, D_f, has proven to be extremely useful in indicating failure when cutting conditions are considered unacceptable(e.g., chatter, built-up edges, etc.). When the tool is worn and tool vibration is serious, the force rate shows large fluctuations quite consistently and can be easily detected. The acoustic emission count signal is, however, not very useful. The output of the AE counter depends strongly on the adjustment of the gain of the processor (Emel,1983), and due to such adjustment problems this signal would probably not be very practical on the shop floor.

TOOL BREAKAGE DETECTION STRATEGY

The Idea of Voting As mentioned before, no system using a single signal or feature is reliable. Those signals or features that might indicate failure sometimes occur when there is no failure event; and when there is a failure event they do not necessarily occur.

A signal feature that may indicate failure can be considered as an indication of a probable failure event. Thus, two or more features mean an increased probability of failure. From this point of view, we devised a detection system which monitors multiple signal features that might indicate failure. It is based on a very simple decision making logic. That is, each signal feature is given a right to vote; if the majority vote for failure, then the system decides a "true" failure event has occurred.

Based upon our experimental work, we include possible failure event signal features as:
i) The average feed force may increase or

decrease when the cutting tool wear progresses, if it changes a certain percentage, e.g.,

$$f_f > \alpha_1 * \underline{F}_f \text{ or } f_f < \alpha_2 * \underline{F}_f \quad (2)$$

a possibility of failure is indicated.
ii) A substantial increase in the force derivative signal also indicates a possibility of a failure event,

$$d_f > \beta * \underline{D}_f \quad (3)$$

iii) Similarly for the acoustic emission count slope, a possible failure is indicated when,

$$s > \gamma_1 * \underline{S} \text{ or } s < \gamma_2 * \underline{S} \quad (4)$$

A note should be given here that all the above values of force, AE count slope, etc. are averages over some time period. While $\underline{F}_f$ is an average over a long time, f_f is an average over a small time interval, and is used because the instantaneous value of a cutting process signal shows large fluctuations, as reported by Kegg(1984). Sometimes a change of less than 20% in force can result in failure, while sometimes the change can be as much as 300% without failure. By averaging those values to be tested over a small time interval, the large random signal change range is reduced, hence the reliability increased and false alarms reduced. This proves important in the process of criteria testing. The equations used to calculate $\underline{F}_f$(current long term average) and f_f (short term average) are,

$$\underline{F}_f(t) = [\, f_f(t) + N * \underline{F}_f(t-\Delta T)]/(N+1) \quad (5)$$

$$f_f(t) = [\sum_{i=0}^{m-1} F_f(t - i*\Delta t)]/M \quad (6)$$

where N and M range from 20 to 30 in our tests. F_f (t) is an instantaneous value , Δt is the sampling interval and

$$\Delta T \approx \Delta t * M \quad (7)$$

d_f, D_f and s, $\underline{S}$ are obtained with the same formulas as (5) and (6).

Fixed clock interrupts were not used for sampling. Thus, the sampling rate was determined by the computation speed and was approximately 50-100 ms for our system. This sampling rate can be increased, e.g., by using a fast processor, for the emergency limit testing described below.

The Concept of Limits Despite the merits of the democratic method of voting based upon a number of signal features, the method should be used with caution. While the above criteria are well suited to gradually developing types of failure events, a decision based upon this method is not made until all features have been searched and analyzed. However, in case of tool breakage(e.g., as a result of impact), even a fraction of a second is important for reducing the degree of damage, but no pre-event signal or features are provided. Therefore, a quick decision-making ability against such emergencies is desired.

A signal feature(e.g., the increase of force) represents a probability of a failure event. The probability is proportional to the magnitude of the feature. By this it is meant that if the force becomes extremely large then there definitely is failure, and the probability is 100 per cent. Thus the concept of a limit is introduced: if a signal feature is at its extreme condition(e.g., either too small or too large), then a quick decision can be made based upon this signal feature. We use limits to test for these extreme conditions. In contrast to criteria testing, values tested by limits are truly instantaneous values.

1) Process Limits These limits indicate extreme changes in the process and are related to the signal averages:

$$F_f > P_1 * \underline{F}_f \text{ or } F_f < P_2 * \underline{F}_f \quad (8)$$

$$D_f > P_d * \underline{D}_f \quad (9)$$

$$S > P_3 * \underline{S} \text{ or } S < P_4 * \underline{S} \quad (10)$$

2) Physical Limits These limits are based upon physical constraints. The upper limit for feed force, L1, is based upon cutting tool strength as well as tool holder and machine characteristics. The lower limit of force, L2, is the value such that the cutting tool and workpiece might be considered to have lost contact. A lower limit for AE count slope, L3, is also used to indicate a loss of contact via a breakage event.

$$F_f > L_2 \text{ or } F_f < L \quad (11)$$

$$S > L \quad (12)$$

The essential logic of the tool breakage detection strategy is voting based on criteria testing and quick response base on limit testing.

The large number of different criteria and limits necessitated a procedure for tuning the system or determining these criteria and limits. To expedite the process, a "flag" is set for each criteria or limit in the program. The corresponding "flag" signals as each criteria or limit is exceeded. We choose those limits and criteria such that all "flags" are initially set. Then we relax each one to have its flag set at an appropriate instant during the tests. The system as tuned using series C test 1 or 2 then works for both groups of tests satisfactorily.

We feel such a system is promising as it not only has the function of detecting breakage, but also can use simple intelligence to analyze for a developing failure event. Such a system is also flexible as it can be easily changed and adapted to different conditions, and incorporate other signal detection techniques. While the tuning process is tedious, dedicated tuning software could be developed to automate the process.

CONCLUSIONS

Here we make the following conclusions:
A. By building basic intelligence into a detection system which monitors multiple signals, the reliability of tool breakage detection is enhanced and false alarms are reduced. Our system is robust enough to quickly detect a breakage event.
B. The system has the ability to forecast breakage but cannot be considered to be robust because the acoustic emission count signal is not adequate.
C. The system is flexible enough to incorporate other signal detection techniques.
D. The force rate signal has proven to be very useful, but the acoustic emission count is not. Other AE processing techniques, such as RMS average should be investigated.

ACKNOWLEDGMENTS

The authors are pleased to acknowledge the financial support of the Ministry of Education of the People's Republic of China for E. Han, as well as support from the following companies: Borg-Warner, Caterpillar, Deere and Company, Eaton, Ex-Cell-O Corp, General Electric, Giddings and Lewis, Kennametal, Lodge and Shipley, and TRW Inc..

REFERENCES

Dornfeld, D.A.(1982). Investigation of Metal Cutting and Forming Process Fundamentals and Control Using Acoustic Emission. ***Proceedings of 10th NSF Grantee's Conference on Production Research and Technology***.

Emel, E. and E. Kannatey-Asibu, Jr.(1983). Acoustic Emission Sensig of Tool Wear and Breakage. ***Report to the Industry-UM Consortium on In-process Sensing and Control For CAM.***

Han, E., and A.G. Ulsoy(1984). Tool Breakage Detection in Turning Using Force and Acoustic Emission Signals. ***Report to the Industry-UM Consortium on In-process Sensing and Control for CAM***.

Jetly, S.(1984). Measuring Cutting Tool Wear On-line: Some Practical Considerations. ***Manufacturing Engineering*** , July.

Kakino, Y.(1981). In Process Detection of Tool Breakage by Monitoring Acoustic Emissions. ***Cutting Tool Materials***, pp 25-40.

Kegg, R (1984). On-line Machine and Process Diagnostics. ***CIRP Conference***, Madison, WI.

Matsushima, K., P. Bertok and T. Sata(1982). In Process Detection of Tool Breakage by Monitoring the Spindle Current of a Machine Tool. ***Measurement and Control for Batch Manufacturing***, ASME, pp145-154.

Micheletti, G.F., A De Filippi, R. Ippolito(1968). Tool Wear and Cutting Forces in Steel Turning. ***Annals of CIRP, Vol.16***

Lee, Y.M.(1984). Tool Fracture Probability of Cutting Tools underDifferent Cutting Conditions. ***Journal of Engineering for Industry, Vol.16***, pp168-170.

Mouri, N. and T. Sata(1981). Automatic Tool Breakage Detection Using Kalman Filter. ***Bulletin of Japanese Society of Precision Engineering, Vol.15, No.4***.

See, H.Y.(1982). Detection and Prevention of Tool Breakage. ***Report to the Industry-UM Consortium on In-process Sensing and Control for CAM***.

Tlusty, D.J. and G.C. Andrews (1983). A Critical Review of Sensors for Unmanned Manufacturing. ***Annals of CIRP, Vol.32, No. 2.***

Uhlman, W.T. and M.J. Schmenk(1978). Torque Controlled Machining for Numerical Control Machining Centers. ***IEEE paper CH1707-9/81/0000-0055***.

Ulsoy, A.G. and H.Y. See(1983). Tool Breakage Detection in Turning. ***Report to the Industry-UM Consortium on In-process Sensing and Control for CAM***.

Xu, H.J.(1978). Investigations On High Heating Rate and Precision Calibration of Thermal Electric Characteristics of the Tool and

CONTROL SYSTEMS OF COKELESS CUPOLA

O. M. Abdel Wahab* and M. N. Ali**

**Technical Department, El Nasr Castings Co., Cairo, Egypt*
***Production Department, El Nasr Castings Co., Cairo, Egypt*

Abstract. The cokeless cupola, where coke is entirely replaced by liquid fuel, is more controllable and economical than the conventional coke cupola. It has a firm place in the foundry industry for its present and future melting requirements.

Control systems for the operating parameters, namely, combustion air, atomizing air, fuel oil, and water cooling, have been designed for and applied at the first two cokeless cupola furnaces erected in Egypt.

INTRODUCTION

The cokeless cupola is a new trend as opposed to the conventional coke cupola operation for melting cast iron. The low cost of fuel oil in some countries, as well as natural gas, compared with the expensive foundry coke, has motivated these countries to replace their coke cupolas with cokeless cupolas.

Coke is replaced in the cokeless cupola by the use of the following:

a. Fuel oil or natural gas, to provide energy required for melting the load;
b. Refractory bed, as a heat exchanger for superheating the molten metal;
c. Injection of carbon, to obtain the required carbon content in the tapped iron.

The cokeless cupola is more controllable than the conventional coke cupola in its operation and metal specification. The cokeless cupola operator can easily control the melting process using the control system which will be discussed later.

Figure 1 shows the diagramatic arrangement of a gas-fired cupola.

Foundries which produce high amounts of ductile iron can use cokeless cupolas for that purpose. Ductile iron can be produced directly from the cokeless when using natural gas or low sulphur fuel oil, without the need for additional treatment.

In areas where pollution is a serious problem, the cokeless cupola is recommended without the application of expensive air-cleaning equipment.

OPERATING PARAMETERS AND THE CONTROL SYSTEM

The control variables which govern the melting process in a cokeless cupola are the following:

a. Fuel to air ratio;
b. Carbon injection rate;
c. Refractory bed height;
d. Coating of the grate tubes.

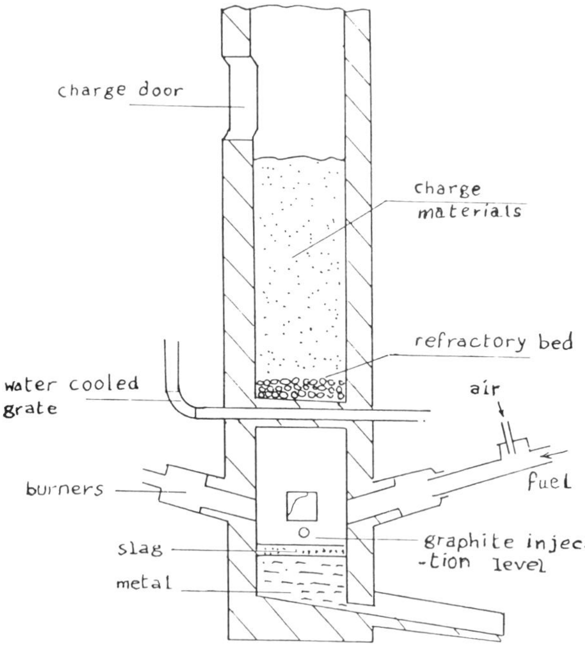

Fig. 1. Diagramatic Arrangement of Gas-Fired Cupola

Fuel to Air Ratio:

As there is no coke below the grate in the cokeless cupola, the combustion of fuel and air must be strictly controlled to avoid any oxygen, which will oxidize the iron falling into the well. Excess fuel is necessary to get a reducing atmosphere inside the cokeless, to preheat the charge under reducing conditions, and also to keep the sphere consumption controllable. Therefore, a certain ratio of fuel to air has to be predetermined to achieve these conditions (85%).

Increasing fuel and air amounts mean higher energy input to the furnace with a constant fuel-to-air ratio.

Carbon Injection :

Carbon injection is necessary, because of the absence of coke, to obtain the correct carbon content in the molten metal. Using good quality of carburizer allows high carbon recovery with low consumption and also affects innoculation process.

Refractory bed height :

The refractory bed superheats the iron droplets as they pass through it into the well. The iron temperature will be affected by controlling the depth of the bed. For temperatures in the range of 1450 ^{0}C to 1480° C. the bed height will be between 45 and 60 cm. For temperatures in the rangesof 1350 ^{0}C to 1450 ^{0}C. the bed height will range between 30 to 45 cm.

Coating of the Grate Tubes :

Coating the grate tubes will decrease heat loss into the cooling water passing through the tubes, consequently iron temperature, melting rate, carbon pick-up, and desulfurization will be improved more over higher ratios of steel scrap can be charged.

CONTROLLING INSTRUMENTS AND CONTROL SYSTEMS OF COKELESS CUPOLA

Controlling Instruments on the Panel :

The control panel contains the following instruments which control the operating parameters and detect any failure or defect during the melting operation :

a. Automatic and manual control for fuel air ratio by means of a flexel ratio station. It's computing element has in output signal which should be maintained in a definite ratio to the input signal.

b. Continuous recorder which has contiuous strip chart recorders containing 1,2 and 3 channels respectively. Each of the input channels in three open recorder for fuel, air, and pressure inside the cokeless, isolated from the others and from earth.

c. Carbon injection system for injecting the choosen recarburizer into the well where the metal and slag are held. This is done by means of a variable speed motor putting the fine grained carbon into a stream of compressed air being blown into the well. The speed of injection can be controlled according to the required carbon analysis in iron.

d. Carbon and carbon equivalent device which is connected to the control panel and has a thermocuple to measure iron temperature.

e. Alarm system for failure of water, air, fuel, electricity and compressed air.

Moreover, there is a gas analizer CO, CO_2 and O_2 in the flow gas coming out from the furnace. This device is separated from the panel and it uses chemical solutions for analysing.

Control Systems :

The following figures 2,3,4 and 5 and tables 1,2,3, and 4 show the different control systems for both two cokeless cupolas each 3.5 ton/hr in Egypt. These systems are :

a. Combustion air system.
b. Atomizing air system.
c. Fuel oil system.
d. Water cooling system.

The control systems contain two main instruments :

a. Actuators for fuel and air, in which the feedback signal is compared electrically with the input signal which represents the desired actuator position.

   Facilities are provided within the instrument to manual control the actuator when required (see fig. 2,4). Then actuators can be either automatic or manual controlled.

b. Transmitters for fuel, air, and compressed air, which they are electric differential pressure transmitters. They operate on the force balance and comprise two basic units, measuring unit, and a transmission unit (see fig. 2,4).

TABLE 1 Combustion Air Items

Combustion Air Items	Quantity No. of existing item X series no.
* Combustion air flow fans 2100 CfM at 50 WG with starter	(2 X 1)
* Pressure gauges 0-250 millibars with isolating cocks	(3 X 2)
* 12" Butterfly valves	(4 X 3)
* 12" Orifice plate assembly	(1 X 4)
* Differential pressure transmitter with equalizing manifold	(2 X 5,6)
* Main air control valve with actuater	(1 X 7)
* Positioner for actuater type AP 200/10	(1 X 8)
* Low pressure switch for combustion air failure type 9/02	(1 X 9)
* 6" Butterfly valves	(6 X 10)
* 6" Orifice plate assemblies	(6 X 11)
* L. ported change-over cocks ¼"	(12 X 12)
* L. ported change-over cupola press. ½"	(4 X 13)
* Isolating valves ½"	(2 X 14)
* Differential pressure cocks ½"	(12 X 15)

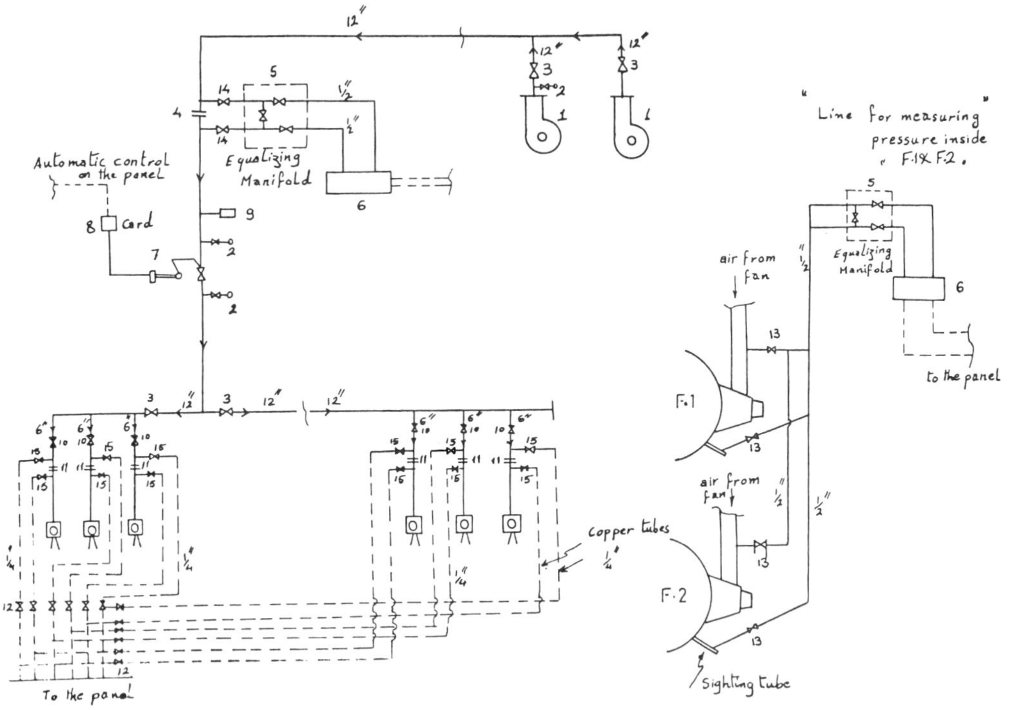

Fig. 2. Combustion Air System

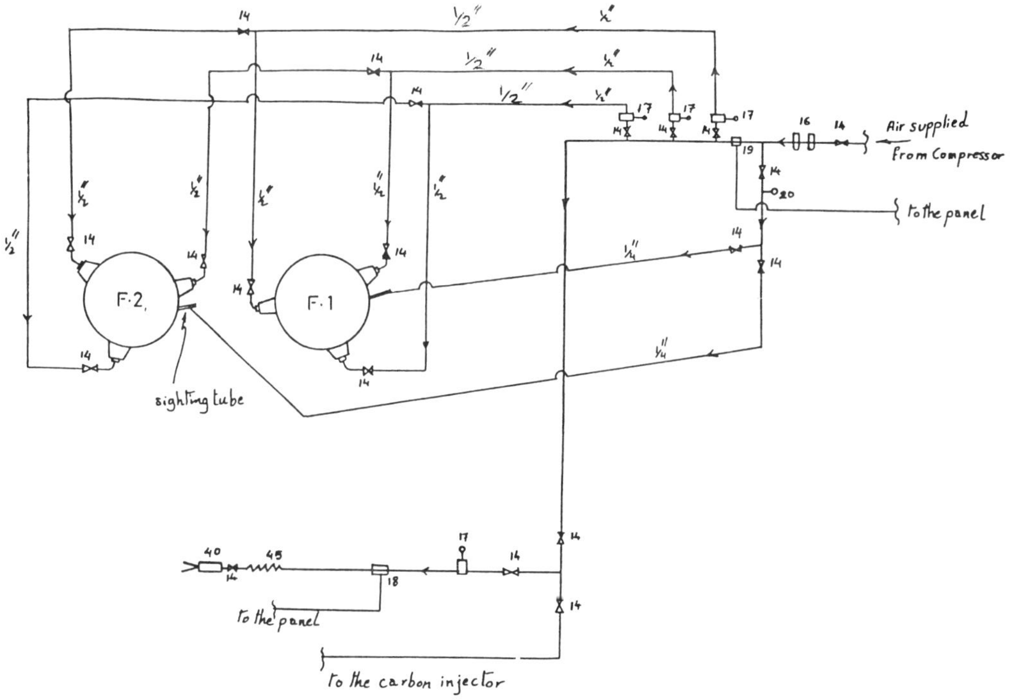

Fig. 3. Atomising Air System

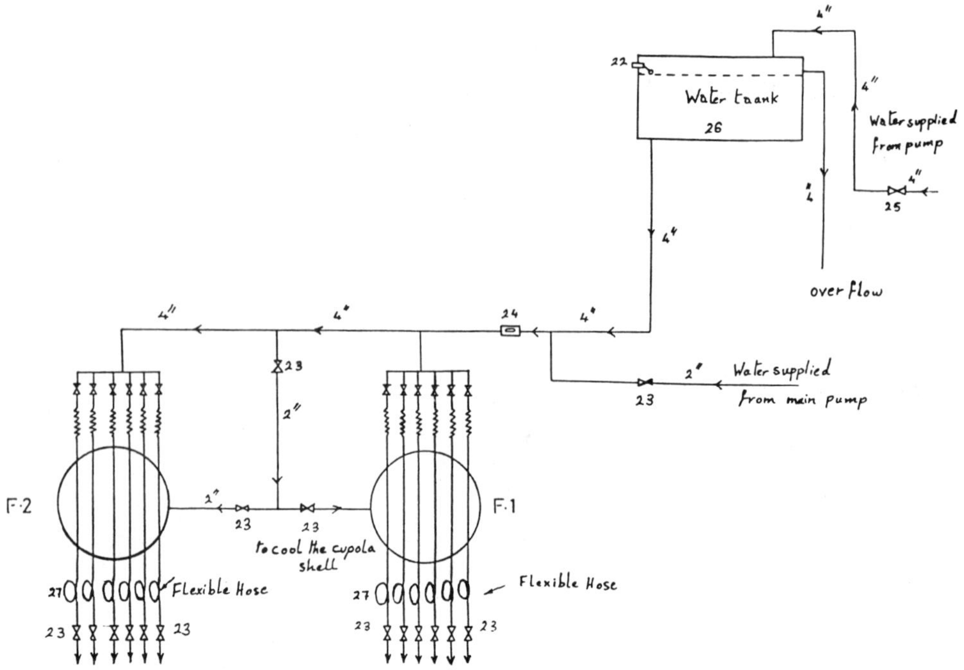

Fig. 4. Water Cooling System

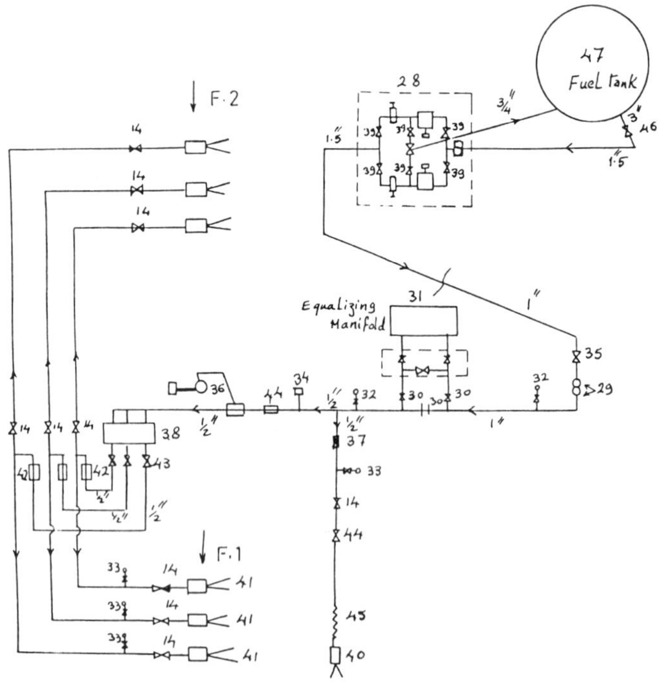

Fig. 5. Fuel Oil System

TABLE 2 Atomising Air Items

Atomising air Items	Quantity No. of existing Item X series No.
* Isolating valves ½"	(20 X 14)
* Auto drain filter	(2 X 16)
* Atomising air pressure regulators with gauges 0/100 PSI	(4 X 17)
* Atomising air failure pressure switches type 8406	(1 X 18)
* Pressure gauges with isolating cocks 0/150 PSI	(2Xin injector line)
* Atomising air isolating valves ¾"	(2Xin injector line)
* Compressed air pressure failure switch type 8406	(1 X 19)
* Air pressure regulators with gauges 0/130 PSI	((1 X 20) (and 1 X in injector line))
* Pressure gauges 0/10 PSI	((1 X 21))

TABLE 3 Water Cooling Items

Water Cooling Items	Quantity No. of existing item X Series no.
* Water level switch mobrey magnetic type SO 1" FO 4/1	(1 X 22)
* 2" Isolating and regulating valves for grate bars and spry rings.	(28 X 23)
* Water flow failure switches	(1 X 24)
* Water pumps rated at 250 gall/min. at 50 foot head with starters	(------)
* 4" Isolating valves	(2 X 25)
* 3" Isolating valves	(-------)
* Water tank	(1 X 26)
* Flexible hose ½"	(- X 27)

TABLE 4 Fuel Oil Items

Fuel Oil Items	Quantity No. of existing Item X Series no.
* Fuel oil pumping and straining unit rated to handle 85 gallons per hour at 120 PSIG	(1 X 28)
* 1 in line filter	(2 X 29)
* Orifice plate assembly with isolating valves	(1 X 30)
* Pressure transmitter with equalising manifold	(1 X 31)
* Pressure gauges with isolating cocks 1"	(2 X 32)

Table 4 (Conti.)

Fuel Oil Items	Quantity No. of existing Item X Series no.
* Pressure gauges with isolating valves ½"	(4 X 32)
* Oil failure pressure switch type 8406	(1 X 34)
* Oilsafety shut-off valves	(1 X 35)
* Fuel control valve assembly with preeons	(1 X 36)
* Pressure reducing valve	(1 X 37)
* Oil flow transmitters mounted in commen Box	(3 X 38)
* Isolating valves ½"	(10 X 14)
* Ball Valves 1"	(6 X 39)
* Heat-up burner (516 GPH)	(1 X 40)
* Burners equipment (HV 600)	(3 X 41)
* Flow meters	(3 X 42)
* Needle valves	(3 X 43)
* Solenoid shut-off valve	(2 X 44)
* Flexible hose	(1 X 45)
* Storage tank isolating valve 3"	(1 X 46)
* Oil storage tank	(1 X 47)
* Positioner for actuator type AP 200/W	(1 X --)

INDUSTRIAL APPLICATION OF COKELESS CUPOLA

It seems likely that, the cokeless cupola will be more widely used in countries where the over all fuel situation favoured it. This is particularly applicable where there are quantities of natural gas, or other cheap fuel, available, for example, in England, Germany, SU., Mexcco, Iran, and Egypt. In Egypt, two cokeless cupolas have been installed, as mentioned, another two are under erection and will be commissioned soon.

CONCLUSION

By means of controlling systems on cokeless cupola, the controlling parameters of operation fuel to air ratio, carbon injection rate, refractory bed height, and plain or coated grate tubes have been controlled. And ductile iron can be produced when using natural gas or low sulfur fuel. Therefore, melting process and metal specifications have also been controlled.Cokeless cupola can help to solve pollution problem.

REFERENCES

Progress and Development Report. (1976). Foundry Trade J., 140, 234.

Nofal, Abdel Wahab, Ali. (1983). 50th International Foundry Congress. 25, Cairo, Nov. 6-11.

Taft, R.T. (1979). British Foundrymen. 72, June 105.

PROFILE AND FULL-FACE PICTURE ANALYSIS FOR AUTOMATIC RECOGNITION OF PEOPLE

B. El Hadj Amor, Dai Mo*, P. Baylou and G. Bousseau

Laboratoire d'Automatique, de Reconnaissance des Formes et de Robotique Agricole (LARFRA) de l'Ecole Nationale Supériure d'Electronique et de Radioélectricité de Bordeaux (ENSERB), 351 cours de la Libération, 33405 Talence Cedex, France

ABSTRACT

The signicance of characterizing a person using anthropological data is examined and the work done in this area is then presented.
A new algorithm for picture processing is applied to the pictures of people. On the pictures obtained, different features are calculated and marked to form a vector of characteristic parameters of the person who has been photographed.

KEYWORDS

Image processing, Filtering, Pattern recognition, People recognition, Computer applications.

GENERAL POINTS

The characterization of a person with the help of particular features of a face photographed full face and in profile is very useful in a number of disciplines. In ethnology, anthropology or in morphopsychology such parameters are used (in this way) in an attempt to study the character of each individual or a group of individuals.

In medecine all problems concerning facial malformation, reconstitution of particular soft or hard facial parts after an accident or the following of bone development in children, require the help of a computer to establish precisely the different facial features, refering to the soft or bony parts, as each case requires.

It goes without saying that this type of characterization will be very useful in resolving problems to do with the security of premises and property.

Admittance into reserve places such as banks or nuclear power stations, and so on, can be filtered by comparing the data magnetically recorded onto a personal card with the data taken from the direct analysis of the pictures of the person.

We are not trying to give an exhaustive list, but to show the relevance of computerized characterization of a person.

BIBLIOGRAPHICAL STUDY

The first works on automatic recognition of people through pictures analysis go back to about twenty years ago. However the following are those which seemed to us to be the must complete :

- Kelly [1] was interested, at the end of the 1960's, in identifying a person standing in front of a camera in a standard position. Measures were then made on the silhouette and face : the width of shoulders, of face, size of waist, the distance between the line of the eyes and the highest point of the head ; the distance between the eyes and the nose and between the eyes and the mouth. Classification was then made by the distance from the nearest neighbor.

- Harmon [2] was interested in the characterization of and individual through the analysis of the line of the profile. The system is in fact not totally automatic since a human operator is needed to trace this line on the photograph under study.

- In a more recent work, Sakaï, Nagao and Kanade[3] developed an identification system through the analysis of pictures seen from the front. Several features are extracted to characterise the individual in the photograph.

The analysis works well for facial types which lend themselves to this type of characterization (i.e. people who do not have hair falling low across the forehead, or a beard or moustache and who do not wear glasses). Despite its limitations this is however the first complete work on automatic analyse of pictures of people.

- We also quote the works developed in our laboratory which in the early stages were concerned with the recognition of an individual by analyzing the line of profile [4, 5, 6]. The procedure, somewhat similar to Harmon's, is made automatic by taking the picture against the light (using a translucid screen) which allows the contrast of the line of profile to be accentuated. The work was complete by studying several methods of classification.

In most of the works on individual identification from anthropological data, automatization is not total. When it is, the study is most often limited to using the line of profile or to a few simple morpholoogical measurements. To our knowledge no work has been undertaken which combines the analysis of both full-face and profile views.

We esteem that such a study is necessary to characterize an individual completely. If studying the line of profile offers certain advantages, the most important being its simplicity, the information remains none the less limited and excludes the use of certain parts which are just as, if not more, pertinent.

DESCRIPTION OF THE SYSTEM UNDER STUDY

Figure 1 presents an outline of the system. It is made up two cameras with good resolution, one of which is used for the full face picture and the other for the profile. These cameras are connected to a 4-way analogic multiplex system, so that the image to be processed can be selected, followed by a 4-bit digitaliser. The image which is digitalised and sampled for 256 x 256 pixels will be stored in an image memory to be processed.

* Computing Center, Northwestern University, Xi'an, China.

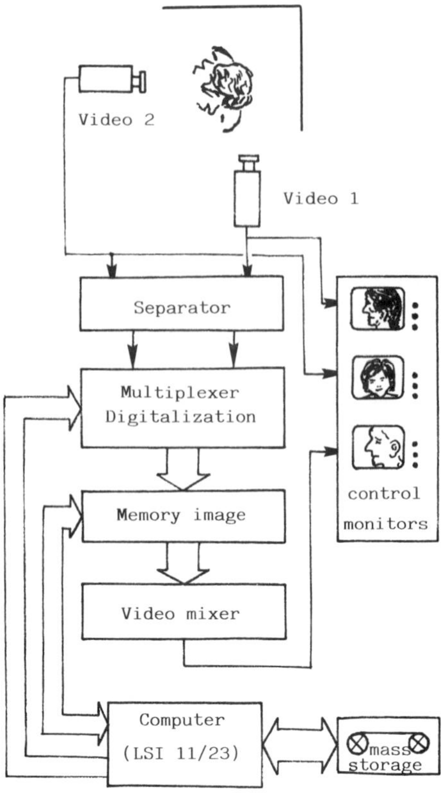

Figure 1

A calculator built around the DEC LSI 11/23 controls the different operations : selection, digitalization, memorization, visualization and processing.

The image memory has a capacity of 16 K bytes which can be taken up to 128 K bytes. The calculator sees the image memory as a peripheral controlled by a status word and by an address word. This avoid the restriction not to say the appropriation of the calculators field of addresses. The duration for image capture is a few milliseconds.

Figures 2a and 2b show what the pictures look like after digitalization.

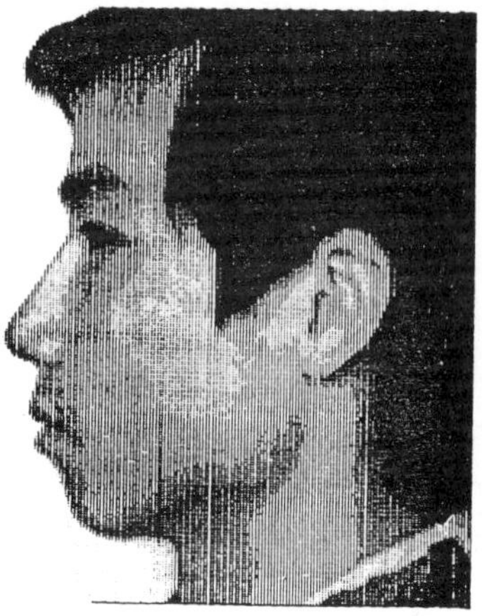

Figure 2a
Profile view after digitalization

Figure 2b
Full-face view after digitalization

PROCESS

From the pictures in fig. 2, the process will give simple images on which the detection of the edges and the localization of the different required curves will be carried out. The features then have to be extracted.

Extracting the image edges can be done in two different ways :

- by using a binary image where the transitions are detected by a simple specialized operation.
- by directly using local operators which make the strongly contrasting points stand out. These operations are generally based on the calculations of the gradient or of the Laplacian.

Let L(x,y) be the function representing the luminance at the point of the coordinates x and y ; the gradient at point (x,y) is represented by :

$$\nabla L(x,y) = [\partial L(x,y)/\partial x \ , \ \partial L(x,y)/\partial y]$$

whose amplitude is given by

$$|\nabla L(x,y)| = \sqrt{[\partial L(x,y)/\partial x]^2 + [\partial L(x,y)/\partial y]^2}$$

The Laplacian at point (x,y) is :

$$\nabla^2 L(x,y) = \partial^2 L(x,y)/\partial x^2 + \partial^2 L(x,y)/\partial y^2$$

in the case where L is defined on a discrete domain, several approximations of ∇ and ∇^2 have been made [7,8].

We have developed a logical for image processing, exploiting the two different approaches. We have also tested a new approximated Laplacian mask.

$\partial^2 L(x,y)/\partial x^2$ is approximated by :

$$2\ L(i,j) - L(i,j-3) - L(i,j+3)$$

and

$\partial^2 L(x,y)/\partial y^2$ by : $2\ L(i,j) - L(i-3,j) - L(i+3,j)$

i being equivalent to x and j to y.

When preceded by a calculation of average in a vicinity of 9 points, this filter proved to be efficient.

Given a simple binary unidimensional image representing a black/white transition (fig. 3a) the filter defined above gives the result 3b.

It can be seen that a simple binarization of the Laplacian allows the edges to be detected but two problems arise :

- the first is due to the size of mask which creates an error on the position of the edges.
- the second is to do with the binarization threshold.

The first problem was solved using a gradient operator on the Laplacian image. When approximated by Prewitt it gives the result on fig. 3c.

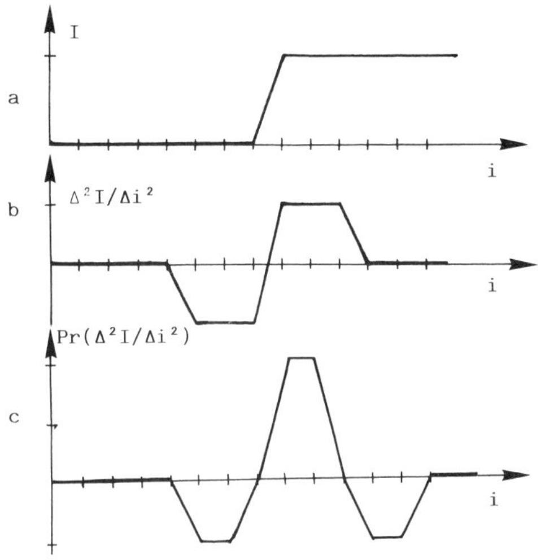

Figure 3

The binarization threshold was fixed from the analysis of the histogram showing the levels of grey. The representation of the number N of pixels selected by the binarization threshold S in function of S shows that for a domain of S included between S_1 and S_2, there is a relative stability of N for 10 % of the total number of pixels (fig. 4) and that the image obtained is satisfactory.

After binarization, a last filter operation is performed. It eliminates picture parts that presents no possibility of connexion with nearly edges.

The processing organigram can therefore be shown as follows (fig. 5).

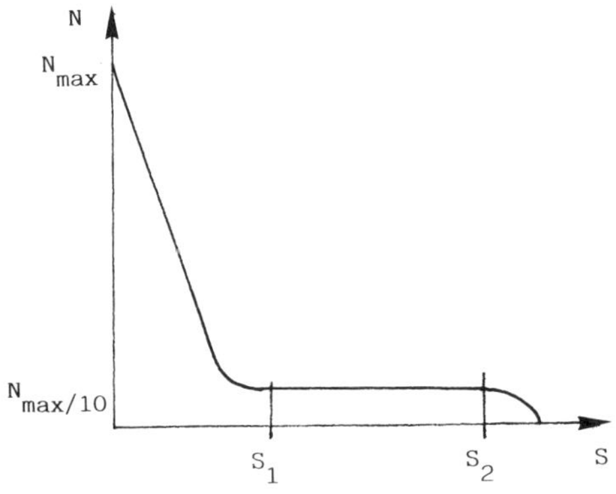

Figure 4

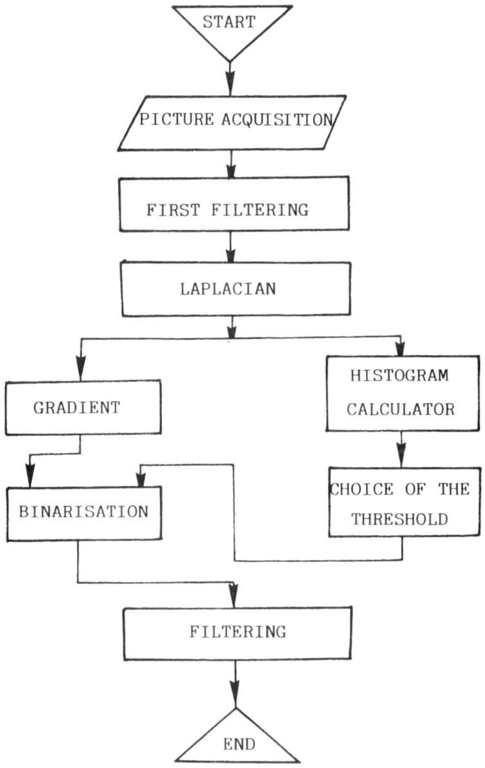

Figure 5

Figure 6 a-b show what the obtained images look like :

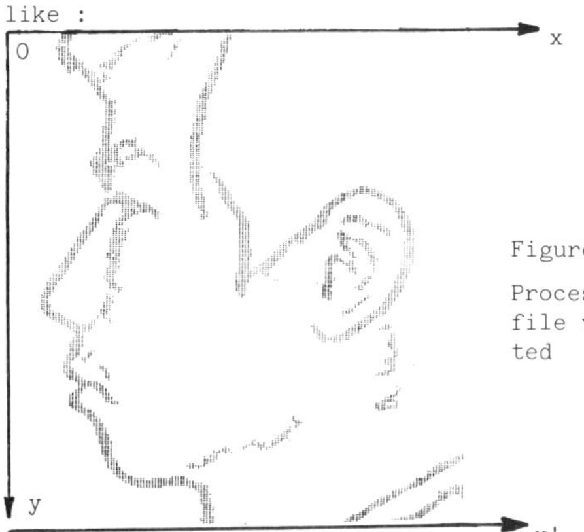

Figure 6a

Processed profile view treated

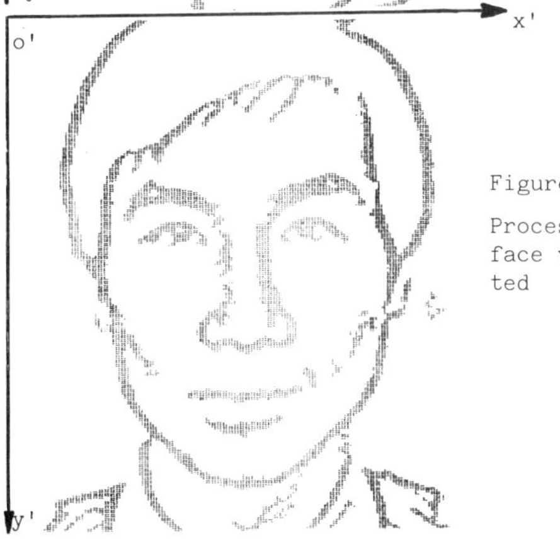

Figure 6b

Processed full face view treated

Marks xoy and x'o'y' are linked the two pictures. The different coordinates vary from Ø to 255.

PICTURE ANALYSIS

The points and the required curves are marked in 9.

It is supposed that we maintain that marks caused by background irregularities are small in size and do not present many possibilities for interconnection.

Moreover it is supposed (the outline detection mask was closen for this) that on the useful edges only the following gaps can appear :
- on the back of the nose (because of light relections)
- in the non regular zones : hair, beard
- at level B.

Several basic programmes have been developed of which the most important are :
- p1 : outline development and calculation of parameters
- p2 : connectivity search
- p3 : curvature calculation from Freeman coding.

The description of the different algorithms is quite long and will be presented in another publication.

With these programmes the profile organigram is as follows (fig. 7).

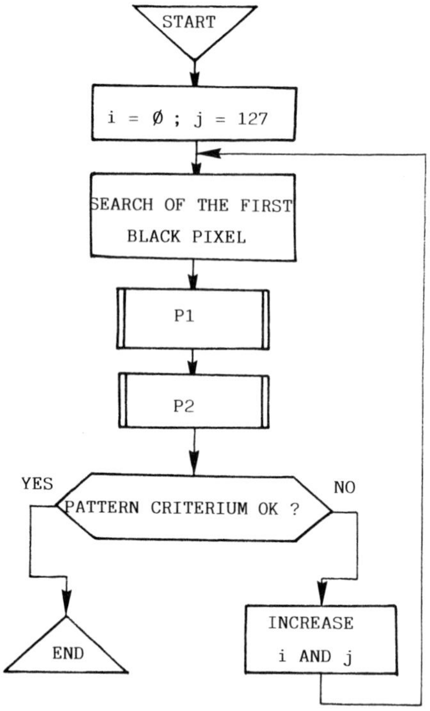

Figure 7

The decision criteria used are linked to the structural description of the line of profile.

On the detected line of profile an approximate lack of extremum curve points N and M is carried out using p3. N and M cut up the profile into three zones. Then the different points of the profile are found by an analog analysis of each zone.

A second analyse is then done in motion to position each point more exactly.

The ear pinna is then traced. It is defined by the contour line and by DP and FP. Finally points T and C are placed.

Figure 9a shows the result obtained.

ANALYSIS OF THE FULL-FACE PHOTO

The analysis is carried out in the following order:

1. The vertical median line is established approximately. This line exists for every one whose hairstyle does not present an apparent dissymetry. The operation can be carried out by studying different lines on the picture of the abscissas of the first and last lines.

2. On each part of the picture ($x' \geqslant x_1$ and $x' < x_1$) the "center" of the cheeks A and A' is situated by detecting the biggest white rectangular surfaces which are practically symmetrical in relation to the median axis (fig. 8a).

3. The width SS' can be found by extending the line AA'. Any eventual gap in the areas S and S' are detected by using a window image. The height of the window is bigger than the maximum gap size observed during research (fig. 8a).

4. The projection on the vertical axis of the points of the picture where $Y' > Y_S$ gives the curve shown in fig. 8b.

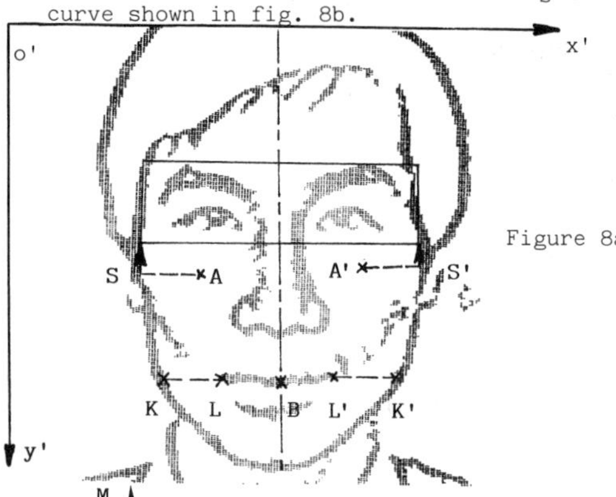

Figure 8a

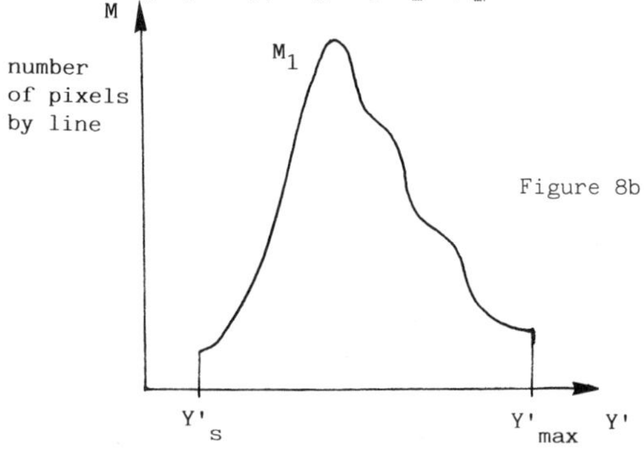

Figure 8b

Establishing the maximum M_1 of the curve allows K, L, L' and K' to be situated (fig. 8a). The intersections between the picture axis $X' = X_1$ and the straight LL' is represented by B'. B' corresponds to point B on the profile.

Any eventual gap in the areas of K and K' are detected in the same way as in § 3.

5. Moving along Y' a window image whose lenght is equal to $X'_S - X_S$ and whose height h allows $G_{L'}$ to be found by searching the density maximum. $G_{L'}$ corresponds to G_L on the profile (fig. 8b).

6. The lenght $Y_{B'} - Y_{GL'}$ is taken as the normalizing lenght on the picture and by correlation between the full-face picture and the profile picture, those differents points can be situated:
   Y_T allows Z and Z' to be obtained
   Y_C allows F and F' to be obtained
   $(Y_N + Y_{SN})/2$ allows J and J' to be obtained.

7. From F and F', the centers of gravity of the apples E, E' and I, I' are obtained.

   In the figure 9b we give example of the results obtained.

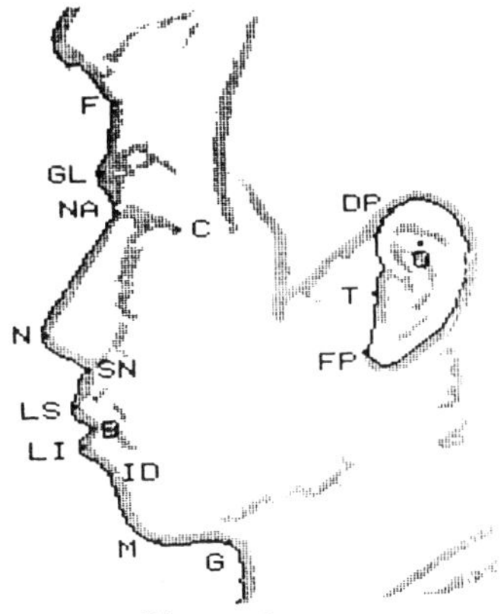

Figure 9a

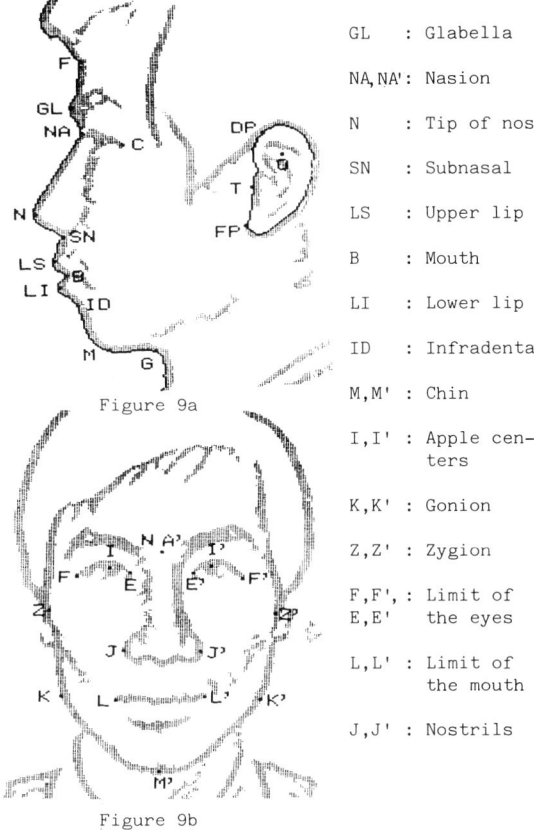

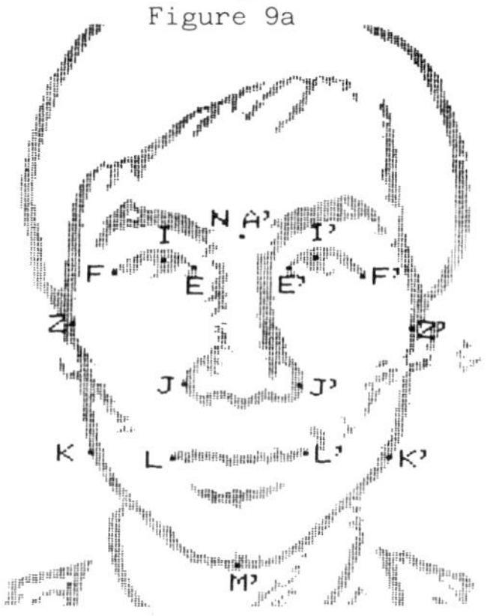

Figure 9b

GL : Glabella

NA, NA' : Nasion

N : Tip of nose

SN : Subnasal

LS : Upper lip

B : Mouth

LI : Lower lip

ID : Infradental

M, M' : Chin

I, I' : Apple centers

K, K' : Gonion

Z, Z' : Zygion

F, F', E, E' : Limit of the eyes

L, L' : Limit of the mouth

J, J' : Nostrils

CONCLUSION

The work presented allows the facial features of a person seen from the front or in profile to be localized, using suitable picture processing algorithms.

From these different points a vector will be established which will allow the recognition of a person. This work is being implemented.

In this study presented we have not mentioned the problems posed by a beard or a moustache. In this case localizing certain points will be debatable (localizing ID, M, G, K and K') but we think that the large number of localized features (about thirty) will give good characterization. An interactive implementation would eventually remove these ambiguities.

BIBLIOGRAPHY

[1] M.D. KELLY
Visual identification of people by computer
Ph. D. Thesis 1970 - Department of Computer Sciences, Stanford University.

[2] L.D. HARMON and W.F. HUNT
Automatic recognition human face profiles
Computer graphics and image processing 6-1977, p. 135-136.

[3] T. KANADE
Computer recognition of human faces
Basel, Stuttgart, 1977.

[4] P. BAYLOU, E.H. BOUYAKHF
Identification des individus par l'analyse automatique du visage.
Rapport interne - E.N.S.E.R.Bordeaux 1978.

[5] P. BAYLOU, E.H. BOUYAKHF, G. BOUSSEAU, A. MORA
Analyse automatique de profils de visages
2ème Congrès AFCET-IRIA - Toulouse France 1979, p. 99-106.

[6] E.H. BOUYAKHF
Etude et réalisation d'un système d'extraction automatique des paramètres anthropométriques du profil du visage
Thèse de Docteur-Ingénieur - U.P.S. Toulouse, 1980.

[7] W.K. PRATT
Digital image processing, p. 770-782, University of Southern California, March 1977.

[8] S. LEVIALDI
Digital Image Processing, p. 105-145, edited by Simon and Haralick, D. Reidel. Publishing company, Dordrecht, Holland.

MINTEST-EXPERT SYSTEM FOR MINERAL IDENTIFICATION

G. L. Kovács

Computer and Automation Institute, Hungarian Academy of Sciences, Budapest, Hungary

Abstract. A microcomputer based expert system is presented for fast and cheap mineral identification. In most cases only some fundamental data are needed to identify a mineral sample (or precious stone), if we know how to use these data. The important, characterizing data of the well-known minerals are stored in the data-base of our program-system. The visible and measured optical and physical properties of the sample are compared with the stored data to get the result, the name of the examined mineral. The system works in a simple, friendly way - asking questions, understanding answers - so that no special computer knowledge is needed to use it; one only has to type in the answers. An advantage of the system is that only relatively simple and fast measurements are required. Based on these measurements, a fast, reliable identification is given by a cheap computer running a sophisticated program-system. The MINTEST system handles even those cases when the tested mineral cannot be determined from the given input data.

Keywords. Microcomputer; mineral identification; expert system

1. INTRODUCTION

The revolutionary fast decrease of microcomputer prices made it possible to use computers not only in big factories, universities and research institutes, but even in households. Small capacity microcomputers (e.g. Commodore 64) are now available for 100-400 $ or high capacity ones (e.g. IBM PC) for 1000-10000 $. The prices depend mostly on disc, printer, other peripherals and software support.

These computer prices are lower than the prices of most of the special mineral identification equipment of big laboratories (for example an X-ray equipment). We could even say that such microcomputers are cheaper than one or two pieces of high quality gemstones, the precise identification of which should be assisted by means of our proposed system.

A microcomputer based program-system is now being developed in the Computer and Automation Institute of the Hungarian Academy of Sciences to help fast and reliable mineral/gemstone identification.

The MINTEST system has some learning facilities to handle new, unexpected data, so its capacity should always be increasing. Based on its knowledge base and learning facility and some other sophisticated features the system is a really expert one.

As most small mineralogical laboratories, jewellers and even mineral collectors, and hobby polishers could afford to purchase an appropriate microcomputer, it seems reasonable to supply the above mentioned program-system.

2. ENUMERATION AND VALUE LIMITS OF MINERAL PROPERTIES

The following section enumerates the main properties, that should be determined in order to identify a polished precious stone or any mineral sample. Some of these properties may be determined by visual observations, others by simple, fast and not too expensive measurements. Some of the data are mostly used in the case of gemstones (g), others give information on non-polished minerals and grown-up crystals (m) only, but most of them are commonly used.

- The knowledge of all the possible values (for transparency, colour, streak, fracture, cleavage, cleavage direction, crystal system and habit, luster, pleochroism and UV sensibility) and the lower and upper limits (for hardness, specific gravity, light refraction, birefrigence, and absorption spectrum, dispersion) is necessary for input data checking when the data base is generated. (See sections 3.1.1 and 3.1.2)

- When a sample is tested, there is only one colour, streak or hardness value, - which has to correspond to one of the valid values found in the data-base during the identification process. (See sections 3.2 and 4.)

The main properties are as follows:

- Transparency:	transparent, semitransparent (translucent), non-transparent (opaque);
- Colour, streak:	there are about 15-20 different, possible colours;
- Hardness:	full or half numbers between 1 and 10;
- Fracture (m):	conchoidal (shell-like), subconchoidal, splintery, hackly, smooth, irregular, uneven, earthy;
- Cleavage (m):	very good (perfect), good, poor, none; along twin boundaries;
- Cleavage direction:	basal, octahedral, prismatic, pinacoidal, cubic;
- Crystal system:	cubic (isometric), trigonal, tetragonal, hexagonal, orthorombic, monoclinic, triclinic, amorphous, (with about 32 subclasses);
- Crystal habit (m):	columnar, prismatic, tabular, bladed, foliated,

botryoidal, reniform, granular, massive;

- Specific gravity: between 1 and 22;
- Luster: metallic, adamantine, greasy, resinous, silky, pearly, non-metallic;
- Light refraction: between 1 and 4;
- Birefrigence: between -0,3 and +0,3; or none;
- Dispersion: between 0 and 0,3;
- Pleochroism (g): none, weak, strong in some colours;
- UV sensibility: none, weak, strong in some colours;
- Absorption spectrum: between 400 Å and 7000 Å.
- Rare light phenomena: asterism, adularescence, fluorescence, thermoluminescence, etc.;
- Other properties: radioactivity, magnetic and electrical properties, etc.

The exact values (or their lower and upper limits) are well-known for all minerals, thus if both the appropriate program and the measured data of an examined sample are available, the sample can be identified.

It can be seen that the numbers of all the posssible data and value ranges are rather limited. That is of great assistance in organizing our program-system. It stands for the Data Input part and for the Interrogation System, as well.

It has to be mentioned that the MINTEST system so far doesn't deal with absorption spectrum measurements, with rare light phenomena and with any "other properties". For gemstone identification an additional microscope test is suggested.

3. THE MINTEST PROGRAM-SYSTEM

The system consists of two main parts:

- Data Input of Minerals
- Interrogation System for Identification.

3.1 Data Input

This program part makes it possible to load the computer with all the necessary data that characterize the minerals/gemstones that we want to identify. The data should be read in easily with validity tests.

The input program is organized in the form of a dialogue and it consists of two parts:

3.1.1 Input of all possible values and value limits. (See blocks 2. and 4. of fig. 1.) The operator should answer the questions of the computer. The questions ask for data as given in the previous section ("Enumeration ...") of this paper. For transparency, colour, streak, fracture, cleavage, crystal system and habit, fluorescence and pleochroism, the operator has to enumerate all the valid answers (e.g. all the colours) and for hardness, specific gravity, light, refraction, birefrigence, dispersion, the possible maximum and minimum values should be typed in.

The system checks some characteristics of the input data. For example if three hardness limits are given, or if the lower limit is higher than the higher one or in the case of some other inconsistencies an error message appears and the question is repeated.

The enumerations of possible values and value limits are used in the following program parts not only for input data checking but as menus as well.

3.1.2 Input of specific mineral data. (See blocks 3. and 5. of fig. 1.) This part of the input system puts questions to the operator, starting with the name and synonyms of the mineral, and then the same questions are asked as for the first input part. The given answers should be based on a good handbook. (In our case the "BLV Bestimmungsbuch 17. Edelsteine und Schmucksteine") Typing in all the specific data of all important minerals is a rather tedious and time consuming activity. However, it would be worth-while doing it.

To make the input process faster and easier, the operator doesn't have to type in all possible values of enumerative properties (e.g. colour, streak) but the program gives assistance offering the previously (3.1.1) learnt data in the form of menus.

All the possible answers to all questions appear on the monitor of the computer and the operator has only to choose the valid ones by using the cursor.

When exact values should be typed in (e.g. refraction, SG), the operator has to do it carefully, as these data will be compared only with the previously defined limit values (e.g. a hardness larger than 10 or smaller than 1 is not accepted).

Recently the program has been built up for 200 minerals (mostly gemstones), but it still can be enlarged. All the data of all minerals can easily be checked and corrected or completed if it is necessary. Further minerals, with all their data can be added to the system and even new properties can be built in.

The Data Input parts of the system should be used only once. All the data of all minerals are stored in an appropriate data-base, which is copied to and stored on disc or magnetic tape when data input is completed.

3.2 Interrogation System for Mineral Identification

This part of the program contains search and compare procedures - looking for the matching of the measured data of the sample in the data-base.

The interrogation program is organized in such a way that after each question and answer the range of possible resulting minerals is getting smaller, and finally the sample is identified. The program structure can be enlarged by asking for further properties, too, which are not used at the moment (e.g. absorption spectrum).

The simplified structure of the system is in Fig. 1. The figure doesn't show the disc store and load processes; they should be logically understood after blocks 2, 3, 4, 5 and 1 respectively.

Any of the main program parts may be skipped or interrupted at any time, and continued or started later (e.g. the operator may input the mineral data in several consecutive sessions, or identification may be stopped and continued 2-3 days later on; intermediate data are always stored).

If any inconsistency appears, the system gives a warning and rejects the required operation (e.g. when mineral identification /block 1/ is initiated, and there is no mineral data in the data-base).

Blocks 4 and 5 stand for data correction. The operator may correct value limits or change or correct enumerations of possible values (e.g.

colours) or may correct mineral data or may add data of new minerals into the data-base.

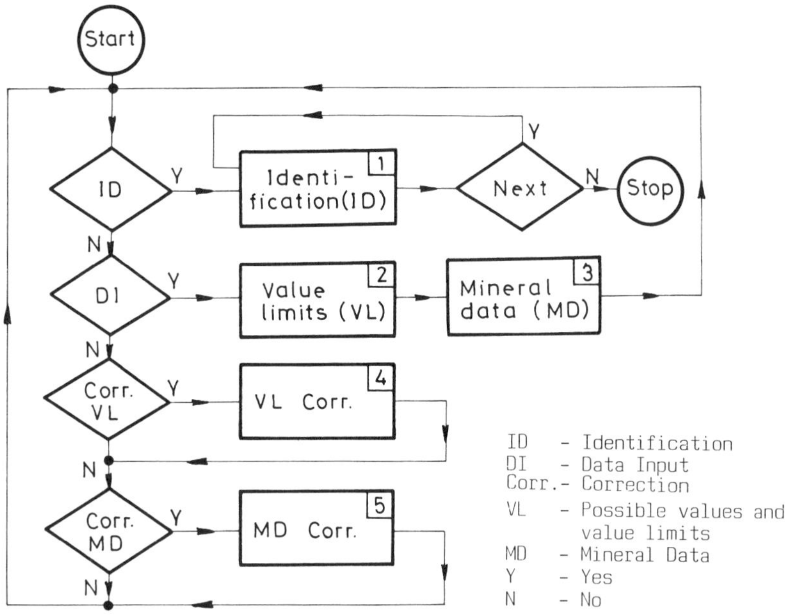

Fig. 1. MINTEST - program-flowchart

4. THE IDENTIFICATION PROCESS USING THE MINTEST SYSTEM

The work of the Identification System starts with loading in the data-base. This data-base contains all the possible values and value limits (section 3.1.1) and all specific mineral data (section 3.1.2). It is followed by the question-answer part (3.2).

The computer puts questions to the user of the system and waits for exact numbers as the answer in the case of S.G., light refraction, etc., or the system provides menus to choose one possibility in the case of colour, crystal form etc. The questions are the same as they were at Data Input. If a measurement result is missing or uncertain, only a Return should be typed in.

It is supposed that in most cases the mineral tests were previously done, however the operator may process the test just when the appropriate question appears.

The validity of all answers is checked and, if an error appears the question is repeated with the message: "invalid answer".

Once a valid answer is given, the next question appears. The sequence of questions is meant to assist a logical testing procedure, starting with easily defined properties, however the operator may choose any other order he wants.

As it was previously mentioned, the range of possible resulting minerals is getting smaller after each answer. If some questions are not yet answered and the range is so narrow that the mineral is already identified the process could be finished.

However it must be underlined that the skipped questions might contain contradictionary answers, thus we suggest no questions to be omitted if answers are available. This way some errors may be avoided.

Some measurements are rarely done (e.g. absorption spectrum), some properties are impossible or hard to define in certain cases (e.g. crystal form for massive, opaque minerals), or some values are missing, even if they could be determined (e.g. the hardness of gemstones, as examination may damage the stone).

If there are too many missing data for a given sample, it may happen that the system doesn't give the exact solution, but 2-3 or more minerals are chosen as possible results. In most of these cases further tests are suggested automatically, to arrive to a correct, unambiguos conclusion - as the system knows which missing data will be decisive. Sometimes human intelligence may help select the only possible solution.

For precious stones examination by microscope almost always helps as the characteristics of microscopic inclusion are assisting to lead to correct diagnosis. This test is the most important in distinguishing between natural and synthetic gemstones.

It is possible that, even if all questions are answered, the system results in no solution. This might be for the following reasons:

- failure in measurement
- failure in data input
- failure in the data-base
- appearence of an unknown mineral.

Having checked that none of the first three reasons exist, the learning feature of the system steps into action. The new data will be incorporated into the system - after their validity is checked in a mineralogy handbook. We suppose that in most cases there will be a rare mineral, the data of which were not coded, and now it is added to the system. However it may actually happen that a new mineral is identified in this way.

If a mineral is identified, the system asks whether the next identification should start or not.

5. CONCLUSION

A system is proposed for mineral and precious stone identification. The system consists of a good mineralogy handbook, of measurement instruments, (as polariscope, refractometer, microscope, etc.) and of a microcomputer equipped with a sophisticated program, called MINTEST.

By means of visual and manual observations and using the measurement instruments most optical and physical properties of a sample can be determined. Based on these data it takes a long time to decide which kind of mineral is present if only a handbook is used as assistance.

Having a computer and our appropriate, friendly useable program-system, the identification may be carried out easily in a couple of minutes. (It is often necessary when precious stones are examined.)

This system can effectively be used when new mineral resources are investigated for future mining and fast identification results are needed.

The cheapest computer on which the MINTEST program runs is a Sinclair Spectrum or a Commodore 64 but on these computers only a restricted number (approx. 150) of minerals can be identified. On an IBM PC category computer, there is no restriction on the number of minerals to be identified.

ACKNOWLEDGEMENT

I should like to express my thanks to Dr.I.Gatter of the Mineralogy Department for his assistance by discussing some problems and correcting the manuscript.

REFERENCES

D.Hawkins (1983). An analysis of expert thinking Int.J.Man-Machine Studies 18, pp.1-47

W. Schumann (1981). Edelsteine und Schmucksteine BLV Bestimmungsbuch 17

USER DATABASE SYSTEM GENERATION

Jiao-jin Xu

Automation Research Institute, The Ministry of Metallurgical Industry, Beijing, China

ABSTRACT. When dBASE is used for solving an office automation problem, the problem can be seen as a tree. The nodes of the tree are classified into several groups. A software called UDBSG (user database system generator) is developed. With this software computer programs for most office automation projects can be generated within half an hour without coding, typing and debugging.

KEYWORDS. Computer software; computer-aided design; database; program generator; office automation.

INTRODUCTION

Let us consider a office automation problem. Micro computers are widely used in this field. The well known software, dBASE II (Ashton-Tate), is very powerful to solve the problem. However, no matter how easy the programming of dBASE II is, the programmers have to code, type and debug the program. Normally it takes a long time to complete a system.

A new idea is presented to get rid of such tedious work. The job is done with a computer as in computer-aided design. As a result a software named User Database System Generator (UDBSG) is developed. With this software all your work to develop a user database system is as simple as follows:

1. Sit at your terminal to answer the questions prompted by a computer.

2. Get your disk with a dozen of programs generated by UDBSG or type the program listings generated if you wish.

The generating process takes you less than half an hour. You can run your generated system immediately.

A TREELIKE PROBLEM

In the real world, many problems have a treelike structure. Let us investigate systems using dBASE II. Many designers would like to elect the menu-driven operation. We call the system with this operation the menu structured system. In this case it is a treelike problem. Let us see a simple system with two database files (PERSONNEL and PROJECT) as an example. Fig. 1 shows the system structure. Obviously it is a treelike structure. Each node represents a procedure (or in dBASE terminology, a command file). It is possible that we classify the nodes into several groups in term of its function. Each group has the common characteristics. For example, each row in Fig. 1 forms a group. Then, we have two procedures (ENTER PERSONNEL and ENTER PROJECT) with the function for entering the data into database files, two (RETRIEVE PERSONNEL and RETRIEVE PROJECT) with the function for retrieving and two (REPORT PERSONNEL and REPORT PROJECT) with the function for reporting. In

```
                  ----------------------------------------
                  |                 MENU                 |
                  ----------------------------------------
                  /                   |                 \
         -----------            ------------        ----------
         | ENTER  |             | RETRIEVE |        | REPORT |
         -----------            ------------        ----------
            |    |                 |      |            |    |
 ------------  ----------          |      |            |    |
 | ENTER    |  | ENTER  |          |      |            |    |
 | PERSONNEL|  | PROJECT|          |      |            |    |
 ------------  ----------          |      |            |    |
                         ------------  -----------     |    |
                         | RETRIEVE |  | RETRIEVE|     |    |
                         | PERSONNEL|  | PROJECT |     |    |
                         ------------  -----------     |    |
                                            ------------  ----------
                                            | REPORT   |  | REPORT |
                                            | PERSONNEL|  | PROJECT|
                                            ------------  ----------
```

Fig. 1 A simple user database system

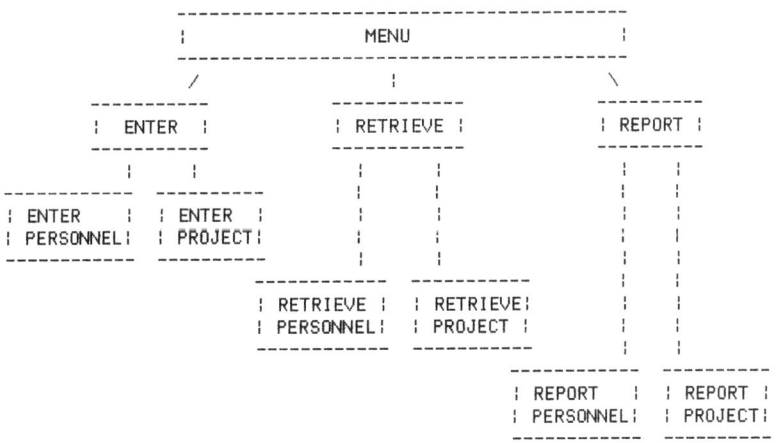

addition, we have a procedure (MENU) and three procedures (ENTER, RETRIEVE and REPORT) for menu and submenu functions respectively. Although different users have different requirements, we can specify several common specifications for each group in addition to those dependent on the user. Based on this idea we have constructed several modules for solving office automation problems. With these modules a lot of user systems can be generated successfully.

THE STRUCTURE OF GENERATED SYSTEMS

There are many ways to explain the generating process. The best way to do it is to use the so-called Backus-Naur Form (BNF) to see the structure of generated systems. BNF was applied in defining all kinds of programming languages (Organick, Forsythe and Plummer, 1978). But it is also useful, with simple amplifications, for describing the structure of generated systems.

The set of BNF syntax definitions for the generated systems is as follows:

```
(a).<generated system>:=<generated tree>
(b).<generated tree>:=<node>|<generated tree>
                       <node>
(c).<node>:=<runnable program>
(d).<runnable program>:=<standard block>|
                          <block comb>
(e).<standard block>:=dBASE command file
(f).<block comb>:=<block>|<block comb><block>
(g).<block>:=<pure block>|<var block>|
             <recurs block>
(h).<pure block>:=<legal statement>|<pure block>
                  <legal statment>
(i).<legal statement>:=dBASE statement
(j).<var block>:=<wide sense statement>|
                 <var block><wide sense statement>
(k).<wide sense statement>:=<legal statement>|
                             <string comb>
(l).<string comb>:=<key word string>
                   <wide sense var>|<string comb>
                   <string comb>
(m).<wide sense var>:=lib var|input var
(n).<key word string>:=store|replace|append|...
                        (i.e. all dBASE key words)
```

The above definitions are easy to understand, but the following definitions require additional explanation. In (j) above, we know that <var block> is a set. If we want to get from the set a definite subset within which <key word string>s are same (determined and constant) while <wide sense var>s are different (non-determined and variable), we can write

```
<def subset E <var block>>
```

to denote the definite subset.

With this amplification to BNF, we can define the recursion block as follows :

```
(o).<recurs block>:=<var block 1><var block 1>
                    <var block 2><var block 2>|
                    <var block 1><recurs block>
                    <var block 2>
(p).<var block 1>:=<def subset E <var block>>
(q).<var block 2>:=<def subset E <var block>>
```

As a general rule in BNF, any object with the absence of angle brackets means a terminal (or atom). From the previous definitions, our atoms include the followings:

(a). dBASE command file.
(b). dBASE statement.
(c). dBASE key word.
(d). lib var/input var.

THE CHARACTERISTICS OF ATOM

The problem of generation can be stated as below. We have a atom pool containing such atoms as files, statements, key words and variables. we want to take the atoms from the pool to form a user system following the BNF syntax definition listed above.

The atom in UDBSG is quite different with that in programming languages. In programming languages, the atom is the simplest element such as digit (1, 2, ..), letter (a, b, ...), arithmetic operator (+, -, ...), and logical operator (.and., .or. ...). But in UDBSG, we even take a dBASE command file as an atom. As well know, a command file is complex in syntax and has never been an atom in programming languages. In UDBSG, to make problem easy and realizable we use it as an atom.

Now let us see the characteristics of atoms.

The first atom is the dBASE command file. It is a syntaxtically no error command file and doesn't crack the system when it enters the user system.

The second atom is the dBASE statement. It is a syntaxtically no error statement. Some examples are as follows :

(a). STORE T TO LOOP
(b). GO TOP

The third atom is dBASE key word. It is the element in dBASE key word set. Some examples are as follows :

(a). STORE
(b). REPLACE
(c). APPEND

The fourth atom is lib var (library variable) or input var (input varible). It can be numeric or string. It doesn't make the statement syntax error when it enters the statement.

THE CHARACTERISTICS OF SYSTEM

In generating, we have a dozen of modules to generate the user system. In general a specific node is generated by a specific module. Fig. 1 shows a very simple example. With this example, we need 5 modules to generate the user system. These 5 modules are MENU MODULE, SUBMENU MODULE, ENTER MODULE, RETRIEVE MODULE and REPORT MODULE. Each module has its special characteristics. In generating process, each module requires the different user data. Let us use Fig. 1 as our example again. MENU MODULE needs to know the son functions to generate procedure MENU. SUBMENU MODULE needs to know the son database file names to generate procedures ENTER, RETRIEVE AND REPORT. ENTER MODULE needs to know the number of fields of the database file and field names, types, widthes and decimal places to generate procedures ENTER PERSONNEL and ENTER PROJECT. RETRIEVE MODULE needs to know the most frequently retrieve items for providing the fast and convenience retrieval to generate procedures RETRIEVE PERSONNEL and RETRIEVE PROJECT. REPORT MODULE needs to know the report formats to generate precedures REPORT PERSONNEL and REPORT PROJECT. When these user dependent data are entered, each module will generate a specified node.

In advanced application, The generation of a more complex node needs not only a specific module, but also some other modules. This is possible with the BNF definition (c), (d), (e) and (f) above.

The above five modules are the basic modules to generate a system. In real applications, it is far from enough. Obviously we need MERGE MODULE to merge a number of database files, CALCULATION MODULE to do all kind of arithmatics and much more.

CONCLUSIONS

UDBSG has been used to generate a couple of user database systems. Typical examples are personnel database, warehouse database, research project administration, finacial acounting and etc. Most systems are generated within half an hour. UDBSG is so powerful that we might give it a nickname "Robot Programmer".

REFERENCES

Ashton-Tate. dBASE II Assembly-Language Relational Database Management System Vol.1 User Manual. Culver City, California.

Organick, E. I., Forsythe, A. I., and Plummer, R. P. (1978). Programming Language Structures. Academic Press, New York.

A LINEAR ELECTRONIC CIRCUIT ANALYSIS PROGAM

E. Leelarasmee

Department of Electrical Engineering, Chulalongkorn University, Bangkok 10500, Thailand

ABSTRACT. A general purpose program for analysing arbitrary linear electronic circuits is described. The program is capable of performing frequency domain analysis of any circuit consisting of linear components, such as resistors capacitors inductors transistors and operational amplifiers, and independent sources, i.e. voltage and current sources. This program is developed on a widely used 8 bits microcomputer. The ease of using the program and various useful features of the program make the program useful as a teaching aid for beginners such as university students and as a design tool for experts in analog circuit design. The program is highly suitable for users who cannot afford the relatively high cost of minicomputers.

Keywords. Computer-aided design; circuit analysis program; frequency response; filters; linear circuits; modified nodal approach; numerical methods.

I. INTRODUCTION

It has been widely accepted that the use of the computer as an aid or tool for engineers in various deciplines is essential for coping with the fast pace of today technologies. With the introduction of low cost and good performance mini-computers and micro-computers, the area of computer-aided design (CAD) is growing very rapidly, both in developed and developing countries. Unfortunately, as the cost of computer hardware goes down, the cost of computer software does not. Therefore in some case the high cost of CAD software may be too prohibitive to be used by universities and small industries, especially in developing countries.

This paper describes an effort taken by the Department of Electrical Engineering, Chulalongkorn University to provide a CAD software in electrical engineering for both local industries and universities. In particular, a program called LEC for analysing arbitrary linear eletronic circuits will be described. Currently the program can perform frequency domain analysis of any circuit consisting of the following components: resistor capacitor inductor transistor operational-amplifier voltage-source and current-source. LEC runs on any APPLE II+ compatible microcomputer which is one of the most widely used and low cost microcomputer. The program is carefully designed to be easy to be used by beginners but yet powerful enough to be a good design tool for experts.

This paper is organized as follows. The next section gives a general overview of the program and shows how easy it can be used. In section III, two practical design cases are demonstrated to show that the program has a potential to be a good design tool for serious circuit designers. A brief technical characteristic of the program is given in section IV. Finally, the conclusion is given in section V.

II. LEC ORGANIZATION AND USER GUIDE.

We first describe the organization of LEC version 2.0. This program is stored on a floppy diskette. It is automatically booted into the microcomputer upon starting the machine. From the user's viewpoint, LEC 2.0 is divided into 4 major parts or modes. The program will automatically print the mode prompter to let the user know which mode he is currently using. The user can always switch to any mode by simply typing in its corresponding mode prompter. These 4 modes are:

1) EDIT mode. This mode allows the user to specify the details of the circuit to be analysed. Its behaviour is similar to a text editor except that the contents to be editted are the components of the circuit and their connectivities. The user can modify or change the contents easily as he wishes.

2) LEC mode. This mode allows the user to issue commands to set the frequency range, start the analysis, print out the results, modify the component values and perform diskette input/output processes.

3) SIM mode. This is the pure analysis mode. It contains various subroutines that are written to set up the matrix equation and solve it. This mode is currently non-interruptible by the user. It will automatically switch to other mode upon finishing.

4} PLOT mode. This mode allows the user to display the results of the analysis obtained from the SIM mode in a high resolution graphic plot on the screen. Any nodal voltage or some circuit currents can be plot in many graphic options the can be defined by the user. The plot can also be saved into the diskette for later comparisons or dumped on to a dot-matrix printer to obtain a hard copy.

As for illustrations, two examples are shown to demonstrate how to use the program. The first example shown in Fig. 2.1 is a simple AC circuit which the user wants to analyse at the frequency 50Hz to see the voltages at various nodes and the power factor of the voltage source, i.e. the phase of I(VI). From this example, one can see that each input line enterred in the EDIT mode describes the type of each component, the nodes to which it is connected and the component value. This input syntax is quite similar to that of SPICE (Nagel, 1975). However, unlike SPICE, LEC's input is interactive. The user can list what have been enterred and corrects any mistake by simply retyping it.

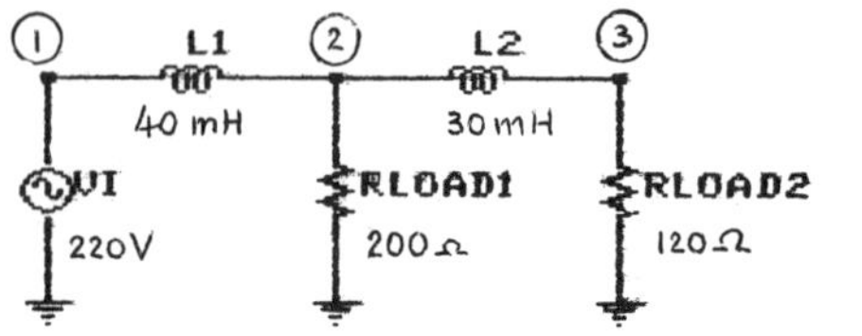


```
EDIT>VI 1 0 220VOLT-AC
EDIT>L1 1 2 40mHENRY
EDIT>L2 2 3 30mHENRY
EDIT>RLOAD1 2 0 200ohm
EDIT>RLOAD2 3 0 120oHm
EDIT>SIM AT 50HZ

VARIABLE  MAGNITUDE    PHASE(deg)
V(1)      220          0
V(2)      215.285818   -9.4004115
V(3)      214.624878   -13.891192
I(L1)     2.86290595   -12.204180
I(L2)     1.78854065   -13.891192
I(VI)     2.86290595   -12.204180
entering.. LEC

LEC>
```

Fig. 2.1 An example to show how to use LEC 2.0 to input the circuit. All underlined characters are those that were input by the user. The rest are outputs from the program or the mode prompter: LEC> EDIT> and PLOT>.

The second example shown in Fig. 2.2 is an electronic low pass filter consisting of an active device, i.e. an operational amplifier. This example is devised to show that LEC 2.0 has a built-in or default model for each active device. The user can edit this model so as to make the active device characteristics as closed to the practical values as possible. This feature will be shown in the next section. The model parameters for an operartional amplifier and their default values are:

1) DC open loop gain : A0 (infinity)
2) DC input resistance: RI (infinity)
3) Input capacitance: CI (zero)
4) Output resistance: RO (zero)
5) Open loop poles of the gain: P1,P2,P3 (infinity)

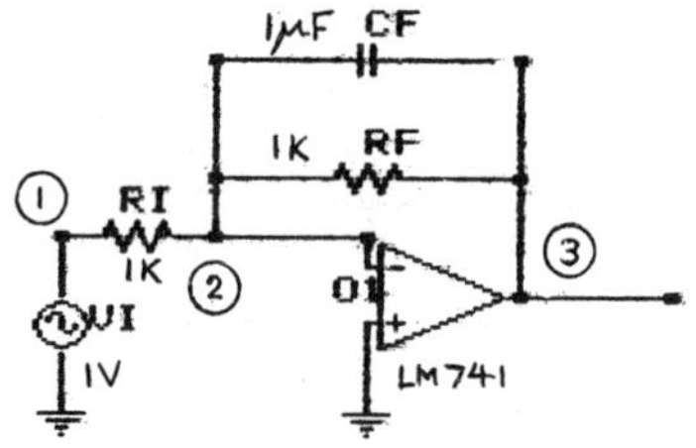


```
LEC>LIST
    MODEL LM741 OPAMP
A0 =1E+32              RI =1E+32ohm
CI =0farad             RO =0ohm
P1 =1E+32hz            P2 =1E+32hz
P3 =1E+32hz
    O1 0 2 3 LM741
    RI 1 2 1000ohm
    RF 2 3 1000ohm
    CF 2 3 1E-06farad
    VI 1 0 1ac-volt 0deg
LEC>SWEEP FREQ
    BEGIN FREQ(hz) = 10 >20HZ
    END FREQ(hz) = 10000 >1KHZ
    SCALE <LOG LIN> = LOG >
    #INTERVALS(<=24) = 10 >
LEC>SIM

    0   AT FREQ(hz) = 20
    1   AT FREQ(hz) = 29.5751527
    2   AT FREQ(hz) = 43.734483
    3   AT FREQ(hz) = 64.6727007
    4   AT FREQ(hz) = 95.63525
    5   AT FREQ(hz) = 141.421356
    6   AT FREQ(hz) = 209.127911
    7   AT FREQ(hz) = 309.249495
    8   AT FREQ(hz) = 457.305052
    9   AT FREQ(hz) = 676.243338
    10  AT FREQ(hz) = 1000
entering.. LEC
```

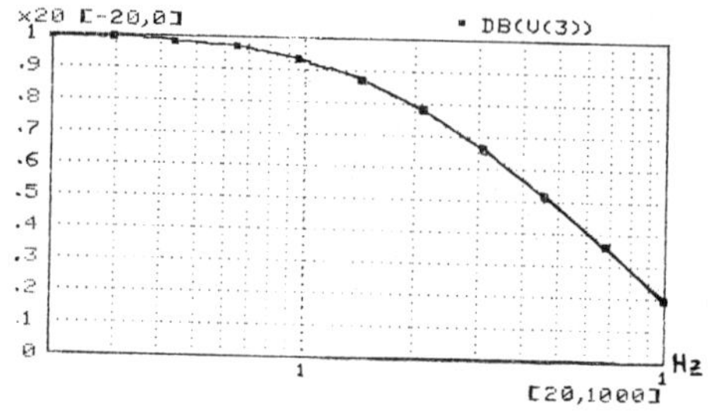

Fig. 2.2 An example to show how to use LEC 2.0 to compute the frequency response of a low pass filter.

III. PRACTICAL CIRCUIT DESIGN USING LEC

In this section, we will show that LEC 2.0 has been carefully designed to be able to deal with practical circuit design cases. Referring back to the first example (Fig. 2.1) of the previous section, we now want to add a capacitor across RLOAD1 and find a good value so as to correct the power factor of the AC voltage source, i.e. to make the phase of I(VI) as close to zero as possible. This is done in Fig. 3.1. From the figure we see that LEC 2.0 has a useful utility that enable the user to add new components easily in the EDIT mode and adjust the component values in the LEC mode (the SEE command). After a few iterations of adjusting the capacitor values, the user can easily get the corresponding capacitor value as shown in Fig. 3.1.

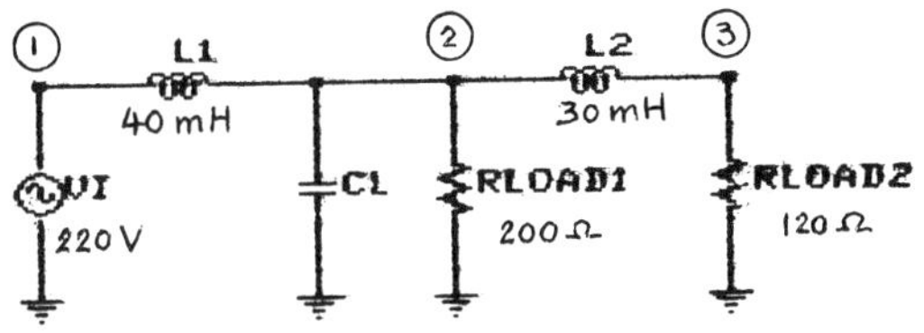


```
LEC>EDIT  (CONTINUE FROM FIG. 2.1)

EDIT>CL 2 0 10UF  (ADD CL=10UF)

EDIT>LIST
      RLOAD1 2 0 200ohm
      RLOAD2 3 0 120ohm
      L1 1 2 .04henry
      L2 2 3 .03henry
      CL 2 0 1E-05farad
      VI 1 0 220ac-volt 0deg

EDIT>SIM AT 50HZ

VARIABLE    MAGNITUDE       PHASE(deg)
V(1)        220             9.3078E-09
V(2)        223.811463      -9.7762655
V(3)        223.12435       -14.267046
I(L1)       3.02455027      .846247988
I(L2)       1.85936958      -14.267046
I(VI)       3.02455027      .846247988
entering.. LEC

 LEC>SEE CL  (CHANGE COMMAND)
     CL = 1E-05farad >8UF    CHANGE TO 8UF
 LEC>SIM AT 50HZ

VARIABLE    MAGNITUDE       PHASE(deg)
V(1)        220             0
V(2)        222.05352       -9.6987322
V(3)        221.371804      -14.189513
I(L1)       2.97823355      -1.7152708
I(L2)       1.84476503      -14.189513
I(VI)       2.97823355      -1.7152708
entering.. LEC
```

Fig. 3.1 An example to show how to add and adjust the value of a capacitor to correct the power factor of the voltage source.

Referring back to the second example (Fig. 2.2) of the previous section, we now want to see the effect of nonideal characteristics of the actual operational amplifier on the frequency response of the filter. This is shown in Fig. 3.2 in which we enter the actual characteristics of the LM741 operational amplifier in the EDIT mode as follows:

A0=100db, RI=1.0Megaohm, CI=20pf, RO=50ohm, P1=10Hz, P2=1.0MHz and P3=10MHz.

Comparing the results with the previous section, we see that, in this case, the effect of nonideal characteristics of the operational amplifier is insignificant. However, this effect is of particular important to serious circuit designers, especially when they have to check the circuit to be sure of its reliability.

```
LEC>EDIT  (CONTINUE FROM FIG.2.2)

EDIT>LIST MODEL
     MODEL LM741 OPAMP
 A0  =1E+32              RI  =1E+32ohm
 CI  =0farad             RO  =0ohm
 P1  =1E+32hz            P2  =1E+32hz
 P3  =1E+32hz
EDIT>MODEL LM741
            A0 =1E+32 >100DB
            RI(ohm) =1E+32 >1MEG
            CI(farad) =0 >20PF
            RO(ohm) =0 >50
            P1(hz) =1E+32 >10HZ
            P2(hz) =1E+32 >1MEGHZ
            P3(hz) =1E+32 >10MEGHZ
```

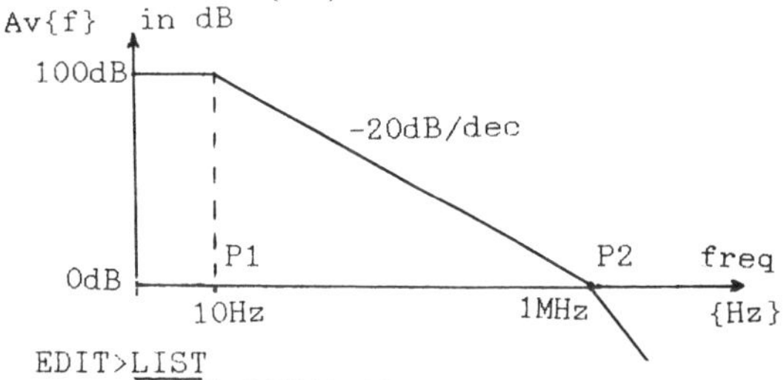


```
EDIT>LIST
     MODEL LM741 OPAMP
 A0  =100000             RI  =1000000ohm
 CI  =2E-11farad         RO  =50ohm
 P1  =10hz               P2  =1000000hz
 P3  =10000000hz
     RI  1 2  1000ohm
     RF  2 3  1000ohm
     CF  2 3  1E-06farad
     VI  1 0  1ac-volt 0deg
     01  0 2 3 LM741
EDIT>SIM
 LEC>PLOT
PLOT>DB V 3 •
       [-16.0634516 , -.0683545868]
PLOT>DUMP
```

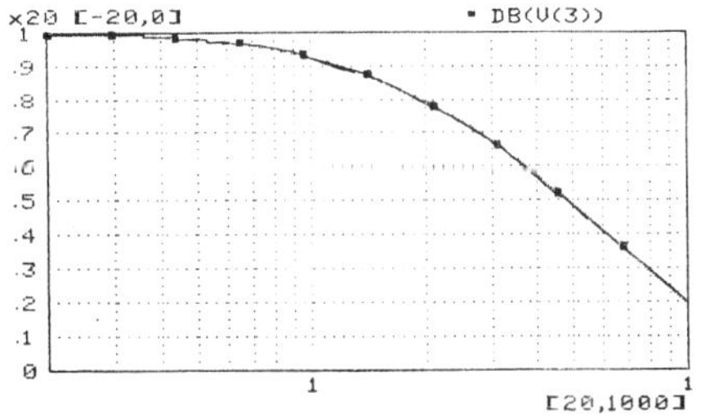

Fig. 3.2 An example to show how to enter the desired operational amplifier characteristics and observe these effects on the frequency response.

IV. TECHNICAL INFORMATION OF LEC

LEC 2.0 is developed in APPLESOFT BASIC language on an APPLE II+ compatible microcomputer. The language is used because the program can be easily debugged and there are many program developing tools existing under APPLESOFT BASIC. It is then compiled into machine language upon released to the general users. With only 48Kbytes of RAM memory in the microcomputer, the program can analysed a circuit up to 28 variables, corresponding to approximately 25 nodes, with as much as 100 components. The maximum number of frequency values that can be analysed simutaneously each time is 25.

The numerical methods used in the program are fairly standard. The circuit description is stored in a linked list form that is easily managable and can facilitate both the frequency and time domain analysis (to be implemented later on). The circuit equation are formed automatically in terms of matrix and vectors by using the Modified Nodal Approach (Ho et al, 1971). The resulting matrix equation is then solved by using the Gaussian Elimination method with partial threshold pivoting strategy (Chua and Lin, 1975). Since LEC 2.0 is not intended to handle large circuits, the equation formulation and solution are carried out in full matrix forms. Therefore the speed of the analysis phase depends on the cube of the size of the matrix. However the speed is not a critical factor for developing LEC 2.0 since it is not run on a fast computer anyway.

V. CONCLUSION.

We have described a general purpose program on a microcomputer to analyse arbitrary linear electronic circuits. The program is called LEC and is now being used and tested by the students of the Department of Electrical Engineering, Chulalongkorn University. The low cost of the microcomputer, the ease of program uses and the power to handle practical cases of the program make LEC highly suitable as a teaching aid and as a design aid for electrical engineers. This program is just only a first step towards the development of more useful and more powerful CAD software which is now going on at the department.

VI. REFERENCES.

Nagel L.N. {1975}. SPICE 2: A computer program to simulate semiconductor circuits. U.C. Berkeley Memorandum No. ERL-M520, May 1975.

Ho C.W. et al {1971}. The modified nodal approach to network analysis and design. IEEE Transaction on Circuit Theory, Vol. CT-18, No. 1, January 1971, page 101-113.

Chua L.O. and P.M. Lin {1975}. Computer-aided analysis of electronic circuits: algorithms and computational techniques. Prentice Hall Inc., 1975.

A CAD LANGUAGE FOR LINEAR CONTROL SYSTEMS

M. Jamshidi, T. C. Yenn and G. Schotik*

CAD Laboratory for Systems and Robotics, Department of Electrical and Computer Engineering, The University of New Mexico, Albuquerque, NM 87131, USA

ABSTRACT

Computer-aided design (CAD) has found a new home in the control systems theory. Many control system's related problems such as analysis, design, estimation, filtering and simulation are now handled through CAD packages and languages. Through these software tools, even a novice computer programmer can take advantage of powerful computational and numerical algorithms for control systems problems. In this paper a CAD language for control and Kalman filtering will be presented. The language, called **CONTROL.lab** is written in FORTRAN/77 and can run under UNIX or VMS on a DEC 11/780 VAX computer system.

Keywords. Computer-aided design, control systems, Kalman filtering.

INTRODUCTION

Computer-aided design (CAD) of control systems is emerging as an indispensible tool for the control system engineer. The CAD capability of control engineers would free them from routine and tiring tasks, but perhaps more importantly, would make powerful numerical and complex algorithms available to them to handle challenging design and analysis problems. From the management point of view, CAD can be a very cost-effective tool for control system engineering. A good and reliable CAD language for control systems would need to be tested extensively by several users through a number of circumstances.

In this paper a CAD language for design (synthesis) and analysis of control systems is presented. The language called **CONTROL.lab**, can handle many challenging problems for control systems. Several application areas of the language is illustrated through actual computer session. Using the capabilities of MATLAB [1] a great number of functions have been added to make **CONTROL.lab** a very powerful, user-friendly and computationally efficient language.

STRUCTURE

CONTROL.lab is designed to handle both linear continuous-time and discrete-time systems defined by a set of four matrices (A, B, C, D), i.e.

State:

$$\dot{x} = Ax + Bu \text{ or } x(k+1) = Ax(k) + Bu(k) \qquad (1)$$

*Los Alamos National Labortory, Los Alamos, NM, U.S.A.

Output:

$$y = Cx + Du \text{ or } y(k) = Cx(k) + Du(k) \qquad (2)$$

Once the system is defined, its description and all the subsequent calculations are stored and are on standby for future use. Figure 1 shows a pictorial representation of the structure of **CONTROL.lab**.

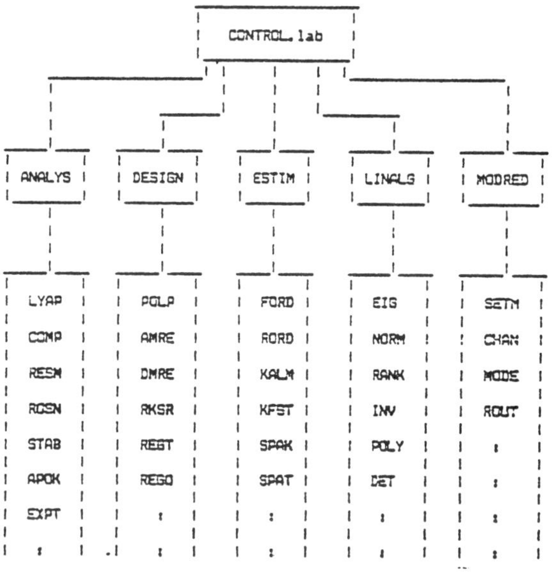

Figure 1. A pictorial structure of CONTROL.lab.

Among the list of functions currently supported by **CONTROL.lab**, "LINALG" is basically MATLAB [1] which includes over seventy seven functions, six of which are:

EIG - matrix eigenvalue/eigenvector
NORM - matrix norm
RANK - matrix rank
INV - matrix inverse
POLY - matrix characteristic polynomial
DET - matrix determinant
etc.

as shown in Figure 1. Other sections of **CONTROL.lab** shown would provide the following:

(i) Analysis (ANALYS):

LYAP - Solution of a continuous-time matrix Lyapunov equation.
COMP - Companion form transformation
RESM - Analytical solution of the resolvent matrix.

ROSN - Rosenbrock's functional minimization.
STAB - Stability of the origin as an equilibrium point.
CON - Complete Controllability Check
OBS - Complete Observability Check
APOK - Analytical solution of the transition matrix of a linear time-invariant discrete-time system, i.e. A k.
EXPT - Analytical solution of the transition matrix of a linear time-invariant continuous-time system
CSCS - Complete solution (simulation) of multivariable continuous-time systems.
CSDS - Complete solution (simulation) of multivariable discrete-time systems.
etc.

(ii) Design (DESIGN):

POLP - Pole placement, via full state feedback.
AMRE - Algebraic matrix Riccati equation.
DMRE - Differential matrix Riccati equation solution via integration.
RKSR - Solution of algebraic matrix Riccati equation by Schur transformation.
FSF - Pole Placement via full State feedback MIMO.
PROP - Pole Placement via Proportional output feedback, MIMO.
PRDR - Pole Placement via PD output feedback, MIMO.
PRPI - Pole Placement via PID output feedback, MIMO.
REGT - Optimal State regulator problem.
REGO - Optimal Output regulator problem.
etc.

(iii) Estimation/Filtering (ESTIM):

FORD - Full-order observer design.
RORD - Reduced-order observer design.
KALM - Standard Kalman filter design.
KFST - Discrete-time standard Kalman filter with zero mean Gaussian noise.
SPAK - Discrete-time decentralized Kalman filter via partitioned two-subsystems with zero mean Gaussian noise.
etc.

(iv) Model Reduction (MODRED):

SETM - Separation of singularly perturbed time sclaes.
CHAN - Chained aggregation.
MODE - Modal aggregation.
ROUT - Aggregation via Routh approximation.
PADE - Aggregation via Pade approximation.
BALN - Aggregation via balanced approach.
etc.

In yet another schematic form, Figure 2 shows a diagram showing the relation of MATLAB, [1] LINPAK [2], EISPAK [3] and PLOTV [4], a graphical package developed at the University of New Mexico which runs with Digital's GIGI terminal and supports **CONTROL.lab**. The dashed box in the figure represents the future incorporation of frequency-domain pakages such as FREDOM/FPC [5,6]. A second graphics interface for CONTROL. lab based on Tektronix's IGL [7] is now implemented.

EXAMPLES

Below are several examples of use of **CONTROL.lab**:

<u>Example 1: LYAP</u> Consider the Lyapunov equation

$$A'P + PA + Q = 0$$

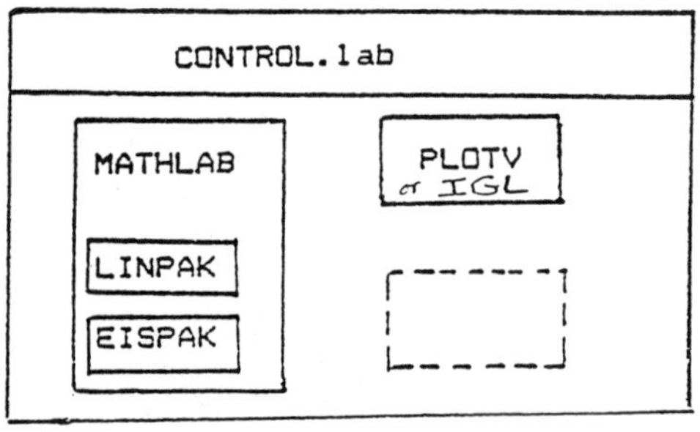

Figure 2. A schematic of <u>CONTROL.lab</u> and its relations with other packages.

Then, in **CONTROL.lab** LYAP(A,Q) provides a solution for it:

```
<> A = <-10 0 0;0 -0.5 0;0 0 -8>


<> A
         -10   0     0
   A =    0   -0.5   0
          0    0    -8

<> Q = <3 2 0.6;2 5 0.2;0.6 0.2 10>

<> Q
          3     2     0.6
   Q =    2     5     0.2
          0.6   0.2   10

<> P = LYAP(A,Q)

<> P
          0.1500  0.1905  0.0333
   P    = 0.1905  5.0000  0.0235
          0.0333  0.0235  0.6250
```

LYAP is based on an iterative scheme described by an algorithm in reference [8].

<u>Example 2: RKSR</u> Consider the continuous-time algebraic Riccati equation:

$$A'K + KA - KBR^{-1}B'K + Q = 0 \qquad (3)$$

then in **CONTROL.lab**, RKSR(A,B,Q,R) will give the results:

```
<> A = <0 1 0;0 0 1;-.5 .5 -.5>

<> A
          0     1     0
   A =    0     0     1
         -0.5   0.5  -0.5

<> B = <0 1;1 1;1 0>

<> B
          0  1
   B =    1  1
          1  0

<> Q = EYE (3,3)

<> Q
          1  0  0
   Q =    0  1  0
          0  0  1

<> R = <1 0;0 1>

<> R

   R =    1  0
          0  1
```

```
<> K = RKSR (A,B,Q,R)

<> K
          0.9858  0.1099 -0.2108
     K =  0.1099  0.6899  0.1845
         -0.2108  0.1845  0.6582
```

To verify the answer, one can calculate the left-hand side of (3) and compare it with zero, i.e. the right-hand side of (3). Hence,

```
<> KDOT = (A')*K + K*A -K*B*INV(R)*(B')*K + Q

<> KDOT

     ANS = 1.0d - 15*

           0.0      -0.1527   0.0173
          -0.0516   -0.9278  -0.0139
          -0.0312   -0.0933  -0.0555
```

RKSR is based on the Schur transformation method due to Laub [9].

<u>Example 3: COMP</u> Consider a multivariable system (A,B,C), where all matrices are general in shape. Then in **CONTROL.lab**, COMP (A,B,C) would transform then into controllable (or observable) companion forms, i.e.

```
<> a = diag (<-1 -2 -3>)

     -1  0  0
A =   0 -2  0
      0  0 -3

<> b = <4 5 6>

B =  1 2 3

<> c = <4 5 6>

C =  4 5 6

<Bc, Cc, Ac> = COMP (a, b', c)

      0   1  0
Ac =  0   0  1
     -6 -11 -6

Cc =  90 114 32

      0
Bc =  0
      1
```

<u>Example 4: POLP</u> Consider a SISO linear time-invariant system

$$\dot{x} = Ax + bu$$

then function POLP ( ) would design a state feedback control law

$$u = kx$$

such that the eigenvalues of the closed-loop system matrix (A + bk) would be assigned to pre-specified locations. In sequel, the nx2 matrix LAMB contains the real and imaginary parts of the desired eigenvalues. Let these desired eigenvalues for the following third order system be -1, -2 $\pm$ j2, i.e.

```
<> a = diag (<1 1>, 1) + <0 0 0; 0 0 0; 1 2 3>

     0 1 0
A =  0 0 1
     1 2 3

<> b = <0 0 1>

     0
B =  0
     1

<> lamb = <-1 0; -2 2; -2 -2>

        -1  0
LAMB =  -2  2
        -2 -2
```

then

```
<> k = POLP (a,b,lamb)

K =  -9 -14 -8
```

to check that the right poles were actually placed, we compute the A + BK matrix and find its eigenvalues:

```
<> Acl = A + B*K

        0   1  0
ACL =   0   0  1
       -8 -12 -5

<> eig (Acl)

       -2.0000 + 2.0000i
ANS =  -2.0000 - 2.0000i
       -1.0000 + 0.0000i
```

which checks with the desired eigenvalues.

<u>Example 5: PID</u> This primitive designs an output feedback control for a MIMO system via a PID controller, i.e. given an nth order, m-input, r-output linear system, for a specified set of closed-loop poles.

$$\dot{x} = Ax + Bu, \; y = cx$$

determine a control,

$$u = -Kpy - Ki \int e \, dt - k_d \, \dot{y}$$

where e=v-y, v is the command input and Kp, K_i and K_d are the desired, m x r proportional, integral and derivative gain matrices. The following "help" documentation on PID describes how to use it.

Find the feedback matrices Kp & Ki & Kd for an observable & controllable MIMO linear time-invariant system as shown:

```
ẋ = Ax + Bu:  y = Cx
u = Kp*y - Ki*INT(e) - Kd*(dy/dt):  e = v - y

PID (A,B,C,PCL,Qhat,ql or kl)
        A      (nxn)
        B      (nxm)
        C      (rxn)
        PDL    ((3m+r-1)x2)
        Qhat   (mxr)
        ql     (mxl)    m>=1
        kl     (rxl)    m<1
```

Where A and B are the system matrices, POL contains the 3m+r-1<=n+1 desired nonrepeated poles; s.t. the REAL and IMAGINARY Parts are in the first and second columns, respectivley. Qhat, ql, and kl are arbitrarily chosen by the user. If m>=1 then input column vector ql, otherwise input row vector kl.

The following 5th order, system with two inputs and two outputs will illustrate the use of PID:

```
<> LOAD ('PID')

<> A

A =

      0      1      0      0      0
      0      0      1      0      0
      0      0      0      1      0
      0      0      0      0      1
    -12     -4     15      5     -3

<> B

B =

      0      1
      0      0
      0      2
      0      0
      1      1

<> C

C =

      1      0      0      0      0
      0      1      0      0      0

POL =

     -1      0
     -2      0
     -3      0
     -4      0
     -5      0
     -1      1
     -1     -1

<> QHAT

QHAT =

      1      0
      0      1

<> Q1

Q1 =

      1
      1

<> <KP,KI,KD>=PID(A,B,C,POL,QHAT,Q1)

KD =

 -4.2262   4.2262
 -0.9973   0.9973

KI =

 -1.3856  -0.6144
 -1.7061  -0.2939

KP =

 -6.2404   6.2404
 -1.9573   1.9573
```

Example 6: AMRE This is an alternative function to RKSR (see Example 2) to find the solution to algebraic matrix Riccati equation (3) using Runge-Kutta integration

```
<> a = <0 1 0; 0 0 1; -1 -2 -3>

A =
     0  1  0
     0  0  1
    -1 -2 -3

<> b = <1 0; 0 1; 1 1>

B =
    1 0
    0 1
    1 1

<> q = EYE (3,3)

Q =
    1 0 0
    0 1 0
    0 0 1

<> r = 2*EYE (2,2)

R =
    2 0
    0 2

<> Kf = <0 0 0 0 0 0>

KF =
     0 0 0 0 0 0

<> e = <0 1 .1 .0001>

E =
    0. 1.0000 0.1000 0.0001
```

where Kf is a 3(3 + 1)/2 = 6-dimensional vector representing the final value of the upper-diagonal Riccati matrix $K(t_f)$ and the vector e represents e = (t_o, t_f, dt, Eps); where t_o and t_f are the initial and final times, respectively dt is the step size and eps is a tolerance constant. Then the following **CONTROL.lab** sentence would give,

```
<> amre (a, b, q, r, kf, e)

ANS =
       0.8708 0.3646 -0.0003
       0.3646 0.9862  0.0910
      -0.0003 0.0910  0.1804
```

Example 7: REGT Below is a test run of REGT of **CONTROL.lab** in solving an optimal linear state regulator problem:

Given a linear time-invariant system with a quadratic cost function,

$$\dot{x} = Ax + Bu, \quad x(t_o) = x_o$$

$$y = cx$$

$$J = 1/2 \int_{o}^{2} (x'Qx + u'Ru)dt$$

where A, B, Q, and R are nxn, nxm, nxn constant matrices with additional constraint that Q and R matrices be positive semi-definate and positive definite, respectively. For this example run we have:

```
<> A = <0 1 0;0 0 1;-.5 .5 -.5>

A =
      0   1   0
      0   0   1
    -.5  .5 -.5

<> B = <0 1;1 1;1 0>
```

```
B =
      0  1
      1  1
      1  0

<> Q = EYE (3,3)

Q =
      1  0  0
      0  1  0
      0  0  1

<> C = <1 1 1>

C =
      1  1  1

<> R = <1 0;0 1>

R =
      1  0
      0  1

<> XO = <1 1 1>

XO =
      1 1 1

<> KF = <1 0 0 1 0 1>

KF =
      1  0 0 1 0 1

<> COND = <0 2 0.1 0.00001>

COND =
        0.   2.    0.1000   0.00001

<> REGT (A,B,C,Q,R,XO,XF,COND)
```

In the above formulation, XO is the initial state vector, KF is an $n(n+1)/2$ -dimensional vector representing the upper triangular value of the nxn final Riccati matrix $K(t_f)$, and the vector COND is defined by

$$COND = (t_o, t_f, \Delta t, Tol) \quad (4)$$

where in (4), t_o = initial time of simulation, t_f = final time of simulation, Δt = step size and Tol = a tolerance constant for integration.

Figures 3-5 show a graphical representation of the optimal states, optimal outputs and optimal controls for this sample use of REGT.

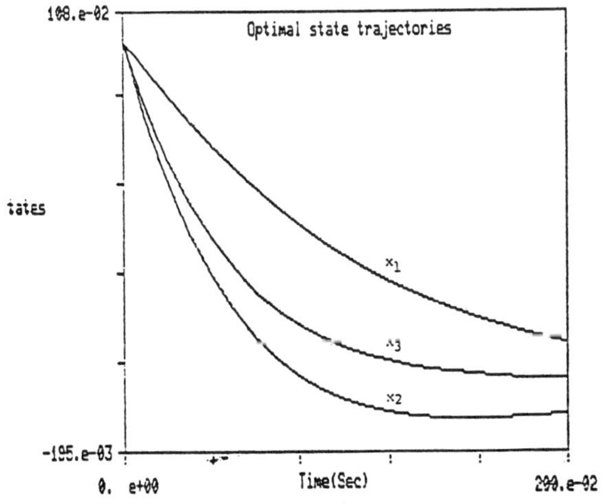

Figure 3. Graphical presentation of the optimal states in the sample use of REGT.

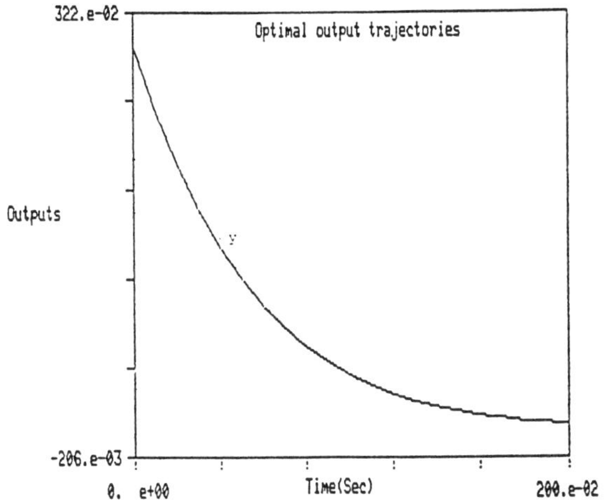

Figure 4. Graphical presentation of the optimal output in the sample use of REGT.

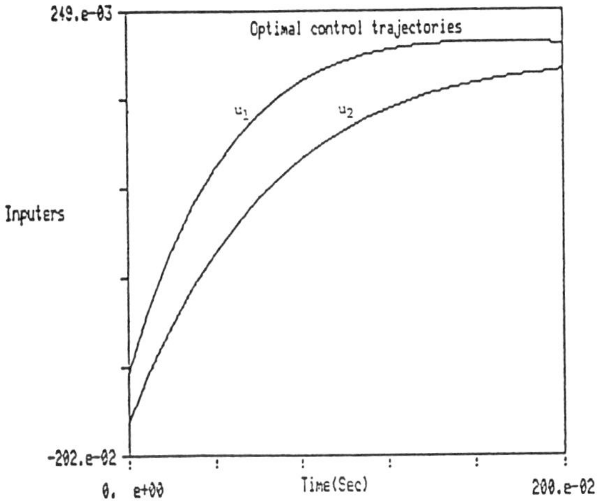

Figure 5. Graphical presentation of the optimal controls in the sample use of REGT.

HELP FACILITY

In an actual situation, in addition to the User's Guide, the user has access to the **CONTROL.lab** is "HELP" facilities which would familiarize him (or her) about such details. In Example 5 a typical use of this facility was illustrated.

Below are two additional HELP menus for two of the 108 functions in **CONTROL.lab:**

POLY: Characteristic polynomial - if A is an nxn matrix, POLY(A) ia an (n+1)-dimensional column vector whose elements are the coefficients of the characteristic polynomial, |sI-A| or in **CONTROL.lab's** expression: DET(S*EYE-A). If z is a vector, POLY(z) is a vector whose elements are the coefficients of the polynomial whose roots are the roots of z, or eigenvalues of A.

RESM: Resolvent matrix - if A is an nxn matrix, RESM(A) provides an analytical solution ot the reslvent matrix $(sI-A)^{-1}$ or INV(s*I-A), explained below:

$$(sI-A)^{-1} = Adj(sI-A)/det(sI-A)$$

$$= \sum_{i=D}^{n-1} Q_i s^i / \sum_{j=D}^{n} p_j s^j$$

where Q_o, Q_1, ..., Q_{n-1} are nxn coefficients of the charactgeristic polynomial (i.e. elements of POLY(A).

CONCLUSIONS

In this paper a brief description of a CAD language for control systems is presented. This CAD language called **CONTROL.lab**, is designed for mini computers such as DEC 11/750/780 VAX systems under UNIX 4.2 or VMS operating systems. A version of this language is also being implemented on the SUN Microsystem workstations which would take advantage of split screens and graphic shells. For future efforts, the present software would be continuously improved and new functions and graphics would be added to it.

REFERENCES

1. C.B. Moler, "MATLAB User's Guide", Department of Computer Science, University of New Mexico, Albuquerque, NM, June 1981.

2. J.J. Dongarra, C.B. Moler, J.R. Bunch and G.W. Stewart, "LINPAK USER'S GUIDE", SIAM, Philadelphia, PA, 1979.

3. EISPAK, "Eigensystem Workshop", Argone National Laboratory, Argone, IL, 1973.

4. D.M. Etter, J.K. McDowell and L.J. White, "A Multi-Purpose Plot Package for GIGI Color Terminals", Technical Report No. ECE 2-83 NSF-0271-1, Department of Electrical and Computer Engineering, the University of New Mexico, Albuquerque, NM, 1981.

5. M. Jamshidi and M. Malek-Zavarei, "LINEAR CONTROL SYSTEMS -- a Computer-Aided Approach", Pergamon Press, Ltd., Oxford, England, 1985.

6. M. Jamshidi, R. Morel, T.C. Yenn and G. Shotik, "Computer-Aided Design of Systems and Networks - Packages and Languages", COMPUTER-AIDED CONTROL SYSTEMS ENGINEERING, M. Jamshidi and C.J. Herget (eds.), North-Holland, Amsterdam, the Netherlands, 1985.

7. Tektronix Corp., Interactive Graphics Language (IGL) User's Guide, Tektronix Dorp., Beaverton, OR, 1978.

8. M. Jamshidi, LARGE-SCALE SYSTEMS: MODELING AND CONTROL, Elsevier North-Holland, New York, 1983.

9. A.J. Laub, "A Schur method for solving algebraic Riccatid equations," IEEE Trans. Auto. Contr., Vol. AC-24, pp. 913-921, 1979.

COMPUTER-AIDED DESIGN OF DECENTRALIZED REGULATORS FOR INDUSTRIAL PROCESSES

G. Guardabassi*, A. Locatelli*, N. Schiavoni*, Y. I. Peng and J. Xiao****

**Dipartimento di Elettronica, Politecnico di Milano, Milano, Italy*
***Department of Electrical Engineering, Hunan University, Changsha, China*

Abstract. The paper briefly illustrates a computer-aided design technique for multivariable regulators and presents the application of such a design procedure to the decentralized control of a plate binary distillation column.

Keywords. Computer-aided design, decentralized control, distillation column.

INTRODUCTION

As is well known, a number of nontrivial obstacles prevent process control engineers from exploiting, in a direct straightforward way, most of the nicest results of the so called "modern control theory". As a matter of fact, almost all control problems to be faced within an industrial context (including those appearing, at a first glance, as the very simplest ones) are characterized by the presence of a set of significant unavoidable constraints, which are in open conflict with some of the fundamental assumptions inherent to almost any result in "modern control theory". In particular, the following constraints are worth mentioning: 1) Decentralized or distributed control structure; 2) Bounds on the complexity (e.g. order) of the local regulators; 3) Use of standard local regulators (e.g. three terms industrial regulators); 4) Bounds on the amplitude of the control effort.

Whenever requirements of the kind (1-4) have to be faced, it is advisable to adopt a somewhat heuristic but more flexible approach which basically consists of two subsequent (possibly interacting) phases: i) selecting, out of a given class, the "structure" of the regulation system: namely, the internal structure (parametrized control algorithm) of each local regulator, plus the topology of the communication network connecting them; ii) tuning, in an optimal way, the parameters of each regulator.

Indeed, this is a demanding task. Up to the present state-of-the-art, phase (i) remains the most difficult and challenging problem. Even more so if such important requirements as system robustness against plant or operation uncertainties must be taken into consideration. Since there is no way to identify, a priori, the 'best' (simplest, cheapest, safest, etc.) structure able to comply with all the aforementioned constraints, the only way out left, in general, is to start with an overdimensioned system and then improve it, progressively, by intelligent computer-aided analysis.

As for phase (ii), the situation is certainly more confortable, due to the power and flexibility of many nonlinear programming codes presently available for parameter optimization. Yet a few points are worth paying particular attention.

1) One possible way to deal with constraints difficult to check at each step of the optimization procedure is to devise a clever parametrization of the entire admissible search domain, such as to ensure feasibility, relative to the 'difficult' constraints, for all values of the parameters (the important word, here, is: 'entire'). In computer-aided control system design, this approach has been taken for instance in Guardabassi and Colleagues (1979), (1983a) relative to the important constraint of robust asymptotic zero-error regulation.

2) A crucial role in any control system design is played by the uncertain variables (initial state, disturbances, set points). Fixing them a priori at a nominal value, before starting the parameter optimization process, may result in a poor result due to overtuning. On the other hand, finding a design which is best in the average may readily become a tremendous task, from a computational point of view, unless some special feature of the problem at hand is suitably exploited.

The parameter-optimization-based design technique presented in this paper is the result of several research efforts developped over a period of approximately fifteen years (see, e.g. Guardabassi and Colleagues (1979); (1983b), Davison and Chaing (1982), Davison and Ferguson (1981) and the references quoted there). It is however only in the last few years that some hidden difficulties connected with the need of meeting with (some of) the crucial constraints (1-4) recalled above have been sufficiently clarified so as to make the development of CAD tools capable of dealing with real size problems possible and rewarding.

The paper is organized in sections as follows. In Section 2, a basic Linear-Quadratic-Gaussian (LQG) problem suited to be solved by a parameter optimization approach is stated and briefly discussed. As an illustration, the design of a decentralized regulator for a binary plate-type distillation column is presented in Section 3. The example does also show how the technique discussed in this paper can (at least to some extent) be heuristically used to deal with nonlinear problems subject to somewhat untractable state and/or control path constraints. The paper ends with some concluding remarks.

STATEMENT OF THE BASIC PROBLEM

Consider a plant described by

$$\dot{x}_P = A_P x_P + B_P u + M_P d, \quad x_P(0) = x_{PO} \tag{1a}$$

$$y_R = C_{PR} x_P + N_{PR} d \quad , \quad y_F = C_{PF} x_P + N_{PF} d \tag{1b}$$

where $x \in R^n$, $u \in R^m$, and $d \in R^1$ are the state, control and disturbance vectors, respectively. As for the output variables, $y_R \in R^p$, $p \leq m$, is the vector of variables which should track a given reference signal, say y°_R, whereas $y_F \in R^r$ is the vector of other measured variables to be possibly exploited for the regulation purposes. The reference signal y°_R and the disturbance d are assumed to be constant for positive time (step signals).

The control objectives considered here are typical of most practical applications and consist in zeroing in a robust way (see Davison, 1976) the error

$$e = y^\circ_R - y_R \tag{1c}$$

at the steady-state, by producing sufficiently fast and smooth transients, while taking possible bounds on the control efforts into adequate consideration. We might therefore assume that these objectives are achieved if the condition

$$\bar{e} = \lim_{t \to \infty} e(t) = 0, \ \forall \ y^\circ_R \text{ and } d \tag{2}$$

is satisfied in a robust way, and a performance index of the form

$$\bar{J} = \int_0^\infty \left(\|x_P(t) - \bar{x}_P\|^2_T + \|u(t) - \bar{u}(t)\|^2_U + \|\dot{u}(t)\|^2_V + \|e(t)\|^2_W \right) dt \tag{3a}$$

is minimized, where T, U, V and W are given positive semidefinite matrices, while

$$\bar{u} = \lim_{t \to \infty} u(t) \quad , \quad \bar{x}_P = \lim_{t \to \infty} x_P(t) . \tag{3b}$$

A parameter optimization approach to the solution of the above control problem consists in determining the parameters of a suitably chosen (and parameterized) regulation system so as to minimize $\tilde{J}$ while satisfying eqn.(2). In what follows, we consider a class of linear time-invariant regulation systems made out of two main parts, which are in charge of the feedback ($\mathcal{S}_1$) and the feedforward ($\mathcal{S}_2$) actions, respectively (fig. 1). Specifically, each block $\mathcal{R}_i$, i=1,2,3, represents any linear time-invariant system depending on a parameter vector q while the block I.M. simply consists of p noninteracting integrators.

On the other hand, each block $\mathcal{R}_i$ may be thought of as constituted by a set of subsystems interconnected according to a loop-free signal-flow diagram suited to fit with a nonclassical information pattern.

In any case, such a regulation system guarantees that eq.(2) is satisfied in a robust way whenever the overall feedback system is asymptotically stable (see Guardabassi and Colleagues, 1983a). Therefore explicit mention to eq.(2) is no more needed.

As a whole, any parametrized regulation system is described by

$$\dot{x}_R = A_R x_R + B_{RE} e + B_{BF} y_F + B_{RS} y^\circ_R + M_R d, \quad x_R(0) = x_{RO} \tag{4a}$$

$$u = C_R x_R + D_{RE} e + D_{RF} y_F + D_{RS} y^\circ_R + N_R d , \tag{4b}$$

where $x \in R^\nu$, ν being a given constant, and all matrices A_R, B_{RE}, etc. are known functions of a design parameter vector q ranging into an admissible parameter set $\Omega \in R^\rho$.

Note that the optimal value q* of q does generally depend on the exogenous variables (y°_R,d) and the initial state of both the plant (y_{PO}) and the regulation system (x_{RO}). Since it is either impractical or plainly meaningless to make q* to depend in real time on this set of variables, it is expedient to think of y°_R, d, x_{PO} and x_{RO} as random variables and seek the value of q which is optimal in the average, i.e. the q minimizing

$$J = E\{\tilde{J}\} , \tag{5}$$

where E{.} is the expected value operator.

As for the characterization of y°_R, d, x_{PO}, x_{RO}, a conceivable choice is to assume that they are zero-mean, mutually independent, random vectors with given covariance matrices Y°_R , D, X_P, X_R respectively; namely:

$$x_{PO} \sim (0, X_P) \ , \quad x_{RO} \sim (0, X_R) \tag{6a}$$

$$y^\circ_R \sim (0, Y^\circ_R) \quad , \quad d \sim (0, D). \tag{6b}$$

In conclusion, the optimization problem we really need to solve can finally be stated as follows, where $Q_S \subset Q$ is the set of values of q such that the overall closed-loop system is asymptotically stable.

The Basic Problem can readily be recasted as a suitable Mathematical Programming problem endowed with a number of interesting and peculiar properties which make fairly easy its solution (see e.g. Guardabassi and Colleagues, 1983a).

AN ILLUSTRATIVE EXAMPLE: CONTROL OF A DISTILLATION COLUMN

The value of the approach outlined in the preceding sections is difficult to appreciate if reference is not made to some specific case, to which other methods or approaches may possibly be applied. In this section, a simulation study is presented, based on a fairly standard model of distillation column dynamics. On the other hand, several applications have already been reported in the literature, to which the interested reader may also refer (see, for instance, Guardabassi and Colleagues (1983b) and the papers quoted therein).

Given a plate-type distillation column for binary systems, consider the problem of controlling the top and bottom compositions (X_T and X_B) in such a way that:

i) the steady-state offsets (under constant set-points and disturbances) do not exceed the (systematic) errors of the composition transducers, whatever the value of both the set-points and the disturbances may be; neglecting then the contribution due to the sensors uncertainty, the steady-state errors have to be zero;

ii) the transients produced by step changes of the feed flow-rate F or composition X_F, or even by step changes of the set-points X_T° and X_B°, be as short as possible, in keeping with the fact that the control variables (reflux flow-rate and vapour flow-rate V) should not exceed, at any time, a priori specified limits;

iii) the overall controller be totally decentralized (fig. 2) and make up of "linear" industrial regulators only.

As for the dynamic model of the distillation column, a set of mass balance equations, one for each plate plus one for the reboiler and one for the condenser, is assumed to provide a description of the plant dynamics accurate enough for all purposes of the present simulation study. More precisely, the adopted simulation model is taken from Rosenbrock (1958).

Let $\bar{R},\bar{V},\bar{F},\bar{X}_F$ be the nominal (constant) values of the input variables, and $\bar{X}_T$, $\bar{X}_B$ be the corresponding (nominal) equilibrium outputs. By linearization around the nominal equilibrium a model in the form of system (1) is obtained, where:

$$u = \begin{vmatrix} R - \bar{R} \\ V - \bar{V} \end{vmatrix}, \quad d = \begin{vmatrix} X_F - \bar{X}_F \\ F - \bar{F} \end{vmatrix}, \quad y_R = \begin{vmatrix} X_T - \bar{X}_T \\ X_B - \bar{X}_B \end{vmatrix}, \quad r=0.$$

We may rephrase now the requirements listed at the beginning of this section in the following way. The multivariable regulator we are looking for

A) must be made up of acyclically interconnected single-input single-output linear industrial regulating units (P, or PI, or PID, etc.).

B) must be fitting with the prescribed information structure (decentralized control), in the sense that each control variable u_i must depend upon the corresponding error variable e_i only (plus, possibly, some disturbances, in case a feed-forward action is also included),

C) when combined with the linear time-invariant model of the plant dynamics previously referred to as system (1), must produce an asymptotically stable overall system S,

D) must be such that the outputs y_{R1} and y_{R2} corresponding to the (asymptotically stable) equilibrium state of S, produced by any constant values of the set points and the disturbances, be exactly equal to the set points (e=0),

E) must be such that the transients produced by any admissible constant values of the set points and the disturbances, starting from zero initial state (step responses), settle in as short a time as possible, without showing exceedingly high deviations of the control variables u from their steady-state value $\bar{u}$, which obviously depends upon the set points (y_R°), the disturbances (d), and the particular value given to the regulator parameters.

Note that, in view of Section 2, condition (D) is automatically met with. Condition (E) may be (partially) expressed as the need that q be chosen in Q_S (condition (C)) so as to minimize a performance index of the form:

$$J(q) = E\left\{\int_0^\infty (||u-\bar{u}(y_R^\circ,d)||_U^2 + ||e||_W^2)\, dt\right\},$$

where y_R° and d are thought of as zero-mean random variables with known covariance matrices Y_R° and D, respectively; U and W are usually positive definite and diagonal. Unfortunately, the setting above does not incorporate any requirement of the form: ($|e_i(t)| \leqslant e_{imax}$ and $|u_i(t)| \leqslant u_{imax}$, for all t) in response to any step variation of either the disturbances or the set points not exceeding, in amplitude, a priori specified limits. However, if an efficient package is available to solve interactively the Basic Problem, even this last "difficult" constraint can easily be incorporated (at least to a certain extent) in the design procedure. The following heuristic "algorithm" may in fact be expedient.

Step 1. Set the weighting matrices U and W to a conceivable initial value;

Step 2. Find the corresponding optimal value q* of the regulator parameters;

Step 3. By an accurate (nonlinear) simulation code, test the most critical step responses of the overall system S;

Step 4. If, no constraint being violated, both control variables (during the worst transient) approach the limits of their admissible regions, accept q* as final design;
if, no constraint being violated, a control variable is poorly active, decrease the corresponding entry on the main diagonal of U, and go to Step 2;
if all error and control variables exceed the prescribed limits, enrich (if possible) the regulator subsystems $\mathcal{R}_i$'s and go to Step 1; otherwise, stop;
if an error or control variable exceeds the prescribed limits, increase the corresponding entry on the main diagonal of W or U, respectively, and go to Step 2.

The column considered for this illustration is taken from Takamatsu and Colleagues (1979). Both the stripping and the enriching sections consist of 4 plates. The relative volatility of the considered system is α=2, while the liquid hold-ups in the column are

H_0 = 200 mols for the reboiler; H_i = 25 mols, for each of the 9 plates; H_{10} = 100 mols for the condenser.

The nominal values of the input (control and disturbance) and output (controlled) variables are:

$\bar{R} = 13$ mols/min; $\bar{V} = 17.5$ mols/min; $\bar{X}_F = 45\%$;

$\bar{F} = 10$ mols/min; $\bar{X}_F = 90\%$; $\bar{X}_B = 10.4\%$.

The maximum constant variations of the set points and the disturbances have been set as follows:

$$|\Delta X^\circ_T|_{max} = |y^\circ_{R1}|_{max} = 5\%, \quad |\Delta X^\circ_B|_{max} = |y^\circ_{R2}|_{max} = 5\%,$$

$$|\Delta X_F|_{max} = |d_1|_{max} = 22\%, \quad |\Delta F|_{max} = |d_2|_{max} = 2.5 \text{ mols/min},$$

while $0 < R(t) < V(t) < 30$ mols/min is the admissible region for the control variables.

Recalling that the variance of a random variable having constant probability density 1/2b, over the interval (-b,b), is $b^2/3$, the above constraints on y°_R and d have been carried into the respective covariance matrices in the following way:

$$Y^\circ_R = \begin{vmatrix} 8.3 & 0 \\ 0 & 8.3 \end{vmatrix}, \qquad D = \begin{vmatrix} 160 & 0 \\ 0 & 2.1 \end{vmatrix}.$$

A first goal has been to examine the best performance achievable by a pair of feedback loops using two term (PI) regulating units only. Referring to the description of the regulation system given by eq.(4), we set

$$A_{RE} = 0, \; B_{RE} = I, \; B_{RS} = 0, \; M_R = 0,$$

$$C_R = \begin{vmatrix} q_1 & 0 \\ 0 & q_2 \end{vmatrix}, \; D_{RE} = \begin{vmatrix} q_3 & 0 \\ 0 & q_4 \end{vmatrix}, \; D_{RS} = 0, \; N_R = 0,$$

so that, referring to fig. 1,

$$\text{I.M.} \quad : \quad \dot{x}_R = B_{RE}\, e,$$

$$\mathcal{R}_2(q) \quad : \quad u = C_R(q)\, x_R + D_{RE}(q)\, e,$$

$\mathcal{R}_1(q)$ is trivial, and no feedforward unit ($\mathcal{R}_3(q)$) has been introduced. Note that q_1 and q_2 are the coefficients of the integral actions (q_3 and q_4 are the coefficients of the proportional actions) of the top and bottom regulating units, respectively.

Since the performance index (3) is of quadratic type and, in view of the adopted units, the range of the control variables is about ten times as large as the admissible range of the errors, the afore mentioned interactive "algorithm" has been initialized by setting U = 5 I and W = 0.05 I. By sequentially applying (bottom-up) empirical rules (Ziegler and Nichols; Aikman, Ream and Rutherford) for the design of industrial regulators, the value $q_1 = 1.4$, $q_2 = -0.59$, $q_3 = 2.61$, $q_4 = -3.7$ have first been obtained. Then the interactive parameter optimization "algorithm" readly provides the following design: $q_1 = 0.2$, $q_2 = -0.28$, $q_3 = 2.5$, $q_4 = -3$. A comparison between the empirical and the optimal choice of the parameter vector q is shown in fig. 3. Despite the simplicity of this regulator (4 parameters), it has to be stressed that achieving this result (good transients and feasible controls) by any simulation-based operator-driven trial-and-error procedure does not appear a trivial task at all.

Moreover, a moderate additional effort enables ascertaining that the performance of the overall control system is not significantly improved by adding either a derivative action to the local regulating units or any feedforward action, from the set points to the control variables. A feedforward purely proportional action, from one of the disturbances (feed flow-rate) to both the control variables, does instead greatly improve the system performance. In particular, referring to eq.(4) and fig. 1, the optimal value found for the two feedforward parameters involved is

$$N_R = \begin{vmatrix} 0 & 1.3 \\ 0 & 1.8 \end{vmatrix}.$$

Far less significant is finally the improvement produced by a feedforward purely proportional action driven by the perturbations of the feed composition.

More details on the multivariable regulator design for plate-type distillation columns may be found in Cattaneo and Colleagues (1981), (1982).

CONCLUDING REMARKS

The main purpose of this paper has been to illustrate the main features of an optimization-based computer-aided design method, suited to handle a variety of control problems of the kind commonly encountered in an industrial environment and not easily solvable by different, however sophisticated techniques.

ACKNOWLEDGMENT

This work has been partially supported by MPI and CNR, Centro di Teoria dei Sistemi.

REFERENCES

Cattaneo, F., P. Colaneri, and R. Colcerasa (1981). Progettazione assistita da calcolatore di sistemi di controllo decentralizzati per una colonna di distillazione a piatti. Tesi di laurea, Politecnico di Milano.

Cattaneo, F., P. Colaneri, R. Colcerasa, G. Guardabassi, Y.I. Peng, and J. Xiao (1982). Computer-aided design of a decentralized regulator for plate distillation columns. Automazione e Strumentazione, 12, 85-92, in Italian.

Davison, E.J. (1976a). The robust decentralized control of a general servomechanism problem. IEEE Trans. Automat. Contr., AC-21, 14-24.

Davison, E.J., and T. Chang (1982). The design of decentralized controllers for the robust servo mechanism problem using parameter optimization methods. Proc. American Control Conference, III, 905-909.

Davison, E.J. and I. Ferguson (1981). The design of controllers for the multivariable robust servomechanism problem using parameter optimization methods. IEEE Trans Automat.Contr.,

AC-26, 93-100.

Guardabassi, G., A. Locatelli, C. Maffezzoni, and N. Schiavoni (1979). Parameter optimization approach in decentralized process control: a unified setting for multivariable industrial regulator design. M.A. Cuenod (Ed.) Proc. IFAC Symposium on Computer-Aided Design of Control Systems, Pergamon Press, Oxford, 87-92.

Guardabassi, G., A. Locatelli, C. Maffezzoni, and N. Schiavoni (1983a). Computer-aided design of structurally constrained multivariable regulators. Part. I: Problem statement, analysis and solution. Proc. IEE-Part D, 130, 155-164.

Guardabassi, G., A. Locatelli, C. Maffezzoni, and N. Schiavoni (1983b). Computer-aided design of structurally constrained multivariable regulators. Part II: Applications. Proc. IEE-Part D, 130, 165-172.

Rosenbrock, H.H. (1958). Calculation of the transient behaviour of distillation columns. British Chemical Engineering, 364-367, 432-435, 491-494.

Takamatsu, T., I. Hashimoto, and Y. Nakai (1979). A geometric approach to multivariable control system design of a distillation column. Automatica, 15, 387-402.

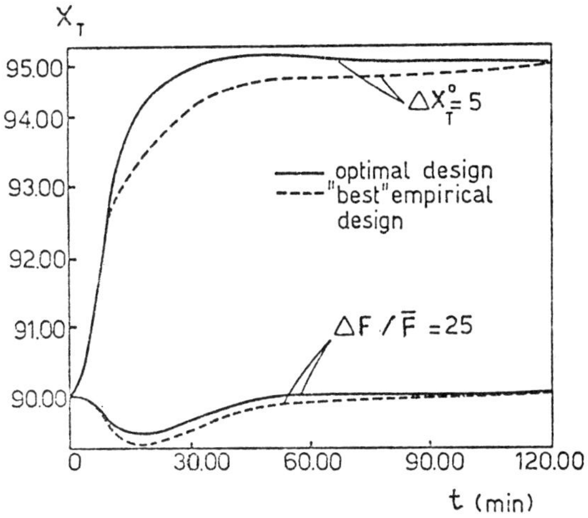

Fig.3 Top concentration transients with an "empirical" or an "optimal" regulator

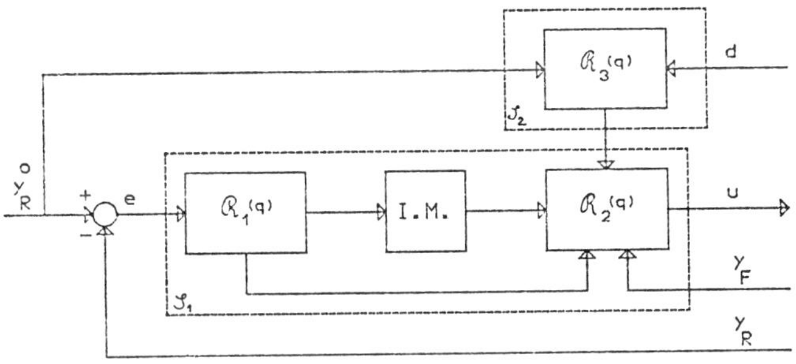

Fig.1 The regulation system.

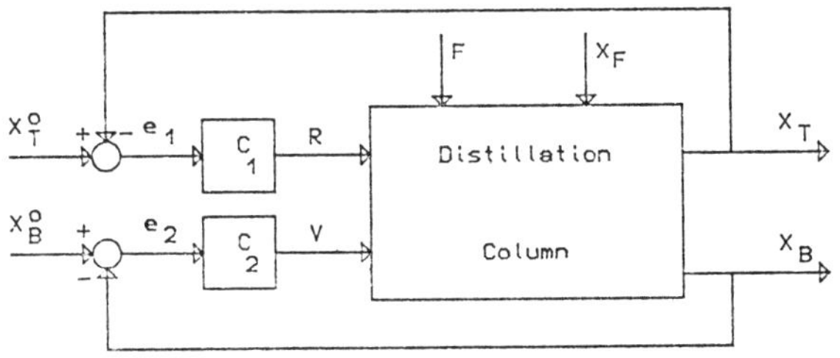

Fig.2 The block-scheme of the regulation system

AN IMPROVED ALGORITHM FOR GENERALISING PSEUDODIAGONALISATION

Y. L. Bao, G. Z. Pang and S. F. Li

Department of System & Management Science, University of Science and Technology of China, Hefei, Anhui, China

Abstract. The multivariable frequency domain design method is playing an increasingly important role in computer-aided control system design and it depends on achieving a diagonally-dominant system to complete the design with success. The pseudodomination proposed in this paper enables us to design dynamic compensator to improve the dominance of the compensated plant over an extended range of frequencies. The procedure arises from Hawkins's generalising pseudodiagonalisation in a number of directions; in particular, the calculation is reduced and some important messages that provide a real facilitation to the design can be easily gotten at in computational process. Three examples designed interactively by computer are included to illustrate the effectiveness of the improved algorithm.

Keywords. Multivariable control system; linear systems; dynamic compensator; computer-aided design; pseudodiagonalisation.

INTRODUCTION

A major difficulty which often exists with several multivariable frequency domain design techniques, is the necessity that the system appears sufficiently noninteractive that it may be considered as a set of single loops. As a means of allowing single loop design, the concept of diagonal dominance was introduced by Rosenbrock (1969, 1974). Hawkins (1972a, 1972b) proposed the pseudodiagonalisation procedure which is an effective method for producing a compensator to improve the dominance of compensated plant.

The notation adopted here follows that previously established. The open loop transfer-function matrix is written as

$$Q(s) = G(s)K(s) \tag{1}$$

where $G(s)$ is nonsingular mxm rational matrix of a given plant transfer functions and $K(s)$ is an mxm polynomal precompensator matrix denoted by

$$K(s) = K^{(0)} + K^{(1)}s + \cdots + K^{(\alpha)}s^{\alpha} \tag{2}$$

$Q(s)$ is said diagonally column dominant if for $\forall s = i\omega \in D$

$$NDj(\omega) = \sum{}' |q_{kj}(i\omega)| / |q_{jj}(i\omega)| < 1 \qquad j=1,2,\cdots,m \tag{3}$$

where $\sum'$ denotes $\sum_{\substack{k=1\\k\neq j}}^{m}$ and D is Nyquist contour without the zero or pole of $G(s)$ on it. The $NDj(\omega)$ is called Normalized Dominance of $Q_{*j}(s)$ (the j-th column of Q) at the point $s=i\omega$.

Without loss of generality the $K(s)$ is always restricted to a polynomal matrix because that any rational matrix $Kp(s)$ can be decomposed to $Kp(s) = K(s)D(s)$, where $D(s)$ is a diagonal rational matrix which will be considered in the design process of the main controller, and the $NDj(\omega)$ of $G(s)Kp(s)$ and that of $G(s)K(s)$ are exactly the same.

Pose the problem of choosing a dynamic precompensator matrix $K(s)$ to make a given $G(s)$ dominant at extended range of frequencies $s = i\omega_t$ $(t=1,2,\cdots,r)$. This can be reduced to m independent subproblems by choosing to obtain column dominance (Note that if a postcompensator is used, namely $Q(s) = K(s)G(s)$, then the consideration of row dominance yields a similar decomposition), in which case

$$Q_{*j}(i\omega_t) = G(i\omega_t)K_{*j}(i\omega_t)$$

$$j=1,2,\cdots,m \quad t=1,2,\cdots,r$$

and the associated dominance $NDj(\omega_t)$ is denoted in eqn. (3). The j-th subproblem is to choose a precompensator column $K_{*j}(s)$ so that $NDj(\omega_t) < 1$, or possibly improve the dominance which already exists at $s = i\omega_t$ $(t=1,2,\cdots,r)$. Write

$$G(i\omega) = M(\omega) + iN(\omega)$$

where $M = Re(G)$ and $N = Im(G)$. Let e_j be j-th column of I_m and

$$E_j = \begin{bmatrix} e_j & 0 \\ 0 & e_j \end{bmatrix} \qquad W(\omega) = \begin{bmatrix} M(\omega) & -N(\omega) \\ N(\omega) & M(\omega) \end{bmatrix}$$

CSTD-H*

$$V(\omega) = \begin{bmatrix} I_m & 0 & -I_m\omega^2 & \cdots & 0 \\ 0 & I_m\omega & 0 & \cdots & (-1)^{\frac{1}{2}(\alpha-1)} I_m\omega^\alpha \end{bmatrix}$$

$$k_j = \begin{bmatrix} K_{*j}^{(0)} \\ K_{*j}^{(1)} \\ \vdots \\ K_{*j}^{(\alpha)} \end{bmatrix}$$ be correspond to $K_{*j}(s)$.

Let $W_t = W(\omega_t)$, $V_t = V(\omega_t)$, and $B_{tk} = V_t^T W_t^T E_k$ be an $(\alpha+1)m \times 2$ matrix, where V^T denotes transpose matrix of V.

To improve column dominance over an extended range of frequencies, by generalising pseudodiagonalisation method (Rosenbrock 1974), each column $K_{*j}(s)$ is calculated by minimising

$$J_j = \sum_{t=1}^{r} c_t \Sigma' \left| q_{kj}(i\omega_t) \right|^2 \quad (4)$$

subject to $k_j^T k_j = 1$. It leads to an $(\alpha+1)m \times (\alpha+1)m$ eigenvalue/vector problem (See Appendix 1):

$$\begin{cases} Rk_j = \theta_{min} k_j \\ k_j^T k_j = 1 \end{cases} \quad (5)$$

$$\text{where} \quad R = \sum_{t=1}^{r} c_t \Sigma' B_{tk} B_{tk}^T \quad (6)$$

Furthermore, Ford and Daly (1979), Mason, Daly and Lee (1979) extended this procedure by minimising

$$J_j = \int_{\omega_1}^{\omega_r} c(\omega) \Sigma' \left| q_{kj}(i\omega) \right|^2 d\omega$$

subject to

$$\int_{\omega_1}^{\omega_r} c(\omega) \left| q_{jj}(i\omega) \right|^2 d\omega = 1$$

It leads to an $(\alpha+1)m \times (\alpha+1)m$ generalised eigenvalue/vector problem (Its deductive process is similar to that given in appendix 1):

$$\begin{cases} Rk_j = \theta_{min} Lk_j \\ k_j^T Lk_j = 1 \end{cases} \quad (7)$$

$$\text{where} \quad R = \int_{\omega_1}^{\omega_r} c(\omega) \Sigma' B_{\omega k} B_{\omega k}^T d\omega$$

$$L = \int_{\omega_1}^{\omega_r} c(\omega) B_{\omega j} B_{\omega j}^T d\omega$$

An improved algorithm presented in this paper reduces the computation. It has no need to solve the above-mentioned large order eigenvalue/vector equation for any m (input/output number) and any order α of K(s). For all examples concerned (distillation column, boiler furnace, aircraft autostabilisation system, turbofan engine, generator set, paper machine, unstable batch process), using the algorithm we have computed compensators with either same or better results.

PSEUDODOMINATION ALGORITHM

We also calculate $K_{*j}(s)$ by minimising J_j as in eqn. (4) but suggest

$$\left| q_{jj}(i\omega_0) \right| = 1 \quad (8)$$

where $\omega_0 \in [\omega_1, \omega_2, \cdots, \omega_r]$. An analysis similar to that given in Appendix 1 shows that k_j is given again by an eigenvalue/vector problem:

$$\begin{cases} Rk_j = \theta_{min} Lk_j & (9) \\ k_j^T Lk_j = 1 & (10) \end{cases}$$

where R is the same as that given in Eq.(6) and

$$L = B_{oj} B_{oj}^T$$

The condition of optimality gives that $J_j = \theta_{min} \geqslant 0$.

While each matrix $\Sigma' B_{tk} B_{tk}^T$ is only positive semidefinite, as well as the $B_{oj} B_{oj}^T$, the matrix R+L will usually be positive definite whenever $r \geqslant \alpha+1$. Then the generalised eigenvalue/vector relationship eqn. (9) may be rewritten as

$$\lambda_{max} k_j = (R+L)^{-1} Lk_j \quad (11)$$

where $\lambda_{max} = 1/(1+\theta_{min})$. Obviously, we have $0 < \lambda_{max} \leqslant 1$.

Appendix 2 shows the solution k_j of eqns. (11) and (10) can be obtained equivalently from eqns. (12-14) as follows:

$$\begin{cases} Ax = \lambda_{max} x & (12) \\ x^T x = 1 & (13) \end{cases}$$

$$k_j = \lambda_{max}^{-1} Dx \quad (14)$$

$$\text{where} \quad D = (R+L)^{-1} B_{oj} \quad \text{and} \quad A = B_{oj}^T D \quad (15)$$

the λ_{max} is the larger eigenvalue of the 2×2 symmetrical matrix A and the 2 dimension column vector x is the corresponding eigenvector.

It is extremely easy to solve the 2×2 symmetric matrix eigenvalue/vector eqns. (12-13) and obtain k_j from eqn. (14).

As compared with pseudodiagonalisation the improved algorithm makes what better as follows:

1. Having no need to solve the $(\alpha+1)m \times (\alpha+1)m$ eigenvalue/vector problem, it reduces the calculation.

2. If the $\lambda_{max} = 1$, then the $Q_{*j}(s)$ is "decoupled", that means $q_{jj}(s) \neq 0$ and $q_{kj}(s) = 0$ for $k \neq j$. In fact, if there exists a β-order $K_{*j}(s)$ which can "decouple" $Q_{*j}(s)$, then such one will be given

by using the pseudodomination algorithm provided we take the $\alpha=\beta$ and $r \geqslant \alpha+1$.

3. If we take $\alpha > \beta$ (assume that $Q_{*j}(s)$ can be decoupled by a β-order compensator) then the two eigenvalues of 2×2 symmetric matrix A, $\lambda_{max} = \lambda_{min} = 1$. In this case the computer will give designer a message to reduce the order α of $K_{*j}(s)$.

4. In order to have the general trend of $NDj(\omega)$ on $\omega \in [\omega_1, \omega_r]$ under control we can choose the weighted coefficients c_t $(t=1, 2, \cdots, r)$ as follows:

(A) Take $c_t = \omega_t^n$. If we let $n>0$ then the $NDj(\omega)$ at high frequencies will be even more improved; similarly, if we let that $n<0$ then the $NDj(\omega)$ at low frequencies will be more improved. Often we let that $n = 0$, that gets $c_t = 1$.

(B) When we desire that the $NDj(\omega)$ at all $\omega \in [\omega_1, \omega_r]$ should be approximately the same, we may twice calculate k_j from eqns. (12-14), for the first time taking that $c_t = 1$ $(t=1,2,\cdots,r)$ and obtaining $NDj'(\omega_t)$, and for the last time taking that $c_t = NDj'(\omega_t)$. (See examples 1 and 2, we call it "self-weighting".)

Remark. From the point of view of obtaining the same normalised dominance $NDj(\omega)$, the k_j is equivalent to ck_j for any real $c \neq 0$. So we may substitute $k_j = Dx$ for eqn. (14). Furthermore, we may substitute $c(s)K_{*j}(s)$ for $K_{*j}(s)$, where $c(s)$ is any real rational function, $c(s) \neq 0$.

DESIGN EXAMPLES

Example 1. The example is chosen to demonstrate that the algorithm can deal with the unstable system considered by Rosenbrock (1974) and Mason, Daly and Lee (1979). It is the model of a chemical reactor given by

$$G(s) = N(s)/d(s)$$

with

$$N_{11} = 263.3+29.2s$$
$$N_{12} = -31.8-89.83s-32.62s^2-3.146s^3$$
$$N_{21} = -106.8-68.84s+42.67s^2+5.679s^3$$
$$N_{22} = 15.15+9.43s$$
$$d(s) = 5.514-88.31s+15.75s^2+11.67s^3+s^4$$

Having chosen the order of $K(s)$, $\alpha = 1$, for calculating k_1 we take $r=2$, $\omega_1=.0316$, $\omega_0=\omega_2=100$; for calculating k_2 we take $r=25$, $\omega_1=.001$, $\omega_{t+1}=10^{\frac{1}{4}}\omega_t$ $(t=1,2,\cdots,24)$ and $\omega_0=1$, obtain:

$$K(s) = \begin{bmatrix} -.0882-.0005s & .0877+.2599s \\ -.6217-.3185s & .7275-.0169s \end{bmatrix} \tag{16}$$

The plots of $\mathrm{Log}_{10}NDj(\omega)$ $(j=1,2)$ vs. frequencies ω are given in Fig. 1.

For comparing, the results obtained from eqn. (7) by Mason, Daly and Lee (1979) listed below:

$$\overline{K}(s) = \begin{bmatrix} -.0869+.0168s & .1058+.3904s \\ -.6335+.7686s & .9098-.0925s \end{bmatrix} \tag{17}$$

The corresponding $\mathrm{Log}_{10}\overline{ND}j$ (ω) are also drawn in Fig. 1. with dotted line.

In order to obtain approximately same $NDj(\omega)$ for all $i\omega \in D$, we use the "self-weighting" calculation (the results obtained at first time were as above and also drawn in Fig. 2. with dotted line) which gives:

$$K^*(s) = \begin{bmatrix} .1449-.0176s & .1639+.4765s \\ 1.032-.5848s & 1.368-.1345s \end{bmatrix} \tag{18}$$

The corresponding $\mathrm{Log}_{10}NDj(\omega)$ are given in Fig. 2.

Example 2. This example concerns the following model of boiler considered by Ford and Daly (1979). The elements are complex. Having chosen $\alpha = 2$ and frequencies $\omega \in [.001, 10]$, Ford and Daly (1979) obtained (from eqn. (7)):

$$\overline{K}_{*1} = \begin{bmatrix} 52.97(s+3.892\ 10^{-3})(s+0.1438) \\ 7.683\ 10^{-3}(s-7.994)(s-0.8228) \end{bmatrix} \tag{19}$$

The corresponding $\mathrm{Log}_{10}\overline{ND}_1(\omega)$ plot is drawn with dotted line in Fig. 3.a. It is obvious that $\overline{Q}_{*1}(s)$ is not dominant. Having tried to remove the two zeroes from right plane, they gave:

$$\overline{K}^*_{*1} = \begin{bmatrix} (s+3.892\ 10^{-3})(s+0.1438) \\ 9.541\ 10^{-4} \end{bmatrix} \tag{20}$$

The corresponding $\mathrm{Log}_{10}\overline{ND}^*_1(\omega)$ is given in Fig. 3.b. with dotted line.

Now we used the self-weighting of the pseudodomination algorithm with eqns. (12-14) and obtained

$$K_{*1}(s) = \begin{bmatrix} 1.6841+417.87s+3976.9s^2 \\ 2.8382-0.4457s+0.0504s^2 \end{bmatrix} \tag{21}$$

The corresponding $\mathrm{Log}_{10}ND_1(\omega)$ is given in Fig. 3.a. It is obvious that the $Q_{*1}(s)$ is already dominant. As we try to remove the two zeroes from right plane we obtain

$$K^*_{*1}(s) = \begin{bmatrix} 1.6841+417.87s+3977s^2 \\ 2.8382 \end{bmatrix} \tag{22}$$

The corresponding plot of $\mathrm{Log}_{10}ND^*_1(\omega)$ is given in Fig. 3.b.

The calculation of K_{*2} in this example is trivial. Taking order $\alpha = 0$, the results obtained from eqns. (12-14) or from eqn.(5) or eqn. (7) are just the identical.

Example 3. This example concerns the model of a 30-plate distillation column in the UMIST pilot plant considered by

Hawkins (1972a) and Rosenbrock (1974). Taking the order $\alpha = 0$, we obtained the result as the same as that given by them. Taking $\alpha = 1$, we discover $\lambda_{max}=1$ ($\lambda_{min} \neq 1$), and obtained

$$K(s) = \begin{bmatrix} \frac{-1.461(1+10s)}{1+15s} & \frac{-2.0739(1+75s)}{1+15s} \end{bmatrix}$$

The compensated system $Q(s) = G(s)K(s)$ has been decoupled. Having taken $\alpha = 2$, the computer with pseudodomination CAD program gives that $\lambda_{max} = \lambda_{min} = 1$, by which we are warned that there is no need for taking so high order of precompensator.

CONCLUSION

As discussed above, the improved algorithm presented in this paper can be used to design dynamic compensator for reducing open loop interaction in multivariable systems. The method gives a systematic guide to the design of dynamic precompensator (postcompensator) that provide a high degree of column (row) dominance improvement over a wide range of frequencies. Furthermore, the algorithm can also be applied to the systems with time delays. Its advantage is reduction of calculation. Moreover, it is easy to give some important information through the use of the algorithm. The examples presented clearly demonstrate the computational efficiency of the proposed algorithm.

REFERENCES

Ford, M.P. and K.C. Daly (1979). Dominance improvement by pseudodecoupling. Proc. IEE, 126, 1316-1320.

Hawkins, D.J. (1972a). Pseudodiagonalisation and the inverse-Nyquist-array method. Proc. IEE, 119, 337-342

Hawkins, D.J. (1972b). Multifrequency version of pseudodiagonalisation. Electron. Lett., 8, 473-474

Mason, P.J., K.C. Daly and K.B. Lee (1979). The design of dynamic compensators for reducing open loop interaction in multivariable systems. In M.A. Cuenod (Ed.), IFAC Symp. Computer Aided Design of Control Systems, Zurich, Switzerland. pp. 331-336

Rosenbrock, H.H. (1969). Design of multivariable control systems using the inverse Nyquist array. Proc. IEE, 116, 1929-1936.

Rosenbrock, H.H. (1974). Computer-Aided Control Systems Design. Academic Press, London.

APPENDICES

Appendix 1. Deduce eqn. (5).
In the notation given in this paper it is obvious that

$$K_{*j}(i\omega) = [I_m,\ iI_m]\cdot V(\omega)k_j$$

and $G_{k*}(i\omega) = M_{k*}+iN_{k*} = e_k^TM + ie_k^TN$

Hence $q_{kj}(i\omega) = G_{k*}K_{*j} =$

$$= (e_k^TM + ie_k^TN)\cdot[I_m,\ iI_m]Vk_j =$$

$$= [e_k^TM+ie_k^TN,\ -e_k^TN+ie_k^TM]\cdot Vk_j =$$

$$= [1,\ i]\cdot E_k^TWVk_j$$

Thus $q_{kj}(i\omega_t) = [1,\ i]\cdot B_{tk}^Tk_j$

$$|q_{kj}(i\omega_t)|^2 = q_{kj}^+q_{kj} = k_j^TB_{tk}\begin{bmatrix}1\\-i\end{bmatrix}\cdot[1,i]\cdot B_{tk}^Tk_j$$

$$= k_j^TB_{tk}B_{tk}^Tk_j$$

Therefore the index of eqn. (4) may be written as

$$J_j = \sum_{t=1}^{r} c_t \Sigma' k_j^TB_{tk}B_{tk}^Tk_j = k_j^TRk_j$$

where the $(\alpha+1)m \times (\alpha+1)m$ matrix R is denoted by eqn. (6). For choosing k_j to minimize J_j subject to constraint that $k_j^Tk_j=1$, on introducing a Lagrange multiplier θ this requires us to minimize

$$L_j = k_j^TRk_j + \theta(1-k_j^Tk_j)$$

Under the condition of optimality $\nabla L_j = 0$ we obtain

$$\frac{\partial L_j}{\partial k_j} = 2Rk_j-2\theta k_j=0 \text{ and } \frac{\partial L_j}{\partial \theta} = k_j^Tk_j-1 = 0$$

The condition of optimality gives $J_j = \theta \geq 0$. It follows that to minimize J_j in eqn. (4) we must choose that k_j corresponding to, the smallest eigenvalue of R, θ_{min}. Therefore the eqn. (5) yields.

Appendix 2. Show the solution k_j of eqns. (10-11) can be obtained equivalently from eqns. (12-14).
Proof. The proof is in two parts. First we show that if k_j satisfies eqns. (10-11) then it will satisfy the eqns. (12-14). Note that in the notation of equation $L = B_{oj}B_{oj}^T$, eqns. (11) and (10) are respectively

$$\lambda_{max}k_j = (R+L)^{-1}B_{oj}B_{oj}^Tk_j \qquad (23)$$

$$(B_{oj}^Tk_j)^T(B_{oj}^Tk_j) = 1 \qquad (24)$$

Multiplying B_{oj}^T to both sides of eqn. (23) yields

$$\lambda_{max}B_{oj}^Tk_j = B_{oj}^T(R+L)^{-1}B_{oj}B_{oj}^Tk_j \qquad (25)$$

Denote $x = B_{oj}^Tk_j$, the eigenvalue/vector relationship eqns. (24) and (25) are simply eqns. (13) and (12) respectively. From eqn. (23), noting that $\lambda_{max} > 0$, we insure that the k_j satisfies eqn. (14).
Secondly we show that if k_j satisfies eqns. (12-14) then it will satisfy eqns. (10-11). Note: if the largest eigenvalue of the 2 X 2 real symmetric matrix A in eqn. (12) then A=0 $\Rightarrow$ $B_{oj}=0$ $\Rightarrow$ $V_o^TW_o^TE_j=0$ $\Rightarrow$

$$\Rightarrow \quad [I_m, 0] W_o^T E_j = 0 \;\Rightarrow\; [I_m, 0] \begin{bmatrix} M^T & N^T \\ -N^T & M^T \end{bmatrix} \begin{bmatrix} e_j & 0 \\ 0 & e_j \end{bmatrix} = 0$$

$$\Rightarrow \quad M_{j*} = N_{j*} = 0 \quad \Rightarrow \quad G_{j*}(i\omega_o) = 0 \;,$$

it is contradictory to that $G(s)$ is non-singular at any $s = i\omega_o \in D$. Therefore $\lambda_{max} > 0$ in eqn. (12). Multiplying $\lambda_{max} B_{oj}^T$ to both sides of eqn. (14) yields

$$\lambda_{max} B_{oj}^T k_j = B_{oj}^T (R+L)^{-1} B_{oj} x \qquad (26)$$

The relationship eqn. (12), with the notation of eqn. (15), implies

$$B_{oj}^T (R+L)^{-1} B_{oj} x = Ax = \lambda_{max} x \qquad (27)$$

Clearly that eqns. (26) and (27) together give

$$x = B_{oj}^T k_j$$

Substitute it in eqns. (13) and (14) yields eqns. (10) and (11) respectively.

The transfer-function matrix of the plant in Example 2 is given as follows:

$$g_{11} = 2.617\times10^{-3}/(s^4 + 11.112s^3 + 11.229s^2 + \\ +1.1195s + 2.266\times10^{-3})$$

$$g_{12} = -7.88\times10^{-5}/(s^3 + 1.112s^2 + 0.112s + \\ +2.266\times10^{-4})$$

$$g_{21} = (-2\times10^{-4}s - 1.997\times10^{-5})/(s^6 + 11.442s^5 + \\ +14.9058s^4 + 4.9362s^3 + 0.484s^2 + 0.0127s + \\ +2.266\times10^{-5})$$

$$g_{22} = (0.025s^4 + .0303s^3 + .00557s^2 + 2.909\times10^{-4}s \\ +1.1633\times10^{-6})/(s^5 + 1.442s^4 + .4888s^3 + \\ +.0482s^2 + 1.1942\times10^{-3}s + 2.266\times10^{-6})$$

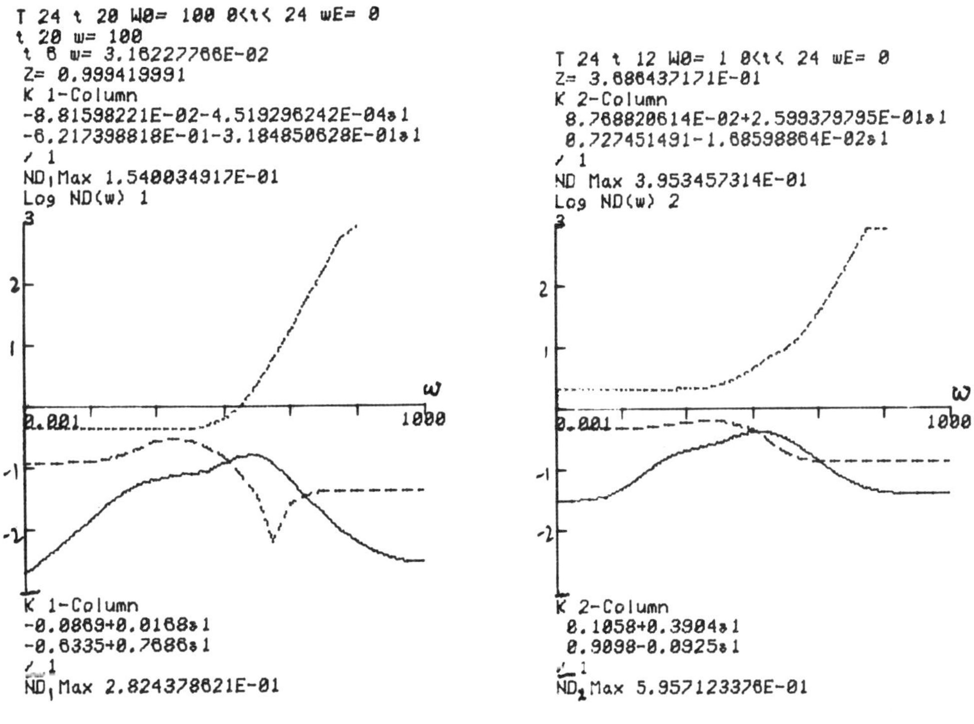

.... uncompensated
- - - - compensated with $\bar{K}(s)$ of Eq. (17)
——— compensated with $K(s)$ of Eq. (16)

Fig. 1. $\text{Log}_{10} ND_j(\omega)$ against frequency

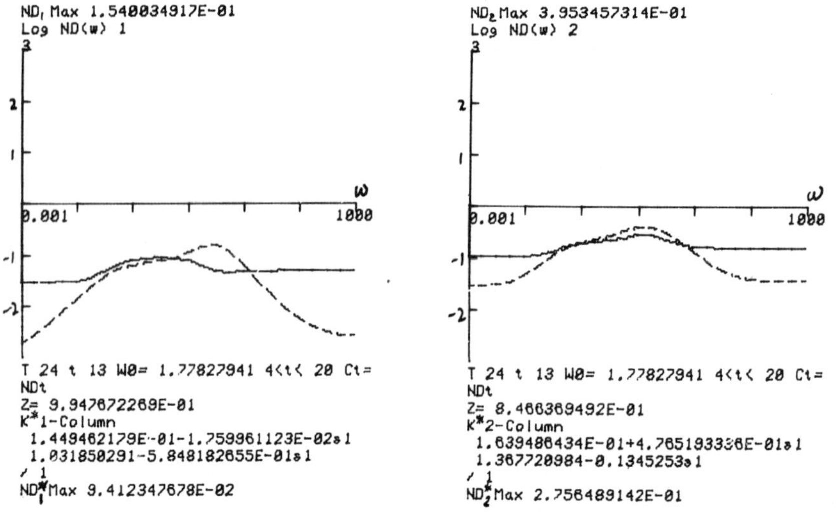

-------- compensated for the first time
———— compensated with self-weighting

Fig. 2. $\mathrm{Log}_{10}\mathrm{ND}_j^*(\omega)$ of Q(s) with K*(s) of Eq. (18)

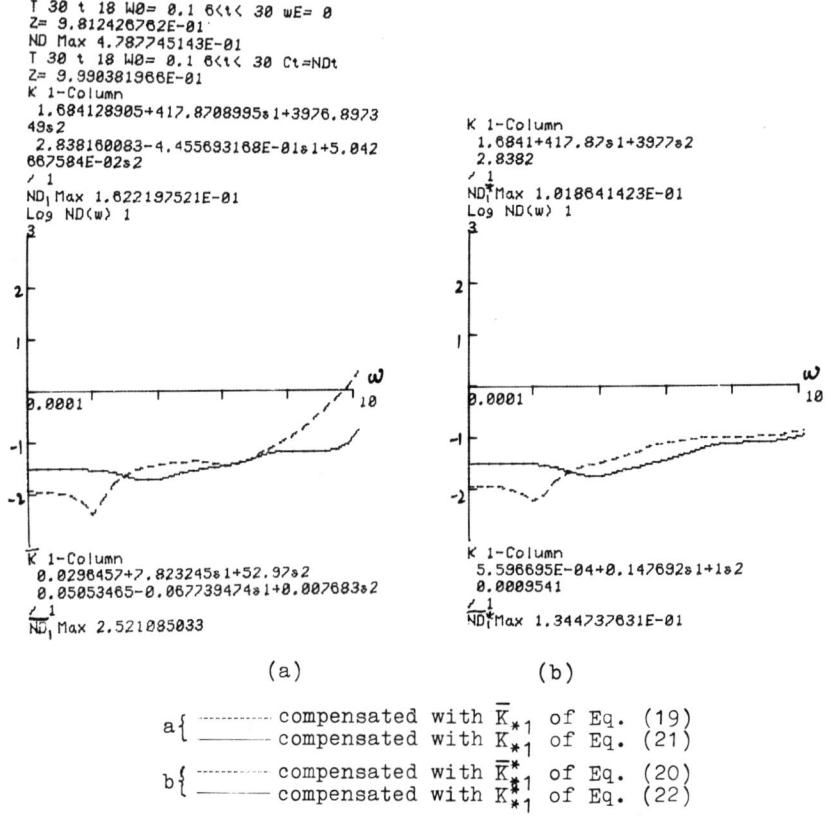

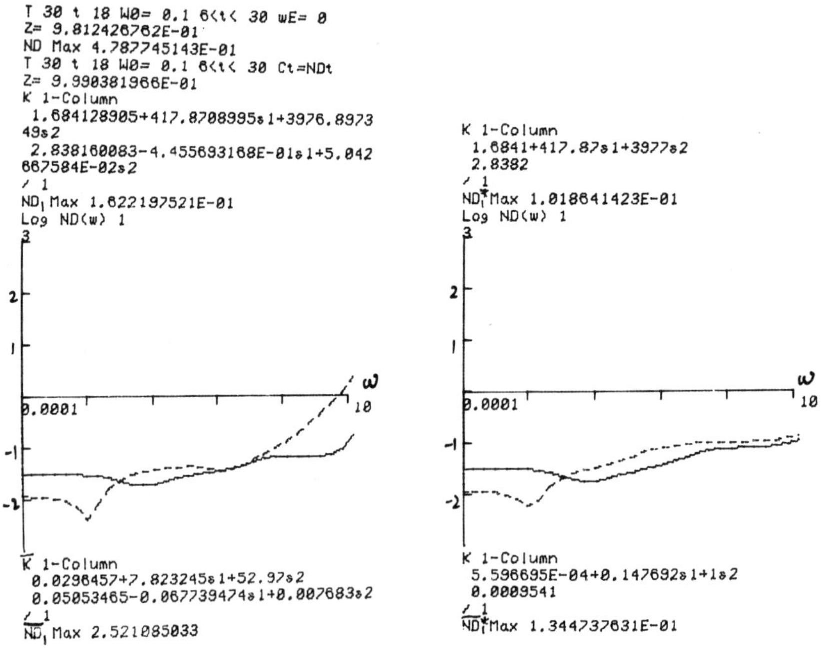

(a) (b)

a{ -------- compensated with $\bar{K}_{*1}$ of Eq. (19)
a{ ———— compensated with K_{*1} of Eq. (21)

b{ -------- compensated with $\bar{K}^*_{*1}$ of Eq. (20)
b{ ———— compensated with K^*_{*1} of Eq. (22)

Fig. 3. $\mathrm{Log}_{10}\mathrm{ND}_1(\omega)$ plots in Example 2.

3D SURFACE DESIGN ON PROFESSIONAL PERSONAL COMPUTER

G. L. Kovács and F. Fenyves

Computer and Automation Institute, Hungarian Academy of Sciences, Budapest, Hungary

Abstract. A wide range of machine parts have boundaries which can be described as translational surfaces. To illustrate their characteristics we use the following definition: a generator curve travels along a directrix curve to scrape the covering of the object. This way only two plane curves (directrix and generator) should be given to define the exact shape. The available 3D systems use volumetric modeling or sculptured (free-form) surfaces with all their advantages and disadvantages, namely the simple definition and restricted possibilities of the first, and the rather tedious definition of the second method.
In our CAD/CAM system translational surfaces may be combined with basic volumetric modeling elements (plane, sphere, cone, cylinder) by means of set operations (union, intersection, difference).
Having the appropriate technological data (type of milling, tool parameters, etc.) even an automatic tool-path generation of the rough-, final-and intermediate surfaces is supplied for a CNC milling machine.
The parameters of technology and tolerences should be given as input by the operator. The resulting tool path may be zig-zag (for roughing) or contour following (for finishing).
The paper describes the mathematics, software modules of the system and the design procedure with some examples to show what kind of industrial applications (dies, moulds, etc.) make it reasonable to have such a relatively cheap and powerful system for surface design.

Keywords. Microcomputer; computer aided design; 3D modelling

1. INTRODUCTION

One of the most challanging tasks of using professional personal computers is the computer aided design and manufacturing (CAD/CAM) of complicated, three-dimensional (3D) objects, as these kind of problems are solved only on high-performance, interactive minicomputers (e.g. VAX or PR1ME) with expensive, sophisticated graphic devices.

3D surfaces which can not be defined by means of basic volumetric modeling elements (plane, cube, prism, sphere, cone and cylinder) are called free-form (or sculptured) surfaces.

There are several methods and systems (Böhm, 1984) to define the geometry of sculptured surfaces with the common problem of being forced to input a great amount of data. (Points and relations among them.)

The introduction of the so-called translational surfaces (Fenyves, 1983) decreased the freedom of 3D forms, but on the other hand they may easily be defined, and the amount of data involved in the modeling is not extremely large. These facts made it possible to develop a system for translational surface design and manufacturing on a 16 bit microcomputer.

2. TRANSLATIONAL SURFACES

Our 3D processor has a surface-based model. A surface may be a plane or a translation surface.

The basic elements of geometry of the 3D processor are: point, vector, plane curve and surface.

Plane curves may be: a straight line, circle, plane curve interpolated by line- and circle segments or interpolated by quadratic splines, or plane curve composed by concatenation of plane curves.

Plane curves may be closed or open. The open plane curve should be finite.

Description of surfaces can be done by means of:

- two plane curves
- transformations of given surfaces
- set operations of given surfaces or planes.

Translational surfaces are defined by means of two plane curves. The generator curve is moved along the directrix curve. The recent version of our 3D processor has some restrictions concerning these surfaces:

- both the directrix and generator should be loop-free plane curves, in different planes,
- the crossing or common point of generator with directrix is the same point of the generator all the time,
- the plane of the generator is always perpendicular to the plane of the directrix,
- the angle between the tangent of the generator in the common point and the plane of the directrix is constant (if there is no tangent then a line can substitute it with the same relation with the directrix).

The geometry of the row part and of final part is given by planes and translational surfaces and by set operations among them. To have an easier input description, some basic surfaces (sphere, cone, cylinder) can be defined by their names and parameters. However, even in these cases they are represented in the computer by means of directrix and generator, i.e. as translational surfaces.

Depending on the type of generator and directrix (open or closed) different kind of surfaces (open, closed) may be generated (Fig. 1-5).

If the generator intersects itself when moved along the directrix it results in a surface having (one or more) intersection lines or points with itself. Cutting the surface along all the intersection lines and points, more surfaces are produced Then the surface which contains the directrix is defined.

The set operations (union, intersection, difference) on translation surfaces and planes make it possible to generate complex translational surfaces.

Set operations are processed on the inside of the surfaces, as on subsets of the Cartesian 3D space. The resulting surface is the boundary surface of the resulting point set.

The general parametric equations of translational surfaces can be given as follows:

$\underline{f}(s,t)=\underline{d}(s)+a(s)\underline{g}(t)$, where

$\underline{d}(s)=(d_x(s),d_y(s),d_z(s))$ is the directrix curve

$\underline{g}(t)=(g_x(t),g_y(t),g_z(t))$ is the generator curve

in parametric form, and a(s) is a scalar function which should be chosen properly.

3. THE STRUCTURE OF THE 3D PROCESSOR

The machining of the most complicated parts can be designed by the 3D processor. It serves to design the geometry and technology of translational surfaces. The processor automatically defines the tool paths for cutting consecutive operations, i.e. no input language level description is necessary. For automatic tool-path generation of intermediate surfaces between the rough and final surfaces it is enough to define the following on the input language level: the type of cutting (surface milling, contour milling, etc.), tool parameters, machining parameters, final surfaces (goal surface), and rough surfaces.

The processor consists of functional modules organized as program sections. The input and output of the consecutive sections are sequential files, which are handled by a monitor system.

The part-program is converted by the Input-Decode section to a file where each record corresponds to a defintion of the part program. This file is the input of the geometry design.

The main sections are represented in Fig. 6.

As the previous section gave a detailed description of the geometry, we now deal briefly with the remaining program sections.

Checking of machinability
This section decides whether to continue processing or not, using all necessary data from data-bases and from the partprogram.

Technology
This section designs the surface for elementary operations and the tool path cycles for them. The definitions of tools and technology with all the parameters and tolerances should be given as input by the designer.

We suppose that in the part program all translational surfaces have directrices, which are in parallel planes and cutting is going to be in these planes and in planes parallel to them.

There is only one rough and one final part in every part program and in this way all surfaces, including intermediate ones should be assigned to each other by means of set operations.

The type of tool path may be: zig-zag (for roughing) or contour following (for finishing).

Having all the necessary technological data (parameters of tools, speeds, depths, etc.) the technology section computes tool paths and prepares data for the CL redactor. If the surfaces cannot be machined with the given tools, error messages appear.

The rough, final and intermediate surfaces which are defined as input, should be planes or translational surfaces or surfaces built up from the previous ones by means of set operations (union, intersection, difference) and transformations.

The logical structure of technology module for translational surfaces is based on two simple theses:

- All the plane cut-segments of a translational surface, which are parallel to the plane of the directrix are offsets of the directrix. The same way a tool-path is an offset with respect to the directrix in the case of contour following.
- The plane cut-segment of the union (or intersection or difference) of two translational surfaces equals to the union (or intersection or difference) of the plane segments.

The main steps of technology design were determined based on the above given theses: transformation of the surface, calculation of offsets, filtering of intersections, resolving of set operations, inverse transformation of surfaces.

In case of zig-zag movement the resulting curves are the limits of the motion, these are intersected with lines parallel to the x axis (routing width), the intersection points are ordered as necessary, and the tool paths are given by calculating the normals of the surface.

For contour following the calculated curves are the tool paths, only the surface normals should be calculated at the ends of the curve segments.

CL redactor
This section serves as input data for the post-processors which produce the control information of the available NC machine tools.

NC tape checking
The raster display of the personal computer makes it possible to check the NC information prior to machining to avoid unnecessary tool breaks and other collisions. Checking the information may be done at CL file level, or after post-processing the CNC file may be displayed and checked. An error free post-processing should be assumed if only the CL file is checked.

The recent version needs the CNC tape as input (Kovács, 1982). The input program is expected a) to be syntactically correct, b) to contain path elements (that is, the path, possibly indirectly defined - e.g. contour and tool-size conditions -, has been processed to its explicit form), c) calls and jumps have been resolved, d) there are no conditional statements, nor workshop-time value assignments. (Otherwise, they will be neglected.)

g : generator
d : directrix

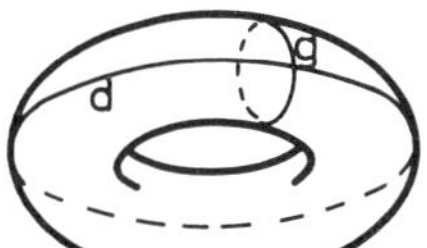

Fig. 1.

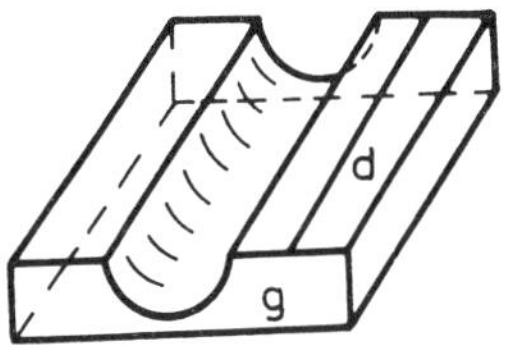

Fig. 2.

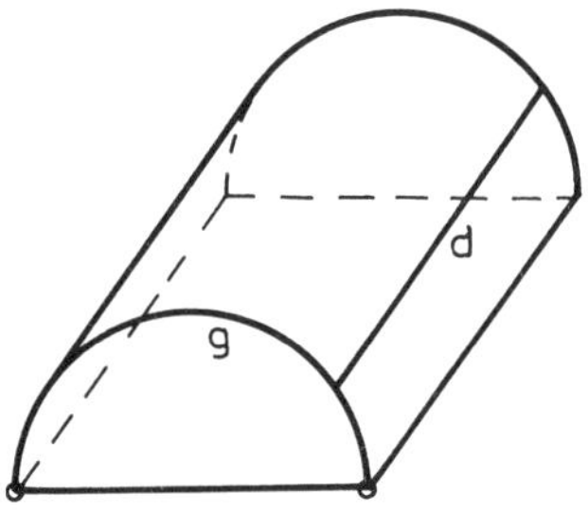

Fig. 3.

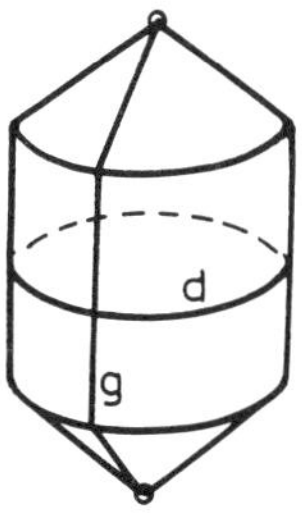

Fig. 4.

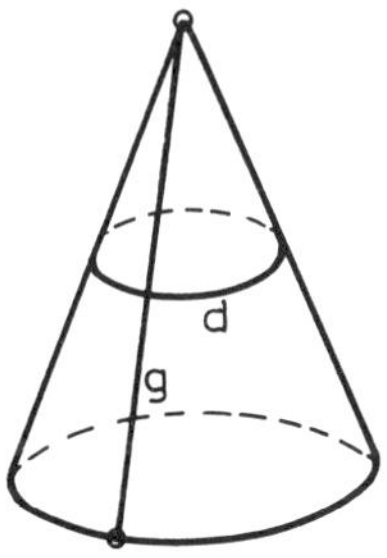

Fig. 5a.

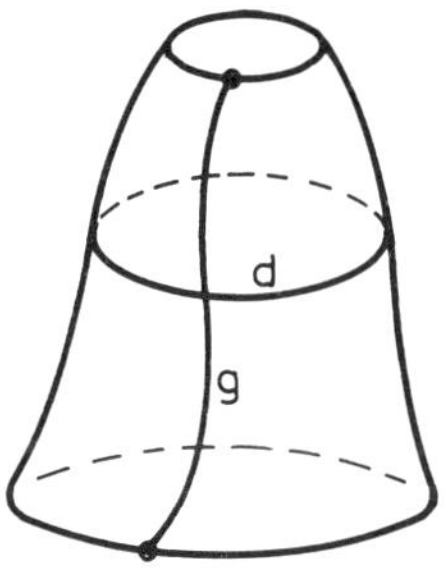

Fig. 5b.

Fig. 1÷5. Translational surfaces

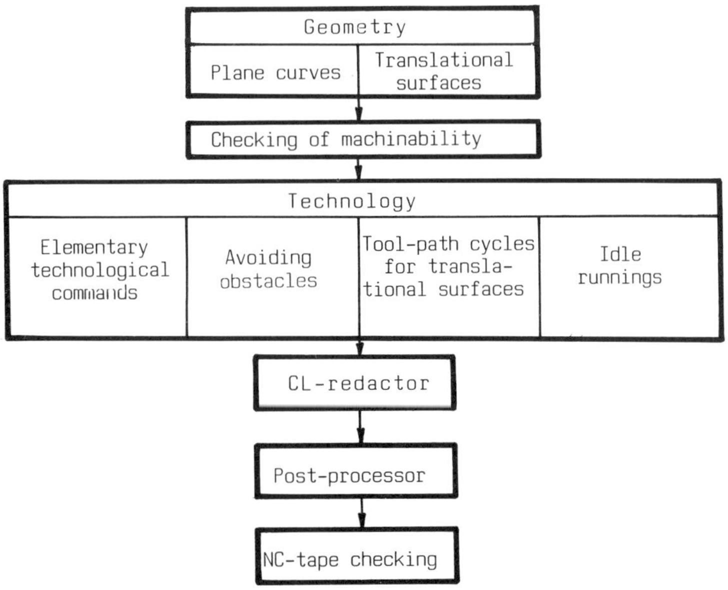

Fig. 6. Structure of the 3D processor

The only operation available is (for the time being) "2 1/2 dimensional" path display. The plane is to be specified XY, YZ, or XZ. The path display consists of pictures: one contains motions performed by the same tool in the same height (at the same distance from the specified plane). Some or all of the pictures can be superposed.

The path display appears on the screen in a coordinate square, scaling is such so as to fit the path into that square. Coordinate values are in the machine's system.

By means of the display pictures some errors of surface definition, tool selection, tool-path strategy, etc. can be detected. Having a small dialogue-system, detailed information of any picture element can be checked and minor corrections may be processed. Brute errors should be corrected on definition level, and the complete processor has to be run again.

4. CONCLUSION

A microcomputer based CAD/CAM system was discussed in detail for designing complicated 3D surfaces, the so called translational surfaces.

However, translational surfaces are less general than the real sculptured surfaces, there are many applications which made the development of our system possible. For example there are different dies, moulds, tools, etc. which may effectively be designed and produced as translational surfaces.

Finally Fig. 7. shows the part program and the drawing of a relatively simple mould which is defined as a translational surface. It should be mentioned that any other type of definition of the same piece would definitely be more complicated.

5. REFERENCES

Böhm, W., G.Farin and J.Kahmann (1984). A survey of Curve and Surface Methods in CAGD. Computer Aided Geometric Design, Vol.1. pp.1-60.

Fenyves, F., G.Kovács and I.Licskó (1983). Translation Surfaces in the 3D Subsystem of the COMECON NC Programming System. ICED'83 København, 15-18.Aug.Vol.1.pp.105-112

Kovács, G. (1982). Displaying Tool Paths IKM'82 Vortrag 39, Leipzig, 10-12 March

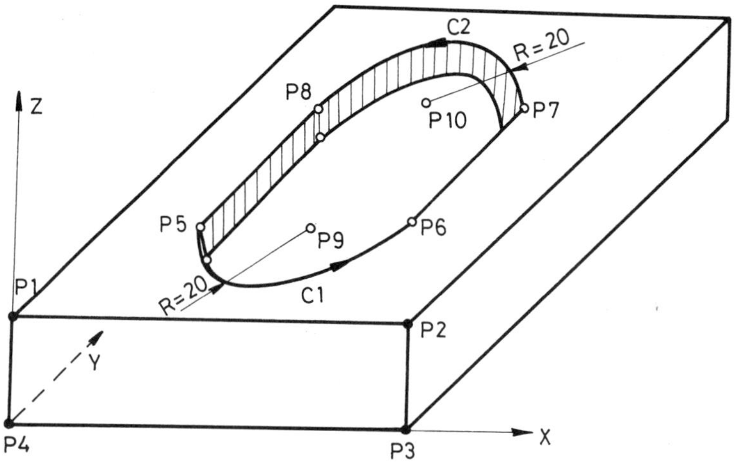


```
PARTNO='POCKET';
P1=POINT/0,0,20;
P2=POINT/80,0,20;
P3=POINT/80,0,0;
P4=POINT/0,0,0;
P5=POINT/20,40,20;
P6=POINT/60,40,20;
P7=POINT/60,80,20;
P8=POINT/20,80,20;
P9=POINT/40,40,20;
P10=POINT/40,80,20;
C1=CIRCLE/NORMAL,V1=VECTOR(0,0,1), CENTRE,P9,
     RADIUS,20;
C2=CIRCLE/NORMAL,V1,CENTRE,P10,RADIUS,20;
L1=LINE/P6,P7;
L2=LINE/P8,P5;
CURV1=CURVE/P5,C1,CCLW,P6,L1,P7,C2,CCLW,P8,L2,P5;
CURV2=CURVE/P1,P2,P3,P4,P1;
GEN1=CURVE/P5,P30=POINT(20,40,14);
CYL1=TRSURF/DIRECT,CURV1,GENER,GEN1;
P11=POINT/0,80,100;
CURV3=CURVE/P1,P11;
CYL2=TRSURF/DIRECT,CURV3,GENER,CURV2;
FNPIEC=TRSURF/DIFFER,CYL2,CYL1;
PL1=PLANE/XYPAR,21;
PL2=PLANE/XYPAR,14;TRANS/0,0,0;
FROM/PFROM=POINT(-40,-40,80);
SAFPOS/PSAFE=POINT(-40,-40,60);
CLRSRF/PL1;
CLDIST/2;
TLDATA/MILL,113001,113,94111,DIAMET,4,LENGTH,40,
   LENG1,0,LENG2,12,LENG3,24,RADIUS,1,ANGL,90;
WORK/F1=FACMIL(SO,TOOL,113001,FEED,0,2,SPEED,30,
   PLGFED,0,1,CUTDEP,5,DELTA,15,ROUGH,1.5,DRCOM,
   DLCOM,ZIGZAG,V2=VECTOR(0,1,0));
CUT/RPIECE,CYL2,NPIECE,FNPIEC,WSURF,CYL1;
WORK/F2=FACMIL(F1,FEED,0.1,SPEED,40,FIN,CONTUR,
   OMIT,CUTDEP,V2);
CUT/RPIECE,CYL2,NPIECE,FNPIEC,WSURF,CYL1;
FINI;
```

Fig. 7. Example of a 3D part program

COMPUTER AIDED DESIGN OF HYDRAULIC SYSTEM

D. Y. Chen, H. Q. Huo and X. Q. Ma

Northwestern Polytechnical University, Xian, Shaanxi, China

Abstract. This artical deals with the application of CAD method in the design of hydraulic system. The system here used is a domestic complete set of CAD equipment. A set of CAHSD1 application software is compiled, which can display images of various hydraulic elements, units and systems and the primary technical parameters on the screen. The transient response to a step input or a pulse input could be obtained with input of structural parameters of hydraulic servo system. Images and datum obtained could be modified at real-time and finally transfered to the plotter. According to the basic demand of dynamic character and working precision of hydraulic profiling cutter saddle, the objective function is made and a set of application software CAHSD2 of optimization design is compiled by optimization method. The dynamic target is based on Time Dommain Analysis. Flexible Polygon Search method is used to optimize structural parameters of the saddle. The results so far obtained are within reasonable range. The parameters of a high-precision hydraulic profiling cuttle saddle are also in this range. So the optimization design has produced a practical reference value, it also can be used in similar hydraulic servo systems.
Keywords. Computer-aided design; hydraulic systems; software; optimization.

COMPUTER AIDED DESIGN OF HYDRAULIC TRANSMISSION SYSTEM

Basic contents and procedures of design about hydraulic transmission system can be classified as follows:

1. To set a plan of hydraulic transmission system;
2. To draw up charts of hydraulic transmission system;
3. To calculaté hydraulic transmission system and modification parameters;
4. To select standard hydraulic units and design nonstandard hydraulic units;
5. To draw up formal working charts and compile technique files.

Among them, the second and the third are the most important. Normally, designer has to revise and calculate many times to make a decision. So work is heavy and design period is long. In order to improve quality and efficiency of the design, we make use of CAD to design hydraulic transmission system.

CAD system used in design is composed of domestic equipments completely, as shown in Fig. 1

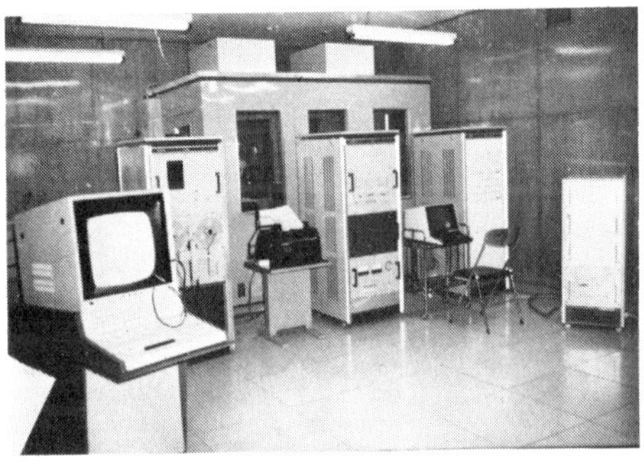

Fig. 1. CAD system

Software CAHSD1

1. It can display common charts of hydraulic elements and units. For constructing desired charts of hydraulic transmission system, designer can use light-pen and key board to call corresponding subroutines.
2. It can input the required feature parameters of the hydraulic system such as load, speed, etc. by communication between man and computer. After calculating, CAD system can display primary technical parameters of the system such as pressure, size of cylinder, etc. immediately.
3. Designer can modify at real-time, delete or recompile charts and parameters. If design is satisfied, he can call output subroutine to obtain related charts and files, as shown in Fig 2
4. It can desplay curves of transient response to

a step or a pulse input for related parameters of transfer function for hydraulic servo system and display overshoot, settling time, vertextime and vibrating times. It can also display several sets of curves of transient response on the same screen for various parameters of one system or of different systems, which can be used to analyse and compare. Response to a step and a pulse input of the same system are shown in Fig 3 and Fig 4. The

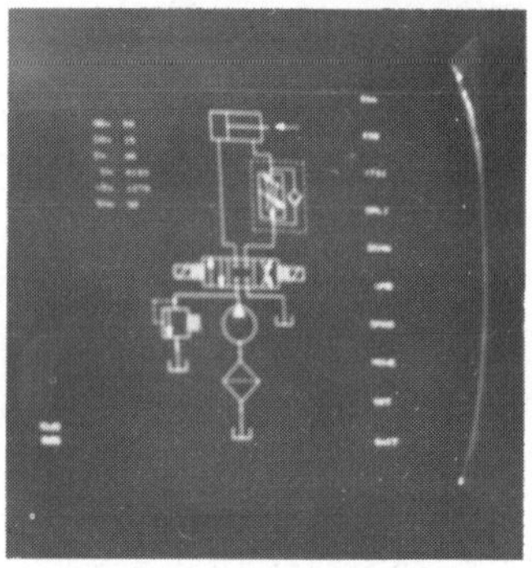

Fig. 2. Hydraulic transmission system on CRT

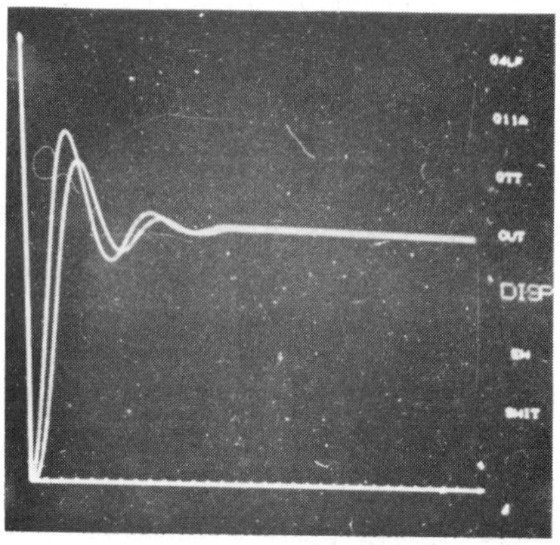

Fig. 3. Response to a step input with different parameters

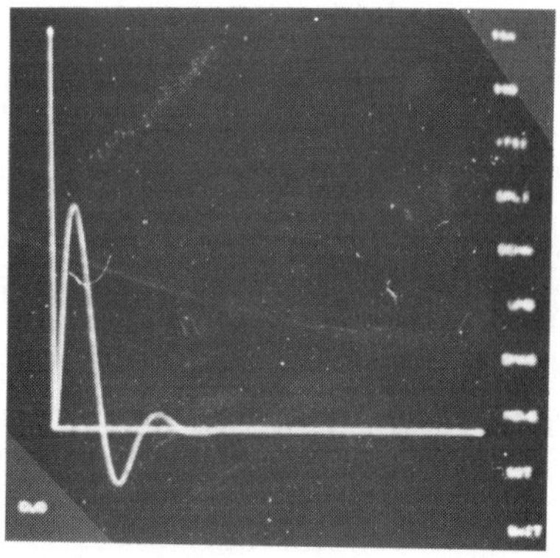

Fig. 4. Response to a pulse input

transfer function of the system is:

$$(S)= \frac{156.25}{S^2+7.5S+156.25} \qquad (1)$$

COMPUTER AIDED OPTIMIZATION DESIGN OF HYDRAULIC PROFILING CUTTER SADDLE

Hydraulic profiling cutter saddle is an efficient attachment of machine tool in serial production. In recent years, optimization design has become one of primary contents of CAD in mechanical engineering. Only using CAD can we accomplish optimization design, for hydraulic system such as hydraulic profiling cutter saddle. Preliminary work is done in this field and promising results are obtained. Software CAHSD2 used in optimization design of hydraulic profiling cutter saddle has been compiled. Hydraulic profiling cutter saddle prossessing four-way control and differential cylinder is shown in Fig 6.

The saddle belongs to hydraulic servo system.

The static character of system is connected with structural parameters of the system. In consideration of static character, there are contradictions among selections of the system parameters.

Dynamic differential equation is:

$$\frac{K'M}{F_1^2(1+m)^2}\frac{d^3y}{dt^3}+M\left|\frac{\partial V}{\partial P}\right|\frac{d^2y}{dt^2}+\frac{dy}{dt}+\frac{\partial V}{\partial \delta}y=\frac{\partial V}{\partial \delta}X-\left|\frac{\partial V}{\partial P}\right|R \qquad (2)$$

where:

K' = coefficient of elasticity
M = the mass of moving parts in system
R = external load force
Y = coordinate of cylinder (saddle)
X = coordinate of feeler
F_1 = the piston area of the large chamber
m = ratio of area
v = cylinder velocity
p = applying force of cylinder
δ = profiling error

Therefore the transfer functions of the system are:

$$\phi_x(S)=\frac{Y(S)}{X(S)}=\frac{C}{S(T^2S^2+2hTS+1)+C} \qquad (3)$$

$$\phi_R(S)=\frac{Y(S)}{R(S)}=\frac{-\frac{C}{E}}{S(T^2S^2+2hTS+1)+C} \qquad (4)$$

The block diagram of the system is shown in Fig 7.

The open-loop transfer function of the system is:

$$G(S)=\frac{Y(S)}{\delta(S)}=\frac{C}{S(T^2S^2+2hTS+1)} \qquad (5)$$

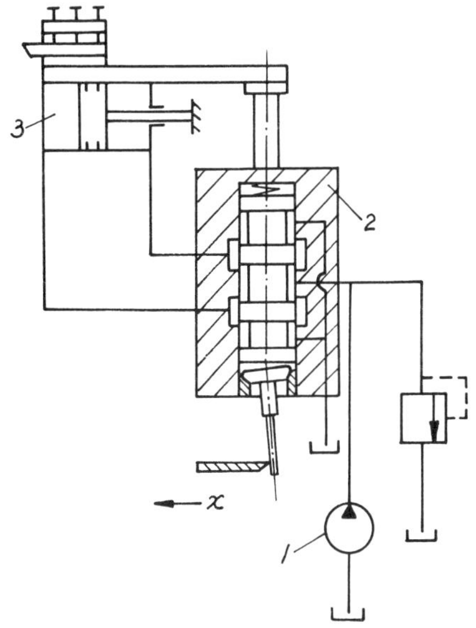

Fig. 6. Hydraulic Profile Cutter Saddle

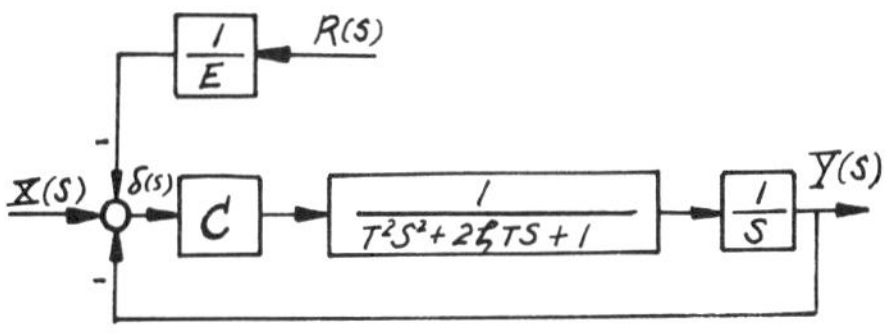

Fig. 7. Block diagram of the system

Every coefficient of the transfer function is closely related with structural parameters of the system. Transfer function and differential equation correspond one to one. They describe the dynamic characters of the system. For obtaining better follower effect, we must pay attention to the determination of the structural parameters.Dynamic targets are based on Time Domain Analysis and error criterion is used to make optimizing selection. Because targets are function of the structural parameters of the system, we can adjust the system parameters in the design to make minimum error of targets. Considering good selective property of criterion and stability accuracy and speed demand of the hydraulic follower system, ITAE criterion of Graham and Lathrop is used. The fine degree of the system is assessed by $\int_0^\infty t\left|e(t)\right| dt$.

Considering the practical work range, the selection range is given and restraint is thus formed.Through change of complex domain, the objective function of the structural optimization is:

$$\Phi(y)=\left\{\frac{1}{(1+\sqrt{m})^2 F_1}\left[\frac{E_u 4kd}{Hm}\right]^{\frac{2}{3}}\left[(A+\delta_o)(1+m)\right]^{\frac{5}{3}}\left[\frac{M}{P_o i}\right]^{\frac{1}{3}} \frac{1}{m}^{\frac{1}{6}} - 1.75\right\}^2+\left\{F_1(1+m^2)\left[\frac{E_u(1+m)}{MHm}\right]^{\frac{1}{3}} \left[\frac{1}{4kdP_o i(A+\delta_o)}\right]^{\frac{2}{3}}\left[\frac{1}{m}\right]^{\frac{1}{3}}-2.15\right\}^2 \qquad (6)$$

If the combination of P_o , F_1 , d , m , etc. could make $\Phi(y)$ minimum, the system would prossess optimal dynamic property.

where:

- H = length of oil column
- E_u = modulus of oil elasticity
- k = coefficient of flow
- d = diameter of valve
- $A+\delta_o$ = preparing edge
- P_o = supply pressure
- i = link ratio of output to input

In this way, we can obtain optimal systems in engineering field, which is close to ideal system of small overshoot and short settling time. With satisfactory properties, the system has maximum rapidity and then the static error could also be minimized.

Flexble Polygon Search is used as optimization method.

Search is started from many points.

The searched combinations of structural parameters P_o, d, F_1, m are concentrated to a due degree and the parameters of a high-precision hydraulic profiling cutter saddle made by a certain research institute of our country are also in this range. So the optimization method is feasible. Because the method of optimization design uses advanced computer technique, it is an innovation in design method. As it can improve quality and shorten period of design, it has important practical value.

CONCLUSION

1. Computer Aided Design of Hydraulic Transmission System has advantages of Analog and Digital computer, such as rapid speed, strong visual property, high work efficiency, high copying precision, easy change of structural parameters and communication between man and computer. It is an advanced method in design.

2. Design of hydraulic profiling cutter saddle uses optimization method and CAD technique. After

objective function is determined, computer is used to optimize. Design speed is rapid, calculation correct, quality improved and period shortened.
3. The design method in this article is a new try. The work done can be used for reference to engineering designer in Computer Aided Design of hydraulic systems.

REFERENCE

Blackburn, J.F. (1960). Fluid Power Control. The Technology Press of M.I.T. and John Wiley & Sons, Inc.
Himmelblau, D.M. (1972). Applied Nonlinear Programming. McGraw-Hill Book Company.
Stringer, J. (1976). Hydraulic System Design. The Macmillan Press Ltd.

CAD FOR PNEUMATIC CIRCUIT DESIGN IN LOW COST AUTOMATION

T. P. Leung* and W. K. Chick**

**Department of Mechanical and Marine Engineering, Hong Kong Polytechnic, Hong Kong*
***Welltec Industrial Equipment Co. Ltd., Hong Kong*

Abstract. This paper describes the design of a CAD system for pneumatic sequential control. A special language for describing the sequence is developed. The procedure of memory designation and function generation, switching function minimization, and circuit construction is described in detail. Referring to a simple design problem, the computer process from the high level language statement to the pneumatic circuitry is illustrated. Finally, the application of the design package to other control problems is outlined.

Keywords. Computer-aided circuit design; pneumatic control equipment; Boolean functions; Quine-McCluskey algorithm.

INTRODUCTION

Pneumatic sequential control has played an important role in production and factory automation. Various methods, from the intuitive method to Karnaugh map method, have been used for pneumatic circuitry design. The former is limited to relatively simple control problems, and its success depends on the designer's experience. The latter is a systematic approach for complicated circuit design. It also provides a solution that makes use of the minimum number of logic elements.

Computer Aided Design (CAD) of pneumatic sequential circuit is based on the Karnaugh mapping method. Software had been developed to facilitate the design of large, complicated problems with numbers of input, output variables and the control steps were only limited by the memory size of the computer. The CAD design package is written in FORTRAN-77 and runs on a DEC VAX 11/750 computer.

GENERAL DESCRIPTION

An Application Oriented Language is first developed which allows the designer to use declaration, statements and symbols to describe the operating sequence. These statements are subsequently converted into data compatible with the system synthesis program by a language processor program.

Sequential control basically consists of combinational circuit and memory elements. The memory designation program MEMGEN is used to make each state unique by addition of auxiliary memory variables. The program FUNGEN then formulates truth tables for each memory variable depending on the type of power device. The resulting truth table represents a combinational network for sequential control.

The output switching functions defined by the raw truth table is minimized by the program MINIM. The minimization alogrithm used is a derivative of the Quine-McCluskey algorithm which bases on the containment properties of cellular structure. The minimized functions are represented by two integer arrays defining a two-level switching network. The program BOOLEAN converts these data into a set of Boolean functions which specify an abstract model of the control system. Before plotting the circuit, functions having same product terms are sorted out and combined together to eliminate surplus logic elements. The circuit layout program LAYOUT places the symbol of logic elements to produce a well-presented diagram. A graphic library is written to support the drawing of the standard pneumatic symbols on a bench type plotter. Another program LINK plots all the interconnection according to the output of program MINIM.

APPLICATION ORIENTED LANGUAGE

The procedure of defining the problem may be formalized by the application oriented language, which allows machine specifications in the manner familiar to the designer, using nomenclature and syntax defined by the rules of the language. The designer labels each device and then makes use of these labels to specify the machine cycle. The language is designed for easily-learned, and user-oriented programming. The machine cycle can be thoroughly expressed by statements which enable the designer to handle at ease.

Program statements can be broken down into three categories: output control statements, sequence statements and declaration statements. Statements begin with an identifying name, starting in the first column of the line. Declaration statement pertains to entities such as:

```
DEVICE A,B:BISTABLE, C:MONOSTABLE
EXTERN D,E
```

which assign labels to power or external devices. The label must be a single character, followed by the type of device. INIT statement determines the de-actuated rest position of power device, eg.

```
INIT    A-, B+, C-
```

Sequence description statement is the heart of the program in that the machine cycle is here defined. The language uses special characters to indicate the sequences:

- \+ power device actuate (or traverse forward)
- \- power device de-actuate (or traverse backward)

:	external signal
/	step delimiter
//	end of machine cycle

These characters must follow the label of power device. More than one power device can traverse simultaneously within one step, eg.

SEQUEN A+/B+C+/D:B-/E:B+/C-/B-A-//

The form of output is specified with a OUTPUT statement. The output options are:

BOOLE	Boolean Expression
LADDER	Electric Ladder Diagram
PCPGM	Programmable Controller Control Programme
LOGIC	Logic Control Circuit
PNEUM	Pneumatic Control Circuit
HYDRA	Hydraulic Control Circuit

If more than one output form is required, the options are separated by commas.

LANGUAGE PROCESSOR

The main purpose of the language processor is to produce a flow table for control system synthesis. The flow table is an integer array; each entry corresponds to one step of the machine sequence. The binary notation of the integer defines the status of all power devices at each step, one binary digit for each power device. Thus, the maximum number of power devices possible is limited by the world length of the computer. After inputting the declaration statement to the computer, the processor assigns a bit address for every power device and the content of the bit then denotes the status of that power device. A 'one' in the corresponding bit indicates the power device in its actuated state and a 'zero' for its de-actuated state. The labels and their bit addresses are stored into a Device Label table. The processor scans through the sequence statement to set or reset the corresponding bit of the flow table, leaving only the external device for later processing.

Language processor also provides thorough checking and lists detailed, self-explanatory error messages. The processor checks both the syntax and logic errors, such as undeclared labels, invalid symbols, illegal identifying names, and unreasonable operating sequence, for instance, the sequence A+/A+/.

MEMORY DESIGNATION

CAD of sequential control system is carried out in two steps; the memory designation, followed by the synthesis of the resulting combinational network. Memory designation can begin when the flow table is read in. The program designates a state for all memories in every step of the machine cycle. For the memory of the power devices, the states are just the same as those in the flow table. However, under certain conditions, additional states must be added. These conditions are:

1. If a memory designation has been used, an auxiliary memory must be set or reset for the purpose of discrimination.

2. Prior to the simultaneous movement of two or more power devices, an auxiliary memory must be set or reset.

3. When a memory designation is immediately succeeded by its complementary designation, an auxiliary memory must be set or reset to avoid premature signal cancellation.

The computer continuously compares the present state value with other state values. If it has happened before, then the state of auxiliary memory will change. The second and third conditions are checked by bit Exclusive OR operation. The computer will check whether there is more than one bit difference in any two adjacent entries. Any bit component changing state in both preceding and succeeding entries is identified. In either of these two cases, the computer will toggle the bit components corresponding to the auxiliary memory and put this extra state in between. The state table will be outputted in the form of integer array in order.

SWITCHING FUNCTION GENERATION

The synthesis of combinational network requires the control problem specified by a truth table. For each power device, there will be one or two excitation truth tables depending on the nature of the device. For a monostable device, only one truth table is required while bi-stable device has two tables, one for either traverse direction. For the device to traverse forward, it is referred as 'Set' and for backward traverse 'Reset'. The state table resulting from the memory designation program MEMGEN is now treated as a two dimensional bit array. Each row represents the status of all memories at that step and each column denotes the status of a memory throughout the machine cycle. Each column of data is scanned step by step for the whole column to generate a Karnaugh map for the combinational network of that memory. Each row of the state table defines the cell address of the Karnaugh map and the cell content is determined by the excitation map. For monostable device, the excitation map is as follows:

State		Excitation
Present	Next	Set
0	0	0
0	1	1
1	0	0
1	1	1

The first two columns denote the present/next state relationship while the last column is the excitation function. Excitation map for a bistable device is shown below:

State		Excitation	
Present	Next	Set	Reset
0	0	0	d
0	1	1	0
1	0	0	1
1	1	d	0

where 'd' represents don't care state, other cells in the Karnaught map that are not addressed are also don't cares. Every column of the state table, which corresponds to one memory, will generate its own Karnaugh map accordingly.

The resulting truth tables represent a combinational network, and the problem of synthesis is reduced to minimize the output switching functions defined by the truth table.

FUNCTION MINIMIZATION

Computer aided minimization of output logic functions is usually based on the Quine-McCluskey algorithm, or its derivatives. Quine-McCluskey algorithm accepts inputs in the form of a fully specified expanded truth table and the resulting Boolean expressions define a two level (OR-AND or AND-OR) switching network. This network is minimal (or nearly so) in terms of the selected cost function.

The minimization process, as mechanized by computer, starts with the truth table in minterm form. The minterms and don't cares are combined and arranged in ascending order. The computer examines the minterms in pairs and checks if a pair is a prime implicant. The prime implicants are then checked to see if they are essential. The final minimized function is derived from the minimal cover -- a process that involves selection of the fewest prime implicants that contain, or cover, all the minterms of the Boolean function. All essential prime implicants must also be included.

The computerisation of Quine-McCluskey algorithm for CAD involves the cellular structure of Boolean functions. Each product term of a Boolean function is a cell or a vertex of the cellular n-cube. After decimal transformation the cell is represented by two integers. The integers correspond to the maximum and the minimum vertices of the cell. The containment properties of the decimal representation for cells of the n-cube enable the computer to determine the following conditions by simple logical AND operations.

1. A pair of minterms of a Boolean function can be combined to form a cell.

2. A cell contains another cell.

3. A cell contains or covers a vertex.

The first condition is the same as looping the minterms on a Karnaugh map, because prime implicants are represented by cells. The second condition determines if one product term of Boolean function contains another. This is equivalent to the task of generating the minimizing table and a list of prime implicants for the Quine-McCluskey algorithm. The third condition is a test for minterm coverage when trying to find the essential prime implicants from a prime implicant table. In addition, in order for one vertex to contain another vertex, the decimal transform of the former must be greater than that of the latter. These properties allow the Quine-McCluskey algorithm to be implemented easily on a digital computer. The computer can use comparison as a "screening' test of containment for a pair of cells before finding the prime implicants of a Boolean function using the logical AND operation. The logical AND and less than or equal comparisons greatly increase the speed of solution over the conventional Quine-McCluskey approach.

The minimization program must run once for each memory. The result is two arrays of integers, denoting the product terms of the final Boolean function.

CIRCUIT CONSTRUCTION

Of the six different output options, only Boolean equation, pneumatic and hydraulic circuitry are now supported by the program. Boolean function can easily be obtained by inverse decimal transform of the resulting integer pairs, with each pair of integers corresponding to one product term. The relationship between the integer pair and minterm is illustrated as follows:

Integer	Binary form	Minterm
3	000000011	0------11
255	011111111	

The BOOLEAN program then finds the names of the variables that are in complemented and uncomplemented form from the Device Label table. The Boolean expression is obtained by converting integer pairs to produce terms until the array of data is exhausted and then all the product terms are summed together.

The circuit construction program comprises four steps, namely, element reduction, component placement, wire connection and plotter output.

The number of logic elements required to implement a two-level network can be reduced if the Boolean expressions of different memory have the same minterm, so that they can share the same combinational network. This condition can be spotted by inspecting the integer array pair and checking whether any two pairs of integer have the same value.

The plot of circuit is divided across the page into five horizontal regions, each for one category of components. At the top is the power device, under which are memory elements, OR valves the third, and next to it is the AND valves, with all signal lines grouped and laid horizontally at the bottom. The component placement program first determines the number of AND valves by summing the total number of AND function for all minterms. The number of minterms minus one for each boolean expression will be the number of OR valves for that memory. Once the number of valves is known, the AND valves is first placed evenly on a horizontal row, the OR valves is located afterwards. On the top of the combinational network are the memory elements and power devices. For a monostable power device, a single-acting spring-return two-position valve is used as memory element whilst bistable power device employs double-acting two-position valve. All the logic elements are then numbered and stored in a table together with the positions of elements.

A connection program uses output data from the minimization program to produce a table specifying the interconnection between logic elements. The computer then divides the spaces between the logic elements on the plot into certain number of orthogonal grid. As both ends of every signal line are fully stated by the interconnection table and element location table, the routing program LINK can look for shortest path from the start point to the terminate node along the unoccupied grid.

In order to produce a well-presented circuit, all symbols are drawn in a horizontal position and lines are attached to the right-hand block of two-position valves. Valves are drawn in their deactuated positions, while cylinders are in their rest positions.

The outcome is plotted on an A3 size bench type intelligent plotter. Lines, circles, and dotted lines of the circuit can be drawn by simple character commands. A graphic subroutine library is developed to plot the OR valve, AND valve, cylinder, and directional control valves. The PLOT program draws the symbols by calling the library and connects all the signal lines. As hydraulic symbols are similar to those of pneumatic components, hydraulic circuit can be easily drawn with another graphic library.

APPLICATION EXAMPLE

This section illustrates the package use for the design of a pneumatic control circuit referring to a simple machine cycle as shown in Fig. 1. The pneumatic system consists of three cylinders entitled A, B, and C. The program for describing the machine sequence is as follows:

```
DEVICE A,B,C:BISTABLE
INIT A+,B-,C-
```

```
SEQUEN A-/B+C+/A+C-/B-/A-/C+/C-/A+//
OUTPUT BOOLE, PNEUMA
```

The flow table generated by the language processor, the state table as a result of memory designation, and the device label table are shown respectively in Tables 1, 2 and 3. Memory designation has inserted six extra states and has added two auxiliary memories X and Y.

TABLE 1 Flow Table

C B A	Decimal
0 0 1	1
0 0 0	0
1 1 0	6
0 1 1	3
0 0 1	1
0 0 0	0
1 0 0	4
0 0 0	0
0 0 1	1

TABLE 2 State Table

Y X C B A	Decimal
0 0 0 0 1	1
0 0 0 0 0	0
0 1 0 0 0	8
0 1 1 1 0	14
0 0 1 1 0	6
0 0 0 1 1	3
0 1 0 1 1	11
0 1 0 0 1	9
1 1 0 0 1	25
1 1 0 0 0	24
1 1 1 0 0	28
1 0 1 0 0	20
1 0 0 0 0	16
1 0 0 0 1	17
0 0 0 0 1	1

TABLE 3 Device Label Table

Label	Bit No.
A	0
B	1
C	2
X	3
Y	4

Ten truth tables are produced by the function generation program; each is minimized using the modified Quine-McCluskey algorithm. The package generates Boolean functions for the combinational network of memories A, B, C and auxiliary memories X, Y are listed below:

$$A1 = C_1.\bar{y}.\bar{x} + c_0.\bar{x}.y$$
$$A0 = b_0.\bar{y}.\bar{x} + x.y$$
$$B1 = a_0.x.\bar{y}$$
$$B0 = a_1.x$$
$$C1 = a_0.x$$
$$C0 = \bar{x}$$
$$X1 = a_0.b_0.\bar{y} + c_0.b_1$$
$$X0 = c_1.b_1 + c_1.\bar{y}$$
$$Y1 = a_1.x.b_0$$
$$Y0 = a_1.\bar{x}$$

The output of the circuit construction program is the pneumatic circuit as shown in Fig. 2.

CONCLUSION

This CAD package was developed using a DEC VAX 11/750 computer. It can be easily adapted to use on a personal computer that can run FORTRAN-77. The package saves a lot of time to obtain optimum pneumatic circuit design especially for complicated circuits.

The methodology adopted in the development of the present package can be extended to the CAD of electrical ladder diagram. This will be very useful for the application of programmable controllers.

ACKNOWLEDGEMENT

The authors would like to acknowledge Mr. Y.M. To for his help in the preparation of the graphic library for this piece of work.

REFERENCES

Rohner, P. (1979). Fluid Power Logic Circuit Design, Macmillan Press. Chap. 7, pp. 46-111.

Nagle, H.T., B.D. Carroll, and J.D. Irwin (1975). An Introduction to Computer Logic, Prentice Hall. Chap. 4, pp. 158-178.

Stecki, J.S., and O.A. Reddecliffe (1980). Designing complex control systems with the computer. Hydraulic and Pneumatic, 33, 107-109.

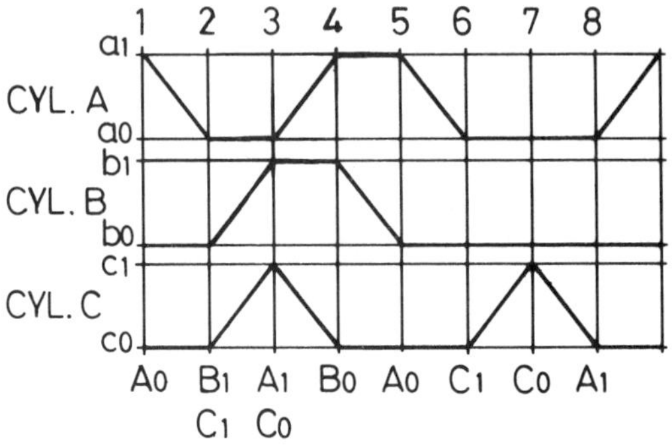

Fig. 1. Traverse-time diagram

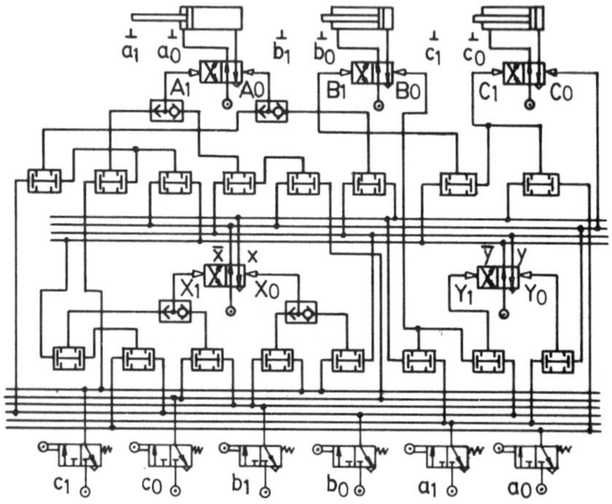

Fig. 2. Pneumatic circuit

A DATA BASE MANAGEMENT FOR DIE CAD/CAM

Jia Minghua and Weng Shixiu

Department of Mechanical Engineering, Shanghai Jiao Tong University, Shanghai 200030, China

Abstract. CADB is a network model data base management system (DBMS) developed for die CAD/CAM systems and it is based on a simplified DBTG data base model. Besides having the characteristics which are common to the conventional DBMS, it has its own features to meet the special requirements of die CAD/CAM. The structure of this system, the access methods, the representations of the data relationships and their physical storage structures, and the system's main commands are described in this paper. This system was implemented on a Burroughs 1955 computer, and has been transferred to microcomputers and used in die CAD/CAM.

Keywords. Data base management system; computer aided design; data aquisition; die design; data handling.

INTRODUCTION

CADB, which is developed to meet the needs of an assembling die CAD/CAM system, is a network model data base management system. In designing the system, the normalization theory, the schema decomposition method and the entity-relationship approach (Wiederhold,1983; Sa, 1984) are applied to analyse the die CAD/CAM system and to establish the data models of the subsystems of the die CAD/CAM system. By synthesizing these data models, we sum up the conceptual schema of the die CAD/CAM special purpose data base system and draw the basic data models which exist widely in the conceptual schema. According to the basic data models and the characteristics of the data access requirements in the die CAD/CAM process, a network model data base management system CADB was designed and implemented on a Burroughs 1955 computer, and it has been transferred to microcomputers and used in developing CAD/CAM systems of dies and moulds.

CADB is based on the proposal of the CODASYL DBTG report (1971), but here the proposal has been simplified and modified considerably. In the system, the concept "set" is used to represent the relationships among data, and two methods, the pointer array and the directory, are used for the storage of "set". Several random access methods are used, which include hash method and index method. In the organization of hash files, the division method with prime numbers as dividers, together with a new overflow processing algorithm, is used; in index files, B^+ tree is used to organize indexes.

So far, the system contains the Data Description Language (DDL), the Data Physical Storage Description Language (PSDL) and the Data Manipulation Language (DML).

THE ORGANIZATION OF CADB

The Architecture of the Data Base System

The architecture of the die CAD/CAM data base system to be set up can be illustrated with Fig. 1, which is based on the ANSI/SPARC 3-schema structure (Jardine, 1977) having a conceptual schema, a subschema (or external schema) and an internal schema. The conceptual model of the die CAD/CAM data base is defined with DDL as the conceptual schema; the internal schema (the data storage structure) is described with PSDL; the subschema represents a subset of the schema which is needed for a particular application.

The Working Area in Secondary Memory

The secondary memory is divided into two parts, the system's working area (SWA) and the data base area. SWA includes the data dictionary and the index area, shown in Fig. 2 and Fig. 3. The data dictionary is a special data base used for storing the information of the descriptions of schema, subschema and internal schema, and the mappings between schema and subschema, and between schema and internal schema. The index area is used for storing indexes and directories of data files. The data files are stored in pages (or blocks) so that the storage density can be raised and the performance and efficiency of the file organizations can be improved.

The data dictionary, the index and the data are all stored on magnetic disc, and the volumn of the data base is only limited by the disc space available. The system uses soft disc as backup in case of possible failures in the software and hardware of the computer system.

The Working Area in the Main Memory

The user's schema description area. In applying CADB to set up a data base, every user can define his own subschema at the stage of data description. When operating on the data base, the user must present the name of his subschema which will be transferred into the user's schema description area in main memory by DBMS for inquiring whenever necessary. The area contains a central subschema which is always in the main memory and a user's subschema which is currently used by the application program and may be changed at the different stages of die CAD/CAM.

The system's communication units (SCU). SCU is divided into two parts. One is the currency indication table which is used for the communications inside the DBMS and includes the currency indications of record types, set types and the running unit, and the other is the data base running status unit which is used for the communications between DBMS and users to indicate whether the manipulations on the data base are successful or not.

THE ACCESS METHODS

The access methods used in the system are mainly random ones instead of sequential ones so as to raise the response speed to users' inquiries. The key of a record type can be an attribute or a set of attributes. For example, the specifications of die sets and typical assemblies in the die design are generally determined by the dimensions of the cavity dies, (L, B) or (L, B, H), so it is necessary that the attribute group (L, B) or (L, B, H) be defined as key, and the records can be inquired by a set of values (L, B) or (L, B, H). The relation symbols, such as $>, \geqslant, =, \leqslant, <$, can be used to inquire about records. For example, in selecting the press, we require that the force of the press be equal to or greater than the working force. The system can also offer multiple access paths.

The file organizations in this system are hash files and index files. The division method with prime numbers as dividers is used in hashing and the handling of overflow has its own features. When a record overflows, it is stored in a temporary unit at first, and after all the records are hashed, the overflowed records will then be handled. Instead of setting up an overflow area separately, the empty space of the main data storing area is used to store these records so that the efficiency of the storage space and the performance of the hash files can be enhanced and improved. In the index files, the distributed free space technique is used to handle the overflows caused by inserting new records into the files. In index organizations, B^+ tree is used which is different from those used in the usual index files. Because the sequential access to data is not needed in die CAD/CAM, the data blocks which are the leaves of the tree are not linked together sequentially so that the system can be simplified and the maintenance be made easier.

THE REPRESENTATIONS OF DATA RELATIONSHIPS

The relationships among data are represented by the concept "set" which is not completely the same as the CODASYL-set. This system offers two types of set to construct complicated network data models.

The single-direction multiple-member set type. This set type consists of one record type called owner and a number of record types called members, shown in Fig. 4, and it can be expressed as

$$S=(O;\ M^{(1)},\ M^{(2)},\ \cdots,\ M^{(n)})$$

where O is the owner and $M^{(i)}$ is the member ($i=1, 2, \cdots, n.\ n \geqslant 0$). In this set type, only such queries as from the owner to members are permitted, that is to say, the data relationship is single-directed. Such set type can represent the relationship between an assembly and the parts which constitute the assembly. In Fig. 5, the set type SA represents the data relationship between the diagonal guide pillars die set (type A) and the parts(punch holder, die holder, guide pillar and guide bushing) which compose this type of die set. An occurrence of set SA is shown in Fig. 6, which explains a certain specification of the die set and the dimensions of the parts. The pointer array method is used in the storage structure of this set type, as shown in Fig. 7, in which every owner record is followed by a pointer array to indicate the addresses of all the member records subordinate to the owner record.

The double-direction single-member set type. This set type consists of one owner record type O and one member record type M, as shown in Fig. 8. In this set type, we can inquire about not only the member record from the owner record but also the owner record from the member records. But it is limited that a member record cannot belong to the different set occurrences of the same set type. The directory method is used in the storage structure of this set type. There are two directories, the owner to member and the member to owner directory, as shown in Fig. 9 and Fig. 10.

THE COMPOSITION OF CADB

The whole system can be divided into three parts, i.e. DDL, PSDL and DML. DDL is used to define schema, subschema, record types and set types, and to describe the total logical structure of the data base; PSDL is used to describe the physical storage structure of the data base, the access path and the storing position, and also to load the data on the data base and to organize hash files and index files; DML is the interface between users and the data base, and is the means offered to the application programmer by the system to retrieve, to modify, to insert and to delete the data in data base. The main commands of DDL, PSDL and DML are shown in Table 1.

THE RUNNING OF THE SYSTEM

In the trial running of the system, we implemented a small data base system according to the practical requirements in the application environment. A number of subschema are defined in the system. The subschema DIESET is about the data of die structure, and a part of it is shown in Fig. 11; the subschema TECHVA is about the technological variables as well as the material and the press parameters; and the subschema TESTDB is designed for debugging and used to check the performances of CADB. The debugging and the trial running process is as follows:

Data definition. The subschemas DIESET, TECHVA, TESTDB and the record types and the set types which are contained in these subschemas are defined. For the model in Fig. 11, eight record types and two set types are defined. The sets are

SA=(GB2851.1; GB2855.1, GB2855.2,
GB2861.1, GB2861.6)
SA=(GB2851.2; GB2855.3, GB2855.4,
GB2861.1, GB2861.6)

where GB2851.1, GB2851.2, etc. are the names of the record types.

The description of the storage structure of the data. The storage structures of the record types in DIESET and TECHVA are determined according to the practical requirements of the die CAD/CAM process. For example, the primary keys of the record types GB2851.1 and GB2851.2 are attribute group (L, B, H), and hashing is used in organizing the files because the values of (L,B,H) are standardized so that they can be easily handled by hashing. As for the data in TESTDB, every storage structure is used and multiple access paths are defined in this subschema.

The retrieval, modification, insertion and deletion of records. In the process, the records in different storage structures are retrieved not only according to primary keys and secondary keys and with every relation symbol ($>$, $\geqslant$, $=$, $\leqslant$, $<$), but also through data relationships, i.e. via set. A series of operations are also made, which include the modification, the insertion and the deletion of records and the corresponding indexes of B^+ tree structure and include the spliting and the merging of the tree nodes, the increasing and the decreasing of the height of B^+ trees.

CONCLUSIONS

The development of CADB supplies a powerful means for the information exchange and the data handling in die CAD/CAM systems. The system CADB has such advantages as follows:

Higher independence between data and application programs. When come some changes in the logical structure or in the physical storage structure of data, it is not necessary to change the application programs because there are the mappings both between subschema and schema, and between schema and internal schema, that is to say, the logical independence and the physical independence of the data are achieved.

Lower data redundance. Applying data base techniques to manage the data in CAD/CAM systems makes it possible that different users can use the same data set in their traditional views and according to their own application requirements, that is to say, the data can be commonly shared, so the data redundance can be greatly decreased. Besides, as the primitive data are often rearranged, conbined and compressed, the practical storage of the data can be made much smaller.

The general application. Because of some common characteristics existing in the mechanical products, CADB can be used not only in die CAD/CAM systems, but also in CAD/CAM systems of other mechanical products, such as modular machine tools. It is general purpose to some extent.

Rapid access speed. In the development of CADB, much consideration and emphasis have been put on the response speed of the system to users' access requirements. Compared with file systems, this system has faster response speed.

Through this research work, we try to find

the main features and the general rules of the DBMS used in the CAD/CAM systems of mechanical products, to accumulate some useful experiences and to lay a good foundation for developing a more powerful and general purpose CAD/CAM DBMS used in designing mechanical products.

REFERENCES

Data Base Task Group of CODASYL Programming Language Committee. (1971). Report. ACM, New York, N.Y.

Jardine, D. (1977). The ANSI/SPARC DBMS Model. North-Holland Press, New York.

Sa, S. X., and S. Wang (1984). Introduction to Data Base Systems (in Chinese) High Education Press, Beijing. pp. 335-350.

Wiederhold, G. (1983). In E. M. Munson and J. F. Murphy (Ed.), Data Base Design. McGraw-Hill Book Company, New York. Chap. 7, pp. 345-401.

TABLE 1 The Main Commands of CADB

DDL-Data Description Language
CALL SCHEMA(schema-name, IST) CALL RECORD(record-type-name, length, data-field-number, field-name, field-type, field-length, field-position, primary-key-name, IST) CALL SET(set-type-name, owner-record-name, member-record-name, type, IST)
PSDL-Data Physical Storage Description Language
CALL HASHFL(parameter list) CALL INDXFL(parameter list) CALL KEYDES(record-type-name, key-name, type, IST)
DML-Data Manipulation Language
CALL OPEN(subschema-name, IST) CALL FIND(record-type-name, primary-key-name, relation-symbol, key-value, IST) CALL FINDSC(set-type-name, IST) *Find the addresses of the member records in the current occurrence of a set type and put them into Set Currency Indication Table. CALL FINDVS(set-type-name, query-sequence, IST) *Find the owner record or the member records of the current occurrence of a set-type. CALL SKFIND(record-type-name, secondary-key-name, relation-symbol, key-value, query-sequence, IST) CALL GET(user-working-area, IST) CALL MODIFY(parameters) CALL INSERT(parameters) CALL DELETE(parameters)

*IST is the running status unit.
Parameter list includes record-type-name, disc-file-number, the parameters of physical structures, the number of records, the values of records, etc.

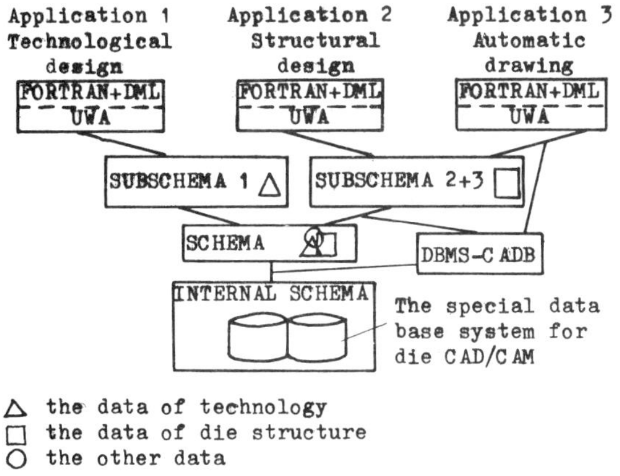

Fig. 1. The architecture of the special data base system for die CAD/CAM

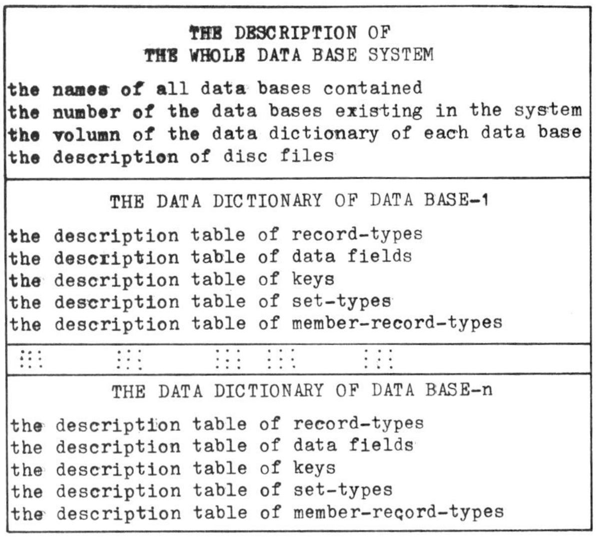

Fig. 2. The system's working area in secondary memory

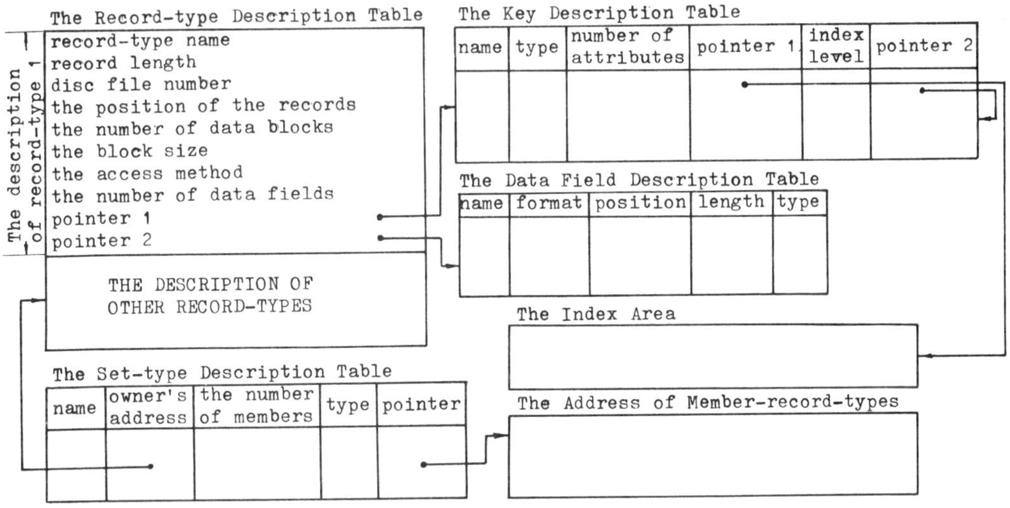

Fig. 3. The description tables and the index area

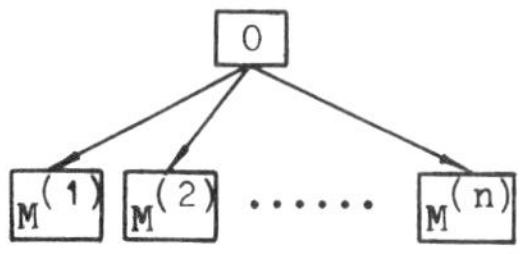

Fig. 4. The single-direction multiple-member set-type

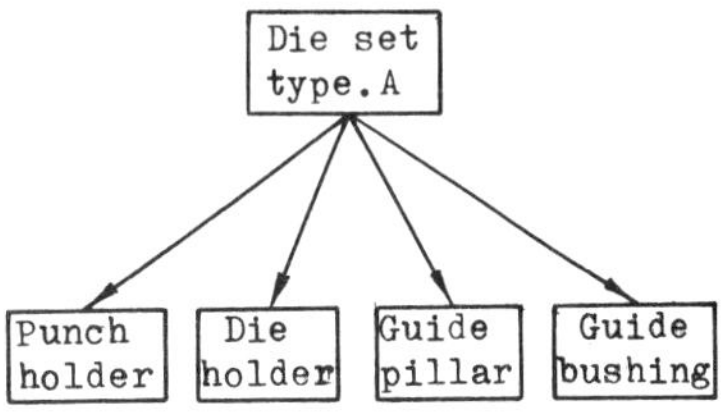

Fig. 5. Set-type SA

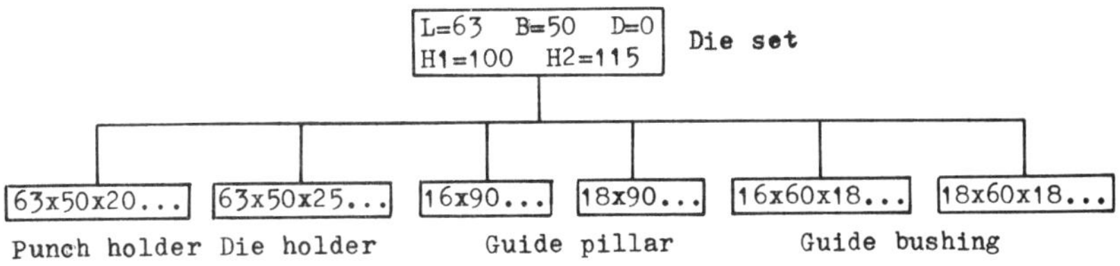

Fig. 6. An occurrence of set-type SA

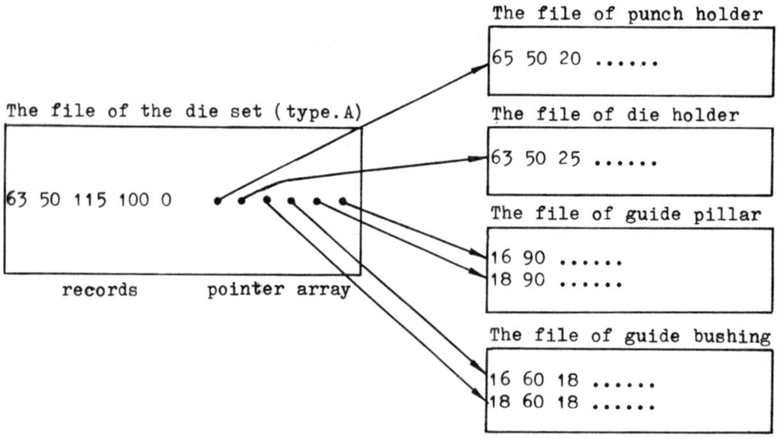

Fig. 7. Pointer array method

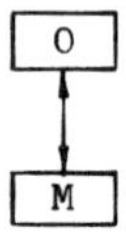

Fig. 8. Double-direction single-member set-type

primary key values of owner records	addresses of member records	pointer

Fig. 9. Owner-member directory

primary key values of member records	addresses of owner records

Fig. 10. Member-owner directory

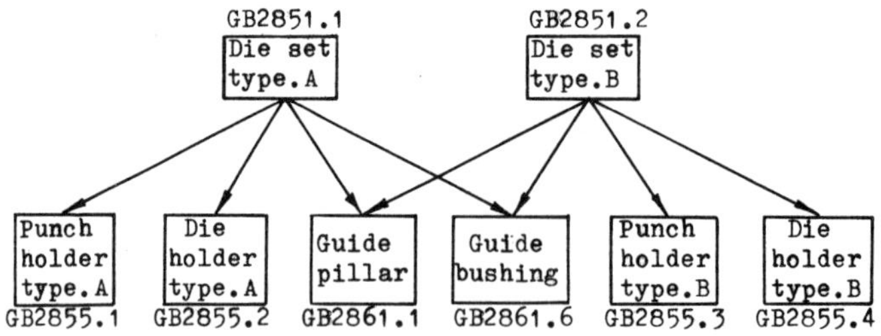

Fig. 11. A simple data model

DESIGNING A RAILWAY STATION: THE DEVELOPMENT OF AN INTERACTIVE COMPUTER SYSTEM TO EVALUATE ALTERNATIVES

R. C. Oliveira and L. V. Tavares

CESUR, Technical University of Lisbon, IST, Av. Rovisco Pais, 1000 Lisbon, Portugal

Abstract. This paper describes a project aiming to develop an interactive computer-based system to support the design studies of Campanhã Railway Station the main station serving Oporto area (the second largest Portuguese city). This computer system includes facilities to generate lay-out solutions and to define future traffic scenarios and a simulation model that enables designers to assess the performance of the solution.

The paper reports the development stages of the project, namely the role played by a prototype in the early stages of specifying the computer system.

Keywords: Railways; modelling; computer software; discrete event simulation; lay-out design

PROBLEM BACKGROUND

Transport is a critical factor for development in most developing countries not only because it is responsible for a large fraction of the energy consumption (in Portugal, this fraction is about 1/3) but also because it is a prime production factor for the commercialization of agricultural goods, for the growth of most industries and for a more balanced regional distribution of population.

Land transportation includes two major modes (rail and road) with rather distinct features: the unit consumption of land or energy for the latter is much greater than for the former but, unfortunately, the investment required by railways exceeds the initial cost of a road system. Therefore, the preference for a rail solution for adequate problems (intensive passenger traffic or freight transport with a high index of ton x km) implies special attention to the key task of optimising the required investment.

The problem of investment optimisation can be formulated in general terms using the basic model of System Theory:

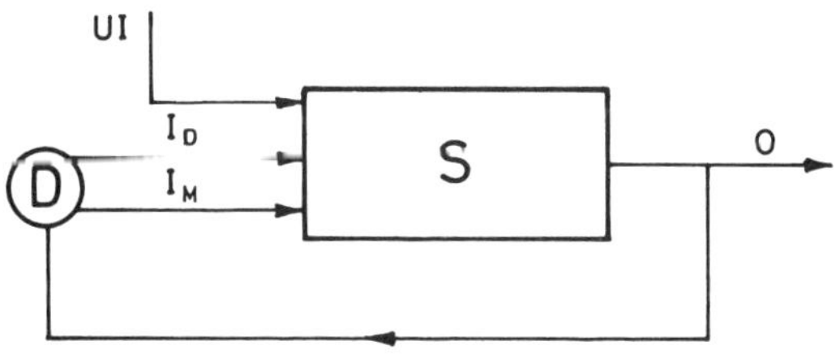

where

S - studied system (e.g., a new railway station)

UI - uncontrolled inputs acting on S (e.g., timetables and random delays of trains arriving to or leaving from the studied station due to exogeneous factors)

I_D - controled inputs decided by the decision maker D about the design of S (e.g., the graph describing the lay-out of the station which is responsible for the matrix of internal routes)

I_M - controlled inputs decided by D about the management of a specific design (e.g., allocation of trains to platforms)

O - Output measuring the performance of S (e.g., measures of the railway station capacity to deal with the adopted timetable of train arrivals and departures, of the delays induced by the station, of facilities utilization, etc).

This problem can be then reduced to solving one of these models:

1) Max $|U(O) - C(ID) - C(IM)|$
with

$I_D \in F_D$

$I_M \in F_M(I_D)$

$O = f(I_D, I_M)$

2) Max $U(O) \;/\; |Cx(I_D) + C(I_M)|$
with

$I_D \subset \Gamma_D$

$I_M \in F_M(I_D)$

$O = f(I_D, I_M)$

3) Max $|U(O)|$
with

$C(I_D) \leq L_D$

$C(I_M) \leq L_M$

$$I_M \in F_M(I_D)$$

$$O = f(I_D, I_M)$$

4) Min $|C(I_D + C(I_M)|$
with

$$U(O) \geqq L$$

$$I_D \in F_D$$

$$I_M \in F_M(I_D)$$

$$O = f(I_D, I_M)$$

where $C(I_D)$, $C(I_M)$ are appropriate cost functions, F_D is the feasible set of alternative designs, $F_M(I_D)$ is the feasible set of alternative operational policies defined in terms of the adopted design (I_D), L, L_D or L_M are bounds to be respected and, the last but not the least, f is the function relating the performance of the system with the controlled inputs.

This last function is particularly critical to any of these models and it is not easily estimated by traditional engineering techniques (e.g., Zuquete, 1984). The major aim of the project reported in this paper is the development of an interactive computer system oriented to the estimation of f for a crucial and complex component of any railway system: a railway station.

This choice is also motivated by an important real case taking place in the northern part of Portugal: the renewal of the railway node of Oporto. Oporto is the main town of the northern region of Portugal which generates more than 1/3 of the national industrial product and a large renewal program is being developed to improve its railway facilities including (Fig.1):

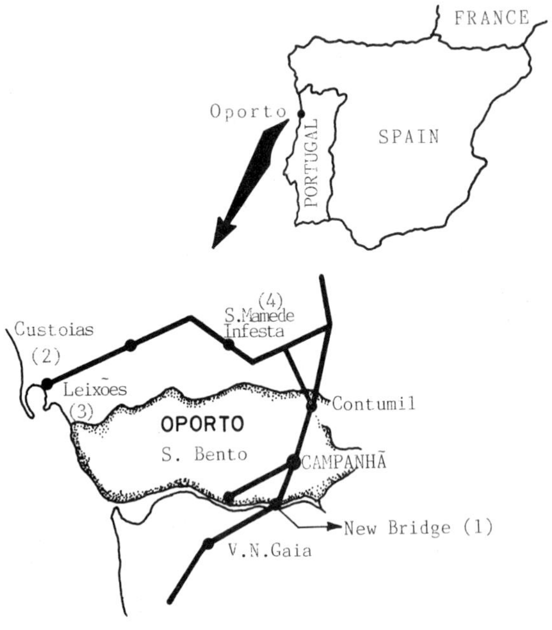

Fig. 1
Oporto area railway network

- a new bridgge (1) to substitute the single-track old D.Maria bridge built over river Douro last century by the french engineer Eiffel;
- a new railway workshop (2) at Custoias;
- renewal of the track connecting the important harbour Leixões with Oporto (3);
- a new freight terminal to be located at S.Mamede de Infesta (4);
- new railway stations at Campanhã, Vila Nova de Gaia, General Torres, S.Bento and Contumil.

A large fraction of the total investment is due to the necessity of renewing Campanhã which is the main station of Oporto connecting its central urban area (S.Bento) with the north (Minho), the northeast (Douro), the harbour of Leixões (Leixões) and the south (Oporto-Lisbon). Campanhã is a complex commuting station from and to several origins and destinations but acting also as a terminal station.

No need to emphasize that any improvement on Campanhã lay-out is expressed by the generation of substantial economies due to the large cost of any component. As an example, if the utility of a switch point is found negligible, its supression corresponds to a cost reduction of about U.S.$50000 plus other benefits due to the reduction of maintenance charges and operational complexity. The high investment costs and the large life-time of a railway station (usually, about 50 years) calls for extensive investigation of alternative lay-outs and it is critical that planners can assess the solutions performance under different (uncertain) future scenarios.

Having realized this, the Governmental agency responsible for the renewal programme of Oporto railway node (GPFD, Ministry of Public Works) decided to sponsor this research project. The Portuguese Railway Company (CP) appointed a working group that provided railway technical expertise and all the relevant data, assisting the authors in all development stages of the computer-system.

THE INITIAL STAGES OF THE PROJECT

In face of the problem caracterization presented above, it was clear that a fully analytical model aiming at defining clear-out solutions and optimizing decisions would certainly be inappropriate. Due to the complexity of the problem, such approach would require over-simplifications, and validity questions would ineviatably be arosen. Furthermore, managers of the railway company would favour an approach where their experience and judgement could play an important role in the decision-making process, whereas they were reluctant to accept solutions derived by an "optimizing procedure" of which they had little understanding and in which they had no intervention. This is a key issue, since it is believed that users motivation and involvement is a crucial condition for a project to be successful.

Hence, we were aiming at a computer system that:

i) would not require over-simplification of the problem;
ii) decision-makers could understand and trust;
iii) would allow decision-makers to carry out extensive experimentations and use their experience and judgement in generating and evaluting alternative solutions.

The idea of experimenting solutions using a computer model that is a replica of the station and its operation appealed very much to the railway company management. Therefore, computer simulation appeared to be an atractive approach that could meet requirements of the problem. However, managers had not been exposed to simulation concepts and felt some difficulties in specifying the model. To overcome this difficulty, a prototype model was developed.

This prototype was a rough representation of the existing station and its operating rules and, though a simplified version, it was already a reasonable approaximation of the final model. The prototype played a very important role in two ways. First, managers could understand how the model worked and became aware of the potentialities of the approach. Second, having realized that this would be a powerful decision aid tool, they felt motivated to get involved in the project and cooperate in the development of the computer system. The visual and interactive characteristics of the prototype, also included in the final model, contributed significantly to this initial understanding and enthusiasm.

The first specifications for the model were then produced, namely based on criticisms about the prototype. Several modifications and additions were made as the project progressed and managers became more familiar with the method.

COMPUTER SYSTEM OVERVIEW

The core of this computer-based system is a simulation model developed to enable designers to evaluate alternative lay-out solutions under various traffic load conditions. This is a visual interactive model: it produces a dynamic representation of the station and all train movements and it enables the user to interrupt the simulation run namely to obtain additional information or to alter the state of the simulation system. After each simulation run the model provides a series of performance measures (such as train delay statistics, platform occupation data, routes utilization, etc.) which enable the user to assess each lay-out solution.

The main inputs for the simulation model are:

a) Traffic load: the system can support up to 20 different train timetable versions tipifying future traffic scenarios. Timetable editing facilities are provided.

b) Lay-out solutions: the system can support any number of lay-out solutions. For Campanhã railway station it was possible to define a "macro-solution" (CP - Serviço de Estudos, 1983) so that any lay-out one can reasonably envisage is a sub set of the former. With this assumption in mind, a routine was developed to generate lay-out solutions and to prepare the associated data used as input for the simulation model.
This routine starts by depicting on the screen a diagram of the station with the tracks, platforms and switch points corresponding to the "macro-solution" (any other previously defined lay-out solution can also be chosen as starting point). Simple comands enable the user to eliminate (or to put back) any of these elements. When the user chooses to do so, an automatic analysis is carried out to determine incompatibilities between routes. The results of this analysis are recorded on a file that will be used as input for the simulation model.

In addition to this, the system includes a series of menu-driven routines that enable the user to edit other basic data necessary to run the simulation model, namely travel times, alternative routes, train/platform assignment preferences, etc.

THE SIMULATION MODEL

The model represents Campanhã railway station with a previously defined lay-out solution, the incoming lines and connections with the neighbouring station. A detailed description of the model will be presented elsewhere, the remainer of this section being a short overview.

The model comprises:
- traffic generation
- operating rules
- statistics and performance measures gathering
- dynamic visual representation
- interaction facilities

1) Traffic Generation

A train entres the (simulated) system once it is in a neighbouring station ready to depart to Campanhã railway station. This event is controled by the timetable version identified at the beginning of the simulation run. However, if specified by the user, randomly generated delays can be added to the times specified in the timetable in order to evaluate how the station responds to these random disturbances.

2) Operation Rules

The model includes a set of operation rules governing all train movements and the assignment of trains to platforms and routes. The model cheques the relevant conditions for train movement and, once this takes place, updates accordingly the state of the station (like, for instance, when a route is assigned to a train, the relevant incompatible routes are identified and their use by other trains is prevented).

3) Statistics and Performance Measures Gathering

As simulation evolves, several statistics and performance measures are computed. These include:

a) Train delays statistics, for arrival and departure times and for several levels of aggregation (by origin, or destination, or origin/destination, or origin/destination and type of train)

b) Train/platform assignment statistics

c) Platform track utilization: frequency of utilization, occupancy and reservation times

d) Routes utilization: frequency of utilization, occupancy and blockage times

e) Tables displaying data on track and platform occupancy by each train
f) List of all train movements

4) Dynamic Visual Representation

The model depicts a dynamic representation of the system on a colour screen showing the state of the station and all train movements. This facility proved to be very useful in making end-users understand the model, gaining their confidence and involvement. It is also a useful tool for validation purposes and it allows for a qualitative assessment of the solution performance.

Figures 2 to 4 illustrate a sequence of events as it is represented by the model. In Fig. 2, train no. 719 is in Gaia station ready to leave. Since it has a free platform (platform IV) and route, these are assigned to it (Fig. 3). In Fig. 4, the train stops at platform IV and the route is released.

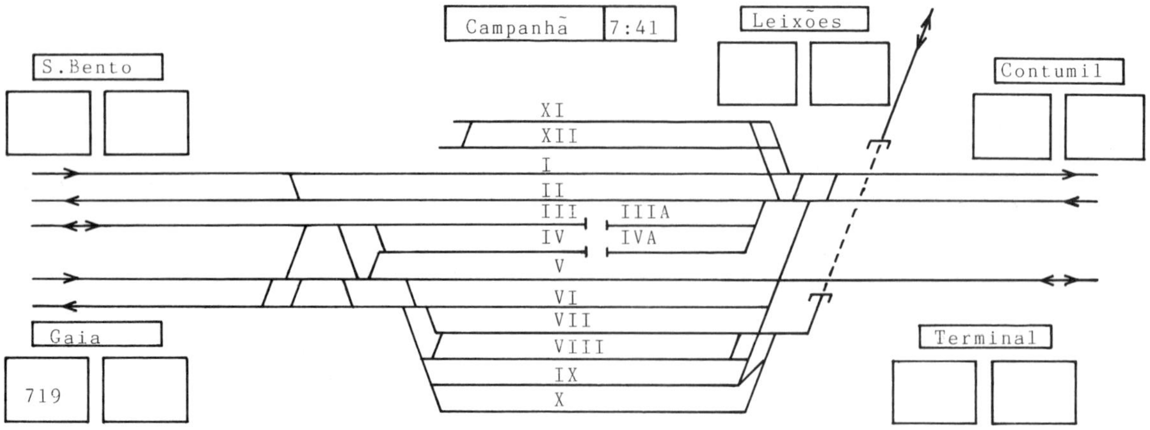

Fig. 2 - Train no. 719 in Gaia Station

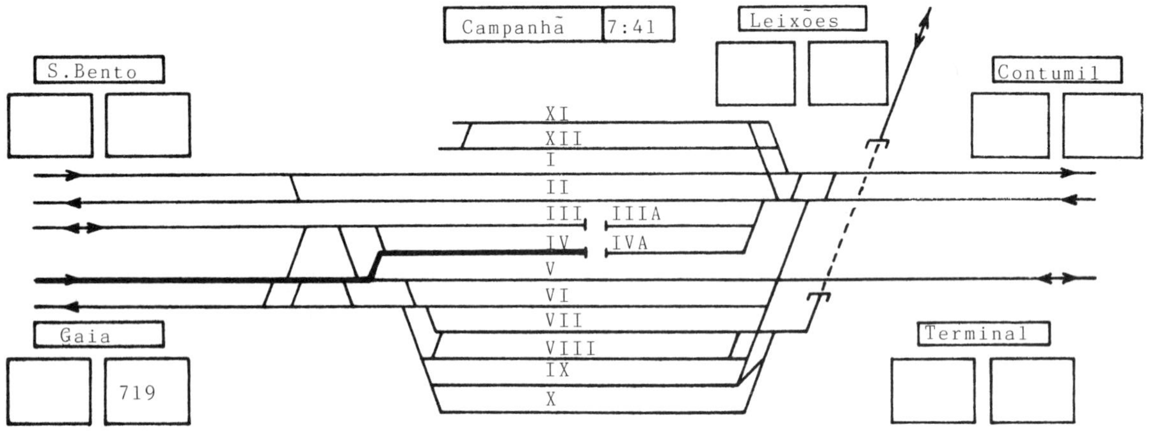

Fig. 3 - Train no. 719 Advances to Campanhã Station

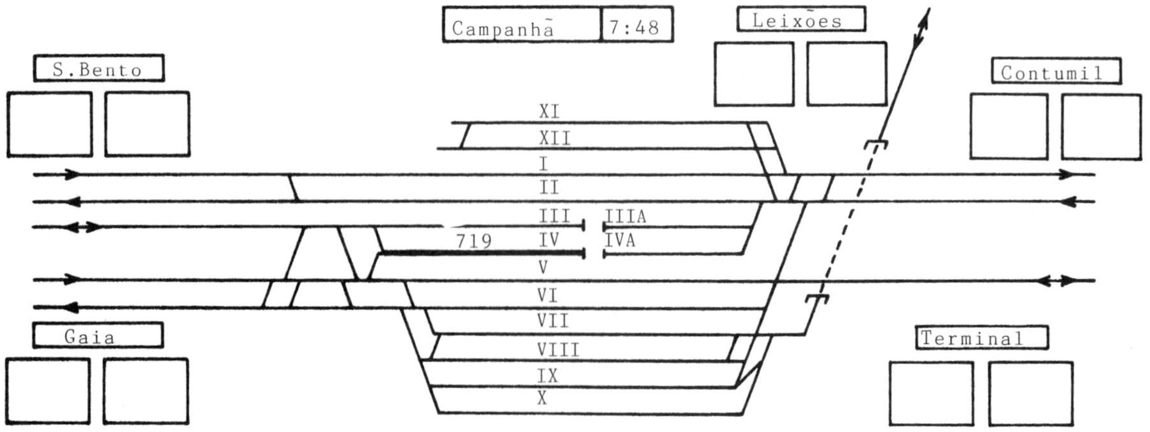

Fig. 4 - Train no. 719 Stops at Platform IV

5) Interaction Facilities

The model includes interaction facilities that enable the user to interrupt the simulation run with three main purposes

a) To ask for additional information, such as about trains (origin, destination, type, scheduled and/or actual arrival and departure times, etc), platforms or routes (status, scheduled time of released, etc) or scheduled future events;

b) To alter the state of the simulated system, either by altering the status of a component of the instalation (like disabling platforms or routes) or afecting the train movements (such as forcing an early departure of the train).

c) To make a dump (saving the present situation) and to move forward or backwards in (simulation) time to a previously saved situation.

CONCLUSIONS

The computer system described above is now implemented and the Railway Company staff responsible for the design of Campanhã station have started experimentations using it. Hence, the impact of the system on the decision process is still to be determined. Nevertheless, the model is already regarded as a powerful instrument of analysis and an important decision support tool that decision-makers can trust and use.
For this end-users satisfaction and confidence contributes not only the final product - the computer system itself - but also the way the whole project was conducted. From the very early stages of the project, it was the authors concern to gain end-users involvement and motivation, and the developed computer system is the result of a strong and mutually enriching cooperation with the CP working group. Other similar projects have failed not because of the poor quality of final product (in purely technical grounds), but because it was not taken in due account the importance of these motivational aspects that might undermine end--users confidence and acceptance.

Though developed for planning purposes (to support design studies of Campanhã railway station), the model can easily be adappted to cover other fields of application, namely at the operational level and with staff training purposes. As for the former, possible applications include the optimization of a number of operation rules (e.g., train/platform assignment) and the refinement of operation programmes (timetables). The visual and interactive characteristics of the model are certainly valuable assets for these purposes and create opportunity for synergetic decisions envolving the operational staff, though other approaches (e.g., mathematical programming and scheduling techniques) can also be envisaged. Further work will be done along these lines.

AKNOWLEDGEMENTS

This project was sponsored by GPFD (Ministry of Public Works) to whom the authors are indebted. The cooperation, involvment and valuable suggestion of several persons from the Portuguese Railway Company (in particular, the working group apointed to assist the project led by Mr. Martins de Brito) greatly contributed to the results achieved and are therefore acknowledged. Mário Baptista and José Carlos Branco programmed part of the computer system.

REFERENCES

Zuquete, E. (1984), Algumas Contribuições para o Estudo de Estações Terminais de Vias Duplas. MSc. dissertation, IST, Technical University of Lisbon.

CP - Serviço de Estudos (1983). "Remodelação da Estação Ferroviária de Campanhã - Memória Justificativa", CP - Serviço de Estudos.

AN LSI MASK ARTWORK VERIFICATION AND PROCESSING SYSTEM JC–81

Hong Xianlong*, Xue Shu*, Xu Qinglin and Zhong Longbao****

**Department of Computer Science and Technique, Tsinghua University, Beijing, China*
***Beijing Research Institute of Automation for Machine-Building Industry, Ministry of Machine-Building Industry, Beijing, China*

Abstract. This paper describes a general purpose LSI mask artwork verification and processing system JC-81, which provides the designers with functions of logical, topological, and geometrical operations to eliminate most false and unchecked errors. Some algorithms, an analysis language used in JC-81 and several application examples are also presented.

Keywords. Computer-aided circuit design; computer software; computer debugging; LSI mask artwork verification; design rule checking.

INTRODUCTION

Based on the design automation level, layout design methods can be classified into three categories: full-automatic, semi-automatic, and manual. It is obvious that a full-automatic design method does not need a design verification tool. But most of today's LSI chips, especially mass production LSI chips, are unfortunately designed manually or semi-automatically, processes which inevitably cause errors due to human intervention. It is extremely difficult to eliminate all these errors from the LSI's by hand; thus, there is an intense need for a design verification tool.

To solve these problems, several verification programs have been developed, most of which only provide users with a set of separated checking commands. Though easy to use, the lack of flexibility restricts the checking ability of such programs.

JC-81 described in this paper is a general purpose system without the above-mentioned disadvantage. It provides a wide range of functions including logical operation, topology analysis, geometric calculations, geometric checking, graphic process etc., and provides an artwork analysis language (AAL) which enables the user to perform the desired complicated checking by different combinations of system functions. By means of this, both common geometric checking as width, space and area of figures, and some special design rules checking, including overlap of figures between different mask layers, MOS channel length or width etc., can be easily performed.

The input of JC-81 is an artwork data file. After this kind of data file has been provided on the disc, JC-81 can interactively accept the source program of AAL in process mode, then interpret and execute it to implement the checking request.

The output of JC-81 can be used for either generating masks or for producing an error file.

JC-81 has different operating modes, such as automatic batch process or interactive, real-time process, and has different types of output, such as list, display or plot.

Finally, AAL is a language independent of the IC technology, design method and artwork editor. This is another advantage over many other checking systems.

THE FEATURES OF JC-81

The major features of JC-81 are as follows:

1. Universality

JC-81 can be used for checking various types of LSIs' artwork as its verification method is independent of the IC's production procedure. The only condition necessary for checking is the corresponding statements given in AAL following the design rules and the user's data file.

The input data file is an artwork data file, edited by various artwork graphic editors (AGE) according to the data structure format of JC-81.

2. Accuracy

One of the most difficult tasks in design rule checking has been to reduce the number of false and unchecked errors. JC-81 provides an effective method for overcoming this problem. It is capable of not only processing normal figures used in LSI artwork, but also performing logical operations, topological analysis and geometric expansion; therefore, accurate checking can be obtained if the user properly organizes the relative statements of AAL into a necessary command file.

3. High Efficiency

Reducing the computer time to an acceptable

level is another important problem. As a result of adopting several advanced algorithms the time complexity in JC-81 is only $O(n*\log_2 n)$ and even $O(m)$ for width checking where $m<<n$ and n is the number of figures in the artwork. From this users will get an efficiant and economic checking performance with a mini-computer.

4. Flexibility
In order to meet requirements from different user, JC-81 supplies several checking modes, such as automatic or self-difining by user, batch or interactive etc.. Users are allowed to select any of them to suit his own application problem.

5. Explicitness
The AAL's statements are simple, explicit and easy to use. The result of the check can be output in three types: list, display and plot. For the list output the information includes the position coordinates of error figures, the layer number and the block number. For the output from display and plotter it includes the violation position with intersected lines.

6. As JC-81 adopts block structure and programing techniques of overlap, swap etc. it can be easily expanded in the furture when it is neede. The system can also be composed in different size with different functions to meet the requirements of the user.

THE STRUCTURE OF JC-81

The system consists of 46 blocks which produce 6 objective programs with techniques of swap and overlap. It is shown as Fig. 1.

ARTWORK ANALYSIS LANGUAGE (AAL)

The statement format in AAL is as follows:

output file name=COMMAND (option, parameter, input file name)

There are 6 types of statements in AAL, they are shown as follows:

1. GET (get and transform)

OF = GET (m , n , IF)

where OF is an output filename and IF is an input filename.

This statement is used for getting graphic data from the given input file and parameter n determines the layer number of the artwork. According to m the graphic data file will be selected from 5 different options in a way more suitable to the user's request.

In summary GET is used for obtaining data file which is the process object for all the other statements. Therefore it must be contained in every source program of AAL.

2. Logical operations

A. "AND" statement

OF = AND (IF1, IF2, ... IFn)

B. "OR" statement

OF = OR (IF1, IF2, ... IFn)

C. "SUB" statement

OF = SUB (IF1, IF2)

These three operations can be in a form of combination, for example,

OF = AND (IF1, OR (IF2, IF3))
OF = OR (SUB (IF1, AND (IF2,IF3),IF4)

The results of logical operations are shown in Fig. 2.

3. Topology analysis

OF = TOPO (IF1, IF2)

where TOPO is one of the following operations.

A. CONTAIN: From IF2 are selected such figures that completely contain one or more figures in IF1.

B. UNCONTAIN: From IF2 are selected such figures that do not contain any parts of the figures in IF1.

C. CONTAINED: From IF1 are selected such figures that are completely contained by some figures in IF2.

D. INTERSECT: First, make a judgement whether two figures, which belong to IF1 and IF2 respectively, are intersected, that is, whether each one is divided into two parts by another. If so, take the AND of IF1 and IF2 as intersect result.

4. Geometric process

A. OF = EXPAND (d, IF)

B. OF = CONTRACT (d, IF)

All figures in IF are expanded outward (or contracted inward) by d.

C. OF = INSIDE (X0, Y0, X1, Y1, IF)

where X0, Y0, X1 and Y1 determine the position of the window. It picks out all the figures inside the indicated window from IF. If a figure crosses the edge of window, it will be cut and inside parts are taken.

5. Geometric check

A. OF = WCHEK (d, IF)

B. OF = AREA (d, IF)

C. OF = SPCHEK(d, m, IF1, IF2,...)

where d is a given checking value and m is an option which determines the type of space checking from 7 different choices. These statements are used to pick out error figures that violated the design rules.

6. Output

A. OF = LIST (IF1, IF2, ...)

B. OF = PLOT (m, IF1, IF2, ...)

C. OF = DISPLAY (m, IF1, IF2, ...)

where the option m is an optimization mark. When JC-81 executes DISPLAY or PLOT, it offers a set of interactive commands.

SOME MAIN ALGORITHMS

1. Partial ordered vector method

This algorithm is used for implementing the logical operation and topological analysis. It consists of two parts, a preprocessor and a scanning processor. The preprocessor transforms all figures in the input file into edges producing a X-coordinate file XF and a vector file VF, which consists of horizontal and slopping vector. Then the entities in these files are sorted.

A double scanning method is adopted in scanning processor. At first, the scanning processor scans the X-coordinates in the XF from less to larger, and then scans the vectors in the working list WL from less Y to larger Y at every X. During the course of vector scanning following works are completed: calculating the intersection, cutting the vector, inserting newly generated vector, determining result vector and forming the result figures according to different conditions.

2. Multiple index touching method

This algorithm is used for checking space between figures. Because of the block nesting structure of the artwork data file in JC-81 system, it is easy to form a index file for a data file. Every building block has a corresponding outline index in index file and so has every figure. An index consists of the length, width, left-bottom X and Y coordinate and pointer. It is obvious that index records have fixed length so they can be easily scanned.

At begining, these indexes are sorted according to left-bottom X coordinate (or Y coordinate). For space checking, if two indexes do not touch each other, their corresponding blocks (or figures) need not be checked, so the computing time is saved.

3. The scanning algorithm for checking width

This algorithm has two features. First, the same block or same figure is checked only once, no matter how many times they appear, as the algorithm uses the block nesting structure. Second, all figures are classified as rectangles and non-rectangles , the latter are checked by means of ordered vector method.

4. Other algorithms

Besides the above algorithms JC-81 system has adopted many other algorithms as well, such as sorting, merging, clipping, oversize, undersize and computing etc..

APPLICATION EXAMPLE

Here is an example of N-channel Si-gate EDMOS artwork shown in Fig. 4. Its layer number, material and type of outline are shown in Table 1. Assuming that an artwork data file names MASKFILE exists in the disk. The design rules of the third layer of this artwork are as follows, in which the values are minimum tolerance.

C1 - N^+ covering area with poly-Si= 25
C2 - N^+ overlap of above area = 1
C3 - contact over poly-Si in the direction of N^+ = 3
C4 - contact overlap in the direction of N^+ = 2
C5 - contact to edge of unrelative Si-gate spacing = 4

The source program with AAL is as follows.

```
CHECK CONTACT BETWEEN N+ AND POLY-SI
 (comment line )
NPLUS = GET (0,1,MASKFILE)
 (produce the figures of N+)
CONTACTNP = GET (0,3,MASKFILE)
 (produce the contacts between N+ and
  poly-Si)
POLY = GET (0,4,MASKFILE)
 (produce the figures of poly-Si)
CONTACT = AND (NPLUS,CONTACTNP,POLY)
 (produce the covering area)
ERRC1 = AREA (25,CONTACT)
 (checking rule C1)
OVERLAPNP = SUB (NPLUS,CONTACT)
 (produce the overlap figures)
ERRC2 = WCHEK (1,OVERLAPNP)
 (checking rule C2)
OVERPN = SUB (CONTACTNP,POLY)
 (produce the figures over poly-Si in
  the direction of N+)
ERRC3 = WCHEK (3,OVERPN)
 (checking rule C3)
OVERLAPC = SUB (CONTACTNP,NPLUS)
 (produce the overlap figures of
  contact in the direction of N+)
ERRC4 = WCHEK (2,OVERLAPC)
 (checking rule C4)
CONPOLY = OR (CONTACTNP,POLY)
 (combine the CONTACTNP and POLY)
ERRC5 = SPCHEK (4,0,CONPOLY)
 (checking rule C5)
```

The last two statement can also be rewriten as follows.

```
SGRID = INTERSECT (NPLUS,POLY)
 (produce the figures of Si-gate)
ERRC5 = SPCHEK (4,1,SGRID,CONTACTNP)
 (checking rule C5)
```

CONCLUSION

The JC-81 system has successfully been applied for checking several LSI artwork. Here are some data showing its effectiveness of checking bipolar and MOS artwork. When program is run on NOVA minicomputer, the logical operation for 1088 figures takes 35 sec., the logical operation for 5493 fiqures consumes 441 sec., where most of the computer time are spent on accessing the disk.

REFERENCE

Xue, S., and X. L. Hong (1984). The boolean operation and topological analysis algorithm for LSI mask artwork and its implementation. Journal of Tsinghua University, Vol. 24. Beijing. China. pp. 11-22.

Mitsuhashi, T., T. Chiba, and M. Yoshida (1980). An integrated mask artwork analysis system. Proceedings of 17th Design Automation Conference. U.S.A. pp. 227-284.

Lawrence, M. R., etc. (1974). CRITIC: An integrated circuit design rule checking program. Proceedings of 11th Design Automation Conference. U.S.A.

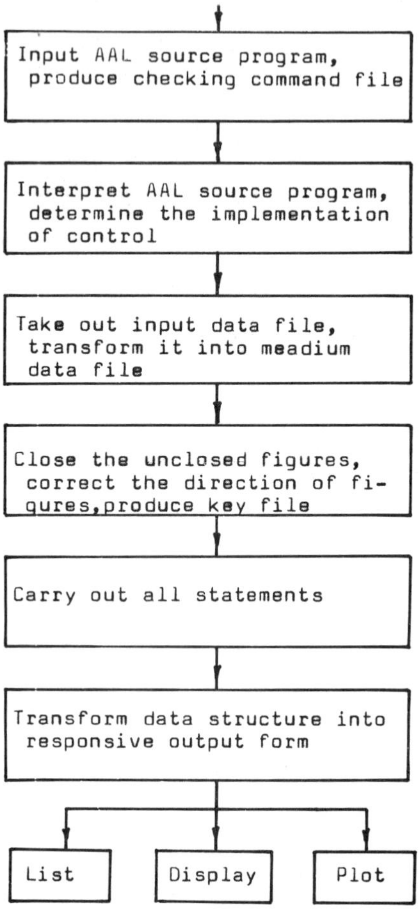

Fig. 1. System structure.

TABLE 1

layer number	material	type of outline
1	N^+	solid line
2	depletion-implantation	
3	the contacts between N^+ and poly-Si	dash line
4	poly-Si	thin line with dots

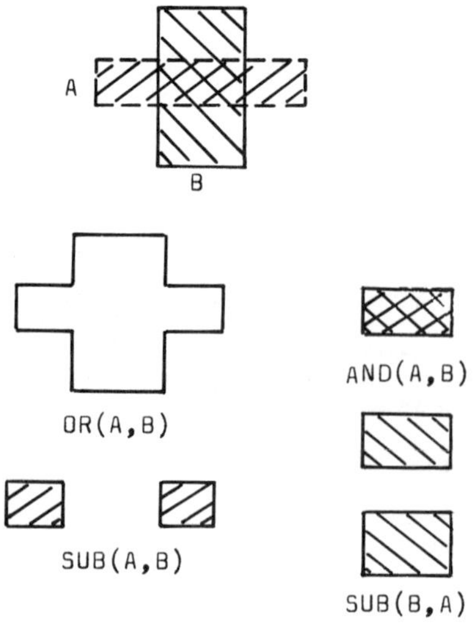

Fig. 2. Logical operation sketch.

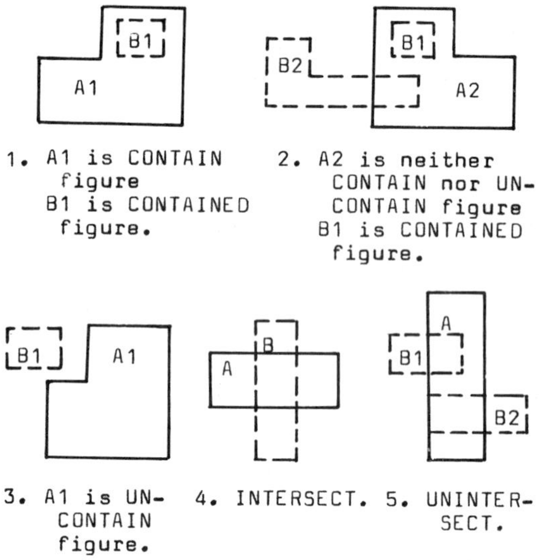

Fig. 3. Topological analysis.

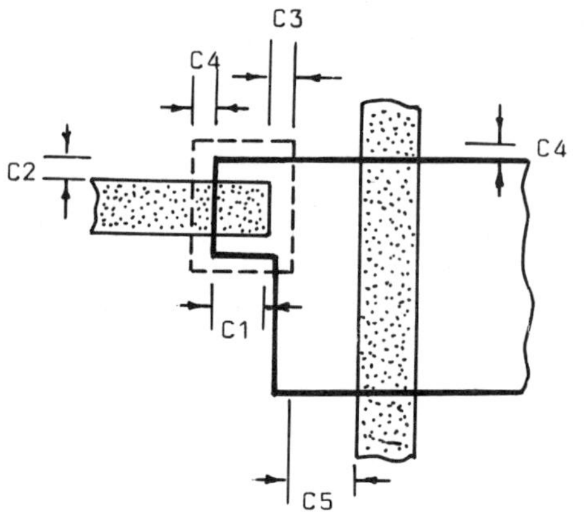

Fig. 4. The example of design rule checking.

COMPUTER AIDED OPTIMIZATION OF MACHINING CONDITIONS

Tong-jian Chen* and N. Fabris**

**The First Department of Mechanical Engineering, South China Institute of Technology, Guangzhou, China*

***Department of Mechanical Engineering, California State University, Los Angeles, USA*

Abstract. The paper proposes a new optimization method of machining conditions. The target of optimization is getting the performance index, the unit cost of workpiece, to be minimum. A proper expression has been suggested to describe the relation between the unit cost and the cutting conditions, it becomes a complete mathematical model by means of Stepwise Regression Method. The process of optimization is divided in two steps, widespread search and partial search. Uses Embedding Method and Golden Section Method for two steps individually to search the real optimal point within the constrained range with one run of computer easily and certainly.

Keywords. Manufacturing processes; optimization; modelling; computer selection & evaluation; nonlinear equations; convergence of numerical methods; embedding.

MODELLING

A Real Performance Index of machining

Usually a performance index (PI) can be any measurable response of the machining operation. If the purpose of optimization is to improve the economics of machining, the logical choise of index would be the unit cost of workpiece, it has been selected as a real performance index in this paper.

The Relation Between PI and Machining Conditions

The unit cost C_u in a machining operation is determined by:

$$C_u = C_l \cdot T_m + C_t \cdot N_t \quad (1)$$

where C_u is the unit cost($), C_l is the labor and overhead rate($/min), C_t is the tooling cost($/edge), N_t is the number of tool changes for one piece, T_m is the time period of machining(min), It includes

$$T_m = T_h + T_c + T_t \quad (2)$$

where T_h, T_c, and T_t are the machine handling time, cutting time and tool changing time, in which

$$T_t = t_t \cdot N_t, \quad (3)$$

$$N_t = T_c / T \quad (4)$$

where t_t is tool changing time for one operation, T is tool life. Substituting Eq.(2), (3), (4) into (1), the cost is

$$C_u = C_l \cdot T_h + C_l \cdot T_c + C_l \cdot t_t \cdot T_c/T + C_t \cdot T_c/T \quad (5)$$

In order to find the relation between Cu and cutting conditions, cutting speed and feed rate, the famous Taylor's tool life Eq. can be used, it is

$$V \cdot T^{\alpha} \cdot f^{\beta} = C$$

where v is cutting speed(fpm), T is tool life(min), f is feed rate(ipr), C is a constant which depends on the materials of workpiece and tool, depth of cut, cutting fluid, etc, α, β are exponents. Rewrite the tool life Eq. in another pattern:

$$T = C^{\frac{1}{\alpha}} / (v^{\frac{1}{\alpha}} \cdot f^{\frac{\beta}{\alpha}}) \quad (6)$$

Besides that, the cutting time T_c is concerned with the cutting conditions also, their relation is

$$T_c = C_c / (v \cdot f) \quad (7)$$

in which C_c is constant concerned in the size of workpiece and depth of cut. Substituting Eqs.(6),(7) into Eq.(5) yields

$$C_u = A_0 + A_1/(f \cdot v) + A_2 \cdot v^{\frac{1-\alpha}{\alpha}} \cdot f^{\frac{\beta-\alpha}{\alpha}} \quad (8)$$

where $A_0 = C_l \cdot T_h$, $A_1 = C_l \cdot C_c$, $A_2 = (C_l \cdot t_t + C_t) C_c/C$. For lathe work, usually the exponent α equals 0.2-0.4, β equals 0.2-0.6, selecting $\alpha = 0.3$, $\beta = 0.4$, the Eq.(8) becomes

$$C_u = A_0 + A_1/(f \cdot v) + A_2 \cdot v^{2.3} \cdot f^{0.33} \quad (9)$$

From Eq.(9), it can be seen that the cost C_u is proportional to power $f^{-1} \cdot v^{-1}$, $f^{0.33} \cdot v^{2.3}$ In order to improve the versatility of the model, the model was set up as a polynomial included the terms of Eq.(9), it is

$$C_u = B_0 + B_1 \cdot v + B_2 \cdot f + B_3 \cdot v^3 + B_4 \cdot f^{0.3} + B_5 \cdot v^{-1} \cdot f^{-1} + B_6 \cdot v^{2.3} \cdot f^{0.33} \quad (10)$$

While the processing of Stepwise Regression, if some terms of the model are redundant, they will be removed from the model by the program automatically. After the process, the remainder terms will be the significant correlative ones and their regression coefficients are ready simultaneously.

Estimate in Stepwise Regression Analysis

For each step of Stepwise Regression, it is necessary to do a F test for variables that are either inside the Eq. or outside it. If the value F of a variable outside the Eq. is larger than the given level FA_{out}, the variable is significant enough, it will be led into the Eq.. If the value F of a variable inside the Eq. is less than the given level FA_{in}, the variable will be eliminated from the Eq.. The process of regression will not stop until there is not any variable can be led in or removed out the Eq.. Rewrite Eq.(10) with some common symbols, there is

$$Y=B_0+B_1\cdot X_1+B_2\cdot X_2+B_3\cdot X_3+\dots\dots+B_6\cdot X_6 \quad (11)$$

where Y is a dependent variable(C_u), X_1-X_6 are independent variables($v, f, v^3, f^{0.3}, v^{-1}\cdot f^{-1}, v^{2.3}\cdot f^{0.33}$), B_0-B_6 are regressors. In order to estimate the regressors, it is necessary to arrange a group of test in production with different values of parameters v and f to obtain a series of results C_u as follows:

$$\left.\begin{array}{l} Y_1=B_0+B_1\cdot X_{11}+B_2\cdot X_{21}+\dots\dots+B_6\cdot X_{61} \\ Y_2=B_0+B_1\cdot X_{12}+B_2\cdot X_{22}+\dots\dots+B_6\cdot X_{62} \\ \vdots \qquad\qquad\qquad \vdots \\ Y_m=B_0+B_1\cdot X_{1m}+B_2\cdot X_{2m}+\dots\dots+B_6\cdot X_{6m} \end{array}\right\} \quad (12)$$

In which m is the number of test, for X=6, m=9 should be reasonable. The procedure of Stepwise Regression is to apply linear transformations to the partitioned matrix

$$\begin{pmatrix} S & T' \\ T & Z \end{pmatrix} \quad \text{(Efroymson's(1960)works)} \quad (13)$$

where

$$(S)_{ij}=\sum_{t=1}^{m}(X_{it}-\bar{X}_i)(X_{jt}-\bar{X}_j), \quad \text{(6*6 matrix)}$$

$$(T)_{1j}=\sum_{t=1}^{m}(X_{jt}-\bar{X}_j)(Y_t-\bar{Y}), \quad \text{(1*6 vector)}$$

$$Z=\sum_{t=1}^{m}(Y_t-\bar{Y})(Y_t-\bar{Y}), \quad \text{(scalar)}$$

The elements of the matrix are called normalized correlation coefficients. At the end of the variables selecting, the selected correlation coefficients are in the last column of above matrix and their subscripts in row are corresponding with the numbers that have been selected as a real regressor. After determining the correlative regression coeffs., the next procedure is to convert them into the natural ones B_0 to B_6. The summary flow chart of regression is shown in Fig. 1.

Machining Test for Estimate of Regressors

1. Design a complete machining process that will be adopted in actual production. Select the number of test points considering the requirement of statistics.

2. Locate the testing points on a reasonable places within the constraint of cutting conditions.

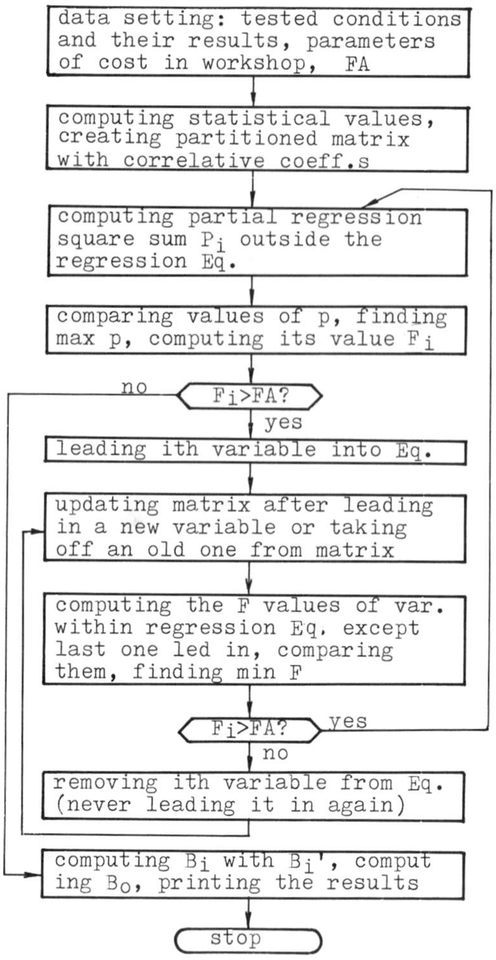

Fig. 1. Flow chart of regression

3. Run the test one by one. It is necessary to produce several pieces in an identical pair of cutting conditions for getting the even value of results so that it could improve the reliability of data.

4. Following information must be recorded from each test: a. number of tool change (N_{tb}), b. time period of test(T_e, min), c. number of parts produced(N_p, in a batch), d. feed rate(f, ipr), e. cutting speed (v, fpm). After getting above information, the unit cost of each testing point C_u can be determined by formula(14). It needs to arrange the data to agree with the requirement of Eqs.(12) for regression analysis. The tester only needs to record the original data such as v, f, N_p, N_{tb}, Te and inputs them to computer, all the calculations will be done by computer in an united program of optimization. The formula for calculating the unit cost in each test is

$$C_u = (C_l\cdot T_e+C_t\cdot N_{tb})/N_p \quad (14)$$

STRATEGY OF OPTIMIZATION IN CONSTRAINT

Fig. 2. shows the constraint of cutting conditions. The real cutting parameters only can be selected inside the spoted area M including the boundary. The strategy of optimizing is as follows:

1. Searching the optimum point in wide-spread range covered the first quadrant quickly, then identifying its position to see where it is. If the optimum point falls inside the area M, it is just what we want, otherwise, searching the effective point again with next steps.

2. If the optimum point falls into area U, going on the partial search along the line 1-2 with one-dimension search.

3. If the point falls into area R, going on the partial search along the line 3-4 with the same way as above.

4. If the point falls into area C, searching along the line 2-3.

5. If the point falls into area 01, searching along the line 5-4.

6. If the point falls into area 02, searching along the line 1-5.

7. If the point falls into area 00, the point 5 is just the ultimate choice. The flow chart of criteria with above logic is shown in Fig. 3..

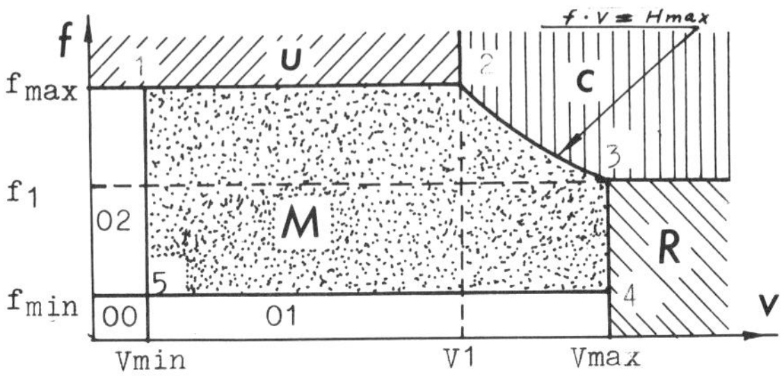

Fig. 2. Division of region for partial search step.

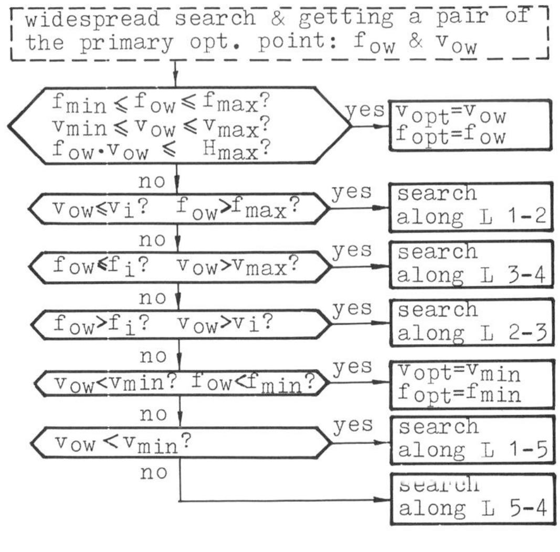

Fig. 3. Logic for selecting the line of partial search.

WIDESPREAD SEARCH

For widespread searching, an efficacious method, Embedding Method, is used. It could get the search to have solution under all circumstances.

Embedding Method

Rewrite Eq.(10) to be a common form: $C_u = f(X)$, X is a vector. Assume the function f(X) has only one pole and is derivable respect to X, the most common way for searching the pole is solving the equations: Gradiant(f(X))=0, that is

$$DF(X) = \frac{f(X)}{X} = 0 \qquad (15)$$

In Embedding Method, for evaluating, it doesn't evaluate the solution of Eqs.(15) directly. For an identical target, it creates a similar Eqs.

$$DF(X) - W \cdot DF(X_o) = 0 \qquad (16)$$

where X_o is a starting vector and it can be given optionally, W is a parameter variable, its values are between 1 and 0. When the value of W tends to zero, the solution of Eqs.(16)(X(W)) is just the solution of Eqs.(15)(X*), that is

$$\lim_{w \to 0} X(W) = X^* \qquad (17)$$

There are different methods can be used to solve Eqs.(16). The paper offers a method of integral. Rewrite Eqs. (16) in compound functions

$$DF(X(W)) - W \cdot DF(X_o) = 0$$

Evaluate the derivative of DF with respect to W, there is

$$\frac{\partial DF}{\partial X} \cdot \frac{dX}{dW} - DF(X_o) = 0$$

$$\frac{dX}{dW} = \left[\frac{\partial DF}{\partial X}\right]^{-1} \cdot DF(X_o) \qquad (18)$$

Integrating Eqs.(18) with respect to W

$$X^* = \int_1^0 \left[\frac{\partial DF}{\partial X}\right]^{-1} \cdot DF(X_o) \cdot dW + X(1) \qquad (19)$$

in which X(1) is initial condition, it equals the value of X when W equals 1, it is just X_o mentioned above. It shows that the solution of Eqs.(15) is just the solution of differential Eqs.(18). If the integral value in Eqs.(19) can be calculated, the solution should be definite.

Real Algorithm

In order to compute the integral by computer, we may use the numerical integral. Rewrite Eqs.(18) in increment form:

$$\frac{\partial DF}{\partial X} \frac{\Delta X}{\Delta W} = DF(X_o) \qquad (20)$$

$$\Delta X_i = \left[\frac{\partial DF(X_{i-1})}{\partial X}\right]^{-1} \cdot DF(X_o) \cdot \Delta W$$

$$X = \sum_{i=1}^{n} \left[\left[\frac{DF(X_{i-1})}{X}\right]^{-1} \cdot DF(X_o) \cdot \Delta W\right] + X_o \qquad (21)$$

$$\text{or} \quad X = \sum_{i=1}^{n} X_i + X_o \qquad (22)$$

in which $n = 1/|\Delta W|$, ΔW is the length of integral step, usually ΔW are divided to be equal value between 1 and 0, their sign are negative when the sequence of integrating is from 1 to 0. As it is known, the smaller the $|\Delta W|$, the more precise the result of integration. X_o is initial condition. It is in correspondance with the initial value of W, which equals 1. Actually, for simplifying the algorithm, we use another procedure to get each value ΔX. From Eqs.(20), we have

$$\frac{\partial DF(X_{i-1})}{\partial X} \cdot \Delta X_i = \Delta W \cdot DF(X_o) \qquad (23)$$

This is just a standard linear equation system in matrix form, we can evaluate the solution ΔX_i by any sophisticated method such as Gauss or Crout and so on. Our target is to obtain the integral result, so it needs to accumulate every value ΔX_i and initial value X_o as showing in Eqs.(22).

Consideration for Improving the Algorithm

In order to get the result with a high precision, it is necessary to give n a big value. Thus, it will waste so much computer time. For instance, if ΔW equals -0.001, the n will be 1000, that means it needs to accumulate ΔX_i at 1000 times. For making good the defect, this paper has suggested a skilful algorithm. In this method it needs to fix ΔW at 0.1 only, in other words the accumulating times of ΔX_i equals 10 merely. As far as it goes, the precision of the result will be poor indeed, but the process doesn't stop here, the Newton's iteration is arranged to go on the process with the starting-point that is the rough result of numerical integrating. Since the starting-point of iteration will close the real solution very much after the integral process, it should be easy to get the ultimate result with high precision in a little iterated steps that are just several times usually. The flow chart of this algorithm is shown in Fig. 4..

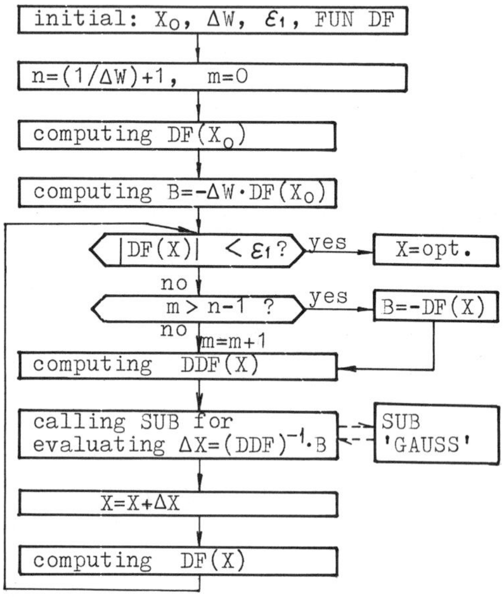

Fig. 4. Flow chart of widespread optimizing

PARTIAL SEARCH

If the result in widespread search process is inside the constraint area (including boundary), it is just a final solution, otherwise, searching the optimum point along the boundary of constraint again. There are three configurations of partial search, they are a. partial search going along a horizontal boundary, b. partial search going along a vertical boundary, c. partial search going along a hyperbola boundary.

Horizontal Partial Search

There are two possible searches on this type, they are going along line 1-2 or 5-4 shown in Fig. 2. . For both lines, the variable f will be fixed in a constant f_{max} or f_{min}, the fn. C_u becomes monovariant:

$$C_u = f(v)$$

As the existance of the pole within the searching range is uncertain, most methods concerned with derivative of the function will be unavailable. Golden Section Method would be useful for any configuration and it is chosen for this project.

Vertical Partial Search

There are two possible searches on this type too, they are going along line 4-3 or 5-1 (Fig. 2.). For both lines, the variable v will be fixed in a constant v_{max} or v_{min}. The fn. of cost becomes

$$C_u = f(f)$$

The subroutine is the same as using at horizontal partial search.

Hyperbolic Partial Search

The hyperbolic part is defined by

$$f \cdot v = H_{max} \qquad (24)$$

in which H_{max} is constant concerned with max. horsepower offered by the machine tool, the subroutine using at horizontal search is avaliable also because f can be substituted by H_{max}/v, it gets function C_u to become one-dimension. the complete process of optimization for cutting conditions is shown in the overall flow chart (Fig. 5.). The detail of the program has been written in 377 statements with FORTRAN language as a software package.

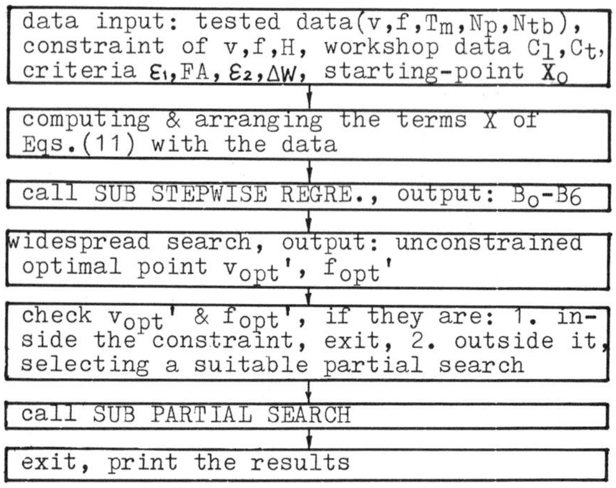

Fig. 5. Overall flow chart of optimization.

EXAMPLE

As the emphasis of this paper is on efficient algorithm, the informations of production in the example were obtained from Ham's (1972) data listed in TABLE 1 . The others of parameter are:

a. parameters of cost in workshop
   C_l = 0.1 \$/min(laber & overhead rate)
   C_t = 0.2 \$/edge (cutting tool cost)
b. constraint of conditions
   v_{max} = 900 fpm
   v_{min} = 10 fpm
   f_{max} = 0.0102 ipr
   f_{min} = 0.001 ipr
   $H_{max} = (f \cdot v)_{max} = 5$
c. parameter for regression
   FA = 4.11 (consulted from Handbook)
d. parameters for widespread search
   starting-point: adopted the first test point
   v_s = 251 fpm
   f_s = 0.0068 ipr
   convergent accuracy: $\varepsilon_1 = 10^{-2}$
   length of integral step: ΔW = 0.1
e. parameter for partial search
   convergent accuracy: $\varepsilon_2 = (B-A) \cdot 10^{-3}$, where (B-A) is the range of search

TABLE 1 Record of Production

test condition		test result		
speed fpm	feed ipr	time min	no. of pieces	no. of tool change
251	0.0068	240	11	1
251	0.0092	240	15	2
251	0.0102	240	16	2
346	0.0068	240	15	7
346	0.0092	240	20	9
346	0.0102	240	22	10
408	0.0068	240	18	16
408	0.0092	240	24	21
408	0.0102	240	27	23

After inputing the initial as a data statement to to the program, the computer outputs the complete result in one time of run. The explanation of the results is as follows.

1. Regression result:
   B_1=-0.01128259938 B_4=-24.8784852119
   B_6=0.000008735714 B_o=10.00135995175
   $B_2=B_3=B_5=0$

The function of cost should be:

$$C_u = 10.0014 - 0.0113v - 24.8785f^{0.3} + 0.0000087v^{2.3} \cdot f^{0.33} \quad (25)$$

In order to check the precision of optimization, we have graphod an Equal-Cost-Curves based on the cost function by TEKTRONIX Graphical Computer model 4051 (Fig. 6). Both estimate and practice of cost are listed together as TABLE 2 for comparision.

2. Optimizing search:
The times of accumulation in integral process are 10 and the times of iteration in last step are 10 for the widespread search (unconstraint). The first opt. point is
   v_{opt}' = 22.2fpm
   f_{opt}' = 598.5 ipr

This is a false one because it falls outside the constraint. The ultimate (real) optimum point in partial search is
   v_{opt} = 417.1 fpm
   f_{opt} = 0.0102 ipr

The minimum unit cost of workpiece is
   $C_{u,min}$ = \$ 1.05394

The total exhaustion of computer time is 8 unit (about 4 second)

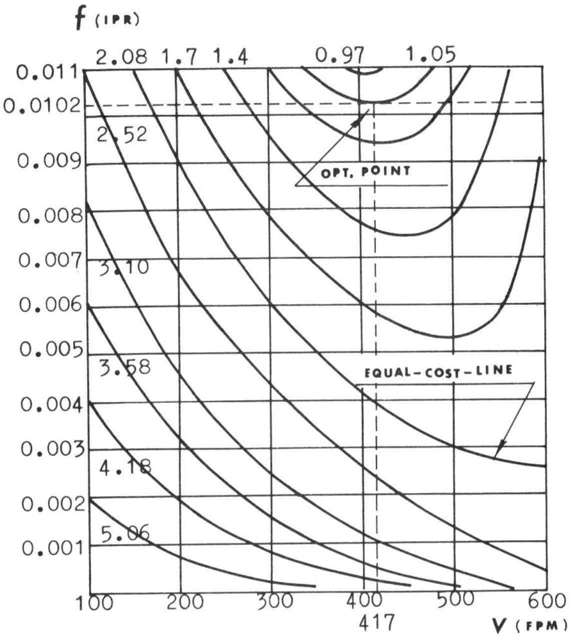

Fig. 6. The Equal-Height-Line of unit cost graphed by TEKTRONIX 4050 computer

TABLE 2 The comparision of unit costs btween practical ones computed with tested data and estimated ones computed with the model

v fpm	f ipr	unit cost \$ practical	unit cost \$ estimate	error
251	0.0068	2.2	2.159	-0.041
251	0.0092	1.627	1.689	0.063
251	0.0102	1.525	1.519	-0.006
346	0.0068	1.693	1.695	0.002
346	0.0092	1.290	1.289	-0.001
346	0.0102	1.182	1.142	-0.040
408	0.0068	1.511	1.532	0.021
408	0.0092	1.175	1.182	0.007
408	0.0102	1.059	1.055	-0.004

CONCLUSION

1. The primary data for optimization can be obtained from the record of production easily and directly. During the normal process of production, change the cutting conditions according to the expected distribution, record the corresponding results, the primary data that the program

require are ready. The processing and computing for the data, processing for regression and optimizing are accomplished by computer with one sequence.

2. As a result of efficacious creation of model structure and the using of effective Stepwise Regression, the estimate has fitted the test data very much in one fitting sequence.

3. The skilful strategy of optimization gets the optimizing process having solution certainly, exactly and saving a lot of computer running time. The result of optimization agrees with the graph (showing as Fig. 6.) completely.

4. The result of optimization can be adjusted in any accuracy expected. The algorithm should be available for any kind of machine tool particularly for the machines that have precise transmission of stageless adjustment.

5. As an algorithm, the dimension of variable in optimized function can be set up in any size. So this method could be used in optimization problems of optional multivariable.

REFERENCES

Efroymson, M.A. (1960). Multiple regression analysis. In Mathods for Digital Computers, John Wiley & Sons, New York. Chap. 17, pp. 193-203.

Ham, T. (1972). Computer optimization of machining conditions for shop production. Engineering Research Bulletin, B-105, Pennsylvania State University.

DIRECT SPLINE INTERPOLATION ON CNC-MACHINE TOOL

Ji Huan

Beijing University of Astronautics and Aeronautics, Beijing, China

Abstract. This paper introduces a method of direct cubic spline interpolation on CNC-machines based on a single board microcomputer, and the corresponding format of NC instructions. It can be used to control directly the machine tool to generate the required movements defined by the spline, according to given parameters. This results in a simplification of continuous path NC programming and a drastic contraction of the NC-program, and provides a more efficient way for NC data transmission.

Keywords. Computer control; function generators; machine tools; numerical control; splines (mathematics).

INTRODUCTION

When free-curve contours or sculptured surfaces are machined on numerically controlled machine tools, it is usually required to describe the part contour or the tool path by a cubic spline. Because the interpolating ability of common NC-controllers is generally limited to linear and circular processing between input points, the tool motion cannot be directly produced with the spline information. Instead, straight line segments must first be used to approximate the cubic spline and then transferred as input data to the NC-controller which, in turn, produces, through linear interpolation, the tool motion, as illustrated in Fig. 1a. The path resulting from the linear interpolation is not a perfect spline; but if the length of each individual segment is short enough, a close approximation may be obtained. Since each programmed line segment must be input to the NC-controller, a path with small tolerances would require enormous NC-data blocks, and a large NC-program would result. It is evident that the inconsistency between the interpolation function of the NC-controller and the mathematical description of the tool path would give rise to remarkable increase or expansion of the volume of data transmitted between the programming system and the machine tool. The NC-tape used for machining a large surface may be a few hundred or even a thousand meters in length, giving more opportunities for errors to occur during tape punching and reading. Though some CNC-control systems have the ability to store the entire NC-program that can eliminate tape-reading errors during machining, an extension of memory capacity for storing long NC-tapes will greatly increase the cost of the NC-controller (e.g., every 80 meter tape requires a capacity of 32 k bytes RAM).

Recently, much attention has been paid to efficient NC-data transfer. The key point is to implement a sophisticated software interpolation on the NC-machine tool--for example, spline interpolation--by the use of a micro-computer. In this case, the interpolation function of the NC-machine is consistent with the mathematical description of the tool path. Under this condition, there is no need to transmit large volumes of NC-instructions for the line segments, and the spline parameters defined by given points can be directly transmitted to the NC-controller. A spline interpolator built in the NC-system will calculate the coordinates on the spline, and the machine is controlled to generate the required movements, as shown in Fig. 1b. This will simplify NC-programming and reduce NC data.

Theoretically it is possible to accomplish software interpolation of any mathematically expressed curve. But the working speed of software interpolation is limited by the calculation speed of the computer. In general, it is preferred to apply a combination of software and hardware interpolation facilities. The software performs complex and time-noncritical calculations, and the hardware the simple and time-critical ones. Another possible solution is based on a multi-microprocessor configuration to share the interpolation task. However, owing to the limitation of interpolation speed, the available interpolation function of currently used industrial NC-controllers is still confined to linear and cir-

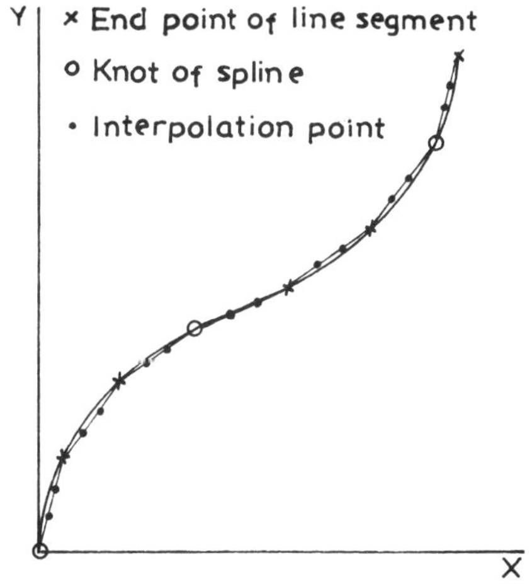

Fig. 1a. Spline approximated with line segments

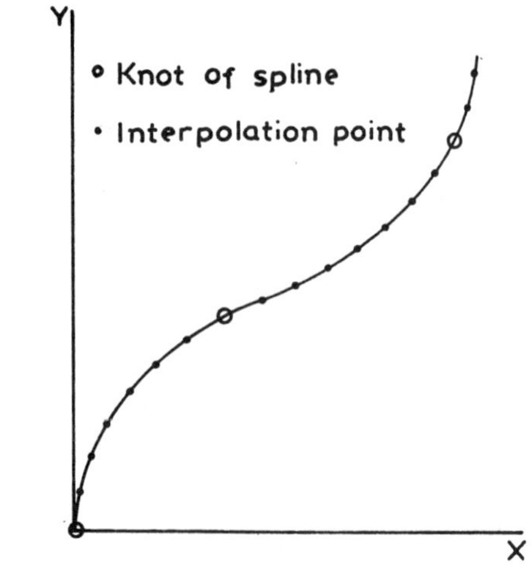

Fig. 1b. Spline directly produced by spline interpolator

We are trying to develop a more simple method to meet the demand for time-critical cubic spline interpolation. It consists of software configuration and the corresponding interpolation algorithm suitable for this configuration based on a 16 bit microcomputer. The aim is to provide a low cost control system.

THE PARAMETRIC SPLINE

A spline is a smooth curve through a set of predefined points (knots) in the space, as illustrated in Fig. 3. Referring to Fig. 2, let A and B be two vectors with end points a and b, and vectors T_a and T_b tangent to the curve at a and b respectively. A space curve through points a and b, tangent to T_a and T_b can be defined in the following form:

$$P(u) = R_3 u^3 + R_2 u^2 + R_1 u + R_0, \tag{1}$$

where u is a nondimensional parameter taking the values of $0 \leqslant u \leqslant 1$ and considering:

$$\begin{aligned} R_o &= A \\ R_1 &= T_a \\ R_2 &= 3\ (B-A) - 2T_a - T_b \\ R_3 &= 2\ (A-B) + T_a + T_b \end{aligned} \tag{2}$$

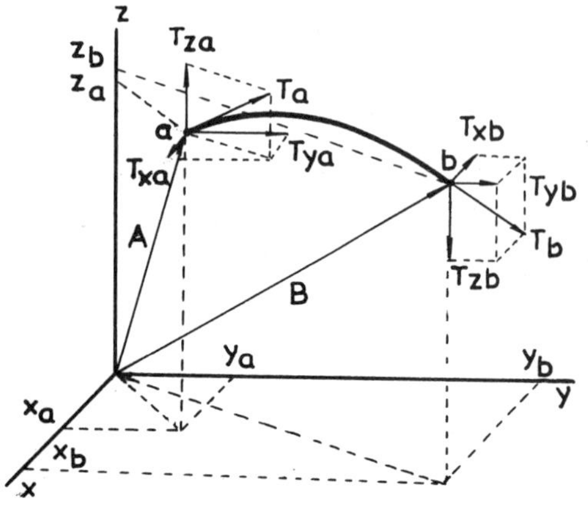

Fig. 2. Three dimensional tangent vector components

Since the above expressions determine a general interval between two knots, it can easily be adapted to n knots with n-1 intervals. Referring to Fig. 3, let $p_i(x_i, y_i, z_i)$ be n vectors and T_i be the corresponding tangent vectors and assuming zero curvature at the end points, we have a solvable system of simultaneous equations from which all tangent vectors $T_i(T_{xi}, T_{yi}, T_{zi})$ can be obtained:

$$\begin{aligned}
2T_1 + T_2 + 0 + 0 + 0 + 0 + \ldots\ldots &= 3(P_2 - P_1) \\
T_1 + 4T_2 + T_3 + 0 + 0 + 0 + \ldots\ldots &= 3(P_3 - P_1) \\
0 + T_2 + 4T_3 + T_4 + 0 + 0 + \ldots\ldots &= 3(P_4 - P_2) \\
0 + 0 + T_3 + 4T_4 + T_5 + 0 + \ldots\ldots &= 3(P_5 - P_3) \\
&\vdots \\
0 + 0 + 0 \quad \ldots\ldots\ T_{n-2} + 4T_{n-1} + T_n &= 3(P_n - P_{n-2}) \\
0 + 0 + 0 + 0 \ \ldots\ldots\ T_{n-1} + 2T_n &= 3(P_n - P_{n-1})
\end{aligned} \tag{3}$$

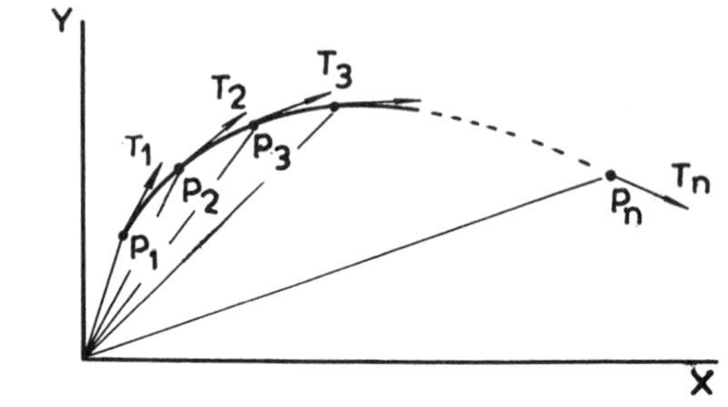

Fig. 3. Representation of parametric spline

Once the knots of the spline are given, the tangent components T_{xi}, T_{yi}, T_{zi} of T_i can be determined by (3), and the tool path can be interpolated using (1) for each interval.

NC-INPUT DATA FORMAT FOR SPLINE INTERPOLATION

The function of the NC-programming system is to find the tangent vectors of the knots by (3) and the coordinate increments between them. These data make up the input instruction of the NC-machine. According to these instructions the spline interpolator accomplishes the spline interpolation and controls the motion of machine tool. Since there is no specification for spline interpolation in the strandard NC-instructions, a new input data format must be introduced and satisfy the following requirements:

- easy for interpretation and calculation in CNC-control system,
- conform with the convention of the standard NC input data format,
- capable of being integrated into the standard NC-blocks.

With the introduction of spline interpolation, the format of the NC-instruction is as follows:

```
:
standard NC-data format
:

sp DI---- DJ---- DK----
   X----- Y----- Z----- DX---- DY---- DZ---- ;
   X----- Y----- Z----- DX---- DY---- DZ---- ;
   :
   X----- Y----- Z----- DX---- DY---- DZ----CR
   :
standard NC-data format
   :
```

Where "sp" indicates that the following NC-data are used to carry out spline interpolation defined by points $P_1, P_2, \ldots, P_n$, as shown in Fig. 3. A special symbol ";" is used to separate NC blocks each of which describes a spline interval between two points P_i and P_{i+1}, $i = 1,2,\ldots,n-1$. The data and symbol in the NC-block are defined in the following order:

- X,Y,Z axis increments along the interval between two knots; $X=X_{i+1}-X_i$, $Y=Y_{i+1}-Y_i$, $Z=Z_{i+1}-Z_i$, as shown in Fig. 3.
- DI, DJ, DK tangent components at first point P_1; $DI=T_{x1}$, $DJ=T_{y1}$, $DK=T_{z1}$, referring to Figs. 2 and 3. They are adopted from the standard NC-data format as I, J, K of the circular interpolation.
- DX, DY, DZ tangent components at the successive points $P_2, P_3, \ldots, P_n$; $DX=T_{xi}$, $DY=T_{yi}$, $DZ=T_{zi}$, $i=2, 3, \ldots, n$.
- "CR" ending of the spline interpolation NC block started at the symbol "sp".

THE CONFIGURATION OF THE SPLINE INTERPOLATION

The spline interpolator is a software program in the control microcomputer of an NC-machine. It must possess the ability to calculate, with sufficient speed and accuracy, the intermediate points within the spline interval, to control the simultaneous motion of the machine axes. A practically usable spline interpolator should meet the following requirements:

- capable of interpolating space curve,
- the interpolation error should be less than or equal to the resolution value of the axis movement,
- the maximum feedrate should not be less than 3 meters per minute,
- when interpolations are performed on the three principle plans, XY, ZX and YZ for machining continuous contours, the tool compensation can be automatically accomplished.

Spline interpolation with equation (1) will not generate theoretical and cumulative errors. But it is very time-consuming. It is desirable to divide the spline interpolator into two stages, the coarse interpolator is used to carry out spline interpolation and the fine one the segment interpolation, as shown in Fig. 4. For simplicity, we consider only 2-D spline.

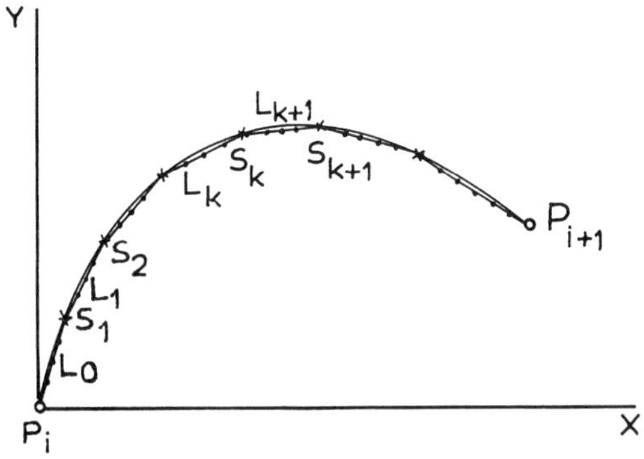

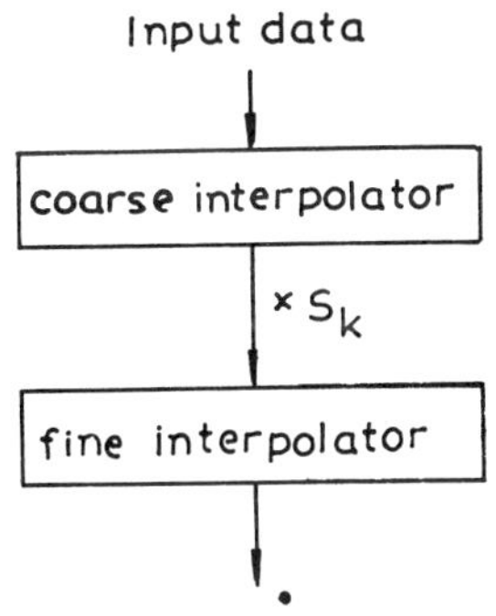

Fig. 4. Two stages software spline interpolation

Firstly a parametric increment Δu can be obtained from the axis increment X and Y. The parameter u is changed by Δu:

$$u_k = u_{k-1} + \Delta u, \quad u_0 = 0, \quad k = 1,2,\ldots \tag{4}$$

The value of Δu must be chosen that the path error will not exceed the resolution value of the axis movement. The coarse interpolator finds out points S_1, S_2, ... on the spline with equation (1). Then the fine interpolator generates the coordinates of intermediate points on the line segments. As the algorithm for fine interpolator is very simple, and the calculation for most points is accomplished through linear interpolation so the interpolation time can be drastically shortened. It makes the software spline interpolator suitable for industrial use. For the feedrate of the machine tool is determined by the fine interpolation period, a constant feedrate can easily be obtained. Figure 5 illustrates the time behaviour of two-stage spline interpolation.

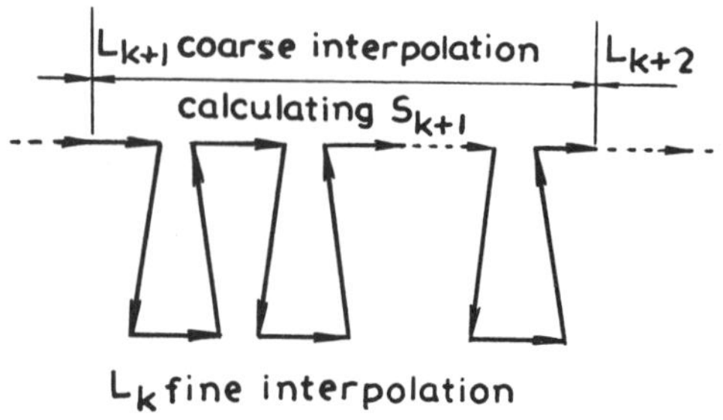

Fig. 5. Time behaviour of two-stage software interpolation

Computer is constantly under the condition of coarse interpolation (or waiting). According to the required feedrate, through every time interval of T, an interrupt request for fine interpolation is produced. The computer suspends the current calculating process for coarse interpolation, and the fine interpolation is carried out to control the movement of the machine. The computer then returns to the main program to perform coarse interpolation. As shown in Fig. 5, the calculating operation of coarse interpolation should lead the fine. When the fine interpolator is working on the line segment L_k, the end point S_{k+1} of next line segment L_{k+1} is determined by coarse interpolator using time remained after the fine interpolation. This interpolation method can be used independently of the actuation system of machine to produce simultaneous axis movements. Therefore it is applicable to both an open and closed loop system.

In order to prove the validity of the proposed method, an advanced numerical control system based on MC68000 single board microcomputer has been developed as shown in Fig. 6. In this system, in addition to the common linear (4D) and circular (on principle plans) interpolation, three axes spline interpolation without tool compensation and two axes with tool compensation are also available. The CNC-controller possesses the following features:

- . CRT display and keyboard for inputing, editing and testing of NC-program,
- . interface connecting machine actuation system,
- . tape reader interface,
- . DNC-interface,
- . cassette interface,
- . RAM with 32 k byts capacity for storage of NC-program corresponding to paper tape of 80 meters in length.

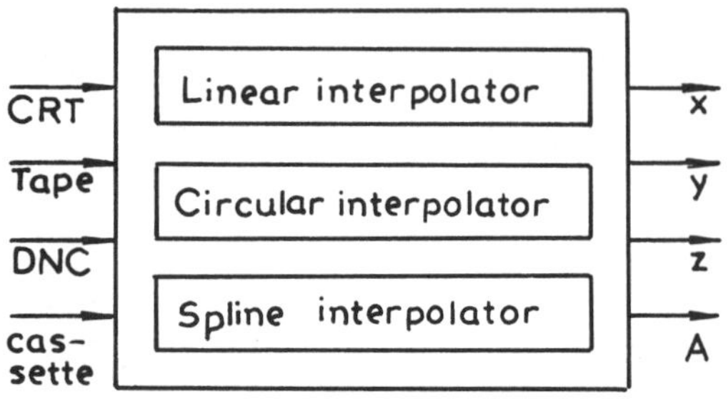

Fig. 6. A CNC-controller with spline interpolator

APPLICATION OF THE SPLINE INTERPOLATOR AND EFFICIENT NC-DATA TRANSFER

The application of CNC-controller with spline interpolator offers an easier solution for NC-programming and efficient NC-data transfer.

If the NC-machined contours are defined by a set of points (e.g. by spline), then the spline interpolating instruction can be used directly for programming and machining without linear approximation as in general.

Spline interpolation can also be used efficiently for analytic curve elements such as ellipse, parabola and hyperbola. For example, as shown in Fig. 7, an ellipse can be approximated with a contour tolerance of 0.08 mm by the spline, when 20 points are taken on it. Using the conventional method it would need 60 points to satisfy the same accuracy requirement. Figure 8 shows the workpiece machined on the NC-machine with tool compensation.

Direct spline interpolation is especially suitable for machining 3-dimensional sculptured surface. Its outstanding advantage is enormous NC-data compression. The surface is usually defined by mesh of patches. Using the basic data an object surface, and then a drive surface on which tool runs can be produced, referring to Fig. 9. It consists of a set of drive curves, the tool path made up of spline. For conventional NC-controller with 3-D linear interpolation drive curves must be converted into line segments as NC-data referring to Fig. 10.

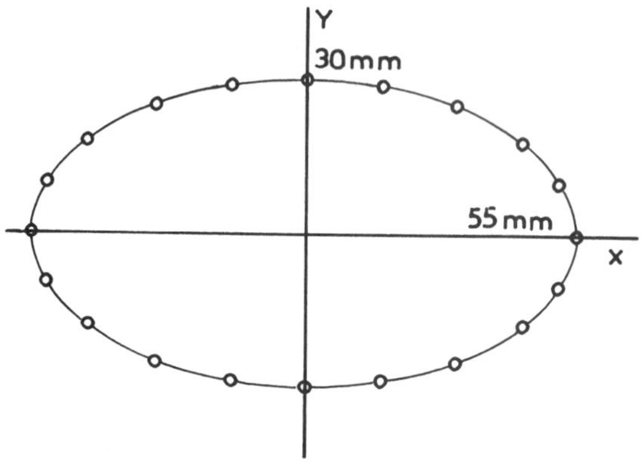

Fig. 7. An ellipse approximated by parametric spline

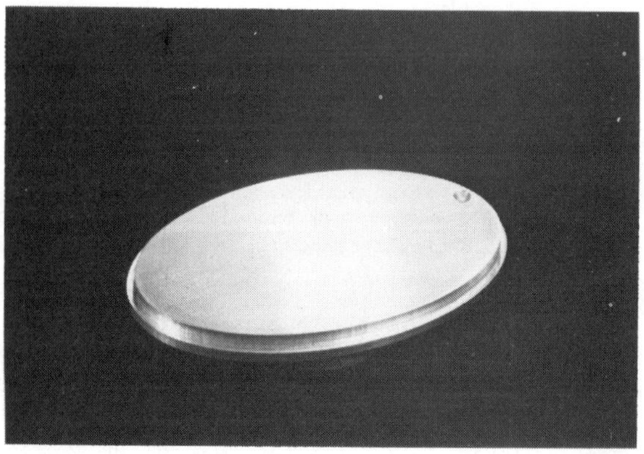

Fig. 8. Machining result of analytic contour

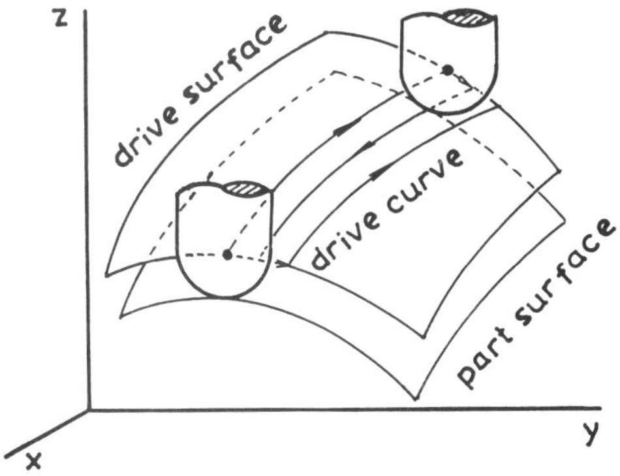

Fig. 9. The tool path for machining sculptured surface

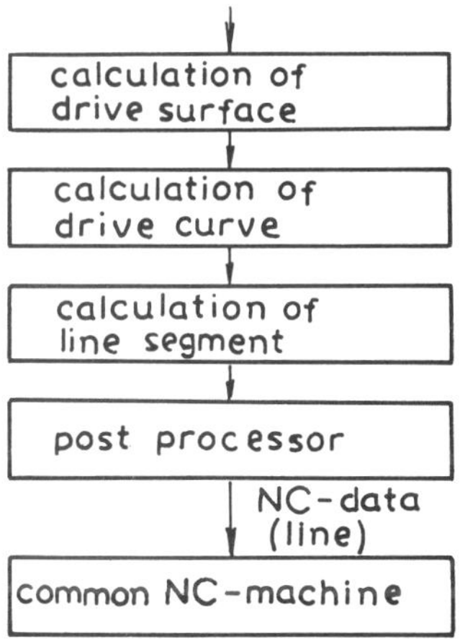

a) on common NC-machine

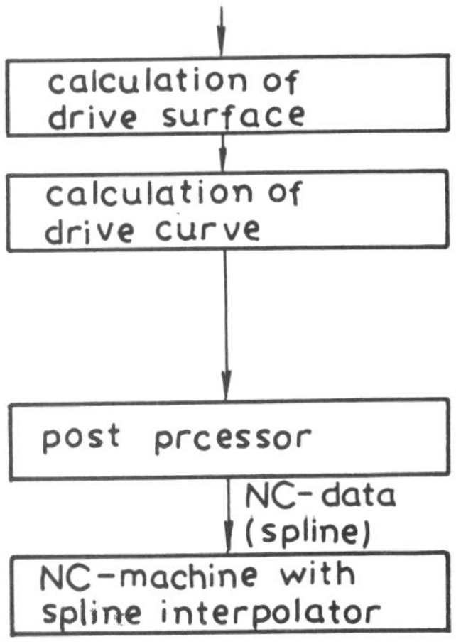

b) on the NC-machine with spline interpolator

Fig. 10. NC-programming and NC-data transfer between NC-programming system and NC-machine

However, to machine a large surface, enormous NC-data e.g. long tape are needed. The NC-controller described in the previous section requires only spline parameters like coordinate increments and tangent vectors as NC-data. The conversion of drive curve to line segments is not necessary, referring to Fig. 10.

Passing spline parameter from NC-programming system to NC-machine results in more efficient transfer of NC-data. NC-input data can be contracted on average by 80%. It will be of great importance for machining large and complex sculptured surfaces. In a test as shown in Fig. 11, machining a die for turbine blade, only 200 NC-data blocks corresponding 30 meters of paper tape must be prepared. This presents a striking contrast between the two.

Fig. 11. Sculptured surface machined with spline interpolator on NC-machine

REFERENCES

Kochan, D. and K.-H. Zehe (1984). Effective Datenfluss durch software-mässige Funktionserweiterung der CNC-Steuerung für Probleme der NC-Fräsbearbeitung. Messen. Steuern. Regeln., No.2.

Pressman, R.S. and William, J.E. (1977). Numerical control and computer-aided manufacturing. John Wiley and Sons, Inc.

Pritty, D.W. (1976). A continuous path microprocessor numerical control system. Proceedins of the 3rd International IFIP/IFAC Conference on Programming Languages for Machine Tool.

Sata, T., Kimura, F., Okada, N. and Hosaka, M. (1981). A new method of NC-interpolation for machining the sculptured surface. Annals of CIRP, Vol. 30/1.

THREE-DIMENSION AUTOMATIC TRACING SYSTEM FOR ROBOTIC GUIDANCE

Chao-zhen Hou

Department of Automatic Control Engineering, Beijing Institute of Technology, Beijing, China

Abstract. This paper describes the development and experimental results of the three-dimension(3-D) tracing system with tactile sensors. The analog tracing probe can sense the 3-D deflection of the probe stylus. According to the deflection vector and tracing objectives, a microcomputer implements the control algorithm and drives the probe to trace a space curve which is an intersection of two unknown physical surfaces automatically. The tracing scheme and algorithm have been developed and the system dynamics has also been analyzed. As the application examples, the test structures having been traced are a 3-D lap-joint seam, a butt-joint seam and a V-butt-joint seam. This system can be used for the robotic automatic guidance as a kind of self-learning industrial robots.

Keywords. Robots; computer control; tracking system; welding; corrdinate measurement

INTRODUCTION

The rapid development of the technology in computer and control field is able to increase the intelligence of industrial robots. The tracing system developed by us using tactile sensors can guide the motion of a robot according to the force feedback. There is no need of the information of the part's figure and teaching of the operator for this system. It is a sophisticated automated tracing and self-learning system.

Recently, some of the research work have been developed in the field of 2-D tracing with tactile sensors (Kusic, 1980 and Lau, 1981). Reference (Bollinger, 1979) described a kind of 3-D tracing system, but the motion along one axis is externally driven, and only the motion along other two axes are controlled in real-time tracing mode.

The 3-D tracing discussed here is tracing along two physical surfaces. The tracjectory of the tracing probe is a space curve which is the intersection of two unknown physical surfaces. The concept of the 3-D tracing is shown in Fig. 1. The surface A intersects the surface B at space curve CD. The control system is required to drive the probe stylus P moving along the space curve CD automactically. In this case, both two surfaces are physical constrained surfaces to the tracing system. The 3-D tracing scheme would require the probe stylus to keep contact with both the surface A and the surface B, while at the same time, moving it along their intersection — the space curve CD at the desired velocity.

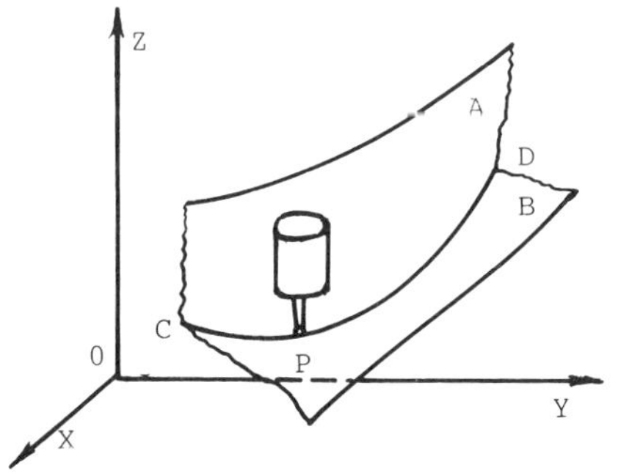

Fig. 1. 3-D tracing with 2-surface constraint

A direct application example of 3-D tracing along two surfaces is robotic welding guidance. Most welded seam are constrained by two surfaces. As shown in Fig. 2, the tracing system is driving the probe stylus to move along the space lap-joint seam to be welded. The test parts are the truck structural members. The ultimate purpose of 3-D tracing in this case is to create a data base that specially describes the seam to be welded along with other information necessory for welding.

Fig. 2. 3-D tracing along a lap-joint seam

MICROCOMPUTER-CONTROLLED 3-D TRACING SYSTEM

Fig. 3. illustrates a simplified block diagram for 3-D tracing system with tactile sensors. The transducer of the system is the 3-D probe. The probe deflections($\Delta X, \Delta Y, \Delta Z$) are the difference between the coordinates(X_P, Y_P, and Z_P) of the center of the probe ball as constrained by two surfaces and the actual origin of the probe positioned by the measuring machine(X_C, Y_C, and Z_C). These deflections can be sensed through three LVDTs mounted in the probe and transfered into the microcomputer through an A/D converter. According to these deflections and the control objectives, the microcomputer implements an algorithm to obtain the velocity commands V_X,V_Y and V_Z, and issues them to the measuring machine through a D/A converter.

The basic requirement to the 3-D tracing system are keeping desired travelling speed and maintaining contact with two constraint surfaces. Therefore, it is convenient to select the contact force and the travelling speed as the control objectives. The contact force can be decomposed in one plane and other axis of the measuring machine. In many welding applications, e.g., the lap-joint welding, one constraint surface is perpendicular to X-Y plane of the measuring machine. Then the contact force can be decomposed into X-Y plane and Z-axis. Thus the control objectives may be selected as follows:

- reference tracing velocity V_R
- reference contact force in the X-Y plane F_{RXY}
- reference contact force in the Z-axis F_{RZ}

In order to implement these control objectives, the microcomputer should at first calculate the contact force in terms of the deflections and the spring constants of the probe. Then, according to the contact force and the control objectives, the computer can obtain the general guidance velocity $\vec{V}_G$ and the force-correcting velocity $\vec{V}_F$ and issue the resultant velocity commands to realize the multiple control objectives.

For the experimental system, as shown in Fig. 4, the probe is a 3-D analog probe; the computer is Giddings & Lewis 868 microcomputer; the measuring machine is the Giddings & Lewis 10V Numericenter; and the laser system for coordinates measurement is HP 5501A.

PRINCIPLES OF RECURSIVE TRACING

How can one obtain the velocity commands in terms of the contact force and the control objectives? This is the basic problem which should be considered in the control of tracing system with tactile sensors.

Fig.4. Experimental 3-D tracing system

In 2-D tracing, the velocity commands can directly be determined by means of the contact force and the control objectives. In this case all velocity commands are constrained to lie in one plane —— the X-Y plane. As shown in Fig.5, the general-guidance command velocity $\vec{V}_G$ lies in the X-Y plane and is perpendicular to the contact force $\vec{F}$ due to the probe deflection. If there was no friction, the direction of the contact force $\vec{F}$ should be normal to the part surface at the point of contact and $\vec{V}_G$ would be parallel to the tangent to the part surface TT at the point of contact M. In the presence of friction force, $\vec{F}$ is not normal to the surface and $\vec{V}_G$ is not parallel to $\overline{TT}$, then a force-correcting velocity command $\vec{V}_F$ can be generated along the direction of $\vec{F}$ according to the control objectives. The resultant command velocity $\vec{V}$ is the sum of $\vec{V}_G$ and $\vec{V}_F$ and parallel to the tangent to the part surface TT. What is important to note here is that $\vec{V}_F$ and $\vec{V}_G$ lie in the X-Y plane and are uniquely determined by the contact force for a given tracing speed and trend.

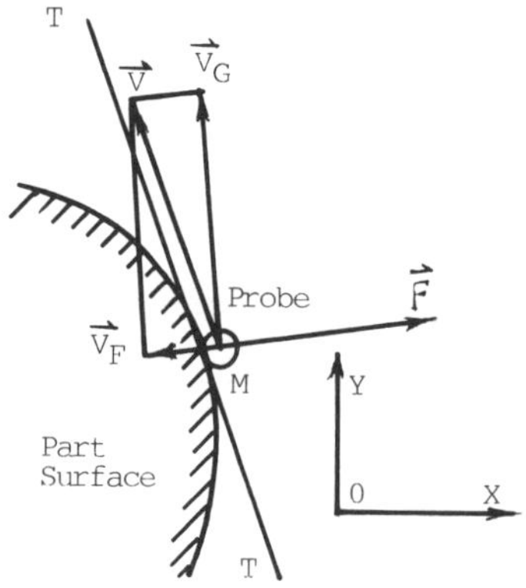

Fig.5. The case of 2-D tracing

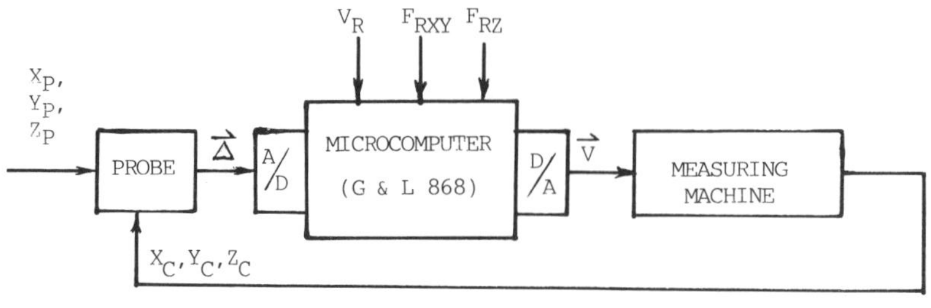

Fig. 3. Block diagram of 3-D tracing with 2-surface constraint

In 3-D tracing along two surfaces, the situation is different. As shown in Fig.6, the general-guidance command velocity $\vec{V_G}$ is still perpendicular to the contact force $\vec{F}$, i.e., $\vec{V_G}$ should lie on the plane H perpendicular to the contact force $\vec{F}$ at the center of the probe ball P. But the direction of $\vec{V_G}$ on the plane H can not be determined without further information about the parts. In other words, the direction of the general-guidance velocity in 3-D case is uncertain only according to the contact force $\vec{F}$.

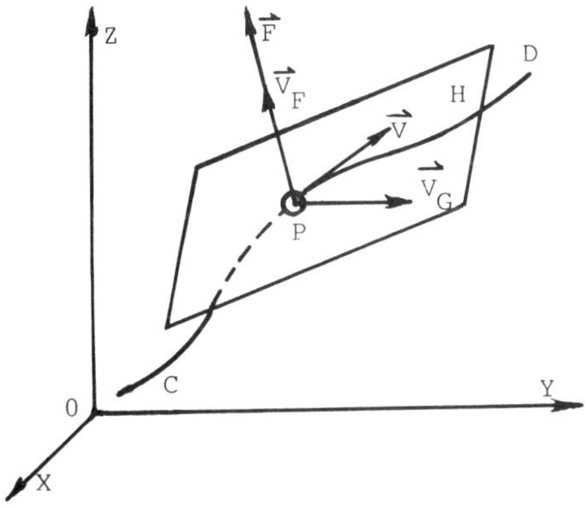

Fig.6. The uncertainty of general-guidance velocity in 3-D

In order to solve this problem, a technique called "recursive tracing" or "inertial tracing" has been developed. The basic idea of recursive tracing is to use the direction of previous tracing velocity to determine the direction of the current general-guidance velocity.

The strategy for determining $\vec{V_G}$ in 3-D recursive tracing is shown in Fig.7. P(n) is the position of the center of the probe ball at the current sampling instant t(n), and P(n-1) is the position of the probe ball at the previous sampling instant t(n-1). $\vec{V}(n-1)$ is the velocity of the probe at t(n-1). In the recursive tracing scheme, the direction of the general-guidance velocity $\vec{V_G}(n)$ is assigned the same direction as the previous resultant velocity $\vec{V}(n-1)$.

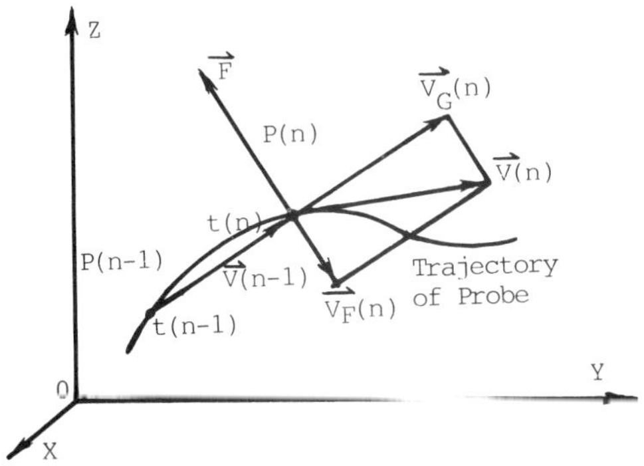

Fig. 7. Recursive tracing

Thast is

$$\vec{V_G}(n) = \frac{|\vec{V_G}(n)|}{|\vec{V}(n-1)|}\vec{V}(n-1) \qquad (1)$$

where $\vec{V_G}$ is the vector of the general-guidance velocity at the current sampling instant t(n); $\vec{V}(n-1)$ is the vector of the resultant velocity at the previous sampling instant t(n-1); $|\vec{V_G}(n)|$ and $|\vec{V}(n-1)|$ are the magnitude of the vector $\vec{V_G}(n)$ and $\vec{V}(n-1)$ respectively.

At the same time, a force-correcting velocity command $\vec{V_F}(n)$ is determined according to the diviation of the contact force $\vec{F}$ from that desired during tracing. By combining $\vec{V_G}(n)$ and $\vec{V_F}(n)$, a resultant tracing command velocity $\vec{V}(n)$ is obtained.

CONTROL ALGORITHM OF 3-D TRACING SYSTEM AND SOFTWARE ORGANIZATION

The first step of control algorithms of 3-D tracing system is the initialization. In the recursive tracing scheme, the present velocity commands are very much dependent upon the previous velocity. It is therefore important to establish the proper initial velocity when tracing is starting.

The second step is to modify the contact force. In 3-D tracing, the contact forces F_X, F_Y and F_Z measured by the probe result from deflection of the probe while in contact with two surfaces. If the friction force is not considered, F_Z may be the contact force with one surface, and F_X and F_Y may be the contact forces with another surface. But while the slope of the space curve being traced is not horizontal, F_X and F_Y would be influenced by two surfaces. In order to keep contact with two surfaces while tracing and simplify the control strategy, the control forces should be modified. Theoretical analysis and experimental test have shown the modified contact forces should be

$$F_X' = F_X + \frac{V_X V_Z F_Z}{V_{XY}^2} \qquad (2)$$

$$F_Y' = F_Y + \frac{V_Y V_Z F_Z}{V_{XY}^2} \qquad (3)$$

$$F_{XY} = \sqrt{(F_X')^2 + (F_Y')^2} \qquad (4)$$

where F_X', F_Y' and F_{XY}' are the modified contact forces along X, Y, and X-Y plane respectively; V_X, V_Y, V_Z and V_{XY} are the velocity components along X,Y,Z axes and in X-Y plane respectively.

Then using PD control algorithm to obtain the force-correcting command velocities

$$V_{FX}(n) = -\left[K_P(F_{RXY} - F_{XY}'(n)) + \frac{K_D}{T}(F_{XY}'(n-1) - F_{XY}'(n))\right]\frac{F_X'(n)}{F_{XY}'(n)} \qquad (5)$$

$$V_{FY}(n) = -\left[K_P(F_{RXY} - F_{XY}'(n)) + \frac{K_D}{T}(F_{XY}'(n-1) - F_{XY}'(n))\right]\frac{F_Y'(n)}{F_{XY}'(n)} \qquad (6)$$

$$V_{FZ}(n) = -\left[K_{PZ}(F_{RZ} - F_Z(n)) + \frac{K_{DZ}}{T}(F_Z(n-1) - FZ(n))\right] \qquad (7)$$

where V_{FX}, V_{FY} and V_{FZ} are the force-correcting velocities along X,Y and Z axes respectively; (n) and (n-1) are the low marks which represent the nth sampling instant and (n-1)th sampling instant respectively; K_P and K_{PZ} are the constants of proportional controller in the X-Y plane and along the Z axis respectively; K_D and K_{DZ} are the constants of derivative controller in the X-Y plane and along the Z axis respectively; F_{RXY} and F_{RZ} are the desired contact forces in the X-Y plane and along Z axis respectively; F_Z is the actual contact force along the Z axis; T is the sampling interval.

For the recursive tracing, the force-correcting velocity has an accumulative character to the resultant command velocity due to the recursive algorithm. The digital PD controller which is something similar to the PI controller in the non-recursive tracing system is resonable here. The magnitude of the resultant force-correcting command velocity is then

$$V_F = \sqrt{V_{FX}^2 + V_{FY}^2 + V_{FZ}^2} \tag{8}$$

The general-guidance command velocity V_G should be perpendicular to the force-correcting command velocity V_F, and the magnitude of the resultant velocity should be equal to the reference velocity V_R, i.e.

$$V_R^2 = V_G^2 + V_F^2 \tag{9}$$

or

$$V_G = \sqrt{V_R^2 - V_F^2} = \sqrt{V_R^2 - (V_{FX}^2 + V_{FY}^2 + V_{FZ}^2)} \tag{10}$$

The components of V_G in X,Y and Z axes can be obtained by projecting Eq.(1) to X,Y and Z axes

$$V_{GX}(n) = V_G(n)\frac{V_X(n-1)}{V(n-1)} \tag{11}$$

$$V_{GY}(n) = V_G(n)\frac{V_Y(n-1)}{V(n-1)} \tag{12}$$

$$V_{GZ}(n) = V_G(n)\frac{V_Z(n-1)}{V(n-1)} \tag{13}$$

where $V_X(n-1)$, $V_Y(n-1)$ and $V_Z(n-1)$ are the X,y and Z components of V(n-1) (the resultant velocity at (n-1)th sampling instant) respectively; $V_{GX}(n)$, $V_{GY}(n)$ and $V_{GZ}(n)$ are the X,Y and Z components of $V_G(n)$ (the general guidance command velocity at nth sampling instant) respectively.

At last, the command of machine velocity is the result of the vector sum of the velocities

$$\vec{V}(n) = \vec{V}_G(n) + \vec{V}_F(n) \tag{14}$$

The components of $\vec{V}(n)$ in the X,Y and Z axes can be found from

$$V_X(n) = V_{GX}(n) + V_{FX}(n) \tag{15}$$

$$V_Y(n) = V_{GY}(n) + V_{FY}(n) \tag{16}$$

$$V_Z(n) = V_{GZ}(n) + V_{FZ}(n) \tag{17}$$

where $V_X(n)$, $V_Y(n)$ and $V_Z(n)$ are the X,Y and Z components of $\vec{V}(n)$ respectively.

It is should be noted that since the vibration of the machine and the probe exists in the tracing system, and the roughness of the surfaces of the traced parts may increase the vibration, the digital filters are necessary for this tracing system. If the surfaces of the traced parts are rather rough, an additional algorithm for the compensation of the friction force should be used in the system. After tracing the parts, a data base about the space curve can be created according to the geometrical relationship between the coordinats and the deflection of the probe and the tracing parameters; and a set of equations can be developed for the data processing.

The total program flowcharts including 3-D tracing control and data processing are shown in Fig. 8. The offline program excutive initializes the system and inputs tracing parameters such as the reference tracing velocity, the desired contact forces, the direction of the searching velocity and the initial velocity. While waiting for the control interrupts the program processes, stores, and displays the data. Once a control interrupt is acknowledged, the computer excutes the control routine. This includes sampling the probe deflections, implementing the recursive tracing scheme, calculating the velocity commands and output them.

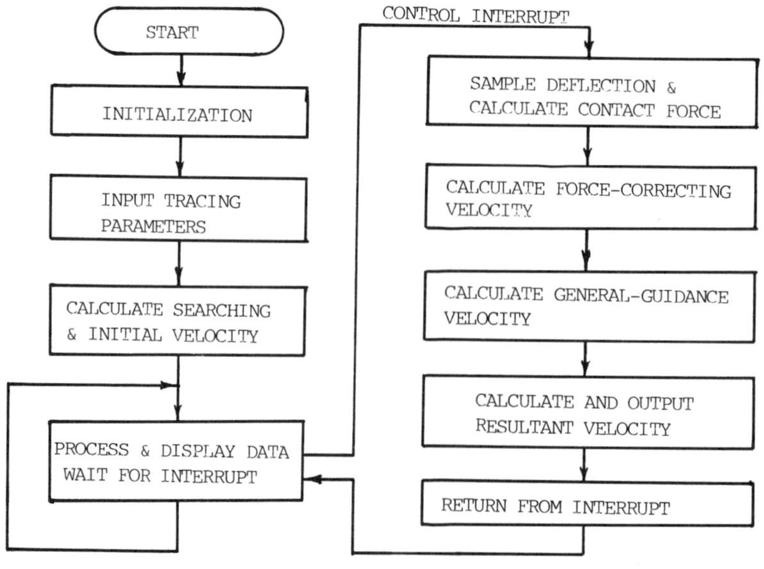

Fig. 8. Flow chart of program

ANALYSIS OF THE SYSTEM DYNAMICS

The control block diagram of 3-D tracing system can be shown in Fig. 9. There are six dynamic elements in this system: $P_X(S)$, $P_Y(S)$ and $P_Z(S)$ are the transfer functions of the probe elements in X,Y and Z axes respectively; $G_X(S)$, $G_Y(S)$ and $G_Z(S)$ are the transfer functions of the measuring machine system in X,Y and Z axes respectively.

The system is a multiple control objectives, coupling multivariables, and non-linear system. Three control objectives are the reference tracing velocity V_R, the reference contact force in X-Y plane F_{RXY}, and the reference contact force in Z axis F_{RZ}. The control variables may be the origin coordinates of the probe positioned by the measuring machine X_C, Y_C and Z_C.

From Eq. (2) to Eq. (17), one has three equations of the velocity commands

$$V_X = F_1(\Delta X, \Delta Y, \Delta Z, V_R, F_{RXY}) \tag{18}$$

$$V_Y = F_2(\Delta X, \Delta Y, \Delta Z, V_R, F_{RXY}) \tag{19}$$

$$V_Z = F_3(\Delta Z, V_R, F_{RZ}) \tag{20}$$

In general case, the functions F_1 and F_2 are non-linear functions. It is very difficult to analyze this kind of multivariables and non-linear system. Since the tracing speed is not very fast, the linearization method can be used to simplify the system.

Expanding the functions F_1, F_2 and F_3 into the Taylor series at Δ (ΔX, ΔY and ΔZ) and neglecting the higher than second order parts, one has

$$V_X = \frac{\partial F_1}{\partial(\Delta X)}\Delta X + \frac{\partial F_1}{\partial(\Delta Y)}\Delta Y + \frac{\partial F_1}{\partial(\Delta Z)}\Delta Z \tag{21}$$

$$V_Y = \frac{\partial F_2}{\partial(\Delta X)}\Delta X + \frac{\partial F_2}{\partial(\Delta Y)}\Delta Y + \frac{\partial F_2}{\partial(\Delta Z)}\Delta Z \tag{22}$$

$$V_Z = \frac{\partial F_3}{\partial(\Delta Z)}\Delta Z \tag{23}$$

or

$$\begin{pmatrix} V_X \\ V_Y \\ V_Z \end{pmatrix} = \begin{pmatrix} \frac{\partial F_1}{\partial(\Delta X)} & \frac{\partial F_1}{\partial(\Delta Y)} & \frac{\partial F_1}{\partial(\Delta Z)} \\ \frac{\partial F_2}{\partial(\Delta X)} & \frac{\partial F_2}{\partial(\Delta Y)} & \frac{\partial F_2}{\partial(\Delta Z)} \\ 0 & 0 & \frac{\partial F_3}{\partial(\Delta Z)} \end{pmatrix} \begin{pmatrix} \Delta X \\ \Delta Y \\ \Delta Z \end{pmatrix} \tag{24}$$

$$= \begin{pmatrix} j_{11} & j_{12} & j_{13} \\ j_{21} & j_{22} & j_{23} \\ 0 & 0 & j_{33} \end{pmatrix} \begin{pmatrix} \Delta X \\ \Delta Y \\ \Delta Z \end{pmatrix} \tag{24}$$

or

$$V = J\Delta \tag{25}$$

where J is the Jacobian matrix.

Then one can fix the coefficients at every tracing interval and get the linear system approximately. The Z-transform method and the state variable technique can be used for the system analysis.

The performance of the experimental system has regulated by changing the form and the parameters of the digital controller. If there is only a proportional controller, the system would be very osillatory even unstable due to the accumulation of the force-correcting velocity in the recursive algorithm; if there is only a derivative controller, the reference contact force could not be kept and the probe might lose contact. Only using the PD controller and selecting the suitable parameters, the better results, faster response and less vibration, can be obtained.

Fig.10 is the typical output of the sensors of the probe in the 3-D tracing process. The deflection in Z direction is almost constant due to the requirement of the contact force in the Z direction. The deflection in X and Y direction are changed according to the geometry of the traced parts, but they should satisfy the requirement of the contact force in the X-Y plane. Because the roughness of the surfaces of the traced parts and the dynamics of the measuring machine and the probe, the response is not very smooth. But all the data collection are finished almost at the same sampling instant, and the error of the data is within ±0.025 mm in the experimental system. It is accurate for the kind of welding applications.

TEST RESULTS AND CONCLUSIONS

The results of the experimental tests have shown that the scheme of the 3-D tracing along two surfaces can be used in some of the practical case.

As shown in Fig. 2, the lap-joint seam can be traced. In this case, the tracing parameters

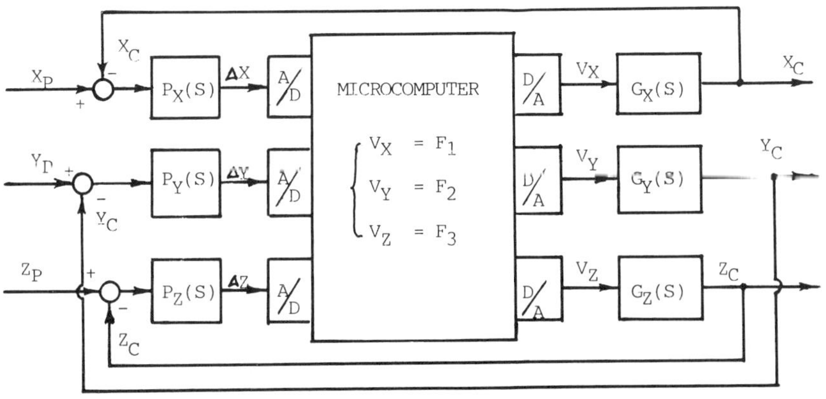

Fig. 9. Control block diagram of 3-D tracing

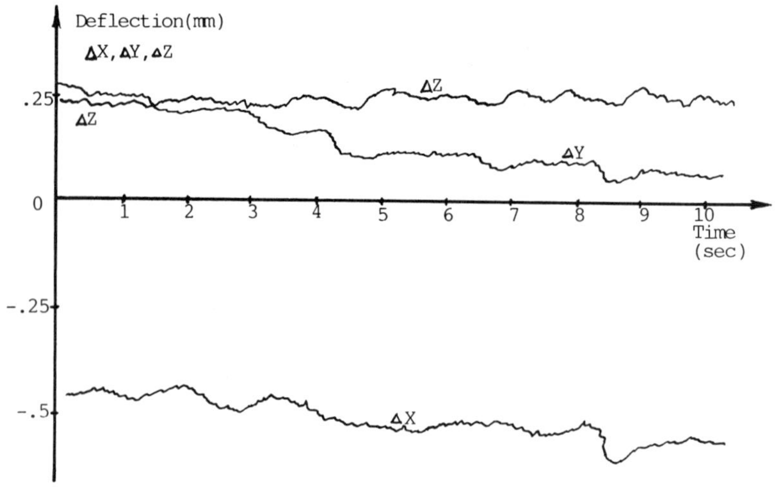

Fig. 10. Deflection response of the probe in 3-D tracing

are as follows

- reference tracing velocity V_R = 300 mm/min
- reference contact force in the X-Y plane F_{RXY} = 55g
- reference contact force in the Z direction F_{RZ} = 55g

In this situation, one surface is roughly perpendicular to the X-Y plane, and the resultant contact force is approximately at 45° from the Z axis. Other applications can be found in the V-butt-joint welding and the butt-joint welding. The experimental parts are two parts with the V-channel and a space gap respectively. In these cases, the ideal contact force in the X-Y plane is zero while maintaining reference contact force in the Z direction.

From the results of the experiments and analysis, it is shown that:

1. The recursive tracing scheme and the digital PD controller are convenient for the 3-D tracing along two intersecting surfaces.

2. The 3-D tracing system is a non-linear, multiple control objectives and coupling multi-variables system. The linearization method can be used for the analysis of the system dynamics.

3. This tracing should require some information about the traced surfaces. In the case of the lap-joint welding, one surface is roughly vertical to the X-Y plane, and another surface is almost perpendicular to the first surface. In butt-joint and V-butt-joint welding cases, the bisector of the two surfaces is roughly perpendicular to the X-Y plane.

If the traced surfaces are twisty, the tracing scheme would fail due to the selected reference contact forces can not be suitable for this situation. In other words, for this scheme, the only information of the probe deflection in three directions is not enough for the general 3-D tracing along two unknown surfaces. In the situation the controller can not distinguish whether the probe is in contact with both surfaces or only one surface.

In the general case of the 3-D tracing along two unknown surfaces, the more complicated probe might be required for sensing the changing surfaces and the more complex control strategy might be used for determining the changing of the surfaces.

ACKNOWLEDGMENTS

The author acknowledges Prof. J. G. Bollinger, Prof. N. Duffie and Dr. Kam C. Lau of University of Wisconsin-Madison (U.S.A.) for their help and support in the research work.

REFERENCES

Bollinger, J. G. and P. W. Ramsey, (1979). Computer controlled self programming welding machine.Weld Journal May 1979.

Kusic,George L.(1980).Feedback design of closed contour servomechanisms. IEEE Transaction on Industry Applications. Vol. IA-14, No.4, July/Aug.

Hou, Chao-zhen. (1982).Microcomputer controlled three-dimension tracing and measurement system. (in Chinese). The first conference of micro computer application of China.

Lau, Kam C. (1981). Development of an automated guidance and dynamic measurement system for coordinate measuring machines and robotic devices, Ph.D thesis of University of Wisconsin-Wadison, U.S.A.

COOPERATIVE CONTROL OF TWO MANIPULATORS

Joonhong Lim and Dong H. Chyung

Department of Electrical and Computer Engineering, University of Iowa, IA 52242, USA

Abstract. The problem of controlling two manipulators in a cooperative manner is investigated. The basic task is to move an object from one place to another by grasping it at two different points using two manipulators. A position control method for two cooperating manipulators is proposed. The Cartesian trajectories of each manipulator are determined from the planned path of the object, and then the cooresponding joint trajectories of the two manipulators are derived. The movement of each manipulator is controlled in small increments. An experiment was performed to demonstrate the proposed method.

Keywords. Robots; cooperation; position control; inverse kinematics; path planning; joint variables

INTRODUCTION

Although many tasks can be performed satisfactorily by a single manipulator, there are cases where it is either necessary or more economical to use more than one manipulator. Moving a large box or a long flexible bar is a typical example.

The problem of controlling two manipulators in a cooperative manner is considered. The basic task under consideration is to transfer an object from one place to another by holding it at two different points using two manipulators. In a single manipulator system, it is usually sufficient to control the manipulator hand so that it follows a predetermined trajectory, and the exact time at which the hand passes through a particular point on the trajectory is irrelevant. The orientation of the hand during the movement may also be irrelevant. In a two manipulator system, however, once the two hands grasp an object, their relative positions and orientations with respect to each other must remain invariant during the entire movement. Therefore, each hand must pass through a particular point on its trajectory at exactly the right moment, and the orientations of the hands also must assume the correct values.

In this paper, a position control method for two cooperating manipulators is proposed. In order to perform a task, the position and orientation of each hand must first be determined as a function of time, and then the control command must be given to each manipulator so that the manipulator hands will follow their respective positions and orientations.

The path of the object is planned first. Then, the positions and orientations of the hands are determined so that the motion of the object follows the preplanned path. The corresponding joint trajectories are obtained by the inverse kinematics solution. The joints for each manipulator are position-controlled to follow the trajectories by using a conventional servo controller[1].

The path of the object is planned in such a way that a predetermined point of the object is moved along the straight line, and the required orientation change is achieved by a rotation about an axis through that point. The straight line path is easily visualized and predictable.

TASK REPRESENTATION

In describing the cooperative task of a two-manipulator system, it is necessary to represent the position and orientation of each manipulator hand. Also, it is necessary to represent the position and orientation of the object to be manipulated. Homogeneous transformation matrices[2] are used to describe positions and orientations. The structure of the cooperative task is defined in terms of homogeneous transformation matrices.

Consider the task of transferring an object from one place to another by holding it with two hands. The system configuration together with the coordinate systems is given in Fig.1.

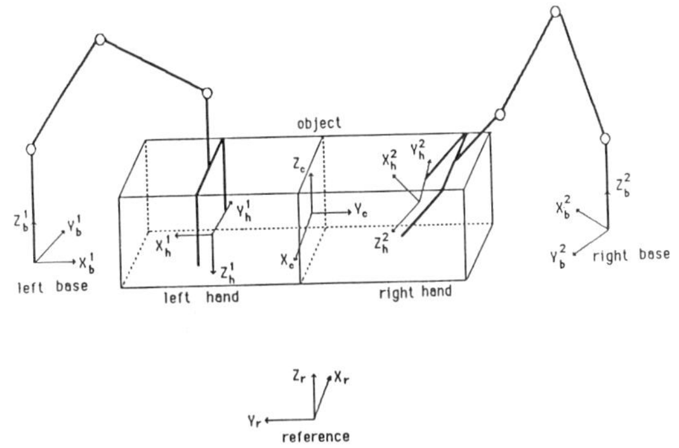

Fig.1. Task description

Let $X(t)$ be the homogeneous transformation matrix of the object coordinate system with respect to the reference coordinate system at time t. Let $Y^i(t)$ be the homogeneous transformation matrix of the hand i (i=1 for the left hand, i=2 for the right hand) with respect to its own base coordinate system at time t. Also, let T_r^i be the homogeneous transformation matrix of the base coordinate system of the manipulator i with respect to the reference coordinate system, and let $T_o^i(t)$ be the homogeneous transformation matrix of the hand of the manipulator i with respect to the object coordinate system at time t. Then, the cooperative task is to move the object from its initial position and orientation $X(0)$ to a final place $X(t_f)$, where t_f is the time at which the task is completed. Once the task is given, the values of $X(0)$ and $X(t_f)$ are known. The initial grasping configurations of each manipulator hand $Y^i(0)$, i=1,2, are also known.

In order to perform a task, the position and orientation of each hand $Y^i(t)$, i=1,2, must be determined for the given $X(t)$. The major difficulty in executing the given task, of course, is the fact that the positions and orientations of the two hands must be such that their relative positions and orientations with respect to each other must be invariant during the execution of the task.

The constraints imposed on the relative position and orientation of the two hands during the execution of the task imply that $T_o^i(t)$, i=1,2, remain constant for the entire duration. Thus,

$$T_o^i(t) = T_o^i(0) = T_o^i, \qquad i=1,2, \quad \text{for all } t \ (0 \leq t \leq t_f). \tag{1}$$

From the property of the homogeneous transformation matrix[2], it is easy to see that

$$T_r^i Y^i(t) = X(t) T_o^i(t), \qquad i=1,2, \quad \text{for all } t \ (0 \leq t \leq t_f). \tag{2}$$

Therefore, from the known initial configurations, T_o^i, i=1,2, are given by

$$T_o^i = X(0)^{-1} T_r^i Y^i(0), \qquad i=1,2. \tag{3}$$

A POSITION CONTROL METHOD

The object motion can be decomposed into a line translation of the origin of the object coordinate system and a rotation about an axis through the origin. The translation is along the vector $p=(x,y,z)^t$ representing the vector between the origins of the initial and final object coordinate system. Also, the orientation change is represented by the rotation angle θ about an axis whose unit vector is $k=(k_x,k_y,k_z)^t$. The values of x,y,z,θ and the unit vector k can be determined by the initial and final configurations of the object as follows: Let

$$X(t) = \begin{bmatrix} n(t) & o(t) & a(t) & p(t) \\ 0 & 0 & 0 & 1 \end{bmatrix} \tag{4}$$

where $n(t)$, $o(t)$, $a(t)$ and $p(t)$ are 3x1 vectors. Then it can be shown that

$$x = n(0)^t \cdot (p(t_f) - p(0)) \tag{5}$$

$$y = o(0)^t \cdot (p(t_f) - p(0)) \tag{6}$$

$$z = a(0)^t \cdot (p(t_f) - p(0)) \tag{7}$$

$$\theta = \cos^{-1}\left\{ \frac{1}{2}\left(n(0)^t \cdot n(t_f) + o(0)^t \cdot o(t_f) + a(0)^t \cdot a(t_f) - 1\right)\right\},$$

$$0 \leq \theta < \pi \tag{8}$$

$$k = \frac{1}{2\sin\theta}\begin{bmatrix} a(0)^t \cdot o(t_f) - o(0)^t \cdot a(t_f) \\ n(0)^t \cdot a(t_f) - a(0)^t \cdot n(t_f) \\ o(0)^t \cdot n(t_f) - n(0)^t \cdot o(t_f) \end{bmatrix}. \tag{9}$$

It is proposed that the object motion between two places be planned so that the origin of the object coordinate system is moving along a straight line represented by the vector p and the required change in the orientation is achieved by a single rotation about the axis whose unit vector is k. Thus, the position and orientation of the object $X(t)$ is given by

$$X(t) = X(0) D(t) \tag{10}$$

where $D(t)$ is a drive transformation matrix[3] representing a straight line translation and a single rotation of the object. If we let $h = t/t_f$, $D(t)$ is given by

$$D(t) = \begin{bmatrix} k_x k_x V(h\cdot\theta) + C(h\cdot\theta) & k_y k_x V(h\cdot\theta) - k_z S(h\cdot\theta) & k_z k_x V(h\cdot\theta) + k_y S(h\cdot\theta) & h\cdot x \\ k_x k_y V(h\cdot\theta) + k_z S(h\cdot\theta) & k_y k_y V(h\cdot\theta) + C(h\cdot\theta) & k_z k_y V(h\cdot\theta) - k_x S(h\cdot\theta) & h\cdot y \\ k_x k_z V(h\cdot\theta) - k_y S(h\cdot\theta) & k_y k_z V(h\cdot\theta) + k_x S(h\cdot\theta) & k_z k_z V(h\cdot\theta) + C(h\cdot\theta) & h\cdot z \\ 0 & 0 & 0 & 1 \end{bmatrix} \tag{11}$$

where

$$V(h\cdot\theta) = 1 - \cos(h\cdot\theta) \tag{12}$$

$$C(h\cdot\theta) = \cos(h\cdot\theta) \tag{13}$$

$$S(h\cdot\theta) = \sin(h\cdot\theta). \tag{14}$$

Now the position and orientation of each hand can be uniquely determined from $X(t)$ and the task constraints given by Eq.(1). From Eq.(2), the motion of each hand can be expressed by

$$Y^i(t) = (T_r^i)^{-1} X(t) T_o^i(t), \qquad i = 1,2. \tag{15}$$

Therefore, from Eqs.(1) and (10), $Y^i(t)$ is given by

$$Y^i(t) = (T_r^i)^{-1} X(0) D(t) T_o^i, \qquad i = 1,2. \tag{16}$$

Let $q^i(t)$ be the m^i dimensional vector representing the joint variables of the manipulator i. Then the inverse kinematic equations can be expressed by

$$q^i(t) = G^i(Y^i(t)), \qquad i = 1,2, \tag{17}$$

where G^i are m^i dimensional vector function of the elements of $Y^i(t)$. There is no general procedure for finding the inverse. However, 6 degrees of freedom manipulators, which are kinematically simple[4], the inverse can be determined by either an algebraic or a geometric approach[5,6]. It was shown that most of manipulators commercially available fall into this category[7]. A manipulator with fewer than 6 degrees of freedom can be treated as a special case of a 6 degrees of freedom manipulator.

Once the joint variables are determined, a control command for each joint is now given to the manipulators 1 and 2 so that each manipulator follows the joint trajectories. Two manipulators are position-controlled in small increments by using a conventinal servo controller.

EXPERIMENTAL TASK

In this study, two RHINO robot manipulators are used. The manipulator has 5 degrees of freedom, and each joint is controlled by a d.c. motor. It also has a control for opening and closing the fingers. Figure 2 shows the kinematic structure of the manipulator.

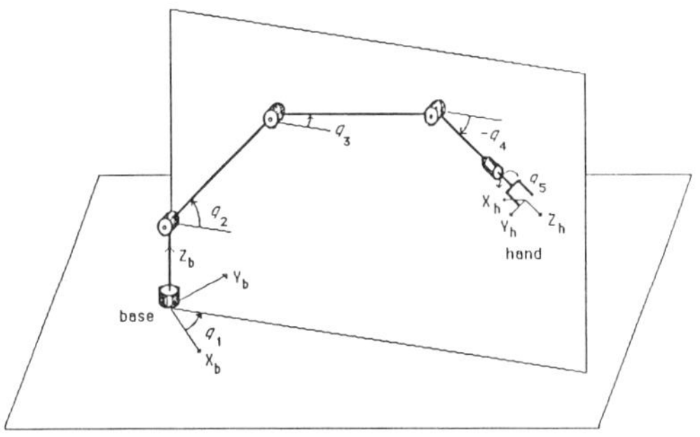

Fig.2. The RHINO robot

The homogeneous transformation matrix Y which represents the position and orientation of a hand with respect to the base coordinate system is given by

$$Y=\begin{bmatrix} C_1S_4C_5-S_1S_5 & -C_1S_4S_5-S_1C_5 & C_1C_4 & C_1(9C_2+9C_3+4C_4) \\ S_1S_4C_5+C_1S_5 & -S_1S_4S_5+C_1C_5 & S_1C_4 & S_1(9C_2+9C_3+4C_4) \\ -C_4C_5 & C_4S_5 & S_4 & 9S_2+9S_3+4S_4+10.5 \\ 0 & 0 & 0 & 1 \end{bmatrix} \qquad (18)$$

where $C_j=\cos(q_j)$ and $S_j=\sin(q_j)$, $j=1,2,...,5$.

The inverse kinematics solutions can be obtained as follows: When a homogeneous transformation matrix Y is given by

$$Y=\begin{bmatrix} n_x & o_x & a_x & p_x \\ n_y & o_y & a_y & p_y \\ n_z & o_z & a_z & p_z \\ 0 & 0 & 0 & 1 \end{bmatrix}, \qquad (19)$$

it is easy to see that

$$q_1=ATAN2(p_y, p_x) \qquad (20)$$

$$q_4=ATAN2(a_z, C_1a_x+S_1a_y) \qquad (21)$$

$$q_5=ATAN2(-S_1n_x+C_1n_y, -S_1o_x+C_1o_y) \qquad (22)$$

where ATAN2(y,x) is the four-quadrant version of $\tan^{-1}$ (y/x). Also, if we let

$$\alpha=(C_1p_x+S_1p_y-4C_4)/9 \qquad (23)$$

$$\beta=(p_z-10.5-4S_4)/9 \qquad (24)$$

then it follows that

$$\cos(q_2)+\cos(q_3)=\alpha \qquad (25)$$

$$\sin(q_2)+\sin(q_3)=\beta. \qquad (26)$$

Eqs.(25) and (26) have two solutions. They are given by

$$q_2=ATAN2(\beta, \alpha)-ATAN2(\pm\sqrt{4\gamma-\gamma^2}, \gamma) \qquad (27)$$

$$q_3=q_2+ATAN2(\pm\sqrt{4\gamma-\gamma^2}, \gamma-2) \qquad (28)$$

where $\gamma=\alpha^2+\beta^2$. Two solutions are identified as elbow-down configuration ($q_3-q_2>0$) when the + sign is taken in the second term in Eqs.(27) and (28), and elbow-up configuration ($q_3-q_2<0$) when the – sign is taken.

An experimental task using two RHINO robot manipulators was performed to verify the proposed control method. The experimental setup is shown in Fig.3.(a), where the objective is to pick up the bar and move it onto the stack of books. The object motions involved in executing the task are confined to a translation in space and a rotation about an axis parallel to the z-axis of the reference coordinate system, for the RHINO robot manipulator has only five degrees of freedom.

The reference coordinate system is defined at the middle point of the two manipulators so that the location of two manipulators are given by

$$T_r^{\,1}=\begin{bmatrix} 1 & 0 & 0 & 0 \\ 0 & 1 & 0 & 8 \\ 0 & 0 & 1 & 0 \\ 0 & 0 & 0 & 1 \end{bmatrix}, \qquad T_r^{\,2}=\begin{bmatrix} 1 & 0 & 0 & -.15 \\ 0 & 1 & 0 & -8 \\ 0 & 0 & 1 & 0 \\ 0 & 0 & 0 & 1 \end{bmatrix}. \qquad (29)$$

The initial object position and orientation is given by

$$X=\begin{bmatrix} .866 & -.5 & 0 & 12.16 \\ .5 & .866 & 0 & -5.3 \\ 0 & 0 & 1 & .8 \\ 0 & 0 & 0 & 1 \end{bmatrix}. \qquad (30)$$

The initial grasping configurations give the following constraints:

$$T_o^{\,1}=\begin{bmatrix} 0 & 1 & 0 & 0 \\ 1 & 0 & 0 & 3.12 \\ 0 & 0 & -1 & 0 \\ 0 & 0 & 0 & 1 \end{bmatrix}, \qquad T_o^{\,2}=\begin{bmatrix} 0 & 1 & 0 & 0 \\ 1 & 0 & 0 & -3.12 \\ 0 & 0 & -1 & 0 \\ 0 & 0 & 0 & 1 \end{bmatrix}. \qquad (31)$$

The desired object location is given by

$$X=\begin{bmatrix} .866 & .5 & 0 & 9.37 \\ -.5 & .866 & 0 & 8.23 \\ 0 & 0 & 1 & 4 \\ 0 & 0 & 0 & 1 \end{bmatrix}. \qquad (32)$$

The initial and final configurations are shown in Figs.3.(a) and (f). The required rotation is equivalent to rotating the bar by 60 degrees about an axis parallel to the z-axis of the reference coordinate system.

The task was divided into 3 subtasks in order to avoid possible collisions with the books. The task was accomplished by first moving up the bar from the initial point by 5.2 inches without any rotation, then moving to the point above the final place while rotating the bar and finally moving to the desired point. Figures 3.(b),(c),(d) and (e) show some of the intermediate points.

The object motion of each subtask was planned as explained previously. The elbow-up configuration was chosen for the inverse solutions. The entire task was performed by making small incremental motions. The resulting motion of manipulators was slow but quite smooth.

CONCLUSIONS

The cooperative motion of two manipulators is analyzed, and a position control method is discussed. The basic cooperative task is investigated in a systematic way by using the homogeneous transformation matrices.

The cooperative control is achieved by first defining the task constraint for given task configurations and then by transforming the motion of the object into that of each manipulator. The corresponding joint trajectories are obtained by solving the kinematic equations.

The joint trajectories are generated in such a way that a predetermined point of the object is moved along the straight line and the required orientation change is achieved by a rotation about an axis through that point.

REFERENCES

[1] J. Y. S. Luh, "Conventional controller design of industrial robots - a tutorial", IEEE Trans. Syst., Man, Cybern., vol.SMC-13, no.3, pp.298-316, May/June, 1983

[2] R. P. Paul, Robot manipulators : Mathemathics, Programming and Control, The MIT Press, 1981

[3] R. P. Paul, "Manipulator Cartesian path control", IEEE Trans. Syst., Man, Cybern., vol.SMC-9, no.11, pp.702-711, Nov., 1979

[4] D. L. Pieper, The kinematics of manipulators under computer control, Ph.D. Thesis, Department of Computer Science, Stanford University, 1968

[5] R. P. Paul, B. Shimano and G. E. Mayer, "Kinematic control equations for simple manipulators", IEEE Trans. Syst., Man, Cybern., vol.SMC-11, no.6, pp.449-455, June, 1981

[6] C. S. G. Lee and M. Ziegler, "A geometric approach in solving the inverse kinematics of PUMA robots", Proc. of the 13th Int'l Symp. on Ind. Robots, pp.16-1 to 16-18, 1983

[7] J. C. Colson and N. D. Perreira, "Kinematic arrangements used in industrial robots", Proc. of the 13th Int'l Symp. on Ind. Robots, pp.20-1 to 20-18, 1983

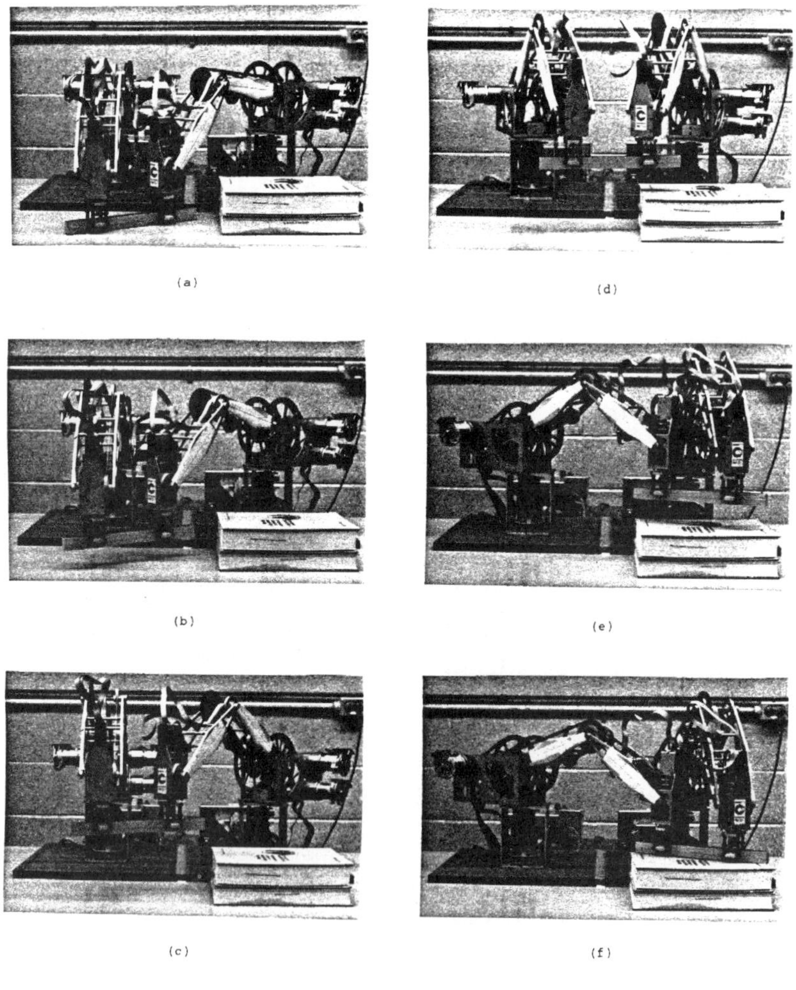

Fig.3. Task execution

APPLICATION OF THREE-DIMENSIONAL STATE OBSERVERS IN THE MANIPULATOR

Liu Bochun

Department of Automatic Control Engineering, Nanjing Institute of Technology, Nanjing, China

Abstract. In this paper, a new method is suggested for designing three-dimensional observers, namely, load disturbance, speed, and integrating component output observers for a speed regulator. These observers are directly applied to a manipulator control system. This paper presents the design procedures and experimental circuits of the observers. The experimental results show not only that the design of the observers is correct, but also that the results are satisfactory from a practical engineering point of view.

INTRODUCTION

Since D. G. Luenberger [1] introduced the theory of the state observer, it has gained wide application in control systems. In China, Long [2] and his colleagues proposed a design method for load disturbance and speed observers and applied these observers in a speed regulator system with Silicon Controlled Rectifier (SCR) double closed loops. This experiment showed improved results over previous approaches. This method of constructing two-dimensional observers must utilize the voltage across a d.c. motor armature. However, this voltage is generated by an SCR device and is generally noisy, especially as the waveform becomes noncontinuous. Therefore, the observer output signal will inevitably contain measurement noise. In an actual system, in order to make use of such a state variable, it is necessary to attach a twin T network filter. This filter circuit increases the complexity of the system and the time delay.

In this paper, a new method is suggested for designing the three-dimensional observers. This method eliminates the use of the voltage on the d.c. motor armature. A new state variable, the input voltage is applied instead. This method is advantageous in two respects. First, this method is practically quite convenient and simple. Secondly, because the input voltage is stabilized by the d.c. stabilized voltage supply, the output of the three-dimensional observers also possesses high stability. The output waveforms of these observers are completely smooth. These advantages are demonstrated by experimental results.

In this paper, the design procedures and experimental circuits of these observers are given. Also, the observers are applied to a manipulator control system. Experimental data obtained with the manipulators coincided with the theoretical analysis.

A speed control system with a Pulse-Width Modulated (PWM) amplifier was applied to the internal loop of the manipulator control system. In this paper, we show how the three-dimensional state observers facilitate more effective control in the manipulator control system.

SPEED CONTROL SYSTEM WITH PWM

The block diagram of the speed control with PWM is shown in Fig. 1.

Since the output waveform of the tacho-generator in Fig. 1 is inferior in quality, we therefore replace the tacho-generator by a speed observer.

The Pulse-Width Modulated amplifier consists of four distinct elements: comparator, trigonometric waveform generator, double-action delay circuit, and power amplifier with PWM. The block diagram of the PWM amplifier is shown in Fig. 2.

A complete circuit of the Pulse-Width Modulated power amplifier is shown in Fig. 3. A double-end control in the diagonals Ax and Bx is used in the four arms of the bridge in Fig. 3. Note in particular that the circuit in Fig. 2 provides logical safeguard; thus, the reliability and safety of the PWM amplifier in Fig. 3 is improved significantly.

We now choose some experimental values for the various parameters: d.c. motor S--569, 160W, 2.2A, 110V, 3300 RPM, $R=4.32\Omega$, $L=35mH$, $C_e=0.029$, $C_m=0.028$, $T_m=0.18s$, $K_{PWM}=38$, $\beta=1V/2.2A=0.45V/A$, $T_t=8ms$, $T_{oi} = 2ms$, $T_{on}= 4.7ms$, $\alpha=0.003$, $T_{\Sigma i}= T_{oi}+ 0.5T = 2.3ms$,

$T_{\Sigma n}=2T_{\Sigma i}+T_{on}=9.3\text{ms}$

The transfer function of the current loop in Fig. 1 is given by

$$W_L(S)=I(S)/U_{gi}(S)=1/\beta(2T_{\Sigma i}S+1) \quad (1)$$

Quite clearly, the current loop is simply considered as an one-order inertia loop.

A state feedback structure of the speed control system is shown in Fig. 4. Clearly, the system in Fig. 4. is a single-input single-output system. In addition. the open-loop transfer function contains no zeros. In this paper, the state feedback coefficients for this system may be obtained by the optimal frequency-domain solution approach[3].

In terms of Fig. 4, we have seen that the transfer function of the plant $G(s)$ is

$$G(s)=\frac{R/\beta C_eT_m}{S^2(T_{\Sigma n}S+1)} \quad (2)$$

while the return- path transfer function $H(s)$ is

$$H(s)=K_1+\frac{C_eT_mS}{R}K_2+\frac{\beta C_eT_m(T_{\Sigma n}S+1)S}{R}K_3 \quad (3)$$

The optimal frequency-domain solution approach is based on the known $G(S)$ and the form of $H(S)$, as given in Eq. (3). And determine an exact expression for $1+G(s)H(s)$. Equate the exact and approximate expressions for $1+G(s)H(s)$ and solve for the unknown values of K. Since $G(s)$ in Eq. (2) has three more poles than zeros, the appropriate Butterworth filter is third-order. Therefore, the approximate expression for $1+G(s)H(s)$ can be accomplished by multiplying $G(s)$ by a Butterworth polynomial with a characteristic frequency ω_c of the order of the pole-zero excess of $G(s)$.

$$1+G(S)H(S)=\frac{R/\beta C_eT_m}{S^2(T_{\Sigma n}S+1)}\left[\left(\frac{S}{\omega_c}\right)^3+2\left(\frac{S}{\omega_c}\right)^2+2\frac{S}{\omega_c}+1\right]$$

$$=\frac{(R/\beta C_eT_m\omega_c^3)(S^3+2\omega_cS^2+2\omega_c^2S+\omega_c^3)}{S^2(T_{\Sigma n}S+1)} \quad (4)$$

The exact expression for $1+G(s)H(s)$ is

$$1+G(S)H(S)=\frac{T_{\Sigma n}S^3+(1+T_{\Sigma n}K_3)S^2+(K_2/\beta+K_3)S+K_1R/\beta C_eT_m}{S^2(T_{\Sigma n}S+1)} \quad (5)$$

The exact expression for $1+G(s)H(s)$ is equal to the approximate expression. We obtain the state feedback coefficients.

$$K=[K_1\ \ K_2\ \ K_3]=[1\ \ 24.35\ \ 9.03] \quad (6)$$

A fact that should be stressed in regard to the use of optimal frequency-domain solution method is that the technique works only if the two equations for the exact and approximate expressions for $1+G(s)H(s)$ have the same form. The exact and approximate expression for $1+G(s)H(s)$ must agree in form in both numerator and denominator. Note that these conditions can be held in this paper.

In order to realize the return-path transfer function $H(s)$ in the equation (3), it is necessary that the speed regulator in Fig. 4. is obtained by using the "proportional + integral" controllers and the "tachometrical + differential" feedbacks. As shown in Fig. 5.

Comparing the structure in Fig. 4. with the system in Fig. 5. we have

$$K_1+\frac{C_eT_m}{R}(K_2+\beta K_3)S+\frac{\beta C_eT_mT_{\Sigma n}}{R}K_3S^2=\frac{K_na}{\tau_n}+\frac{K_na}{\tau_n}(\tau_n+\tau_d)S+S+aK_n\tau_dS^2 \quad (7)$$

The following answer is obtained:

$$K_n=11;\qquad \tau_n=0.033S;\qquad \tau_d=0.0014S\,. \quad (8)$$

In a practice control system, this as a fortunate state of affairs, as the construction of the "tachometrical + differential" feedbacks to realize it can be considered as separate problems. They satisfy the separation property. The tachometrical feedback only has been used here in order to keep the derivation on an intuitive level. As shown in Fig. 6. The structure of the speed control system in Fig. 6. can be equivalently changed into Fig. 7.

In accordance with Fig. 7, the state equation and the output equation of the speed control system can be put in a standard form as

$$\dot{x}=Ax+Bu+B_0I_L \quad (9)$$

$$y=cx$$

The dimensions of x, u and y are n, r and P, respectively.

$$\text{where}\quad A=\begin{bmatrix}0 & \frac{R}{C_eT_m} & 0\\ -\frac{ak_n}{\beta T_{\Sigma n}} & -\frac{1}{T_{\Sigma n}} & \frac{1}{\beta T_{\Sigma n}}\\ -\frac{ak_n}{\tau_n} & 0 & 0\end{bmatrix},$$

$$B=\begin{bmatrix}0\\ \frac{k_n}{\beta T_{\Sigma n}}\\ \frac{k_n}{\tau_n}\end{bmatrix},\qquad B_0=\begin{bmatrix}-\frac{R}{C_eT_m}\\ 0\\ 0\end{bmatrix},$$

$$C=\begin{bmatrix}1&0&0\\0&1&0\\0&0&1\end{bmatrix},\qquad x=\begin{bmatrix}n\\ I\\ V\end{bmatrix}.$$

The constant load disturbance equation is given by

$$\dot{I}_L=0 \quad (10)$$

REDUCED ORDER OBSERVERS

It is easily included by augmenting the original system (9). (10) as follows

$$\begin{bmatrix}\dot{I}_L\\ \dot{x}\end{bmatrix}=\begin{bmatrix}0&0\\B_0&A\end{bmatrix}\begin{bmatrix}I_L\\ x\end{bmatrix}+\begin{bmatrix}0\\B\end{bmatrix}u \quad (11)$$

$$y=[0\quad I]\begin{bmatrix}I_L\\ x\end{bmatrix} \quad (12)$$

Note that the system (11), (12) is completely observable. Now we shall consider the reduced-order observers of the system. Equations (11) and (12) can be put in a partitioned form. We lose no generality in choosing this form, since any system can be put it with a simple state transformation.

$$\begin{bmatrix}\dot{x}_1\\ \dot{x}_2\end{bmatrix}=\begin{bmatrix}A_{11}&A_{12}\\A_{21}&A_{22}\end{bmatrix}\begin{bmatrix}x_1\\ x_2\end{bmatrix}+\begin{bmatrix}B_1\\B_2\end{bmatrix}u \quad (13)$$

$$y=[0\quad I]\begin{bmatrix}x_1\\ x_2\end{bmatrix}=x_2 \quad (14)$$

Where the x_2 are available as outputs for direct measurement. There is clearly no need to estimate the m states. We may estimate only the $(n-m)$ states which make up x_1, as these are not directly avilable for measurement. This observer is called an $(n-m)$ dimensional reduced-order observers.

By manipulating the equation in (13), x_1 may be viewed as the states of an $(n-m)$ dimensional subsystem.

$$\dot{x}_1=A_{11}x_1+A_{12}x_2+B_1u \quad (15)$$

Similarly, the $A_{21}x_1$ in (13) may be expressed as

$$A_{21}x_1 = \dot{y} - A_{22}y - B_2u \tag{16}$$

In the equation (15), we denote the estimate of x_1 by $\hat{x}_1$. And Eq.(15) may be written as

$$\dot{\hat{x}}_1 = (A_{11} + LA_{21})\hat{x}_1 + A_{12}x_2 + B_1u - LA_{21}\hat{x}_1 \tag{17}$$

The poles of (17) can be placed in any desired locations by choosing L appropriately. Substituting (16) into the equation (17), we have

$$\dot{\hat{x}}_1 = (A_{11} + LA_{21})\hat{x}_1 + (B_1 + LB_2)u + (A_{12} + LA_{22})y - L\dot{y} \tag{18}$$

The only apparent difficulty in imlementing the observer (18) is that differentiation of the output y is required. This can easily be avoided, however, by redefining the state of the observer to be

$$\hat{x}_1 = W - Ly \tag{19}$$

Substituting (19) into (18), we find

$$\dot{W} = (A_{11} + LA_{21})W + (B_1 + LB_2)u + [A_{12} + LA_{22} - (A_{11} + LA_{21}L)]y \tag{20}$$

Finally, equations (19) and (20) denote a reduced-order observers. A method deriving a reduced-order observers in this paper is relatively more simple than that in [5].

DESIGN FOR THE LOAD DISTURBANCE OBSERVER

By use of Eq. (19) and (20), choosing separately these matricex as

$$L = [L_{11}\ 0\ 0],\quad x_1 = I_L,\quad x_2 = [n\ \ I\ \ V]^T,\quad A_{11} = 0,$$
$$A_{12} = [0\ 0\ 0],\quad A_{21} = B_0,\quad A_{22} = A,\quad B_1 = 0,\quad B_2 = B$$

Substituting the above these matricex into the Eq. (19) and (20), we have

$$\dot{W}_1 = -\frac{RL_{11}}{C_eT_m}W_1 + \frac{RL_{11}^2}{C_eT_m}n + \frac{RL_{11}}{C_eT_m}I \tag{21}$$

$$\hat{I}_L = W_1 - L_{11}n \tag{22}$$

Equations (21) and (22) denote a load disturbance observer. Let us choose the pole of the load disturbance observer as

$$\lambda_{11} = -\frac{R}{C_eT_m}L_{11} = -\frac{1}{\tau_1} \tag{23}$$

DESIGN FOR THE SPEED OBSERVER

By use of Eq. (19) and (20), choosing separately these matricex as

$$L = [0\ \ L_{12}\ \ 0],\ x_1 = n,\ x_2 = [I_L\ I\ V]^T,\ A_{11} = 0,$$
$$A_{12} = \left[-\frac{R}{C_eT_m}\ \ \frac{R}{C_eT_m}\ \ 0\right],$$

$$A_{21} = \begin{bmatrix} 0 \\ -\frac{ak_n}{\beta T_{\Sigma n}} \\ -\frac{ak_n}{\tau_n} \end{bmatrix},\quad A_{22} = \begin{bmatrix} 0 & 0 & 0 \\ 0 & -\frac{1}{T_{\Sigma n}} & \frac{1}{\beta T_{\Sigma n}} \\ 0 & 0 & 0 \end{bmatrix},$$

$$B_1 = 0,\quad B_2 = B.$$

Substituting the above these matricex into the Eq. (19) and (20), we find

$$\dot{W}_2 = -\frac{ak_nL_{12}}{\beta T_{\Sigma n}}W_2 + \left[\frac{R}{C_eT_m} - \frac{L_{12}}{T_{\Sigma n}} + \frac{ak_n}{\beta T_{\Sigma n}}L_{12}^2\right]I + \frac{L_{12}}{\beta T_{\Sigma n}}V - \frac{R}{C_eT_m}I_L + \frac{k_nL_{12}}{\beta T_{\Sigma n}}u \tag{24}$$

$$\hat{n} = W_2 - L_{12}I \tag{25}$$

Equations (24) and (25) denote a speed observer. Choosing the pole of the speed observer as

$$\lambda_{12} = -\frac{ak_nL_{12}}{\beta T_{\Sigma n}} = -\frac{1}{\tau_2} \tag{26}$$

DESIGN FOR INTEGRATING COMPONENT OUTPUT OBSERVER OF THE SPEED REGULATOR

The state variable V in Fig. 7 is not directly available for measurement.

We note that it is neccessary to construct the speed observer in Eq. (24). Therefore, it is neccessary to estimate V with an observer. Choosing separately these matricex as

$$L = [0\ 0\ -L_{13}],\ x_1 = V,\ x_2 = [I_L\ n\ I]^T,\ A_{11} = 0,$$
$$A_{12} = \left[0\ \ -\frac{ak_n}{\tau_n}\ \ 0\right],\quad A_{21} = \left[0\ \ 0\ \ \frac{1}{\beta T_{\Sigma n}}\right]^T,$$

$$A_{22} = \begin{bmatrix} 0 & 0 & 0 \\ -\frac{R}{C_eT_m} & 0 & \frac{R}{C_eT_m} \\ 0 & -\frac{ak_n}{\beta T_{\Sigma n}} & -\frac{1}{T_{\Sigma n}} \end{bmatrix},$$

$$B_1 = \frac{k_n}{\tau_n}\qquad B_2 = \begin{bmatrix} 0 \\ 0 \\ \frac{k_n}{\beta T_{\Sigma n}} \end{bmatrix}$$

The last of these two equations was obtained and then substiuting the above these matricex into the equations (19) and (20).

$$\dot{W}_3 = -\frac{L_{13}}{\beta T_{\Sigma n}}W_3 + \left(\frac{ak_n}{\beta T_{\Sigma n}}L_{13} - \frac{ak_n}{\tau_n}\right)n + \left(\frac{L_{13}}{T_{\Sigma n}} - \frac{L_{13}^2}{\beta T_{\Sigma n}}\right)I + \left(\frac{k_n}{\tau_n} - \frac{k_nL_{13}}{\beta T_{\Sigma n}}\right)u \tag{27}$$

$$\hat{V} = W_3 + L_{13}I \tag{28}$$

Choosing the pole of the observer $\hat{V}$ as

$$\lambda_{13} = -\frac{L_{13}}{\beta T_{\Sigma n}} = -\frac{1}{\tau_3} \tag{29}$$

COMPOSITE DESIGN OF THREE ORDER OBSERVERS OF $\hat{I}_L$, $\hat{n}$ AND $\hat{V}$

We now note that the three-order observers of $\hat{I}_L$, $\hat{n}$ and $\hat{V}$ can be combined into a composite observer in Fig. 8

As state above, we have seen that the three-order observers of $\hat{I}_L$, $\hat{n}$ and $\hat{V}$ are separately developed by means of the equations(19)and (20) deriving a reduced order observers. Based on these conditions, it would be an easy matter to construct a composite observer in Fig. 8.

The last was obtained by substituting the equation (25) into the equation (21).

$$\dot{W}_1 = -\frac{RL_{11}}{C_eT_m}W_1 + \frac{RL_{11}^2}{C_eT_m}W_2 + \left(\frac{RL_{111}}{C_eT_m} - \frac{RL_{11}^2L_{12}}{C_eT_m}\right)I \tag{30}$$

The last was obtained by substiuting the equations (22) and (28) into the equation (24).

$$\dot{W}_2 = -\frac{R}{C_eT_m}W_1 + \left(\frac{R}{C_eT_m}L_{11} - \frac{ak_nL_{12}}{\beta T_{\Sigma n}}\right)W_2$$

$$+\frac{L_{12}}{\beta T_{\Sigma n}}W_3+\Big(\frac{R}{C_eT_m}-\frac{L_{12}}{T_{\Sigma n}}+\frac{\alpha k_n L_{12}^2}{\beta T_{\Sigma n}}$$

$$-\frac{RL_{11}L_{12}}{C_eT_m}+\frac{L_{12}L_{13}}{\beta T_{\Sigma n}}\Big)I+\frac{k_nL_{12}}{\beta T_{\Sigma n}}u \qquad (31)$$

The last was obtained by substituting the equation (25) into the equation (27).

$$\dot{W}_3=\Big(\frac{\alpha k_n L_{13}}{\beta T_{\Sigma n}}-\frac{\alpha k_n}{\tau_n}\Big)W_2-\frac{L_{13}}{\beta T_{\Sigma n}}W_3$$

$$+\Big(\frac{L_{13}}{T_{\Sigma n}}-\frac{L_{13}^2}{\beta T_{\Sigma n}}+\frac{\alpha k_n}{\tau_n}L_{12}-\frac{\alpha k_n L_{12}L_{13}}{\beta T_{\Sigma n}}\Big)I$$

$$+\Big(\frac{k_n}{\tau_n}-\frac{k_nL_{13}}{\beta T_{\Sigma n}}\Big)u \qquad (32)$$

The combination of Eqs. (30)、(31) and (32) can be put in a standard form as

$$\dot{W}=A_rW+B_ry_r \qquad (33)$$

$$x_1=C_rW+D_ry_r \qquad (34)$$

where $\dot{W}=[\dot{W}_1 \dot{W}_3 \dot{W}_3]^T$, $y_r=[I \quad u]^T$.

$$A_r=\begin{pmatrix} -\frac{RL_{11}}{C_eT_m} & \frac{RL_{11}^2}{C_eT_m} & 0 \\ -\frac{R}{C_eT_m} & \frac{RL_{11}}{C_eT_m}-\frac{\alpha k_n}{\beta T_{\Sigma n}}L_{12} & \frac{L_{12}}{\beta T_{\Sigma n}} \\ 0 & \frac{\alpha k_n L_{13}}{\beta T_{\Sigma n}}-\frac{\alpha k_n}{\tau_n} & -\frac{L_{13}}{\beta T_{\Sigma n}} \end{pmatrix} \qquad (35)$$

$$B_r=\begin{pmatrix} \frac{RL_{11}}{C_eT_m}-\frac{RL_{11}^2L_{12}}{C_eT_m} & 0 \\ -\frac{R}{C_eT_m}-\frac{L_{12}}{T_{\Sigma n}}+\frac{k_nL_{12}^2}{\beta T_{\Sigma n}}-\frac{RL_{11}L_{12}}{C_eT_m}+\frac{L_{12}L_{13}}{\beta T_{\Sigma n}} & \frac{k_nL_{12}}{\beta T_{\Sigma n}} \\ \frac{L_{13}}{T_{\Sigma n}}-\frac{L_{13}^2}{\beta T_{\Sigma n}}+\frac{\alpha k_n}{\tau_n}L_{12}-\frac{\alpha k_n L_{12}L_{13}}{\beta T_{\Sigma n}} & \frac{k_n}{\tau_n}-\frac{k_nL_{13}}{\beta T_{\Sigma n}} \end{pmatrix}$$

(36)

$$C_r=\begin{pmatrix} 1 & -L_{11} & 0 \\ 0 & 1 & 0 \\ 0 & 0 & 1 \end{pmatrix}, \qquad D_r=\begin{pmatrix} L_{11}L_{12} & 0 \\ -L_{12} & 0 \\ L_{13} & 0 \end{pmatrix} \qquad (37)$$

The choice of observer poles is completely arbitrary in principle. In practice, however, it is very important that these observers of $\hat{I}_L$、$\hat{n}$ and $\widehat{V}$ ought to be stable and able to be physically realized. To do so requires that all the poles λ_{11}, λ_{12} and λ_{13} of these observers have negative real parts.According to (23)、(26) and (29), the uumbers of L_{11}、L_{12} and L_{13} must be positive real ones. Generally, it is necessary to make the transient response of these observers faster than the original system. Accordingly, the pole values of these observers ought to be greater than system itself by a factor of 3~5. Another practical consideration is that the pole values of these observers ought not to be too great. The pass-band of these abservers will be widened, and the output of these observers will inevitably contain a large amount of the noise interference if the observer is too fast. The value of a dominant pole in a original system is equal to −30.3. Therefore, in this paper, the pole values of these observers are chosen as

$$\lambda_{11}=\lambda_{12}=\lambda_{13}=-100$$

then $L_{11}=0.12$, $L_{12}=12.68$ and $L_{13}=0.42$

TEST RESULTS

We now note, the transient response of the original system with PWM amplifiers in response to a unit step input can be unsatisfactory for an overshoot of 5%. Figure 9(a) shows the dynamic behaviour of the original speed control system. The dynamic response with the construction of an observers $\hat{I}_L$ $\hat{n}$ and $\widehat{V}$ to realize speed control system can be satisfactory, as shown in Fig. 9(b)

The impact drop of the original system is shown in Fig. 10 (a). The impact drop of the system with the observers is shown in Fig. 10 (b). Clearly, the impact drop is obtained successfully improved.

REFERENCES

1. Luenberger D G, Observers of multivariable system, IEEE Trans on automatic control. Vol. AC-11 No. 2 pp190~197 (1966)
2. Gan Long. Xiong Guangleng. Liang Dequan: Engineering design method for a two-order state observer in electrical drive systems. Acta Automatica sinica, No. 2 (1982)
3. Schultz D G and Melsa J L, State functions and linear control system. New York (1967)
4. Kalman R E, When is a linear control system optimal? Trans ASME Basic Eng Vol. 11. No. 1 (1966) PP51−60.
5. Fortmann T E and Hitz K L, An introduction to linear control system. Marcel Dekker INC (1977)
6. Morecki A and Kedzioz K, Theory and practice of Robots and manipulators. (1977)

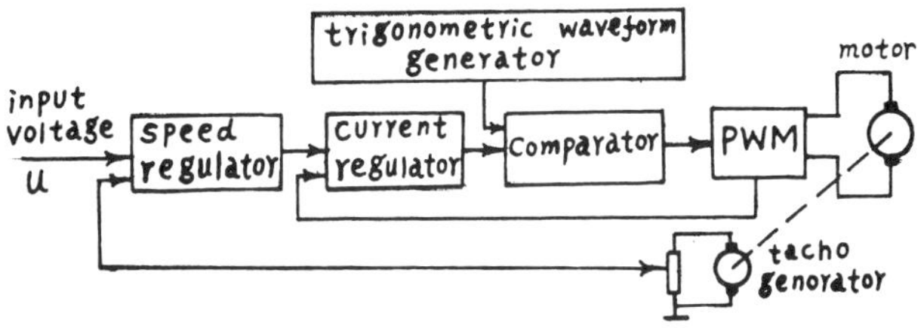

Fig. 1. Block diagram of speed control system with PWM

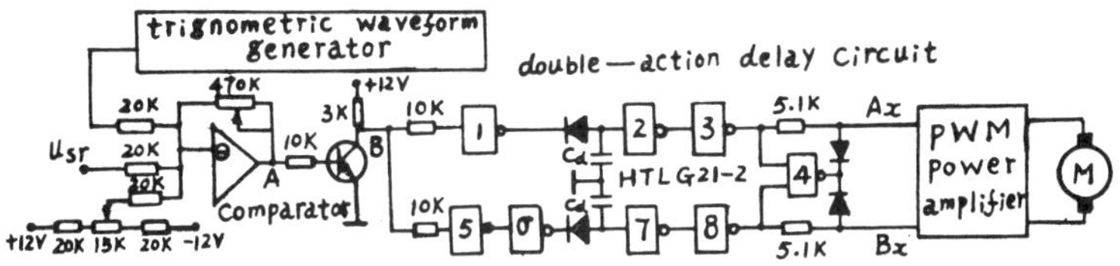

Fig. 2. Block diagram of PWM amplifier

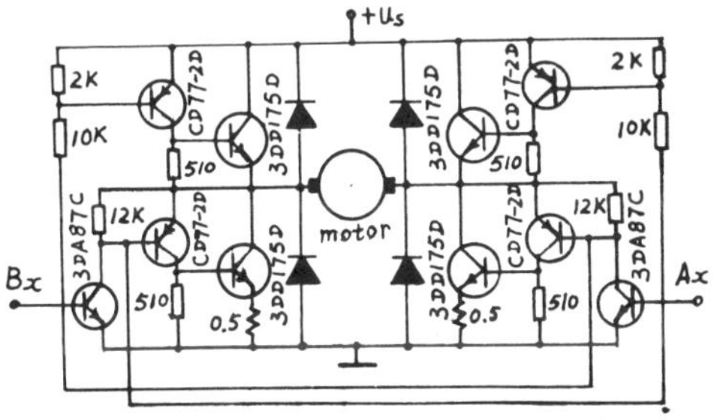

Fig. 3. Circuit of PWM power amplifier

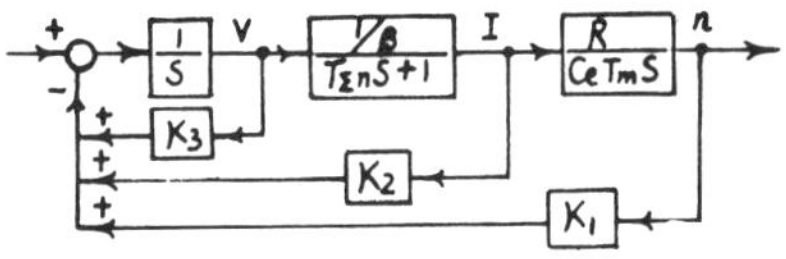

Fig. 4. State feedback structure

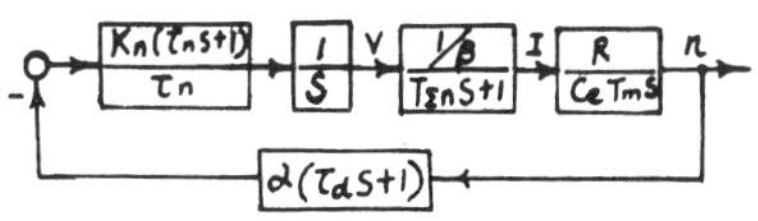

Fig. 5. System structure

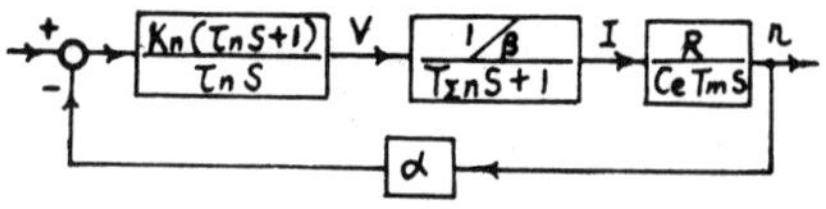

Fig. 6. Speed control system

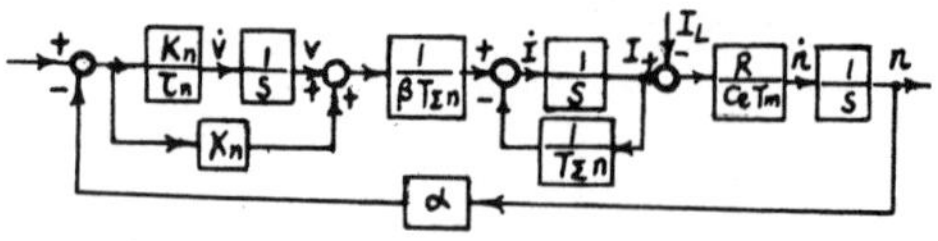

Fig. 7. Equivalent structure

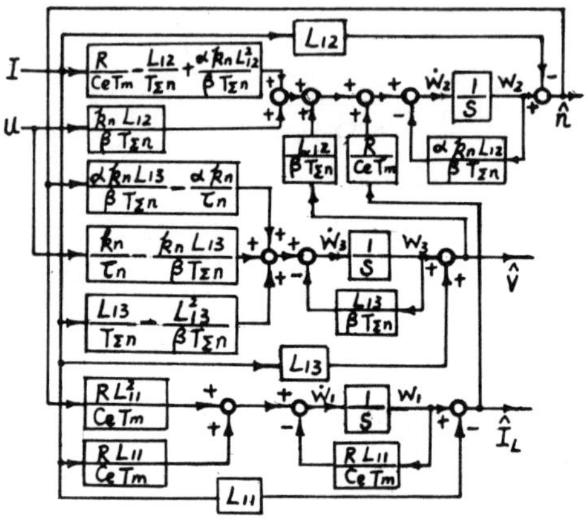

Fig. 8. Composite observer

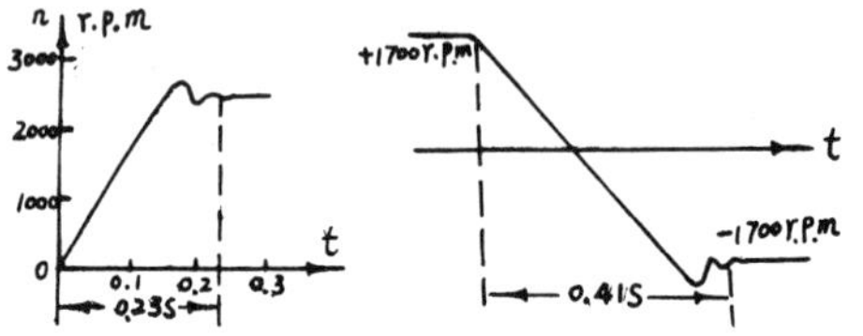

(a) original system

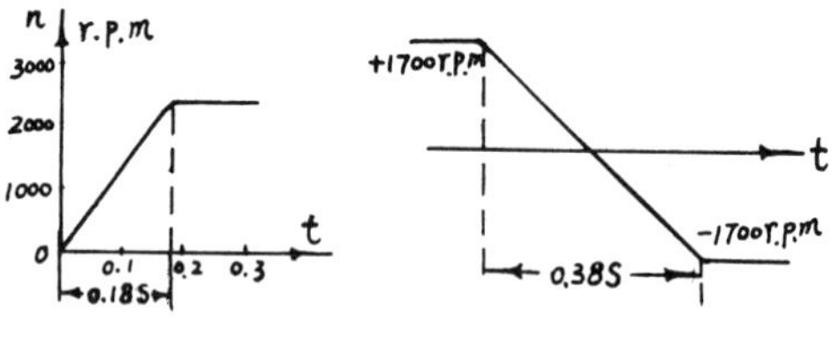

(b) system with the observers
Fig. 9. Dynamic responses of the speed control system

(a) Original system (b) System with the observers
Fig. 10. Impact drop waveform

TRACKING CONTROLLER DESIGN FOR A ROBOTIC MANIPULATOR

S. C. Won and D. H. Chyung

Department of Electrical and Computer Engineering, University of Iowa, IA 52242, USA

Abstract. In this paper, a tracking controller design method for a dc motor driven robotic manipulator is presented. The reference input to be followed is a ramp function. The main difficulties are caused by the Coulomb frictions, disturbance due to the gravitational pull and the rather stiff spring effect of the link between the drive motor and the manipulator arm. The proposed design method is based on the feeding back of the delayed values of the observable variables and augmenting the system by additional integrators.

Keywords. Tracking control; manipulator; stiff nonlinearities; delayed feedback control; system augmentation; limit cycles

INTRODUCTION

The problem of designing a feedback controller for a dc motor driven robotic manipulator is investigated. It is desired that the system output follows a ramp reference input with zero steady state error. This kind of tracking is essential for an accurate control of a manipulator, especially when more than one manipulator is used in a cooperative manner. Although a tracking controller can be designed routinely for linear systems, there are three major problems with dc motor driven manipulators. The first is the fact that not all the state variables are measurable for the feedback control. The stiff spring effect, which is due to the high ratio gear train and the linking shaft between the motor and the manipulator arm, increases the order of the control system by 2. This, in turn, requires feeding back the angle and velocity variables of the motor shaft as well as those of the load. The angle at the motor shaft, however, is difficult to measure, for it must make multiple revolutions due to the high gear ratio. The second is the stiff system nonlinearities which are mainly due to the Coulomb frictions. The Coulomb friction, in certain cases, causes the system response to go into a limit cycle. In order to eliminate the limit cycle, the Describing function method is applied. Lastly, some manipulator joints receive external disturbances because of the gravitational pull. This disturbance must be canceled out to force the system response to follow the reference input. The proposed method is based on augmenting the system with additional integrators at the output side and applying the delayed feedback method to the augmented system.

The delayed feedback parameters are obtained by the following method. First, the nonlinear part is removed from the system. Then a delayed feedback controller is designed for the resulting linear system such that the closed loop system is sufficiently stable. Finally, the feedback parameters are readjusted using the Describing function method so that potential limit cycles are eliminated from the closed loop nonlinear system.

The Coulomb friction at the motor shaft has a benificial effect as well. Without the friction, it is necessary to supply a constant current to the motor to cancel the stationary load torque due to the gravitational pull. Otherwise, a mechanical brake must be employed. If the Coulomb friction is larger than the gravitational pull when it is reflected at the motor shaft, then the position will be maintained even if the current to the motor is turned off. This feature is especially attractive in small manipulators.

The controller presented in this paper is for the particular manipulator which is under investigation in our laboratory. However, the same scheme can be applied to other manipulators as well.

MODEL

The manipulator under consideration can be represented by the diagram shown in Fig. 1.

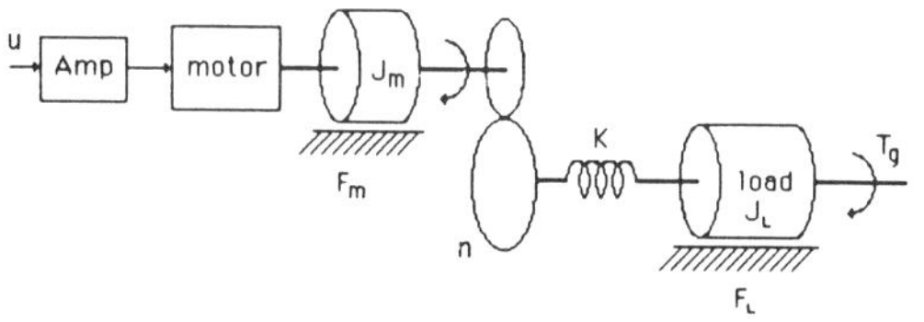

Fig. 1. Robot drive system.

In the figure, F represents the Coulomb friction and T_g represents the gravitational pull. The latter is regarded as an external disturbance to the system. u is the control input to the amplifier for activating the motor.

The above system can be modeled by the block diagram shown in Fig. 2.

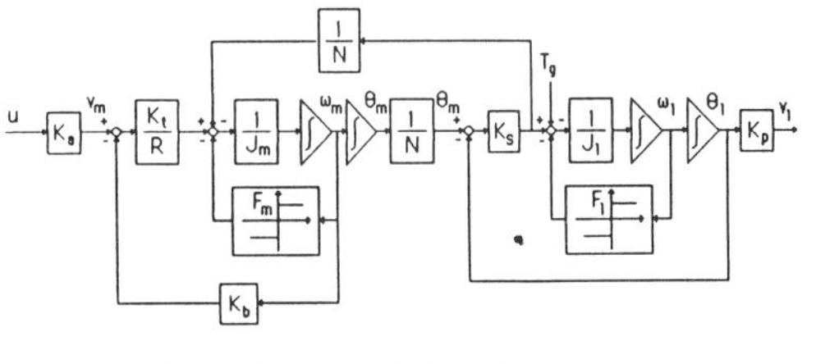

Fig. 2. Block diagram of the robot driv

The motor inductance and the gear backlash is found, experimentally, to be insignificant. Hence they are ignored in the block diagram. The parameters for the system under consideration are:

u : input voltage to the amplifier
v_m : the voltage applied to the motor
T_g : disturbance torque due to the gravitational pull
$K_p=6.5$: potentiometer constant
$R=8.19$: motor winding resistance
$K_b=0.0388$: motor back emf constant
$K_t=0.0388$: motor torque constant
$K_a=6$: power amplifier constant
$N=436.7$: gear ratio
$K_s=3000$: spring constant
$F_m=0.004264$: motor Coulomb friction
$J_m=0.000004$: motor inertia
$F_l=0.207$: load side Coulomb friction
$J_l=1.8$: norminal load inertia
v_l : potentiometer output for position measurement
θ_l : load position angle
ω_l : load angular velocity
θ_m : motor shaft angle
ω_m : motor shaft angular velocity
T_g : torque due to gravitational pull

Let

$x_1=v_l$: potentiometer output for the angular position of the load
$x_2=\omega_l$: load angular velocity
$x_3=\theta_m/N$: motor position divided by the gear ratio
$x_4=\omega_m/N$: motor velocity divided by the gear ratio
$y_1=x_1$: observable variable
$y_2=x_2$: observable variable
$x=(x_1,x_2,x_3,x_4)^t$
$y=(y_1,y_2)^t$

Then the system equation in vector form is given by

$$\dot{x} = \begin{bmatrix} 0 & 6.5 & 0 & 0 \\ -256 & 0 & 1167 & 0 \\ 0 & 0 & 0 & 1 \\ 605 & 0 & -3933 & -46 \end{bmatrix} x - \begin{bmatrix} 0 \\ .556 f_l(x_2) \\ 0 \\ 573 f_m(x_4) \end{bmatrix} + \begin{bmatrix} 0 \\ 0 \\ 0 \\ 16.27 \end{bmatrix} u + \begin{bmatrix} 0 \\ .556 T_g \\ 0 \\ 0 \end{bmatrix} \quad (S)$$

$$y = \begin{bmatrix} 1 & 0 & 0 & 0 \\ 0 & 1 & 0 & 0 \end{bmatrix} x$$

CONTROLLER DESIGN

Let z be the ramp reference input, and let T_g be the constant disturbance torque due to the gravitational pull. The objective is to design a feedback control u of the observable variables only such that $x_1=z$ in steady state under the constant torque disturbance T_g.

To follow a ramp reference input with zero steady state error, the system is now augmented by introducing additional state variables x_5, x_6 and observable variables y_3, y_4 which are defined by

$$\dot{x}_5 = x_6$$
$$\dot{x}_6 = x_1 - z$$
$$y_3 = x_5$$
$$y_4 = x_6$$

The variables y_3 and y_4 are observable because the state variables x_5 and x_6 are externaly introduced to the system. The augmented system by additional integrators at the output side is shown in Fig. 3.

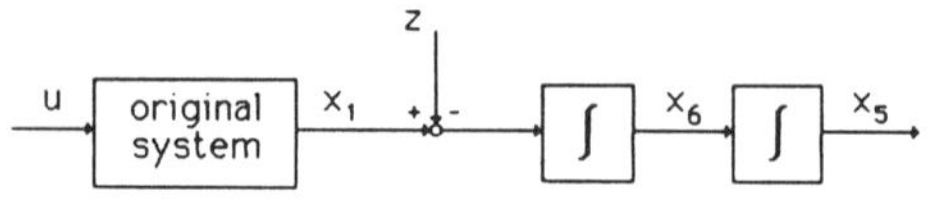

Fig. 3. Augmented system.

Let $\hat{x}=(x_1,x_2,x_3,x_4,x_5,x_6)^t$ and $\hat{y}=(y_1,y_2,y_3,y_4)^t$. Then the augmented system $(\hat{S})$ is given by

$$\dot{\hat{x}} = \begin{bmatrix} 0 & 6.5 & 0 & 0 & 0 \\ -256 & 0 & 1167 & 0 & 0 \\ 0 & 0 & 0 & 1 & 0 \\ 605 & 0 & -3933 & -46 & 0 \\ 0 & 0 & 0 & 0 & 1 \\ 1 & 0 & 0 & 0 & 0 \end{bmatrix} \hat{x} - \begin{bmatrix} 0 \\ 0.566 f_l(x_2) \\ 0 \\ 573 f_m(x_4) \\ 0 \\ 0 \end{bmatrix} + \begin{bmatrix} 0 \\ 0 \\ 0 \\ 16.27 \\ 0 \\ 0 \end{bmatrix} u + \begin{bmatrix} 0 \\ 0.566 T_g \\ 0 \\ 0 \\ 0 \\ 0 \end{bmatrix} - \begin{bmatrix} 0 \\ 0 \\ 0 \\ 0 \\ 0 \\ z \end{bmatrix} \quad (\hat{S})$$

$$\hat{y} = \begin{bmatrix} 1 & 0 & 0 & 0 & 0 & 0 \\ 0 & 1 & 0 & 0 & 0 & 0 \\ 0 & 0 & 0 & 0 & 1 & 0 \\ 0 & 0 & 0 & 0 & 0 & 1 \end{bmatrix} \hat{x}$$

In general it is necessary to feedback the entire state variables to synthesize a satisfactory controller. In the present case, however, x_3 and x_4 are not measurable and hence are not available for feedback. Therefore the delayed feedback controller of the form

$$u(t) = -(K_{10}y_1(t)+K_{11}y_1(t-h)+K_{20}y_2(t)+K_{21}y_2(t-h) + K_{22}y_2(t-2h)+K_3y_3(t)+K_4y_4(t))$$

is considered. Here h is the basic time delay. It is shown [2] that there always exists such a controller which stabilizes the given system $(\hat{S})$ when $z=0$ and $T_g=0$. Furthermore, if the controller stabilizes the system, then for the stabilizing control u, $x_1=z$ for all T_g in steady state. Therefore the task now is to find a stabilizing feedback control u for the system $(\hat{S})$. Let $T_g=0$ and $z=0$. Define the cost functional J by

$$J = \int_0^\infty (\hat{x}^T Q \hat{x} + u^2) dt$$

Then we find the feedback control

$$u(t) = -(K_{10}y_1(t)+K_{11}y_1(t-h)+K_{20}y_2(t)+K_{21}y_2(t-h) + K_{22}y_2(t-2h)+K_3y_3(t)+K_4y_4(t))$$

which minimizes the cost functional J. If the optimal controller stabilizes the system ($\hat{S}$), then the controller satisfies the given requirement. The overshoot and settling time of the output response are determined by choosing an appropriate matrix Q. If the controller causes a limit cycle, then the feedback parameters are readjusted to eliminate the limit cycle by applying the Describing function method (for example, see Ref. 1).

Let h be 0.01 then for the following weighting matrix Q

$$Q = \begin{bmatrix} 0 & 0 & 0 & 0 & 0 & 0 \\ 0 & 0 & 0 & 0 & 0 & 0 \\ 0 & 0 & 0 & 0 & 0 & 0 \\ 0 & 0 & 0 & 0 & 0 & 0 \\ 0 & 0 & 0 & 0 & 0.1 & 0 \\ 0 & 0 & 0 & 0 & 0 & 100 \end{bmatrix}$$

the delayed feedback control u is found to be

$$u(t) = -(9.92y_1(t) - 6.31y_1(t-0.01) + 0.72y_2(t) - 0.943y_2(t-0.01) + 0.38y_2(t-0.02) + 0.004y_3(t) + 9.6y_4(t))$$

The resulting feedback control system is shown in Fig. 4.

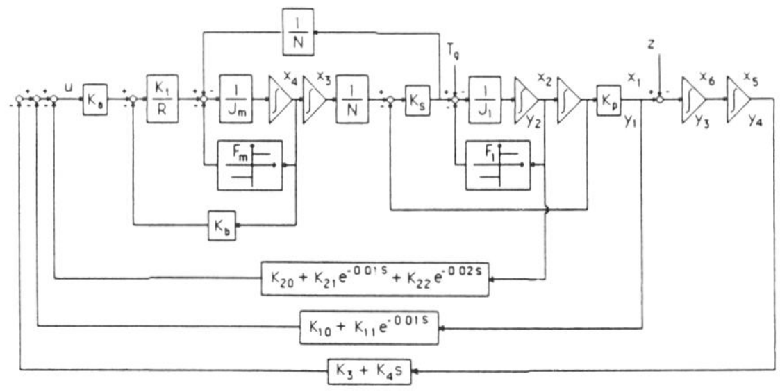

Fig. 4. Delayed feedback control system

Without going into details, the plots of Describing function and the transfer function are as shown in Fig. 5. Since there are no intersections it is unlikely that a limit cycle exists for the controller.

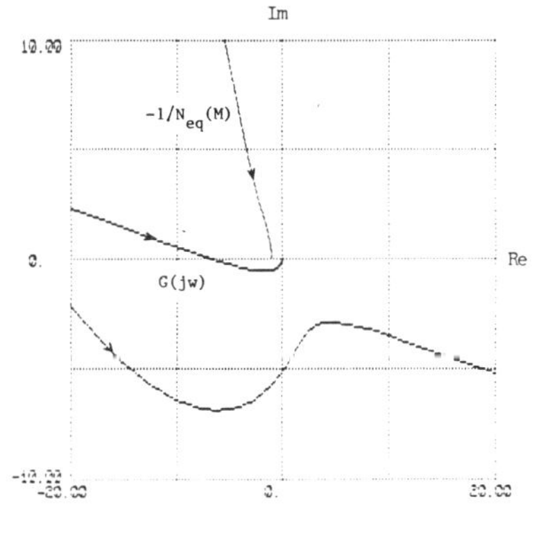

Fig.5. Loci of G($j\omega$) and -1/N_{eq} (M)

When T_g=0, the unit step input response and the ramp input response for the above controller are shown in Fig. 6.

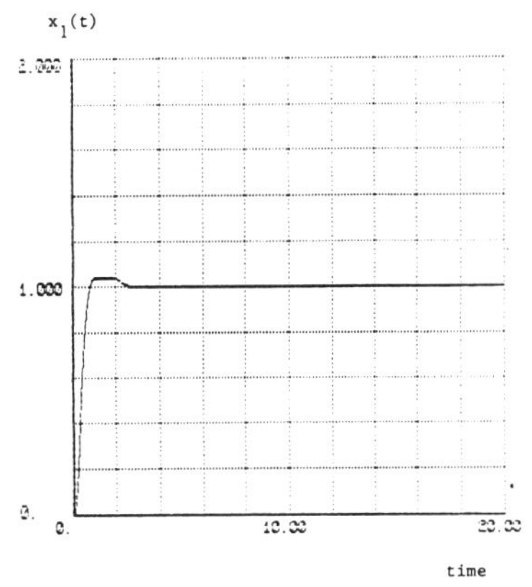

a) z(t) = 1

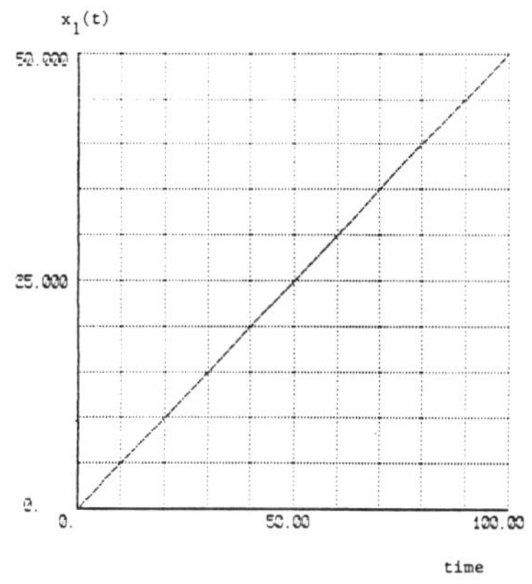

b) z(t) = 0.5t

Fig. 6. Output responses of the system

In the figure one can see that the proposed delayed feedback controller indeed gives very satisfactory results. To evaluate the effect of a nonzero gravitational disturbance, a constant disturbance of T_g=3 is applied to the system with the unit step reference at t=6. The corresponding response is shown in Fig. 7.

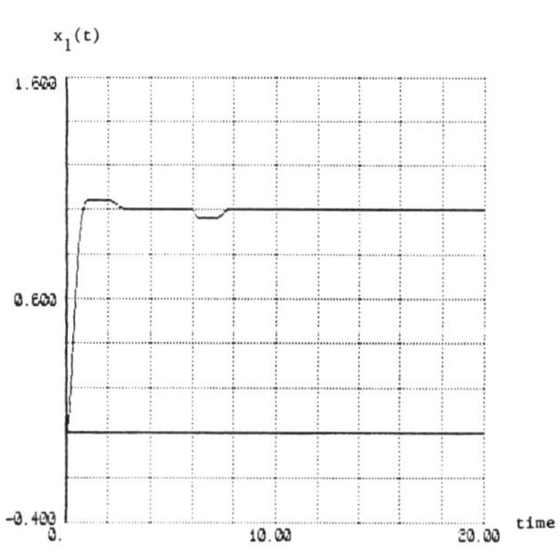

Fig.7. System response with disturbance

As expected, after an initial disturbance, the response returns to the original state. It should be noted that, in this case, the transient responses are slightly different when the manipulator is moving in the same direction as the direction of the gravitational pull and when it is moving in the opposite direction.

CONCLUSIONS

A systematic method for designing a tracking controller for a dc motor driven robotic manipulator is presented. The objective is to maintain zero steady state error for ramp reference inputs when the system contains Coulomb frictions and also when a gravitational pull is applied to the system. The method is based on augmenting the system with integrators at the output side and feeding back the delayed values of the observable variables to the augmented system. The feedback parameters are found by applying the optimal control theory without the nonlinearities, and the existence of possible limit cycles are examined by the Desribing function method.

REFERENCE

[1] S. C. Won and D. H. Chyung, "Delayed Feedback Tracking Controller for Single-input Single-output Nonlinear Systems", Proceedings of 23rd CDC, December, 1984.

[2] J. H. Kim and D. H. Chyung, "Delayed Feedback Optimal Controller", IFAC World Congress IX, Budapest, Hungary, July 1984.

[3] R.P. Paul, "Robotic manipulators: mathematics, programming and control", Chapter 5, The MIT Press, 1981.

MATHEMATICAL MODEL OF SEQUENCING HYDROELECTRIC STATIONS IN CASCADED DEVELOPMENT OF A RIVER BASIN

Li Mi-an and Wu Xianglin

Department of Automatic Control Engineering, Huazhong University of Science and Technology, Wuhan, Hubei, China

Abstract. The capacity expansion of a regional electric power system is considered in this paper. A nonlinear integer programming model with logical expressions is constructed for this problem, and a quasi-gradient screening algorithm is suggested. Case studies show that the mathematical model is adequate, and the algorithm suggested is more effective than the ones used in the past.

Keywords. Water resources; Sequencing model; Electric power system planning; Quasi-gradient screening algorithm; Capacity expansion; Decision theory.

INTRODUCTION

Sequencing or timing an engineering project is a kind of problem often encountered in China. Here a capacity expansion problem will be studied. In particular, the sequence of putting into operation hydro-electric and thermo-electric power stations will be investigated.

We assumed that the following conditions are satisfied: The location of all hydro-electric and thermo-electric power stations to be built and their characteristic parameters are given. In the process of construction of electric power stations, the network of transmission lines will be constructed so that the electric power can be supplied to assigned customers. There is no inflation, so that cost flow may be estimated at an unchanged price.

The mathematical model for sequencing projects and then the algorithm for solving the model will be discussed. A quasi-gradient screening approach is suggested. At the end, a case study is presented.

COST FLOW AND ELECTRIC POWER FLOW OF THE PLANNED ELECTRIC POWER STATIONS

The cost flow of the i-th electric power station is shown in Fig. 1, where TB_i is the time to the start of building, TG_i is the time of the start of operation beyond TB_i, and TP_i is the time interval of the trial run.The cost flow can be expressed as:

$$C_i(n) = C_i^\circ(n-TB_i), \quad (1)$$

$$C_i^\circ(n) = \sum_{j=1}^{TG_i+TP_i+1} C_{ij}\,\delta(n-j) \quad (2)$$

where $\delta(x)$ is unit delta function, i.e.

$$\delta(x) = \begin{cases} 1 & \text{for } x = 0, \\ 0 & \text{for } x \neq 0. \end{cases} \quad (3)$$

The present value in the year TB_i+TG_i+1 of cost flow of the i-th electric power station is shown as:

$$CEQ_i = \sum_{j=1}^{TG_i+TP_i+1} C_{ij}\,q^{TG_i+1-j}, \quad (4)$$

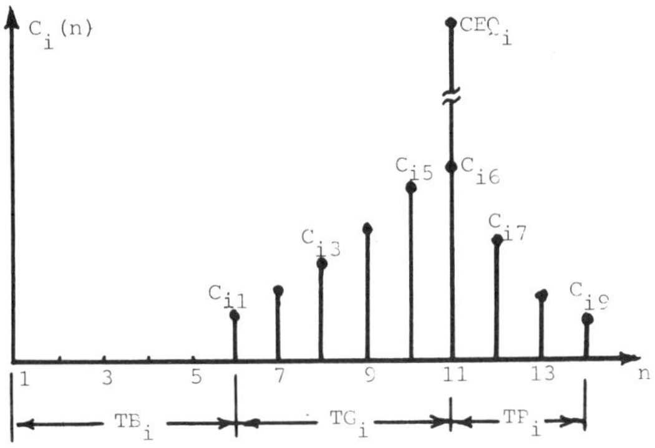

Fig. 1. Cost flow of the i-th electric power station.

where q=1+r, and r is discounted rate.

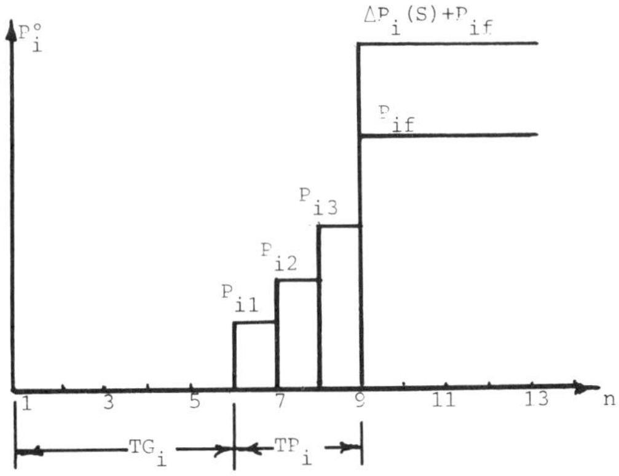

Fig. 2. Power flow of the i-th electric power station.

The power output flow of the i-th electric power station when $TB_i=0$ (Fig. 2) can be expressed as:

$$P_i^\circ(n) = \sum_{j=1}^{TP_i} P_{ij}\,\delta(n-TG_i-j)+P_{if}STP(n-TG_i-TP_i-1) \quad (5)$$

where STP is step function, i.e.

$$STP(x) = \begin{cases} 1 & \text{for } x \geq 0, \\ 0 & \text{for } x < 0, \end{cases}$$

and P_{if} is the firm power or rated power capacity. When $TB_i \neq 0$, the total power of the i-th station including both generated from and caused by itself would be

$$P_i(n) = P_i^{\circ}(n-TB_i) + \Delta P_i(S)\,STP(n-TB_i-TG_i-TP_i-1), \quad (6)$$

where $\Delta P_i(S)$ is the increment of firm power of cascaded hydro-electric power stations put into operation.

BALANCE EQUATIONS OF SUPPLY AND DEMAND

The balance condition occured in the worst hydrologic situation would be considered and attention would be focused on that the demand should be satisfied and the normal operating condition should be ensured in any worst situation. Assume that the electric power system has been operated normally before the planning period and the increases of demand are met with new power stations to be builded. For every power station, the utilization factor f_u of installed capacity is given. In any year, capacity or electric power balance equations can be expressed as

$$\sum_{i=1}^{K} f_{ui}^{-1} P_i(j) x_{ij} \geq P_{max}(j) \qquad \text{for } j \leq N \qquad (7)$$

where x_{ij} is the operating state of i-th station in the j-th year, with $x_{ij}=1$ means that the station has been operated, otherwise, $x_{ij}=0$; $P_{max}(j)$ is the maximum of average power demand within one hour in the j-th year; and N denotes the number of years for the planning period.

Electric energy balance equations are expressed as:

$$\sum_{i=1}^{K} P_i(j) x_{ij} \geq K_e D(j) \ , \qquad \text{for } 1 \leq j \leq N \qquad (8)$$

where D(j) is load average power in j-th year, and K_e is a modified factor to take care of the disagreement between the curves of the annual power delivered by stations and the load power. For the studied regional system, $K_e=1.015$. For the integer variables x_{ij} we can write down the following constraints:

$$\sum_{j=1}^{N} x_{ij} \geq 1, \qquad \text{for } 1 \leq i \leq K, \qquad (9a)$$

$$x_{ij+1} \geq x_{ij} \ , \qquad \text{for all i, j}, \qquad (9b)$$

$$TB_i = N - \sum_{j=1}^{N} x_{ij} - TG_i \qquad \text{for all i.} \qquad (9c)$$

For each station, the eariest possible time of beginning to build is TBL_i+1, so there are constraints:

$$TB_i \geq TBL_i \ , \qquad \text{for all i.} \qquad (10)$$

OBJECTIVE FUNCTION AND NET-BENEFITS OF POWER STATIONS

The benefits of an electric power station consists of the following five components:

1. B_{if}°, it is the present value of income corresponding to firm power P_{if} in the year (TB_i+TG_i+1). The final year of computing benefits is set in the year after the beginning of planning period for 70 years. B_{if}° is expressed as:

$$B_{if}^{\circ} = P_{if}\, k_t\, p\, ALF_i \ , \qquad (11)$$

$$ALF_i = q^{-TP_i}(1-q^{-SPAN_i})/(1-q^{-1}) \ , \qquad (11a)$$

$$SPAN_i = 70 - TB_i - TG_i - TP_i \ , \qquad (11b)$$

where k_t is the annual operating hours for the firm power, p is the income minus the fuel cost per KWH, and $SPAN_i$ is the number of continuous operating year of the i-th station.

2. Supplementary benefits SB_i°, it consists of the benefits due to seasonal electric energy, irrigation, flood control and navigation. If the tatal benefits per year is SB_i then

$$SB_i^{\circ} = SB_i\, ALF_i. \qquad (12)$$

3. Benefit corresponding to the increment of firm power $\Delta P_i(S)$ due to sequential interdependence is

$$\Delta B_i(S) = \Delta P_i(S) k_t\, p\, ALF_i \qquad (13)$$

4. Benefit B_{gi}° of electric energy during trial run time TP_i is expressed as

$$B_{gi}^{\circ} = \sum_{j=1}^{TP_i} P_{ij}\, k_t p q^{-j+1} \qquad (14)$$

5. Benefit loss C_{cgsi}°, resulted from that the supply exceeds the demand in m years can be estimated by

$$C_{cgsi}^{\circ} = \sum_{j=1}^{m} (P_{ij} - k_e d_{ij}) k_t p q^{-j+1} \qquad (15)$$

where j=1 is the year TG_i+TB_i+1 and d_{ij} is load demand increment blanced by the i-th station.

With respect to the final year for computing benefits, the cost CEQ_i of power station should be revised by replacement cost, therefore the equivalent cost could be expressed as:

$$RCEQ_i = CEQ_i \alpha_i \qquad (16)$$

where α_i is a modified coefficient of the i-th power station.

In some literatures, the objective function of sequencing models are the minimization of the total present value of cost without considering benefits. Such a objective function takes no regard of the advantageous condition of some hydro-electric power stations. Now we introduce the benefits terms into the objective function, so that the objective function is the maximization of the total present value of net-benefits. Net-benefits of the i-th power station can be expressed as:

$$PNB_i = [B_{if}^{\circ} + \Delta B_i(S) + SB_i^{\circ} + \Delta C_i(S) + \Delta B_{gi}^{\circ} - C_{cgso}^{\circ} - CEQ_i \alpha_i] q^{-TB_i - TG_i} \qquad (17)$$

where $\Delta C_i(S)$ is decrement of cost of the i-th station due to interdependent relation between hydro-electric power stations. With capacity expansion problems, the congestion cost should be taken into account when operating possible capacity were greater than load demand. The following formula is suggested for estimating the congestion cost:

$$\Delta C_{cgsi} = \pi C_{cgsi}^{\circ} \ , \qquad \pi > 1.$$

Then the objective function would be:

$$\max PNB = \max \sum_{i=1}^{K} [B_{if}^{\circ} + SB_i^{\circ} + \Delta C_i(S) + B_{gi}^{\circ} + \Delta B_i(S) - \Delta C_{cgsi} - CEQ_i \alpha_i] q^{-TB_i - TG_i} \qquad (18)$$

CONSTRAINTS ON TOTAL COST FLOW PATTERN

When the solution of the optimal sequence has been found, then the total cost flow can be expressed as:

$$SUM(n) = \sum_{i=1}^{K} C_i^o(n-TB_i)+C_T(n) \quad (19)$$

where $C_T(n)$ is the cost flow of common transmission lines. Budget constraints can be transformed into constraints on total cost flow pattern. Constraints on the peak of annual cost and the maximum increment of annual cost should be taken, so that we have:

$$\max_{1<n\leq N} SUMC(n) \leq (2 \sim 2.5)C_{av} \quad , \quad (20)$$

$$\max_{1\leq n\leq N+15} [SUM(n+1)-SUM(n)] \leq 1.5C_{av} \quad , \quad (21)$$

$$C_{av} = \frac{1}{N_x} \sum_{n=1}^{N_x} SUM(n) \quad , \quad (22a)$$

$$N_x = \max_{SUMC(n) \neq 0} n. \quad (22b)$$

when the optimal solution violated the constraints on total cost flow pattern, we should inform decision maker and designing group of power stations.

CHANGE OF COST AND OUTPUT POWER RESULTED SEQUENCING DECISION

Sequencing decision about hydro-electric stations in cascaded development and thermal electric power stations can result eccoomic effects in three sides: (a) the cost of the hydro-electric power station to be builded may be decreased, since the upward reservoir had been constructed; (b) the data of inflow of downward hydro-electric power stations can be varied because of the stream regulation at upward reservoirs, then the firm power of the downward stations can be increased; (c) the share of electric transmision lines and equipments amony power stations is determined by sequence to put into operation of power stations.

Let 1,2,..., m denote the codes of hydro-electric power stations and m+1, m+2,...,K denote the codes of thermoelectric power stations. The codes of hydro-electric power stations with annual regulation reservoirs is denoted by x,y, and z. Now, the vector $S=[TB_1, TB_2,...TB_K]$ is used to express sequence variables. The decrement of cost of the i-th station is expressed as

$$\Delta C_i(S)=\Delta C_i^x L_i^x(S)+\Delta C_i^y L_i^y(S)+\Delta C_i^z L_i^z(S), \quad \text{for all } i \quad (24)$$

where ΔC_i^x is the cost decrement resulted from the x-th station in the year TB_i+TG_i+1, and $L_i^x(S)$ is a logic function of S, as $L_i^x(S)=1$, the cost decrement can be implemented, otherwise $L_i^x(S)=0$, cannot be implemented, similarly for other symbols.

The increment of firm power of the i-th power station can be expressed as

$$\Delta P_{ii}(S)=\Delta P_i^x L_{pi}^x(S)+\Delta P_i^y L_{pi}^y(S)+\Delta P_i^z L_{pi}^z(S)+\Delta P_i^{xy}(S) +\Delta P_i^{yz} L_{pi}^{yz}(S)+\Delta P_i^{zx} L_{pi}^{zx}(S)+\Delta P_i^{xyz} L_{pi}^{xyz}(S) \quad , \quad (25)$$

where ΔP_i^x, ΔP_i^{xy}, and ΔP_i^{xyz} are the increment of firm power of the i-th power station, the upper indexes x, xy, and xyz show the codes of the power stations which have been builded up, $L_{pi}^x(S)$, $L_{pi}^{xy}(S)$ and $L_{pi}^{xyz}(S)$ denote corresponding logic function respectively, and similarly for other symbols. The second index i of $\Delta P_{ii}(S)$ expressed that the power increment was delivered in the year $TB_i+TG_i+TP_i+1$.

Because the i-th power station has been builded up, when x power station put into production and there exits the increment of firm power that can be expressed as:

$$\Delta P_{ix}(S)=\Delta P_i^{ix} L_{px}^i(S)+\Delta P_i^{zix} L_{px}^{zi}(S)+\Delta P_i^{yix} L_{px}^{yi}(S)+ \Delta P_i^{yzix} L_{px}^{yzi}(S) \quad (26a)$$

where the upper indexes denote the sequence of power stations i, y and z having been put into opeation. The total firm power increment is shown as:

$$\Delta P_x(S)= \sum_{i=1}^{m} \Delta P_{ix}(S) \quad . \quad (26)$$

Similarly we can write down:

$$\Delta P_y(S)= \sum_{i=1}^{m} \Delta P_{iy}(S) \quad , \quad (27)$$

$$\Delta P_z(S)= \sum_{i=1}^{m} \Delta P_{iz}(S) \quad , \quad (28)$$

where $P_{iy}(S)$ and $P_{iz}(S)$ can be expressed similarly as (26a).

There are ℓ sections of common transmission lines and equivalent cost in single year are $C^1, C^2,...C^\ell$, then the increment of transmission cost in the year TB_i+TG_i+1 can be expressed as:

$$\Delta C_i^T = \sum_{h=1}^{\ell} C^h L_{Ti}^h(S) \quad , \quad i=1,2,...,K, \quad (29)$$

where $L_{Ti}^h(S)$ is a logic function, as $L_{Ti}^h(S)=1$, it means that the cost of h section of transmission network belongs to the i-th power station.

Now, the mathematical model for sequencing hydroelectric and thermal electric power stations to be builded up in the planning period have been constructed. This is a nonlinear integer programming with logic expressions.

SCREENING FOR THE OPTIMAL SEQUENCE

The order or sequence to put into operation for K projects evidently was a permutation of K elements and the number of possible permutation was K!. As K=17, $K!=3.56\times10^{14}$. It is impossible to select the optimal sequence from such a large number of permutation by enumerable approach.

In the Permutation theory, a basic theory said:"All the n! permutations of n elements can be arranged into a sequence of permutations so that the sequence is beginning from any permutation and any other permutation in the sequence may be obtained by once interchange on two elements of the nearest permutation in the front". Any interchange can be instead of finitive interchanges on neighbouring elements, called neighbouring interchange, therefore the neighbouring interchange is the smallest difference between two permutations.

According to the definition of optimal permutation or sequence, the necessary condition of the optimal permutation can be written as:

NECESSARY CONDITION: Any interchange on one permutation leads to the objective value of that permutation decreasing or nonincreasing, this is the necessary condition of the optimal permutation.

On the basis of the necessary condition, a method for screening off non-optimal permutation and searching the optimal permutation can be constructed. A sequence P_1ABP_2, in which P_1 and P_2 are subsequence contained m_1 and m_2 power stations respectively and A and B expressed two power stations, is given. The order of A and B can be interchanged then a new sequence P_1BAP_2 will be yielded. Let $NB(P_1ABP_2)$ and $NB(P_1BAP_2)$ denote the present value of net-benefits of this two sequences, the following definition can be written:

DEFINITION: The increment of objective value obtained by once interchange in the given sequence is called quasi-gradient and denoted by $QG(P_1\bar{A}\bar{B}P_2)$, i.e.

$$QC(P_1\bar{A}\bar{B}P_2)=NB(P_1ABP_2)-NB(P_1BAP) \quad . \tag{30}$$

Sequence P_1ACBP_2 has been become sequence P_1BCAP_2 with once interchange on projects A and B. Similarly, the quasi gradient $QG(P_1\bar{A}C\bar{B}P_2)$ is expressed as:

$$QG(P_1\bar{A}C\bar{B}P_2)=NB(P_1ACBP_2)-NB(P_1BCAP_2) \quad . \tag{31}$$

Then we obtain:

COROLLARY 1: If quasi-gradient $QG(P_1\bar{A}\bar{B}P_2)>0$, then sequence P_1BAP_2 is nonoptimal, for maximizing problems, similarly if $QG(P_1\bar{A}C\bar{B}P_2)>0$, then sequence P_1BCAP_2 is non-optimal.

Total capacity and firm power of the electric power stations contained in subsequences P_1AB and P_1BA, according to the balance constraints (7) and (8) can supply load until time τ_{AB} and τ_{BA} respectively. Because P_1AB and P_1BA have equal total capacity and firm power, from the constraints (7) and (8) we have $\tau_{AB}=\tau_{BA}$. And then we can deduct a property of the quasi-gradient as follows:

PROPERTY1: Quasi-gradient $QG(G_1\bar{A}\bar{B}P_2)$ is independent of the sub-permutations P_2, for the determinate load demand increasing curves.

With respects to subsequence P_2, the beginning supplied power to electric power system is the year $\tau_{BA}+1$ which is called the deficient year of P_1AB or P_1BA and is independent of order of A and B. PNB_i expressed in (17) is really the net-benefits occured by the decisions making after the time $\tau_{AB}+1$, therefore the net-benefits occured by any determined subsequence P_2 can be canceled as calculating quasi-gradient. The above property is true. For permutation P_1ABP_2, if P_1 does not containt any station, then calculated once quasi-gradient, we could determine that the (K-2)! permutations were non-optimal. For the optimal sequence, there is a important property which is similar to Bellman's principle of optimality and can be expressed as following:

PROPERTY2: To bring away $R\geq 2$ projects at the front of the optimal permutation, forming a sequencing problem again with the remained K-R projects in the deficient year of R projects, the optimal permutation included K-R Projects of the new problem must be the sub-permutation of the original optimal permutation. And forming the sequencing problem in the beginning of planning period with R projects, the optimal permutation must be the subpermutation in the front of the original optimal permutation.

The property can be proved by reduction to absurdity. According the property of the optimal permutation, we could pose a problem to find the fore two elements of the optimal permutation. We suggested the following screening algoritrm.

QUASI-GRADIENT SCREENING ALGORITHM

A real value of the objective function can be calculated for each permutation of K elements or electric power stations or projects, where assume all permutations are feasible solutions, therefore all permutation of K elements become a partially ordered set. In that set, can obtain non-equal complete ordered subsets, and the supremum of any complete ordered subset can be found. According to Zorn's lemma, there exist the maximum elements in the partially ordered set. For the partially ordered set, the maximum elements is nonounique, otherwise for complete ordered set, unique. The maximum element is just the optimal permutation.

For K elements, the number of combinations contained two elements is C_K^2. Concerning given combination e.g. A and B, if $QG(\bar{A}\bar{B}P_2)>0$, then BAP_2 is non-optimal permutation and ABP_2 may be optimal, so we define A prior to B, or A preceded B, and denots as $A \succ B$. With this definition, K elements become a complet ordered set, denoted by P. In set P, let the index of the most precedent element be K-1 and the index of the other elements be K-2, ...,1,0, according the precedence respectively. From the necessary condition of the optimal permutation we have two corollaries:

COROLLARY 2: The element with zero index is not the first element of optimal permutation.

COROLLARY 3: The element with index K-1 is not the second element of the optimal permutation.

Generally, the first element of the optimal permutation cannot be determined according to the index number of elements. We suppose the problem to select the first two elements of the optimal permutation. The number of possible decisions is $2C_K^2$, and calculation the quasigradient for each pair of element, we can detemine that the C_K^2 decisions are non-optimal decisions. The remainded C_K^2 decisions can be divided into K-1 decision subsets, denoted by D(i), i=1,2,..., K-1, where i is the index of first elements of the decisions. Now, the possible decision set is

$$D=D(K-1) \cup D(K-2) \cup \dots \cup D(2) \cup D(1)$$

Let $C_{K-1}, C_{K-2}, \dots C_1, C_0$ express K elements and the subscript equals the index of the element in set P. The possible decisions in set D can be denoted by C_iC_{i-j}, j=1,2,...,i, i=1,2,...,K-1. For two decisions C_iC_{i-j} and $C_\ell C_{\ell-m}$, when $i=\ell$, calculate $QG(C_i\bar{C}_{i-j}\bar{C}_{i-m})$; when $i-j=\ell-m$, calculate $QG(\bar{C}_iC_{i-j}\bar{C}_\ell)$; when $i-j=\ell$, calculate $QG(\bar{C}_i\overline{C_\ell C_{\ell-m}})$; otherwise, calculate $QG(\overline{C_iC_{i-j}}\,\overline{C_\ell C_{\ell-m}})$, if $QG>0$, we define decision C_iC_{i-j} is prior to $C_\ell C_{\ell-m}$, denoted by $C_iC_{i-j} \succ C_\ell C_{\ell-m}$. According to this definition, as $QG\neq 0$, the decision set D is a complete ordered set and there exist the most prior and unique decision, which is prior to all other decisions.

THEOREM: The first two elements sub-permutation of the optimal permutation is the sub-permutation corresponding to the most prior decision.

The necessity of the theorem have been proved by mathematical induction. A permutation in which the first two elements is not the most prior decision, is non-optimal,evidently the sufficiency is true.

With above theorem as a guide we can form a screening procedure for searching the most prior decision. Within D(i) subset, there are i decisions formed as C_iX_1, C_iX_2,...,C_iX_i. Calculating $QG(C_i\bar{X}_\ell\bar{X}_j)$, for ℓ, j and i, but ℓ≠j,if QG≠0, a decision $d_i^*=C_iX_{mi}$ which is prior to other decisions in D(i) can be found. In the similar way, for i=K-1, K-2,...,2, we do so-and-so, and obtain the possible optimal decisions $d^*_{K-1}=C_{K-1}X_{mK-1}$,..., $d_2^*=C_2X_{m2}$ and $d_1=C_1C_0$. For these K-1 decisions, calculating quasi-gradient and comparing two decisions, when QG≠0, one and only one decision $C_{jK-1}X_{mjK-1}$ can be found, and this decision is just the optimal decision at the present step. The screening and searching procedure can be expressed as Fig.3. May call the diagram as a screening tree.

When QG≠0, with the above algorithm, the first two elements of the optimal permutation can be determined. The property of the optimal permutation shows that we can form a new sequencing problem as the first two elements of the optimal permutation have been determined. The first two elements of new problem can be found similarly. Carry on such searching procedure until the solution of the optimal permutation is obtained.

Selected the optimal sub-permutation of two elements, the calculation times of quasi-gradient are $2C_K^2-1$. For searching the optimal permutation, the calculation times of quasi-gradient are:

$$N=C^3_{K+2}-(K^2+4K-1)/4, \quad \text{for } K=2m+1 ,$$

$$N=C^3_{K+2}-(K^2+4K+2)/4, \quad \text{for } K=2m.$$

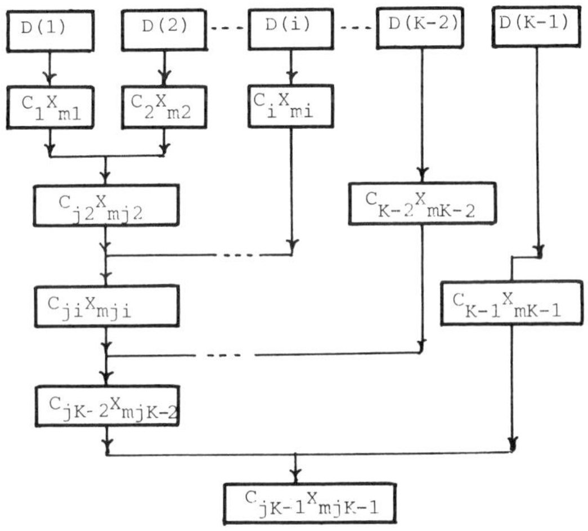

Fig. 3. Screening procedure of the first two elements of the optimal permutation.

For K=17, N=880, we should say that the quasi-gradient screening and searching algorithm was feasible and effective.

CASE STUDY

In order to demonstrate the effectiveness of the quasi-gradient screening algorithm, the algorithm is applied to a situation described by the U.S.Army Corps of Engineers. This situation is similar to the description of the Columbia River System in some reference. The lay out of potential hydro-electric projects in the river basin is given in Fig.4. The capacity and cost data of the nine

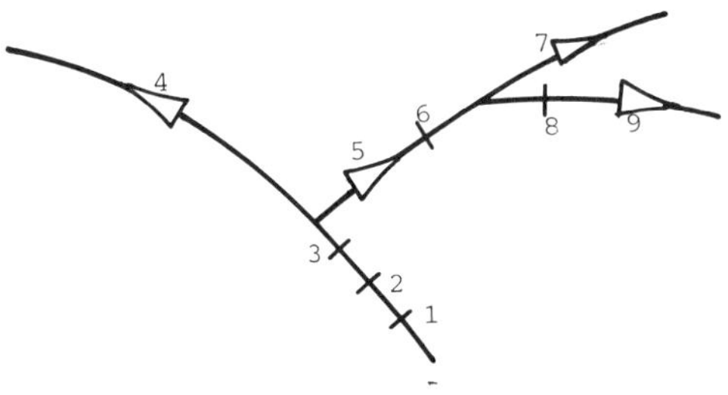

Fig. 4. Hydro-electric projects for a river basin

projects are given in Table 1. Base capacity is the prime power delivery available from a site in the absence of additional upstream storage projects. Additional capacity is the additional prime power available at the site if all storage projects above the site have been established. If a subset of those storage projects has been established, a corresponding fraction of the additional prime power is avaiable. In addition to the investment for establishing base capacity, supplimental investment is required to realize the additional capacity at each site. Table 1 gives the investment total for the full additional capacity; for intermediate levels of additional capacity the investment is adjusted proportionally. Assume the increasing law of power demand which is supplied from the hydro-electric projects is $D(t)=500e^{0.07t}$, here 500MW is the primary capacity and load of electric power systems, and the discounted law of cost is $PC=Ce^{-0.05}$, here, unit of t is in year. Determine the sequence putting into operation of the nine projects in order that the total present value of cost is minimum. D. Erlenkotter(1973) has solved this problem but applied dynamic programming approach. Now, applies the quasi-gradient screening algorithm to solve the same problem. The time is considered as continuous variable.

Select possible optimal decisions consisted of two projects sub-permutation and partition into subsets D(i), i=8,...,1, according to the same first project of the sub-permutations. The results are shown in table 2. The possible optimal decisions d_i^* selected from subsets D(i), are given in the last row of this table. Among this eight posible

Table 1 Capacity and Cost Data of the Nine Projects

Projects code	Capacity MW		Investment 10^6 $		Fraction of additional capacity for K upstream storage			
	Base	Add.	Base	Add.	K=1	K=2	K=3	K=4
1	200	200	60.6	16.0	0.40	0.75	0.90	1.00
2	260	260	70.0	20.8	0.40	0.75	0.90	1.00
3	180	180	80.0	14.4	0.40	0.75	0.90	1.00
4	270		80.0					
5	315	105	131.6	8.4	0.70	1.00		
6	140	210	108.0	16.8	0.75	1.00		
7	120		96.0					
8	80	120	36.0	9.6				
9	220		100.0					

Table 3 Calculating Results

Sequency	cose	Capacity MW	Total Capacity MW	Deficient year	Cost 10^6\$	Present cost 10^6\$	Present cost*10^6\$
0		500	500	0	0	0	0
1	2	260	760	5.9815	70.0	70.0	70.0
2	4	370	1134	11.6985	88.32	65.4896	65.49
3	1	280	1414	14.850	66.4	36.9964	37.00
4	9	381	1795	18.2593	112.88	53.7221	53.72
5	8	200	1995	19.7684	45.60	18.3008	18.30
6	3	315	2310	21.8627	90.80	33.7925	33.79
7	5	484.5	2794.5	24.5828	145.16	48.6524	48.65
8	6	297.5	3092	26.028	120.60	35.2808	35.28
9	7	268.0	3360	27.22	107.84	29.3487	29.35
Total present value of cost of optimal sequency						391.5815	391.58

*This column shows the results obtained by Erlenkotter on IBM 360/91.

Table 2 Intermediate Results

Present cost 10^6\$	Subsets							
	D(8)	D(7)	D(6)	D(5)	D(4)	D(3)	D(2)	D(1)
114.490	21	13	43	83	35	95	56	67
129.320	23	14	45	85	36	96	57	
135.490	24	15	46	86	37	97		
173.751	25	16	47	87	39			
150.082	26	17	48	89				
147.354	27	18	49					
96.694	28	19						
150.320	29							
Possible optimal decision	24	14	48	89	39	95	57	67

optimal decisions, can screen out the first two projects of the optimal sequence that are projects 24. The screening procedure may be shown as Fig.5.

Thereafter we would order the remained seven projects so as to obtain the optimal subsequence. It is unnecessary to describe the procedure in detail. At the end, we find that the optimal sequence is 241983567. All the results have been shown as Table 3. In the last column of this table, the results which had been obtained by D. Erlenkotter (1973), is shown. It is evidently that the two solutions are consistent each other.

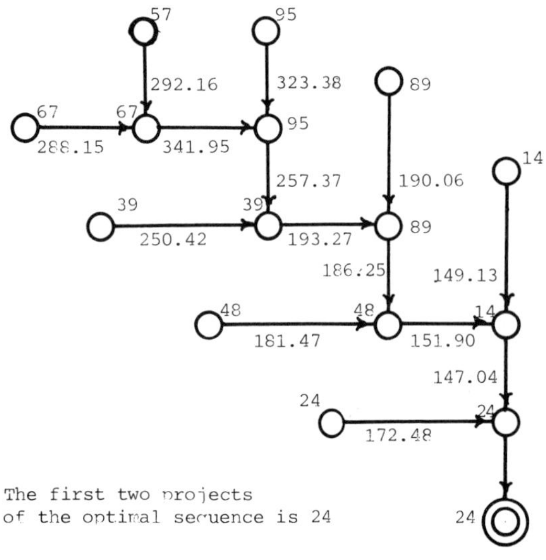

Fig. 5. The screening procedure of the first two projects of the optimal sequence, $QG(\overline{675})$= 288.15-292.16, similarly for other.

We have applied this screening algorithm to search the optimal sequence putting into operation of ten hydro-electric projects and seven thermal electric power projects for the development planning of the regional electric power system of China. Obtained solution are not contradictory with judgement from engineering experience. Among the given permutations by designers and a large number of permutations yielded by stochastic programming, we have not found any permutation is better than obtained solution.

CONCLUSION

Suggested sequencing model is nonlinear mixed integer programming contained logical expressions. In the model, the interdependence of cost and power output between power stations have been considered. The dynamic characteristic is expressed by acted order of the balance constraints between supply and demand, so that the model is a non-classical dynamic programming.

On the basis of property of the optimal sequence, a quasi-gradient screening algorithm have been proposed, applied this algorithm to situation of Columbia River System, the solution obtained consistents with the result obtained by Erlenkotter. But for Erlenkotter's algorithm, the calculation times of objective value of sub-permutation is $K2^{K-1}$, whereas for the quasi-gradient screening algorithm, the calculation times is $2C^3_{K+2}-(K^2+4K-1)/2$, for K= 2m+1, which is a polynomial of degree three, so that the quasi-gradient screening algorithm is more efficient than the former.

REFERENCES

Kurosh, A.G. (1952). Course of Higher Algebra. In Russian, translated into Chinese. Higher-Education Press, Beijing. p.81.
Erlenkotter, D. (1973). Sequencing expansion projects Opns. Res.,21, 542553.
Erlenkotter, D. (1973). Sequencing of interdependent hydroelectric projects. Water Resources Research, 9, 21-27.
Erlenkotter, D. and J.S. Rogers (1977). Sequencing competitive expansion projects. Opns. Res., 25, 937-951.
Cai Yao Zhong (1982). The study of sequencing hydroelectric stations in cascaded development of Hongshui River. Hydroelectric Power, No.4, In Chinese.
Topkis, D.N. (1978). Minimizing a submodular function on a lattice. Opns. Res., 26, 305-321.
Lucks, D.D., J.R. Stedinger and D.A.Haith (1981). Water Resource System Planning and Analysis. Prentice-Hall Inc..
Luss, H. (1982). Operations research and capacity expansion problems: A Survey. Opns.Res., 30, 907-947.

STRUCTURAL APPROACH APPLIED TO POWER SYSTEMS ANALYSIS

Liu Hsu and E. Kaszkurewicz

Department of Electrical Engineering COPPE/Federal University of Rio de Janeiro, CP 68504, 21944 Rio de Janeiro, RJ, Brazil

Abstract. This paper contributes to the Lyapunov method in transient stability analysis of power systems. By the so called structural approach, a Lyapunov function is constructed taking into account governors, flux decay and automatic voltage regulators (AVRs).

INTRODUCTION

From nearly two decades, the application of Lyapunov's direct method in transient stability analysis of multimachine power systems has been the subject of numerous works (see reference[1] for a comprehensive account). The state-of-art in this field already offers possibilities of practical application by the electric utilities. Two aspects in the method are basic: first, the determination of a Lyapunov function (LF) and second, how it can be efficiently used in transient stability assessment (e.g., determination of the critical clearing time following a fault). Concerning the second problem, the conservativeness of early procedures is now apparently circumvented by the recent breakthrough achieved by Athay and collegues[2] who obtained excellent practical results.

This paper deals with the first problem. Existing LFs, usually refer to very simplified machine and control apparatus models. Here, governors, flux decay and AVR are taken into account. A fairly simple method for determining a LF is obtained by the structural approach outlined in the next section. The result is more general then those obtained in earlier works[3,4,7] where some of these model improvements were also considered.

THE STRUCTURAL APPROACH - BASIC CONCEPTS

The structural approach (STA) consists basically in representing the dynamical system under consideration in a form such that a certain structure becomes apparent. Then, the stability problem can be studied in terms of the mentioned structure. The STA was first proposed[6] for solving absolute stability problems of single-input/single-output systems. Later, it was applied for the stability analysis of interconnected systems. The aim of this section is to give a brief outline of the STA in a somewhat broadened version with respect to the mentioned references.

A fundamental point in the STA is how to define the term "structure". In the context of this paper the definition will be given only for a special class of systems (P-systems). However, the same definition is also useful for more general systems as will be clarified later on.

P-systems. A dynamical system is a P-system if it can be represented as

$$\dot{x} = A\,\phi(x), \qquad (1)$$

where $A=(a_{ij})$ is a real nxn matrix; $x\varepsilon R^n$; $\phi(x) = |\phi_1(x_1), \ldots, \phi_n(x_n)|'$, where $\phi_i(x_i)$ are continuous functions belonging to the positive infinite sector, i.e., $\phi_i(x_i)x_i>0$ if $x_i\neq 0$ and $\phi_i(0)=0$ (this is denote $\phi_i\varepsilon(0,\infty)$). When the system is a P-system in some neighborhood of x=0, then it is called a local P-system.

The structure of a P-system is defined through an equivalence relationship. Two P-system $\dot{x}=A\,\phi(x)$ and $\dot{y}=B\,\psi(x)$ have the same structure if there exists a positive definite diagonal matrix D such that, either A=DB or A=BD. Thus, the matrix A characterizes the structure of system (1).

Class $\mathcal{D}$. The following class of nxn matrices is important in the STA:

$$\mathcal{D} = \{A : PA + A'P = -Q < 0;\ \text{for some } P>0 \text{ and diagonal}\}$$

Favorable structure. A P-system $\dot{x}=A\,\phi(x)$ has favorable structure if $A\,\varepsilon\,\mathcal{D}$. The structure is weakly favorable if there exists a diagonal $D \geq 0$ such that $A - \mu D\,\varepsilon\,\mathcal{D}$ for any $\mu > 0$.

Some important favorable structures correspond to the following class $\mathcal{D}$ matrices: sign-stable matrices[8], quasi-dominant matrices[9], M-matrices with inverted sign[10] and D-skew symmetric matrices[7,11].

Favorable Structure Realizations

Suppose that a certain physical system has a initial model which is neither a P-system nor a P-system with favorable structure. Redefining the state variables by some realization technique opens the possibility of obtaining a new model which is a P-system with a favorable structure. This approach was first used by Kaszkurewicz and Hsu[1] in absolute stability theory. A simple example is given, following Corollary 1.

STABILITY THEOREMS

The following theorem due to Persidskii[12] has wide application in the STA.

Theorem 1. Consider a P-system (1) where ϕ satisfies

$$\lim_{|x_j|\to\infty} \int_0^{x_j} \phi_j(\tau)d\tau=\infty \qquad (j=1,\ldots,n) \qquad (2)$$

Then, for the existence of a positive definite function (a Lyapunov function)

$$V(x) = \sum_{j=1}^{n} \alpha_j \int_0^{x_j} \phi_j(\tau)d\tau;\ \alpha_j>0\ (j=1,2,\dots,n) \quad (3)$$

that has negative definite time-derivative with respect to (1), it is necessary and sufficient that $A \in \mathcal{D}$.

The Lyapunov function (3) can be given explicitly as follows. Since $A \in \mathcal{D}$, there exists $P=\text{diag}(p_i)>0$ such that $PA + A'P=-Q<0$. Then setting $\alpha_i = 2\ p_i$ one has $\dot V(x)=-\phi(x)\ Q\phi(x)$ which is negative definite.

Corollary 1. If (2) is satisfied and $A \in \mathcal{D}$, the P-system (1) is absolutely stable in the positive infinite sector, i.e., $x = 0$ is globally asymptotically stable for all $\phi_i \epsilon (0,\infty)$.

This corollary is, of course, directly related with the definition of favorable structures and it could obviously be restated in terms of the favorable structure of the P-system (1). Further results along the lines of Theorem 1 can be found in reference[7].

Example: Let

$$\dot x = \begin{vmatrix} 0 & 1 & 0 \\ 0 & 0 & 1 \\ 0 & -d & -c \end{vmatrix} x + \begin{vmatrix} 0 \\ 0 \\ -1 \end{vmatrix} f(y);\ f\epsilon(0,\infty) \quad (4)$$

$$y = ax_1 + x_2$$

be the controllable canonical state-space realization of the system $G(p)\ f(y)+y = 0$; $G(p)=(p+a)\ /\ p(p^2+cp+d)$; $p=d/dt$. System (4) is not a P-system. However, from G(p) it is possible to obtain a special realization of the form $\dot x = Jx-bf(y)$; $y=x_1$; where J is a tridiagonal (or Jacobi) matrix and $b' = (0,1,0)$, by means of a simple algorithm[6]. In the present example one has

$$J = \begin{vmatrix} 0 & 1 & 0 \\ 0 & -h & 1 \\ 0 & ah-d & -a \end{vmatrix};\ b = \begin{vmatrix} 0 \\ 1 \\ 0 \end{vmatrix};\ h = c - a \quad (5)$$

In the form of a P-system this realization is given by

$$\dot x = \begin{vmatrix} 0 & 1 & 0 \\ -1 & -h & 1 \\ 0 & ah-d & -a \end{vmatrix} \cdot \begin{vmatrix} f(x_1) \\ x_2 \\ x_3 \end{vmatrix} \quad (6)$$

One easily concludes[11] that the structure of (6) is weakly favorable if $h=c-a>0$; $a,d>0$. It follows, from a slight generalization of Corollary 1 to weakly favorable systems, that the given system is absolutely stable for $f\epsilon(0,\infty)$, f satisfying (2). Note that the result is algebraic in contrast with Popov's criterion of absolute stability.

Consider now a general system $\dot x=f(x)$, $f(0)=0$. It is always possible to rewrite it in the form

$$\dot x = A\ \phi(x) + F(x) \quad (7)$$

where $\phi(x)=|\phi_1(x)\ ,\dots,\phi_n(x)|'$; $F(x)=|F_1(x),\dots,F_n(x)|'$; $\phi(0)=F(0)=0$. The functions ϕ and F and the partial derivatives $\partial\phi_i/\partial x_j$ $(i\neq j)$ are supposed continuous.

Suppose $A\epsilon\mathcal{D}$. Then there exists $P=\text{diag}(p_i)>0$ such that $PA + A'P = -Q < 0$. Consider the function

$$V(x) = \sum_{i=1}^{n} p_i \int_0^{x_i} \phi_i(x)dx_i \quad (8)$$

where $\int_0^{x_i} \phi_i(x)dx_i \triangleq \int_0^{x_i} \phi_i(x_1,\dots,x_{i-1},\sigma_i;x_{i+1},\dots,x_n)d\sigma_i$

and define $\psi(x) = |\psi_1(x),\dots,\psi_n(x)|'$ where

$$\psi_j(x) = \sum_{\substack{i=1 \\ i\neq j}}^{n} p_i \frac{\partial\phi_i(x)}{\partial x_j} dx_i\ . \quad (9)$$

Note that $\text{grad } V(x)=P\phi(x) + \psi(x)$.

Now, a new result similar to Theorem 1 can be stated for system (7).

Theorem 2. If in some domain H containing x=0

(a) $|\text{grad } V(x)|'\ |F(x)-AP^{-1}\psi(x)|\leq 0$;

(b) $\text{grad } V(x) \neq 0$ if $x \neq 0$;

then, $\dot V(x)$ along the trajectories of (7) is negative definite.

Proof. It is possible to write $\dot V(x)$ in the form

$$\dot V(t) = -|P^{-1}\ \text{grad } V|'\ \frac{Q}{2}\ |P^{-1}\ \text{grad } V| + |\text{grad } V|'\ |F - AP^{-1}\ \chi| \quad (10)$$

Then, the theorem follows easily from (a) and (b) since $Q > 0$.

Corollary 2. With the assumptions of Theorem 2

(c) if V(x) is positive definite the trivial solution of (7) is asymptotical stable, and

(d) if V(x) is also radially unbounded and H is the whole state-space the trivial solution of (7) is globally asymptotically stable

Comments on Theorem 2 and Corollary 2. Condition (b) is not always easy to check. However, if $F(x)-AP^{-1}\ \psi(x) \equiv 0$ then (b) can be replaced by the following condition: "the only equilibrium point of (7) is x=0". When conditions (a) and (b) are verified then the positive definiteness condition in (c) can be replaced by the requirement that x=0 be (locally) asymptotically stable.

Reference Systems. The decomposition of a system $\dot x=f(x)$ into the form (7) should be made so that Theorem 2 is applicable. The following considerations give some guidelines for the decomposition. For system (7) the reference system is defined as

$$\dot x = A\ \Phi_0(x) \quad (11)$$

where, $\Phi_0(x)=|\Phi_{01}(x_1),\dots,\Phi_{0n}(x_n)|'$ with $\Phi_{01}(x_1) = \Phi_1(x_1,0,\dots,0)$; $\Phi_{02}(x_2)=\Phi_2(0,x_2,0,\dots,0)$; and so on.

It can be easily shown that a necessary condition for Theorem 2 to be satisfied is that the reference system (11) be a P-system with favorable structure, at least locally.

System (7) can be viewed as a "perturbation" of the reference system (11), and the term $F - AP^{-1}\ \psi$ in condition (b) of Theorem 2, as the effect of the perturbation. This term can have undesirable effects in the sign of $\dot V(x)$. Thus the decomposition should be made so that the sign-definiteness of $\dot V(x)$ is retained, e.g., making $F - AP^{-1}\ \psi = 0$.

Furthermore, realization techniques can help in obtaining favorable structures for the reference system, as explained earlier.

LYAPUNOV FUNCTIONS FOR POWER SYSTEMS

In this section a Lyapunov function is constructed for a n-machine power system using the STA. Initially, only governors with load-frequency control structure is given in Figure 1. The mathematical model is derived under same usual assumptions, e. g., synchronous machines are represented as constant voltages behind their transient reactances; transmission line conductances, are neglected, ect.

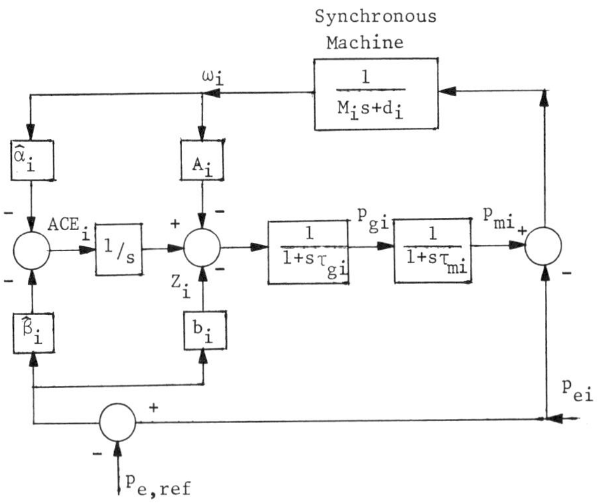

Fig.1 - Load-frequency control structure; $\hat{\alpha}_i$, $\hat{\beta}_i$, a_i, b_i, are the controller parameters; p_{mi} is the turbine mechanical output; P_{ei} is the net tie-line power; $P_{e,ref}$ is the scheduled net power interchange; ACE_i is the area control error.

The following state variables are used

(I) $x' = |x_1', x_2', \ldots, x_n'|$

$x_i' = |w_i, p_{mi}, p_{gi}, z_i|;\ i=1,\ldots,n$

where w_i is the i-th machine speed relative to a synchronous reference; the states p_{mi}, p_{gi} and z_i are defined in Fig. 1.

(II) $\sigma' = |\sigma_1, \sigma_2, \ldots, \sigma_m|;\ m = (n-1)\,n/2$

where σ_k are the power angle differences defined as follows. Let $\Theta_{ij} = \Theta_i - \Theta_j$ denote the power angle difference between the i-th and j-th machines. Then

$$\sigma_1 = \Theta_{12};\ \sigma_2 = \Theta_{13};\ \ldots;\ \sigma_{n-1} = \Theta_{1n}$$

$$\sigma_n = \Theta_{23};\ \sigma_{n+1} = \Theta_{24};\ \ldots;\ \sigma_m = \Theta_{n-1,n}$$

Let x^* and σ^* correspond to the operating equilibrium point. Then the incremental state variables with respect to be operating point are denoted

$$\xi_i = \Delta x_i = x_i - x_i^*;\ \xi' = |\xi_1', \xi_2', \ldots, \xi_n'|;$$

$$\delta = \Delta\sigma = \sigma - \sigma^*$$

Then it can be shown that the system equations are of the form[7]

$$\begin{vmatrix} \dot{\xi} \\ \dot{\delta} \end{vmatrix} = S^{-1} A \begin{vmatrix} \xi \\ \phi(\delta) \end{vmatrix};\quad A = \begin{vmatrix} K & \Gamma \\ H & 0 \end{vmatrix} \tag{12}$$

with matrices S, K, Γ, H defined in the Appendix. Neglecting machine saliency the function $\phi(\delta) = |\phi_1(\delta_1), \ldots, \phi_m(\delta_m)|'$ is given by

$$\phi_k(\delta_k) = F_{ij}\ \sin(\delta_k + \sigma_k^*) - \sin\sigma_k^*|$$

where F_{ij} is a constant and the index k is related to the pair (i, j) (i.e., $\sigma_k = \Theta_i - \Theta_j$).

Thus system (12) is a local P-system when $|\sigma_k^*| < \pi/2$ for all k. However, its structure is not favorable. Then, in order to obtain a favorable structure the realization technique, outlined in the example of Sec. 3, is used. Consider from system (12) the equation (see Appendix for notation)

$$\dot{\xi}_i = S_i^{-1} K_i\ \xi_i + S_i^{-1}\ \Gamma_i \phi(\delta) \tag{13}$$

Then it can be shown by using the Jacobi (tridiagonal) realization algorithm[6,7] that there exists a transformation

$$\eta_i = \begin{vmatrix} 1 & \vdots & 0 \\ \cdots & \vdots & \cdots \\ L_i & \vdots & N_i \end{vmatrix} \xi_i \tag{14}$$

such that equation (13) takes the form

$$\dot{\eta}_i = J_i \cdot \eta_i + \tilde{\Gamma}_i\ \phi(\delta) \tag{15}$$

where J_i is tridiagonal and $\tilde{\Gamma}_i' = T_i'|-1/M_i\ 0\ 0\ 0|$.

The matrix J_i is given by the mentioned realization algorithm. Then, transformation (14) can also be determined. The important point is that neither w_i or δ_k are changed by this tranformation. It follows that the overall system in the new variables has the form

$$\begin{vmatrix} \dot{\eta} \\ \dot{\delta} \end{vmatrix} = \bar{A} \begin{vmatrix} \eta \\ \phi(\delta) \end{vmatrix};\quad \bar{A} = D_M \begin{vmatrix} J & \vdots & -H' \\ \cdots & \vdots & \cdots \\ H & \vdots & 0 \end{vmatrix} \tag{16}$$

where J = block diag$|J_i|$, J_i is Jacobi, D_M = block diag$|M_1^{-1}, I_3, M_2^{-1}, I_3, \ldots, M_n^{-1}, I_3, I_m|$; I_3 and I_m are respectively the 3x3 and mxm identity matrices. To verify if (16) has favorable structure is quite simple. If all J_i's are sign-stable then $\bar{A}$ is D-skew symmetric[7,11] and therefore the structure is weakly favorable. There is no loss of generality in supposing that J+J' is diagonal (negative definite, because of sign stability). Thus, $D_M^{-1}\ \bar{A} + \bar{A}'\ D_M^{-1} \le 0$.

Identifying D_M^{-1} with P in the definition of class $\mathcal{D}$ (except that $-Q \le 0$) one has from Theorem 1 the Lyapunov functions

$$V(\eta,\delta) = \sum_{i=1}^{n} \frac{1}{2}\ \eta_i' P_i \eta_i + \sum_{k=1}^{m} \int_0^{\delta_k} \psi_k(\tau_k)\, d\tau_k \tag{17}$$

where $P_i = \text{diag}(M_i, I_3)$; $\dot{V}$ is only semidefine negative but asymptotic stability can be proved via La Salle's theorem[7].

When the J_i's are not sign stable it suffices to show that they are of class $\mathcal{D}$. A simple characterization of Jacobi and quasi-Jacobi matrices of class $\mathcal{D}$ can be found in reference[11]. The resulting V function is similar to (17). Of course V can be expressed in terms of (ξ,δ) by means of transformation (14). More details can be found in reference[7].

Hence, a rather simple way of constructing a

Lyapunov function for the power system is possible via the STA. Note that the realization (15) is obtained separately for each machine (or generation area). Thus, the calculations involved are very simple. A noteworthy point is that the local controller parameters are such that if the J_is are of class $\mathcal{D}$ (e.g., sign-stable) the overall system is asymptotically stable. This can be obtained on a local basis by adequate choice of local controller parameters.

Flux Decay and AVR Effects

It is now shown that the STA allows the inclusion of flux decay and AVR (automatic voltage regulator) effects. The flux decay is modelled according to (Kakimoto and Colleagues, 1980) and the AVR action is approximated by a simple exponential representation. However, more realistic AVR models such as the one proposed by Miyagi[4] can also be considered with few modifications.

$$\begin{vmatrix} \dot\eta \\ \dot\delta \\ \cdots \\ \dot e \\ \dot\varepsilon \end{vmatrix} = \left|\begin{array}{c:cc} \tilde A & & \\ & 0 & \\ \cdots & \cdots & \cdots \\ & -D_e & D_eC \\ 0 & & \\ & 0 & -D_\varepsilon \end{array}\right| \begin{vmatrix} \eta \\ \Phi(\delta,e) \\ \cdots \\ e \\ z \end{vmatrix} + \begin{vmatrix} 0 \\ 0 \\ \\ F_e(\delta,e) \\ 0 \end{vmatrix} \qquad (18)$$

where η, δ and $\bar{A}$ are defined as before and $e \in R^n$, $\varepsilon \in R^n$ are respectively the vector of incremental values of the machine internal voltages (e) and exciter voltages (ε) with respect to the steady state values; $\tilde{\Phi}' = |\tilde\Phi_1,\ldots,\tilde\Phi_m|$; $F_e' = |F_{e1},\ldots,F_{en}|$ are defined by

$$\tilde\Phi_k(\delta_k,e) = B_{ij}\,|(E^*_i+e_i)\ (E^*_j+e_j)\ \sin(\delta_k+\sigma^*_k) -$$

$$-\ E^*_i\ E^*_j \text{ sub } \sigma^*_k|; \qquad (19)$$

E^*_i, E^*_j are steady state values of internal voltages and B_{ij} is a positive constant;

$$F_{ei}(\delta,e) = -\beta_i \sum_{\substack{i=1\\ j\neq i}}^{n} B_{ij}(E^*_j+e_j)\,|\cos\sigma^*_k - \cos(\sigma^*_k+\delta_k)|;$$

β_i is a positive constant.

The matrices D_e, D_ε and C are diagonal and, positive definite. System (18) is in the form (17). The structure of the reference system is weakly favorable. Indeed, with $P = \text{block diag}|P_1,P_2,P_3|$ it suffices to take $P_1 = \text{block diag}|M_1,I_3,\ldots,M_n,I_3,I_m|$ and any positive diagonal matrices P_2 and P_3 satisfying

$$\begin{vmatrix} -2P_2D_e & P_2D_eC \\ P_2D_eC & -2P_3D_\varepsilon \end{vmatrix} < 0 \qquad (20)$$

Then, $A'P + PA = -Q \leq 0$. Due to the diagonal machine of D_e, D_ε and C, it can be easily shown that given any $P_2 > 0$ there exists $P_3 > 0$ such that condition (20) holds. Moreover, choosing $P_2 = \text{diag}(\beta_i)\ D_e^{-1}$

$$AP^{-1}\ \psi - F \equiv 0 \qquad (21)$$

where $F = |0\ \ 0\ \ F_e'\ \ 0|'$ and Ψ is defined by (9) where $\Phi = |\eta',\ \Phi',\ e',\ \varepsilon'|$.

The function V(x) defined in (5) is in the present case

$$V(\eta,\delta,e,\varepsilon) = 1/2(\eta'\,P_1\,\eta + e'\,P_2e + \varepsilon' P_3\varepsilon) +$$

$$+ \sum_{k=1}^{m} \int_0^{\delta_k} \tilde\Phi_k(\delta,e)\,d\delta_k \qquad (22)$$

However, Q is only semi-definite negative. It is easy to see that in this case Theorem 2 only gives $\dot V \leq 0$. If V is positive definite asymptotic stability can be proved via La Salle's Theorem. It can be shown that if the trivial solution of (18) is asymptotically stable (which is obviously the practical situation) then V must be positive definite.

APPENDIX

$K = \text{block diag}(K_i)$; $S = \text{block diag}(S_i)$;

$\Gamma' = |\Gamma_1'\ \ \Gamma_2'\ \cdots\ \Gamma_n'|$ and

$$K_i = \begin{vmatrix} -d_i & 1 & 0 & 0 \\ 0 & -1 & 1 & 0 \\ -a_i & 0 & -1 & 1 \\ -\hat\alpha_i & 0 & 0 & 0 \end{vmatrix}; \quad \Gamma_i = \begin{vmatrix} -1 \\ 0 \\ -b_i \\ -\hat\beta_i \end{vmatrix}.T_i'; \quad H = |H_1\ H_2\ \ldots\ H_n|$$

$(i = 1,\ldots,n)$ and T_i, H_i are defined as follows

Let

$$C_2 = |1\ \ -1|$$

$$C_3 = \begin{vmatrix} 1 & -1 & 0 \\ 1 & 0 & -1 \\ 0 & 1 & -1 \end{vmatrix}; \quad C_4 = \begin{vmatrix} 1 & -1 & 0 & 0 \\ 1 & 0 & -1 & 0 \\ 1 & 0 & 0 & -1 \\ 0 & 1 & -1 & 0 \\ 0 & 1 & 0 & -1 \\ 0 & 0 & 1 & -1 \end{vmatrix}; \text{ etc...}$$

(C_j has $\frac{j(j-1)}{2}$ rows and j columns)

T_i is the i-th column of C_n $(n \geq 2)$

$H_i = |T_i\ \ 0\ \ 0\ \ 0|$;

$S_i = \text{diag}(M_i,\ \tau_{mi},\ \tau_{gi},\ 1)$.

REFERENCES

1. Pai MA, Power System Stability, North Holland (1981).
2. Athay T et al, IEEE Trans. PAS, 98 (2) (1979) 573.
3. Pai MA and Rai V, Int. J. Control, 19(4)(1974) 817.
4. Miyagi H, IEEE, Trans. AC. 29(12) (1984), 1120.
5. Kakimoto N et al, IEEE Trans. PAS, 99(5) (1980), 1819.
6. Kaszkurewicz E. and Hsu, L., Automatica, 15 (1979), 609.
7. Hsu L. et al., Proc. II IFAC Symp. Large Scale Systems, Toulouse (1980), 67.
8. Quirk, J. and Ruppert R., Rev. Economic Studies, 32 (1965), 311.
9. Moyian P. and Hill D.J., IEEE AC, 23(2) (1978), 143.
10. Arak. M., IEEE Trans. AC 23(2) (1978), 129.
11. Kaszkurewicz E. and Hsu L., Linear Algebra and its Applications, 59 (1984), 19.
12. Persidskii S.K., Autom. Remote Control, 12 (1969), 1889.

OPTIMAL CONTROL OF HYDROTHERMAL SYSTEMS USING A MIN-MAX DECOMPOSITION APPROACH

P. A. V. Ferreira and J. C. Geromel

Faculdade de Engenharia de Campinas, UNICAMP, CP 6122 Campinas, Brazil

Abstract. A decomposition approach to the operation scheduling of hydrothermal power systems is proposed and numerically applied to a network in the Southest of Brazil. The methodology proposed decomposes the overall system into two decoupled subsystems (hydraulic and thermal) introducing the concept of hydraulic subsystem effectiveness with respect to the demand meeting. This measure is then maximized through a MIN-MAX approach, having as objective to compromise the demand and hydraulic resource cycles. Once this optimal policy has been obtained, the remaining problem relies on an Economic Dispatch of Thermal Units whose solution is already well known. Some results regarding an application to a network in the Southest of Brazil are also included.

Keywords. Optimal control, Hydrothermal systems, Mathematical programming

INTRODUCTION

The long-term economical operating schedule of a power system is usually concerned with the efficient management of two coupled, although distinct means of energy production-hydraulic and thermal. Basically, the problem consists in determining the parcel of the expected demand level for electricity that each one of these power subsystems must fulfill during each stage of the planning horizon, minimizing the overall generation costs of the system. Further, it is known that the thermal subsystem normally plays a secondary role in the demand meeting, only supplying a possible deficit between the market demand and the hydraulic generation, once the latter subsystem has negligible marginal generation costs (Arvanitidis & Rosing, 1970).

Besides the difficulties which naturally arise from deal with large scale problems, the system management represents a difficult task, preliminary, due the cyclical nature of the hydraulic resources that, in general, does not agree with the market demand cycle. In view of this, for instance, the hydraulic subsystem can present a major generation capability during periods of the planning when the demand levels are lower and conversely, a small capability when the demand levels are higher. Having as a coordination and economic objective to minimize the complementary thermal production, it becomes then strictly necessary to regularize the water inflows employment (hydraulic resources) according to the expected demand behavior.

In fact, the hydraulic resources management in this context constitutes without doubt the most important aspect of the problem to be considered and is the basic motivation for many authors (Arvanitidis and Rosing, 1970; Soares and colleagues, 1980).

However, it is possible to accomplish this management in a somewhat indirect manner. Putting aside the classical approaching, we can visualize the following decomposition of the overall problem in to two *decoupled subproblems:*

Subproblem 1 : Determination of the hydraulic subsystem optimal policy taking into account the demand cycle;

Subproblem 2 : Determination of the thermal subsystem optimal policy taking into account the deficit between the hydraulic generation and the market demand.

To perform such a decomposition, the paper introduces the concept of hydraulic subsystem *effectiveness* as the relation at each stage of the planning between the total amount of energy to be generated by the subsystem and the corresponding demand level. This relation is then maximized through a Min-Max approach and, as a consequence, a kind of *equalization effect* of the water inflows and demand cycles is verified. Once the Subproblem 1 has been solved, the thermal optimal policy can be readily obtained (Subproblem 2).

The paper is organized as follows: in the next section we present the mathematical model with detailed representation of the subsystems and then formulate the corresponding optimization problem. Its resolution by the Min-Max approach is proposed in Section 3. An example of hydrothermal operating planning is treated in Section 4 and, finally, in Section 5 we conclude the paper.

PROBLEM STATEMENT

As previously mentioned, a power system is usually composed by two subsystems. The mathematical model of the hydraulic subsystem expresses the interconnected operation of a number of hydroplants. In the model, with n units, the principal quantities are:

$x_i(k)$ - reservoir i storage at the beginning of period k;

$u_i(k)$ - hydroplant i discharge during period k;

$y_i(k)$ - independent water inflows during period k;

J_i - index set of immediate upstream neighbours dams of hydroplant i.

At any period $k \in K = [0, T-1]$, the state of the subsystem is described by $x(k)' = [x_1(k) \ldots x_n(k)]$, where (') means transpose. The vector $u(k)' = [u_1(k) \ldots u_n(k)]$ denotes the controlled discharge of the plants and $y(k)' = [y_1(k) \ldots y_n(k)]$ the independent water inflows reaching the reservoirs at $k \in K$ and represented here only by their mean expected values.

For the transition of period k to $k+1$, the reservoir storages obey the linear dynamic equation

$$x(k+1) = x(k) + Bu(k) + y(k), \quad x(0) = x_o \quad k \in K \qquad (1)$$

where x_o is the initial state of the subsystem supposed to be known. For simplicity, it is considered that the final state $x(T)$ can vary freely. Matrix $B \in R^{nxn}$ establishes the interconnection structure between the hydroplants and is defined as follows:

$$B = \{b_{ij}\} = \begin{cases} -1, & \text{if } i = j \\ 1, & \text{if } j \in J_i \\ 0, & \text{otherwise} \end{cases} \qquad (2)$$

Moreover, the capacities of the reservoirs and turbines cannot be exceeded, that is

$$x(k+1) \in X = \{x(k) \mid \underline{x} \leq x(k+1) \leq \bar{x}\} \qquad (3)$$

$$u(k) \in U = \{u(k) \mid \underline{u} \leq u(k) \leq \bar{u}\}, \quad \forall k \in K \qquad (4)$$

The hydroplant power generation is a function of the water discharge $u_i(k)$ and reservoir water head which is, in turn, a function of storage $x_i(k)$. Thus, the overall power subsystem generation can be represented by the nonconcave differentiable function (Ferreira, 1983; Geromel and colleagues, 1981)

$$\sum_{i=1}^{n} f_i(x_i(k), u_i(k)) = \sum_{i=1}^{n} \rho_i a u_i(k) h_i(x_i(k)) \qquad (5)$$

for any $k \in K$, where ρ_i denotes the equivalent turbine-generator efficiency of hydroplant i and a is the gravity acceleration.

Let the thermal subsystem be composed by m generation units. If $g(k)' = [g_1(k) \ldots g_m(k)]$ denotes the thermal subsystem generation during period k, then

$$g(k) \in G = \{g(k) \mid \underline{g} \leq g(k) \leq \bar{g}\}, \quad \forall k \in K \qquad (6)$$

where $\underline{g}$ and $\bar{g}$ are the lower and upper bounds on the generation capacity of the subsystem. The corresponding production or running cost $C_i(\cdot)$, usually associated with oil consumption is convex and increasing with respect to the production level $g_i(k)$ (Turgeon, 1978). So being, the whole problem can be mathematically posed as:

$$\underset{g(k) \in G}{\text{Minimize}} \sum_{k=0}^{T-1} \sum_{i=1}^{m} C_i(g_i(k)) \qquad (7)$$

$$\text{s.t.} \quad \sum_{i=1}^{m} g_i(k) + \sum_{i=1}^{n} f_i(x_i(k), u_i(k)) = d(k), \quad k \in K \qquad (8)$$

$$x(k+1) = x(k) + Bu(k) + y(k) \qquad (9)$$

$$x(k+1) \in X \; ; \; u(k) \in U \qquad (10)$$

where d(k) is the expected demand for period k. An important aspect of this problem is that for any fixed amount of hydraulic generation, the remaining problem constitutes an Economic Dispatch of Thermal Units (Kirchmayer, 1958). The well-known solution of this problem consists in equalizing the marginal costs of all the units. Thus, if $C_i(\cdot)$, $i = 1, \ldots, m$ are continuously differentiable functions and $\delta(k)$ represents the marginal cost,

$$\left.\frac{dC_i(\zeta)}{d\zeta}\right|_{\zeta = \tilde{g}_i(k)} = \delta(k), \quad i = 1, \ldots, m \qquad (11)$$

the optimal solution g^* being given by (Geromel and Baptistella, 1984)

$$g_i^* = \max\{\underline{g}_i, \min\{\bar{g}_i, \tilde{g}_i(k)\}\} \quad i = 1, \ldots, m \qquad (12)$$

where $\tilde{g}_i(k)$ solves (11). Hence, there is a clear advantage in trying to obtain the optimal operation policy of the hydraulic subsystem, *a priori*. This will be possible through the methodology proposed in next section

A MIN-MAX DECOMPOSITION APPROACH

In order to develop a valid decomposition of the overall system into independent subsystems, we must define carefully some kind of performance index for the hydraulic subsystem with respect to the demand meeting which assures a complementary thermal generation with minimum cost. For this, we introduce a measure of the hydraulic subsystem utilization in any stage $k \in K$ as

$$J(x(k), u(k)) = d(k)^{-1} \sum_{i=1}^{n} f_i(x_i(k), u_i(k)) \qquad (13)$$

trying, next, to make this measure as great as possible in a MIN-MAX sense:

$$\underset{u(k)}{\text{Maximize}} \; \underset{k \in K}{\text{minimum}} \; J(x(k), u(k)) \qquad (14)$$

subject to (9)-(10). Initially, it should be noted that (14) represents a mixed integer optimization problem since involves a decision on the integer set K. Thus, to avoid unnecessary difficulties, we rewrite problem (14) as

$$\underset{u(k)}{\text{Maximize}} \; \underset{\beta}{\text{maximum}} \; \beta \qquad (15)$$

$$\beta \leq J(x(k), u(k)), \quad \forall k \in K \qquad (16)$$

still subject to (9)-(10).

Considering the resolution of this problem by Duality (Geoffrion, 1971; Lasdon, 1970), let $\lambda' = [\lambda(0) \ldots \lambda(T-1)]$ be the Kuhn-Tucker multipliers attached to the constraints (16). Then, the dual problem concerned with only these constraints is

$$\underset{\lambda \geq 0}{\text{Minimize}} \; \underset{u(k), \beta}{\text{maximum}} \; L(x(k), u(k), \lambda(k)) \qquad (17)$$

where $L(\cdot) : R^n \times R^n \times R^T \rightarrow R$ is the Lagrangean function associated to (15)-(16), namely

$$L(x(k), u(k), \lambda(k)) = \beta + \sum_{k \in K} \lambda(k)(J(x(k), u(k)) - \beta) \qquad (18)$$

Some results derived by Geoffrion (1970) will be invocated in what follows. Rearranging expression (18), problem (17) can be stated as

$$\underset{\lambda \geq 0}{\text{Minimize}} \; \underset{u(k),\beta}{\text{maximum}} \; (1 - \sum_{k \in K} \lambda(k))\beta + \sum_{k \in K} \lambda(k) J(x(k), u(k)) \quad (19)$$

or, taking advantage of the separability of the maximum,

$$\underset{\lambda \geq 0}{\text{Minimize}} \; \underset{\beta}{\text{maximum}} \; (1 - \sum_{k \in K} \lambda(k))\beta + \underset{u(k)}{\text{maximum}} \sum_{k \in K} \lambda(k) J(x(k), u(k)) \quad (20)$$

which clearly imposes $\sum_{k \in K} \lambda(k) = 1$ in the first term, assuring that (20) has a finite solution (β is unrestricted in sign). Problem (17) relies then on

$$\underset{\lambda \in \Lambda}{\text{Minimize}} \; \phi(\lambda) \quad (21)$$

where $\Lambda \overset{\Delta}{=} \{\lambda \in R^T \mid \lambda(k) \geq 0, \sum_{k \in K} \lambda(k) = 1\}$ and

$$\phi(\lambda) = \underset{u(k)}{\text{maximum}} \sum_{k \in K} \lambda(k) J(x(k), u(k)) \quad (22)$$

subject to (9)-(10).

Concerned with this problem manipulation, we should point out that it is not possible (at least formally) to guarantee that problems (14) and (21) are absolutely equivalents since problem (14) is generally nonconcave and hence, a *duality gap* can always occur. Even it occurs, however, the proposed decomposition scheme remains valid, considering that no unfeasibility is produced and a suboptimal schedule is readily available.

Thus, supposing that no duality gap is verified, we progress noting that without any additional assumption on $J(x(k), u(k))$, $k \in K$, the function $\phi(\lambda)$ is convex, although not necessarily differentiable everywhere. In view of this, we cannot be confident in applying any gradient descent method (Luemberger, 1973). Methods based on *subgradients* of $\phi(\lambda)$ are more indicated. In order to employ a specific one discussed by Lasdon (1970), let ξ^i, $i = 1, \ldots, p$ be any p subgradients of $\phi(\lambda)$ evaluated at λ^i, $i = 1, \ldots, p$. In particular, we may choose

$$\xi^i = \left[J(x^i(0), u^i(0)) \ldots J(x^i(T-1), u^i(T-1))\right], \quad i = 1, \ldots, p \quad (23)$$

if $\left[u^i(0), \ldots, u^i(t-1)\right]$ solves (22) at $\lambda = \lambda^i$. In this case, an outer linearization of order p to $\phi(\lambda)$ would be expressed as

$$\phi(\lambda) = \underset{1 \leq i \leq p}{\text{maximum}} \{\phi(\lambda^i) + (\lambda - \lambda^i)'\xi^i \quad (24)$$

where $\phi(\lambda^i) + (\lambda - \lambda^i)'\xi^i = \lambda'\xi^i$ is the corresponding supporting function to $\phi(\lambda)$ at $\lambda = \lambda^i$ (Lasdon, 1970). Finally, we may note that (21) can be replaced by the linear program in variables (λ, σ):

$$\underset{\lambda \in \Lambda, \sigma}{\text{Minimize}} \; \sigma \quad (25)$$

$$\sigma \geq \lambda'\xi^i, \quad i = 1, \ldots, p \quad (26)$$

in that, it is possible to consider $\sigma > 0$ since $\xi^i > 0$ (physically, $J(x(k), u(k)) > 0$, $\forall k \in K$).

An iterative procedure to solve (7)-(10) via the MIN-MAX decomposition approach is stated below. It is assumed that a feasible initial solution $[u^o(0), \ldots, u^o(T-1)]$ is available. Choosing any $\lambda^o \in \Lambda$ and setting $\ell = 1$, an arbitrary iteration ℓ of the procedure would be

Step 1 : Solve the nonlinear program (22) for $\lambda = \lambda^\ell$, obtaining $\phi(\lambda^\ell)$ and by (23), the subgradient ξ^ℓ. Initialize the procedure with $(u^{\ell-1}(0), \ldots, u^{\ell-1}(T-1))$

Step 2 : Solve the linear program (25)-(26) for ξ^i, $i = 1, \ldots, \ell$, obtaining σ^ℓ and λ^ℓ

Step 3 : If $(\phi(\lambda^\ell) - \sigma^\ell) < \varepsilon$, being ε an arbitrarily small positive scalar, then stop. The current control policy $[u^\ell(0)), \ldots, u^\ell(T-1)]$ solve (7)-(10). Otherwise, replace ℓ by $\ell + 1$, and return to step 1.

Remarks :

1. Of course, if $\phi(\lambda^*) > \beta^*$ where λ^* solves the dual problem (21)-(23) and β^* is the corresponding primal solution of (15)-(16), a duality gap is detected and only a suboptimal schedule is then available.

2. For a given $\lambda \in \Lambda$, problem (22) can be efficiently solved by a number of nonlinear programming procedure like the reduced gradient method (Luemberger, 1973) or by some of its extensions specially developed to explore, with savings of computer time, the very particular structure of the hydraulic subsystem (Ferreira, 1983).

3. Problem (25)-(26) is simple to be solved. Eventually, if the number of constraints increases too much, nonbinding constraints may be always dropped at the end of each iteration. Only T linearly independent hyperplanes are needed to represent any point in R^T.

APPLICATION

To illustrate the MIN-MAX decomposition approach performance, a real example of hydroelectric generation scheduling is presented in this section. The hydraulic subsystem under consideration is composed by four hydroelectric plants located in two connected valleys of the Paranã River Bassin, Brazil (see Fig. 1)

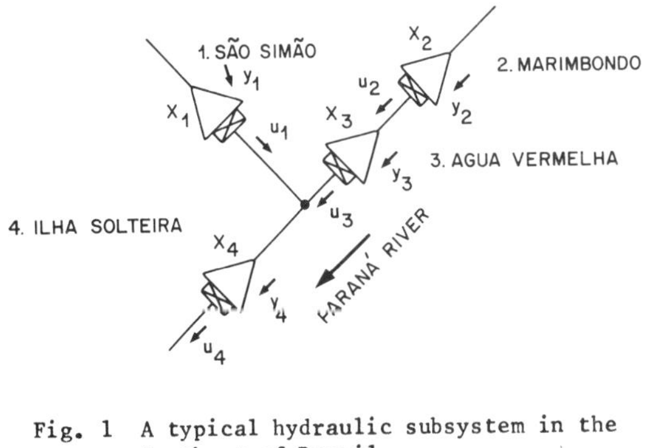

Fig. 1 A typical hydraulic subsystem in the Southest of Brazil.

Our numerical experiments are enterely based on the actual data (hydroplants operational characteristics, independent water inflows, ...) described in detail in (Geromel and colleagues, 1981).

Fig. 2 shows how this subsystem emploies the hydraulic resource availables in order to maximize, without any connection with the demand market, the energy to be generated during the planning horizon of one year, monthly discretized.

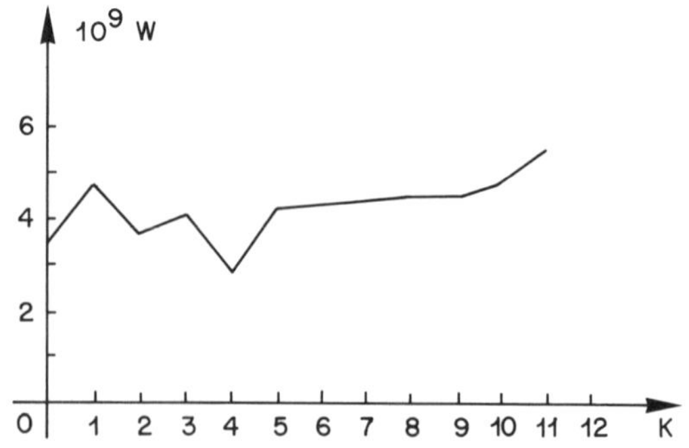

Fig. 2 Hydraulic subsystem monthly production

In a first experiment, the same subsystem is required to meet as much as physically possible, a constant monthly average consumption of 4000 MW. By the methodology outlined in this paper, we obtained an optimal scheduling to this subsystem through 17 iterations between the nonlinear subproblem (96 variables and 240 linear constraints) and the linear master (13 variables), after the total CPU time of 208 s on a Digital PDP-10 Computer System. The stop-criterium ε was made equal to 0.01.

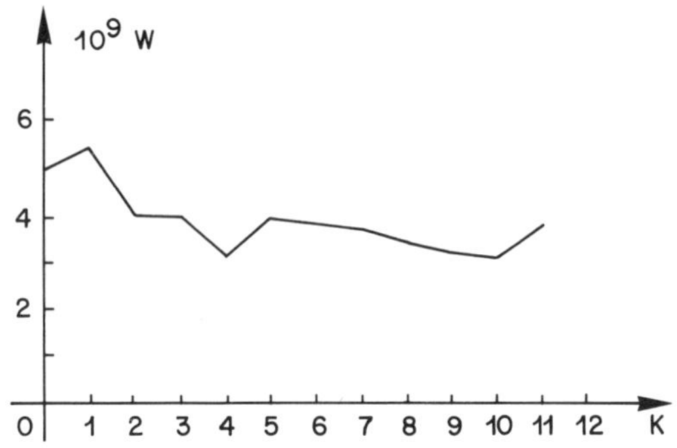

Fig. 3 Production for constant market demand.

An attempt to increase the hydrogeneration in the fifth period and to follow the market demand is apparent. This tendency can be also observed when the energy market to be met is made (artifitially) time-variables as shown in Fig. 4.

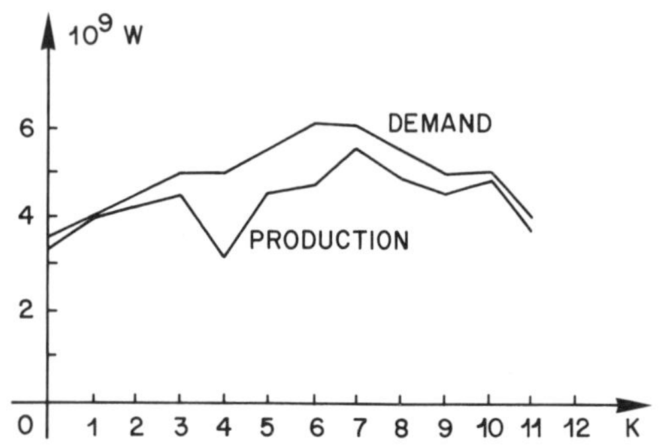

Fig. 4 Production for variable market demand.

This latter experiment took approximately 403 s of CPU time, corresponding to 19 master-subproblem iterations.

CONCLUSIONS

A decomposition based on a MIN-MAX approach of the hydrothermal power system operating schedule has been proposed in this paper. This decomposition becomes possible primary due the introduction of a performance index to the hydraulic subsystem, allowing us to deal hydraulic and thermal schedules as decoupled subproblems. The former is solved by a feasible (although *dual*) MIN-MAX approach, divided into a maximization of a nonlinear objective subject to the hydraulic constraints, for which, specially developed algorithms can be employed, and a linear minimization problem, simple to be solved, that represents the coordination level of the procedure. Taking advantages of the particular characteristics of these problems, the hydraulic schedule is quickly obtained, as illustrated by the examples of the previous section.

For those periods of the planning which present deficits between the hydraulic generation and the market demand, an optimal schedule of thermal units is then required. This latter subproblem is significantly more treatable if the *start-up* costs (Geromel and Baptistella, 1984) of the units are not taking into account. Even they are, however, this subproblem can be solved by some methodologies (Geromel and Baptistella, 1984; Turgeon, 1978).

Finally, in the decomposition scheme developed here, it is interesting to notice that the dual variables $\lambda' = [\lambda(0) \ldots \lambda(T-1)]$ have an obvious economic interpretation. They give us a measure of the relative values of water inflows concerned with the demand levels, being greater in periods of small inflowing. Indeed, the coordination problem tries to adjust the values of $\lambda(k)$, $k \in K$ in order to transfer resources to those periods.

REFERENCES

Arvanitides, N.V., and J. Rosing (1970). Optimal Operation of Multireservoir Systems Using a Composite Representation. IEEE Trans. on Power App. Systems, 89, 327-325.

Ferreira, P.A.V. (1983). A Primal Method for Solving Optimal Control Problems of Large Scale Dynamic Systems (in Portuguese). M.Sc. Dissertation, Universidade Estadual de Campinas, FEC-DEE, Brazil.

Geoffrion, A. (1971). Duality in Nonlinear Programming: A Simplified Applications - Oriented Development. SIAM Review, 13, 1-37.

Geromel, J.C., H.P.L. Luna and L.C.S. Silva (1981). Saving Fuel for Electrical Energy Through Optimal Control of Hydraulic Resources. Proceedings of the IFAC Congress VIII, XX, 23-28.

Geromel, J.C. and L.F.B. Baptistella (1984). Optimal Operation of Thermal Systems with Start-up Costs. Proceedings of the 6th Int. Conf. on Analysis and Optimiz. of Systems, 1, 116-126.

Kirchmayer, L.K. (1958). Economic Operation of Power Systems. Wiley, New York, 8-15.

Lasdon, L.S. (1970). Optimization Theory for Large Systems. The McMillan Co., London, 428-435.

Luemberger, D. (1973). Introduction to Linear and Nonlinear Programming, Addison-Wesley Publish. Co., Massachusetts, 262-265.

Soares, S., C. Lyra and H. Tavares (1980). Optimal Generation Scheduling of Hydro-Thermal Power Systems. IEEE Trans. on Power App. Systems, 99, 1107-1115.

Turgeon, A. (1978). Optimal Scheduling of Thermal Generating Units. IEEE Trans. Automat. Control, 23, 1000-1005.

ACKNOWLEDGMENT

This work was partially supported by Fundação de Amparo à Pesquisa do Estado de São Paulo, FAPESP, and Conselho Nacional de Desenvolvimento Científico e Tecnológico, CNPq.

HIERARCHICAL CONTROL FOR CITY WATER SUPPLY SYSTEMS*

Wang Qingyu

Institute of Automation, Academia Sinica, Beijing, China

Abstract. A "hierarchical operation scheduling model "has been developed to facilitate stable and economical control of water supply systems (WSS) for large and medium-sized cities in North China, where the WSS take their sources from ground water. The model is structured on three levels and based on the current manual dispatching system. Problems involving theory and technology which must be solved in realizing computer control of WSS are also introduced, i.e. predicting the short term water demand (daily and hourly) by the Kalman Filter method, establishing the macro model of the piping network by dividing time zones, as well as developing the optimizing algorithm of the nonlinear programming and dynamic programming. The models have been presented in practical simulation and test in a large city with a population of millions. The results have shown the ability of the models to enable the control system to obtain the set objectives.

Keywords. Water resources; Modelling; Kalman filters;Dynamic programing; Nonlinear programing.

INTRODUCTION

With the growth of economy and population explosion, the city water demand has been increased. In fact, the shortages of water resources become a severe problem, especially in those cities where their sources come from ground water. The researches of estimating the increasement of overall water supply demand, supplying water of good quality and saving consumption of energy and water resource are of great urgency. In china's large and meaium sized cities, water supply company is one of the citie's largest power consumers. Nearly half of cost for water supply takes on power consumption. Therefore, reduction of water supplying power consumption is also necessary for relaxing the power shortages of cities.

Since the city's water supply systems (CWSS) are complicated, including both industrial and domestic water; the fluctuation of water demand is nonstationary stochastic process; the hydraulic models of the piping network are nonlinear; the data transmission and information communication system is quite enormous. For all of those, only realizing an on-line scheduling model for water operations and computer supervisary control can meet the following objectives: [1,2]

1 Reliable and secure operation, maintenance of service pressure to consumers;
2 Economical operation to minimize the total power consumption;
3 Saving consumption of water resources through controlling service pressure strictly and reducing water leakage.
4 Laying a foundation for further reationally exploiting water resources.

In most of large and medium sized cities in North China the WSS adopt multi-sources from groundwater, usually with several water plants, and the piping networks are the type of ring structure shown in Figure 1. There are no distribution reservoirs in the piping network. For all of these conditions, a "hierarchical operation scheduling model" is developed to facilitate stable and economical control of the total WSS from one central station. Because it's based on the current manual scneduling, therefore can be realized easily. Models and methods developed for realizing hierarchical operation scheduling are also introduced. They include predicting short-term demand by means of the Kalman Filter method; getting the mathematical equations of the piping networks by macro model in distribution time zones; and adopting the optimization algorithm of the nonlinear programming and the dynamic programming. All these models and methods have been put into test in the control system of the large scale WSS in one of China's large-sized cities. The results of the test have been shown feasibility of the models and enabling the control system to attain the set objectives.

FORMULATION OF THE PRINCIPAL PROBLEMS

The principal problems which must be solved in large-scale CWSS are: 1) searching for optimal distribution of total water demands among n water plants from time to time; 2) optimal water level of the underground reservoir in every plant; and 3) an optimal

---

* The co-operators with the author are -- Nin Kui-Xi, Bai Zun-Liang and Chen Hua Yang, Technical Research Laboratory, the Beijing Water Supply Company.

arrangement of pumps, including both first- and second-stage pumps, subject to maintenance of service pressure on all control points in the piping network and other constraints, such that the daily total power consumption of n water plants attains minimum. We can formulate these as:

Find a group of feasible solutions: $(Q_i(t))$ $(Li(t))$, i-1,2,......,n,
Such that

$$J=\sum_{i=1}^{n}\int_{t_0}^{t_n} N_i(Q_i(t),L_i(t))dt=\min \quad (1)$$

$$\text{s.t}\quad H_{jmin}\leq H_j(t)\leq H_{jmax}\ \forall t,\quad j=1,2,\ldots,m \quad (2)$$

$$\sum_{i=1}^{n} Q_i(t)=Qd(t) \quad \forall t, \quad (3)$$

$$L_{imin}\leq L(t)\leq L_{imax} \quad \forall t,\ i=1,2,\cdots,n \quad (4)$$

Where: n is the number of water plants; $H_j(t)$, Hjmin, Hjmax are the pressure value at time t, the given minimum and maximum pressure values on the jth control point in the piping network respectively; Qd(t) is the flow of the total water demand; Qi(t), Li(t) are the effluent flow and the water level of the reservoir at time t in the ith water plant respectively; Ni(·,·) is the power consumption of the ith plant at time t, refering to the power consumption of the pumps mainly; to and tn are the initial and final time a day. This is a functional optimization problem to find the optimal control law. Here two questions exist, the first is: the objective function Ni(Qi(t),Li(t)) is a nonlinear function of Qi(t) and Li(t), even usually can't be expressed by mathematical analytic formula. Therefore, it's very difficult to judge its convexity in mathematics. The second is: the constraint function Hj(Q(t)) Hj(Q1(t)...Qn(t)) is inseparable. Because there are not distribution reservoirs in the piping network, and several water plants transport water jointly to the piping network, such that it's difficult to divide the service area into a few of independent pressure zones, the pressure of the control points is the result from the effect made by all plants. The conventional hierarchical dynamic programming is inapplicable. All of those throw a shadow upon proving the existence of the optimum solutions in mathematics. However, we still can estimate the exist of the optimum solutions from the angles of engineering and technology, and further try to find out numerical solutions. In addition, the following three essential problems must be solved for realizing the hierarchical scheduling of a large scale WSS:

1 short-term demand prediction
2 the mathematical models of the piping network
3 modeling for the hierarchical operation scheduling.

STRUCTURE OF HIERARCHICAL OPERATION SCHEDULING MODEL

There are two-stage pumps and a large underground reservoir in every water plant of WSS of large and medium sized cities in North China. In these systems, practically two-level scheduling already exists, and to perform hierarchical operation scheduling is realizable.

Now let's consider the above-mentioned principal problems by the Lagrange multiplier theorem. Suppose Ni is continuously differentiable with respect to Qi(t) and Li(t), and so Hj does with respect to Qi(t); the optimal solutions of the principal problems exist. Firstly, let us find out the feasible solution field,
From the equation (2), it can be got from solving the following problem Pbl.
Pbl: find a group of feasible solution (Qi(t)), i=1,2,......,n, such that

$$J1=\sum_{j=1}^{n} W_j(H_j(Q(t))-H_{jmin})=\min \quad \forall t$$

$$\text{s.t.}\quad \sum_{i=1}^{n} Qi(t)=Qd(t) \quad \forall t$$

$$H_j(t)\geq H_{jmin} \quad j=1,2,\cdots,m\ \forall t$$

Secondly, because the iluctuation of the water consumption is slow, we can approximately approach the dynamical optimization with static optimization such as we can decompose the objective function in time, derive the necessary conditions which the minimum must be satisfied at every time, i.e.

$$J_t=\sum_{i=1}^{n} N_i(Q(t),L_i(t))=\min \quad (6)$$

According to the Kuhn-Tucker conditions, we have:

$$F(Q,L,\lambda,v,u)=J_t+\lambda(\sum_{i=1}^{n} Q_i(t)-Q_d(t))-v^T(L(t)-L_{min})+u^T(L(t)-L_{max})$$

Where F is the Langrange functional, $v=(v1,v2,\ldots,vn)T$, $u=(u1,u2,\ldots,un)T$ and λ all are Lagrange multipliers related to time t. The necessary conditions are:

$$\frac{\partial F}{\partial Q_i(t)}=\left(\frac{\partial N_i}{\partial Q_i(t)}+\lambda\right)=0 \quad \forall t \quad (8)$$

$$\sum_{i=1}^{n} Q_i(t)=Q_d(t) \quad \forall t \quad (9)$$

$$\frac{\partial F}{\partial L_i(t)}=\frac{\partial N_i}{\partial L_i(t)}-v_i+u_i=0 \quad \forall t \quad (10)$$

$$v_i(L_i(t)-L_{imin})=0 \quad v_i\geq 0,i=1,2,\cdots,n,\forall t \quad (11)$$

$$u_i(L_i(t)-L_{imax})=0 \quad u_i\geq 0,i=1.2,\cdots,n,\forall t \quad (12)$$

In addition, (Qi(t)) is a group of the feasible solution from the field θ, i.e. (Qi(t)) or $Q(t)\subset\theta(t)$

And the solution which satisfy equations (10),(11) and (12) can be got from solving following problem Pb2: find a group of feasible solutions (Li(t)), i=1,2,...n, Such that

$$J2=\sum_{i=1}^{n}\int_{t_0}^{t_n} N_i(Q_i^*(t),L_i(t))dt=\min$$

$$\text{s.t.} \quad L_{imin} \leqslant L_i(t) \leqslant L_{imax} \qquad \forall t,\ i=1,2,\cdots,n$$

Where suppose Qi(t) take the optimal value Qi(t). Obviously, Pb2 can be decomposed in space as Pb2'

$$J_i=\int_{t_0}^{t_n} N_i(Q_i^*(t),L_i(t))dt=\min$$

$$\text{s.t.} \quad L_{imin} \leqslant L_i(t) \leqslant L_{imax} \qquad \forall t,\ i=1,2,\cdots,n$$

Thus we can optimize pb2 plant by plant. Now the last necessary condition left for us is to make the equation (8) equal to zero i.e.

$$\frac{\partial N_i}{\partial Q_i(t)} + \lambda(t)=0 \qquad \text{or}$$

$$\frac{\partial N_i}{\partial Q_i(t)} = -\lambda(t) \qquad \forall t,\ i=1,2,\cdots,n$$

The solutions (Qi(t)) which satisfy the equations (8),(9) and Pb1 are usually called "Optimal hourly flow distribution among all plants". From the above Pbl and the condition (14), like the discussion in water and wastewater control engineering, we can give the physical illustration of optimal (Qi(t)) i.e. the optimal (Qi(t)) should make:

1 the pressure value of every control point be close to the set index as possible.

2 the marginal cost (or instantaneous unit cost, i.e. $\frac{\partial N_i}{\partial Q_i(t)}$) of all water plants approximate to be equal.

For fulfilling the condition (14), let's examine the Pbl. In fact, it's a scalar optimization problem instead of seeking for the Pareto-optimal set. Thus we can obtain the condition (14) by adjusting the weight coefficient Wj, like defining:

$$W_j^{(k)} = \alpha_j\left(\frac{\partial N_i}{\partial Q_i(t)} - \frac{1}{n}\sum_{i=1}^{n}\frac{\partial N_i}{\partial Q_i(t)}\right)^{(k-1)}$$

in the kth iteration where the α_j is an undefined coefficient.

A rough proof follows.

We can choose m=n, n control points, and make the pressure value of every control point reflect approximately the changes of the effluent flow of a corresponding water plant. In the kth iteration, if

$$W_j(k) \geqslant W_j^{(k-1)}$$

then

$$\left(\frac{\partial N_i}{\partial Q_i(t)}\right)^{(k-1)} \geqslant \left(\frac{1}{n}\sum_{i=1}^{n}\frac{\partial N_i}{\partial Q_i(t)}\right)^{(k-1)}$$

according to the conception of scalar optimization, we have: $Q_i^{(k)} \leqslant Q_i^{(k-1)}$

Since Ni is the nondecreased function with respect to Qi(t) and so is the instantaneous unit cost $\frac{\partial N_i}{\partial Q_i(t)}$, therefore we have:

$$\left(\frac{\partial N_i}{\partial Q_i(t)}\right)^{(k)} \leqslant \left(\frac{\partial N_i}{\partial Q_i(t)}\right)^{(k-1)}$$

hence

$$\left(\frac{\partial N_i}{\partial Q_i(t)} - \frac{1}{n}\sum_{i=1}^{n}\frac{\partial N_i}{\partial Q_i(t)}\right)^{(k)} \leqslant \left(\frac{N_i}{Q_i(t)} - \frac{1}{n}\sum_{i=1}^{n}\frac{\partial N_i}{\partial Q_i(t)}\right)^{(k-1)}$$

to iterate repeatly until $|W_j(k)-W_j^{(k-1)}| \leqslant \varepsilon$

Then we can make the instantaneous unit cost of all water plants approximate to be equal.

So the structure of the hierarchical operation scheduling model for the total control sytem can be sketched in Figure 2.

MODELING, TESTING AND VERIFYING

The essential model and algorithm developed for the control system include water demand prediction, the mathematical models of the piping network, optimization algorithm, and so on.

1. Demand Prediction

Although the fluctuation of the water demand in cities is a nonstationary stochastic process, we still can predict the water demand by the Kalman Filter method in several duration divided in the light of the changes of climate and season in a year. In the replace time of season or climate a modifying way on-line is adopted for transition.

We define the state vector as: Xk=(X(1)), X(2),X(3)) . Its elements represent the ratio of the daily water demand in two adjacent days; the ratio of the periodic fluctuation in a week and the ratio of having a effect on the daily water demand in response to changes in temperature respectively. The prediction equation of the daily water demand is

$$\hat{Q}_{k+1}=H_{k+1}\hat{X}_{k+1}$$

Where Qk+1 is the water demand of the k+1th day. And the observation vector Hk+1 is:

$$H_{k+1}=(Qk,((Qk-6-Qk-7)-B(T_k-6-T_k-7)), T_k+1-T_k)$$

Where Qk-i is actual water demand of the k-1th day, Tk is the highest temperature of the kth day, B is the ratio of the water demand fluctuation induced only by the difference of temperature in the corresponding days of the last week. The hourly water demand prediction is completely similar. There are many advantages to predict the water demand by the Kalman Filter method like high accuracy, small memory requirement, calculating easily in a computer, and so on. The results of practically predicting in about three years, from 1981 to 1983, in this large sized city of China have shown that the normal average prediction errors are within ±5%, and within ±3% in the peak demand periods.

2. Mathematical Models of the Piping Networks

It's possible for developing computer control of WSS, only if the mathematical models of the piping networks which actually reflect the state of the piping networks is found out. It's the key of optimization scheduling. The conception of a macro model was put forward by Robert Demoyer for reducing calculation of the piping networks. But it only can be applied in a proportional load condition, i.e. the flow of all nodes in a piping networks must be direct proportion to the total water demand flow of cities. Because CWSS supply water for both industrial usage and domestic usage in the cities of China, it's difficult to satisfy this proportional load condition. For this reason, carefully examming the actual situation of water demand, we may divide the whole 24 hours a day into several different time zones, such that within each time zone we may view the load condition in the light of being approximatelly proportional, and find the macro models. They include two types of models.

A. System curve
It shows the relationship of the effluent pressure of a plant with the flow of other plants and total water demand. With it and the characteristic curves of parallelizing equivalent pumps, we can calculate the work-points of pumps, then make out the power consumption or the power cost of pumps. Normally, the equations can be obtained by the piece-wise regressing method as

$$P_i=P_{i0}+S_{iQ_i}{}^2$$

With

$$P_{i0}=C_{i,1}+C_{i,2}Q_d{}^2+\sum_{j=1}^{n} C_{i,j+2}Q_j{}^2$$

$$S_i=C_{i,n+3}+\sum_{j=1}^{n} C_{i,j+n+3}Q_j/Q_i$$

Where Pi, Pio are the effluent pressure and the static head of the ith plant respectively; Si is the equivalent hydraulic friction coefficient; Qd is the flow of the total water demand; Q and Qj are the effluent flow of the ith and jth plant respectively; and C , is the regressive coefficient. For ex., we have obtained the equation of a plant in the time zone on 8:00-22:00 like:

$$P_7=28.57-1.16843\times10^{-7}\times Q_1^2 \quad -4.85458\times10^{-8}\,Q_2^2$$
$$+9.28635\times10^{-10}\,Q_3^2 \quad +4.4144\times10^{-8}\,Q_4^2$$
$$+7.98795\times10^{-7}\,Q_5^2 \quad +1.38863\times10^{-8}\,Q_8^2$$
$$-2.76428\times10^{-9}\,Q_d^2 \quad +7.37261\times10^{-7}\,Q_7^2$$

complex related coefficient	r=0.925014
variance	s=1.48m
value of the F-Function	f=6.24

B. Pressure curve of the control points:
It shows the relationship of the control point pressure in the piping network with the effluent flow of all plants and the total demand. That is just the constraint function in the principal problem. That can be obtained by the similar method just mentioned above.

The results of calculation and tests with a vast amount of data have shown that the errors of the models campared with the actual measuring values have been controlled within 2 metres in about 90 percent samples. There is no problem in accuracy for computer control of WSS.

3. Optimal Scheduling
As mentioned above, we must solve the Pb1 and Pb2' on level 2 and level 3 respectively. The former is a nonlinear programming problem with inequality constraints for getting optimal hourly flow distribution among all water plants. The latter is a typical optimization problem in time process. It's clearly to use the dynamic programming with the continuous transition equation like:

$$S_i(X_i(k-1))\times X_i(k-1)=S_i(X(k))\times X(k)-Q'(k)-Q_i(k)$$

$$k=1,2,\cdots\cdots 24$$

Where: K is the kth hour, S(.) is the effective area of the ith water plant reservoir, which is the function of water level state Xi(K); Qi(K) and Qi(K) are the intake flow of the first-stage pumps from the wells to the reservoir and the effluent flow of the second-stage pumps from the reservoir to the piping network in the ith plant respectively. Its iteration formula of the cost function is:

$$I_i(X_i(k),k)=F_i(X_i(k),k)-I_i(X_i(k-1),k-1)$$

Where: Fi(Xi(k),k) is the ith plant total operation power cost at the state X (k) in the kth hour.

The results of the initial experiment in the piping network by the nonlinear programming have shown that the average pressure values of the control points in the piping network have been reduced 1-2 meters, by estimating, water consumption can be cut down 2-5%. Further, The simulating of the dispatching in a water plant of this city by the dynamic programming has shown that the amount of power consumption can be reduced 4.3%. The dispatching table printed by computer is shown in Figure 3. All of the computer programm have been written by the FORTRAN-IV and BASIC Languages.

For optimal control of WSS. We also have developed a series problems in control as well as in water and wastewater control engineering including the algorithm of the dynamic programming on the 3th level; robust control, conception of the equavalent pump; parallelization of characteristic curves of the intake pumps in the complicated pipelines and so on.

CONCLUSION

The preceeding sections describe the main theoretical and practically problems which may be encountered in the optimal control (or the economical scheduling) of the WSS. For the WSS of large and medium sized cities in North China, a hierarchical operation Scheduling model has been developed. The results of the practice tests have shown its feasibility and enabling the control system to realize the set objectives.

ACKNOWLEDGEMENT

The author would like to express his gratitude to the participators of the Beijing Water Supply Company for their support in this joint project and for contributing much valuable data. Many thanks are also due to professor Wu Jong-Ming and Yang Zhi-Jian of the Institution of Automation, Academia Sinica for offering much useful technical advices.

REFERENCE

1. Patrick F. Perry "Decentralized optimum control for water distribution system optimization" Proc. of the 1977 J. Auto. Control Conference TP28/352, Vol.2.
2. Yoshihiko Sato "Study on Water Distribution Control for Pressure Equalijing" Journal of Japan Water Works Association, No.446, 1971.
3. F.Fallside, P.F. Perry "On line prediction of consumption for water supply network control", 1975 6th IFAC Congress, Part IIIC.
4. F.Fallside, P.F. Perry "Hierarchical optimization of a water supply network", Proc. IEE, Vol.122, No.2, 1975.
5. The Working Conference on Water Supply Control of Japan. by FUJI Electrical Company, 15 Oct. 1980.
6. Robert Demoyer Jr. and Lawrence B. Horwitz "Macro Scopic Distribution System Modeling", JAWWA, July, 1975.
7. Robert V. Hogg "Robust in Statistics - An Introduction to Robust Estimation", "Robusthess in Statistics", Edited by Robert L. Launer, Granam N. Wilkinson.

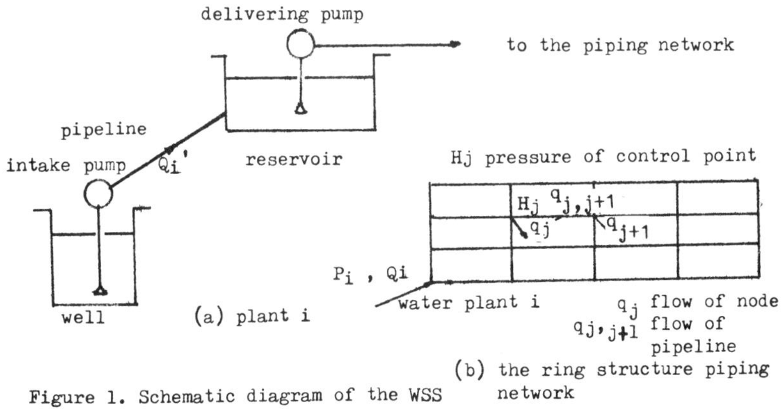

Figure 1. Schematic diagram of the WSS

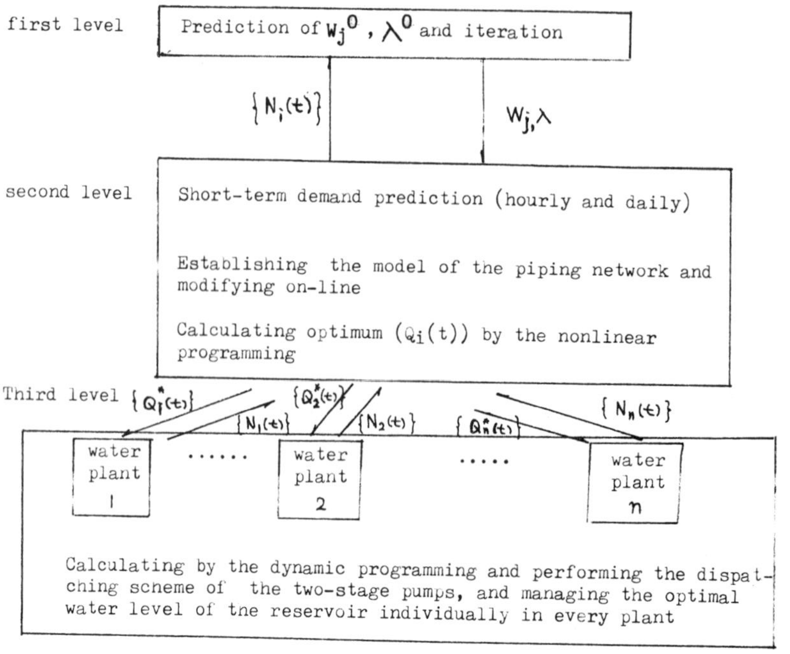

Figure 2 Structure of hierarchical scheduling model

PUMP STATION NO.7 1983.8.28

hour	cm.of WL	total £.	£.of WP	t/h of WP	£.of DP	t/h of DP	£/t of WP	£/t of DP	total £/t
1	240	25	11	1852	14	1647	.0059	.0085	.0152
2	240	25	11	1853	14	1647	.0059	.0085	.0152
3	240	32	18	2381	14	1647	.0076	.0085	.0194
4	260	32	18	2381	14	1649	.0076	.0085	.0194
5	280	32	18	2381	14	1655	.0076	.0085	.0193
6	300	33	18	2381	15	1607	.0076	.0093	.0205
7	320	41	26	2750	15	1609	.0095	.0093	.0255
8	360	47	18	2381	29	3173	.0076	.0091	.0148
9	340	47	18	2381	29	3169	.0076	.0092	.0148
10	320	47	18	2381	29	3109	.0076	.0093	.0151
11	300	40	11	1846	29	3159	.0060	.0092	.0127
12	260	48	18	2381	30	3150	.0076	.0095	.0152
13	240	41	11	1852	30	3146	.0059	.0095	.0130
14	200	33	18	2381	15	1596	.0076	.0094	.0207
15	220	34	18	2381	16	1598	.0076	.0100	.0213
16	240	48	18	2381	30	3146	.0076	.0095	.0153
17	220	4[illegible]	18	2381	30	3141	.0076	.0096	.0153
18	200	48	18	2381	30	3136	.0076	.0096	.0153
19	180	43	13	1954	30	3131	.0067	.0096	.0137
20	140	33	18	2381	15	1590	.0076	.0094	.0208
21	160	26	11	1861	15	1592	.0059	.0094	.0163
22	160	33	18	2381	15	1592	.0076	.0094	.0207
23	180	40	26	2750	14	1646	.0095	.0085	.0243
24	220	32	18	2381	14	1645	.0076	.0085	.0195

the daily total £. = 901 the daily total delivering water tx1000.=54.1800

the daily total £/t.=.0166

xx THE DISPATCHING TABLE OF DP XX

1	0----2400	
4	700----1300	1500----1900

xx THE DISPATCHING TABLE OF WP xx

772	0----24	
771	0----24	
783	0----24	
780	0----24	
778	0----24	
779	2----24	
781	6----9	22----23
775	2----24	
773	6---- 9	22----23
782	8----9	

Figure 3.
Dispatching table

SITING AND DIMENSIONING OF HYDROELECTRIC POWER PLANTS

A. Turgeon

Institut de recherche d'Hydro-Québec (IREQ), 1800, montée Sainte-Julie, Varennes, Québec, Canada JOL 2PO

Abstract. This paper presents a new approach to the problem of selecting a development scheme for a river valley from preliminary surveys on candidate sites. Rather than proceeding by elimination, this approach identifies the scheme which minimizes the investment and operating costs. This minimization problem is decomposed using the Benders technique into a master problem covering the site selection and sizing aspects and a subproblem covering the production aspect. The former is solved by mixed-integer linear programming whereas the latter is solved by a nonlinear network flow technique. Subsequently, a study is made to determine whether demand will be satisfied with the desired probability if worse sequences of river flows than those registered in the past occur in the future with a relatively high probability; if not, the development scheme is revised.

INTRODUCTION

A virgin river valley can contain literally dozens of possible sites for the construction of a hydroelectric complex while each site, in turn, can offer a host of possibilities with respect to the type and size of the installations built. The result is that an almost infinite number of potential development schemes exists for each valley. The purpose of the work described in this paper was to find a method for identifying the best scheme.

Detailed study of a hydroelectric development scheme comprises several stages. The first involves topological surveys of the land upstream of each site to determine the locations where retaining dikes will be required to hold back the water for the reservoir. Meanwhile, at the site itself, surface and underground geological studies are required to determine the type, strength and stability of the ground on which the dam and power station will be built. Hydrological surveys are also undertaken at each site to establish the characteristics of the water flow. Once the various surveys have been completed, of course, the data has to be analyzed and validated while hydraulic and other studies must be carried out before a specific project can be drawn up for the site selected. The final stage in this process is the cost evaluation.

Such a study is obviously time-consuming, strenuous and costly and can therefore not be undertaken blindly. One efficient and rigorous way of determining the best scheme is outlined below:

1) First, estimate the development cost for each site, based on the preliminary surveys, as a function of the reservoir capacity and power-plant capacity. (It is assumed that this estimate will not exceed the real cost). Then fix ITER = 1.
2) Determine the development scheme that minimizes the combined investment and operating costs using the method described below. If ITER = 1, go directly to step 4.
3) If the development scheme just identified is identical to the one found at the previous iteration, stop the search because the scheme is an optimal one. If not, go the next step.
4) Make a detailed study of the last development scheme to determine the exact cost. If this is greater than the estimate, replace the latter by the exact cost, make ITER = ITER + 1 and return to step 2.

A development scheme determined in this way is really optimal for three reasons: 1) it minimizes the investment and operating costs; 2) its cost is exact; 3) because of the assumption made in step 1, the cost of the other schemes is not overestimated. As for the number of iterations required and , consequently, the number of development schemes to be studied in detail, these depend on the original estimate of the investment cost. If this was exact, a single iteration is required whereas if not, several iterations may be needed. It should not be forgotten, however, that even if the development scheme changes at each iteration, the sites involved do not necessarily change. Only the size of some of the installations might be altered.

The remainder of this paper is devoted to describing the method used in step 2 of the above procedure to determine the development scheme that minimizes the combined investment and operating costs. The main difficulty with this minimization problem stems from the fact that river flows vary randomly, so that the production of the powerplants, and hence the operating costs, cannot be determined with certainty. The difficulty can be avoided by supposing that future river flows will behave exactly as those in the past and using past hydrological data to solve the problem. But if worse sequences of river flows can occur in the

CSTD-K

future with a relatively high probability, then the new installations may prove inadequate to meet demand. In fact, the only way to be sure that demand will be satisfied with the desired probability is to take into consideration all possible flow sequences. However, this would require an enormous amount of time especially if step 2 has to be solved several times. This is why the following compromise solution has been adopted:

1) Find a development scheme for the river using the above procedure and past hydrological data as river flows.

2) Determine whether the dimensions of the installations in the development scheme should be increased to take into account the stochasticity of the river flows.

The content of the paper has been arranged as follows. The next section formulates the problem of determining the development scheme which minimizes the sum of investment and operating costs in mathematical form. It is shown in the following section that the problem can be decomposed using the Benders technique into a planning problem and a production subproblem; the former, which covers site selection and sizing of the installations, is solved by mixed-integer linear programming while the latter can be solved with a nonlinear network flow algorithm or if the production functions are piecewise-linearized, by linear programming. The final section shows how to adjust the dimensions of the installations to account for the stochasticity of river flows.

PROBLEM FORMULATION

The system

The problem is to select a development scheme for a river that has n possible sites for the construction of a hydroelectric complex (Fig. 1).

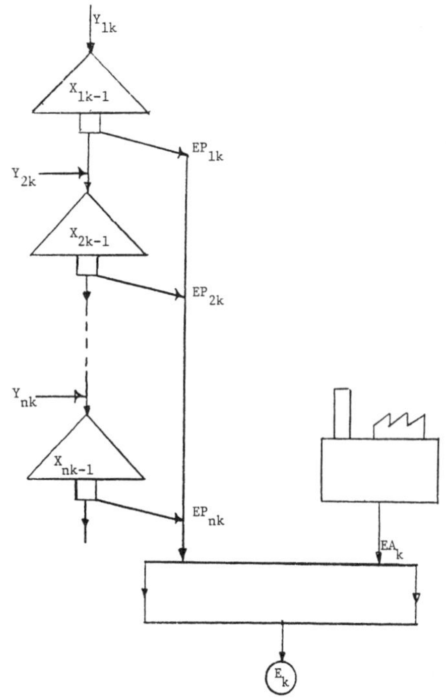

Fig. 1: The system

It is assumed that:

- K weeks of hydrological data are available;
- the probability distributions of the weekly river flows are known;
- perfect correlation exists between the river flows at two different sites;
- there are no competing uses for the water in the reservoir;
- the electricity demand is known;
- the difference between the demand and the hydroelectric generation of the river will be produced by the alternative energy source shown in Figure 1;
- the transmission line losses are negligeable;
- the installations will all be commissioned at the same time and have a lifetime of K weeks.

The objective function

The objective is to find the development scheme that minimizes the sum of the investment and operating costs or, in mathematical terms, minimizes

$$\sum_{i=1}^{n} (crf_i \cdot IR_i + crv_i \cdot HR_i + cpf_i \cdot IP_i + cpv_i \cdot PP_i)$$

$$+ \text{cav PA} + \sum_{k=1}^{K} cap \cdot \gamma_k \cdot EA_k \qquad (1)$$

where:

IR_i = binary variable which is 1 if a reservoir is built at site i and 0 otherwise;
crf_i = fixed cost of building a reservoir at site i (\$);
HR_i = maximum level of reservoir i (m);
crv_i = variable cost of reservoir construction at site i (\$/m);
IP_i = binary variable which is 1 if a powerplant is built at site i and 0 if not;
cpf_i = fixed cost of building a powerplant at site i (\$);
PP_i = installed capacity at powerplant i (MW);
cpv_i = variable cost of building a powerplant at site i (\$/MW);
PA = capacity of the alternative energy source (MW);
cav = variable cost of building an alternative source (\$/MW);
EA_k = energy generated by the alternative source in week k (MWh);
γ_k = discounting factor;
cap = production cost for alternative energy (\$/MWh);

Although the cost of reservoir i is assumed to increase linearly with size in (1), this may not be the case in practice. For example, when a reservoir is expanded, new dikes are often required to prevent water from spilling into other valleys, thus occasioning new fixed costs as well as higher variable costs. Therefore, the cost of reservoir i may follow the curve shown in Figure 2.

However, since such a curve can be represented by a sum of linear functions, there is consequently no loss of generality in using (1) in our presentation. The same reasoning applies to the powerhouse cost.

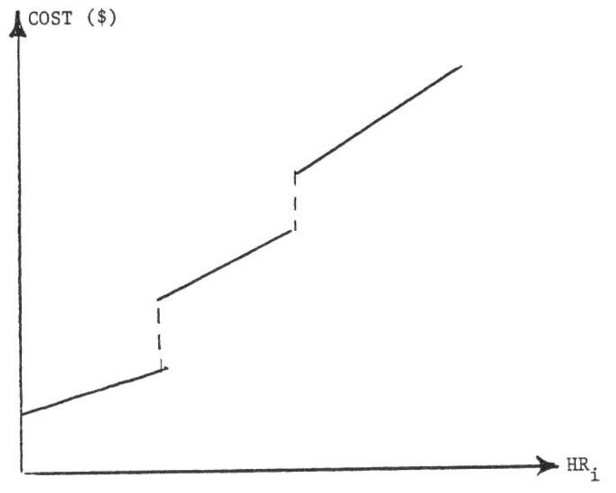

Fig. 2 Cost of reservoir i

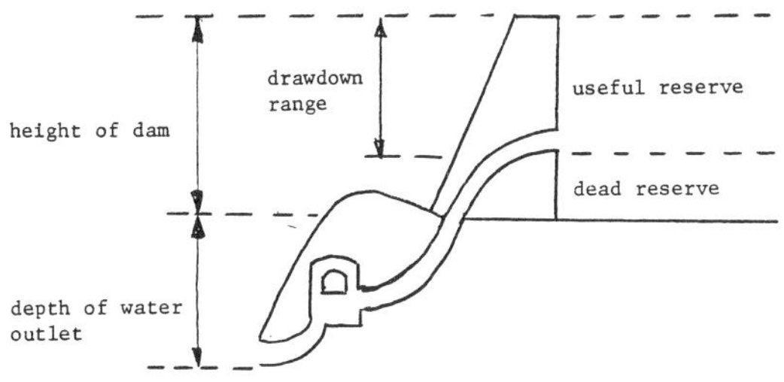

Fig. 3 Hydroelectric scheme

The constraints

The minimization of function (1) must be done under a large number of constraints, which can be grouped as follows: reservoir operation constraints, physical constraints and production constraints.

The operating constraints for reservoir i can be expressed mathematically as follows:

$$X_{ik} = X_{ik-1} + A_{ik} - U_{ik} - V_{ik} \qquad (2)$$

$$X_{ik} \leq \bar{X}_i \qquad (3)$$

$$U_{ik} \leq \bar{U}_i \qquad (4)$$

$$X_{io} = X_{ik} = \bar{X}_i \qquad (5)$$

$$\text{where } A_{ik} = Y_{ik} + \sum_{j\epsilon\Gamma_i} (U_{jk} + V_{jk}) \qquad (6)$$

Equation (2), the state equation for reservoir i, specifies that X_{ik}, the reservoir content at the beginning of period k, is equal to the content at the beginning of period k-1 plus A_{ik}, the total inflow in period k, less U_{ik}, the effective draft in period k, less V_{ik}, the spillage during period k. The total inflow in period k, as specified in Eq. (6), is equal to Y_{ik}, the natural inflow during period k, plus the sum of the effective discharges and spillages from the reservoirs immediately upstream. Constraint (3) shows that the content of reservoir i at the beginning of period k cannot exceed $\bar{X}_i$, the reservoir storage capacity. Constraint (4) limits the effective draft from reservoir i during period k to the value $\bar{U}_i$, which represents the discharge capacity of the generating station. Constraint (5) fixes the content of reservoir i at the beginning and end of the period studied to the value X_i.

As seen in Fig. 3, the active reserve $\bar{X}_i$ depends on the height of the dam HR_i and the height of the water inlet HD_i. The water stored below the inlet, the so-called dead reserve, is denoted by XD_i. The head of water at powerhouse i is equal to the depth of the powerhouse HP_i plus the height of the water level in the reservoir. It is important that the head not vary too much, in order that the generating station can operate at more or less maximum capacity. One means of doing this is to fix the drawdown range (which is equal to $HR_i - HD_i$) to a fraction (0.10 to 0.15) of the maximum head. This fraction, or drawdown coefficient, is represented by the letters "ac" in the constraints given below

$$HR_i - HD_i \leq ac\,(HR_i + HP_i) \qquad (7)$$

$$HD_i = H_i(XD_i) \qquad (8)$$

$$HR_i = H_i(XD_i + \bar{X}_i) \qquad (9)$$

$$HP_i \leq \bar{HP}_i \cdot IP_i \qquad (10)$$

$$HR_i \leq \bar{HR}_i \cdot IR_i \qquad (11)$$

H_i in Eqs. (8) and (9) represents an increasing concave function which establishes the relationship between the reservoir content and the height of the water level above the foot of the dam. The symbol $\bar{HP}_i$ in Eq. (10) represents the maximum depth to which the generating station at site i can be located while $\bar{HR}_i$ in Eq. (11) denotes the maximum height the dam at site i can attain.

It should be emphasized that only the location of the dam is fixed a priori. For dam i, for example, this location will be assumed to be at an altitude of "$foot_i$" metres. As for generating station i, it could be located anywhere between altitude "$foot_i - \bar{HP}_i$" and "$foot_i$". Therefore, denoting the altitude of generating station i as AP_i and the maximum height the water level in reservoir i can attain as AR_i, we see that

$$AR_i - AP_i = HR_i + HP_i \qquad (12)$$

$$AR_i \geq foot_i \qquad (13)$$

$$AP_i \geq foot_i - \bar{HP}_i \qquad (14)$$

If sites i and i+1 are very close together, "$foot_i - \bar{HP}_i$" could be lower than "$foot_{i+1} + \bar{HR}_i$", with the result that the reservoir at site i+1 could flood the powerhouse at site i. To avoid this, the following constraint must be added

$$AP_i \geq AR_{i+1} \qquad (15)$$

Furthermore, care must be taken to ensure that the space between the crest of the dam and its foot is full, otherwise the project selected by the computer program risks comprising no more than a section of dam at maximum altitude, a powerhouse at minimum altitude and nothing between the two (which incidentally would provide a high head at little cost!). This situation can also be avoided by adding four linear constraints to the model:

$$AR_i - HR_i \leq altm - (altm - foot_i).IR_i \quad (16)$$

$$AR_i - HR_i \geq foot_i.IR_i \quad (17)$$

$$AP_i + HP_i \leq altm - (altm - foot_i).IP_i \quad (18)$$

$$AP_i + HP_i \geq foot_i.IP_i \quad (19)$$

where $altm = foot_1 + \overline{HP}_1$

The final constraints to be taken into consideration concern the electrical demand to be satisfied. First there is the maximum load P that the new facilities must be able to provide. Therefore,

$$\sum_{i=1}^{n} PP_i + PA \geq P \quad (20)$$

where PP_i represents the peak power generated by power station i. Also to be satisfied is the electrical demand in period k, E_k. Consequently,

$$\sum_{i=1}^{n} EP_{ik} + EA_k \geq E_k \quad (21)$$

where EP_{ik} denotes the output from power station i in period k. Naturally the production of the alternative power station cannot exceed its capacity, with the result that

$$EA_k \leq m_k.PA \quad (22)$$

where m_k represents the number of hours in period k.

Both the power and the energy generated at power station i are functions of the head and the amount of water flowing through the turbines. Several assumptions have been formulated for these functions in the past including:

$$PP_i = \alpha.[HP_i + H_i(XD_i + 0.5\ \overline{X}_i)].\overline{U}_i \quad (23)$$

$$EP_{ik} = \alpha.m_k.[HP_i + H_i(XD_i + X_{ik-1})].U_{ik} \quad (24)$$

where α is a conversion factor. To calculate the head in Eq. (23), it is asusmed that reservoir i will be half full at the time the peak occurs.

Before concluding this description of the problem, it should be mentioned that although not explicitly noted above, all the variables involved can take only non-negative values.

METHOD OF SOLUTION

Generalized Benders method

The Benders decomposition method, generalized by Geoffrion [1], is used for solving problems of the following form:

$$\min_{w,z} F(w,z) \text{ subject to } G(w,z) \leq 0;\ w \in W;\ z \in Z \quad (25)$$

where z is a vector of complicating variables in the sense that (25) is a much easier problem in w when z is temporarily held fixed. This method is interesting for our problem because once the development scheme has been selected, and, therefore, the vector z fixed, the problem becomes an operating problem for which various methods of solution have been put forward in the past.

The idea to Benders decomposition is to project problem (25) into the z space, which can be shown to yield the following problem:

$$\min_{z \in Z \cap R} q \quad (26)$$

subject to:

$$q \geq \operatorname{infinium}_{w \in W} \{F(w,z) + \lambda G(w,z)\}\ \forall \lambda \geq 0 \quad (27)$$

where $R = \{z:\ G(w,z) \leq 0 \text{ for some } w\}$

Each of the constraints in (27) is called a Benders cut. In view of the very large number of such constraints, the most natural strategy for solving the problem is relaxation [2]. This strategy served as inspiration for the following algorithm. Denote by z_p, q_p, w_p and λ_p the values of z, q, w and λ found at iteration p.

Step 1: Set $p = 1$, $q_p \leq -\infty$, $UBD = \infty$, and z_1 to any value of $z \in Z \quad R$. Go to step 2

Step 2: Solve the subproblem

$$\min_{w \in W} F(w,z_p) \text{ subject to } G(w,z_p) \leq 0$$

Let w_p be an optimal solution. If $q_p \geq F(w_p, z_p) - \varepsilon$, where ε is the desired precision, stop. Otherwise, find λ_p, the value of λ corresponding to the solution, set $UBD = F(w_p, z_p)$, $p = p + 1$, and go to step 3.

Step 3: Solve the master problem

$$\min_{z \in Z \cap R} q$$

subject to

$$q \geq \operatorname{infinium}_{w \in W} \{F(w,z) + \lambda_i\ G(w,z)\};\ i=1, 2,\ldots, p-1 \quad (28)$$

Let (z_p, q_p) be the optimal solution. If $q_p \geq UBD - \varepsilon$, stop. Otherwise go to step 2.

Two problems hamper the use of the algorithm just described. The first is to find the z's that form part of the set R or, which comes to the same thing, to ensure that the z_p selected in Step 3 will be such that a solution will be obtained to the subproblem in Step 2. When applied to our problem, this amounts to wondering whether the development scheme selected by the master will satisfy the energy demand, i.e.constraint (21), at all times. In fact the only way to do so is to solve the operating subproblem, which could be very time-consuming if there are several schemes under investigation. This difficulty, can be avoided by allowing generation shedding if constraint (21) cannot be fulfilled, although at a cost so high that shedding will not in fact take place with the optimal scheme.

The second problem with the above algorithm is that constraint (28) must be evaluated, which means solving p-1 minimization problems at iteration p for each value of z considered. For our development problem, for instance, this amounts to solving p-1 operating problems for each development scheme under study. In view of the possible enormity of such a task and the extensive CPU time involved, it seemed wiser to replace (28) by the following heuristic cuts:

$$q \geq F(w_i, z) + \lambda_i G(w_i, z); \quad i=1, 2, \ldots, p-1 \qquad (29)$$

where w_i denotes the solution to the subproblem found at iteration i. However, it is important to remember that the right-hand side of (29) is bigger than or equal to that of (28), with the result that the value of q_p obtained with this constraint could be higher than that obtained with (28). Moreover, if z_p is very different from z_i, $F(w_i, z) + \lambda_i G(w_i, z)$ could be a poor approximation of the exact cut, with the risk that the solution selected may not be the right one. To prevent this, the old cuts, i.e. those obtained with a z very different from z_p, should be set aside and only the most recent ones applied.

The master problem

In addition to constraint (29), the master includes constraints (7) to (20) and (23). As mentioned earlier, the problem is solved by linear programming with mixed-integer variables (MPSX) but only after constraints (8), (9) and (23) have been piecewise-linearized.

The subproblem

The operating subproblem consists in determining the values of U_{ik}, EA_k and ES_k, i=1, 2,...,n, k=1, k, which minimize $\sum_{k=1}^{k} (cap.EA_k + cs.ES_k)$ under constraints (2)-(6), (22) and the following:

$$\sum_{i=1}^{n} EP_{ik} + EA_k + ES_k \geq E_k, \qquad (30)$$

where ES_k denotes the energy shed during period k and cs the cost of the shedding operation in \$/MWh.

This problem can also be solved by linear programming if constraint (24) is piecewise-linearized. However, should this approach not be satisfactory because it introduces too many new variables, it might be worthwhile solving the following non differentiable network flow problem instead. Let

$$C(DEF_k) = \begin{cases} 0 & \text{if } DEF_k \leq 0 \\ cap.DEF_k & \text{if } 0 \leq DEF_k \leq m_k PA \\ cap.m_k.PA + cs\,(DEF_k - m_k PA) & \text{otherwise} \end{cases}$$

The subproblem can their be rewritten

$$\text{minimize} \sum_{k=1}^{k} C(DEF_k)$$

under constraints (2)-(6) with $DEF_k = E_k - \sum_i EP_{ik}$.

Bissonnette, Lafond and Côté in a recent article [3] have shown how this problem can be solved using a smoothing technique developed by Bertsekas [4].

Optimal sizing of the installations

As mentioned in the introduction, the development scheme determined by the above method may fail to meet demand with the desired probability if worse sequences of river flows have a relatively high probability of occurring in the future. There are two ways of finding out this. The first is to determine the probability distribution of $\sum EP_{ik}$ but, since this calls for the solution of a Markov chain of as many dimensions as there are reservoirs in the valley, needless to say this is a difficult approach. The second way, which is much easier, is to perform Monte-Carlo simulations, i.e. simulations of the operation of hydroelectric plants in which the river flows are randomly generated, as they are in reality, and count the number of times constraint (21) is not satisfied. If the number is larger than desired, then the development scheme determined above with past river flows is not adequate and must consequently be revised.

Two types of revision are possible, if we exclude the possibility of developing other sites: to increase the capacity of the alternative energy source or to increase the capacity of some reservoirs. The first allows the value of EA_k to be increased in critical weeks and consequently reduces the number of times demand is not satisfied. The second allows more water to be stored in periods of high flows and hence reduces the number of periods in which there are shortages. Whether it is better to increase the capacity of a reservoir or that of the alternative source depends on the costs involved. For instance, it will be better to increase the size of reservoir i if the following inequality holds:

$$crv_i - \sum_{k=1}^{k} cap.\gamma_k.\Delta G_k \leq (cav - cpv_j)\frac{\partial PA}{\partial HR_i} \qquad (31)$$

where ΔG_k denotes the expected increase in the hydroelectric production in period k. The left-hand side of inequality (31) represents the increase in the total cost if the height of dam i is increased by one meter. On the one hand, the construction cost will increase by crv_i, but, on the other hand, the operating cost will decrease by $\sum cap.\gamma_k.\Delta G_k$ since there will be less spillage and hence more hydroelectric energy produced. The right-hand side of (31) represents the cost of increasing the capacity of the alternative energy source by $\partial PA/\partial HR_i$ MW, i.e. an amount that will produce the same results as increasing the height of dam i by one meter. However, if the capacity of the alternative source is increased, the capacity of some hydroelectric plants, for instance plant j, can be reduced by the same amount without violating constraint (20), which explains the presence of the term $- cpv_j$ in (31).

The additional hydroelectric production, ΔG_k, can be determined by performing Monte-Carlo simulations first with the original height of dam i, then with the height increased by one foot, and by computing the differences in the hydroelectric energy produced. However, before performing Monte-Carlo simulations, the operating policy of the installations has to be determined. This can be done by straightforward stochastic dynamic programming if the number of reservoirs is smaller than five and by dynamic programming combined with a decomposition method [5] otherwise.

The Monte-Carlo simulations allow not only ΔG_k to be determined but also the number of times constraint (21) will be violated. Subsequently, it is easy to find $\partial PA/\partial HR_i$, that is the amount PA must be increased to obtain the same results.

The procedure of raising the height of a dam or increasing the capacity of the alternative source should be repeated until constraint (21) is satisfied with the desired probability.

REFERENCES

A.M. Geoffrion, Journal of Optimization Theory and Applications, 10(1972), 237.
A.M. Geoffrion, Management Science, 16(1970).
V. Bissonnette, L. Lafond and G. Côté, 1985, PICA Conference.
D.P. Bertsekas, Mathematical Programming, 3(1975), 1.
A. Turgeon, Water Resources Research, 16(2) (1980), 275.

OPTIMAL ENERGY EXCHANGE IN A DECENTRALIZED POWER POOL

S. H. Wan*

Department of Engineering-Economic Systems, Stanford University, Stanford, CA 94305, USA, Energy Systems Division, Systems Control, Inc., Palo Alto, CA 94303, USA

ABSTRACT

Due to the natural diversity of incremental generating costs among interconnected power systems, electric utility companies have been "pooling" together to improve economics of operation by exchanging energy. In decentralized power pools, each utility still retains the control over its generating facilities. They report to a computerized "broker" the hourly buy and sell bids for block energy based on their local economic dispatch computations. The broker then matches the buyers and sellers to determine the flow of energy exchange.

The main drawback of the "greedy" brokerage method is presented in this paper. The mathematical model for the decentralized power pool is presented as a generalized minimum cost flow problem with convex node production costs. A new combinatorial optimization algorithm, Maximum Path Return method, (MPR) is shown suitable as a substitute for the brokerage method. At each iteration, the method finds the maximum return (cost reduction) of a unit flow augmentation. This return is the buying price of the receiving node less the sum of the selling price of the sending node and the transmission cost of the cheapest path connecting the two nodes.

This method takes the network topology into consideration. It combines the marginal analysis algorithm for discrete convex production allocation problems and the path algorithm embodied in many LP network flow methods. A special scaling technique for the MPR method is presented. The scaling MPR method solves the problem in $O(n^4)$ polynomial time, where n is the number of utilities in the network.

1. Power Pooling

Throughout its history, the electric power industry has become more and more interconnected as utilities have built transmission links to their neighbors. Due to natural diversities of load distributions, types of fuels and generating units, a utility may have its incremental generation cost significantly different from that of other interconnected utilities. In such cases, it will be cheaper for the higher cost utility to reduce generation on its own system and purchase energy from the lower cost utility. Such transactions are known as economy energy exchanges. It is estimated that the cost savings from bulk energy exchange are in the range of 3 to 5 percent, as compared to isolated operation (Power Pooling in the United States, 1981). In order to capture the full economic benefits, utilities have been pooling together to have some exchange arrangements and a center to coordinate, which are referred to power pools.

The Florida Energy Broker System represents the type of decentralized power pools in the U.S. A minicomputer, as an electronic broker, matches the lowest incremental cost utility with the highest decremental cost utility, based on their hourly sell and buy bids. It then matches the next-to-lowest-cost and the next-to-highest-cost utilities, and so on. The calculation determines the transfer amounts and the transaction costs and makes this information available to utilities for decision. The utilities have bilateral agreements between each other and transmission arrangements which allow the exchanges to take place after each match. If a proposed match violates transmission constraints, it is omitted and the next match is determined. In 1980, $43 million cost savings was reported by the Florida Broker System (Power Pooling in the United States 1981). At the end of 1980, over 80 utilities in 23 states of the U.S. were participating in energy-broker arrangements.

2. The Broker Algorithm and Marginal Analysis

Assume in a power pool, each utility determines its current buy and sell bids as follows:

$$BP_i\,(G_i) = \theta_i\,(G_i) - \theta_i\,(G_i - 1)$$

$$SP_i\,(G_i) = \theta_i\,(G_i + 1) - \theta_i\,(G_i)$$

where G_i is the current generation level and θ_i, the cost function of a utility, is assumed to be convex.

The broker actually solves the following problem.

(Problem 1)

$$\text{Min } \Sigma\, \theta_i\,(G_i)$$

$$\text{s.t. } \Sigma\, G_i = \Sigma\, L_i$$

$$\underline{G}_i \leqslant G_i \leqslant \overline{G}_i \qquad i = 1,\ldots,n$$

$$G_i \in \{0,1,2,\ \ldots.\}$$

where L_i is the local load of utility i. L_i, $\underline{G}_i$ and $\overline{G}_i$ are all integers.

For the above discrete convex production allocation problem, Marginal Analysis is an efficient algorithm (Fox, 1966). It considers two related

* A citizen of the People's Republic of China.

questions about a feasible allocation $G = (G_i,\ldots,G_n)$. Is it profitable to increase production of some i by 1 unit and decrease production of some other j by 1 unit? If so, which switch increases profit (decreases cost) the most?

$$\text{Let } I(G) = \min_{G_i < \overline{G}_i} \{SP_i\ (G_i)\}$$

$$D(G) = \max_{G_j > \underline{G}_j} \{BP_j\ (G_j)\}$$

Marginal analysis will stop if $I(G) \geq D(G)$. Otherwise, it switches 1 unit from i to j if $I(G) = SP_i\ (G_i)$ and $D(G) = BP_j\ (G_j)$.

An important property of the marginal analysis method, due to the monotonicity of marginal costs in convex programming, is stated by the following theorem (Denardo, 1983).

Theorem 2.1 (Important Property of Marginal Analysis)

When marginal analysis algorithm is applied to a convex allocation problem, if a variable G_i is increased (alternatively, decreased) at any given iteration, it will not be decreased (alternatively, increased) at any subsequent iteration.

In an Energy Broker System, successive bids are required from each matched pairs by the center and the broker actually performs a marginal analysis algorithm. The major drawback of this algorithm is that it ignores the network topology of the energy exchange problem. As a result, although the algorithm can correctly solve problem 1, it does not necessarily provide optimal solution for the energy exchange. Doty and McEntire have given an example to show this point (Doty, 1982).

3. Generalized Minimum Cost Flow Problem

In order to improve the efficiency of energy exchange in a broker system, lets formulate the problem as follows:

(Problem 2)

$$\text{Min } \Sigma\ \theta_i\ (G_i) + \Sigma\ C_{ij}\ (x_{ij})$$

$$\text{s.t. } G_i + \Sigma\ x_{ji} - \Sigma\ x_{ij} = L_i$$

$$0 \leq x_{ij} \leq K_{ij}$$

$$\underline{G}_i \leq G_i \leq \overline{G}_i$$

L_i, K_{ij}, $\underline{G}_i$, $\overline{G}_i$ are all integers and the decision variables G_i and x_{ij} are required to be integers.

The transmission cost C_{ij} is mainly caused by transmission losses. We shall assume $C_{ij}\ (\cdot)$ to be linear first. Our algorithm, however, will apply to any convex functional forms of $C_{ij}\ (\cdot)$.

This problem is a generalization of the linear minimum cost flow problem. For if $C_{ij}\ (\cdot)$ is linear and $\overline{G}_i = \underline{G}_i$, then the constant term of the node costs, $\Sigma\ \theta_i\ (G_i)$, can be dropped and the problem degenerates to a standard minimum cost flow problem.

4. Path Algorithm

The generalized minimum cost flow problem can be transformed to a convex cost network flow problem, which has been studied by Beake (Beale, 1955), Hu (Hu, 1968), Klein (Klein, 1967), and Minty (Minty, 1961). All their algorithms are essentially extensions of some LP algorithms.

Since the network topology is another essential structure of the problem in addition to the discrete convexity, we should explore this aspect by first giving the following concept of incremental network.

Definition 4.1

In a linear network flow problem, the incremental network associated with a flow X is defined by assigning the following arc lengths.

$$a_{ij} = \begin{cases} C_{ij} & \text{if } 0 \leq x_{ij} < K_{ij},\ x_{ji} = 0 \\ -\ C_{ij} & \text{if } x_{ij} = 0, \quad x_{ji} > 0 \\ \infty & \text{if } x_{ij} = K_{ij}, \quad x_{ji} = 0 \end{cases}$$

The following two facts are well-known in combinatorial optimization literatures:

1) For a LP network flow problem, a flow X is optimal if and only if there is no cycle of negative length in the incremental network.

2) Flow augmentations along shortest paths in the incremental network do not generate any negative cycle.

Zadeh (Zadeh, 1979) discovered that in all network flow methods, including simplex, out-of-Kilter, dual simplex, Lemke's complementary pivot algorithm etc., a shortest path tree in the incremental network is computed by mimicking the Dijkstra's algorithm (Lawler, 1976) and flows are then sent along a sequence of minimum cost paths.

5. Maximum Path Return Method (MPR)

With the concept developed in the last section, the maximum return achievable by sending 1 unit energy from i to j is given by the following path return.

$$PR_{ij} = BP_j\ (G_j) - SP_i\ (G_i) - U_{ij}.$$

where U_{ij} is the length of a shortest path from i to j in the incremental network.

Thus, the broker algorithm becomes the following Maximum Path Return algorithm (MPR).

MPR Method

Start with a feasible solution without negative cycle in the incremental network. Augment a unit flow from node i to node j along a shortest path if PR_{ij} is the maximum positive path return. Iterate until all PR's become non-positive.

In general, $\underline{G}_i \leq L_i \leq \overline{G}_i$ for all i, thus $(G,X) = (L,0)$ is an initial feasible solution without negative cycle in the incremental network. Otherwise, some procedures can be used to obtain such an initial feasible solution (Wan, 1985).

The following is a numerical example of an eight utility network. The link costs and capacities are specified in the figure, e.g., $C_{12} = 0.3$ and $K_{12} = 5$.

TABLE 1 Generation Parameters for MPR Example

Utility	Load	Cost Function	$\overline{G}$	$\underline{G}$
1	26	$0.2G^2 - G$	40	25
2	3	$3G^2 - 1.5G$	5	1
3	2	$8G^2 - 5.5G$	3	1
4	2	$G^2 + 9G$	5	1
5	7	$1.6G^2 + G$	10	4
6	3	$3.5G^2 + 4G$	8	1
7	16	$0.4G^2$	30	10
8	25	$0.22G^2 - G$	40	10

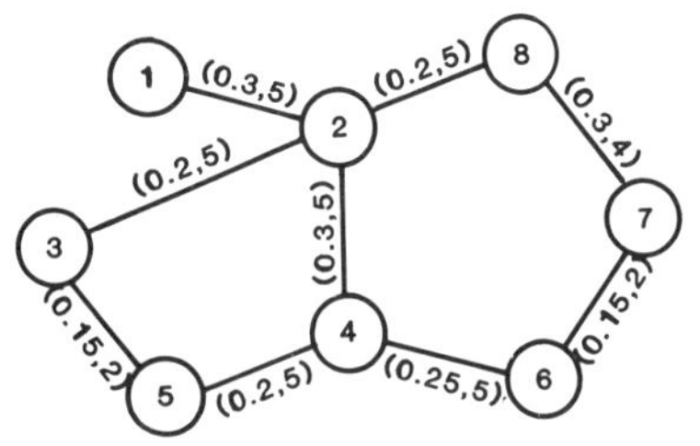

Figure 1

TABLE 2 MPR Iterations

Step	Augmenting Path	MPR (Cost Reduction)
1	1 - 2 - 3 - 5	11.55
2	8 - 7 - 6	10.83
3	1 - 2 - 3	8.0
4	1 - 2 - 3 - 5	7.55
5	8 - 2 - 4 - 5	4.04
6	1 - 2 - 8 - 7 - 6	3.15
7	8 - 2	2.2
8	1 - 2 - 8 - 7	0.8

The total cost is reduced from 518.5 to 470.38.

From the above example, we see that

1) nodes in a subset S = {1,8} only send flows out and nodes in another subset R = {2,3,5,6,7} only receive inflows;

2) the path returns (cost reductions) at successive iterations are monotonically decreasing.

6. An $O(n^2)$ Algorithm for Each MPR Iteration

In order to find the maximum path return at each MPR iteration, a mechanical application of the MPR principle would require a large amount of computations in both determining shortest paths between all the node pairs and the crucial comparisons. Therefore, we want to apply a combinatorial optimization technique and to take advantage of the fundamental property of the marginal analysis for developing an efficient algorithm.

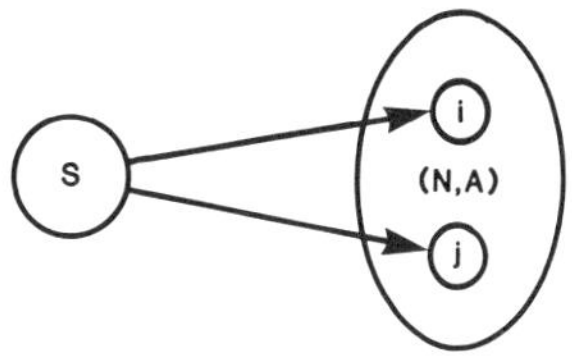

Fig. 2

Let N be the set of nodes and A be the set of arcs. In Figure 2, we append a source node s to N. Let the length of a directed arc (s,i) be $a_{si} = SP_i\ (G_i)$ (the sell price of node i) for each node i in N. Since there is no negative cycle in (N,A) at each iteration, there exist shortest paths from s to all nodes in N. Let $\lambda_i^{(k)}$ be the length of a shortest path from s to node i before a MPR flow augmentation at iteration k. Since $\lambda_i^{(k)} = \min\ (a_{si}^{(k)} + U_{ij}^{(k)})$, the maximum path return can be determined by

$$PR_{i^*j^*} = \max PR_{ij}^{(k)} = \max\ (BP_j^{(k)} - \lambda_j^{(k)}) \qquad (1)$$

and the augmenting path from i* to j* together with the arc (s,i*) form a shortest path from s to j*. Therefore, the computation for shortest paths from s to all nodes in N is the central part of a MPR iteration. In general, this computation requires $O(n^3)$ running time where n is the number of nodes in N, i.e., the number of additions and comparisons of the computation is bounded by a polynomial of order n^3. Zadeh's analysis of the path algorithm suggests that the $O(n^2)$ Dijkstra's shortest path algorithm is the most efficient way in all different versions of the solution methods. Therefore, we shall apply Dijkstra's computation by extending Edmonds & Karp's idea (Emands, 1972) to achieve an $O(n^2)$ algorithm for each MPR iteration.

By the definition of λ, we have for each arc (i,j),

$$\lambda_i^{(k)} + a_{ij}^{(k)} - \lambda_j^{(k)} \geq 0 \qquad (2)$$

Lemma 6.1

Let $a_{ij}^{(k+1)}$ be the arc length after a MPR flow augmentation at iteration k.

Then $\lambda_i^{(k)} + a_{ij}^{(k+1)} - \lambda_j^{(k)} \geq 0. \qquad (3)$

Proof

Let the sending and receiving nodes at iteration k be i* and j* respectively. We only need to consider arcs (s,i*), (s,j*) and the arcs along the augmenting path. Since for all other arcs, $a_{ij}^{(k)} = a_{ij}^{(k+1)}$ and (3) is equivalent to (2).

For the sending node i*, $G_{i*}^{(k+1)} = G_{i*}^{(k)} + 1$. By convexity, $a_{si*}^{(k+1)} = SP_{i*}(G_{i*}^{(k+1)}) \geq SP_{i*}(G_{i*}^{(k)}) = a_{si*}^{(k)}$. Therefore, (3) holds for (s,i*).

For the receiving node j*, $G_{j*}^{(k+1)} = G_{j*}^{(k)} - 1$ and $a_{sj*}^{(k+1)} = SP_{j*}(G_{j*}^{(k+1)}) = BP_{j*}(G_{j*}^{(k)})$. From (1), $PR_{i*j*} = BP_{j*}(G_{j*}(G_{j*}^{(k)}) - \lambda_{j*}^{(k)}$. Since $PR_{i*j*} > 0$ and $\lambda_s = 0$, we have $\lambda_s^{(k)} + a_{sj*}^{(k+1)} - \lambda_{j*}^{(k)} > 0$.

For any arc (i,j) associated with the augmenting path from i* to j*, $a_{ij}^{(k+1)} < a_{ij}^{(k)}$ only if $x_{ij}^{(k)} = x_{ji}^{(k)} = 0$ and the unit flow traversed from j to i such that $x_{ij}^{(k)} = 0$ and $x_{ji}^{(k)} = 1$. In this case, $a_{ij}^{(k)} = C_{ij}$ and $a_{ij}^{(k+1)} = - C_{ij}$. However, since arc (j,i) is on the shortest path from s to i, we have $\lambda_i^{(k)} = \lambda_j^{(k)} + a_{ji}^{(k)} = \lambda_j^{(k)} + C_{ij}$. Therefore, we still have $\lambda_i^{(k)} + a_{ij}^{(k+1)} - \lambda_j^{(k)} = 0$.

Definition 6.1

Let $\underline{a}_{ij} = \lambda_i^{(k)} + a_{ij}^{(k+1)} - \lambda_j^{(k)}$ be the modified arc length.

Lemma 6.2

Any cycle length with respect to $a_{ij}^{(k+1)}$ is equivalent to the length with respect to $\underline{a}_{ij}$. A shortest path from p to q with respect to modified arc lengths $\underline{a}_{ij}$ is also a shortest path with respect to the regular length $a_{ij}^{(k+1)}$, differing in length by $\lambda_p^{(k)} - \lambda_q^{(k)}$, i.e.,

$$\underline{U}_{pq} = \lambda_p^{(k)} + U_{pq}^{(k+1)} - \lambda_q^{(k)}.$$

Proof

Conclusions hold trivially by definition 6.1.

Lemma 6.3

Let $\underline{U}_{sj}$ be the length of a shortest path with respect to modified arc lengths.

Then $\lambda_j^{(k+1)} = \lambda_j^{(k)} + \underline{U}_{sj}$ (4)

Proof

It is a consequence of lemma 2 since $\lambda_j^{(k+1)} = U_{sj}^{(k+1)}$ and $\lambda_s = 0$.

By lemma 6.1, all modified arc lengths are non-negative, therefore, the $O(n^2)$ Dijkstra's algorithm (Lawler, 1976) can be applied to calculate $\underline{U}_{sj}$. According to lemma 6.3, λ_j can be updated by adding $\underline{U}_{sj}$ after each MPR flow augmentation.

Theorem 6.1

Assuming the evaluations for node incremental and decremental costs do not take any time, the computational bound for each MPR iteration is $O(n^2)$.

Proof

The computational bound of Dijkstra's algorithm has a higher polynomial order than other computations in a MPR iteration. Therefore, $O(n^2)$ is also a bound for each MPR iteration.

7. Fundamental Theorem of MPR Method

From the numerical example of Section 5, we made two observations. The first observation leads to the following theorem.

Theorem 7.1 (Fundamental Theorem)

In the MPR method, if a node i sends 1 unit flow out, then the path return from any other node q to i, PR_{qi}, will always be non-positive. And node i will never receive any inflow. If a node j receives 1 unit inflow, then $PR_{jq} \leq 0$ and j never sends any flow out.

Proof

For sending node i:

From lemma 6.3, $\lambda_i^{(k+1)} = \lambda_i^{(k)} + \underline{U}_{si} \geq \lambda_i^{(k)}$ since $\underline{U}_{si} \geq 0$. Suppose node i sends 1 unit flow out at iteration k, then

$$\lambda_i^{(k)} = a_{si}^{(k)} = SP_i(G_i^{(k)}) = BP_i(G_i^{(k)} + 1) = BP_i(G_i^{(k+1)}).$$

For any other node q in N,

$$PR_{qi}^{(k+1)} = BP_i^{(k+1)} - U_{qi}^{(k+1)} - SP_q^{(k+1)}$$

$$\leq BP_i^{(k+1)} - \lambda_i^{(k+1)} = \lambda_i^{(k)} - \lambda_i^{(k+1)} \leq 0$$

Therefore, i will not receive inflow at iteration k+1. The conclusion follows an induction.

Similarly, we can prove that a receiving node will not send flow out.

The second observation is justified by the following corollary.

Collary 7.1

The maximum path returns are monotonically decreasing in MPR iterations.

Proof

According to the above theorem, all the receiving nodes are in a subset R. Thus equation (1) can be rewritten as

$$\max PR_{ij}^{(k)} = \max \{BP_j^{(k)} - \lambda_j^{(k)} \mid j \in R\}$$

Since $\lambda_j^{(k)}$ is increasing and $BP_j^{(k)}$ is decreasing in k (since G_j is decreasing) respectively, the conclusion follows.

Corollary 7.2

The MPR method terminates in finite iterations. The number of iterations is equal to the number of units transferred from a sending subset S to a receiving subset R.

8. Scaling

If we regard a unit augmenting path in the MPR method as a feasible direction to move the current solution (G,X), then it is easy to show that the maximum path return corresponds to the most negative directional difference with respect to a fixed step size of 1 unit. The MPR method is thus a feasible algorithm of steepest descent directions. Moreover, the descent magnitudes are monotonically decreasing in the number of iterations according to Corollary 7.1.

The above interpretation naturally leads to the scheme of scaling the "unit" size. Initially, we choose two smallest integers L_1 and L_2 such that $2^{L1} \geq \max (\overline{G}_i - \underline{G}_i)$ and $2^{L2} \geq \max K_{ij}$. Let $L = \min (L_1, L_2)$. The "unit" sizes at scaling steps 1,2,..., L are $2^{L-1}, 2^{L-2},..., 1$ respectively. Thus at scaling step L, we achieve the desired solution.

The above scaling scheme is different from those proposed by Edmonds & Karp (Edmonds, 1972), and Rock (Rock, 1980). It does not change the arc capacities at each step, neither the arc costs. The flow solution at each scaling step is feasible for the original problem as well as for the problem of the next scaling step.

In reality, the scaling factor does not have to be 2. For example, in a power pool, all the utility companies may agree to exchange energy with a "unit" (or a block as called in the industry) size of 10 MW first, then 1 MW, etc. The scaling factor 2 is suitable for the analysis of computational complexity. This scaling technique is obviously natural and applicable in practice.

9. Computational Complexity

Detailed computational complexity analysis is beyond the scope of this paper. However, we shall present the major results in this section.

Since we can use the $O(n^2)$ Dijkstra's shortest path algorithm for each MPR flow augmentation, the total number of elementary operations in the MPR method without the scaling is bounded by $O(n^2 V)$, where V is the number of units switched in the network. However, V is not limited by the problem size (either the number of nodes or the number of arcs).

The fundamental theorem of the MPR method tells us that all these V units must be switched from a subset of nodes, S, to another subset of nodes, R. The number of the node pairs between S and R is bounded by $n^2/4$, where n is the number of the nodes in the network. It can be shown (Wan, 1985) that with the new scaling procedure, at most 1 "unit" at each scaling step can be transferred between each node pair from S to R if there is no arc capacity constraints. For if it were profitable to send 2 "units" from i to j at scaling step k, then they (equivalent to 1 "unit" at the previous scaling step) should have been transferred from i to j at scaling step k-1. Therefore, the computational bound for problems without arc capacity constraints is given by $(n^4 L/4)$, where L is the number of scaling steps and L is limited by the logrithm of the capacity parameters.

For problems having arc capacity constraints, it is possible to have negative cycles at the beginning of each scaling step. These negative cycles, however, can be removed efficiently (Wan, 1985) and the polynomial bound is not affected.

10. Convex Transmission Costs

As mentioned in Section 3, the transmission cost functions do not have to be linear. Suppose the transmission cost functions are strictly convex. In this situation, we can define the arc lengths at each MPR augmentation as follows:

$$a_{ij} = C_{ij}(x_{ij} + 1) - C_{ij}(x_{ij}) \quad \text{if } x_{ij} < K_{ij} \text{ and } x_{ij} = 0$$

$$= \infty \quad \text{if } x_{ij} = K_{ij}$$

$$a_{ji} = - [C_{ij}(x_{ij}) - C_{ij}(x_{ij} -1)] \text{ if } x_{ij} > 0$$

The computational efficiency of the MPR method will not be affected by the above definition.

11. Power Pool Revisited

The MPR method, although deep and rich in theory and computation, is not a pure mathematical or computational exercise. It is rooted in real problem solving. When it is applied to a decentralized power pool as a substitute for the broker algorithm, several advantages and some further challenges are presented.

As all other primal-feasible algorithms, the MPR method can be started with decision maker's guess for an optimal solution. Dynamic programming,

(Doty, 1982) "build-up" algorithm (Hu, 1966) etc. cannot make use of this information. The explicit cost functions of local utilities are not required. Instead, only successive buy bids of some utilities and sell bids of some other utilities are required by the center. Since there is no negative cycle, the power flow pattern is optimal at each iteration given the generation levels at that iteration. Therefore, the algorithm can be stopped even before the completion of the algorithm to get a sub-optimal solution.

The method can provide a complete set of bilateral transactions within the pool. For each transaction, only the sending utility, receiving utility and the wheeling utilities on the path are involved. A possible procedure for saving allocation would be as follows. Let the buying utility pay the selling utility and wheeling utilities to fully compensate their costs. The net benefit PR is then split in some way between the sending and receiving utilities. The problem is, will this simple billing procedure be "fair" such that there is no incentive for a subset of utilities to form a subcoalition and leave the pool ? many economists and operations researchers have already been interested in this production game problem. An open question is, will the MPR method provide some new insight into this complicated issue?

ACKNOWLEDGEMENTS

The author wishes to thank Professor D.G. Luenberger, D. Dantzig and M. Beale for their helpful discussions on the subject. He is also indebted to his colleagues at Systems Control, Inc. for their support.

REFERENCES

Beale, E.M.L., 1955). "On Minimizing a Convex Function Subject to Linear Inequalities," Journal of the Royal Statistical Society, Series B, Vol. XVII, 173-184.

Dantzig, G.B., (1963). "Linear Programming and Extensions," Princeton University Press, Princeton.

Denardo, E.V., (1983). "Dynamic Programming," Prentice-Hall, Englewood Cliffs.

Doty, K.W., and McEntire, P.L., (1982). "An Analysis of Electric Power Brokerage Systems," IEEE Trans. PAS-101, No. 2.

Emands, J., and R.M. Karp, (1972). "Theoretical Improvements in Algorithmic Efficiency for Network Flow Problems," J.A.C.M. 19, 248-264.

Fox, B.L., (1966). "Discrete Optimization via Marginal Analysis," Management Science, Vol. 13, No. 3, November, 210-215.

Hu, T.C., (1966). "Minimum-Cost Flows in Convex-Cost Networks," Naval Research Logistics Quarterly, 13, 1-9.

Klein, M., (1967). "A Primal Method for Minimal Cost Flows with Applications to the Assignment and Transportation Problems," Management Science, Vol. 14, No. 3, November, 205-220.

Lawler, E., (1976). "Combinatorial Optimization: Networks and Matroids," Holt, Rinehart and Winston, New York.

Minty, G.J., (1961). "Solving Steady-State Nonlinear Networks of 'Monotone' Elements," I.R.E. Trans. Circuit. Theory, CT-8, 99-104.

Papadimidriou, C.H., and Steiglitz, K., (1982). "Combinatorial Optimization," Prentice-Hall, Inc., Englewood.

Power Pooling in the United States, (1982). Federal Energy Regulatory Commission, Office of Electric Power Regulation, Washington, D.C., December.

Rock, H., (1980). "Scaling Techinques for Minimum Cost Network Flows", in Discrete Structures and Algorithms, edited by Uwe Pape, Carl Hanser Verlag Munchen Wien, 181-191.

Wan, S.H., (1985). "Maximum Path Return Method," Technical Report, Dept. of Engineering-Economic Systems, Stanford University.

Wood, A.J. and Wollenberg, B.F., (1984). "Power Generation, Operation, and Control," John Wiley & Sons, New York.

Zadeh, N., (1979). "Near-Equivalence of Network Flow Algorithms," Technical Report No. 26, Dept. of Operations Research, Stanford University.

OPTIMIZATION OF LARGE SCALE TIME-DELAY SYSTEMS: TWO INTERACTION PREDICTION ALGORITHMS WITH CONVERGENCE PROOFS AND APPLICATION

Cai Xiaoqiang and Zhou Junren

Department of Automation, Tsinghua University, Beijing, China

Abstract. Two multilevel algorithms are developed to solve the problem of optimizing a large scale system with multiple delays. It is shown that the algorithms convert the problem, even if its cost function is nonseparable, into a sequence of independent subproblems of optimizing smaller subsystems without delays and a coordination problem of predicting interactions. The convergences of the algorithms are studied. By applying one of the algorithms to the economic operation problem of large hydrothermal power systems, we obtain a new computational method for this problem. A numerical example is solved to compare the computational efficiency of this method with that of a current one. The results show that the former decreases the computation time on a large scale.

Keywords. Large-scale systems; Time lag systems; Delays; Optimization; Computational methods; Power systems; Hydraulic systems; economic operation.

INTRODUCTION

The optimization for a large scale system with multiple delays is computationally very difficult for two reasons: First, the existence of delays makes the computation cumbersome; Second, the high dimensionality causes some problems, in particular those termed "the curse of dimensionality". Therefore, in order to solve the optimization problem effectively, the computational methods used should be of the ability to get arould both of the difficulties.

It seems that many of existing methods were not provided with this ability. For instance, the traditional approach to problems with delays can reduce the delay model of a discrete system to a non-delay one by augmenting the state vector; however, it increases the dimension enormously. Many multilevel methods (Mahmoud, 1977) can replace a single large problem by a sequence of smaller subproblems (and a problem of coordination). However, these methods are generally constructed for non-delay systems.

A two-level method especially for delay problems has been presented by Tamura (1975). Although the lower level of this method is effective, the coordination of this method is inefficient if the problem to be solved is large.

In the case without delays, many papers have pointed out that the multilevel methods based on the interaction prediction principle (Measarovic, 1970) are computationally efficient. One purpose of this paper is to develop two interaction prediction methods for the large time-delay problem. By decomposing the system in space and in time, we can see that the difficulties arising from the high dimensionality and the delays can be avoided if two sets of interactions are coordinated simultaneously. Respectively extending the coordination scheme of Takahara (1965) and that of Cohen (1977) to the coordination of these interactions, we obtain the two algorithms. The convergences of the algorithms are proven.

The second purpose of this paper is to present a new algorithm for the economic operation problem of large hydrothermal power systems. It has been shown by Soares, Lyra and Tavares (1980) that, to solve this problem, a subproblem of optimizing a large scale system with multiple delays must be solved a number of times. Extending one of the interaction prediction methods developed in this paper to the solution of this subproblem, we form the new algorithm for the hydrothermal problem. Numerical results show that this algorithm is very fast. Compared with the algorithm presented by Soares, Lyra and Tavares (1980), which uses Tamura's method to solve the subproblem, this algorithm decreases the computation time enormously.

STATEMENT OF THE PROBLEM

The large time-delay problem considered is as follows:

$$P_1 \begin{cases} \min\ J=\frac{1}{2}X(N)^T Q(N)X(N)+\frac{1}{2}\sum_{K=0}^{N-1}\{X(K)^T Q(K)X(K)+U(K)^T R(K)U(K)\} & (1)\\ \text{s.t. } X(K+1)=\sum_{L=0}^{\theta}\{A^L(K)X(K-L)+B^L(K)U(K-L)\}, \quad K=0,1\ldots N-1 & (2)\\ X(-L)=a(L), \quad L=0,1\ldots\theta & (3a)\\ U(-L)=b(L), \quad L=1,2\ldots\theta & (3b)\end{cases}$$

where (2) are the state equations of the system; θ is the maximum time delay constant; a(L) and b(L) are initial vectors; $X(K)\in R^n$ and $U(K)\in R^r$ are respectively the state and control vectors; other matrices and vectors are real and of appropriate dimensions. It is assumed that $Q(K+1)\geq 0$ and $R(K)>0$ $(0\leq K\leq N-1)$.

DECOMPOSITION

Spatial Decomposition

By using some proper decomposition method (a direct method is to decompose the system according to its natural structure), we can decompose the system into M $(M\geq 1)$ subsystems. The state equation of the ith subsystem is:

$$X_i(K+1)=\sum_{L=0}^{\theta}\sum_{j=1}^{M}\{A_{ij}^L(K)X_j(K-L)+B_{ij}^L(K)U_j(K-L)\}, \quad (4)$$

where $X_j(K)$ and $U_j(K)$ are respectively the n_j $(n_1+\ldots+n_M=n)$ dimensional state vector and r_j $(r_1+\ldots+r_M=r)$ dimensional control vector of the jth subsystem; $A^L_{ij}(K)$ and $B^L_{ij}(K)$ are respectively the corresponding blocks of $A^L(L)$ and $B^L(L)$. (All the matrices and vectors in P_1 are partitioned accordingly.)

From (4) and the optimality conditions for P_1 [see (11)~(13)], it is not hard to see that subsystem i is related to the other subsystems through:

$$\pi_{S1i}(K)=\sum_{j\neq i}\sum_{L=0}^{\theta}\{A^L_{ij}(K)X_j(K-L)+B^L_{ij}(K)U_j(K-L)\},\quad K=0\ldots N-1 \tag{5}$$

and, if the cost function is nonseparable, through:

$$\pi_{S2i}(K)=\sum_{j\neq i}Q_{ij}(K)X_j(K),\qquad K=1\ldots N \tag{6}$$

$$\pi_{S3i}(K)=\sum_{j\neq i}R_{ij}(K)U_j(K),\qquad K=0\ldots N-1 \tag{7}$$

Now let us call π_{S1}, π_{S2} and π_{S3} spatial interactions.

Temporal Decomposition

The multi-stage process of the discrete system may be viewed as the composite of N subsystems. The Kth subsystem is defined by the overall state of the system at the discrete instant K: X(K).

In the case without delays, the interconnection structure of the subsystems may be represented as Fig. 1. Obviously, in this case, subsystem K+1 (K=1...N-1) is affected only by subsystem K. However, in the case that the discrete system has multiple delays, the interactions among the subsystems are increased, subsystem K+1 is affected by subsystem K-L $(L=0\ldots\underline{\theta})$ (if $K>\theta$, then $\underline{\theta}=\theta$; otherwise $\underline{\theta}=K-1$). This tells us that it is the additional interactions that cause delays. These additional interactions can be represented as:

$$\pi_t(K)=\sum_{L=1}^{\theta}\{A^L(K)X(K-L)+B^L(K)U(K-L)\},\quad K=1\ldots N-1 \tag{8}$$

Let us call π_t temporal interactions.

In the first section, we have pointed out that there may be two difficulties in solving the large optimization problem with delays. Now we can say that it is the temporal interactions and the spatial interactions that cause these difficulties respectively. Roughly speaking, if a proper coordination scheme is used to coordinate the spatial interactions, the overall problem P_1 may be converted into M independent subproblems, one for each subsystem of lower dimension. However, in general, the subproblems include delays. On the other hand, if the temporal interactions are coordinated by a proper scheme, the delay problem P_1 may be replaced by a non-delay problem. However, this non-delay problem is generally of high dimensionality. Therefore, to overcome both of the difficulties, the two classes of interactions should be coordinated simultaneously. In this paper, we hope that this may be performed by one coordinator. For this reason, let us combine the spatial interactions in the system model $[\pi_{S1i}(K)]$ with the temporal interactions as:

$$\pi_{Ii}(K)=\sum_{j\neq i}\{A^0_{ij}(K)X_j(K)+B^0_{ij}(K)U_j(K)+\sum_{j=1}^{M}\sum_{L=1}^{\theta}\{A^L_{ij}(K)X_j(K-L)+B^L_{ij}(K)U_j(K-L)\}. \tag{9}$$

P_1 may equivalently be reformulated as follows:

$$P_2\begin{cases}\min\ (1)\\ \text{s.t. } X_i(K+1)=A^0_{ii}(K)X_i(K)+B^0_{ii}(K)U_i(K)+\pi_{Ii}(K), \quad (10)\\ (3),(9),\quad i=1\ldots M;K=0\ldots N-1\end{cases}$$

COORDINATION ALGORITHMS

Algorithm 1

According to nonlinear programming theory, one may derive the following optimality conditions for P_2 (λ_i and β_i are respectively the co-state variables for (10) and the Lagrange multipliers for (9)):

$$S_2\begin{cases}\text{For } i=1\ldots M;\ K=0,1\ldots N-1:\\ R_{ii}(K)U_i(K)+\pi_{S3i}(K)+B^0_{ii}(K)^T\lambda_i(K+1)-\sum_{j\neq i}B^0_{ji}(K)^T\beta_j(K)-\sum_{L=1}^{\theta}\sum_{j=1}^{M}B^L_{ji}(K+L)^T\beta_j(K+L)=0\\ \lambda_i(K)=A^0_{ii}(K)^T\lambda_i(K+1)+Q_{ii}(K)X_i(K)+\pi_{S2i}(K)-\sum_{j\neq i}A^0_{ji}(K)^T\beta_j(K)-\sum_{L=1}^{\theta}\sum_{j=1}^{M}A^L_{ji}(K+L)^T\beta_j(K+L)\\ \lambda_i(N)=Q_{ii}(N)X_i(N)+\pi_{S2i}(N)\\ \beta_i(K)=-\lambda_i(K+1)\\ \lambda_i(K_1)=0 \quad\text{for } K_1>N\\ (3),(6),(7),(9),(10)\end{cases}$$

Obviously, if $\pi_{Ii}(K)$, $\pi_{S2i}(K+1)$, $\pi_{S3i}(K)$ and the terms containing $\beta(K)$ are given, then P_2 is equivalent to M independent subproblems. Naturally, we construct the following algorithm (notice that $\beta_i(K)=-\lambda_i(K+1)$):

ALGORITHM 1 :

STEP 1. Choose initial values for $U^c_i(K)$, $X^c_i(K+1)$ and $\lambda^c_i(K+1)$ $(1\leqslant i\leqslant M;\ 0\leqslant K\leqslant N-1)$. Set c=0, where c is the iteration index. Go to step 3.

THE SECOND LEVEL (COORDINATOR):

STEP 2. If $\|Z^{c+1}(K)-Z^c(K)\|\leqslant\varepsilon$, where $Z^c(K)=[U^c(K)^T\ X^c(K+1)^T\ \lambda^c(K+1)^T]^T$ (K=0...N-1) (ε is a prespecified tolerance), let $U^{c+1}(K)$ be the optimal control and stop. Otherwise, make c=c+1, continue.

STEP 3.a. Compute $\pi^c_{Ii}(K)$, $\pi^c_{S2i}(K+1)$ and $\pi^c_{S3i}(K)$ $(1\leqslant i\leqslant M;\ 0\leqslant K\leqslant N-1)$ by substituting U^c, X^c and (3) into (9), (6) and (7).

STEP 3.b. Compute:

$$\pi^c_{2i}(K)=\sum_{j\neq i}A^0_{ji}(K)^T\lambda^c_j(K+1)+\sum_{L=1}^{\theta}\sum_{j=1}^{M}A^L_{ji}(K+L)^T\lambda^c_j(K+1+L)+\pi^c_{S2i}(K)\qquad K=1\ldots N$$

$$\pi^c_{3i}(K)=\sum_{j\neq i}B^0_{ji}(K)^T\lambda^c_j(K+1)+\sum_{L=1}^{\theta}\sum_{j=1}^{M}B^L_{ji}(K+L)^T\lambda^c_j(K+1+L)+\pi^c_{S3i}(K)\qquad K=0\ldots N-1$$

$\lambda^c_i(K)=0$ for all $K>N$

THE FIRST LEVEL:

STEP 4. For i=1...M, solve the ith subproblem P_{3i}

$$P_{3i}\begin{cases}\min\ J_{3i}=\frac{1}{2}X_i(N)^TQ_{ii}(N)X_i(N)+\pi^c_{2i}(N)^TX_i(N)+\\ +\sum_{K=0}^{N-1}\{\frac{1}{2}X_i(K)^TQ_{ii}(K)X_i(K)+\frac{1}{2}U_i(K)^TR_{ii}(K)U_i(K)+\\ +\pi^c_{2i}(K)^TX_i(K)+\pi^c_{3i}(K)^TU_i(K)\}\end{cases}$$

$$\begin{cases} \text{s.t. } X_i(K+1)=A_{ii}^0(K)X_i(K)+B_{ii}^0(K)U_i(K)+\pi_{li}^c(K), & K=0\ldots N-1 \\ X_i(0)=a_i(0) \end{cases}$$

Let the solution be $U_i^{c+1}(K)$, $X_i^{c+1}(K+1)$ and $\lambda_i^{c+1}(K+1)$ $(0\leqslant K\leqslant N-1)$. Go to step 2. (Notice that the analytical solution of P_{3i} can be obtained by computing a matrix Riccati difference equation and a difference linear equation (Cai, 1984).) ∎

The above algorithm may be viewed as an extension of Takahara's method (1965) to the case with delays.

Algorithm 2

By application of nonlinear programming theory, one may derive the following optimality conditions for P_1 [λ are the co-state variables for (2)]:

$$S_1 \begin{cases} \text{For } i=1\ldots M; K=0,1\ldots N-1: \\ \sum_{j=1}^{M} R_{ij}(K)U_j(K)+\sum_{L=0}^{\theta}\sum_{j=1}^{M} B_{ji}^L(K+L)^T\lambda_j(K+1+L)=0 & (11) \\ \lambda_i(K)=\sum_{j=1}^{M} Q_{ij}(K)X_j(K)+\sum_{L=0}^{\theta}\sum_{j=1}^{M} A_{ji}^L(K+L)^T\lambda_j(K+1+L) & (12a) \\ \lambda_i(N)=\sum_{j=1}^{M} Q_{ij}(N)X_j(N) & (12b) \\ \lambda_i(K_1)=0 \quad \text{for all } K_1>N & (13) \\ (2),\ (3) \end{cases}$$

From S_1 we know that $X_i(K+1)$ and $\lambda_i(K+1)$ can be obtained if $U_i(K)$ $(K=0\ldots N-1)$ are given. Therefore, from S_1 and S_2 we construct the following algorithm:

ALGORITHM 2 :

STEP 1. Choose initial values for $U_i^c(K)$ $(1\leqslant i\leqslant M; 0\leqslant K\leqslant N-1)$. Set $c=0$, where c is the iteration index. Go to step 3.

THE SECOND LEVEL (COORDINATOR):

STEP 2. If $\|U^{c+1}(K)-U^c(K)\|\leqslant\varepsilon$ $(0\leqslant K\leqslant N-1)$, where ε is a prespecified tolerance, let $U^{c+1}(K)$ be the optimal control and stop. Otherwise, make $c=c+1$, continue.

STEP 3. Compute $X^c(K+1)$ $(0\leqslant K\leqslant N-1)$ by substituting U^c and (3) into (2).

STEP 4. Compute $\lambda^c(K+1)$ $(0\leqslant K\leqslant N-1)$ by substituting X^c and (13) into (12).

STEP 5. Same as step 3 of Algorithm 1.

THE FIRST LEVEL:

STEP 6. Obtain $U^{c+1}(K)$ $(0\leqslant K\leqslant N-1)$ by solving the subproblem P_{3i} $(i=1\ldots M)$. Go to step 2. ∎

The above algorithm is an extension of Cohen's method (1977) to the case with delays.

CONVERGENCE PROOFS

Let us define:

$\hat{R}(K)=\text{block diag}(R_{ii}(K)\mid i=1\ldots M)$,

$\hat{Q}(K)=\text{block diag}(Q_{ii}(K)\mid i=1\ldots M)$,

$\hat{A}^0(K)=\text{block diag}(A_{ii}^0(K)\mid i=1\ldots M)$,

$\hat{B}^0(K)=\text{block diag}(B_{ii}^0(K)\mid i=1\ldots M)$,

$R=\text{block diag}(R(K)\mid K=0\ldots N-1)$,

$Q=\text{block diag}(Q(K)\mid K=1\ldots N)$,

$$A=\begin{bmatrix} 0 & & & & & \\ A^0(1) & 0 & & & 0 & \\ A^1(2) & A^0(2) & \ddots & & & \\ \vdots & & \ddots & \ddots & & \\ A^\theta(\theta+1) & \ddots & & \ddots & \ddots & \\ & 0 & & A^\theta(N-1) \ldots\ldots & A^0(N-1) & 0 \end{bmatrix},$$

$$B=\begin{bmatrix} B^0(0) & & & & \\ B^1(1) & B^0(1) & & 0 & \\ \vdots & & \ddots & & \\ B^\theta(\theta) & \ddots & & \ddots & \\ & 0 & & B^\theta(N-1) \ldots\ldots & B^0(N-1) \end{bmatrix}$$

$\hat{R}=\text{block diag}(\hat{R}(K)\mid K=0\ldots N-1)$,

$\hat{Q}=\text{block diag}(\hat{Q}(K)\mid K=1\ldots N)$,

$\hat{B}=\text{block diag}(\hat{B}^0(K)\mid K=0\ldots N-1)$,

$$\hat{A}=\begin{bmatrix} 0 & \ddots & & 0 \\ \hat{A}^0(1) & \ddots & & \\ & 0 & \hat{A}^0(N-1) & 0 \end{bmatrix}, \qquad \tilde{R}=R-\hat{R},\quad \tilde{Q}=Q-\hat{Q},\quad \tilde{A}=A-\hat{A},\quad \tilde{B}=B-\hat{B},$$

$U=[U(0)^T\ldots U(N-1)^T]^T$, $X=[X(1)^T\ldots X(N)^T]^T$,

$\lambda=[\lambda(1)^T\ldots\lambda(N)^T]^T$.

LEMMA 1. For any starting vector Z^0 the iteration method $Z^{c+1}=PZ^c$ converges to $Z^*=0$ if there exits a positive definite matrix H such that the matrix given by:

$$G=H-P^THP \tag{14}$$

is positive definite.

The proof of this lemma may be found in (Young, 1971).

THEOREM 1. Assume that $R(K)>0$ and $Q(K+1)>0$ $(0\leqslant K\leqslant N-1)$. Then, if:

(a) $2\hat{R}(K)>R(K)$, $K=0\ldots N-1$;

(b) $2\hat{Q}(K)>Q(K)$, $K=1\ldots N$;

(c) $(BR^{-1}\hat{B}^T+\hat{B}R^{-1}B^T)+[(A-I)Q^{-1}(\hat{A}-I)^T+(\hat{A}-I)Q^{-1}(A-I)^T]>$
$BR^{-1}(\hat{R}+\frac{1}{2}R)R^{-1}B^T+(A-I)Q^{-1}(\hat{Q}+\frac{1}{2}Q)Q^{-1}(A-I)^T$,

Algorithm 1 starting from any initial values (U^0, X^0, λ^0) converges to the optimal solution of the problem P_1.

Proof. Let U^*, X^* and λ^* be respectively the optimal control, state and co-state of P_1. Then from S_1 we have:

$$RU^*+B^T\lambda^*=0, \tag{15}$$

$$\lambda^*=QX^*+A^T\lambda^*, \tag{16}$$

$$X^*=AX^*+BU^*+X_0, \tag{17}$$

where X_0 is a constant vector related to $a(L)$ and $b(L)$.

From step 3 and the optimality conditions for P_{3i} $(i=1\ldots M)$, we obtain:

$$\hat{R}U^{c+1}+\tilde{R}U^c+\hat{B}^T\lambda^{c+1}+\tilde{B}^T\lambda^c=0, \tag{18}$$

$$\lambda^{c+1}=\hat{Q}X^{c+1}+\tilde{Q}X^c+\hat{A}^T\lambda^{c+1}+\tilde{A}^T\lambda^c, \tag{19}$$

$$X^{c+1}=\hat{A}X^{c+1}+\tilde{A}X^{c}+\hat{B}U^{c+1}+\tilde{B}U^{c}+X_0 \quad . \tag{20}$$

Thus, form (15)~(20), we get: $\hat{S}(Z^{c+1}-Z^{c})=-SZ^{c}$, (21)
where:

$$Z=\begin{bmatrix} U^c-U^* \\ X^c-X^* \\ \lambda^c-\lambda^* \end{bmatrix}, \quad \hat{S}=\begin{bmatrix} \hat{R} & 0 & \hat{B}^T \\ 0 & \hat{Q} & (\hat{A}-I)^T \\ \hat{B} & (\hat{A}-I) & 0 \end{bmatrix}, \quad S=\begin{bmatrix} R & 0 & B^T \\ 0 & Q & (A-I)^T \\ B & (A-I) & 0 \end{bmatrix}.$$

From the assumption of the theorem, we can easily prove that $\det(S)\neq 0$ and $\det(\hat{S})\neq 0$. Therefore, from (21) we get:

$$Z^{c+1}=(I-\hat{S}^{-1}S)Z^{c} \quad . \tag{22}$$

Now we use Lemma 1 to prove the convergence of the above iteration method. Let us select:

$$H=S+\Delta S, \quad \text{where } \Delta S=\begin{bmatrix} 0 & 0 & 0 \\ 0 & 0 & 0 \\ 0 & 0 & 2BR^{-1}B^T+2(A-I)Q^{-1}(A-I)^T \end{bmatrix},$$

It is easy to prove that $H>0$. From (14) and (22), one may get:

$$G=(S+\Delta S)-(I-S\hat{S}^{-T})(S+\Delta S)(I-\hat{S}^{-1}S) \quad .$$

After some calculations, one may prove that $G>0$ under the conditions of the theorem. Therefore, according to Lemma 1, the theorem is proven.

THEOREM 2. Assume that $R(K)>0$ and $Q(K+1)\geqslant 0$ $(0\leqslant K\leqslant N-1)$. Then, if

$$2[\hat{R}+\hat{B}^T(I-\hat{A})^{-T}\hat{Q}(I-\hat{A})^{-1}\hat{B}]>[R+B^T(I-A)^{-T}Q(I-A)^{-1}B],$$

Algorithm 2 starting from any initial prediction U^0 converges to the optimal solution of P_1 .

Proof. After some simple but somewhat lengthy calculations [see (Cai, 1984)], one can obtain the following equation from the steps of Algorithm 2 and the optimality conditions for P_1 (see S_1):

$$[\hat{R}+\hat{B}^T(I-\hat{A})^{-T}\hat{Q}(I-\hat{A})^{-1}\hat{B}](U^{c+1}-U^{c})$$

$$=-[R+B^T(I-A)^{-T}Q(I-A)^{-1}B](U^{c}-U^{*}) \quad .$$

Set: $\hat{F}=\hat{R}+\hat{B}^T(I-\hat{A})^{-T}\hat{Q}(I-\hat{A})^{-1}\hat{B}$, $\quad F=R+B^T(I-A)^{-T}Q(I-A)^{-1}B$

and $Z^c=U^c-U^*$. (Obviously, the assumption of the theorem ensure $F>0$ and $\hat{F}>0$.)

Then, $$Z^{c+1}=(I-\hat{F}^{-1}F)Z^{c} \quad . \tag{23}$$

Now we apply Lemma 1 to (23) to prove its convergence. Let us select $H=F$. From (14) and (23):

$$G=F-(I-F\hat{F}^{-1})F(I-\hat{F}^{-1}F)=F\hat{F}^{-1}(2\hat{F}-F)\hat{F}^{-1}F>0 \ .$$

Therefore, according to Lemma 1, this theorem is proven.

APPLICATION

In this section we apply Algorithm 1 to the economic operation problem of large hydrothermal power systems, based on the development in the paper (Soares, Lyra and Tavares, 1980), where the formulation of the problem can be found.

Some Results Obtained in (Soares and others, 1980)

Suppose there are T thermal plants and M hydro plants in a hydrothermal power system. By applying Lagrange duality theory to the problem decomposition, Soares and others (1980) obtained such a computational structure (see Fig 2.):

1. Upper coodinator fixes $\rho^b(K)>0$ by:

$$\rho^{b+1}(K)=\rho^{b}(K)+S(K)[D^b(K)-\sum_{i=1}^{M}H_i^b(K)-\sum_{j=1}^{T}G_j^b(K)],$$
$$K=0\ldots N-1 \tag{24}$$

where b (b=0,1...) is the iteration index of the upper coordinator, S(K) are step lengthes, $\rho(K)$ are Lagrange multipliers for the power balance constraints, $D^b(K)$, $G_j^b(K)$ and $H_i^b(K)$ are respectively the solutions of the stochastic subproblem, the thermal subproblem and the HS.

2. The solution of the thermal subproblem is given by:

$$G_j^b(K)=\min\{\bar{G}_j,\ \max\{\underline{G}_j,\ \hat{G}_j(K)\}\} \ , \quad K=0\ldots N-1, \ j=1\ldots T \tag{25}$$

where $\bar{G}_j$ and $\underline{G}_j$ are the bounds on power generation of thermal plant j, $\hat{G}_j(K)$ is the solution of:

$$d\psi_j[G_j(K)]/dG_j(K)\Big|_{\hat{G}_j(K)}-\rho^b(K)=0 \ , \tag{26}$$

where $\psi_j[G_j(K)]$ is the thermal operation cost function.

3. The solution of the stochastic subproblem is given by:

$$D^b(K)=F_K^{-1}\{[\mu(K)-\rho^b(K)]/[\mu(K)-\nu(K)]\}, \quad K=0\ldots N-1 \tag{27}$$

under the condition that $\mu(K)\geqslant\rho^b(K)\geqslant\nu(K)>0$, where F_K is the cummulative distribution function of the load demand D(K), μ and ν are given factors.

4. The HS is as follows:

$$\text{HS}\begin{cases} \min \quad J=\sum_{i=1}^{M}\sum_{K=0}^{N-1}\{-\rho^b(K)H_i(K)\} & (28) \\ \text{s.t. for } i=1\ldots M;\ K=0,1\ldots N-1: & \\ X_i(K+1)=X_i(K)-U_i(K)+\sum_{j\in S_i}U_j(K-t_{ji})+Y_i(K) \ , & (29) \\ U_j(-t_{ji})=b_j(t_{ji}) \ , & (30) \\ X_i(0)=X_i^0 \ , \quad X_i(N)=X_i^N \ , & (31) \\ \underline{X}_i\leqslant X_i(K)\leqslant\bar{X}_i \ , \quad \underline{U}_i\leqslant U_i(K)\leqslant\bar{U}_i \ , & (32) \end{cases}$$

where: $H_i(K)=C_{1i}X_i^2(K)+C_{2i}U_i^2(K)+C_{3i}X_i(K)U_i(K)+$

$$+C_{4i}X_i(K)+C_{5i}U_i(K)+C_{6i} \ ,$$

is the power generation function of hydro plant i (it is assumed to be a convex function), S_i is the index set of the immediate upstream neighbouring hydro plants of plant i, t_{ji} is the water time delay from hydro plant j until i, $Y_i(K)$ are independent water inflows given previously, $b_j(t_{ji})$ are delayed water discharges prior to optimization, X_i^0 and X_i^N are respectively the initial and final states of the reservoir storge. For convenience, spills are not considered in the reservoir dynamics.

The Solution of The HS---The Application of Algorithm 1

The HS is a dynamic optimization problem with time delays which must be solved repetitively. Noting that the thermal subproblem and the stochastic subproblem are extremely simple (their solutions are obtained analytically), Lyra, Tavares and Soares (1981) stressed that the great computational effort in solving the hydrothermal problem is concentrated on the HS. Now we use Algorithm 1 in this paper to solve the HS. Although the HS is somewhat different from P_1, the extension is streightforward. (Of course, Algorithm 2 can also be extended to the HS. See the results in (Cai, 1984).)

Naturally, we decompose the HS into M subproblems, one for each hydro plant. Obviously, the interactions between subsystems are:

$$\pi_{1i}(K) = \sum_{j \in S_i} U_j(K - t_{ji}), \quad i=1\ldots M;\ K=0,1\ldots N-1 \quad (33)$$

With such a decomposition, an extension of Algorithm 1 is as follows:

ALGORITHM 3 :

STEP 1. If $b \neq 0$, go to step 3. Otherwise choose initial values for $U_i^c(K)$ and $\lambda_i^c(K+1)$ $(1 \leq i \leq M;\ 0 \leq K \leq N-1)$. Put the iteration index c=0, go to step 3.

STEP 2. If some degree of accuracy is reached, reture (to the upper coordinator). Otherwise make c=c+1, continue.

STEP 3. Compute $\pi_{1i}^c(K)$ for i=1...M and K=0...N-1 by substituting U^c and (30) into (33).

STEP 4. Solve subproblem P_{4i} for i=1...M; then go to step 2.

$$P_{4i} \begin{cases} \min\ J_{4i} = \sum_{K=0}^{N-1} \{-\rho^b(K) H_i(K) + \sum_{j \in R_i} [\lambda_j^c(K+1+t_{ij}) U_i(K)]\}, \\ \text{where } R_i \text{ is the index set of the immediate downstream neighbouring hydro plants of hydro plant } i,\ \lambda_j(K)=0 \text{ for all } K>N. \\ \text{s.t. } X_i(K+1) = X_i(K) - U_i(K) + \pi_{1i}^c(K) + Y_i(K), \quad K=0\ldots N-1 \\ (31),\ (32) \end{cases}$$

As space is limited, in this paper it is impossible to enter into details of the solution of the subproblem P_{4i}. The detailed discussion has been given in (Cai, 1984).

The New Algorithm for the Whole Hydrothermal Problem

Combining Algorith 3 with (24)~(27) forms a new algorithm solving the whole hydrothermal problem. We call this algorithm ALGORITHM 4.

A Numerical Example for Comparison

Soares and others (1980) gave an illustrative hydrothermal problem with 4 hydro and 2 thermal plants over 12 subinternals (i.e. N=12). This example is restated as follows:

Thermal subsystem:

$$d\psi_1/dG_1(K) = 10 + 3G_1(K) \quad ; \quad d\psi_2/dG_2(K) = -20 + 1.66G_2(K)$$

$$10 \leq G_1(K) \leq 80 \quad ; \quad 20 \leq G_2(K) \leq 80$$

Hydro subsystems:
The hydro network is shown in Fig. 3. The water delays are: $t_{13}=2$, $t_{23}=3$, $t_{34}=4$.

The particulars of the hydro plants are listed in Table 1~3. The delayed water charges prior to optimization have been included in $Y_i(K)$.

Table 1 Independent Water Inflows

K	Y_1	Y_2	Y_3	Y_4	K	Y_1	Y_2	Y_3	Y_4
0	10	8	20	20	6	8	6	3	0
1	9	8	20	20	7	9	7	2	0
2	8	9	10	20	0	10	8	1	0
3	7	9	2	18	9	11	9	1	0
4	6	8	3	0	10	12	9	1	0
5	7	7	4	0	11	10	8	2	0

Table 2 Hydro Plant Data

i	$\underline{X}_i$	$\bar{X}_i$	$\underline{U}_i$	$\bar{U}_i$	X_i^0	X_i^N
1	80	150	5	15	100	120
2	60	120	6	15	80	70
3	100	240	10	30	170	170
4	70	160	13	25	120	140

Table 3 Hydro Plant Coefficients

m	1	2	3	4	5	6
C_{m1}	-0.001	-0.1	0.01	0.40	4.0	-30.0
C_{m2}	-0.001	-0.1	0.01	0.38	3.5	-30.0
C_{m3}	-0.001	-0.1	0.01	0.30	3.0	-30.0
C_{m4}	-0.001	-0.1	0.01	0.38	3.8	-30.0

Table 4 Load Demand Data

K	0	1	2	3	4	5	6	7	8	9	10	11
Mean	200	210	220	230	240	250	240	230	220	210	200	190
Var.	5	5	5	5	5	5	5	5	5	5	5	5

Load demand:
F_K is a normal distributed function. The expected values and the variance are given in Table 4. $\mu(K)$ and $\nu(K)$ are chosen to be $\mu(K)=100$, $\nu(K)=10$.

Algorithm 4 was used to solve this hydrothermal problem. Table 5 shows the optimal discharges $U_i(K)$ and the optimal generation $G_j(K)$ (notice that other variables, e.g. $X_i(K)$, can be obtained by using these values). To compare the computational efficiency, we also used the algorithm presented by Soares, Lyra and Tavares (1980) (we call it SLT ALGORITHM), which applies Tamara's method to solve the HS, to solve this hydrothermal problem on the same computer PDP11/23. The optimal trajectories obtained by the two algorithms are identical. Table 6 shows the computation times and the total iteration numbers required by these two algorithms to solve this problem under same stopping conditions and with same initial values.

Table 5 Optimal Results

K	U_1	U_2	U_3	U_4	G_1	G_2
0	5.6	6.4	21.7	13.0	15.5	46.0
1	6.5	7.2	21.1	13.0	15.8	46.7
2	7.4	8.0	20.6	13.5	16.2	47.4
3	8.0	8.9	20.1	15.2	16.4	47.8
4	8.9	10.0	19.7	16.9	16.9	48.6
5	9.9	11.3	19.6	18.7	17.4	49.5
6	9.0	11.1	19.1	19.4	15.5	46.1
7	8.0	11.0	18.7	20.1	13.7	42.8
8	7.0	11.0	17.3	20.9	12.1	39.9
9	6.7	7.6	17.8	21.9	11.1	38.1
10	5.0	7.5	18.0	22.7	10.0	35.1
11	5.0	6.0	17.2	23.2	10.0	30.2

Table 6 Computation Times and Iteration Numbers

	Computation Times	Iteration Numbers
Algorithm 4	1.8 min	40
SLT Algorithm	13.7 min	353

CONCLUSION

Two interaction prediction algorithms have been developed to solve the complex problem: the optimization of large scale systems with multiple delays. The algorithms have several useful properties:

(1) The subproblems are not only smaller (lower dimensions), but also simpler (without delays) than the original problem.
(2) The coordination is extremely simple since mere substitution is required into very simple formulae.
(3) The algorithms are proven to be convergent under reasonable conditions and therefore they are not heurisitic ones.
(4) Like many other multilevel methods, the algorithms need comparatively less computer storage than classical methods; this makes it possible to solve

bigger problems with smaller computers.

(5) The algorithms can easily be extended to more general problems than the linear quadratic one considered in this paper. (see (Cai, 1984).)

One of the interaction prediction algorithms developed in this paper has been applied to the economic operation problem of large hydrothermal power systems. This yields a new algorithm for the whole hydrothermal problem. Numerical results show that the new algorithm has very rapid convergence and it decreases the computation time enormously.

REFERENCES

Cai Xiaoqiang (1984). Multilevel Optimization Methods and Their Applications to Economic Operation Problems of Large Power Systems. M.S.Thesis, Tsinghua University.

Cohen, G. (1977). On an algorithm of decentralized optimal control. Journal of Mathematical Analysis and Applications, 59, 242-259.

Lyra, C., H. Tavares and S. Soares (1981). In A. Titli and M. G. Singh (Ed.), Large Scale Systems Theory and Applications, Pergamon Press, Oxford. 417-452.

Mahmoud, M. S. (1977). Multilevel systems control and application: a survey. IEEE Trans. on Syst., Man, and Cybern., SMC-7, 125-143.

Mesarovic, M. D., D. Macko and Y. Takahara (1970). Two coordination principles and their application in large scale systems control. Automatica, 6, 261-270.

Soares, S., C. Lyra and H. Tavares (1980). Optimal generation scheduling of hydrothermal power systems. IEEE Trans. on PAS, PAS-99, 3, 1107-1115.

Takahara, Y. (1965). Multilevel Approach to Dynamic Optimization. M. S. Thesis, Case Institute of Technology.

Tamura, H. (1975). Decentralied optimization for distributed-lag models of discrete systems. Automatica, 11, 593-602.

Young, D. M. (1971). Iteration Solution of Large Linear Systems, Academic Press, New York.

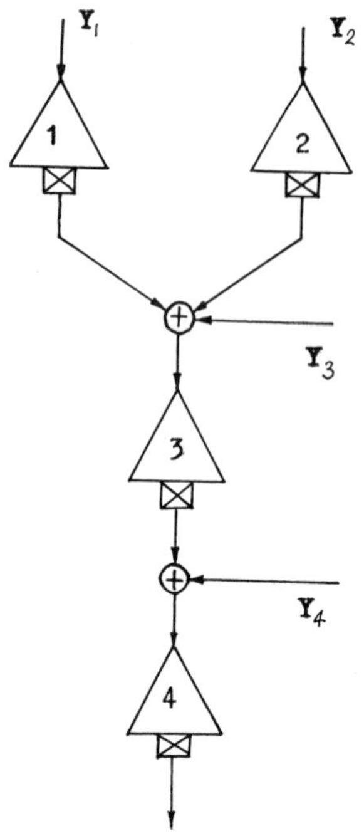

Fig. 3. Hydro network.

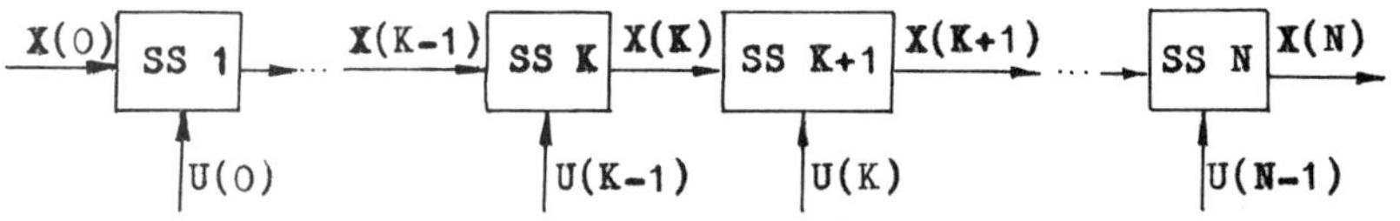

Fig. 1. Structure of discrete systems.

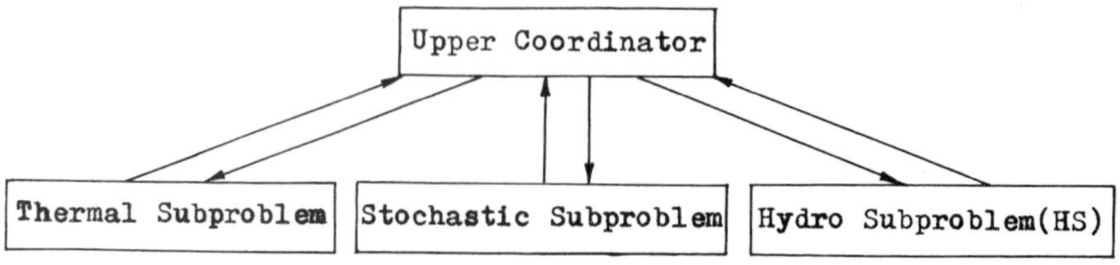

Fig. 2. Computational structre for hydrothermal scheduling.

FORECASTING AND OPTIMAL FLOOD CONTROL FOR RESERVOIRS WITH HYDROELECTRIC PLANTS

D. D. Wang

Institute of Systems Science, Academia Sinica, Beijing, China, East-West Population Institute, 1777 East-West Road, Honolulu, HI 96848, USA

Abstract. Description, identification of a generalized threshold model with integrated threshold elements and its application to forecasting floods for reservoirs with hydroelectric plants are discussed in this paper. Based on this model, a theorem of optimal control of floods for the reservoir aiming on maximum of hydroelectricity output by the plant during the flood period is also given in the paper. This theorem shows the optimal control rule under some assumptions is really independent of the target function. Comparisons of the predicted and the recorded floods, and of the optimal and the common control of floods are graphically showed with the data of real observations.

Keywords. Hydrology; threshold element; optimal control; statistics; time series; prediction; functional analysis; singular optimal control; computational methods; hydroelectric engineering.

INTRODUCTION

The techniques of flood forecasting developed after 1900 are mainly based on statistics and probability theory which, as a whole, is now called "Stochastic Hydrology". There are three important methods evolved from stochastic hydrology during the fifty years from 1930 to 1980. The first is the so-called "Method of Frequency Density", a technique for fitting probability density curve of floods related to precipitations. The second method is based on the exact analysis of phsical procedures between rainfall and runoff, methods like this are called "Physical Simulation". The last method in stochastic hydrology was developed and was applied to hydrology in the 1960s, and was well known as "Method of Time Series", such as (in linear case) AR, ARMA, ARIMA, and Bayesian even Kalman Filter (Chin, 1978).

A comparison of the three methods in actual forecasting shows the advantage of the time series method, especially for the reservoir with a hydroelectric plant. It is important to model for this reservoir that the structure of model fit for floods must be very convenient in readjustment to adapt to a varied environment. This property of model might be called "Structure Adaptability" for being distinguished from that of "Parameter Adaptbility".

TABLE 1 Comparison of Three Methods

	Exact. of forecast.	Compute. time	Paramet. adaptab.	Struct. adaptability
1	low	short	weak	middle
2	high	long	middle	weak
3	middle	middle	strong	strong

Note: "1" stands for method of frequency density, "2" for method of physical simulation, and "3" for method of time series.

The old model used in Xin An River Reservoir for flood forecasting can be classified as a method of physical simulation. This model should be precise enough for application, but this advantage is diminished by its very complicated structure which takes too much computational time to make the decision before the flood peak comes to the reservoir. Any simplification of this method would make the model less exact. Furthermore, the preciseness may be seriously decreased by too many unknown parameters in a physical model which are only estimated by statistics or even by experiences. After all, the weak structural adaptability of the Physical Simulation Model certainly stems the improvement of its exactness.

The first part of this thesis comes from author's actual work in Xin An River Hydroelectric Plant. To meet the needs of flood forecasting in short time and at high enough exactness, the threshold model with integration of threshold elements was established for forecasting floods. A comparison of the predicted and the recorded values of floods shows a good agreement (see Fig. 1 - 12).

The second part is devoted to optimal control of the predicted flood for reservoirs with hydroelectric plants. A singular optimal control problem is discussed in this part, and an optimal control solution might bring in more income if it were adopted as a practical plan.

FORECASTING OF FLOODS

Let $Q_s(t)$ $t \varepsilon I_s$ be discharge of the flood which we want to predict, let $P_s(t)$ $t \varepsilon I_s$ be the average amout of precipitation on the area, $Q_j(t)$ $t \varepsilon I_j$ the discharge of j'th historical flood, $j=1,2,\ldots,s-1$, and $P_j(t)$ $t \varepsilon I_j$ the precipitation of j'th historical flood, $j=1,2,\ldots,s$. Where $I_j=[t_{1j}, t_{2j}]$ if it stands for continuous sampling, and $=\{t_{ij}: 1 \leqslant i \leqslant N_j\}$ if it stands for discrete sampling, and the quantity $t_{2j} - t_{1j}$ or $t_{N_j j} - t_{1j}$ is the lasting time of j'th flood, $j=1,2,\ldots,s$.

If $Q_j(t)$ $t\varepsilon I_j$ can be considered as a function of $P_j(t)$ $t\varepsilon I_j$, then the problem of forecasting s'th flood for the reservoir can be expressed in this mathematical form:

Given $P_j(t)$ $t\varepsilon I_j$ $j=1,2,\ldots,s$ as well as $Q_j(t)$ $t\varepsilon I_j$ $j=1,2,\ldots,s-1$.
To predict $Q_s(t)$ $t\varepsilon I_s$. (1)

This problem has a similar form to the problem of time series forecasting, but there is a crucial difference between the two problems. In fact of hydrology, the group ($P_j(t)$, $Q_j(t)$) $t\varepsilon I_j$ is sometimes basically different (i.e. in probability distribution) from ($P_i(t)$, $Q_i(t)$) $t\varepsilon I_i$ if $j\neq i$. So it is impossible to connect the data

$$P_1(t)\ t\varepsilon I_1,\ P_2(t)\ t\varepsilon I_2,\ldots,\ P_{s-1}(t)\ t\varepsilon I_{s-1}$$

as a single time series and to connect the data

$$Q_1(t)\ t\varepsilon I_1,\ Q_2(t)\ t\varepsilon I_2,\ldots,\ Q_{s-1}(t)\ t\varepsilon I_{s-1}$$

as another time series and then to predict $Q_s(t)$ $t\varepsilon I_s$ with input $P_s(t)$ $t\varepsilon I_s$ according to the model set by the two connected serieses. For this reason, a lot of model of on-time forecasting would not work for our purpose.

To resolve the problem (1), a threshold linear model was considered and the function $Q_s(t)$ $t\varepsilon I_s$ could be treated with technique of functional analysis. The model under consideration is:

$$Q(t)=\begin{cases} a_1^1Q(t-1)+\cdots+a_{n1}^1Q(t-n_1)+b_o^1P(t)+\cdots+b_{m1}^1P(t-m_1) \\ \qquad T_1^1\leqslant \sum_{j=k_1^1}^{k_2^1} P(t-j)\leqslant T_2^1 \\ a_1^2Q(t-1)+\cdots+a_{n2}^2Q(t-n_2)+b_o^2P(t)+\cdots+b_{m2}^2P(t-m_2) \\ \qquad T_1^2\leqslant \sum_{j=k_1^2}^{k_2^2} P(t-j)\leqslant T_2^2 \\ \quad\vdots \\ a_1^\beta Q(t-1)+\cdots+a_{n_\beta}^\beta Q(t-n_\beta)+b_o^\beta P(t)+\cdots+b_{m\beta}^\beta P(t-m_\beta) \\ \qquad T_1^\beta\leqslant \sum_{j=k_1^\beta}^{k_2^\beta} P(t-j)\leqslant T_2^\beta \end{cases} \qquad (2)$$

The vector :

$$Y=\{\beta,n_1,n_2,\ldots,n_\beta,m_1,m_2,\ldots,m_\beta,\ T_1^1,T_2^1,k_1^1,k_2^1,\ldots,T_1^\beta,T_2^\beta,k_1^\beta,k_2^\beta\}$$

is named here "structure vector". We can use usual techniques for identification of linear model such as ARMA or ARX ("X" here means exogenous variable) whenever the structure vector Y has been given. Notifying that the model has integrates of input variables as its threshold elements, we simply write this kind of threshold model "ITARX".

A proof by functional analysis gives us a simple theorem, which implies that if Y is given, then any methods of identifying linear models can be used to identify model (2) without more error than be used to identify a single linear model. In actual prediction, the obvious physical characteristic of floods enables us to determine the structural vector Y.

The comparison of the predicted floods and the recorded floods are showed in Fig. 1 - Fig. 12. It is easily seen that the ITARX can be satisfyingly employed for fitting the nonlinear process of floods.

The computational method used here is so-called "conjugate gradient method", for solving a normal equation. It is better in preciseness than any of recursive computational methods when the sampled data are not great as needed from the point of statistical view.

OPTIMAL CONTROL OF FLOODS

Let $q(t)$ $t\varepsilon I_s$ denote the discharge of water flowing into the reservoir, which is given by the predicted flood $Q_s(t)$ $t\varepsilon I_s$. Let $v(t)$ $t\varepsilon I_s$ be the volume of water filled into the reservoir from original time to time t, and let $h(v(t))$ the height of water level of the reservoir at time t. Let

$$U(t)=\{\ u_1(t),\ u_2(t)\ \}$$

be the control vector in which $u_1(t)$ is the discharge of water flowing out of the hydroelectric plant, and $u_2(t)$ the discharge of water released out from the sluice gate of the reservoir.

The total electric energy produced by the hydroelectric plant during period of s'th flood equals to k×J, where k is a constant and

$$J=\int_{t_o}^{T} h(v(t))\times u_1(t)dt \qquad (3)$$

which is given by hydroelectric engineering.

So the problem of optimal control of floods for reservoir with hydroelectric plant is expressed as

$$\begin{aligned} &\max_U\{\ J=\int_{t_o}^{T} h(v(t))\times u_1(t)dt\ \} \\ &\dot v(t)=q(t)-u_1(t)-u_2(t) \\ &v(t_o)=\bar v \\ \text{s.t.}\quad &\alpha_v\leqslant v(t)\leqslant\beta_v \\ &0\leqslant u_1(t)\leqslant\beta_1 \\ &0\leqslant u_2(t)\leqslant\beta_2 \\ &v(T)=\beta_v \end{aligned} \qquad (4)$$

where the lasting time of the s'th flood is given by $Q_s(t)$ $t\varepsilon I_s$, $I_s=[t_o,\ T]$.

Problem (4) is a singular optimal control problem with a nonlinear goal function. Because of the special characteristics of hydroelectric engineering for a reservoir, Problem (4) has a definite solution given by the main theorem in this part.

The main theorem states that if the optimal solution of (4) exists in class of piece-wise continuous functions, then the optimal control $U^*(t)$ must be the same as:

$$u_1^*(t)\equiv\beta_1$$

$$u_2^*(t)=\begin{cases} 0 & t\varepsilon\,[t_o,\ t_1) \\ \beta_2 & t\varepsilon\,[t_1,\ t_{sup}) \\ q(t)-\beta_2 & t\varepsilon\,[t_{sup},\ T] \end{cases}$$

$$\text{where}\quad t_1\triangleq t_{sup}-\frac{1}{\beta_2}[\bar v-\beta_v+\int_{t_o}^{t_{sup}}(\ q(t)-\beta_1\ )dt\ ]$$

$$t_{sup}\triangleq \operatorname{Sup} I_m$$

$$I_m\triangleq\{t\varepsilon I_s:\ q(t)\geqslant\beta_1+\beta_2\} \qquad (5)$$

Proof of this theorem is rather like the proof of generalized Legender-Clebsch necessary condition for an extremal curve (Bryson, 1978; Bell, 1975;

Fleming, 1975), but this theorem given here supplies a sufficient condition for the optimal curve, so the optimality principle must be used in this proof and some subtle techniques are also needed in proof of this theorem.

The general procedure of optimal control in problem (4) is roughly drew out in Fig. 13.

We show a better effect on utility of water resource than the traditional control method of flood for reservoir (Fig. 14 - Fig. 15), because the latter has never taken the optimal problem like problem (4) into account.

SUMMARY AND ACKNOWLEDGEMENT

In this paper a new method for forecasting of floods and a particular class of singular optimal control problem are discussed. Applications to Xin An River Hydroelectric Plant are also discussed as an example of theory, and as the background of the new model -- ITARX.

The work represented in the first part of this paper was supported by the managers of Xin An River Hydroelectric Plant, and all the work was done under the guidance of Professor Cui Yi.

REFERENCES

Bell, David J. and D.H. Jacobson (1975). Singular optimal control problems. Academic Press, San Diego, CA.

Bryson, A.E. and Y.C. Ho (1978). Applied optimal control. Harvard University Press, Cambridge, MA.

Chin, C.L. (ed.) (1978). Applications of Kalman filter in hydrology, hydrologics and water resources. Stochastic Hydraulics Program, Department of Civil Engineering, University of Pittsburgh, Pittsburgh, PA.

Fleming, W.H. and R.W. Rishel (1975). Deterministic and stochastic optimal control. Springer-Verlag, New York.

GRAPHS

Glossary. In Fig. 1 - Fig. 12, the dotted lines stand for the predicted floods, height of water level and quantity of water. And the real lines stand for the post-recorded floods. All horizontal axes are scales in measuring the flood time, 6 hours per unit, and all vertical axes the quantity of water, 1000 cubic m/sec. per unit, or the height of water level in 0.1 meter per unit with 105 meters base-level.

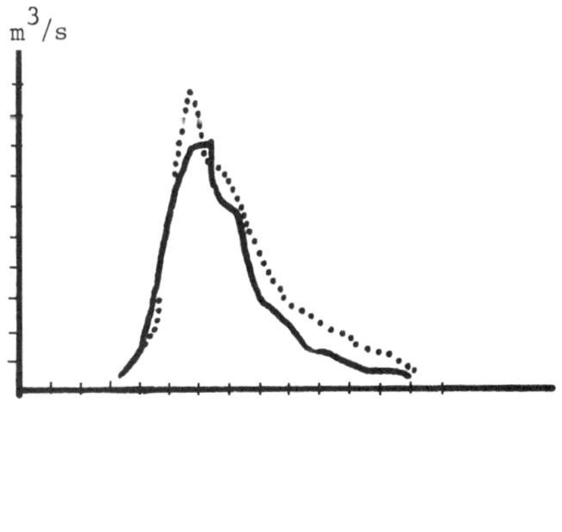

Fig. 1.

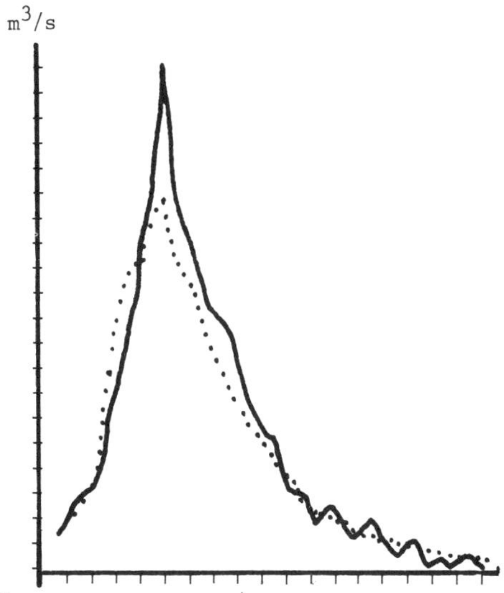

Fig. 2.

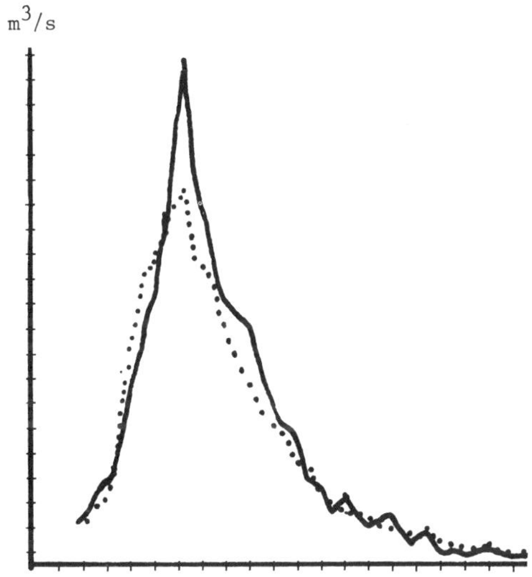

Fig. 3.

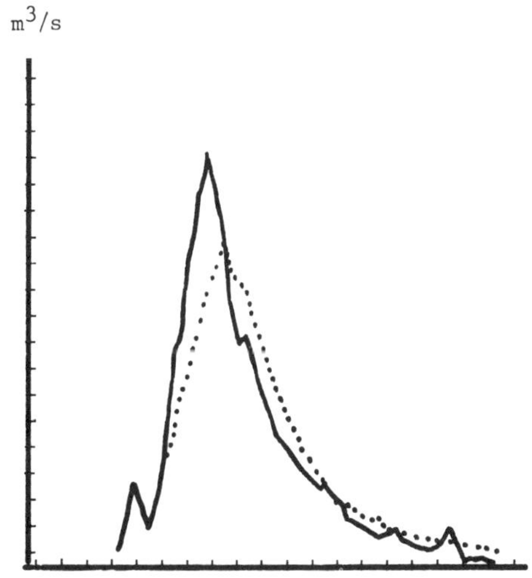

Fig. 4.

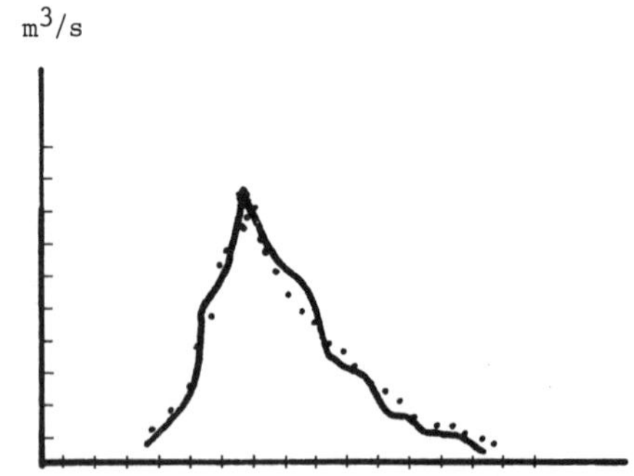

Fig. 5.

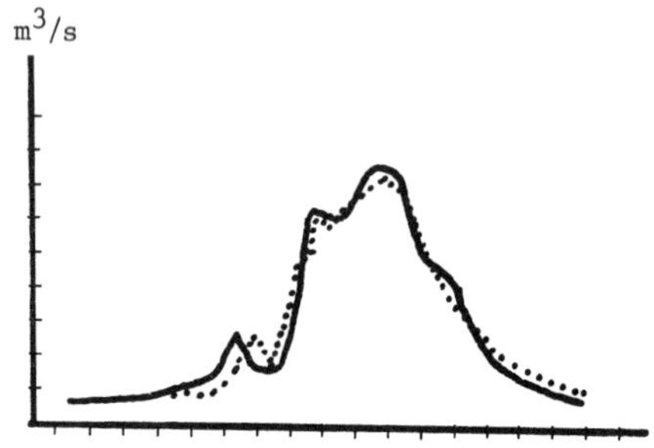

Fig. 6.

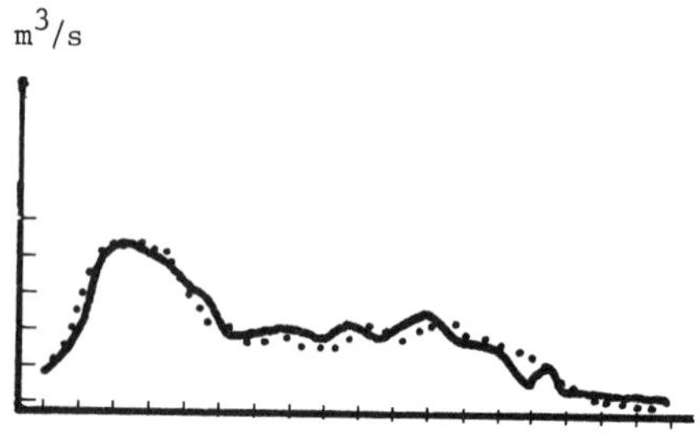

Fig. 7.

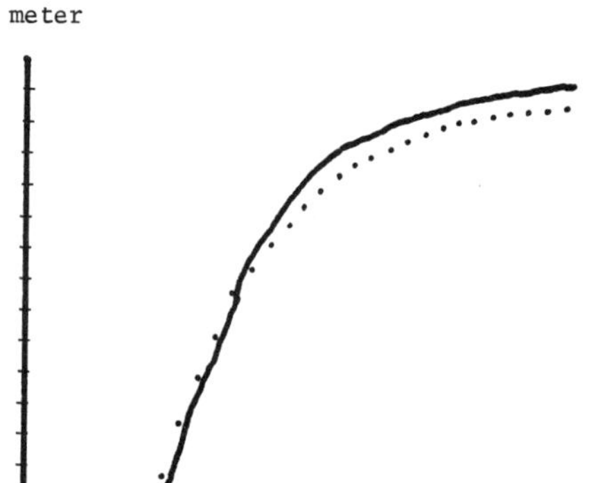

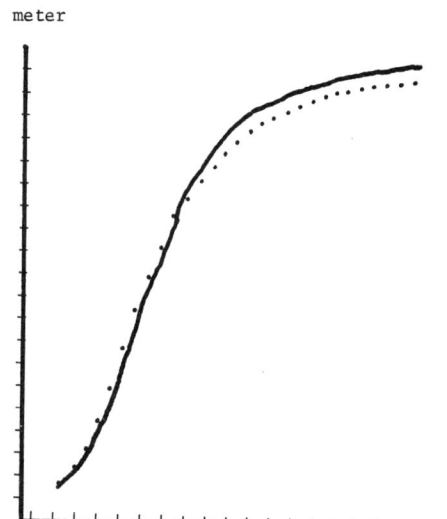

Fig. 8.

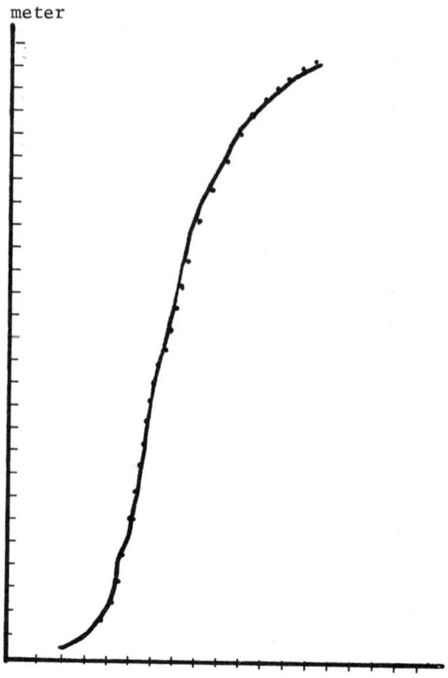

Fig. 9.

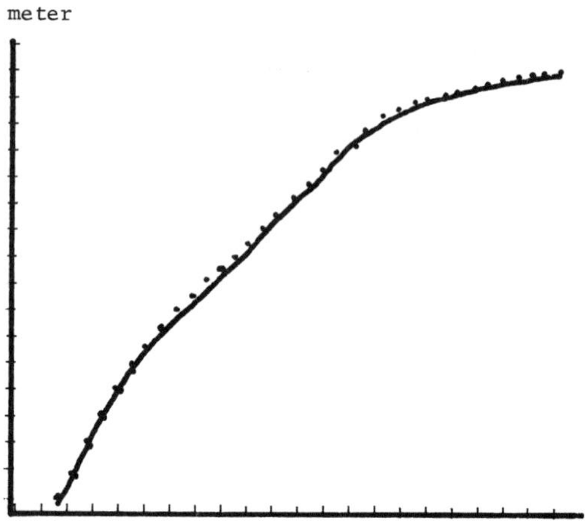

Fig. 10.

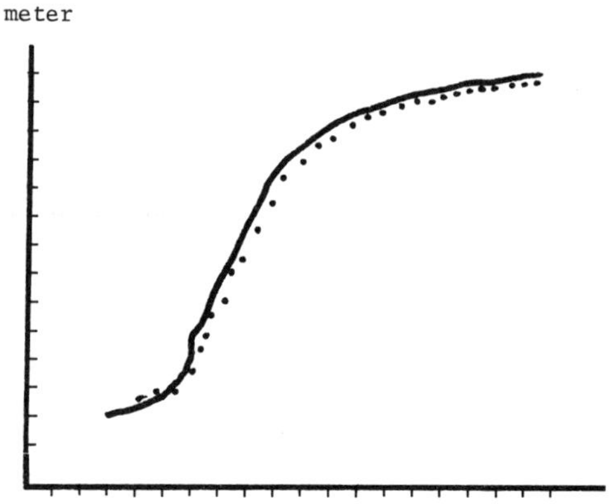

Fig. 11.

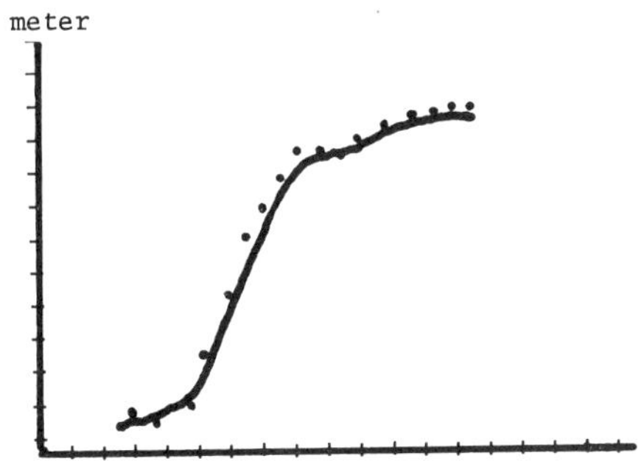

Fig. 12.

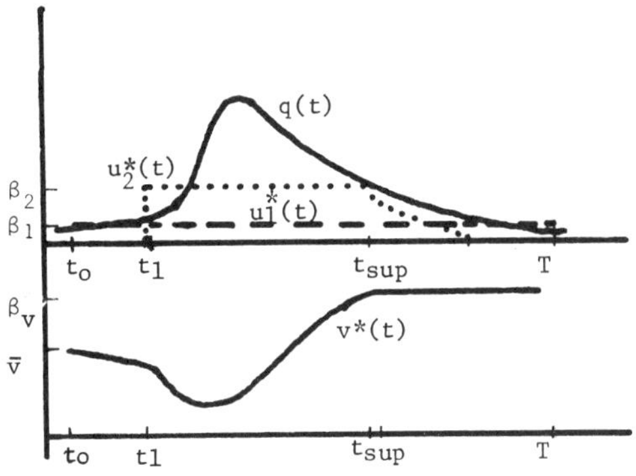

Fig. 13. General procedure of optimal control

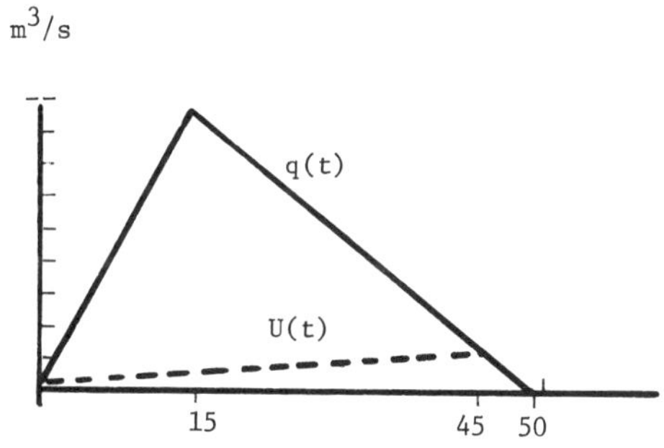

Fig. 14. Traditional control method of flood

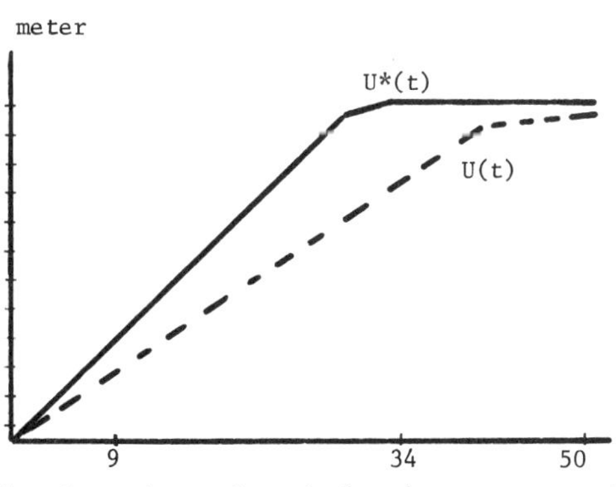

Fig. 15. Comparison of optimal and common control.

A STOCHASTIC MODEL FOR MONTHLY FORECAST OF HYDROELECTRIC ENERGY RESOURCES IN ITALY

R. Anelli*, P. Bonelli, G. Finzi*****

**ENEL-CRA, Cologno Monzese, Milan, Italy,*
***ENEL-CRTN, Via Rubattino 54, Milan, Italy,*
****Centro Teoria dei Sistemi, Politecnico, Via Ponzio n. 34/5, Milan, Italy*

Abstract. The goal of the electric power system is to provide for the most rational utilization of the available energy production and, in critical cases, to prevent uncontrolled blackouts, while taking into account both the security and quality of the service and the economical aspects of the problem. The forecast of monthly hydroelectric energy resources one month in advance requires particular accuracy in order to allow the power dispatchers efficient planning at a national level. The aim of this work is to provide better estimates of future values by means of a prediction model and simultaneously to preserve a certain simplicity in the algorithm so that it can be easily implemented in operation. Two classes of stochastic models are considered: the Exponential Smoothing models and the Cyclostationary Auto Regressive models. After a preliminary statistical analysis of the historical series of data, the parameters of the two models have been estimated for three Italian areas. Finally the performance indices of the predictors applied to a validation period of ten years have been computed and compared.

Keywords. Hydroelectric energy resources; stochastic systems modelling; prediction; power system control.

INTRODUCTION

The monthly energy capability of a hydro installation (or producibility) is conventionally defined as the maximum energy amount that could be produced or accumulated in terms of water storage due to the natural inflows during one month, assuming that the hydro installation is operating at the maximum efficiency.

Due to the fact that hydroelectric energy in Italy still makes up a significant percentage of the total energy production, the interest of the Italian electricity board (ENEL) in the variable just defined is relevant for the economic and secure coordination of generation resources.
In fact, hydroproduction is mainly scheduled to meet the more variable part of the demand diagram; this way of operating well suits the features of the hydroplants with reservoirs and it also allows the most economical scheduling of thermal plants. So, while on one hand hydroelectric energy continually decreases percentagewise in the years to come, on the other its importance as a modulation variable increases more and more.

The aim of this work is to set up a simple stochastic model of the monthly hydroelectric resources to be used for medium term operation (from one month to one year) over three extended areas covering the Italian peninsula.

A good forecast of monthly values could allow the energy system operators to plan for the amount of water to be released from the great seasonal storages, for the energy exchanges with other companies and, finally, for the stocking of fuels.

More specifically, in the following paragraphs, two families of stochastic models will be considered: the "Exponential Smoothing" models and the "Auto Regressive Cyclostationary" models. Both classes take into account the periodicity of the process due to the seasonal climatic component throughout the twelve months of the year. After a preliminary statistical analysis of the historical series of data, the parameters of both models have been estimated for the three Italian regions over a period of eighteen years (1953-1970). The performance indices of the predictors have been computed over a validation period of ten years (1971-1980). The comparison between the two methods shows that quite satisfying results can be obtained in both ways, except for some months of the year for which the variability of meteorological variables, such as rain and temperature, and their correlation with water resources are too strong to be considered solely with the average, as in the following.

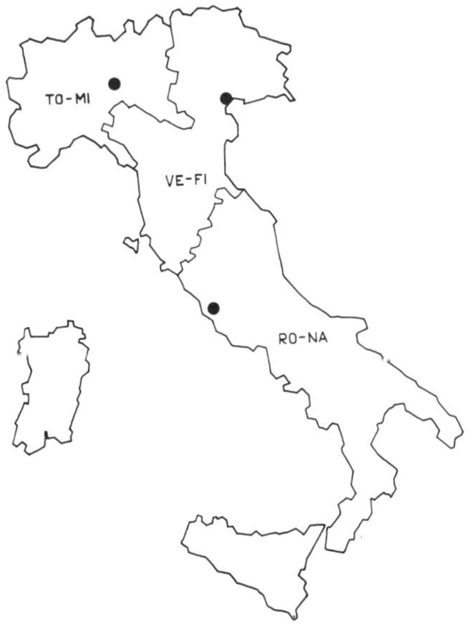

Fig. 1 Italy divided in the three areas

THE DATA SET - A PRELIMINARY STATISTICAL ANALYSIS

Fig. 1 shows how continental Italy has been divided by ENEL into three areas, two of which are characterized mainly by Northern climatology, with high mountains (the Alps) and cold winters, while the third one presents a typical Mediterranean climate with mild winters and hot summers. The three historical series of monthly data of energy capability (GWh) used in this work have been recorded in the years from 1953 to 1980. Fig. 2 shows the series for the first geographical area.

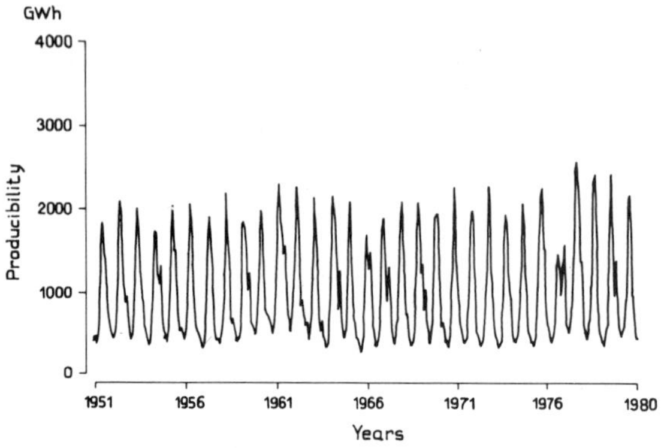

Fig. 2 The observed time series of energy capability for the Area TO-MI

A preliminary statistical analysis of such data is shown in Table 1; the values of monthly means and standard deviations may suggest the following considerations:

A. The yearly cycle (12 months) is quite evident for all the three areas.

B. Looking first at the two Northern zones, it can be noticed that the greatest values of the means are obtained during spring and summer months (that is after the snow melting phenomenon), while the standard deviations are proportionally higher during autumn months, due to the strong variability of rain events from one year to another in that period.

C. As for the third area, the values of the means are generally lower due to the local climatic conditions, characterized by little snow accumulation during winter and dry summers, with a rather narrow range of variability.

TABLE 1 Means and Standard Deviations of the Monthly Producibility for the three Areas.

Month of the year		1	2	3	4	5	6	7	8	9	10	11	12
TO - MI	μ_i	456	402	513	754	1491	2037	1844	1404	1058	913	696	523
	σ_i	79	56	112	162	238	233	281	253	237	311	200	103
VE - FI	μ_i	615	522	684	1030	1554	1616	1322	1070	954	937	877	750
	σ_i	136	117	163	220	237	240	213	232	303	367	364	234
RO-NA	μ_i	837	826	912	906	810	557	386	326	343	448	635	805
	σ_i	241	203	197	176	182	125	73	48	84	138	198	262

IDENTIFICATION OF THE TWO STOCHASTIC MODELS

1. An Exponential Smoothing Model

Exponential Smoothing methods have been developed for about thirty years, mainly in the field of operational research (Brown,1956; Holt, 1957). They have been widely used as predicting filters as a consequence of their simplicity and low operating costs.

The model considered in this paper is an extension of the "simple exponential smoothing", following an approach suggested by Winters (1960).

If $\{e_t\}_{t=1}^{n}$ is the observed time series of the process of the monthly producibilities, three components (the stationary one, the linear trend and the seasonal one) are evidenced by this kind of model in the following way:

$$\begin{aligned} S_t &= \alpha \frac{e_t}{I_{t-T}} + (1-\alpha)(S_{t-1}+b_{t-1}) \\ b_t &= \gamma\,(S_t-S_{t-1})+(1-\gamma)b_{t-1} \qquad (1) \\ I_t &= \beta \frac{e_t}{S_t} + (1-\beta)I_{t-T} \end{aligned}$$

where:

T is the seasonal period (12 months);
b_t is the trend component;
S_t is the deseasonalized component;
I_t is the seasonal factor, taking into account the seasonal cycle of the process;
α, β, γ are the model parameters.

From this model, with the three components combined, the following predicting formula can be obtained for the process:

$$\hat{e}_{t,t+m} = (S_t+mb_t)I_{t+m-T} \qquad (2)$$

where:

$\hat{e}_{t,t+m}$ is the forecast made at time t for m steps (months) ahead.

The structure of the algorithm is clearly recursive; therefore at the first step it is necessary to initialize the values of the components b_t, S_t and I_t and finally find the "best" estimates of the parameters α, β and γ.

For this purpose, the historical series of data used for the identification of the model has been divided into two subsets, the first of which, at least of T (12) terms, has allowed for the definition of the initial components:

$$I_j = \frac{e_j}{\frac{1}{T}\sum_{k=1}^{T} e_k} \qquad j = 1,2,\ldots,T$$

$$S_T = \frac{1}{T}\sum_{k=1}^{T} e_k \qquad (3)$$

$$T_T = 0$$

On the basis of the second set of data up to time T', the parameters of the model have been estimated, minimizing the following function:

$$I = \sum_{k=T}^{T'} (\hat{e}_{k,k+1}-e_{k+1})^2 \qquad (4)$$

This is the formulation of a non-linear optimization problem; to solve it a "quasi-Newton" resolution algorithm has been adopted (Fletcher, 1970). Since the observed values have not shown an evident trend component, only a couple of parameters (α and β) have been estimated for each of the three areas. The model has been subsequently validated over a period of ten successive years, as illustrated in the next paragraph.

2. An Auto Regressive Cyclostationary Model

ARMA Stochastic models (Box and Jenkins, 1970) have been widely used in literature for their simple linear structure and capacity to fit many classes of stochastic processes using but a limited number of parameters, estimated on the basis of the values of the autocorrelation function.

A particular subset of these models is the one considered here. A process is defined as T-cyclostationary if all its statistical moments present a cycle of period T (Franks, 1971).

In the case of the producibility time series examined in this paper, the cycle of twelve months in means values and standard deviations has already been evidenced in Table 1 as a consequence of the different climatic conditions throughout the solar year.

Taking into account also the periodicity of the autocorrelation coefficients between two successive months, the following AutoRegressive model of order one has been considered:

$$e(i+1+sT) = \mu_{i+1} + r_i \frac{\sigma_{i+1}}{\sigma_i} (e(i+sT) - \mu_i) + \varepsilon(i+sT) \quad (5)$$

where:

$T = 12$
$i = 1,2,\dots,12$
$s = 1,2,\dots,n$ (number of years).

The noise term $\varepsilon(t+sT)$ is a T-cyclostationary white process $N(0,\sigma_i)$, while the estimator functions for the means, the standard deviations and the correlation coefficients between one month and the following are given by:

$$\mu_i = \frac{\sum_{s=0}^{n-1} e(i+sT)}{n} \quad (6)$$

$$\sigma_i = \sqrt{\frac{\sum_{s=0}^{n-1} [e(i+sT) - \mu_i]^2}{n-1}} \quad (7)$$

$$r_i = \frac{\frac{1}{n-1} \sum_{s=0}^{n-1} [e(i+sT)e(i+1+sT) - \mu_i \mu_{i+1}]}{\sigma_i \sigma_{i+1}} \quad (8)$$

The predicting model can be easily derived from (5), simply assuming for the noise terms its expected value, that is always zero. So, the one month in advance producibility predictor is:

$$\hat{e}(i+1+sT) = \mu_{i+1} + r_i \frac{\sigma_{i+1}}{\sigma_i} (e(i+sT) - \mu_i) \quad (9)$$

That is, the forecast is obtained through a modulation of the monthly mean value by means of the correlation with the value of energy capability measured during the preceding month.

VALIDATION OF THE TWO MODELS AND COMPARISON OF PREDICTIVE PERFORMANCE

The time series of data from 1953 to 1980 has been divided into two subsets; the first one (from 1953 to 1970) has been used to estimate the parameters of the models, while the second one has been used for their validation. Thus , the predictors (2) and (9) have been applied to the last ten years of data, using for each area the set of parameters optimal with respect to the first period of data, as it should be when actually in operation.

The forecast performance of one month in advance of the two producibility predictors has been measured by two indices:

ρ_i : is the correlation coefficient between predicted and observed data during the i-th month;

$\sigma^2_{\varepsilon_i} / \sigma^2_i$: is the percentage of unexplained variance, namely the ratio between the variance of the forecast error ($\varepsilon = \hat{e} - e$) and the variance of the producibility of the i-th month.

Table 2 and Table 3 show the performance indices respectively for predictor (2) and (9), computed for each one of the three geographical areas.

TABLE 2 Forecast Performance Indices for Exponential Smoothing Predictors (Winters Filters).

Month of the year		1	2	3	4	5	6	7	8	9	10	11	12
TO - MI	ρ	.96	.78	.59	.41	.49	.59	.77	.87	.14	.27	.89	.95
	$\sigma^2_\varepsilon / \sigma^2$	.66	1.06	.66	.90	1.02	.77	.48	.32	1.55	.93	.21	.21
VE - FI	ρ	.85	.73	.78	.37	.80	.77	.83	.83	—	.20	.64	.89
	$\sigma^2_\varepsilon / \sigma^2$	.34	.58	.41	1.43	.55	.44	.32	.31	—	.97	.59	.26
RO-NA	ρ	.52	.57	.66	.34	.44	.71	.86	.84	.63	.69	.15	.65
	$\sigma^2_\varepsilon / \sigma^2$	.75	.78	.92	1.10	.85	.51	.28	.48	.76	.61	1.06	.58

It can be noticed that the results are more or less comparable for the two models and, on the average, satisfying. At any rate, the following more detailed comments can be made:

A. As for the two Northern areas, correlations have high values for both models during winter months (from November to February), and

TABLE 3 Forecast Performance Indices for Auto-Regressive Cyclostationary Predictors.

Month of the year		1	2	3	4	5	6	7	8	9	10	11	12
TO - MI	ρ	.90	.79	.75	.53	.55	.47	.79	.88	.14	.60	.96	.94
	$\sigma^2_\varepsilon / \sigma^2$	.26	.38	.44	.72	.74	.94	.88	.30	1.8	.69	.18	.20
VE - FI	ρ	.81	.85	.95	.52	.76	—	.92	.81	.02	.54	.82	.88
	$\sigma^2_\varepsilon / \sigma^2$	.51	.40	.49	.73	.65		.73	.45	1.76	.77	.34	.25
RO-NA	ρ	.55	.71	.86	.64	.57	.75	.91	.92	.78	.67	.33	.71
	$\sigma^2_\varepsilon / \sigma^2$	.70	.53	.41	.70	.72	.47	.17	.15	.44	.69	.92	.51

during the two hot months (July and August). Otherwise, looking at the percentage of unexplained variance, it can be noticed that the cyclostationary model is often more explicative of the variability of the phenomenon, thanks to its larger set of parameters. Results are instead less satisfying for spring and autumn months, as a consequence of the absence of external meteorological inputs to both models; in spring, in fact, the role of snow melting and its delay effect on producibility is quite significant, while in fall the high intensity and time variability of rain inflows make the forecast more critical.

B. As for the third geographical area, the Southern one, good results are only obtained by both predictors for the months from July to September, the climatology being very regular and dry in those Mediterranean regions. Looking to the other months, again the cyclostationary model seems to offer a better performance and to be satisfactory even on an absolute scale, with the exception of November, usually the first rainy month of the cold season, for which the information about the amount of rain inflow would be a significant explicative input.

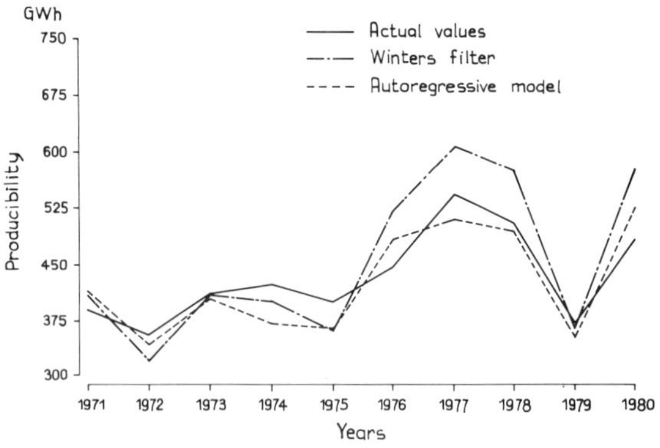

Fig. 3 Observed producibility values vs. the two forecasts during the validation period (month of January).

Finally, Fig. 3 reports, year by year, the behaviour of the historical series of January producibility together with the two different forecasts during the period of validation.

CONCLUDING REMARKS

The comparison between the predictive performance of the two stochastic models identified in this work clearly shows that the class of Cyclostationary Auto Regressive predictors is able to better explain the variability of the monthly producibility features due to the high number of parameters estimated (12 means, 12 standard deviations and 12 correlation coefficiens for every area). Likewise, the Exponential Smoothing predictor also attains results at a good level, just by means of only two parameters (α and β) for every zone.

Moreover, it has been shown that for some critical months (expecially during spring and fall seasons) purely stochastic models such as the ones examined here, are not able to give reliable results because meteorological events are only taken into account as an average and in an indirect way (practically considering only the annual climatological cycle estimated over the identification years).

So, the next step of this research will be to introduce some meteorological external inputs, like temperature and rain. In particular, the good correlation between winter rain and water resources available during spring months is encouraging and supports this further refinement of the model.

REFERENCES

Brown, R.G., (1956). Exponential smoothing for Prediction Demand. Proc. of 10th National Meeting of ORSA, Nov. 1956.

Box, G.E.P. and Jenkins, G.H., (1970). Time Series Analysis, Forecasting and Control. Holden-Day, San Francisco.

Fletcher, L. (1970). A new approach to variable metric algorithms. The Computer Journal, 13, 317.

Franks L. (1971). Signal Theory. Prentice Hall, Henglewood Cliffs, New Jersey.

Holt C.C. (1957). Forecasting Trends and Seasonals by Exponentially Weighted Moving Averages. O.N.R. Memorandum Nr. 53. Carnegie Institute of Technology.

Winters P.R.(1960). Forecasting Sales by Exponential Weighted Averages Moving. Management Science, 6, 324.

A HIERARCHICAL DYNAMIC PROGRAMMING APPROACH FOR SEQUENCING HYDRO POWER PROJECTS

Wang Dingwei* and Fu Minghui**

**Department of Automatic Control, Northeast University of Technology, Shenyang, China*
***Department of Automatic Control, Huazhong University of Science and Technology, Wuhan, China*

Abstract. The sequencing problem is a difficult combinatorial optimization problem. For actual water power resources planning, the projects to be sequenced often have the time constraints, which makes the problem even more complicated and difficult. Referencing to the multiple level hierarchical algorithm for large scale systems, a new hierarchical dynamic programming sequencing approach has been developed from Erlenkotter's sequencing model. This approach can not only find the optimal sequencing, but also select the optimal sizing of the installing capacity from a finite set of feasible sizes. This approach has been applied to actual water power resources planning. The satisfactory results have been achived.

Keywords. Dynamic programming; water resources; hierarchical systems; optimization; sequencing.

INTRODUCTION

Sequencing of hydroelectric power projects and determining of their installing generator capacity are of key importance in water power resources planning. It is well known that there is a close tie between scale and sequencing decisions. But much less attention has been paid to such interaction in conventional water resources planning because of the complexity and difficult in calculation. (Erlenkotter, 1976) The scale of installing capacity is often determined in a static condition, while sequencing is examined under the presumption that the sizes of individual projects are already specified. Becker and Yeh (1974) proposed a dynamic programming model, which is used to find the optimal timing and sequencing of multiple reservoir surface water supply facilities, and the scale of the reservoirs is determined simultaneously. But the complexity of the interdependent projects was not considered enough in their approach, it could result in a nonoptimal overall solution. (Erlenkotter, 1975) Erlenkotter (1976) has given some conclusions useful to coodinating scale and sequencing decisions for water resources projects in the case of linear demand, and pointed out the complexity for more gonoral case.

For the practical problem of water power resources planning, the demand, power requirement, is always nonlinear, Besides, the projects to be sequenced often have some time constraints. They are (a) the constraints of admissible beginning time when the projects can be put into operation, (b) the constraints of necessary interval between two stages of the stage construction projects. In dynamic programming approach, the states generated by the possible capacity combinations don't include time parameter, therefore, these time constraints make the problem even more complicated and difficult.

Referencing to the multiple level hierarchical algorithm for large scale systems, a new hierarchical dynamic programming sequencing approach has been developed from the sequencing algorithm in literatures. (Erlenkotter, 1973; Baker and Schrage, 1978; Morin and Esogbue, 1971) This new approach generates the states via using the hybrid-base number, and transforms the sequencing problem into the shortest path problem. By dividing the state set into several subsets according to the time when the stage construction projects can be put into operation, we constitute some sublevel dynamic programming models. Then the multiple level dynamic programming algorithm is used to find the optimal solution of the shortest path problem. At the same time, the optimal sizing of the installing capacity can be selected from a finite set of feasible sizes when the optimal sequencing is found.

This approach has been applied to determining the optimal sequencing of the hydro power projects on the Hongshui River in South China, and planning the capacity expansion of the Huazhong Power Network based on the construction of the Sanxia Hydro Power Station in the Three Gorges of the Yangtze River. The satisfactory result have been achived.

DESCRIPTION OF THE PROBLEM

In order to determine the installing generator capacity, we consider that each project has several feasible capacity levels. The problem of sequencing multiple projects with time constraints can be described as follows:

Let T be the planning horizon. In the duration $[0,T]$, a finite number of power projects(include thermal plants) are available for construction, index i=1,2, ...,n. Each project i is divided into m_i capacity levels, noted by $Z_i^1, Z_i^2, \ldots Z_i^{mi}$, the investments for m_i capacity levels are $C_i^1, C_i^2, \ldots, C_i^{mi}$, i=1,2,...,n, respectively. Given the pattern of demand over time D(t), D(t) is satisfied with

$$D(t) > 0, \quad \text{for all } t \in [0,T] \quad (1)$$

$$\Delta D(t)/\Delta t \geq 0, \quad \text{for all } t \in [0,T] \quad (2)$$

The admissible beginning year when project i can be put into operation is y_i, i=1,2,...,n. The construction of project n is divided into two stages. The second stage construction is constrainted to be put into operation d years after putting first stage into operation. The objective is to find a sequencing of the capacity selected from n projects to meet the demand during interval $[0,T]$, which should minimize the present worth of total project cost. This is

$$\min J = \sum_{t=0}^{T} (1+r)^{-t} [V(t)+E(t)-B(t)] \quad (3)$$

$$\text{s.t. } Z(t) \geq D(t), \text{ for all } t \in [0,T] \quad (4)$$

$$t_i \geq y_i, \quad i=1,2,\ldots,n \quad (5)$$

$$t_n^2 - t_n^1 = d \quad (6)$$

where, V(t) is the total investment of projects to be put into operation in t-th year.
E(t) is the total operation cost in t-th year.
B(t) is the benefit of the reservoirs except electric power in t-th year.
Z(t) is the total capacity in t-th year.
r is the annual discount rate.
t_i is the year when the project i is put into operation.
t_n^1 is the year when the first stage construction of project n is put into operation.
t_n^2 is the year when the second stage construction of project n is put into operation.

CONSTITUTION OF MODEL

Assumptions. To simplify the problem, the following assumption are made.

1. The steam hydrologies are sufficiently well characterized by their historical monthly flow records.
2. The investment cost C_i^j is incurred at the last year before the capacity Z_i^j is put into operation, and does not vary with time.
3. The project capacity once created has infinite life and does not change over time.
4. The other benifits of each reservoir have been evaluated and are constant.
5. The new capacity is put into operation only when all existing capacity does not meet the demand.
6. The annual discount is constant.

Hybrid-base Number and State Generation. The principle of dynamic programming approach for sequencing expansion projects is to transform the sequencing problem into the shortest path problem via elaborately generating states, then, the optimal solution of the shortest path problem is found by Bellman's (1957) principle of optimality. In order to avoid the emergence of the pathology of dynamic programming sequencing algorithm,(Morin,1973) it is necessary for the state generation to include all possible combinations of capacity selected from the n projects. From the binary representation and the hybrid-base representation,(Erlenkotter, 1973; Baker and Schrage,1978) we develop a new number system, hybrid-base system, and use it to generate the state set, which makes the calculation become very convenient.

Hybrid-base number can be represented as follows (here only deal with the integral)

$$a_n W_n + \ldots + a_2 W_2 + a_1 W_1$$

where, W_i is the weight of the i-th place. If B_i is the base of the i-th place, then

$$W_1 = 1 \quad (7)$$

$$W_i = W_{i-1} \cdot B_{i-1}, \quad \text{for } i > 1 \quad (8)$$

where, a_i is the digital of the i-th place, it can take B_i values $0,1,\ldots,B_i-1$. $B_1, B_2, \ldots, B_n$ have been given, in the definition domain $[0,N)$, arbitrary hybrid-base number can be simply mark as

$$(a_n \ldots a_2 a_1)_H.$$

$$\text{where, } N = \prod_{i=1}^{n} B_i \quad (9)$$

The translation between decimal number and hybrid-base number can be done by following formulas

$$A = \sum_{i=1}^{n} a_i W_i \quad (10)$$

$$a_n = [A/W_n] \quad (11)$$

$$a_i = [(A - \sum_{j=i+1}^{n} a_j W_j)/W_i], \text{ for } i < n \quad (12)$$

where, $[.]$ is to take integral part.
A is the decimal number.
a_i is the i-th place digital of the hybrid-base number.

For the problem of sequencing n projects with multiple capacity levels, we take

$$B_i = m_i + 1, \quad i=1,2,\ldots,n \quad (13)$$

The hybrid-base number $(a_n \ldots a_2 a_1)_H$ represents the state that the capacity of project 1 is $Z_1^{a_1}$, the capacity of project 2 is $Z_2^{a_2}, \ldots$, the capacity of project n is $Z_n^{a_n}$, respectively. Here we add the definition $Z_i^0=0$, represents the project i does not be put into operation, $i=1,2,\ldots,n$.

It is evident that there is an one to one mapping between the definition domain of hybrid-base number and state set consisted of all combinations of the capacity. The total number of possible states is N. The decimal representation of $(a_n \ldots a_2 a_1)_H$, A, can be used as the address label of the state. In following paragraphs, the decimal label A is used to represent the state by $(a_n \ldots a_2 a_1)_H$.

Let $\tau(A)$ be the latest year when the capacity of state A can meet the demand.

$$\tau(A)=\sup\{t \mid Z(A) \geq D(t),\ t \in [0,T]\} \quad (14)$$

where, Z(A) is the sum of the initial capacity, the total capacity of the state A and the compensatory capacity between distinct projects.

The relation between $\tau(A)$, D(t) and Z(A) is shown in Fig.1.

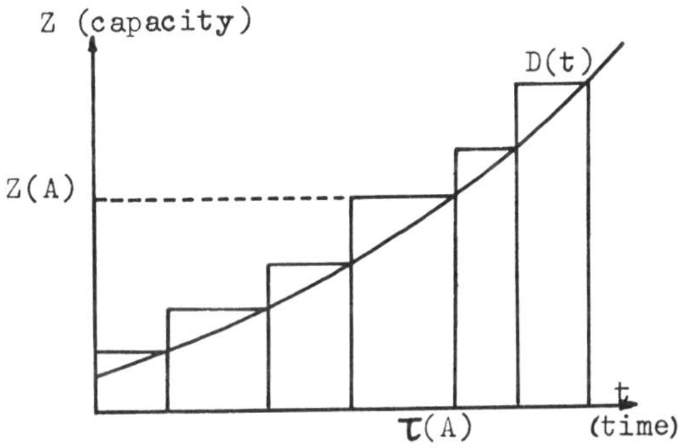

Fig. 1. The relation between $\tau(A)$, D(t) and Z(A).

Let F_A be the set of the state which can reach the state A only one step.

$$F_A=\{f_i \mid f_i=A-W_i,\ \tau(f_i)+1 \geq y,\ i \in I_A\}$$

where, $I_A=\{i \mid a_i>0,\ i=1,2,\ldots,n\}$

$$y=\max\{y_i \mid i \in I_A\}$$

Then, the following forward dynamic programming formulation can be used to find the optimal sequencing.

$$J(0)=C(0) \quad (15)$$

$$J(A)=\min_{f_i \in F_A}\{J(f_i)+[\tau(A)-\tau(f_i)]\cdot C(A)\cdot(1+r)^{-\tau(A)}+\Delta C_i^{a_i}\cdot(1+r)^{-\tau(f_i)}\},$$

$$\text{for } \tau(A) \geq y \quad (16)$$

$$J(A)=\infty \quad \text{for } \tau(A)<y \quad (17)$$

where, C(A) is the annual operation cost of the state A.

$$\Delta C_i^{a_i}=C_i^{a_i}-C_i^{a_i-1}, \quad \text{for } a_i>1$$

$$\Delta C_i^{a_i}=C_i^1, \quad \text{for } a_i=1$$

Because F_A is a finite set for any state A, the optimal partial policy can be found through finite time (<n) comparison. The total number of states N is finite also. Therefore, the optimal solution can be found through less n.N time comparison. Here, we do not consider the increment of the times of comparison caused by the time constraint of the stage construction project n.

The optimal sizing of the installing capacity of n projects in duration 0,T can be found simultaneously. Morin and Esogbue (1971) have proposed that the dynamic programming approach will select the optimal subset of projects to be constructed if the total potential capacity exceeds the maximum demand level.

Let set B be defined as the end state set of dynamic programming approach, that is

$$B=\{x \mid Z(x) \geq D(T)\} \quad \text{or} \quad B=\{x \mid \tau(x) \geq T\}$$

$$\text{If} \quad x^*=\arg\min_{x \in B} J(x), \quad (18)$$

x^* can be translated into the hybrid-base number $(a_n^* \ldots a_2^* a_1^*)_H$, then, $Z_1^{a_1^*}, Z_2^{a_2^*}, \ldots, Z_n^{a_n^*}$ are the optimal installing capacity of project 1, project 2,..., project n in the duration [0,T].

If $Z_j^{a_j^*}<Z_j^{m_j}$, whether or not to install the remainder capacity $Z_j^{m_j}-Z_j^{a_j^*}$ will depend upon the future condition after the duration [0,T].

Hierarchical Dynamic Programming Approach. So far, the time constraints of necessary interval between two stages have not been discussed. As foregoing statement, the states generated by the possible capacity combinations don't include time parameters; therefore, this type of time constraints causes difficulty for the dynamic programming approach. According to Bellman's (1957) principle of optimality, the remaining decisions after state A constitute an optimal policy only in regard to the state A, and should not be influenced by the states before the state A. If we decide the path selection by examining that whether $t_n(x_2)-t_n(x_1)$ equals d, when the recurrence procedure is going to x_2, this must result in a nonoptimal solution. Because the decision at the state x_2 is under the influence of the state x_1, this violates the principle of optimality. Where, x_1 is the state that the first stage construction of project n is put into operation. $t_n(x_1)$ is the year when the first stage construction of project n is put into operation. Clearly,

x_2 and $t_n(x_2)$ are those of the second stage. We have

$$t_n(x_1)=\tau(x_1-W_n)+1 \quad (19)$$

$$t_n(x_2)=\tau(x_2-W_n)+1 \ . \quad (20)$$

To mitigate this difficulty, we propose a new hierarchical dynamic programming approach. First of all, according to the n-th place digitals of the hybrid-base numbers $0,1,\dots,m_n$, the state set can be divided into m_n+1 subsets $S_0,S_1,\dots,S_{m_n}$.

$$S_i=\{x|a_n=i,\ x\varepsilon[0,N)\},\ i=0,1,\dots,m_n.$$

If Z_n^b is the first capacity level of the second stage construction of the project n, $1<b<m_n$, then S_b is the subset of the states that the second stage construction of project n is put into operation, and S_1 is that of the first stage. Let

$$L_1=\max_{x_1\varepsilon S_1}\{t_n(x_1)\},$$

$$L_2=\max_{x_2\varepsilon S_b}\{t_n(x_2)\},$$

$$L=\min\{L_1+d,\ L_2\},$$

and take $l=L-y_n-d$.

Then, according to the year when the first stage construction is put into operation, S_1 is divided into l+1 subsets $G_0,G_1,\dots,G_l$. Where,

$$G_i=\{x_1|t_n(x_1)=y_n+i,\ x_1\varepsilon S_1\},\quad i=0,1,\dots,l.$$

Similarly, S_b is divided into l+1 subsets , $H_0,H_1,\dots,H_l$. Where,

$$H_i=\{x_2|t_n(x_2)=y_n+d+i,\ x_2\varepsilon S_b\},\quad i=0,1,\dots,l.$$

In order to lock the impossible path, let

$$J(x)=\infty,\quad \text{for } x\ \varepsilon\ U_1\bigcup U_2,$$

where,

$$U_1=\{x_1|t_n(x_1)>L-d,\ \text{or } t_n(x_1)<y_n,\ x_1\varepsilon S_1\},$$

$$U_2=\{x_2|t_n(x_2)>L,\ \text{or } t_n(x_2)<y_n+d,\ x_2\varepsilon S_b\}.$$

It is evidend that any path from a state In G_i to a state in H_i satisfies the constraint of necessary interval between two stages of the project n. We compound the i-th sublevel dynamic programming model with G_i and H_i, $i=0,1,\dots,l$. The main dynamic programming model consists of S_0 to S_{m_n}. The solution procedure of the hierarchical dynamic programming is stated as follows.

To simplify the notetion, we mark the $F(x)|_{x\varepsilon S}$ as F(S), where, F(.) is any function, S is a set.

Step 1. Via using the main dynamic programming model calculate $J(S_0)$ by recurrence formulation (15)---(17).

Step 2. Calculate the initial value of $J(S_1)$ from $J(S_0)$ by formule

$$J(S_1)=J(S_0)+[\tau(S_1)-\tau(S_0)]\cdot C(S_1)\cdot(1+r)^{-\tau(S_1)}+C_n^1(1+r)^{-\tau(S_0)}\ .$$

Step 3. Solve the shortest path problem with multiple starting points and multiple end points from G_i to H_i, $i=0,1,\dots,l$ by using the sublevel dynamic programming models.

Step 4. Calculate $J(S_{m_n})$ from $J(S_b)$ by using the main dynamic programming model.

Step 5. Select the optimal end state x*, print the result and stop.

The algorithm construction of the two level hierarchical dynamic programming is shown in Fig.2.

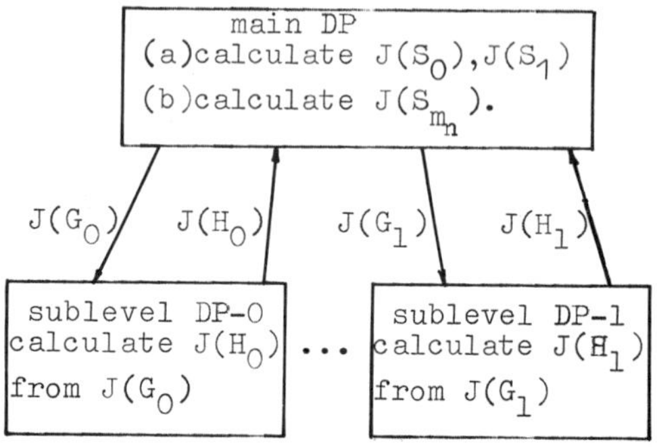

Fig. 2. The algorithm construction of the two level model.

The hierarchical dynamic programming approach does not run counter to Bellman's principle of optimality, and it can find the optimal solution subject to the time constraint.

AN APPLICATION EXAMPLE

The approach is successfuly used in sequencing and sizing of the hydroelectric power projects on the Hongshui (Red Water) River in South China, and the capacity expansion planning of the Huazhong Power Network based on the construction of the Sanxia Hydro Power Station in the Three Gorges of the Yangtze River. Here we only mention the first application in detail.

There are 6 main hydro power projects on the Hongshui River available to construction during the planning horizon 1990 to 2005. The maximum capacity of these projects and thermal power plants to be intended to cooperate with the hydro power projects are shown in Table 1.

TABLE 1 The Maximum Capacity and the Admissible Beginning Year of the Projects on the Hongshui River

main project	maximum capacity (MW)	admissible beginning year
Tianshengqiao (high dam)	6x180	1994
Tianshengqiao (low dam)	2x220[a)]	——
Longtan	8x500	1994
Dahua	2x100[a)]	——
Yantan	4x300	1990
Datengxia	8x150	1991
Thermal plants	10x300	1990

a) The data are additional capacity, 4x 220 MW of capacity of the Tianshengqiao(low dam) project and 4x100 MW of Dahua project will be put into operation before 1990.

Here, the Longtan project is a stage construction project. The second stage construction must be put into operation 5 years after the first stage is. The capacity of the first stage construction may be 4x500 MW or 5x500 MW, it will be determined by the approach.

According to the method of generating states by hybrid-base number, we divide each project into several capacity levels. Because the additional capacity of the Tianshengqiao(low dam) can not begin until the Tianshengqiao(high dam) has begun, we can consider it together with the Tianshengqiao(high dam). For the same reason, the Dahua project can be considered together with the Longtan.

This problem is programmed by using the new approach as mentioned above, calculated by computer PDP 11/70, and following result is achived.

If the annual discount rate is 0.1, the optimal sequencing of the hydro power projects on the Hongshui River is shown in Table 2.

For the scale decision we can give following proposals:

1. Only 4x500 MW is required for the first stage construction of the Longtan project.
2. Before 2005, only 4x180 MW is required for the Tianshengqiao(high dam) project, the additional capacity of the Tianshengqiao(low dam) project is not necessary.

CONCLUSIONS

First of all, we give some illustrations to the approach. For each state x, the latest year, $\tau(x)$, when the capacity of state x can meet the demand, and the annual operation cost of state x, $C(x)$, are calculated by the power and energy balance program. The compensatory capacity between distinct projects is estimated by a simulation model, which is similar to Kuiper and Ortolano's (1973) model.

The study of sequencing and sizing projects leads to following conclusions:

1. Although the objective function is non-convex or non-continuous, the dynamic programming algorithm can always find the optimal solution, therefore, the dynamic programming algorithm is very suitable to the complicated practical problems.
2. The hierarchical dynamic programming approach is effective for sequencing and sizing the projects with time constraints.
3. For the problem of sequencing the projects with multiple levels of capacity there are $(\sum_{i=1}^{n} m_i)!/\prod_{i=1}^{n} m_i!$ possible sequencing. The dynamic programming approach can find the optimal solution through less $n.\prod_{i=1}^{n}(m_i+1)$ comparisons. It is much less than the enumeration algorithm.
4. The computational applicability of this method is limited by "the curse of dimensionality". Both computing time and storage space may exceed the capabilities of the general computer as n increases.

ACKNOWLEDGMENT

We are very grateful to Professor Chen Ting and Professor Li Mian for their direction and help in the preparation of this paper !

REFERENCES

Baker, K. R. and L.E. Schrage (1978). Find an optimal sequencing by dynamic programming: an extension to precedence-related tasks. Opns. Res., 26, 111-120.

Becker, L. and W. W-G. Yeh (1974). Optimal timing, sequencing, and sizing of multiple reservoir surface water supply facilities. Water Resources Res., 10, 57-62

Bellman, R. E. (1957). Dynamic Programming. Princeten, New Jersey.

Erlenkotter, D. (1973). Sequencing expansion projects. Opns. Res., 21, 542-553.

Erlenkotter, D. (1975). Comment on "Optimal timing, sequencing, and sizing of multiple reservoir surface water supply facilities" by L. Becker and W. W-G. Yeh. Water Resources Res., 11, 380-382.

Erlenkotter, D. (1976). Coordinating scale and sequencing decision for water resources projects. In Economic Modelling for Water Policy Evaluation, North-Holand/TIMS Studies in the Management Sciences, Vol.3, 97-112, North-Holand, Amsterdan.

Kuiper, J. and L. Ortolano (1973). A dy-

namic programming-simulation strategy for the capacity expansion of hydroelectric power systems. Water Resources Res., 9, N6.
Morin, T. L. and A. M. O. Esogbue (1971). Some efficient dynamic programming algorithm for the optimal sequencing and scheduling of water supply projects. Water Resources Res., 7, N3.
Morin, T. L. (1973). Pathology of a dynamic programming sequencing algorithm, Water Resources Res., 9, N5.

TABLE 2 The Optimal Sequencing of the Hydro Power Projects on the Hongshui River

(capacity: MW)

project / year	Longtan	Dahua	Tiansheng-qiao	Yantan	Datengxia	Thermal plant
1991						1x300
1992					4x150	
1993					2x150	1x300
1994	2x500					
1995	2x500	1x100				
1996						1x300
1997					1x150	2x300
1998						1x300
1999	1x500 (stage 2)	1x100				
2000	1x500				1x150	
2001						1x300
2002				3x300		
2003	1x500			1x300		1x300
2004	1x500					2x300
2005			4x180 (high dam)			

ENERGY MODELS AS TOOLS FOR POLICY PLANNING IN DEVELOPING COUNTRIES

J. J. Luukkanen* and U. Lehtinen**

*Tampere University of Technology, PO Box 527, SF-33101 Tampere, Finland
**University of Tampere, PO Box 607, SF-33101 Tampere, Finland

Abstract. Energy policy planning in the context of comprehensive development planning is discussed. Development theories, theoretical basis for development planning, are shortly described to create a wider perspective for normative modelling approach. Different groups of energy-economic models, descriptive and normative, and their scope and validity in the policy decision process is reviewed. Problems of modelling noncommercial fuels is discussed.

Keywords. Energy; economics; models; developing countries.

INTRODUCTION

The energy problems of developing countries are often two fold. The rising price of oil has caused severe problems in the balance of payments and financial problems for consumers. In the area of rural energy the fuelwood crisis has cast its shadow over hundreds of millions of people. Fuelwood scarcity has laid severe burden especially on rural women, who mainly collect the wood, the only energy source for cooking in many areas. Also the environmental consequences of the vegetation loss are serious.

The aim of energy policy is to quide energy production and consumption for the development of the society. Energy policy should be a part of the overall development plan of the government, not a distinct sector with its own goals and preferences. The theoretical background to understand and interpret the development of a society is found from the development theories. In the next chapter these theories are discussed and their connections with energy policy are appraised. In the society there are several decision makers in energy production and consumption. The technological energy processes, which are results of these decisions, have effects on the operation of the society thus giving information for the decision making. The structure of this closed-loop system is discussed in third chapter. Further the modelling of this system or part of it is reviewed and the different groups of models, descriptive and normative, are considered. Finally the problems of modelling non-commercial fuels are discussed.

DEVELOPMENT AND ENERGY

When considering the energy problems of the third world the meaning of the energy use must be revealed. Energy use is not a goal in itself but it is a tool for achieving social objects. Usually energy is restricting factor in development, of course excluding the OPEC countries.

When making decisions on energy policy the formulation of the problem is essential. What is the idea of development of society? Who will benefit the decisions? A comprehensive study of the interaction between nature, technology and society must be the basis when trying to solve the energy problems. Because the attemps to solve energy problems are based on the decision maker's ideas of development, the groups of development theories are shortly discussed in the next chapter.

Development theories

(Ganapathy, 1981; Hettne, 1982; Torp and Gould 1980)

Evolutionist theories. The liberalist modernization theories and, according to some researchers also the traditional marxist theories are by nature evolutionist. The idea is that with the help of continuous growth and development the developing countries can achieve the stage of the industrialized western countries. Industrialization and urbanization are key factors in development.

Modernization perpective views development as a linear universal process going through stages ranging from agricultural to post-industrial society. North and South represent different stages in the same process of general development. In modernization theories the attention is paid to the development inside a country and the relations to other countries are thought to be of less importance.

Dependency theories. In these basically marxist theories the comprehensive dependence of developing countries on the capitalist world is reviewed. The dependency perspective views development as a historical process of sustaining polarity, based on unequal relations of interdependence among countries. The affluence of the North is dialectically related to the poverty of the South and its underdevelopment.

Self-Reliance. From the critics of modernization and dependency theories grew alternative theories. These theories are based on the history and culture of developing countries and they emphasize self reliant development. Development is not seen as a linear process, which can be measured only with economic values as in modernization perspective.

The Self-reliance theories consider meaningful development to be contextual and decentralized, oriented towards ecological sustainability and resource conservation, determined and pursued autonomously. Meeting basic needs, it is participatory and empowers people to be autonomously in control of their lives. Development of rural areas is the basis for the development of the whole society.

From the point of view of different development theories the energy problems look also different. Although the target of the development is e.g. rising the standard of living, different development theories will lead to different solutions and so the energy policy will also be different. Modernization approach leads to centralized energy production for centralized industry while self-reliant approach favours decentralized solutions.

MODELLING ENERGY-ECONOMICS

A simplified schematization of the social decision process associated with energy policy is given in Fig. 1. The decision-makers are endowed with resources, which define a feasible set of actions which are open to them at a given point of time (Brock and Nesbitt, 1977). These actions are inputs to the technological-economical energy system. The social actions are e.g. decisions of energy production and consumption. The output of the system can be called "social state", which includes e.g. the prices and quantities of energy used in the society. The system, of course, is stochastic.

Some energy models include this part of the system, from decision making to the output of energy system. These models can be called descriptive. If we include the right-hand box of the figure, operation of society, and the information "feedback" loop we get the so called normative model.

Descriptive models. In descriptive models the "target" for the energy system is given from the outside. In the descriptive (or prospective) philosophy the conclusions about the future are drawn from the study of the past, while in normative approach the objectives to be achieved are defined a priori.

There are two groups of prospective methods used to forecast energy demand (Charpentier, 1983). In purely statistical methods, which are based on regression and extrapolation, the only explicative parameters for energy consumption are time or space (i.e. different countries) or both of them. These methods were used in the "stable" conditions before the first "oil crisis". In these models the growth rate is assumed to be constant, function of time, function of current level of consumption, linear or logarithmic function. The other group is econometric methods. In these models there are more parameters such as price of energy and level of income.

An important group of descriptive models is the supply-demand equilibrium models. These models permit a compact and unified approach to both physical system modelling and behavioral modelling. The behavioral aspect to these models comes from the micro-economic assumption of profit and utility maximization by producers and consumers and the theory of economic equilibrium in the perfectly competitive economy. However equilibrium models are not dynamic, which can be a serious defect, if the changes of the system are studied. From philosophical point of view equilibrium models are positivist.

Dynamic models give information of energy system as a function of time while static models give the equilibrium state at a certain point of time. A rudimentary dynamical character can be achieved by computing the equilibrium states at different points of time and giving the time dependent variables as input variables. In that approach the model in itself doesn't contain any dynamics, but the dynamics is given from the outside.

The dynamic models can be linear. Often in the optimizing linear dynamic models the economic growth is given from the outside. The model can also contain an energy economical and national economical equilibrium model. The system dynamical models, first developed by J. Forrester, generate there own dynamics (Forrester, 1971).

In the paper by Haarasilta, Luukkanen and Majanne (1985) the dynamics of energy consumption and substitution is studied with multivariate signal analysis.

Normative models. In the normative approach the effects of the techno-economical energy process on the society as a whole should be considered. There are several types of criterias for the decision making.

In neoclassical economics the operation of the society can be appraised by using utility function. In that approach social welfare can be assumed to be a function of how well off each member of society thinks he is in each social state. The utilitarian welfare function can be e.g. the arithmetic sum of the utilities of all citizens (Brock and Nesbitt, 1977).

For the selection of technology different lists of criterias have been developed. The criterias include e.g. satisfaction of basic needs, endogeneous self-reliance and harmony with environments (Reddy). These qualitative criterias could be treated as boundary conditions of models, but this can easily lead to non-holistic decisions, which ignore some solutions.
The output of the energy process, the "social state", could also be characterized by different indicators, which can be used for appraisal of the decisions. However, it must be remembered that indicators are atomistic by character, they describe detached parts of a state not the comprehensive process. They are also based on evolutionist theory leading to positivist formulation of the problem.

The operation of society could be also modelled with sectorial "social" models. The energy policy decision-makers should get information of effects that different decisions have e.g. on urban-rural diversity, dependence on foreign countries, social distortions, ecology, agriculture etc. So the box "Operation of society" could contain submodels of these important sectors. One such submodel could be a formal model of dependecia theory developed by Duvall and co-workers (1981). A simplilied flow model of the theory is presented in Fig. 2. This comprehensive dependencia model includes 18 variables and 13 equations.

The effects of fuelwood use on the ecological equilibrium should be modelled so that the decision makers could get information of dynamic consequences of different energy policy decisions. In the same way all important sectors of the operation of society should be submodelled.

Non-commercial energy. In the area of non-commercial fuels the economic equilibrium model cannot be used because there exists no price, which could be measured with monetary units. The consumers of traditional fuels are also producers of such fuels, so the demand and supply are not easily separable. One possible approach to modelling non-commercial fuels is based on allocation of time by the household (Pachauri, 1983). In the model the members of the household are supposed to allocate their time rationally between market wage earning, household consumption and fuel collection, and energy production activities. In the process of time allocation the value on different activities is set by the active rational choices of the people and thus measureable variables, amounts of time, are produced for modelling purposes.

CONCLUSIONS

The formulation of the problem of development is essential for the use of energy-economic models. The ideas of development on which assumptions of the model are based will have a definite effect on the energy policy decisions.

The modernization approach with its trickle down theories has failed in reducing inequalities, on the contrary its solutions have made the situation worse in many places. "Supply" of technology to the poor has not proved to be a solution.

Science and technology are not neutral, they are powerful instruments and the way they are used determines if they are for good or for bad.

Most important is for whom they are good and for whom bad; that is the structure of the society determines the value of different technological solutions. To make sure that it is the poor majority of the people who will benefit the solutions, is to let them decide of the development themselves. To eliminate the obstacles of development and to empower people to be autonomously in control of the decisions and the information for the decisions is the basis for the process of equal development.

REFERENCES

Brock, H.E., and Nesbitt, D.M. (1977). Large-Scale Energy Planning Models: A Methodological Analysis. Stanford Research Institute, California.

Charpentier, J.R. (1983). Basic tools and techniques for energy planning. In H. Neu and D. Bain (Ed.), National Energy Planning and Management in Developing Countries. CEC, D. Reidel Publishing Company, Dordrecht.

Duvall, R.D., Jackson, S., Russet, B.M., Snidal, D. and Sylvan, D., (1981). A formal model of dependencia theory: structure and measurement. In R.L. Merrit and B.M. Russet (Ed.), From National Development to Global Community. George Allen & Unwin Ltd, London.

Forrester, J.W. (1971). World Dynamics. Wright-Allen Press, Cambridge, Massachusetts.

Ganapathy, R.S. (1981). Rural energy and development: A strategic planning analysis. In Southern Perpectives on the Rural Energy Crisis. Nautilus of America Inc. California.

Haarasilta, A., Luukkanen, J. and Majanne, Y. (1985). Signal Analysis in Energy Economic Modelling. Preprints of the CSTD'85, Beijing, China.

Hettne, B. (1982). Development Theory and the Third World. Sarec-report R2:1982, Helsingborg.

Pachauri, R.K. (1983). Third world energy policies, the urban-rural divide. Energy, Policy, 11, 217-224.

Reddy, A.K.N. An alternative pattern of Indian industrialization. In seminar volume: Change and Choice in Indian Industry.

Torp, J.E. and Gould, J. (1980). Development and Development Theories. Intercont no. 18, Helsinki. (in Finnish.)

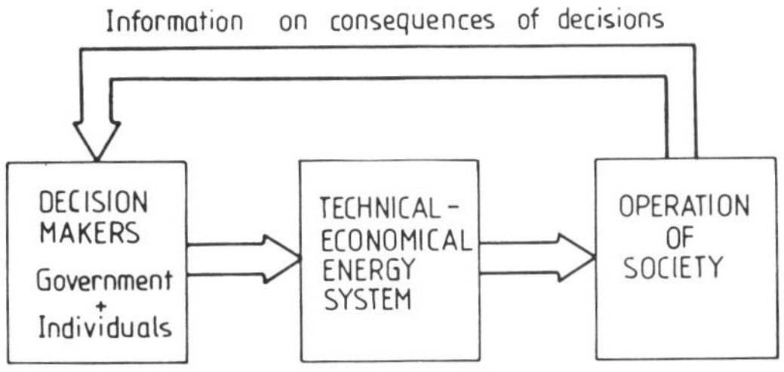

Fig. 1. Schematization of the energy-policy decision process.

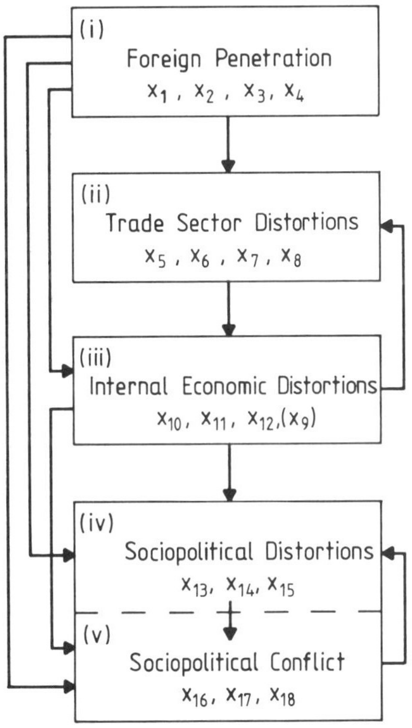

Fig. 2. A simplified flow model of dependencia theory (Duvall and co-workers, 1981).

DECISION ANALYSIS OF THE POLLUTION CONTROL OF THE HUANGPU RIVER

Wei-min Cheng, Yang Chia-ben, Chen Wei-Ji and Cui De-guang

Department of Automation, Tsinghua University, Beijing, China

Abstract. The pollution control of the Huangpu River in Shanghai is a large engineering project which can be benefited by system analysis. A long-term planning is complicated dynamic problem, because there are too many uncertainties, imprecisions, risks as well as intangible impacts to be considered along the time horizon. This paper presents a part of the results of system analysis which have been done jointly with Shanghai EPB and University of Michigan.

Keywords. Decision analysis, water pollution, system analysis, modelling, optimization, robustness.

INTRODUCTION

Shanghai is the largest commercial and industrial city in China (See Fig. 1). The population of Shanghai is 11 millions among which 6 millions are in the urban district. The number of industrial factories is up to 7000.

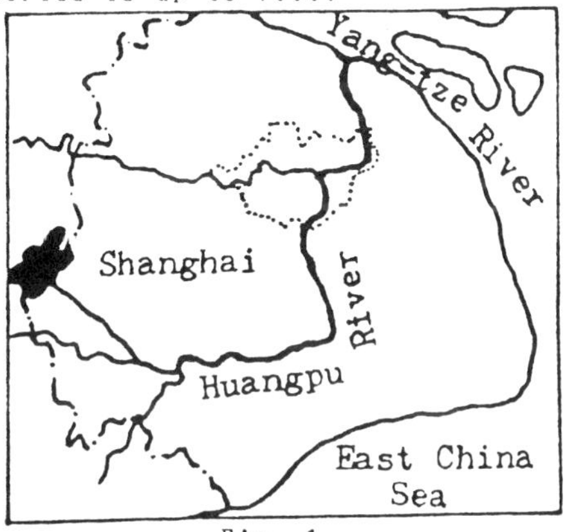

Fig. 1

The Huangpu River transverses the city of Shanghai downwards 25 km from the urban district to Wusongkou where the River converges with the Yangtze River and then empties into the East China Sea. The widths of the Huangpu River are 300 to 500 meters. The depths are 5 to 10 meters. Since the mouth of the river is near the Sea, it is a strong tidal water-way.

According to the field survey in June, 1981, the total amount of wastewater is 5 millions metric tons per day and the total BOD_5 is 500 metric tons per day. Most of the BOD_5 loads are around the urban district where the DO is less than 1 mg/l.The DO along the Huangpu River is given in Fig.2.

The Huangpu River pollution control is a large scale engineering project which needs a long-term planning based on system analysis. In this paper the preliminary analysis, water quality modelling, regional wastewater treatment system planning and dynamic decision analysis are presented.

PRELIMINARY ANALYSIS

The water quality control of the Huangpu River is a large project which may cost billions dollars. The task of system analysis is not just to find a cost-effective way to meet a fixed requirement. It is of utmost importance to do a preliminary analysis on what the objectives are, what the resource constraints are, what the essential problems are to be kept in mind in doing the system analysis.

Since most of the intakes of water supply works are located at the heavily polluted portion of the Huangpu River meanwhile some cancer mortality rates are very high, it is highly suspected that the river pollution may have impacts on cancer mortality. Also the phenomena of black and malodor can be observed in the river due to the degradation of water. As a consequence, the appeal for the improvement of water quality of the Huangpu River is urgent.

On the other hand, the standard of the people's material and cultural life at Shanghai is rather low, for instances, GNP/cap. =$ 2485/year and housing floor space= 5m^2/cap. And although the cancer mortality rate is high, the overall health condition is much improved since 1949.

Therefore the system analysis must answer the following major questions:

- What ia the appropriate water quality?
- How much are the Shanghai people willing to pay?
- What is the cost-effective alternative for the abatement of wastewater?

WATER QUALITY MODELING

Since water quality is the basic objective for river pollution control and is closely related to every part of system analysis, it is of utmost importance to have an appropriate water quality model.

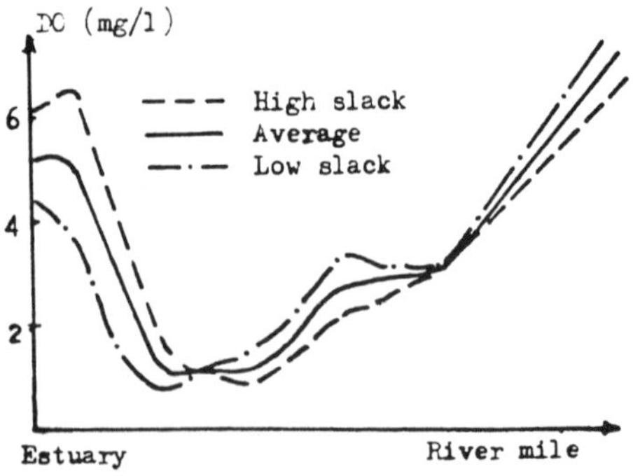

Fig. 2

The Huangpu river is a tidal river, but from field survey as shown in Fig.2, the tidal effect just makes the DO curve swing up and down in the river. From the theoretical study, an equivalent dispersion coefficient E in the static water quality model can be found to take into account the tidal effect. Therefore, static model can be used for planning. Moreover, the net flow is rather low, -35 m^3/s, then in addition to CBOD, NBOD must be considered. Consequently, the mathematical equations of the static water quality model are as follows:

$$DO:\ 0=-u*\partial DO/\partial X+ E*\partial^2 DO/\partial X^2+ k_2(DO_{sat}-DO)-k_1\ CBOD-k_N\ NBOD+ W_{DO}$$

$$CBOD:\ 0=-u*\partial CBOD/\partial X + E*\partial^2 CBOD/\partial X^2-k_1*CBOD+ W_{CBOD}$$

$$NBOD:\ 0=-u*\partial NBOD/\partial X + E*\partial^2 NBOD/\partial X^2 -k_N NBOD+ W_{NBOD}$$

where E is the dispersion coefficient, k_1 is the decay rate of CBOD, k_n is the decay rate of NBOD, k_2 is the aeration rate, u is the velocity of flow and W is the pollution source. In determining the parameters of the water quality model, the major difficulties encountered are the lack of data and the uncertainties of pollution loads. In order to deal with the problem of imprecision, the parameters can not be just calibrated by the method of curve fitting. They should be checked and modified by the following experience and knowledge of experts.

a. Results of laboratory experiments (E=6000 ft^2/s, k_1 = 0.35 1/day, k_n=0.07 1/day, k_2=0.40 1/day).
b. Parameters of similar rivers.
c. Judgements of experts based on the past experiences about the results of some simulation studies on the mathematical model.
d. Results of theoretical analysis and qualitative analysis.

From b,c,d, we have E=1613--6454 ft^2/s, k_1=0.11--0.48 1/day, k_n=0.085--0.535 1/day, k_2=0.05--1.02 1/day. Finally, the calibrated parameters for one set of pollution loads are E=6000 ft^2/s, k_1=0.35 1/day, k_n=0.11 1/day, k_2=0.35 and 0.45 1/day. The calibrated DO curve together with the DO curve of field survey in 1981 are given in Fig. 3. In addition, the mathematical model has been verified by the field survey in 1983, as shown in Fig.4. From the results of the above analysis, the validity of the model is assured.

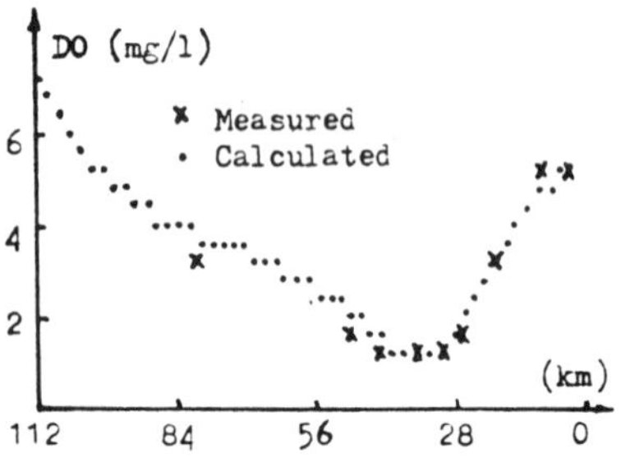

Fig. 3

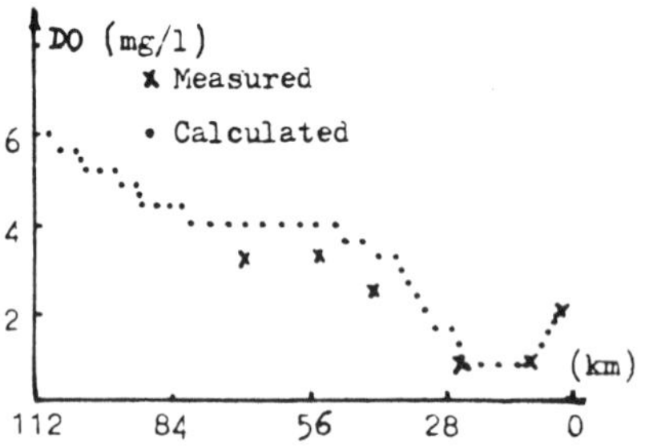

Fig. 4

Also the calibration of parameters of the water quality model depends on the amount of pollution loads. But the pollution loads are very imprecise. In order to assess the credibility of the results of system analysis, three sets of estimated pollution loads and consequently three sets of parameters, as shown in Table 1, have been used throughout the analysis.

Table 1.

	1	2	3
Parameters			
E (ft^2/s)	6000	6000	6000
k_1 (1/day)	.06 -.35	.35	.35
k_n (1/day)	.001-.13	.1	.1
k_2 (1/day)	.22 -.28	.35 -.45	.35 -.45
k_{cs} (1/day)	0.0	0.0	0.0
Loads			
(t/d)			
BOD_5	405	445	485
NH_3-N	77.5	127.5	147.5

REGIONAL WASTEWATER TREATMENT SYSTEM PLANNING

For long-term planning of the pollution control of the Huangpu river, the problems of the optimal layout and the optimal sequence of realization of the wastewater treatment system should be considered. A simplified method that finds out the optimal layout first and then the optimal sequence is shown schematically in Fig.5. A feasible system layout is represented as a network in Fig.6. The feasible water receiving bodies are the East Sea, the Yangtze River, the Huangpu River and its tributaries. Total amount of wastewater is $5*10^6$ m^3/day and total BOD_5 is 500 t/day. Thus the optimization in layout is formulated as a mixed integer programming model of minimizing the total cost.

Optimization in layout
Find the optimal layout of wastewater projects

↓

Optimization in time
Find the optimal sequence of wastewater projects

↓

Multi-objective decision analysis

Fig.5

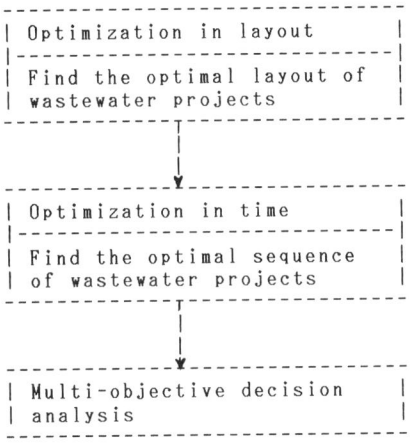

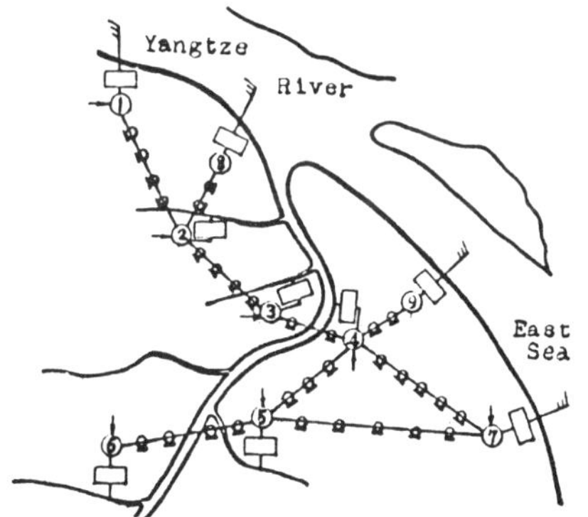

i ith wastewater source
feasible location of treatment plant
feasible pipeline
feasible pumping station

Fig. 6

Model 1

$$\text{Min} \sum_i \sum_K (c_i^K Y_i^K + b_i^K X_i^K) + \sum_i \sum_j \sum_K (c_{ij}^K Y_{ij}^K + b_{ij}^K X_{ij}^K)$$

$$\text{s.t.} \quad \sum_K Y_i^K = s_i + \sum_j \sum_K (Y_{ji}^K - Y_{ij}^K) \quad \text{for all } i \qquad (1)$$

$$Y_i^K \leqslant M_i X_i^K \quad \text{for all } i,k \qquad (2)$$

$$Y_{ij}^K \leqslant m_{ij} X_{ij}^K \quad \text{for all } i,j,k \qquad (3)$$

$$\sum_K X_i^K \leqslant 1 \quad \text{for all } i \qquad (4)$$

$$\sum_K (X_{ij}^K + X_{ji}^K) \leqslant 1 \quad \text{for all } i,j \qquad (5)$$

The constraints (1) are called flow equations. The constraints (2)-(3) are the permissible sizes of the feasible plants and the pipelines. (4)-(5) are restrictions on decision variables. Depending on the economies of scale of wastewater facilities, we have found different optimal layouts under different conditions.

Conditions:
A. It is allowable to discharge directly wastewater into the East Sea and the Yangtze River without any treatment.
B. The treatment plants near the East Sea and the Yangtze River can be primary, others must be secondary.
C. All plants must be secondary.
D. All plants can be primary.
The characteristics of every optimal layout under aforesaid conditions are listed in Table 2.

Table 2

Condi-tion	Total cost (10^8 Y)	DO value increased at 15th segment (mg/l)	Risk
A	1.252	4.41	most
B	1.654	4.41	less
C	2.394	4.38	least
D	1.504	3.02	least

The second problem is the optimal scheduling. The implementation of every alternative is divided into three periods in accordance with three five-year plans as shown in Fig.7.

1	→	2	→	3
1986-1990		1991-1995		1996-2000

Fig.7

At different periods water quality requirements are different. Such a dynamic optimal problem is formulated as a dynamic integer programming model of minimizing the present value of total cost.

Model 2

$$\text{Min} \sum_n \sum_{T_p} \Big(\sum_t C_{t,n} Y_{t,n} + \sum_T M_{T,n} X_{T,n} \Big)$$

$$\text{s.t.} \quad \sum_{T=T_1} \sum_n E_{j,n} X_{T,n} \geqslant D_{j,T} \quad \text{for all } j, T_p \qquad (6)$$

$$X_{T,n} = Y_{(T-Y_n),n} \quad \text{for all } T,n \qquad (7)$$

$$\sum_t Y_{t,n} \leqslant 1, \quad \sum_T X_{T,n} \leqslant 1 \quad \text{for all } n \qquad (8)$$

$$Y_{t,n}=0 \text{ or } 1, \quad X_{T,n}=0 \text{ or } 1 \quad \text{for all } t,T,n \qquad (9)$$

The constraints (6) are water quality specifications. (7)-(9) are constraints on decision variables. Depending on the cost-effectiveness of wastewater facilities and different water quality requirements at different periods, the optimal sequence can be found for every optimal layout. Although there are many uncertainties, but it is found that the abatement of wastewater in Suzhou Creek, which is the major tributary with BOD_5 up to 30% of the Huangpu River and is the nearest to the heavily polluted reach, should be completed at the end of the first period for all alternatives. This conclusion is in accordance with the common sense that the most economical and efficient facility with less risk must be realized first. Therefore by means of optimal sequencing, we can make a robust decision for the first step under uncertainties. For the later periods, new information and experience can be obtained and used to modify the plan, i.e. rolling planning.

Also it is necessary for every alternative to do some sensitivity analysis by varying the pollution loads and cost functions because of their imprecisions. The analysis indicated that every optimal layout and/or optimal sequence owns stronger robustness. For example when the amount of wastewater is varied by 80% including the water quality models with three sets of parameters or one of the coefficients of cost functions is varied by 40%, the optimal layout and sequence remain unchanged for the alternative under the condition B.

In order to find a satisfactory alternative, the multiple objective decision analysis is still needed.

DYNAMIC DECISION ANALYSIS

Altogether fourteen alternatives which appear to be technically feasible, economically reasonable and environmentally acceptable will be assessed. The fourteen alternatives can be classified into four categories.

Category I - Moving the water intakes to the up-stream for the protection of human health- a_{10}.

Category II -In addition to the moving of water intakes, controlling the main industrial pollution sources for the elimination of stink-- a_{11}, a_{12}.

Category III -In addition to the moving of water intakes, conveying all the wastewater into the East China Sea and the Yangtze River without any treatment plant -- a_8 (may violate the regulation)

Category IV --Moving the intakes to the up-stream and establishing wastewater treatment plants -- $a_1, a_2, a_3, a_4, a_5, a_6, a_7, a_9, a_{13}, a_{14}$

For the assessment of alternatives, what the decision makers are considering can be represented by the objectives hierarchy as shown in Fig. 8, in which $u_s(.)$ means the increment or decrement of the utility value of objective s and $w_s(.)$ is its weighting factor. The weighting factors represent the relative worths between objectives and indicate "The willingness to pay".

	$U_s(.)$	$W_s(.)$
Economic factors	Cost	.32/.27
Technical factors	Power	.12/.10
	Land	.06/.06
	Materials	.23/.06
Environmental impacts	Human health	.11/.29
	Aesthetics	.04/.09
Social impacts	Prestige	.07/.07
	Movement of resident	.06/.06

/: 1986 - 1990/ 1991 - 2000

Fig. 8

Considering the discount, all utilities should be refered to the present value. Also the weighting factors may vary with time as shown in the objective-hierarchy of Fig. 8. Therefore the decision analysis should be a dynamic analysis. If $u(a_{ij},k)$ is the over utility function at period k for the construction of facility j of alternative a_i, then the total utility function U of alternative a_i is

$$U \text{ of } a_i = \sum_K \sum_j u\,(a_{ij},k)$$

where

$$u(a_{ij},k)=w_1(k)u_1(a_{ij},k)+\ldots+w_n(k)u_n(a_{ij},k)$$

u_s--increment of the present value of utility

$w_s(k)$-- weighting factor that may vary with time

$s=1,2,\ldots\ldots n$

Here we are not going into the details of utility functions, but only stress the ordinal ranking of alternatives and the robustness of the analysis. The ordinal ranking by dynamic analysis is as follows:

$$a_{10}>a_{12}>a_{14}>a_9>a_{13}>a_{11}>a_6>a_7>a_2>a_1>a_5>a_4>a_8>a_3$$

Since the utility functions and weighting factors can only be obtained very approximately, then the credibility of the analysis is a question to be answered. By means of linear programming, we can get the ranges of the variations of weighting factors by keeping a_{10} being optimal as shown in Fig.9. Such a robustness analysis will help the decision maker to make his\her choice.

CONCLUSIONS

This paper is a delineation of the system analysis of the Huangpu River pollution control. Due to the limited space only the fundamental mathematical models and methods of analysis are given.

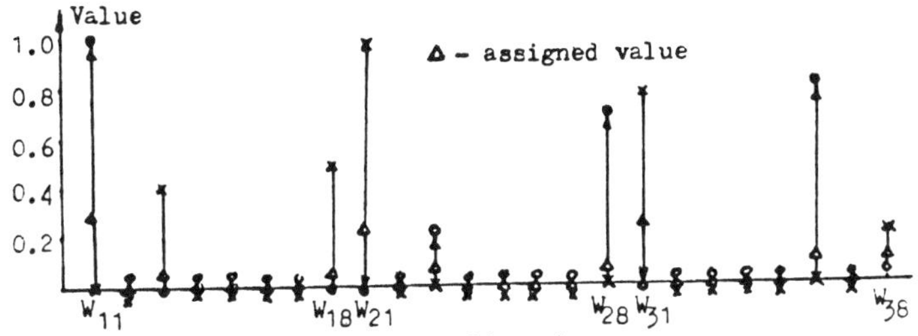

Fig. 9

REFERENCES

Keeney R.L. et al. (1976). Decisions with multiple objectives:preferences and value tradeoffs.

Mandl C.E. (1981). A survey of mathematical optimization models and algorithms for designing and extending irrigation and wastewater networks. Water Resource. Res. 17, 769-775,

Zison S. W. et al. (1978). Rates,Constants and kinetics formulations in surface water quality modeling. (U.S.A. EPA)

DYNAMIC ESTUARY WATER QUALITY MODEL AND ITS APPLICATIONS

Sun Jian-Hua, Chen Yu-Liu and Chen Wei-Ji

Department of Automation, Tsinghua University, Beijing, China

Abstract

In this paper a few problems arisen from the use of dynamic water quality model are discussed. Some special applications are proposed. It shows that under certain situations the dynamic model is needed for giving some more useful information.

Keywords. Environment control; modelling; water pollution; distributed parameter systems; large-scale systems

INTRODUCTION

In 1925, Streeter-Phelps proposed the first water quality model. Later on it was improved and developed according to different situations. For estuaries and tidal influenced rivers, there are two kinds of models used popularly. One is real time dynamic model (simply, "dynamic model"). It can give some more information. Another is the model of tidal cycle average value (simply,"steady-state model").It makes the computation easy. But an equivalent dispersion coefficient should be used to represent the tidal influence. This coefficient is not easy to be determined well.

In this paper, some problems arisen from the applications of dynamic models are discussed. Some special situations are analyzed by using the dynamic model. The rest of this paper is divided into 4 sections. In section 2, the formulation of models is described. And how to choose difference equation form is discussed. In section 3, the numerical analysis of dynamic model and the influence of step size are discussed. The other applications are proposed in section 4 and some brief conclusions are given in section 5.

FORMULATION OF ESTUARY WATER QUALITY MODEL

For investigating suitable difference equation form, an imaginal river is theoretically analyzed first. This imaginal river has uniform rectangular cross-section, and it is deep enough that the change of depth caused by tide can be ignored. Furthermore, suppose that the tidal velocity can be expressed as:

$$u = U_f + U_T \sin \Omega t \qquad (1)$$

where U_f -- tidal cycle average velocity
U_T -- peak value of tidal velocity
$\Omega = 2\Pi/T$
T -- tidal period.

Then water quality model can be presented by using Streeter-Phelps equation as following:

$$\partial B/\partial t = E*\partial^2 B/\partial X - u*\partial B/\partial X - K_d B + W/A \qquad (2)$$

$$\partial C/\partial t = E*\partial^2 C/\partial X - u*\partial C/\partial X - K_a C + K_d B \qquad (3)$$

where B,C: concentrations of BOD and oxygen deficit
K_d, K_a: BOD decay and reaeration coefficients
E: dispersion coefficient
W: pollutant discharge velocity
A: cross-section area.

Under these conditions there exists analytic solution for Streeter-Phelps equation. The further analysis will be concentrated on finding a suitable form for difference calculation to make the computational results close to the analytic solution.

1. Forward difference algorithm is the simplest scheme. But from computational result it has been shown that heavy computational burden will appear.

2. Lax-Wendroff scheme
As shown in Fig. 1 the BOD concentration of location i at time interval t+1 can be considered as transferring from a certain upstream point P, i.e. $B_i^{t+1} = B_p^t$. By using B_{i-1}^t, B_i^t and B_{i+1}^t to do interpolation, B_p^t can be obtained as,

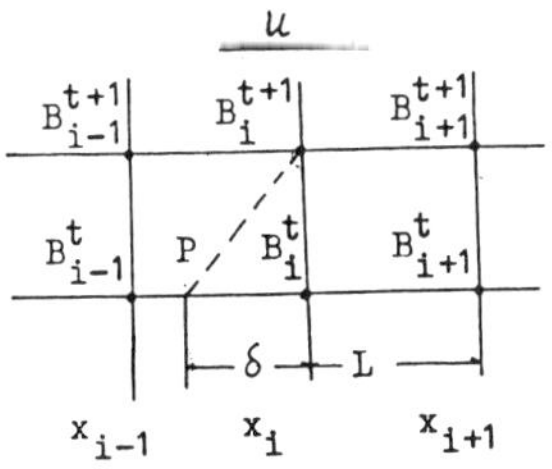

Fig. 1

$$B^t_\rho = B^t_i + \lambda[(1+\lambda)*B^t_{i-1}/2 - \lambda B^t_i - (1-\lambda)*B^t_{i+1}/2] \quad (4)$$

where $\lambda = \delta/L$ and $\delta = u \cdot \Delta t$

Let $r = (1-\lambda)/2$, and take into account of decaying and waste discharge Streeter-Phelps equation can be discretized as

$$\dot{B}^t_i = (B^{t+1}_i - B^t_i)/\Delta t = U[(1-r)B^t_{i-1} - (1-2r)B^t_i - rB^t_{i+1}]/L + E(B^t_{i-1} - 2B_i + B^t_{i+1})/L - K_d B_i + W_i \quad (5)$$

According to this relation, a set of difference streeter-Phelps equations can be established,

when $u \geq 0$

$$B_i = G_1 B_{i-1} - G_2 B_i + G_3 B_{i+1} + W_i \quad (6)$$

when $\overline{u} < 0$

$$B_i = G_3 B_{i-1} - G_2 B_i + G_1 B_{i+1} + W_i \quad (7)$$

where $G_1 = E/L + (1-r)|u|/L$ $\quad G_3 = E/L + r|u|/L$

$G_2 = 2E/L + (1-2r)|u|/L + K_d$

Similar equation can be obtained for oxygen deficit. As shown in Fig. 2 when $r=0.45$, for different section sizes all the results are closed to the analytic one. This means that Lax-Wendroff scheme can be used to do computation without very small step size. In fact, $r \approx 0.37$ can be estimated by using $r=(1-\lambda)/2$. Computational results show that maybe $r=0.45$ is more suitable. Here the error is only within 20%. The chemical reaction part is well known. Here some main expressions for a real river are given briefly.

1. CBOD decaying

$$V^{t+1}_j B^{t+1}_j = V^{t+1}_j B^t_j \beta_4 = V^{t+1}_j B^t_j e^{-k_c \tau} * 1.047^{(T-20)} \quad (8)$$

where V^{t+1}_j -- the volume of jth junction at time interval t+1

B^t_j, B^{t+1}_j -- CBOD concentrations

K_c -- CBOD decay rate

τ -- time step

T -- temperature C

1.047 is the temperature coefficient

2. NH_3-N change

$$V^{t+1}_j N^{t+1}_{1j} = V^{t+1}_j N^t_{1j} \beta_1 + V^{t+1}_j A_j \rho \alpha_1 + A_{sj} \alpha_3 * 1.047^{(T-20)} \quad (9)$$

where N^t_{1j}, N^{t+1}_{1j} -- NH_3 -N concentration

$\beta_1 = e^{-k_{n1}} * 1.020^{(T-20)}$

k_{n1} -- NH_3 - N decay rate

1.020 is the temperature coefficient for β_1

A_j -- algae biomass concentration of jth junction

ρ -- net fractional decrease in algae by respiration and mortality

α_1 -- nitrogen portion of algae biomass

A_{sj} -- surface area of jth junction

α_3 -- rate of release from decaying sediment

1.040 is the temperature coefficient for α_3

3. NO_3-N decaying

$$V^{t+1}_j N^{t+1}_{2j} = V^{t+1}_j N^t_{2j} \beta_2 + V^{t+1}_j N_{1j}(1-\beta_1) \quad (10)$$

where N^t_{2j}, N^{t+1}_{2j} -- NO_2-N concentrations

$\beta_2 = e^{-k_{n2}} * 1.020^{(T-20)}$

k_{n2} -- NO_2-N decay rate

1.020 is the temperature for β_2

4. DO reaeration

To calculate DO saturation concentration first

$$DO_{sat} = 14.553 - 0.382T + 5.426*10^{-3}T^2 - 0.555C_{cl}*(1.665*10 - 5.866*10^{-6}T + 9.796*10^{-8}T^2) \quad (11)$$

where T -- water temperature

C_{cl} -- salinity concentration

Thus we can calculate DO by the following equations

$$V^{t+1}_j O^{t+1}_j = V^{t+1}_j O^t_j + V^{t+1}_j \sqrt{1.037^{(T-20)}}(1 - e^{-k_a \tau})*(O^*_j - O^t_j) - V^{t+1}_j B^t_j(1-\beta_4) - A_{sj} K_B * 1.040^{(T-20)} + V^{t+1}_j * A_j * (\alpha_5 \mu - \alpha_6 \rho) - V^{t+1}_j N_{1j}(1-\beta_1) * \alpha_7 - V^{t+1}_j N_{2j}(1-\beta_2) * \alpha_8 \quad (12)$$

where O^t_j, O^{t+1}_j -- DO concentration

O^*_j -- DO saturation concentration (DO_{sat})

K_B -- rate of dissolved oxygen uptake by decaying sediment

$1.037^{(T-20)}$ -- temperature adjustment coefficient for reaeration

α_5 -- amount of oxygen producrd of algae growth

α_6, α_7, α_8 -- amount of oxygen uptake per unit of algae respired, ammonia nitrogen oxidized, nitrite nitrogen oxidized respectively

Due to space limit, the expressions for other quality components will not be given here.

NUMERICAL ANALYSIS OF THE MODEL

Assume there is an imagine river and there exists only a single point discharge. The tidal velocity is considered as a regular sinusoidal wave with phase lag caused by wave celerity

$$u = U_f + U_T \sin(\Omega t - \phi) \quad (13)$$

where

$\phi \sim x/(\sqrt{gh} * \Omega)$

x -- distance from river mouth

g -- acceleration of gravity
h -- water depth

Then we fix the space step size (distance) first, ex.2 miles,and choose time steps as 7.5,15,30 and 60 min. respectively.The smaller the time step size,the more frequent the advection will be. Thus the effluent components disperse faster. As shown in Fig. 3, the peaks of BOD_5 and oxygen deficit are the smallest for the one of 7.5 min. time step.Next we fix the time step size,e.x. 15 min. , and choose distance steps as 0.5,1.0and 2.0 miles respectively. Their steady state curves of DO and BOD are shown in Fig. 4. Here the physical meaning is also clear.It means after the dispersion frequency in mathematic sense had been given, the bigger space step the bigger volume of junction will be. Thus during the same time interval and equal discharge mass, the bigger volume must have lower concentration.

Combine the above two points together, a third experiment was done.As mentioned in section II, when using Lax-Wendroff scheme a suitable ratio between space step and time step can make the results close to the original differential equation,also close to the actual situation. This make it possible to choose a little bigger step size and save computer time a lot.

As an example,the above theoretical analysis of dynamic water quality model is applied to Huangpu River in Shanghai. This river is 70 miles long. There are almost 200 point sources along the river. In Fig. 5 different time steps are used to do fitting with the DO profile got from the survey data of June, 1981. It can be seen that for 1.4 miles' distance step the suitable time step is 15 min. This is also close to the roughly estimated value explained in section II.

For actual computation there is another important problem, it is the boundary condition. There are two ways taken here. One way is for downstream.The Huangpu River flows into the Yangtze River and exchanges water with it under tidal effect.Because the Yangtze River is a much bigger water body, its quality C_{in} can be seen as constant.We record and store the concentrations, as c_{eb}, of quality components for the water discharged into Yangtze River during ebb tide at each time interval. During flood tide the concentration of quality components will be

$$C_f = C_{in} + (1-\alpha)C_{eb} \qquad (14)$$

Here α =0.3 in our example.Another is for upstream. The Huangpu River is connected with the Dianshan Lake, and the tidal effect influences upto this very end. But its quality is not concerned. Therefore a more roughly approximation is taken. At the terminal junction introduce a pseudo withdraw near the average value of tidal flow there, and an inflow of the same amount with water quality as the same as that in the Dianshan Lake. In this way, both balance of water and exchange of quality are considered.

Finally we have compared the fitting results by using dynamic model with that by steady state model AUTOQUAL (Fig. 6).

OTHER APPLICATIONS OF DYNAMIC MODEL

For planning the steady state model is recommended. But for some quality management and short term prediction steady state model can do nothing, and dynamic model has to be used. Here only a few problems are presented.

1. Prediction of salty tide entering.
The mouth of the Huangpu River is not close to the East Sea. So usually the tidal salinity is not high. During low flow period of the Yangtze River it did have severe influences sometimes. We did computation from the tidal data of June 1981. Assume the salinity at mouth was 2000mg/l, and it lasted for 6 days. The results are shown in Fig. 7.

2. Influence from rainfall runoff.
Rainfall runoff will flush sediment in sewage pipes and foul on street surface into the river. Fig. 8 shows the result from dynamic model and it is compared with the original DO profile. This phenomenon is similar to the survey data of June 1983 after one day raining.

3. Some problems for water quality management.
For water quality management we have investigated the influence of pulse type discharge, e.x. that caused by the failure of a waste water treatment plant. And a control scheme has been studied that a stipulation should be made for allowing discharge only during ebb tide period. Thus more foul will be discharged into the Yangtze River. For this suggestion we have got positive conclusion. Due to space limit, all these problems can not be described in detail here.

CONCLUSION AND ACKNOWLEDGEMENTS

Dynamic model is obviously much more complicated than the steady state model. And it will take 5 even 10 times more computer time than the steady state model. So some researchers negated its usefulness. In our point of view, when doing water resources planning steady state model is enough. But it is necessary to check the equivalent dispersion coefficient of steady state model by using dynamic model. On the other hand for some short term prediction and quality management problems, dynamic model has some functions that steady state model can never do.

The above research work was completed under Prof. Cheng Wei-Ming's supervision. The authors are grateful to our colleagues in the research group of system engineering of Automation Department for their stimulating discussion, and wish to thank Mr. Li Yu-Liang from Hydrolic Engineering Dept. and Prof.Fu Guo-Wei from Environment Protection Dept. for their helpful suggestions.

This work is based on China-America fundamental scientific research cooperation topic "Pollution control Planning and assessment for the Huangpu River in Shanghai ", and supported by Shanghai Environment Protection Bureau. The authors are also wish to acknowledge University of Michigan and Limno-Tech Inc. in U.S.A. for their valuable fundamental materials and excellent experiences.

REFERENCES

Baumgarther, D.J. (1971). Estuarine Modeling.

Chen wei-Ji (1985). Investigation of static water Quality Model of Huangpu River. Technique Reports of Dept. of Automation, Tsinghua University (In Chinese).

Chen Yu-Liu (1984). Simplified Dynamic Water Quality Model and Its Application for Huangpu River. Technique Reports of Dept. of Automation Tsinghua University (In Chinese).

Fu Guo-Wei (1984). Planning of Water Quality Control (In Chinese).

Genet,L.A. D.J Smith and M.B Sonnen (1974) Computer Program Documentation for the Dynamic Estuary An intermediate technical report prepared for SDB of EPA of U.S.

Lu Jin-Fu, and Guan Zhi (1984).Numerical Solution of Partial Differential Equations (in Chinese).

Sun Jian-Hua (1984). Investigation of Water Quality Model of Tidal Influenced River. Technique Reports of Dept. of Automation, Tsinghua University (In Chinese).

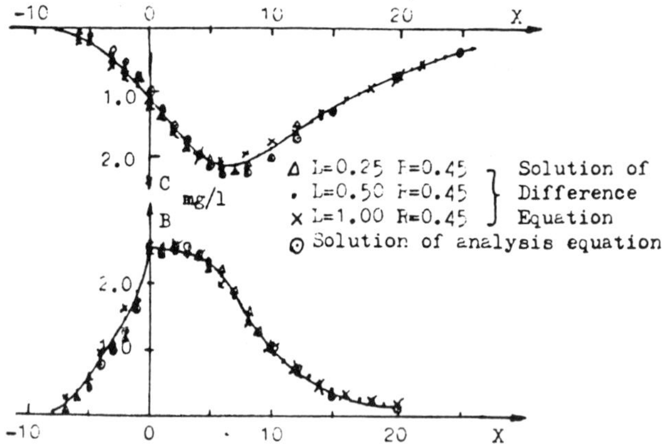

Fig. 2

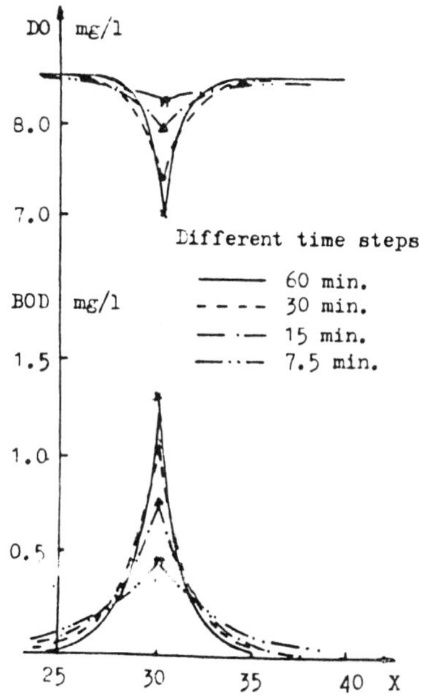

Fig. 3

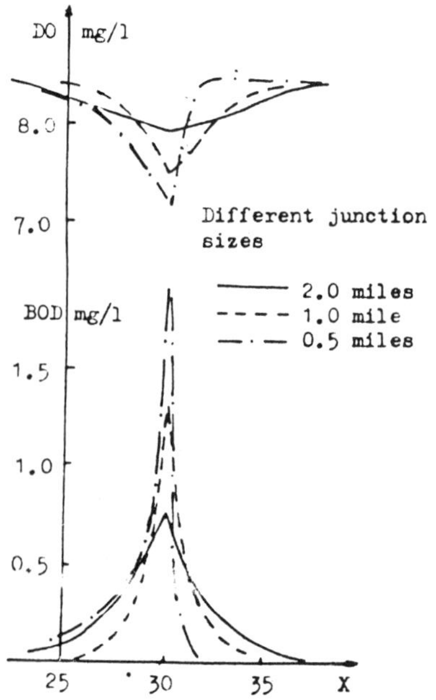

Fig. 4

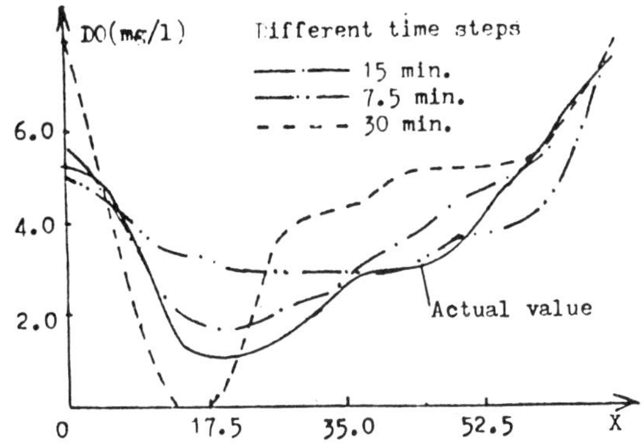

Fig. 5

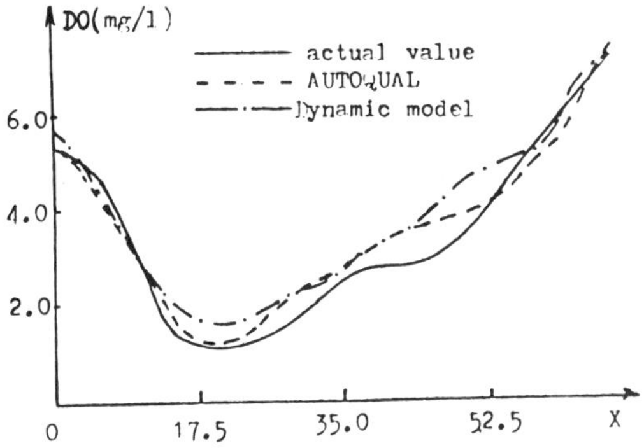

Fig. 6

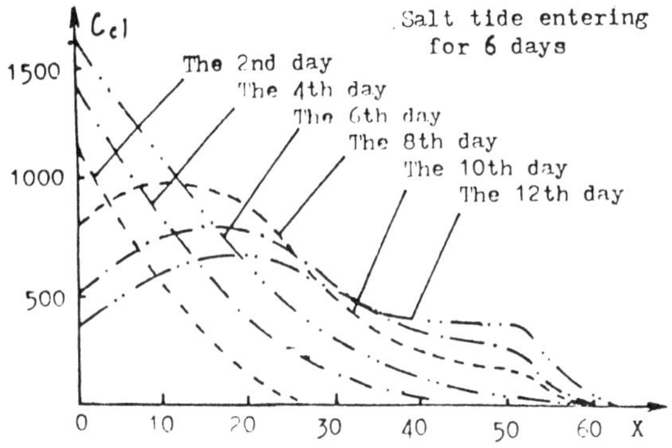

Fig. 7

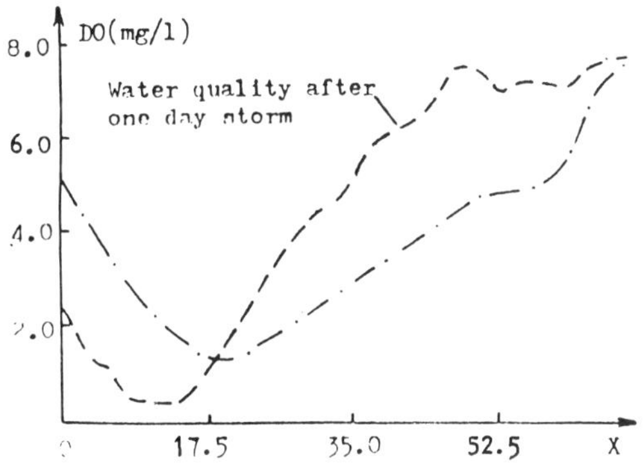

Fig. 8

OPTIMIZATION OF GRAVITY SEWERAGE SYSTEMS

Chen Senfa

Research Institute of Automation, Nanjing Institute of Technology, Nanjing, China

Abstract. This paper presents a method which optimizes both the layout and the pipe design parameters (depth, slope and diameter, etc.) of sewerage systems simultaneously. The procedure of optimization is divided into two steps: First, a Steiner Minimal Tree (called SMT), a tree with the shortest possible total length is found by means of an SI algorithm, which is used as the initial layout of the systems. Secondly, the layout and the pipe design parameters of the systems which minimize the objective function are found by means of the Gauss-Siedel iterative method and the Branch-and-Bound method, under the conditions of flow velocity constraints, depth constraints, etc. A sewerage system with 10 given points and 5 Steiner points has been computed; the optimization results show improvement in economic performance.

Keywords. Optimization; SI algorithm; nonlinear simultaneous equations; Gauss-Siedel iterative method; Branch-and-Bound method.

INTRODUCTION

With the rapid growth of the urban population, expenditures of sewer water disposal and management are increasing continually. Therefore, the optimal design of sewerage systems has economic significance.

In general, the design problem is divided into two aspects: (a) optimization of the system layout, (b) optimization of the pipe design parameters for the known layout. One can obtain better results in optimization by considering these two aspects simultaneously than by considering them as two independent aspects. But this is a complex problem to solve. This is due to the fact that a minimum cost system layout may depend on the various pipe design parameters (depth, slope, diameter, etc.), which, in turn, cannot be determined until the network topology has been defined.

In order to solve this hard problem, Barlow (1972) presented an improved version of the minimum span network algorithm, and Mays et al. (1976) presented the DDDP method. Either of them can determine the layout and the pipe design parameters of systems simultaneously . But these methods can only optimize the systems between the given points, and no extra point can be added later.

The method presented here can optimize systems between the given points and some added extra points (called Steiner points).

The cost implied with the use of the method presented here is lower than that required when the Barlow method is used, under the conditions of disregarding the topography constraints.

MODELS

Assumption. Our method is based on the following assumptions:

A. Sewer flow is steady and uniform.
B. Ground slope is constant. For simplicity, it is assumed to equal zero.
C. The pipe sewerage networks can occupy any position in space, i.e., they are not constrained by hills, depressions and rivers, etc.

Models. The cost of a link in sewerage systems includes cost of pipe, cost of laying (including excavation, jointing, etc.)

The capital cost of a link may be expressed as,

$$C_{ij} = K_1 D_{ij}^{a_1} \; A_{ij}^{a_2} \; L_{ij} \tag{1}$$

where C_{ij} is the capital cost of the link from point i to point j, in RMB(¥). D_{ij}, A_{ij} and L_{ij} are the diameter, average depth and length of the link, respectively, in meter. K_1, a_1 and a_2 are constants, which can be found by means of linear regression technique.

The length of a link, L_{ij}, can be expressed as,

$$L_{ij} = \sqrt{(X_j-X_i)^2 + (Y_j-Y_i)^2 + (H_j-H_i)^2} \tag{2}$$

where X_i and Y_i are the Cartesian coordinates of upstream end of the link, and H_i is its depth. Similarly, X_j, Y_j and H_j are the ones of downstream end of the link.

Slope of a link from i to j, S_{ij}, is taken as,

$$S_{ij} = \frac{H_j-H_i}{L_{ij}} \tag{3}$$

Manning's formula (for a fully filled circular sewage line),

$$Q_{ij} = K_2 D_{ij}^{\frac{8}{3}} S_{ij}^{\frac{1}{2}} \tag{4}$$

where Q_{ij} is the value of load of a link, in m^3/sec, $K_2=0.3115/n$, n is the coefficient of roughness of pipe material.

The average depth of a link, A_{ij}, is expressed as,

$$A_{ij} = \frac{H_i+H_j}{2} \tag{5}$$

From Eq. (4) and Eq. (3), solving for D_{ij}, we can obtain,

$$D_{ij} = \left[\frac{Q_{ij} L_{ij}^{\frac{1}{2}}}{K_2 (H_j-H_i)^{\frac{1}{2}}}\right]^{\frac{3}{8}} \tag{6}$$

Substituting for D_{ij} from Eq. (6), A_{ij} from Eq.(5) and L_{ij} from Eq. (2) in Eq. (1), we can obtain,

$$\begin{aligned} C_{ij} &= K \cdot L_{ij}^{1+\frac{3a_1}{16}} Q_{ij}^{\frac{3a_1}{8}} (H_j+H_i)^{a_2} \cdot (H_j-H_i)^{-\frac{3a_1}{16}} \\ &= K\left[(X_j-X_i)^2+(Y_j-Y_i)^2+(H_j-H_i)^2\right]^{\frac{1}{2}(1+\frac{3a_1}{16})} \\ &\quad \cdot Q^{\frac{3a_1}{8}} \cdot (H_j+H_i)^{a_2} \cdot (H_j-H_i)^{-\frac{3a_1}{16}} \end{aligned} \tag{7}$$

In which $K = K_1 (\frac{n}{0.3115})^{\frac{3a_1}{8}} \cdot 2^{-a_2}$

The total capital cost of systems may be expressed as,

$$C = \sum_K C_{ij} \cdot \phi_{ij} \tag{8}$$

In which, K is the number of link, and

$$\phi_{ij} = \begin{cases} 0, & \text{if i is not connected to j, or } S_{ij} \le 0 \\ 1, & \text{if i is connected to j and } S_{ij} > 0 \end{cases}$$

PROCEDURE OF OPTIMIZATION

Steiner Problem. There are three points A, B and C in the plane (see Fig. 1 (a)). Connecting these points with lines and having the sum of length of all its lines as short as possible, if no extra point is added, we can obtain a tree called Minimal Tree (denoted by MT), as shown in Fig. 1(b); if an extra point is added, we can obtain one called Steiner Minimal Tree (denoted by SMT), as shown in Fig. 1(c). Generally, the length of SMT is shorter than that of MT. In this example, $L_{MT}=400$, and $L_{SMT}=346.5$. How can Steiner Minimal Tree be constructed?

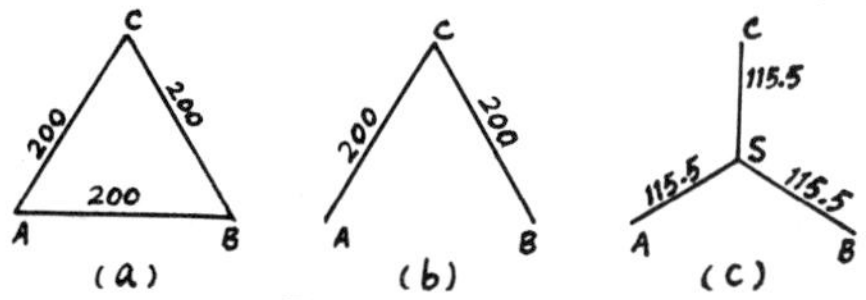

Fig. 1

For 3-point case, there have been many references (Gibert,1968) discussing it in detail.

For general case, n > 3 (n isthe number of given points), SMT is difficult to fine. Garey (1977) has proven that it is a NP-hard problem to find SMT.

Chang (1972) presented a method to find the approximate solutions of SMT, called SI algorithm. This is a fairly good and simple method. So, we use it in this paper.

Determination of the Initial Layout of Systems

From Eq. (8), we can see that it is very complicated to minimize the objective function because ϕ_{ij} is unknown. Nevertheless, it is clear that if the topology of systems is known, ϕ_{ij} and Q_{ij} may be considered as a constant. In practice, in Eq.(7), $a_2 < 1$, $a_1 < 1.5$, $1+\frac{3a_1}{16} > a_2 \gg \frac{3a_1}{16}$; on the other hand, $L_{ij} \gg (H_j + H_i) > (H_j - H_i)$ can be found. Therefore, we can neglect the impact of of $(H_j+H_i)^{a_2}$ and $(H_j-H_i)^{-\frac{3a_1}{16}}$ on objective function when determining the initial feasible solutions of Eq.(8) Thus, the objective function can been seen as nonlinear function of the length of link.

By good fortune, Soukup (1975) has studied the problem and discoveried that the topology of a nonlinear cost function of length is similar to that of a linear one, and that the position of Steiner points are slightly different from that of a linear one.

Accordingly, we can refer to SMT as the initial layout of systems, which can be found by SI algorithm.

Optimization of Objective Function

SMT is referred to as the initial layout of systems. so, ϕ_{ij} in Eq.(8)can be seen to be known. We can observe that there are three dicision variables, ie, X, Y and H in Eq.(7). According to number of variables, we can divide the points on SMT into three kinds as follow:

A. For a given and pendant point, H, X and Y are all known.

B. For a given but not pendant point, X and Y are known, but H is unknown.

C. For a Steiner point, H, X and Y are all unknown.

It is clear to be seen that there are no dicision variable, one dicision variable and three dicision variables for kind A, B and C of points, respectively.

If there are n given points, s Steiner points and n_s pendant points in given points on SMT, then, there will be $3s+n-n_s$ dicision variables in the system.

For simplicity, we denote the dicision variables as $Z_1, Z_2, \dots Z_m$, $(m=3s+n-n_s)$.

It is not difficult to prove that objective function is a strictly convex function and it has a unique minimal value. Therefore, the optimal solutions of Eq. (8) can be obtained by solving Eq. (9).

$$\begin{cases} \dfrac{\partial C}{\partial Z_1} = \sum\limits_K \dfrac{\partial C_{ij}}{\partial Z_1} = 0 \\ \dfrac{\partial C}{\partial Z_2} = \sum\limits_K \dfrac{\partial C_{ij}}{\partial Z_2} = 0 \\ \quad \vdots \\ \dfrac{\partial C}{\partial Z_m} = \sum\limits_K \dfrac{\partial C_{ij}}{\partial Z_m} = 0 \end{cases} \qquad (9)$$

In which, k = 1, 2, ... n+s-1.

This is a multivariate nonlinear simultaneous equations. Its optimal solutions can be found by means of Gauss-Siedel iterative method.

Constraints

In order to ensure effectiveness and structural safety of pipe sewerage systems, the following constraints have been incorporated into this works.

A. Minimum diameter is taken as 0.2m and Maximum one as 1.7m.

B. The diameter of a link on upstream should not be larger than that on downstream.

C. Minimum depth is taken as one metre, and maximum one as eight metres.

D. Flow Velocity Constraint.

The minimum self-cleaning velocity Vmin is taken as,

$$Vmin = \begin{cases} 0.7\text{m/sec} & \text{if } D_{ij} \le 0.5\text{m} \\ 0.8\text{m/sec} & \text{if } D_{ij} > 0.5\text{m} \end{cases}$$

The maximum permitted velocity is taken as 4m/sec. For a fully filled circular pipe, by means of Chezy formula,

$$V = \frac{1}{n} \cdot R^{\frac{2}{3}} \cdot i^{\frac{1}{2}} \qquad (10)$$

Table 1 can be drawn up, giving the minimum and maximum slope for each of the available pipe diameter sizes.

Optimization procedure

The objective in our optimization problem is to minimize the total cost of sewerage systems (Eq. (8)), with performance function delineated in Eq. (7), applicable to each link, subject to the constraints mentioned above from A to D.

Table 1

Pipe diameter m.	Minimum slope Smin ($x10^{-3}$)	Maximum slope Smax
0.2	5.216	0.1703
0.25	3.874	0.1265
0.3	3.038	0.09918
0.35	2.474	0.08075
0.4	2.070	0.06757
0.45	1.769	0.05776
0.5	1.582	0.05164
0.6	1.574	0.03935
0.7	1.281	0.03204
0.8	1.072	0.02681
0.9	0.9164	0.02292

According to the reason mentioned above, we can divide the procedure of optimization into two steps as follow:

A. SMT is found for n given points using SI algorithm, which is referred to as the initial layout of sewerage systems.

B. Optimal solutions are obtained from solving Eq. (9), the pipe diameters are considered as continuity, by means of Gauss-Siedel iterative method, which offer a lower bound value of the solutions of sewerage systems.

Furthermore, slope and commercial pipe diameter for every link, and depth for every point is found using Branch-and-Bound method.

By the way, it should be pointed out that the obtained slope must be between Smin and Smax for every commerical pipe diameter (see Table 1), ie,

$$Smin \le S \le Smax \quad (11)$$

If S > Smax, pipe diameter should be reset; if S < Smin, then let S = Smin.

ILLUSTRATIVE EXAMPLE

Figure 2 is a 10-points system in which point 1 is a sewage treatment plant, point 2 to 10 are discharge sources. The location of Cartesian coordinates and discharge quantity of the given points are listed in Table 2.

Table 2

Point	Location coordinates X(m)	Y(m)	Q (m^3/sec)
1	0	0	-0.256
2	-300	300	0.025
3	-500	300	0.04
4	-400	473.21	0.039
5	-573.21	573.21	0.0156
6	-400	673.21	0.0254
7	300	300	0.035
8	600	300	0.035
9	600	900	0.0156
10	300	900	0.0254

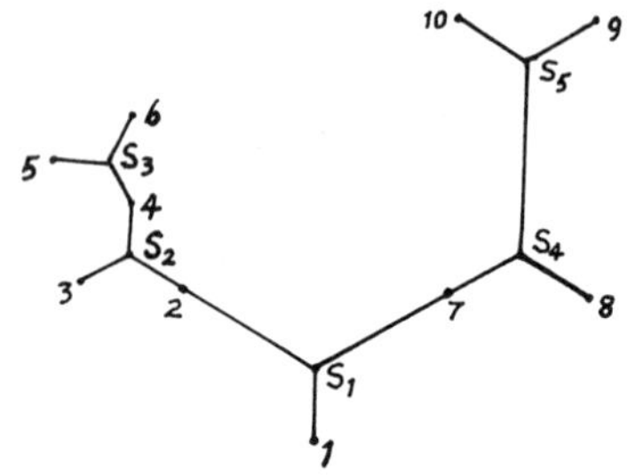

Fig. 2

Circular concrete pipes are used as sewerage pipes in this example. Its coefficient of roughness is taken as 0.014.

According to the available data, we obtained performance function for every link as follow,

$$C_{ij}=77.92\, D_{ij}^{0.58551} \cdot A_{ij}^{0.63448} \cdot L_{ij} \quad (12)$$

Therefore, Eq. (7) can be expressed as,

$$C_{ij}=25.3981\, Q_{ij}^{0.219566}\, L_{ij}^{1.109783} \cdot (H_j+H_i)^{0.63448}\, (H_j-H_i)^{-0.109783} \quad (13)$$

The layout of sewerage systems with minimal length is found using SI algorithm, as shown in Fig. 2, threshould value is taken as 0.01. The location of Cartesian coordinates for every Steiner point is listed in Table 3.

From the continuity equation, we can obtain the load of every link, listed in Table 4.

Table 3

Steiner Poing	Location coordinates X(m.)	Y(m.)
1	0	126.79
2	-400	357.73
3	-457.73	573.21
4	450	386.6
5	450	813.4

We substitute for C_{ij} from Eq. (13) into Eq. (9), taking threshold value as 0.0000001, and solve the nonlinear simultaneous equations by means of Gauss-Siedel iterative method to obtain the lower bound value of solutions of Eq. (8), finally obtain the optimal solutions, listed in Table 4, by means of Branch-and-Bound method.

The total cost is estimated as 185194.44¥.

Also, we optimize the Sewerage systems with the method presented by Barlow (1972), which disregards adding Steiner points into systems. Its solutions is listed in Table 5. Its total cost is estimated as 201412.24¥. Figure 3 is its layout.

The method pressented in this paper needs 8.1 percent lower expenditure than that needs by Barlow.

Table 4 Optimal solutions of sewerage systems

Link	Lent-th(m.)	Dia-meter (m.)	Slope ($x10^{-3}$)	Load (m^3/sec)	up-stream end Depth (m.)
s_1-1	118	0.6	2.07	0.256	3.53
2-s_1	348	0.45	3.038	0.145	2.09
s_2-2	100	0.45	2.057	0.12	1.88
3-s_2	137	0.25	5.245	0.04	1
4-s_2	113	0.4	2.07	0.08	1.65
s_3-4	111	0.35	2.474	0.041	1.37
5-s_3	123	0.3	3.038	0.0156	1
6-s_3	113	0.3	3.038	0.0254	1
7-s_1	355	0.45	1.769	0.111	2.90
s_4-7	140	0.4	2.07	0.076	2.61
8-s_4	210	0.25	4.023	0.035	1
s_5-s_4	419	0.35	2.047	0.041	1.58
10-s_5	165	0.3	3.038	0.0254	1
9-s_5	190	0.3	3.038	0.0156	1

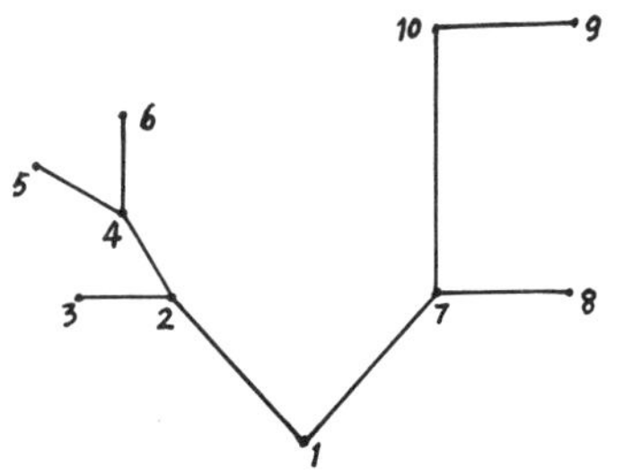

Fig. 3

Table 5

Link	Lent-th(m.)	Dia-meter (m.)	Slope ($x10^{-3}$)	Load (m^3/sec)	up-stream end Depth (m.)
6-4	200	0.35	2.474	0.0254	1
5-4	200	0.35	2.474	0.0156	1
4-2	200	0.4	2.07	0.08	1.50
3-2	200	0.3	3.038	0.04	1
2-1	424	0.5	1.712	0.145	1.91
9-10	300	0.3	3.038	0.0156	1
10-7	600	0.35	2.474	0.041	1.91
7-1	424	0.4	3.299	0.111	3.40
8-7	300	0.25	4.023	0.035	1

CONCLUSIONS

In general, optimization of sewerage systems is divided into two independent aspects: (a) optimization of system layout, (b) optimization of the pipe design parameters for the known layout.

Nevertheless, it can obtain better results of optimization to considering these two aspects simultaneouely. But it is a complex problem to do so. In this paper, according to the characteristics of performance function, we divide the procedure of optimization into two steps presented above. So, the problem can be solved.

Practical results of computing show that it only takes a few iterations to converge to the solutions of Eq. (9), using SMT as the initial feasible solutions, by means of Gauss-Siedel iterative method to solve Eq. (9). Therefore, SMT can offer good initial value to finding the optimal solutions of systems.

The method presented here is little more complex than that presented by Barlow. Nevertheless, it can save much expenditure. Therefore, it still is one of the efficient methods of optimal design of sewerage systems.

REFERENCES

Barlow, J. F. (1972). Cost optimization of pipe sewerage systems. Proceedings of the Institution of Civil Engineers, 53 (Part 2), 57-64.

Chang, S. K. (1972). The generation of minimal trees with a Steiner topology. Journal of the Association for Compting

Machinery, 19, 699-711.

Garey, M. R., R. L. Graham, and D. D. Johnson (1977). The complexity of computing Steiner minimal trees. SIAM Journal on Applied Mathematics, 32, 835-859.

Gibert, E. N., and H. O. Pollak (1968). Steiner minimal trees. SIAM Journal on Applied Mathematics, 16, 1-29.

Mays, L. W., H. G. Wenzel, and J. C. Liebman (1976). Model for layout and design of sewer systems. Journal of the Water Resources Planning and Management Division, ASCE, 102, 385-405.

Soukup, J. (1975). On minimum cost networks with nonlinear cost. SIAM Journal on Applied Mathematics, 29, 571-581.

SOME ANALYSIS METHODS FOR THE PROJECT FEASIBILITY STUDY

Zheng Yingwen

Institute of Automation, Fuzhou University, Fuzhou, Fujian, China

Abstract. Some aspects of the project feasibility study (F/S) are discussed by using the methods of system engineering. F/S is regarded as multilevel optimization under some constraints. The adjacent matrix analysis, decision tree and random test approach (Monte Carlo method) are presented. In the random test approach we discussed specially the stopping problem, the risk and sensitivity analyses for various benefits of the project. These methods are proved useful in practical cases when we are working on.

Keywords. Economics; Developing countries; Decision theory; Optimization; Random processes; Sensitivity analysis; Risk analysis.

INTRODUCTION

The project feasibility study is a very important problem in capital construction either in China or in other developing countries. It has attracted great attention of the government planners. The main concepts, content, procedures, data collecting and processing, various formulas and diagrams can be found in, among others, the "Manual for the Preparation of Industrial Feasibility Studies" edited by United Nations Industrial Development Organization (UNIDO) 1978, the discussion of which will be dispensed with here of. Some system engineering methods are presented as reference for those working at F/S. These methods have been used in the F/S of the following projects:

---- Construction of a cotton printing and dyeing mill. (Case 1)

---- Construction of a power station in Fujian Province. (Case 2)

China is a developing socialist country, in which a large-scale modernization constructions are in progress. So in F/S work some special factors should be noticed:

1. Besides abiding by the laws, it must not contradict the policies and stipulations of the government.

2. Aside from economic profits, the society benefits should also be emphasized, which include:

---- Its effects on ecosystem and environment;

---- Increase of the opportunity in employment;

---- Promotion of the development of science and education;

---- Influence and/or improvement on the relative economic sectors; and

---- Other macroeconomic benefits.

3. Once the social benefits become distinct, the project will be authorized by the government, and the various problems (such as requisition of land, capital investment or employing technicians and workers) can readily be overcome under the government's support.

SYSTEM MODEL

From the viewpoint of system theory, F/S is to decide whether a project is feasible or not, and if it is feasible, what is its optimum alternative. The optimum alternative involves mainly: (a) site location, (b) scale, (c) the method of production and management, (d) the way of carrying out the project. To these points, specialists may propose several alternatives, their number is generally finite. The above four points are denoted by x_1, x_2, x_3, x_4, and each alternative is represented by an evaluated vector $x=(x_1, x_2, x_3, x_4)^T$. We call the set of these alternatives "decision space", denoted by X. Our problem now lies in how to select the optimum one in the decision space.

The decision must first conform to the prevailing laws, decrees, regulations, etc. concerning the project, which is denoted by L.

Then we must investigate the project's states -- the geographical conditions of alternative locations (e.g. earthquake, landform and geologic structure etc.), the source and supply of raw material, water, power and fuels supply, the labour, technicians and leadership status, the channel and amount of investment, the market forecast and so on. All results of these investigation are called "information space", which is denoted by Y. The elements of space Y are often random variables.

An alternative putting into effect will lead to certain economic profits and society benefits. Content of society benefits has been listed in INTRODUCTION. Generally speaking, they are not easy to be evaluated quantitatively, but we can give a grade or an evaluation of the states as "good, medium or bad", and sometimes we can roughly express them by number. There are several ways to describe economic profits. The following economic indices are mainly considered: (a) investment (using present value method), (b) construction period, (c) annual revenues after it is built. Sometimes we have also to consider the useful life and the salvage value of the alternative. Of course, we may also consider other indices (e.g. return on investment, net present value ect.), but most of them can be derived from the preceding data. Both society benefits and economic profits make up a benefit space, which is denoted by Z. Z is a semi-order space, in which a partial order comparison relation can be defined.

To sum up, conforming with L and basing on informations Y, we calculate and compare Z in order to select the optimum point in X. It can be illustrated as in Fig. 1.

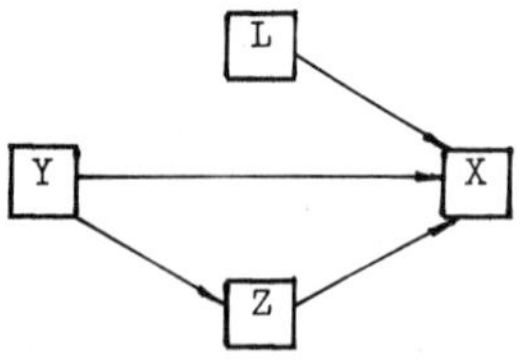

Fig. 1. A system model of F/S.

Therefore, F/S is a multilevel optimization problem under some constraints.

ADJACENT MATRIX

All elements are not necessarily correlative to each other in the spaces. For instance, the geographic condition are closely related to site selection, but not to method of management. The relationship between spaces can be expressed by using adjacent matrices.

Let u, v be m and n dimensional vectors respectively, then adjacent matrix R is a m×n matrix, its entries r_{ij} satisfy $0 \leqslant r_{ij} \leqslant 1$. The relationship between u and v can be expressed as

$$u \sim Rv$$

If $r_{ij} = 0$, the relationship between u_i and v_j is negligible; if $r_{ij}=1$, the relationship is very close; and a real number between 0 and 1 can represent the correlative degree.

Now we discuss the application of adjacent matrices to F/S.

Suppose the relationship between X and Z in Case 1 is:

$$\begin{pmatrix} z_1 \\ z_2 \\ z_3 \\ z_4 \\ z_5 \end{pmatrix} = \begin{pmatrix} 0 & 1 & 0 & 1 \\ 0.5 & 1 & 1 & 1 \\ 0.2 & 1 & 1 & 0 \\ 1 & 1 & 1 & 0 \\ 1 & 0.5 & 0 & 0 \end{pmatrix} \begin{pmatrix} x_1 \\ x_2 \\ x_3 \\ x_4 \end{pmatrix}$$

where z_1 is construction period, z_2 investment, z_3 annual revenue, z_4 ecosystem and environment influence, z_5 society effect, x_1 site location, x_2 scale, x_3 method of production, x_4 way of construction.

It is obvious that the relationship between x and z can also be expressed as :

$$\begin{pmatrix} x_1 \\ x_2 \\ x_3 \\ x_4 \end{pmatrix} = \begin{pmatrix} 0 & 0.5 & 0.2 & 1 & 1 \\ 1 & 1 & 1 & 1 & 0.5 \\ 0 & 1 & 1 & 1 & 0 \\ 1 & 1 & 0 & 0 & 0 \end{pmatrix} \begin{pmatrix} z_1 \\ z_2 \\ z_3 \\ z_4 \\ z_5 \end{pmatrix}$$

Alternative x is regarded as mapping of benefits z. We quantify z with a smaller number if z is better, with a greater number if z is worse. So we can evaluate x from the adjacent matrix and z. For example, for an alternative in Case 1, construction period is short (0.1), investment is small (0.2), annual revenue is not large (0.8), pollution of ecosystem and environment is trifling (0.1), social effect is middling (0.4), then we have

$$\begin{pmatrix} x_1 \\ x_2 \\ x_3 \\ x_4 \end{pmatrix} = \begin{pmatrix} 0 & 0.5 & 0.2 & 1 & 1 \\ 1 & 1 & 1 & 1 & 0.5 \\ 0 & 1 & 1 & 1 & 0 \\ 1 & 1 & 0 & 0 & 0 \end{pmatrix} \begin{pmatrix} 0.1 \\ 0.2 \\ 0.8 \\ 0.1 \\ 0.4 \end{pmatrix}$$

$$= \begin{pmatrix} 0.76 \\ 1.4 \\ 1.1 \\ 0.3 \end{pmatrix}$$

So the feasible degree of this alternative can be roughly determined quantitatively.

DECISION TREE

The method of decision tree can be widely applied for multi-step decision problem. For example, the decision tree for F/S of constructing a power station in Case 2 is shown in the follow figure:

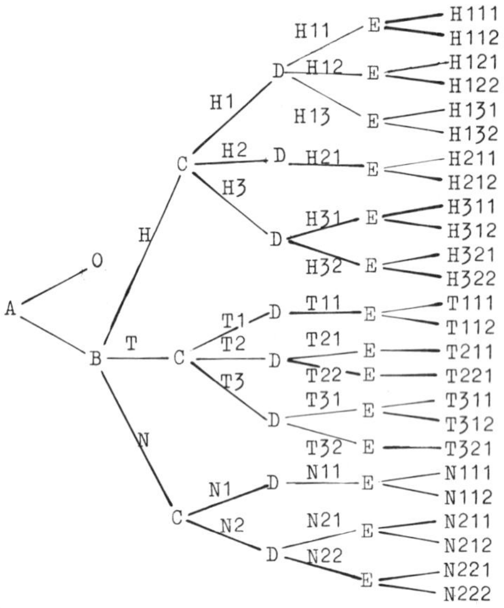

Note:
Decision A: feasible or not. 0 denote "not".
Decision B: type of station.
H: hydroelectric station;
T: thermal station (coal-burning);
N: nuclear station.
Decision C: site location. The first place figures denote different locations.
Decision D: installed capacity. The second figures denote different capacities.
Decision C way of construction. It is denoted by the third figures.
No.1: contract with foreign contractor;
No.2: contract with local corporation.

Fig. 2. Decision tree for construction of a power station (Case 2).

Thus the contents, steps and sequence in the F/S decision are clearly illustrated, where each terminal point (e.g. H111) is an alternative, i.e. an outcome of game between men and system. The benefits of realizing the alternative are called values of decision, which are hardly expressed accurately by numerical figures in general.

After the decision tree is constructed, we can simplify it as follows:

Step 1. Delete infeasible branches.

In this example, suppose N11 is too expensive; T1 is near a scenic spot or residential area, and violates the National Environmental Protection Law; the capacity of H13 is too small to satisfy the predicted requirement. So the branches N11, T1 and H13 are deleted.

Step 2. Eliminate inferior branches.
A branch is called inferior if, in comparison with another, one or more of its "benefits" x_i are worse, and the others are not better. The inferior branches can be eliminated ("inferior" is denoted by "$<$"). In this example, suppose H32$<$H12, T22$<$T32, T31$<$H11, so the branches H32, T22 and T31 are eliminated.

Step 3. Remove weak branches.
If comparing with branch B, the advantages of branch A are much more than that of B (except those benefits that close to each other), then the branch B is weaker than A (denote BwA). The weak branches should also be removed. In this example, we compare T21 with H12, and if realizing T21, we can save one million <u>yuan</u> investment, but have to pay 600 thousand <u>yuan</u> more for annual production cost (other benefits of T21 and H12 are closed), then T21wH12. Furthermore, suppose H21wH31, N21wN22, and construction of a phdroelectric power station by foreign contractor is weaker than that by local corporation. So the branches T21, H21, N21 and H111, H121, H131 are removed.

After these steps, we get a simplified decision tree, which is shown in Fig. 3.

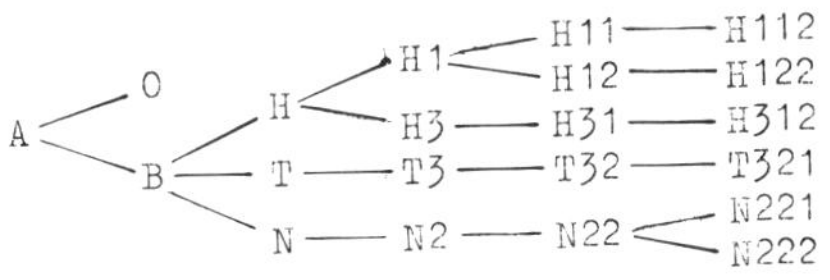

Fig. 3. The simplified decision tree of Case 2.

There are only six alternatives remain.

RANDOM TEST (MONTE CARLO METHOD)

After the decision tree is simplified, we are in a position to analyse the remaining alternatives in order to choose the best one.

We evaluate various benefits of each alternative by using those methods and formulas found in handbooks on F/S, and compare them.

Most of the survey informations are random variables, hence these benefit values are random variables too. The random test technique is applied to express these benefits more comprehensively. We determine firstly the random distributions of various information data, and then generate pseudorandom numbers according to these distributions by a computer. Calculating with these pseudorandom numbers (at least several hundred times), we can get a set of values, which are considered as experimental distribution of various benefits. This is the Monte Carlo method in F/S.

Some analyses for this method are in the following.

Stopping Problem

F/S is to determine whether an alternative is feasible or not. In some cases we don't need evaluate with all pseudorandom variables before we can determine whether the alternative should be given up. There are some permissible scopes in benefit space, e.g. the construction period must be limited to definite time, or the investment to definite amount. Suppose we plan to calculate 500 times, but we find many results go beyond a permissible scope when calculating 100 or 200 times, so we stop calculating and reject the alternative. We present a stopping strategy of binomial distribution here.

Suppose the probability beyond permissible scope is q, we hope that the probability of $q<0.05$ should be more than 90%, i.e. $p(q<0.05)\geq 0.9$. For n times independent random tests, the probability that the overstepped tests (beyond permissible scope) occur less than m times would be:

$$p=(1-q)^n+nq(1-q)^{n-1}+\ldots+C_n^{m-1}q^{m-1}(1-q)^{n-m+1}$$

where

$$C_n^r=\frac{n!}{r!(n-r)!}$$

In this equation, $q=0.05$; we take $n=100, 200, \ldots$ etc. , and calculate term by term. The process stops until $p\geq 0.9$, then we take that m for m_{100}, m_{200}, ... Thus when evaluating 100 times, 200 times, ..., the computer automatically check the number of overstepped tests. If it is less than m_{100}, m_{200}, ..., continue the calculating, otherwise, stop the calculation and give up this alternative. The m_n in preceding example are shown in Table 1.

TABLE 1 <u>Maximum Number of Overstepped Tests</u>

n	100	200	300	400
m_n	8	15	20	25

Risk Analysis

If the results of random test show that the means of an alternative benefits are good, but the values are dispersive, or rather uncertain, the alternative is considered infeasible for its high risk. If the worst benefit values are taken to analyse an alternative, the risk approaches to 0, but this strategy is too conservative to reflect the alternative benefits exactly. Now we apply mainly two methods for risk analysis.

<u>Calculation of mean and deviation.</u> Let the results of random test be $x_1, x_2, \ldots, x_n$, so we calculate mean

$$M=\frac{1}{n}\sum_{i=1}^{n}x_i$$

and deviation

$$S=\sqrt{\frac{1}{n-1}\sum_{j=1}^{n}(x_j-M)^2}\ .$$

The larger is its deviation, the higher its risk is. The mean and deviation will express rough distribution of the alternative benefits; or we express benefit as a single index, e.g. we can consider $M-\mu S$ (μ can be taken $0.5\sim0.8$) as single index if its large mean is better.

<u>Classification and count.</u> We classify the random

test results according to their values, and count the frequencies in each class. For example, to estimate the revenues per year, we classify them in group of loss, different amount of earnings (as 0~1, 1~2, 2~3, ... million yuan), and count the frequencies in each class. So we can find out whether the distribution is concentralized, and how much the risk of loss is.

In calculation it is not necessary to store all the data. We can calculate the means and deviations, one by one, or the frequencies in each class.

Sensitivity Analysis

The various input data lead to different results. Their relation can be studied in sensitivity analysis. For instance, the annual revenue depends on cost of various raw materials, energy cost, way of production, market forecast etc. , but their effects on the annual revenue are different. If a product need less raw material but more energy, the fluctuation of fuel price would cause great effect on the product cost, and the cost of raw material causes little effect. We call the product cost is more sensitive to fuel price and less sensitive to raw material.

To study the sensitivity of a benefit x to input data A in random test calculation, we classify the random numbers of A, which are denoted by $\{a\}$, into groups according to their values. Each group is expressed by $\{a_i\}$, where i= ... -2, -1, 0, 1, 2 ... . The mean of $\{a_i\}$ are A_i, where A_0 should be closed to the mean of $\{a\}$. The calculated benefits from $\{a_i\}$ are $\{x_i\}$, their mean is X_i. The sensitivity of x to A between A_i and A_j can be approximately expressed as

$$s = \left| \frac{X_i - X_j}{A_i - A_j} \right| .$$

We can also describe the sensitivity graphically. Fig. 4 is an example.

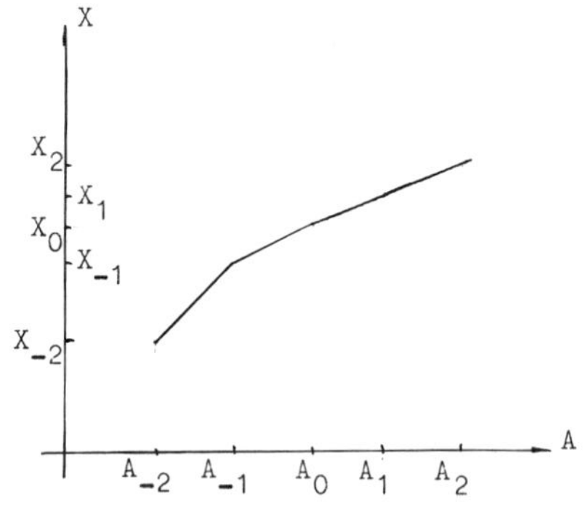

Fig. 4. Sensitivity of x to A.

In Fig. 4, the larger is the gradient, the higher the sensitivity is.

More finance, material and manpower investigation should be spent to improve the sensitive states. For example, if the product cost is sensitive to fuel price, we should predict the consumption and price of fuel as accurate as possible; and try to save energy for improving the benefit. In Fig. 4, it is necessary to avoid the value of A below that of A_{-1}, in order to minimize the sensitive effect on x.

After the random test, several better alternatives (at least three) should be reserved. Then these alternatives are compared by means of an unified cost fuction, or "soft" analyses, i. e. to discuss and debate, so as to select the optimum one.

F/S is a very complex process. It is necessary to investigate, calculate, analyse, reinvestigate ... again and again until a good enough alternative has been got.

CONCLUSION

The project feasibility study is an important procedure in preinvestment phase of capital construction. Compared with conventional method, the approach presented here would improve this work in the following aspects.

1. Verifying the project by means of optimizing over many alternatives, instead of studying the feasibility on a single alternative.

2. Considering all factors as a whole, instead of studying them independently.

3. Regarding their factors as random variables.

4. Paying more attention to the risk and sensitivity of benefits.

We present some system engineering methods to lay F/S on a basis which is more scientific and complete. We think that system engineering are very helpful to F/S, though our work is still in the preliminary stage, and there are a lot of work remained to be done.

SIGNAL ANALYSIS IN ENERGY ECONOMIC MODELLING

J. Luukkanen*, Y. Majanne* and A. Haarasilta**

*Tampere University of Technology, Tampere, Finland
**University of Tampere, Tampere, Finland

Abstract. Multivariable AR-signal analysis is used for energy economic modelling. Three types of energy used in the industrial sector of Finland, coal, oil and electricity, are studied. The dynamic interdependencies of energy prices and consumption are modelled. Based on the MAR-model the dynamic changes in energy consumption due to changing prices are calculated. Spectral contributions and cross-correlations are also computed.

Keywords. Energy; economics; modelling; signal analysis.

INTRODUCTION

Connections between energy and economies were emphasized in 1970s with the rising energy prices. The significant influence of energy on the level of economic activity and structure both in short and long run was noticed. Also interest in the effects of energy price changes and public policy actions has increased.

In Finland energy sector is particularly important because consumption is high and self-sufficiency is fairly low. Cold weather and long distances increase energy demand, but more important is that the most essential manufacturing sectors, forest and metal industries, are both very energy intensive.

For the sensible energy policy decisions a more detailed analysis is needed than only to compare national product and energy demand. In this research our purpose is to study demand for different energy types, coal, oil and electricity in Finnish industry. The method used is the formulation of a Multivariate Autoregressive model (MAR) based on signal analysis. The model reveals among other things the dynamic adjustment process taking place in energy use.

In the second chapter a brief review of economic research that has been done in the area of energy modelling is presented. Next section introduces the MAR-model and data which were used and the results obtained in our research. In the last chapter some problems that have arisen and plans for future research are discussed.

A BRIEF REVIEW OF ENERGY ECONOMIC MODELLING

There has been a wide interest on natural resources and their scarcity since the early days of Malthus and this issue has become popular again in 1970s after the first oil crisis. Already in 1960s Marc Nerlowe studied the substitution possibilities among capital, labour and fuel inputs in the electricity sector of the U.S. economy (Nerlowe, 1963). But the main development of the neoclassical production theory took place in 1970s with the introduction of duality between production and cost functions and the flexible functional forms allowing more than two inputs.

Previous value-edded studies focused on studying dependencies between capital, labour and output. Then it became possible to deal with several inputs, so energy and materials were incorporated. The duality theorem made it possible to derive the factor demand equations from production function through cost function. When we have the information about prices and output, these factor demands and elasticities could be estimated.

Energy modelling in neoclassical production theory has taken place in two main ways, having though many variations (Saariaho, 1984). Firstly there may be aggregative inputs such as capital, labour, energy and materials. Secondly these aggregative inputs may be first solved through sub-models. For example energy sub-model examines the demand for energy input components, coal, electricity etc. (Fuss, 1977). In both cases many problems arise concerning the reality of used structures and required restrictions.

One important issue also in energy economic models is technical change. Often made assumption is that improvments in technology are exogenous and disembodied, that is new inputs are not more efficient than the old ones because of technical advance (Nadiri, 1982). Gradually technical change has been incorporated explicity, but both theoretical and empirical research is still needed.

Earlier most of the empirical studies assumed immediate adjustment to the equilibrium and that long-run estimates can be achieved using cross-sectional data. Time-series data were assumed to be good for short-run estimating. Stapleton has studied the subject and found out that the optained results depend not only on the used data but the underlying model (Stapleton, 1981). Recently the advantage of dynamic empirical models of factor demand has been widely accepted. Also for the Finnish manufacturing the models with slow adjustment process seem to be more suitable than static models (Pikkarainen, 1984). Hartman has classified dynamic econometric energy demand models belonging to three generations (Berndt, Morrison and Watkins, 1981).

(i) Single-equation models using the Koyck partial adjustment or Almon lag procedures. The problem in these models is the limited role of economic theory and neglected interactions between energy and other inputs and externally imposed lag structures.

(ii) In second generation models interrelated factor demands are explicitly incorporated into the firms' short run demand functions. So the short run factor demand equations for variable inputs are solved through static minimisation of a restricted variable cost function relating disequilibrium in one factor market to disequilibrium in other factor market.

(iii) Dynamic optimisation providing short, intermediate and long run price elasticities is explicitly incorporated in the third generation models. The adjustment process of quasi-fixed factors to long-run equilibrium level is endogenously determined and time-varying. The internal costs of adjustment related to the factor prices is used as a constraint of adjustment. For example short run energy demand is determined by energy prices, the capital stock at the beginning of the year and actual output (Mittelstädt, 1983).

In economic energy demand analysis empirical results have been very contradictory. Energy and capital seem to be the most problematic relations from the production factors of K, L, E and M. Between different energy types both substitutability and complementarity properties can be found. Variations in the empirical results may be due to data and definition differences. Also estimation method, function specification, level of aggregation, separability and other assumptions made, static vrs dynamic models all may be possible reasons to cause some problems.

Most of the energy economic studies use econometric methods and concentrate on the demand side. If the substitution possibilities between different energy types or other production factors can be revealed, the information can be used in public policy planning. Recently also many large-scale models have been developed. They are often general equilibrium models taking into consideration all sectors of the economy. Energy sector is one part of the model and also the supply side is more often incorporated.

MAR-MODELLING

The basic idea of our analysis method is to try to model the process under study by means of signal analysis on the basis of the measured signals. The analysis is based on identification of a discrete parameter multivariate signal model and the results derived from the model.

Multivariate autoregressive model can be defined by equation:

$$\underline{x}(k) = \sum_{m=1}^{M} \underline{A}(m)\, \underline{x}(k-m) + \underline{E}(k)$$

$$\underline{x} = [x_1, x_2, x_3, \ldots, x_n]^T$$

$$\underline{A} = \begin{bmatrix} a_{11} & a_{12} & \cdots & a_{1n} \\ a_{21} & a_{22} & \cdots & a_{2n} \\ \vdots & \vdots & & \\ a_{n1} & a_{n2} & & a_{nn} \end{bmatrix}$$

$$\underline{E} = [e_1, e_2, \ldots, e_n]^T \text{ , noise}$$

M is model order.

The calculation of the parameter matrices is based on least-squares optimization and the model order is chosen to minimize the final prediction error. The computer program SIAMS (Signal Identification and Analysis program for Multivariate System) is developed in Tampere University of Technology by R. Suoranta (1984). Applying the MAR-model we can obtain linear relationships defined as inpulse response between all signals.

In our analysis we have studied the energy economics of the Finnish industry. Our research includes coal, oil and electricity consumption and their prices from 1970 to 1980. The prices were deflated with the production price index of the manufacturing industry and the quantities with the industrial production volume index to get a more stationary data. In Fig. 1 oil and electricity prices and consumptions are represented. The data is interpolated with third order quasi-Hermite spline method.

The program computes cross-covariances of the data. In Fig. 2 the covariance of oil price and oil quantity is presented. It shows that after the rise of oil price the decrease in oil consumption is most intense after three years.

For the identification of the MAR-model we selected the fuel prices as inputs and consumptions as outputs. Schematic illustration of model and the interactions between variables is presented in Fig. 3.

First order model was most stable for the data. With the identified model the program computes e.g. step responses for different variables. In Fig. 4 is the open loop step response from oil price to oil consumption. It shows the effect of a 45 % step increase in the price of oil. According to the model after 10 years the oil consumption has fallen 54 %. Similarly in Fig. 5 the effect of 15 % price rise in electricity on the consumption of oil (increase of 22 % after 10 years) is represented.

The effect of the electricity price rise on electricity consumption is shown in Fig. 6. The model expresses that although the price of electricity increases still its consumption increases. This may be due to the fact that the electricity consumption is very sensitive to production changes and during the high conjuncture the consumption increases although the price increases. Electricitv seems not to be price elastic.

The closed loop step responses, that is the relationships between the output variables, can be also computed. The power spectra of the signals and, what is very interesting in our case, input and spectrum contributions of the signals are also available. In Fig. 7 the input contribution to coal consumption is expressed. From the figure it can be seen that in the higher frequences the price of coal has a significant effect on the coal consumption while in the lower frequences the effect is negligible. It means that when the price of coal changes it effects the coal consumption for 2...4 years, but as the time goes on the effect decreases. The electricity price has a very remarkable effect on coal consumption, which is reasonable because coal is mainly used for electricity generation. The importance of the oil price on coal consumption increases in lower frequences, it means after a longer time, which is also reasonable because it takes years to build new power plants for new fuels. Input contribution to oil consumption is presented in Fig. 8. The model is also used to predict the outputs of the system. In Fig. 9 the predicted oil consumption with the residual time series is shown.

In our approach to the modelling problem the energy economic system is assumed to be a "black box". We measure only the input and output signals and make assumptions of the structure of the system by analysing the signals. There is no economical theory in the background, which can cause some problems when interpreting the results. Also the data used for the analysis must satisfy some restrictions (e.g. stationarity).

The greatest advantage of the approach is the inclusion of the dynamics of the systems. The energy consumption cannot be supposed to have an immediate change after the change of the prices as in static models. Also the compactness of the analysis of multivariate stochastic systems with interactions between all variables is an advantage.

CONCLUSIONS

The aim of our research has been to study the dynamics of energy economic systems. In the present study, we have used three energy components, coal, oil and electricity, but we have plans to take into account also other energy sources such as peat, natural gas, wood and waste materials. In Finnish manufacturing large amounts of waste wood and black and sulfite liquors are used. However problems arise with pricing of waste materials and how to get reliable data of quantities used.

Other factors of production could be taken into consideration to make the model more comprehensive Capital, labour and materials can be treated either as aggregate inputs of divided into two or more components. Usually for labour and materials inputs good statistics exist but the pricing of capital is more difficult. Moreover there are many other things than prices influencing the energy decisions. The possibility to model the political and environmental issues should be studied.

The oil crises have taught us that it takes years for economies to adjust for example to a sudden price change of oil. The time constant of the adjustment process is different for different energy sources because it takes longer time to construct e.g. a nuclear power plant than a gas turbine power plant. The information of the dynamics of the energy economical system is essential for the timing of energy political actions.

REFERENCES

Berndt, E.R., Morrison, C.J. and Watkins, G.C. (1981). Dynamic models of energy demand: An assessment and comparison. In E.R. Berndt and Field (Ed.), Modelling and Measuring Natural Resource Substitutions. The MIT Press.

Fuss, M.A. (1977). The demand for energy in Canadian manufacturing. Journal of Economics, 5, 89-116.

Mittelstädt, A. (1983). OECD Economics and Statistics Department, Working Papers, No. 1.

Nadiri, M.I. (1982). Producers Theory. In Arrow and Intriligator (Ed.) Handbook of Economics, Book I, Vol. II. North-Holland.

Nerlowe, M. (1963). Returns to scale in electricity supply. In C. Christ (Ed.) Measurement in Economics. Studies in Mathematical Economics and Econometries in Memory of Yehuda Greenfeld. Stanford Univ. Press, Stanford.

Pikkarainen, P. (1984). The Demand for Energy in Finnish Industry. Bank of Finland. D:57, Helsinki.

Saariaho, M. (1984). Substitutability of Energy. Research Reports in Economics and Statistics, University of Joensuu, No. 1. (in Finnish).

Stapleton, D.C. (1981). Inferring long-term substitution possibilities from cross-section and time-series data. In E.R. Berndt and Field (Ed.) Modelling and Measuring Natural Resource Substitutions. The MIT Press.

Suoranta, R. (1984). SIAMS, Theory of Analysis. Report 23 A, Tampere University of Technology, Tampere, (in Finnish).

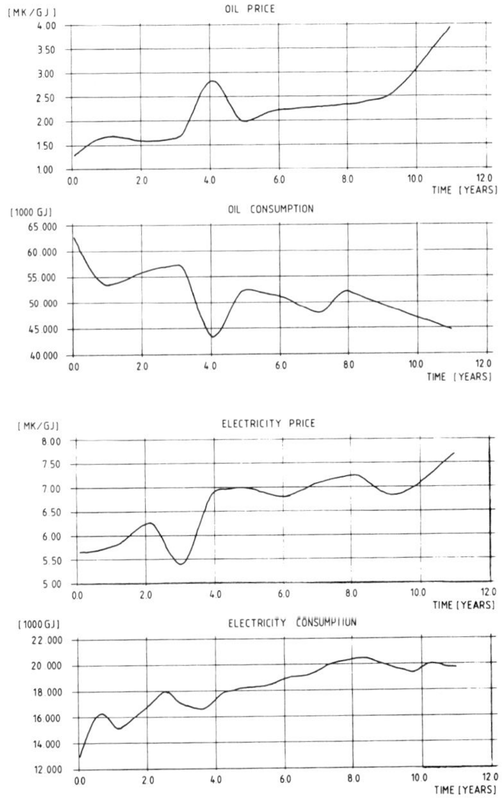

Fig. 1. Oil price and consumption and electricity price and consumption in Finnish industry 1970-80 (deflated).

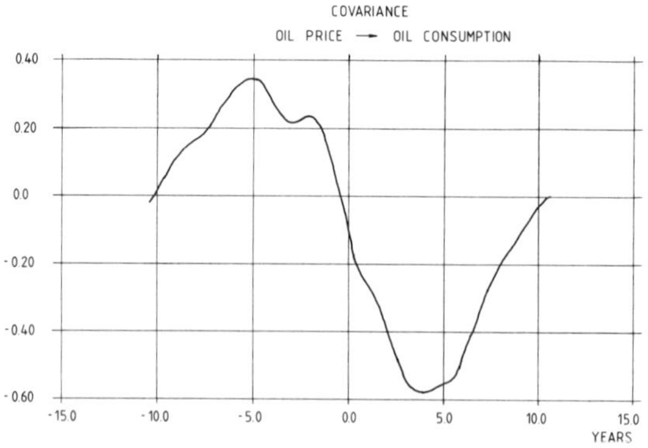

Fig. 2. Oil price → oil consumption covariance.

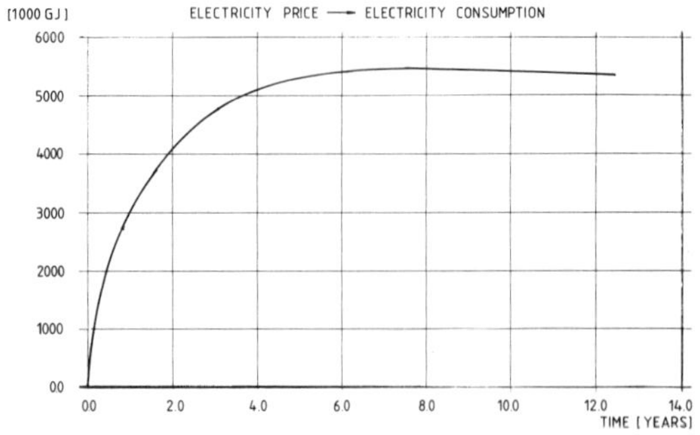

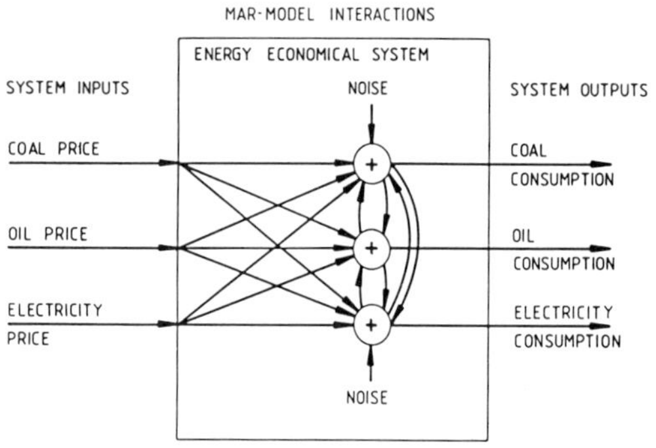

Fig. 3. MAR-model interactions.

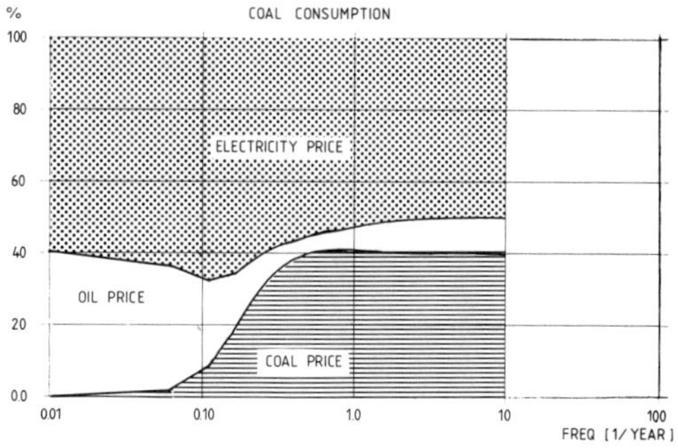

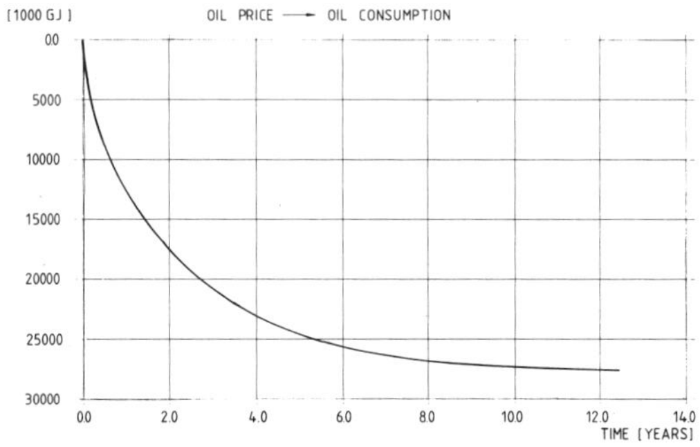

Fig. 4. Step response from oil price to oil consumption.

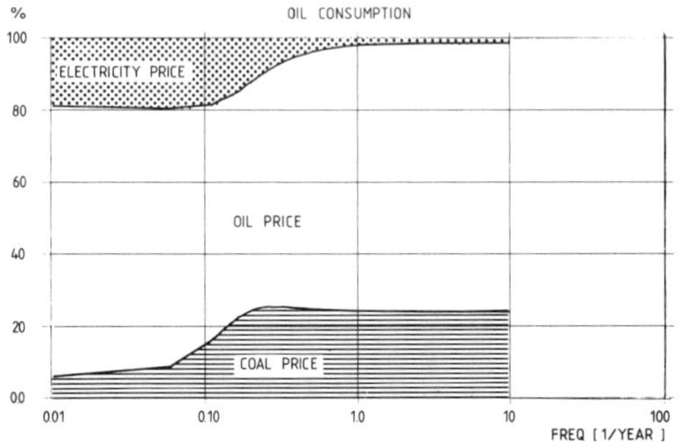

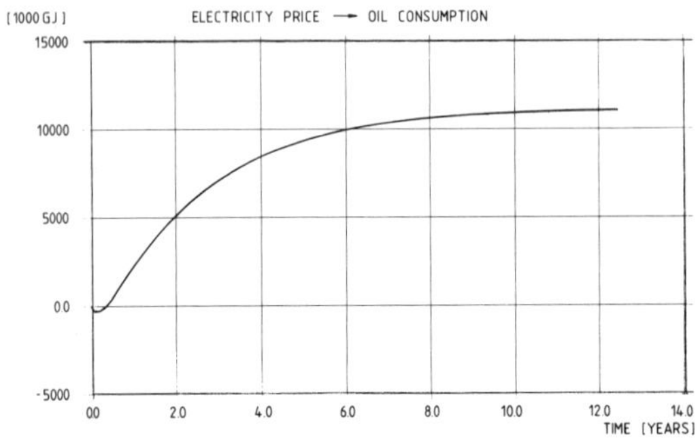

Fig. 5. Step response from electricity price to oil consumption.

Fig. 6. Step response from electricity price to electricity consumption.

Fig. 7. Input contribution to coal consumption.

Fig. 8. Input contribution to oil consumption.

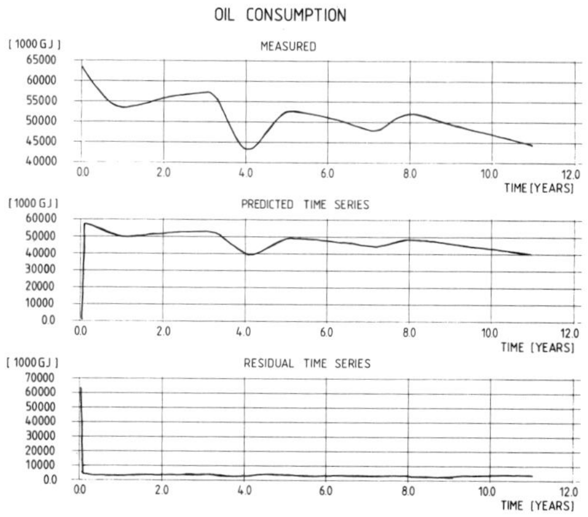

Fig. 9. Measured and predicted (one step) oil consumption and prediction error.

ANALYSIS OF MULTIVARIABLE DYNAMIC ECONOMIC CONTROL SYSTEMS

S. W. Xia

Department of Automation, Tsinghua University, Beijing, China

Abstract. A linear multi-input and multi-output economic control system is introduced based on the dynamic input-output model with multiple year lags. Using the phase variable canonical forms suggested by R.Yokoyama (1973), the dynamic economic input-output discrete transfer function is represented. It is shown that the compact structure of this form is used conveniently to calculate the time sequential values of a multivariable dynamic economic system using the backward recursive method. Some numerical examples are given.

Keywords. Multivariable control systems; canonical forms; control system analysis; dynamic economic control systems; discrete systems; state-space methods; z transforms; time lag systems.

INTRODUCTION

Economic mathematical models are generally used in the strategic analysis of national or regional economics and in making the development planning of a sector. The dynamic economic system is a complicated one with a multi-input and multi-output structure (Chow, 1975). The effect of multiple year lags of investment makes the characteristics of dynamic system high order. Using discrete state space equations to describe such a high order economic system, some difficulties will be encountered in computation. In this paper, based on the dynamic input-output fundamental model with multi-year lags, the computational complexity of the deduced dynamic economic control model is considerably reduced by using the phase variable canonical form suggested by Yokoyama and Kinnen (1973). Then, it can also be used to work out the economical planning and do research work on the control and optimization of dynamic economic discrete systems.

THE DYNAMIC INPUT-OUTPUT DISCRETE CONTROL OF ECONOMIC SYSTEMS

The subject which should be dealt with by the dynamic economic systems is the comprehensive balance among those various sectors to suit the needs of economic development. The technical coefficients of the dynamic models change versus time due to technical progress. Therefore, a time varying multi-input, multi-output economic model should be used. The dynamic input-output models represent the time delay effect of investment. W.Leontief (1977) developed the dynamic input-output model with one year lag and represented its applications. Vogt and Mickle (1975) introduced the gap of supply and demand into dynamic input-output system and examined the time response to the gap of supply and demand of soceity. A one year time lag being considered, the application of these models is limited.

In recent years, the dynamic input-output model with multiple year lags has been applied to medium and long term economical planning of China. The dynamic input-output model with multiple year time delay of various sectors can be divided into two categories: the backward recursive form and the forward recursive form. Because the investment coefficient matrix is always singular and therefore the inverse not be found, in this paper, only the backward recursive form will be discussed (Xia and Zhao, 1984a).

In the dynamic input-output model of this paper, the output vector X_k is the state variable, the final net demand vector YC_k is the control variable and the investment is closely related to the production increment. Here the investment variable I_k, which is implicitly in the output increment $\Delta X_{k+\tau}$, is endogenous (Xia and Zhao, 1984b).

The backward recursive dynamic input-output model with the longest time lag L is

$$G_k X_k = \sum_{\tau=1}^{L} H_k(\tau) X_{k+\tau} + YC_k \tag{1}$$

$$I_k = \sum_{\tau=1}^{L} B_k(\tau)\gamma_k(\tau) \Delta X_{k+\tau} \tag{2}$$

$$G_k = I - A_k + B_k(1)\gamma_k(1) \tag{3}$$

$$H_k(\tau) = B_k(\tau)\gamma_k(\tau) - B_k(\tau+1)\gamma_k(\tau+1) \tag{4}$$

where:

A_k is the input coefficient matrix of year k, (nxn)
n is the number of sectors
$B_k(\tau)$ is the investment coefficient matrix, (nxn), the investment input in year k and will be effective after τ years.
$\gamma_k(\tau)$ is a diagonal matrix, (nxn), which is called decision coefficient matrix. The diagonal element $\gamma_k{}^j(\tau)$ is the ratio of the jth sector's output increment of the year $k+\tau$ caused only by the investment of year k, to the jth sector's total output increment of the year $k+\tau$.
$\gamma_k(L+1)$ is equal to zero.
X_k is the output of year k, (nx1)
YC_k is the final net demand of year k, (nx1)
I_k is the investment of year k, (nx1)
I is the unit matrix.

Using the state space method of dynamic system to describe the dynamic input-output equation mentioned above for an economic system, we have the state space expression in discrete form.

Let

$$\left.\begin{aligned} U_k &= YC_k \\ Z_{k+L-1} &= X_{k+L} \\ Z_{k+L-2} &= X_{k+L-1} \\ &\vdots \\ Z_k &= X_{k+1} \\ X_k &= G_k^{-1}H_k(1)X_{k+1} + G_k^{-1}H_k(2)X_{k+2} + \cdots \\ &\quad + G_k^{-1}H_k(L)X_{k+L} + G_k^{-1}U_k \\ Y_k &= I_k \end{aligned}\right\} \quad (5)$$

Then we have

$$\left.\begin{aligned} \mathcal{X}_k &= D_k\mathcal{X}_{k+1} + E_kU_k \\ Y_k &= F_k\ (\mathcal{X}_{k+1} - \mathcal{X}_k) \end{aligned}\right\} \quad (6)$$

where

$$\mathcal{X}_k = (Z_{k+L-2}\ Z_{k+L-3}\ \cdots\ Z_k\ \ X_k)^T \quad (7)$$

$$\mathcal{X}_{k+1} = (Z_{k+L-1}\ Z_{k+L-2}\ \cdots\ Z_{k+1}\ X_{k+1})^T$$
$$= (X_{k+L}\ \ X_{k+L-1}\ \cdots\ X_{k+2}\ X_{k+1})^T \quad (8)$$

$$D_k = \begin{pmatrix} 0 & I & \cdots & 0 \\ 0 & 0 & \cdots & 0 \\ \cdot & \cdot & \cdots & \cdot \\ \cdot & \cdot & \cdots & \cdot \\ \cdot & \cdot & \cdots & \cdot \\ 0 & 0 & \cdots & I \\ G_k^{-1}H_k(L) & G_k^{-1}H_k(L-1) & \cdots & G_k^{-1}H_k(1) \end{pmatrix} \quad (9)$$

$$E_k = (\ 0\ \ 0\ \cdot\ \cdot\ \cdot\ 0\ \ G_k^{-1}\ )^T \quad (10)$$

$$F_k = (B_k(L)\gamma_k(L)\ B_k(L-1)\gamma_k(L-1)\ \cdots\ B_k(1)\gamma_k(1)) \quad (11)$$

In the dynamic input-output state equation of the backward recursive form, the output of year k is determined by the consumption of year k and the production growth of some years after year k. Therefore, we obtain $\mathcal{X}_k$ recursively from $\mathcal{X}_{k+1}$ and U_k. This expression may be solved by the inverse method.

By transforming equation (6), we have

$$\mathcal{X}_{k+1} = D_k^{-1}\mathcal{X}_k - D_k^{-1}E_k\ U_k \quad (12)$$

$$Y_k = F_k\ (\ \mathcal{X}_{k+1} - \mathcal{X}_k\ ) \quad (13)$$

This state equation of forward recursive form may be solved only when D_k^{-1} is valid. The block diagram of this transformed equation is shown in Fig.1.

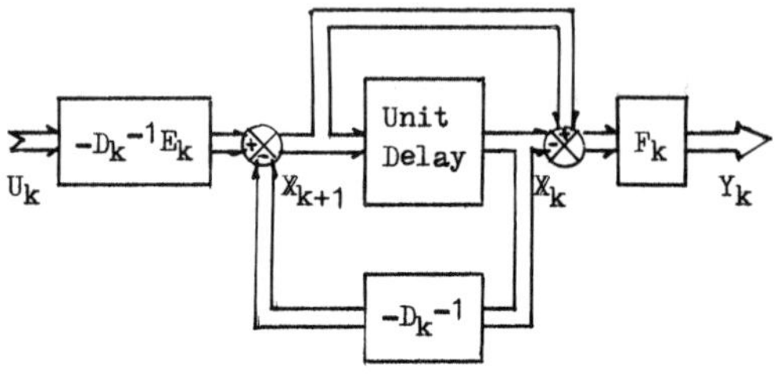

Fig. 1. Block diagram of discrete economic system

It is interesting to compare this block diagram with that of the general forward recursive discrete state systems.

Z TRANSFORMATION OF DYNAMIC INPUT-OUTPUT DISCRETE STATE SYSTEM

After applieing the Z transformation to the set of equation (6) mentioned above, we have

$$\left.\begin{aligned} \tilde{\mathcal{X}}(Z) &= D_kZ\ \tilde{\mathcal{X}}(Z) + E_k\ \tilde{U}(Z) \\ \tilde{Y}(Z) &= F_k(\ Z - 1\ )\ \tilde{\mathcal{X}}(Z) \end{aligned}\right\} \quad (14)$$

Then we obtain the result

$$\tilde{\mathcal{X}}(Z) = (\ I - D_kZ\ )^{-1}E_k\ \tilde{U}(Z)$$
$$= V_k(Z)\ \tilde{U}(Z) \quad (15)$$

$$\tilde{Y}(Z) = F_k(\ Z - 1\ )(\ I - D_kZ\ )^{-1}E_k\ \tilde{U}(Z)$$
$$= W_k(Z)\ \tilde{U}(Z) \quad (16)$$

When the dynamic input-output state equation is used to make planning, the expected value of the economic system in last L years of the planning period should be used as boundary condition. Using equations (15) and (16), the state variable value of every year in the planning period may be obtained recursively. $V_k(Z)$ and $W_k(Z)$ are two transfer function matrices.

As is well known, in the multivariable discrete system of multiple year lags, it is difficult to calculate $V_k(Z)$ and $W_k(Z)$ in equations (15) and (16). The higher the order, the more complicated is the calculation of the inverse $(I-D_kZ)^{-1}$. If the number of planning sectors n is equal to 20, and the longest time lag L is equal to 5, then the D_k is 100×100. It is difficult to find the inverse of a 100X100 matrix with Z variable.

In this paper, the transfer function based on the phase variable canonical form suggested by Yokoyama and Kinnen (1973) is introduced. Their coefficient matrices are the submatrices of dynamic input-output models, and the formulas with simple and compact form may be obtained. Then, the discrete state control method may be used in the complicated economic systems.

CANONICAL TRANSFER FUNCTION OF MULTIVARIABLE ECONOMIC CONTROL SYSTEM

In the phase variable canonical function $V_k(Z)$ of the system mentioned above, we have

$$\text{rank}(E_k,\ D_kE_k,\ \cdots,\ D_k^LE_k)$$
$$= \text{rank}(E_k,\ D_kE_k,\ \cdots,\ D_k^{L-1}E_k) \quad (17)$$

$$l_i = \text{rank}(E_k,\ D_kE_k,\ \cdots,\ D_k^{L-i}E_k)$$
$$- \text{rank}(E_k,\ D_kE_k,\ \cdots,\ D_k^{L-i-1}E_k)$$
$$= l_L \qquad (\ i = 1,\ 2,\ \cdots,\ L-1\ ) \quad (18)$$

$$l_L = \text{rank}\ E_k = n \quad (19)$$

$$l_e = nL - \sum_{i=1}^{L} l_i = 0 \quad (20)$$

Hence, $V_k(Z)$ may be defined as follows

$$V_k(Z) = (I - D_kZ)^{-1}E_k$$
$$= (\Phi_1(Z)\Phi_2(Z)\ \cdots\ \Phi_{L-1}(Z)\Phi_L(Z))^T \quad (21)$$

where

$$\Phi_i(Z) = Z^{L-i}\Phi_L(Z) \quad (22)$$
$$(\ i = 1,\ 2,\ \cdots,\ L-1\ )$$

$$\Phi_L(Z) = (I - G_k^{-1}H_k(1)Z - G_k^{-1}H_k(2)Z^2 - \cdots$$
$$- G_k^{-1}H_k(L-1)Z^{L-1} - G_k^{-1}H_k(L)Z^L)^{-1}G_k \quad (23)$$

Because what we have is a backward recursive state equation , there is a slight difference in form between the set of equation mentioned above and that of Yokoyama.

Similarly, we have

$$W_k(Z) = F_k(Z-1)\ V_k(Z)$$
$$= (\psi_1(Z)\,\psi_2(Z)\cdots\psi_{L-1}(Z)\,\psi_L(Z)) \quad (24)$$

So that the coefficient matrices of $V_k(Z)$ and $W_k(Z)$ may be induced by the research of controllable and observable canonical coefficient matrices. It is possible to find $\Phi_L(Z)$ and then to derive $V_k(Z)$ and $W_k(Z)$. This is beneficial to the calculation and the analysis of this subject.

SOME EXAMPLES AND THE RESPONSE CURVES WITH DISCRETE UNIT INPUT

Example 1. Suppose an economic system consists of three sectors and the longest time lag of investment is one year. When the final demand of a given sector at the end of planning period is increased with one unit, let us find out the time sequence, traced back from k=0, of the desired annual production of each sector. Since L=1, the state equation is quite simple.

Assuming that the three given sectors are industry, agriculture and commerce, the input coefficient matrix A_k and the investment coefficient matrix B_k are

$$A_k = \begin{pmatrix} .328 & .171 & .175 \\ .075 & .171 & .039 \\ .037 & .123 & .018 \end{pmatrix} \quad (25)$$

$$B_k = \begin{pmatrix} .510 & .018 & .102 \\ .157 & .008 & .053 \\ .079 & .003 & .021 \end{pmatrix} \quad (26)$$

Then, we have

$$G_k = I - A_k + B_k$$
$$= \begin{pmatrix} 1.182 & -.153 & -.073 \\ .082 & .837 & .014 \\ .042 & -.120 & 1.003 \end{pmatrix} \quad (27)$$

$$D_k = G_k^{-1}H_k(1) = G_k^{-1}B_k$$
$$= \begin{pmatrix} .455 & .016 & .095 \\ .142 & .008 & .054 \\ .076 & .003 & .003 \end{pmatrix} \quad (28)$$

The characteristic equation may be found out as

$$I - D_kZ = 1 - .486Z + .00467Z^2$$
$$- .000006Z^3 \quad (29)$$

$$(I - D_kZ)^{-1} = \frac{(P_1\ P_2\ P_3)}{|I - D_kZ|} \quad (30)$$

where

$$P_1 = \begin{pmatrix} 1 - .031Z + .00002Z^2 \\ .142Z + .00084Z^2 \\ .076Z - .00018Z^2 \end{pmatrix} \quad (31)$$

$$P_2 = \begin{pmatrix} .016Z - .00008Z^2 \\ 1 - .478Z + .00322Z^2 \\ .003Z - .00015Z^2 \end{pmatrix} \quad (32)$$

$$P_3 = \begin{pmatrix} .095Z + .00011Z^2 \\ .054Z - .01112Z^2 \\ 1 - .463Z + .00141Z^2 \end{pmatrix} \quad (33)$$

If the final demand function U_k is the unit increment of industry in year k, then

$$U_k = (\delta_k\ \ 0\ \ 0)^T \quad (34)$$

$$\tilde{U}(Z) = (\ 1\ \ 0\ \ 0\ )^T \quad (35)$$

Substituting (35) into (16), we obtain

$$G_k^{-1}\tilde{U}(Z) = (\ .833\ -.081\ -.045\ )^T \quad (36)$$

$$\tilde{X}(Z) = (I - D_kZ)^{-1}G_k^{-1}\tilde{U}(Z)$$

$$= \frac{P_4}{|I - D_kZ|} \quad (37)$$

where

$$P_4 = \begin{pmatrix} .833 - .0314Z + .00002Z^2 \\ -.081 + .1541Z + .00094Z^2 \\ -.045 + .0651Z + .00022Z^2 \end{pmatrix} \quad (38)$$

The result can be written as

$$\left.\begin{aligned} \tilde{X}^1(Z) &= .833+.373Z+.185Z^2+.0882Z^3+\cdots \\ \tilde{X}^2(Z) &= -.081+.115Z+.056Z^2+.0269Z^3+\cdots \\ \tilde{X}^3(Z) &= -.045+.043Z+.021Z^2+.0101Z^3+\cdots \end{aligned}\right\} \quad (39)$$

The approprite curves X_k^1, X_k^2, X_k^3 are shown in Fig. 2(a) and Fig. 2(b).

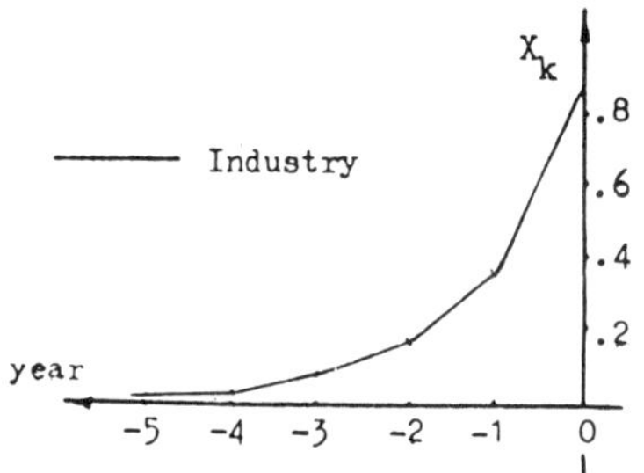

Fig. 2(a)

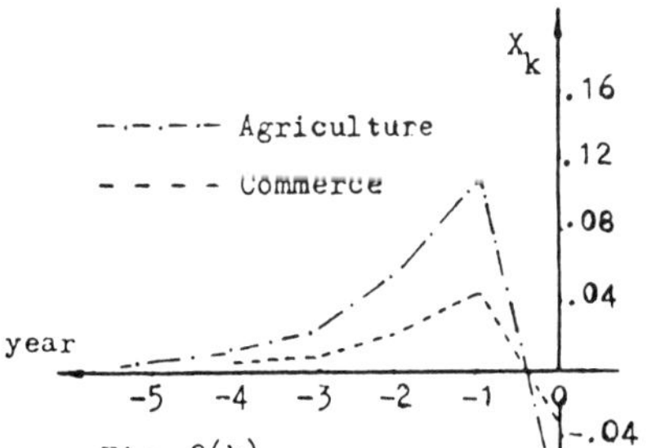

Fig. 2(b)

If all the three sectors (industry, agriculture, and commerce) have one unit final net demand increment in year k, then we have

$$\tilde{U}(Z) = (\ 1 \quad 1 \quad 1\ )^T \tag{40}$$

$$G_k^{-1}\tilde{U}(Z) = (\ 1.051 \quad 1.073 \quad 1.081\ )^T \tag{41}$$

$$\tilde{X}(Z) = \frac{P_5}{|I - D_k Z|} \tag{42}$$

where

$$P_5 = \begin{pmatrix} 1.051 + .0873Z + .00005Z^2 \\ 1.073 - .0305Z - .00768Z^2 \\ 1.081 - .4171Z + .00113Z^2 \end{pmatrix} \tag{43}$$

Then, we obtain

$$\left.\begin{aligned} \tilde{X}^1(Z) &= 1.051+.598Z+.285Z^2+.136Z^3+.0072Z^4+\cdots \\ \tilde{X}^2(Z) &= 1.073+.491Z+.226Z^2+.107Z^3+\cdots \\ \tilde{X}^3(Z) &= 1.081+.108Z+.0485Z^2+.0235Z^3+\cdots \end{aligned}\right\} \tag{44}$$

Hence, the output curve versus time X_k shown in Fig. 3 would be derived from inverse transformation of equation (44).

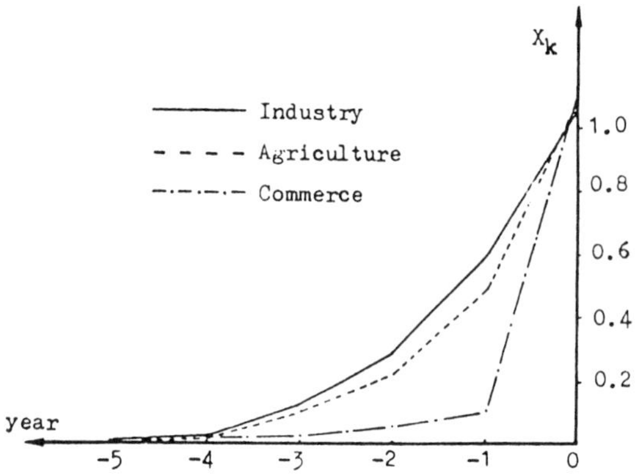

Fig. 3.

Both the two sets of curves shown in Fig. 2(a),2(b) and Fig. 3 are identical with the total output curves of one year time delay developed by Leontief (1977).

It is clear that, while the time delay of investment is only one year, the unit increment final net demand of any sector in year k (such as k = 0) may cause a transient stretching backward more than 5 or 6 years.

Example 2. Dynamic properties of economic system of two sectors with two years time delay.

If the input coefficient matrix and the investment coefficient matrices etc. of this economic system are as follows respectively:

$$A_k = \begin{vmatrix} .328 & .171 \\ .075 & .171 \end{vmatrix} \tag{45}$$

$$B_k(1) = \begin{vmatrix} .511 & .018 \\ .157 & .008 \end{vmatrix} \tag{46}$$

$$B_k(2) = \begin{vmatrix} .621 & .027 \\ .173 & .009 \end{vmatrix} \tag{47}$$

$$k(1) = \begin{vmatrix} .20 & 0 \\ 0 & .60 \end{vmatrix} \tag{48}$$

$$k(2) = \begin{vmatrix} .80 & 0 \\ 0 & .40 \end{vmatrix} \tag{49}$$

Then we have

$$G_k = I - A_k + B_k(1)\ k(1)$$

$$= \begin{vmatrix} .774 & -.016 \\ -.044 & .834 \end{vmatrix} \tag{50}$$

$$H_k(1) = \begin{vmatrix} -.394 & 0 \\ -.107 & .0012 \end{vmatrix} \tag{51}$$

$$H_k(2) = \begin{vmatrix} .496 & .0111 \\ .138 & .0036 \end{vmatrix} \tag{52}$$

$$G_k^{-1}H_k(1) = \begin{vmatrix} -.510 & .00003 \\ -.151 & .00141 \end{vmatrix} \tag{53}$$

$$G_k^{-1}H_k(2) = \begin{vmatrix} .643 & .0143 \\ .199 & .0051 \end{vmatrix} \tag{54}$$

$$\Phi_2(Z) = (I - G_k^{-1}H_k(1)Z - G_k^{-1}H_k(2)Z^2)^{-1}G_k^{-1}$$

$$= \begin{vmatrix} 1+.511Z-.643Z^2 & -.00003Z-.0143Z^2 \\ .155Z-.199Z^2 & 1-.00141Z-.0051Z^2 \end{vmatrix}^{-1} G_k^{-1}$$

$$= \frac{(\ Q_1 \quad Q_2\ )}{1+.511Z-.649Z^2-.0017Z^3+.0033Z^4} \tag{55}$$

where

$$Q_1 = \begin{vmatrix} 1.291 - .012Z + .007Z^2 \\ .068 + .035Z - .025Z^2 \end{vmatrix} \tag{56}$$

$$Q_2 = \begin{vmatrix} .025 - .186Z + .239Z^2 \\ 1.209 + .613Z - .771Z^2 \end{vmatrix} \tag{57}$$

The result may be obtained from

$$\tilde{X}(Z) = \Phi_2(Z)\ \tilde{U}(Z) \tag{58}$$

If there is one unit input used as input function U_k in year k (k = 0) in the industry sector only, then we have

$$\tilde{U}(Z) = (\ 1 \quad 0\ )^T \tag{59}$$

$$\tilde{X}(Z) = \begin{vmatrix} 1.29-.671Z+1.191Z^2-.172Z^3+.681Z^4-\cdots \\ .068+.019Z-.0098Z^2+.007Z^3-.0025Z^4+\cdots \end{vmatrix} \tag{60}$$

Obviously, in this system in which the longest time delay is two years, the response curve of output $\tilde{X}(Z)$ is backward oscillatory attenuately, the oscillatory period T is equal to 2. The curves is shown in Fig. 4(a) and Fig. 4(b).

From the discussion above, the response curves of the dynamic system represent the inherent relationship among systems with different longest time delay and the time sequence relationship, traced back a sufficiently large number of years from year k=0, of comprehensive balance among sectors.

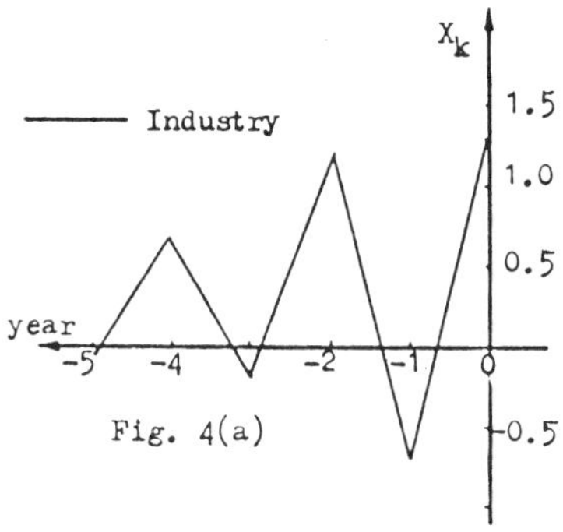

Fig. 4(a)

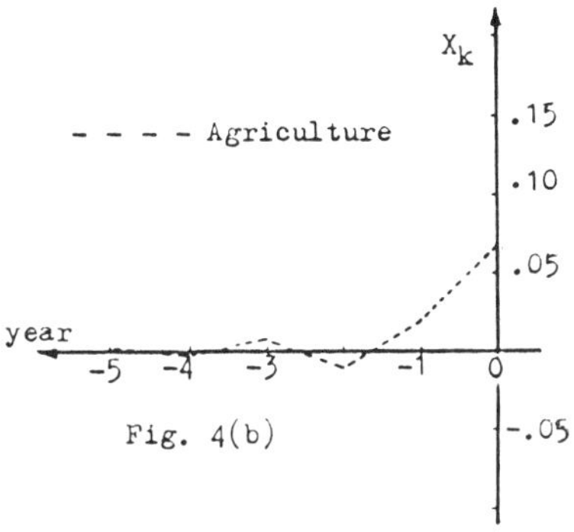

Fig. 4(b)

We have already made the dynamic analysis of macro-economic model for 19 sectors of a province in China and got the reponse curves of this economic system, the longest time delay period is equal to 9, and the order of this system is 171. The curves is shown in Fig. 5.

If the system is time-varying, the calculation of the system will be more complicated and is not within the scope of this paper.

In the previous sections, only the response characteristics of the state vector X_k resulted from the transfer function $V_k(Z)$ has already been discussed. The response curves of investment may easily be obtained. This subject will not be discussed here also.

OPTIMAL CONTROL OF MULTIVARIABLE DYNAMIC ECONOMICAL SYSTEMS

There are two boundary conditions in practical dynamic economical planning. One of them is the expected value at the end of the planning period, such as the value of year 2000, the end of this century, and is called as the end condition, another one is the actual value in the base year of the planning period, such as the value of the base year 1980, and is called the starting condition.

Using dynamic input-output state formulas and the final demand of each year, the magnitude growth of each sector as one goes back in time to starting point may be calculated, but the output value of the starting point may not be match with the existing value in the base year. How to solve this two fixed boundary condition problem? There are many methods, such as goal programming or hierarchical optimization with quadratic criterion (Zhao and Xia, 1984) and etc., may be used.

In this paper, the dynamic input-output descrete state formulas are used as constraints and one attempts to minimize an objective function.

Let

$$\delta X_k = X_k - X_{ko} \tag{61}$$

$$\delta U_k = U_k - U_{ko} = YC_k - YC_{ko} \tag{62}$$

One choice of the objective function is to minimize the deviation. Thus

$$J = \min \sum_{k=1}^{T} (\delta X_k^T Q_k \delta X_k + \delta U_k^T R_k \delta U_k) \tag{63}$$

where

X_{ko}, YC_{ko} are the expected or historical value.

R_k, Q_k are the diagonal weighting matrices.

According to the requirement of the planning system, the values of the element of the matrices R_k and Q_k may be determined.

Alternatively, the objective function may require minimizing the investment also. Thus

$$J = \min \sum_{k=1}^{T} (\delta X_k^T Q_k \delta X_k + \delta U_k^T R_k \delta U_k + Y_k^T S_k Y_k) \tag{64}$$

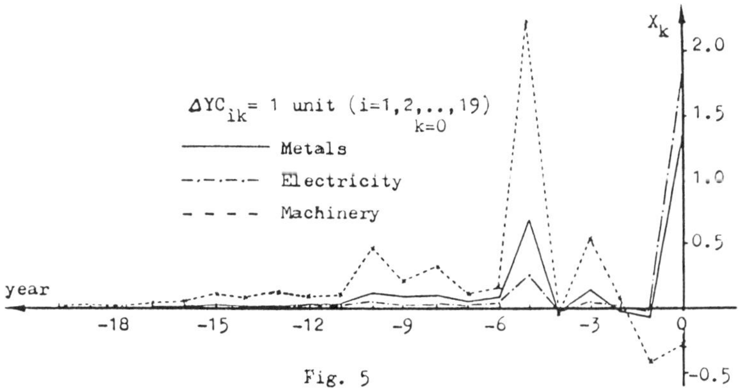

Fig. 5

Using equation (6) and other resource restrictions to be the constraints, this two kinds of optimal equations may be solved through approprite decomposition or other optimal control methods.

CONCLUSION

The transfer function matrices of a linear, multi-input and multi-output control system, based on the dynamic economic input-output model, is described with a compact rational function. The coefficient matrices of this function are submatrices in the dynamic input-output model.

The phase canonical forms introduced by Yokoyama (1973) are used to simplify the calculation. With the simple structure these models may be used in complicated economic system conveniently.

Using the phase canonical foems transfer function, the eigenvalues may be obtained from the coefficient matrices A and B. According to the desired time response of a system, the eigenvalues would be adjusted and the policy analysis and decision making may be accomplished from it.

REFERENCES

Chow, G. C.,(1975). Analysis and Control of Dynamic Economic System, John Wiley & Sons, Chap. 2, pp. 19-37.

Leontief, W., (1977). Essays in Economics, Vol. 2, White Plain Press, pp. 50-77.

Vogt, W. G., M. H. Mickle, and H. Aldermeshian (1975). A dynamic Leontief model for a productive system. Proc. IEEE., 63, 438-443.

Xia, S. W., and C. J. Zhao (1984a). Dynamic input-output models of investment. Journal of Tsinghua University, 24,(3), 103-113.

Xia, S. W., and C. J. Zhao (1984b). The dynamic input-output analysis in economic comprehensive planning. Scientific Methods and Philosophic Topics in System Theory. Tsinghua University Press. pp. 258-285.

Yokoyama, R., (1973), Transfer function matrix of linear, multi-input and multi-output system. INT. J. Control, 18,(2), 369-375.

Zhao, C. J., S. W. Xia, and S. Jiang (1984), Application of hierarchical optimal control method to dynamical input-output analysis, proc. IFAC.

A MACRO ECONOMETRIC MODEL OF SHANGHAI (SHECMOD-A2)

Gu Weiwen, Wu Dinghua and Cai Fuchun

System Engineering Research Institute, Shanghai Jiao Tong University, China

Abstract. In this paper the methodology of econometric model building for Shanghai is proposed. We present a prototype model SHECMOD-A2 based on SHECMOD-A(SHanghai EConometric MODel on Apple II PLUS Microcomputer) using time series economic statistic data of more than twenty years period. SHECMOD-A2 is a nonlinear model, involving thirty-one endogenous variables, and consisting of fifteen stochastic equations and sixteen definitional identities, in total of thirty-one simultaneous equations. Stability study of the model is made based on the computer simulation results. The multiplier analysis and the policy evaluation of six selected economic objectives, i.e. total product of industrial and agricultural sectors, total revenue, total consumption, foreign export, total employment and labor force structure, are inferred from dynamic disturbed simulation results for 1984-1987 years period. Finally, a brief analysis of the results of disturbed multivariable simulation for 1984-1987 years period projected simulation and impact on the six economic objectives has been carried out, and suggestions for Shanghai Municipal economic development are proposed.

Keywords. Economics; Modeling; Econometric model; Optimal system; Computer applications

INTRODUCTION

In this paper we present a prototype macro econometric model of Shanghai. First, the building and features of SHECMOD-A2 based on SHECMOD-A(SHanghai EConometric MODel on Apple II PLUS Microcomputer) are described. Next, we briefly discuss the test results of the model. This is followed by discussion about the performance evaluation and application of SHECOMOD-A2, including stability study, multiplier analysis and policy evaluation. Finally, analysis of policy simulation with multiple variables disturbance is presented and suggestions of Shanghai Municipal economic development are proposed.

THE MODEL

The model is cast in log per capita form for all equations and variables except the variable superscribed with "°" and POIN, which are in arithmetic scale. The logarithm form of Cobb-Douglas eguation with technological progress and constant elastisity is employment for the production function.

The major determinants of local economy, taken into account in this model, are industrial production, agriculatural production, technological progress, physical and human capital input, revenue, consumption, distribution (or wage income) domestic trade, export, commodities transferred, transportation, employment, labor force structure, activity of woman and population growth, etc..

SHECMOD-A2 is a nonlinear model, including fifteen stochastic equations and sixteen definitional identities, thirty-one endogenous variables, eleven exogenous variables, four lagged endogenous variables and five dummy variables, in all thirty-one equations. The flow chart of SHECMOD-A2 is presented in Figure 1. The fifteen stochastic equations include: two sectoral production functions for agriculture and industry to derive per capita product; an employment function to measure

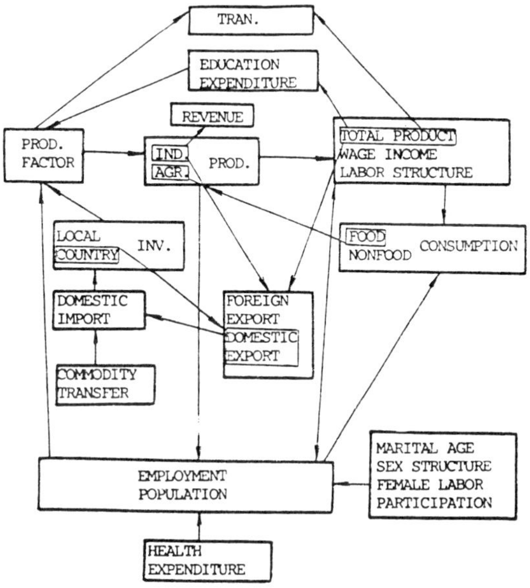

Figure 1 SHECMOD-A2 Flow Chart

the industrial scale; an investment and an accumulative enrollment function to measure physical and human capital formation; two consumption functions for food and nonfood to capture the Engle's effect and the role of food consumption as investment; a revenue function to measure universal municipal per capita income; a foreign export function, a domestic export and a domestic import function to measure the action of foreign and domestic trade; a transportation function to present the capacity of transportation demanded; a fertility and a mortality function to generate population endogenously; an agricultural participation function to reflect the structure of labor force. The model is presented as following equations:

```
 1) Agr. Production
    QAL1=-2.571+.494*NAL1+1.134*FC1+.075*EIR
        (.340) (.213)   (.147)    (.053)
 2) Ind. Production
    YNAN1=8.905+1.246*ENG1-1.211*KNA1+.493*ENP1
         (.482) (.106)   (.127)     (.069)
         for 1957-1969;
    YNAN1=4.005+.381*ENG1+.397*KNA1 (70-83)
         (.903) (.161)   (.046)
 3) Employment
    N1=2.577+.105*QNA1+.124*QA1-.093*WAG+.559*N11
      (.576) (.031)   (.044)   (.043)   (.102)
 4) Physical Investment
    IPC1=-10.220+2.234*MNCRI1-.110*KNA11-.450*D4
         (.546)  (.112)        (.069)    (.067)
         +.330*D5
         (.061)
 5) Human Investment
    ENP1=1.555+.465*EDP1
        (.074) (.015)
 6) Food Consumption
    FC1=-3.665+.151*YPC1+1.026*WAG1+.646*NAN1
       (.788) (.040)    (.109)     (.159)
 7) Public Revenue
    FNC1=1.657+.671*YNA1
        (.230) (.029)
 8) Nonfood Consumption
    NFC1=-5.188+.177*YPC1+1.233*WAG1+.692*(N1-POP1)
         (.509)  (.116)   (.112)     (.303)
 9) Export
    XP1=-2.071+.451*YPC1+.631*XPI1
       (.579)  (.100)   (.172)
10) Domestic Export
    XPI1=3.741+.166*CGT1+(.269+.041*D1)*YNA1
         (.361) (.039)   (.056) (.092)
11) Freight Transportation
    TRAN1=-.880+.294*KNA1+.182*YPC1
          (.386) (.093)   (.065)
12) Domestic Import
    MNCRI1=.079+.621*XPI1+.434*IAA1
          (.173) (.060)   (.061)
13) Crude Birth Rate
    CBR1=10.138-2.229*WOM-.580*ENP1+1.162*FEM
       (1.297) (.554)     (.107)    (.223)
       +.267*D2
        (.098)
14) Crude Death Rate
    CDR1=-.080-.159*HEA+1.084*CDR11+.115*D3
       (.256) (.078)    (.156)      (.065)
15) Structure of Labor Force
    NAN1=.406-.028*YPC1+.950*NAN11
       (.335) (.014)    (.096)
16) Agr Labor-Land Ratio
    NAL1=NAN1+N1-LAND-2*ln(10)
17) Investment Accumulation per Worker
    KNA1=K1-ln[N1°-(NAN1-2*ln(10))°]+4*ln(10)
18) Investment Accumulation
    K1=ln[K11°*95%+IPC1°*POP1°*.01]
19) Investment Accumulation per Capita
    KPC1=K1-POP1+2*ln(10)
20) Consumption of Energy per Nonagr Worker
    ENG1=ENG-ln[N1°-(NAN1+n1-2*ln(10))°]
    +2*ln(10)
21) Population Growth
    POP1=POP11+ln[(CBR1°-CDR1°+POIN)*.001+1]
22) Total Agr.Product
    QA1=QAL1+LAND-2*ln(10)
23) Total Ind.Product
    QNA1=YNAN1+ln[N1°-(NAN1+N1-2*ln(10))°]-4*ln(10)
24) Per Capita Agr-Ind.Product
    YPC1=ln(QA1°+QNA1°)-POP1+2*ln(10)
25) Per Capita Education Expenditure Accumulation
    EDP1=EDR+YPC1-3*ln(10)
26) Per Capita Wage Income of State-owned Persons
    WAG1=WAG-POP1
27) Per Capita Consumption of Food and Nonfood
    C1=ln(FC1°+NFC1°)
28) Per Capita Agr.Product
    YA1=QA1-POP1+2*ln(10)
29) Per Capita Ind.Product
    YNA1=QNA1-POP1+2*ln(10)
30) Per Capita Ind Articles Allocated out
    IAA1=IAA-POP1
31) Per Capita Investment by Government
    CGT1=CGT-pop1
```

Variable difinitions as follows:

C1	:	per capita consumption of food and nonfood, RMB YUAN
CBR1	:	crude birth rate (0/00)
CGT	:	investment by government, RMB Mil. YUAN
D1	:	dummy variable for domestic export
D2	:	dummy variable for new marriage law
D3	:	death dummy variable becuase of hot climate
D4	:	dummy variable for poor investment policy
D5	:	dummy variable for plenty investment policy
EDP1	:	per capita education expenditure cumulated from 1957, RMB YUAN
EDR	:	ratio of EDP1 to YPC1 (0/00)
EIR	:	percentage of electric irrigation field in all farm (0/0)
ENG	:	consumption of coal equivalent in industry, 10,000 tons
ENG1	:	ENG per nonagriculture worker
ENP1	:	ratio of people receiving at least primary education of pop.
FC1	:	food consumption, RMB YUAN per capita
FEM	:	percentage of population of female age 21--35
FNC1	:	public revenue, RMB YUAN per capita
HEA	:	public health expenditure as ratio of YPC1 (0/00)
IAA	:	industrial articles allocated out,RMB Mil. YUAN
IAA1	:	per capita IAA
IPC1	:	productive investment, RMB YUAN per capita
K1	:	accumulation of productive investment, RMB Bil. YUAN
KNA1	:	K1 per nonagr worker, RMB YUAN/nonagr worker
KPC1	:	K1 per capita
LAND	:	farm land. Mil. MOU
MUCRI1	:	domestic imports, RMB YUAN per capita
N1	:	total economically active population (10,000 persons)
NAL1	:	economically active population in agr. per 100 MUO
NAN1	:	percentage of economically active population in agr. (0/00)
NFC1	:	nonfood consumption per capita, RMB YUAN per capita
POIN	:	ratio of population in migration (0/00)
POP1	:	total average population, Mil.
QA1	:	total agr. production, 100 Mil. 1970 constant RMB YUAN
QAL1	:	agr. production per MOU, 1970 constant RMB YUAN per MOU
QNA1	:	total industrial production, 1970 constant RMB 100 Mil. YUAN
TRAN1	:	frieght trans. by land and river, TON per nonagr. worker
WAG	:	total wage income of nonagr. workers in state-owner-ship, RMB Mil. UYAN
WAG1	:	per capita wage income of WAG, RMB YUAN
WOM	:	percentage of women in total labor force (0/0)
XRI1	:	per capita value of domestic exports, RMB YUAN
XR1	:	exports per capita, RMB YUAN
YA1	:	agricultural production per capita in 1970 constant RMB YUAN
YNA1	:	industrial production per capita in 1970 constant RMB YUAN
YNAN1	:	QNA1 per nonagr. worker in 1970 constant RMB YUAN

YPC1 : YA1 plus YNA1

TEST RESULTS OF THE MODEL

The parameters of stochastic equations are estimated by simple OLS (Ordinary Least Square) method using time-series data for twenty-seven years (1957-1983). Table 1 exhibits satisfactory results of the test, which includes standard errors(S) of dependent variables, coefficients of determinents (R^2's), and F-test value ($F_{k,n}$), where: subscript k is the number of independent variables in corresponding equation, and n denotes the number of sample periods. All dependent variables have significant F-test values. All coefficients of determinants R^2 are large enough except that the fourteenth equation is a little below 0.8.

TABLE 1 Standard Error, F-test and Coefficient of Determinants for Stochastic Equations

No. of Equations	S	$F_{k,n}$	R^2
1	.95	$F_{3,27}$=236.5	.969
2 (57-69)	.051	$F_{3,13}$=87.0	.967
2 (70-83)	.040	$F_{2,14}$=57.0	.912
3	.022	$F_{4,26}$=651.2	.992
4	.129	$F_{4,26}$=241.1	.979
5	.080	$F_{1,27}$=902.4	.973
6	.094	$F_{3,27}$=320.1	.977
7	.086	$F_{1,27}$=533.6	.955
8	.066	$F_{3,27}$=471.6	.984
9	.142	$F_{2,27}$=132.8	.917
10	.112	$F_{3,27}$= 70.3	.902
11	.103	$F_{2,27}$= 56.3	.824
12	.048	$F_{2,27}$=756.0	.984
13	.109	$F_{4,26}$= 76.7	.936
14	.063	$F_{3,26}$= 24.6	.770
15	.037	$F_{2,26}$= 49.4	.811

PERFORMANCE EVALUATION AND APPLICATION FOR SHECHMOD-A2

Stability Analysis

Dynamic simulation of SHECMOD-A2 based on SHECMOD-A is made to test for the existance of a stable convergent solution and to evaluate its dynamic performance over the period of 1974-1983 years with given initial values of the lagged endogenous variables and exogenous variables for each year. The mean absolute percentage error (MAPE) and the root of mean squared percentage error (RMSPE) are used to evaluate the stability of the model. For the kth endogenous variable y at year t, MAPE and RMSPE are defined as follows:

$$MAPE_K = \frac{1}{10}\sum_{t=74}^{83} |(\hat{y}_{tk}-y_{tk})/y_{tk}| \quad (1)$$

$$RMSPE_k = \sqrt{\frac{1}{10}\sum_{t=74}^{83} [(\hat{y}_{tk}-y_{tk})/y_{tk}]^2} \quad (2)$$

where y_{tk} and $\hat{y}_{tk}$ are observed and simulated values. Table 2 shows the values of $MAPE_k$ and $RMSPE_k$ for the thirty-one endogenous variables. For the model as a whole, i.e. for all the thirty-one endogenous variables over ten years, MAPE and RMSPE can be expressed in the following formulas:

$$MAPE = \frac{1}{31}\sum_{k=1}^{31} MAPE_k \quad (3)$$

$$RMSPE = \sqrt{\frac{1}{31}\sum_{k=1}^{31}(RMSPE_k)^2} \quad (4)$$

TABLE 2 Stable Analysis--Two Kinds of Error

No.	Variable	MAPE	RMSPE
1	QAL1	.011	.015
2	YNAN1	.004	.005
3	N1	.004	.004
4	IPC1	.022	.026
5	ENP1	.012	.015
6	FC1	.008	.008
7	FNC1	.008	.009
8	NFC1	.006	.009
9	XR1	.015	.019
10	XRI1	.012	.015
11	TRAN1	.021	.025
12	MNCRI1	.007	.008
13	CBR1	.064	.078
14	CDR1	.020	.023
15	NAN1	.002	.003
16	NAL1	.008	.008
17	KNA1	.002	.003
18	K1	.007	.008
19	KPC1	.006	.006
20	ENG1	.004	.004
21	POPP1	.003	.004
22	QA1	.022	.028
23	QNA1	.006	.007
24	YPC1	.005	.006
25	EDP1	.007	.009
26	WAG1	.001	.002
27	C1	.005	.006
28	YA1	.014	.018
29	YNA1	.005	.006
30	IAA1	.001	.001
31	CGT1	.002	.002

The value of MAPE is .0101 and the value of RMSPE is .0187 . Six variables are chosen as economic objectives, which are per capita total product of agriculture and industry YPC1, total economically active population N1, per capita municipal revenue FNC1, per capita food and nonfood consumption C1, per capita export XR1 and percentage of total economically active population in agriculture NAN1. Their average errors are .0065 and .0095 correspondingly, which are lower than that of the whole model. And Fig. 2 also shows that the curves of the simulated values for the six economic objectives fit in well with their observed values.

The above analysis shows that SHECMOD-A2 is stable and can be used as a reliable guide for policy evaluation.

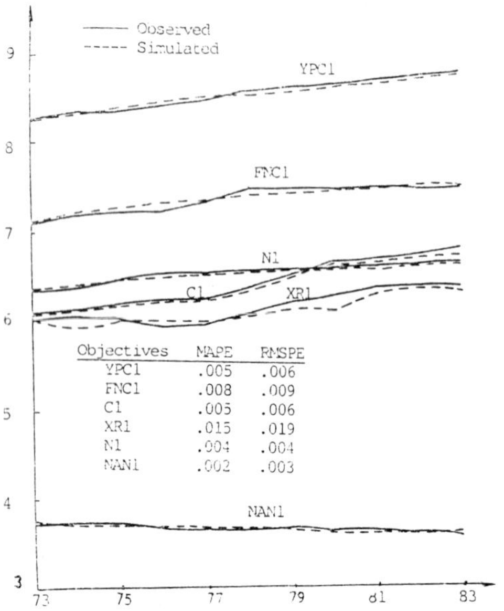

Figure 2 Simulated and Observed Values for the Six Economic Objectives

Structural Analysis and Policy Evaluation

Seven exogenous variables, i.e. energy consumption in coal equivalent ENG, total wage income of nonagriculture workers in state-owner-ship WAG, investment by government CGT, total arable land LAND, percentage of YPC1 spent on public education (accumulated from 1957) EDR, industrial articles allocated out IAA, and percentage of YPC1 spent on public health HEA are chosen as instrumental variables to perform accumulative interim multiplier analysis.

Any change in any of the instrumental variables will affect the values of all endogenous variables. Since our model is in log scale,the multipliers are defined as follows:

$$\frac{\ln y_t^d - \ln y_t^c}{\ln x_t^d - \ln x_t^c} \qquad (5)$$

Where x-- instrumental exogenous variable
y-- endogenous variable
d-- disturbed simulation
c-- control simulation (or original simulation)
t-- a particular year

The cumulative interim multipliers over the period of 1974--1983 years are calculated by the change in endogenous variable of year 1983,i.e. $(\ln y_{83}^d - \ln y_{83}^c)$, if the denominator of (5) is set to be unity.

Policy simulation to measure the impact of the changes in seven instrumental exogenous variables on six objectives have been carried out and the results shown in Table 3.

Analyzing the figures shown in Table 3, we can see the following:

1. Increase in the energy consumption ENG+1 is found to have the largest multipliers. So it will be an attractive policy to increase the ENG, which will lead to increase per capita industrial product directly and then proceed to improve the rest of economic goals.
2. Increase in wage income has the second largest multipliers. It means that the policy of increasing wage income will have positive effect on economic development and will lead to optimistic prospects.
3. Extensive and intensive cultivation through land augmentation and increase in the investment by government will also satisfy the economic goals. Due to land limitation in suburban of Shanghai, land augmentation seems to be not realistic. But increasing investment seems to be realistic.
4. Increase in education expenditure can also contribute to the improvement of all six economic objectives, although not so significant.

Generally speaking, due to the fact that the supply of energy is restricted by planning, the amount of wage increase is restricted by policy and land augmentation is restricted by maximum utilization, therefore increase in the investment and the issue of increase in education expenditure as intellectual investment might be a promising policy.

TABLE 3 The Cumulative Interim Multipliers

Control	YPC1	FNC1	C1	XR1	N1	NAN1
ENG+1	.480	.331	.133	.314	.125	-.101
WAG+1	.128	.030	1.245	.066	.145	-.028
CGT+1	.085	.058	.023	.160	.022	-.014
LAND+1	.076	.032	.088	.043	.177	-.016
EDR+1	.040	.026	.059	.028	.012	-.005
IAA+1	.381	.262	.105	.248	.099	-.064
HEA+1	-.046	-.030	-.071	-.034	-.013	.005

POLICY SIMULATION WITH MULTIPLE VARIABLE DISTURBANCES

We have projected the dynamic simulations beyond the sample period of 1974--1983 years to 1987 with the assumption that observed trends of exogenous variables during past 1974--1983 years would mostly prevail in the future over 1984--1987 years.

In our policy option the disturbances are to increase 0.0618, 0.031, 0.20, 0.10 and 0.05 in log scale for ENG, WAG, CGT, IAA and EDR respectively. The average yearly percentage growth rates of original and the disturbed projected simulations are presented in Table 4.

Table 4 shows that this policy option with multiple variable disturbances can achieve certain improvement on the six objectives in the near future.

TABLE 4 Yearly Growth Rate (in %) of Projected Value for the Six Objectives

Variables	YPC1	FNC1	C1	XR1	N1	NAN1
Original Growth Rate	1.77	1.11	6.13	1.18	.877	-1.71
Option 1 Growth Rate	6.2	4.0	11.4	6.36	2.0	-2.0

CONCLUTIONS

According to the above analysis, a tentative conclusion may be brought up as follows:

1. SHECMOD-A2 is also a small-to-medium size econometric model based on SHECMOD-A. Dynamic simulation over medium term annual periods indicates that SHECMOD-A2 is more reliable in describing the main features of Shanghai Municipal economy during past ten years. Structural analysis shows that the model can be effectively used as a guide and reference to formulate a policy for economic development further.
2. The disturbed projected simulations illustrate that with our policy option it is possible to get an average yearly pecentage growth rate of per capita product of industrial and agricultrial sectors much more than 1.77%, and get an average yearly percentage growth rate of total product more than 7.18% if one percent increase in the population is considered. Thus it is necessary to continue the study of the macro econometrics for Shanghai, which will certainly bring benificial effect.

REFERENCE

Gu Weiwen, Wu Dinghua and Xie Zhiliang (1984). A Macro Econometric Model of Shanghai on Apple II plus Microcomputer (SHECMOD-A), <u>International Conference on Computers, Systems & Signal Processing, Bangalore India,</u> Vol. 2 of 3, 546-549

THE INTELLIGENCE CONTROL FOR PRODUCTION MANAGEMENT SYSTEMS

Song Ji and Chou Hao

Department of Information and Control, Shanghai University of Technology, Shanghai, China

Abstract. Decision-making in production management systems by intelligence retrieval from a knowledge base is presented. A special net graph, ERAI nets, is formulated to specify and resolve the problem. Models represented by ERAI nets have the general characteristics of PETRI net graphs and are more convenient and advantageous in communication between man and models. Problem-solving techniques and algorithms are developed and proved. Application of the method in the planning and control of an electrical equipment manufacturing corporation is taken as an illustrative example.

Keywords. Decision theory; artificial intelligence; production control; man-machine systems; management systems.

INTRODUCTION

The overall economic condition of a corporation depends largely upon the art of production management (PM) of the unit. In the past, the job of PM was inseparable from the intelligence and experience of the manager concerned. So is now in an even faster changing economic, technological and market environment. But the intelligence and experience required for successful management in such a complex environment are beyond the scope of a human alone. Man-made intelligence (artificial intelligence as generally named) and man-made experience (knowledge structure) are helpful assistance for managers in striving for optimum control of their corporation. In this paper, a methodology for a man-machine interactive decision-supported production management system (PMS) with different levels of A.I. is presented.

Production is generally considered a powerful weapon to win a market competition. The concept of PM has been broadened in its scope from narrow internal production scheduling of products being manufactured to the extensive production planning for a corporation as a whole. In terms of systems science, an effective PMS is a large-scale discrete system involving various interrelated subsystems and associated with multi-level and multi-objective decision making. In this paper, we present an interactive decision supported modelling tool, ERAI nets, to facilitate parallel and concurrent simulation of PMS. The net is deadlock-free and can be dynamically executed.

In striving for global control of a corporation, decision making is imperative within its PMS. We categorize decision making activities of PMS according to the level of intelligence it involves into two sorts. One is strategy-oriented decision making for PMS, for which we developed a knowledge-based retrieval system which is adaptable to the high level decision making required by PMS. Where information on states and supports is incomplete, modes of the underlying processes are still in vague concepts. Principles of AI are employed in development of the system [1]. Another decision making structure is operation-oriented PMS, which is adaptable to the lower levels where information on supports and measurements of output and perturbations are sufficient, complete and consistent. Both decision making activities are executed interactively between man and machine; of course, the final decisions are approved by man, not by machine.

Conventional mathematical models are ineffective in specification and resolution of discrete concurrent processing. Various graphic modelling techniques with different capabilities have been developed to represent PM [2,4]. Among them the PETRI net is most commonly used [3]. Yet the PETRI net in its general form is inadequate in explanation capabilities and communication facilities between man and model, which are of primary importance for the execution of PMS decision making. This is the motivation why ERAI nets are developed in this paper for specific task situations generally encountered in PMS.

ERAI-NETS, MODELLING TOOLS FOR PMS

We formulized a system of net graphs for representation, testing and evaluation of PMS. The nets are inherently well-behavioured implying boundedness and safety. Compared with other net generally appeared, ERAI nets have added attribution and facilities for logical retrieval of active information. ERAI net is featured by its characteristics of Evolutionary of Relations with Active Information. The net allows graphical representation of model on different level of abstraction to either top-down or bottom-up execution. Simulation is "token driven" allowing studies of the complex discrete systems. Formal analysis of ERAI net such as boundedness, saftey and invariants will provide necessary structural and behavioural properties of the represented system before its final implementation.

Basic symbols

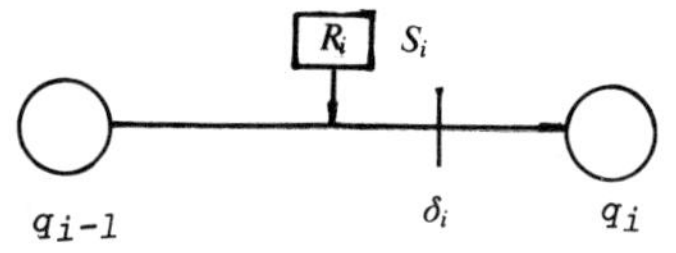

- q — intermediate states and results
- S — support and information required by
- R — the values of support
- δ — event representing knowledge at higher level of decision or production activities at lower level decision and it can be a function, logic formula or time.
- Q — $Q = q \cup S$ total state
- i — positive integer
- P — $P := (F \cup F^{-1}) \cap (Q \times \delta)$
- I — $I := P \cap F$
- O — $O := P \cap F^{-1}$

Definition 1. ERAI net

$$\text{ERAIC}(Q, \delta; I, O) \Leftarrow\Rightarrow$$
$$Q \cap \delta = \phi \wedge Q \cup \delta \neq \phi \wedge I \subseteq Q \times \delta \wedge O \subseteq Q \times \delta \wedge$$
$$\wedge \text{Dom}(I \cup O) \cup \text{Cod}(I \cup O) = Q \cup \delta$$

In order to simulate dynamic behaviour of net, an extra element token q is introduced. q: Q $\rightarrow$ N, N is a set of the positive number.

Definition 2. Marked ERAI net

$$\text{ERAI MC}(Q, \delta; I, O, g) \Leftarrow\Rightarrow \text{ERAIC} \wedge g$$

Definition 3.

ERAI MC with $g = \{g(q_1), g(s_1), \cdots, g(q_n), g(s_n)\}$
$e \in \delta$ is concession if $\forall s_j \in I(e),\ g(s_j) > 0;$
$\forall q_i \in I(e),\ g(q_i) > 0$. That is

$$\text{Concession}(e, g) \Leftarrow\Rightarrow$$
$$\forall s_j \in I(e),\ g(s_j) > 0 \wedge \forall q_i \in I(e),\ g(q_i) > 0$$

Fuzzy knowledge occured in decision making at higher level is represented by weighted arc in ERAI MC. Therefore we have

Definition 4.

$$\text{ERAI FMC}(Q, \delta; I, O, g, \mu) \Leftarrow\Rightarrow \text{ERAI MC} \wedge \mu$$

μ: $F \rightarrow \mathscr{R}$, $\mathscr{R}$ is a set of real number.

In lower level decision making, δ represents production activities and time predicates are introduced in net.

Definition 5.

$$\text{ERAI TMC}(Q, \delta; I, O, T, g) \Leftarrow\Rightarrow \text{ERAI MC} \wedge \text{T}$$

T: $\delta \rightarrow \mathscr{R}$

Conflict may occur in production processes or resources allocation, it is defined as.

Definition 6.

$$\text{Conflict}(e_i, e_j, g) \Leftarrow\Rightarrow$$
$$[(e_i^{\bullet} \cap e_j^{\bullet} \neq \phi) \vee ({}^{\bullet}e_i \cap {}^{\bullet}e_j) \neq \phi)] \wedge \text{Concession}(e_i, g) \wedge$$
$$\wedge \text{Concession}(e_j, g)$$

Other basic elements occured in PMS are: Concurrency, interrelation and series operation. They can be defined in ERAI basic symbols as follows.

Definition 7.

$$\text{Concurrency}(e_i, e_j, g) \Leftarrow\Rightarrow$$
$$\forall t\ \text{Concession}(e_i, g) \wedge \text{Concession}(e_j, g) \wedge$$
$$\wedge \neg \text{Conflict}(e_i, e_j; g)$$

Definition 8.

$$\text{Interrelation}(e_i, e_j, g) \Leftarrow\Rightarrow$$
$$\exists t\ \text{Concurrency}(e_i, e_j; g) \vee [\text{Concession}(e_i, g) \wedge$$
$$\wedge \text{Concession}(e_j, g) \wedge \text{Conflict}(e_i, e_j; g)]$$

Definition 9.

$$\text{Series}(e_i, e_j, g) \Leftarrow\Rightarrow$$
$$\forall t\ \neg \text{Interrelation}(e_i, e_j; g) \wedge \neg \text{Concurrency}(e_i, e_j; g)$$

Using these basic net elements, it is possible to model the whole PMS in any details.

PRODUCTION MANAGEMENT SCHEME

The decision structure of PMS is shown in fig.1. The PMS planning is multi-level (long term, medium term, short term) and multi-branch (technical preparation, part procurement, raw material procurement, sub-product fabricating and product assembly)

LT decision is to decide the type and the number of product to be turned out this year so as to reach the required bonus, quantities and market demands. The MT decision is to assign the LT planning into quarters and months in a reasonable way. The ST scheduling is to realize the higher level decision within the finite resource in PMS. Due to the perturbations arised in PMS, it is impossible to make a successful planning just through one top-down procedure. The static and dynamic balance must be made iteratively within each level and between each level to meet the constraints and criteria. There are two possible perturbations arised during executing the static planning. One is predictable perturbation, which can be found out through simulating each branch of ST planning within deterministic resources. The perturbation can be absorbed through adjusting the net model (in logic form). The other is post perturbation (e.g. machine failure, worker absence). The adjustment can only be made after the perturbation has occured. This is indicated by the feedback arrows in net between each level.

INTELLIGENCE DECISION SYSTEM (IDS)

Cognative psychology has established a problem-solving model (fig.2) for human being. Based on this scheme, a knowledge based IDS (fig.3) is developed. Knowledge base corresponds to the long term memory in human brain. Knowledge base is composed of a net model for PMS in logic form. Moreover, knowledge base consists of several knowledge chunks shown in the figure 4.

The decision data base corresponds to the sensible data and symbolic data in human short term memory. The dynamical data produced during decision in the figure 3 is the dynamical knowledge in short term memory coming from the long term memory temporarily. The data base consists of many data files shown in the left hand of figure 1.

In our IDS, the decision maker (DM) always plays an active role in the whole decision procedure. Before the decision, DM must provide the necessary information. During the decision, DM can use heuristic information to guide the decision. At the end of decision, DM will evaluate and choose the results according to constraints and expected criteria. Next we will show how the IDS works from a simplified decision problem in reality.

DYNAMIC SCHEDULING PROBLEM

1. Hierarchical knowledge processing method.

Due to the enormous volume of knowledge base, a hierarchical knowledge processing method is suggested. The whole knowledge base of PMS has been decomposed into several independent chunks on one hand. On the other hand, the interrelation between them is coordinated in the coordinator (fig.5.). The coordinator has two types: one is the branch coordinator (BC), the other is level coordinator (LC).

BC consists of the interrelated knowledge between each branch and meta-rule which is used to control knowledge flow.

e.g. IF Manufacture technical preparation planning or Adjusting technical preparation planning.

THEN Call for knowledge chunk SDKB1 and STKB.

LC is composed of the meta-rule which switches the knowledge flow from one level to another.

There are three processing modes according to the basic net elements as defined above.

For the concurrency, there exist independent knowledge chunks. So these knowledge chunks can be dealt with concurrently. e.g. the raw material and the part procurement activity.

For the interrelation type of the net structure, the knowledge chunks can be processed in partial concurrency. e.g. the part procurement and the part assembly.

Series net structure has series knowledge chunk. In this case, the knowledge must be processed in a strict order. e.g. LT planning and MT planning.

Having applied the hierarchical knowledge processing method, the decision efficiency can be improved greatly.

2. Dynamic decision in the manufacture

The static planning of the long term has been treated by ERAI FMC net [5]. Here a more difficult dynamic scheduling is taken as an example. The dynamic decision generally consists of the following steps.

- To locate the source of perturbation.
- To suggest the decision results and make decision.
- To evalute the decision selected.

The decision structure in fig.1 has been detailed as figure 6. (in simplified version.) Next a set of converting rules that convert the net model into the first order predicate logic is applied.

In our example, the following predicates have been defined.

$$(\text{LESS} \langle E\rangle\langle A\rangle\langle V_1\rangle\langle V_2\rangle): = \begin{cases} T & V_1 < V_2 \\ \text{NIL} & \text{otherwise} \end{cases}$$

$$(\text{HAVE} \langle E\rangle\langle A\rangle): = \begin{cases} T & \text{the value is "YES" or "HAVE"} \\ \text{NIL} & \text{otherwise} \end{cases}$$

$$(\text{TEST} \langle E\rangle\langle A\rangle\langle V\rangle): = \begin{cases} T & \text{the actual value} = V \\ \text{NIL} & \text{otherwise} \end{cases}$$

$M_1 := M_{11} \wedge M_{12}$

$P_1 := P_{11} \wedge P_{12}$

$M_f := M_{f1} \wedge M_{f2}$

M_2 : = (LE < Raw Material > <Stock > <AN > <PN>)

M_{11}: = (LE < Workshop > <Worker> <AN> <PN>)

M_{12}: = (LE < Workshop > <Machine > <AN> <PN>)

M_3 : = (HAVE < Self-made Parts> <Machining Information>)

P_{11} : = (HAVE < Self-made Parts> <Supplier>)

P_{12} : = (TEST < Raw Material> <Quality> <Good>)

P_2 : = (HAVE < Raw Material> < Technical Data>)

M_{f1} : (LESS < Self-made Parts > < Quantity > <PN > < AN>)
M_{f2} : (LESS < Self-made Parts > < Deadline > < PN> < AN>)
E : Entity ; A : Attribution ; V : Value.
AN : Actual number ; PN : Planning number.

The first step is to locate the perturbation source. This is accomplished by calling for $M_1 \wedge M_2 \wedge M_3 \longrightarrow M_f$, $P_f \rightarrow M_2$, $P_1 \wedge P_2 \longrightarrow P_f$, ... If the predicate P_{12} is false, the perturbation is recognized as the bad quality of the raw material.

Having analysed the perturbation, the machine will give several decision results for selection.

D_1. To warn the supplier to improve quality.
D_2. To change the supplier in next quarter.
D_3. To adjust the original MT production planning.
D_4. To move this product to next quarter.

This final decisions made by decision maker through man-machine interaction mode. Here the D_3 is chosen, the computer will notice the decision maker to adjust the MT planning and give more detailed suggestions about the D_3.

D_{31} : To insert the product at any interval before the deadline.
D_{32} : To increase the equipment and workers in the workshop.

When D_{31} is chosen, the machine will evaluate the D_{31} using net simulation. So more detailed information must be supplied by decision maker or decision data base.

There are three different types of parts P_1 , P_2 , P_3 to be produced before the deadline of P_3 (P_3 is the product to be inserted). The machining procedure for every part is as follows.

$P_{11}\ (M_1:2) \rightarrow P_{12}(M_2:2) \rightarrow$
$P_{13}\ (M_3:2) \rightarrow P_1\ (M_6:2) \rightarrow P_1^*$ 6days deadline
$P_{21}\ (M_4:3) \rightarrow P_{22}(M_1:3) \rightarrow$
$P_2\ (M_5:2) \rightarrow P_2^*$ 6days deadline
$P_{31}\ (M_4:2) \rightarrow P_{32}(M_5:2) \rightarrow$
$P_3\ (M_6:2) \rightarrow P_3^*$ 10days deadline

The ERAI TMC net model for this decision is shown in figure 7.

The inhibitor is used to avoid the conflict when it occures. If the conflict rises, each state of the inhibitor has one token. Which token is to be removed will depend on the external information —— Control function $U_{(x)}$ or decision maker. The production activity relating to the marked state is allowed to be started. Obviously, the control function determines the results of decision whether they are optimum or not.

At the begining of simulation, there is one token in intial state P_0 . The decision will not stop until the final state P_f has one token. The trace record of dynamical token flows is the result of decision. If the planning can not meet the deadline of these products, the control tactic must be changed by DM. The final decision can be obtained through interative man-machine decision. Figure 8 shows the production of these parts can be finished ahead of due-date. So, the D_{31} is the feasible decision. Moreover the accurate time for fabricating P_3 is also obtained.

CONCLUSION

In this article, we have presented an intelligence control methodology for PMS.

A new approach is suggested based on cognative psychology to solve control problem. This methodology emphasize on the fundamental relation among activities rather than the mathematical model.

ERAI net is defined to model the production system. It represents not only the higher level decision knowledge, but also simulates the lower level decision problem.

A hierarchical knowledge processing method is developed to improve the decision efficiency.

An intelligence decision system is suggested by cognative model of human being.

This methodology is also adaptable for the other type of decision planning problem with the same nature.

REFERENCE

[1] Elam, J.J, et al. (1983). Knowledge engineering concepts for decision support system design and implementation. Information and Management, Vol.6, p109-110.

[2] Genrich, H.J. and K. Lautenbach (1981). System modelling with high level PETRI NETS Theoretical Computer Science, Vol.13.

[3] Peterson, J.L.(1982). PETRI NET theory and modelling systems. Prentice Hall.

[4] Pun, L.(1982). GRAI NETS and application. IFAC Symposium on Development Ankara.

[5] Song, J. and Chou Hao. (1985). Interactive decision support system for a production management scheme. The 2nd IFAC/IFIP/IFORS/IEA Conference on Analysis, Design and Evaluation of Man-Manchine System. Italy (to appear)

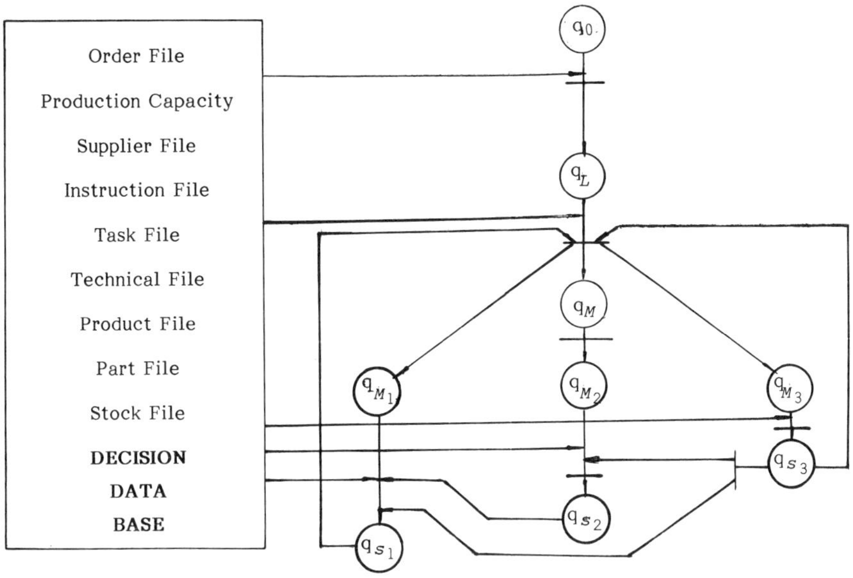

Fig.1. ERAI for the PMS.

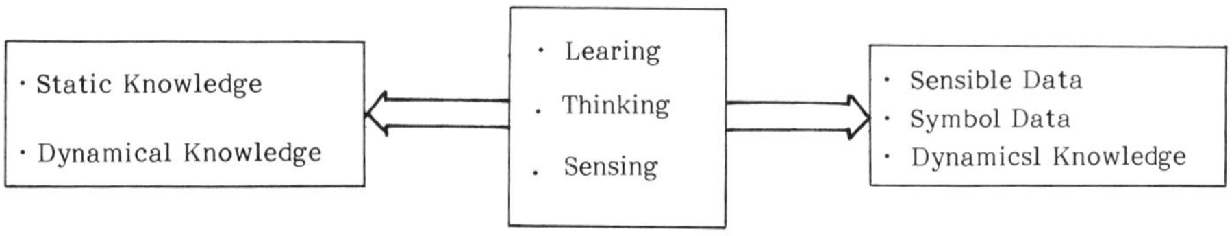

Fig.2. The conceptual model for human problem-solving

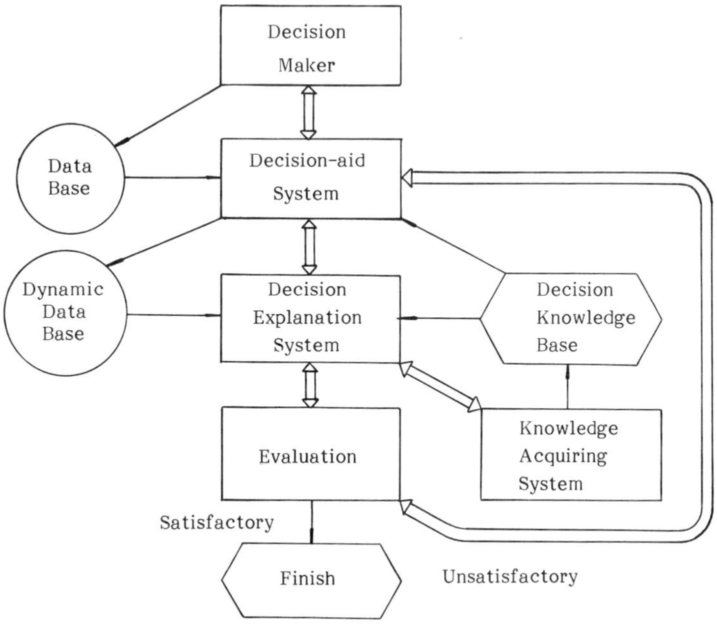

Fig.3. Software Structure for IDS

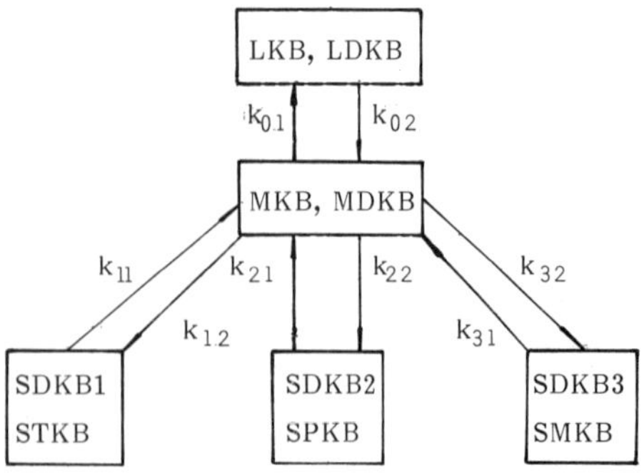

Fig.4. Knowledge structure for PMS.

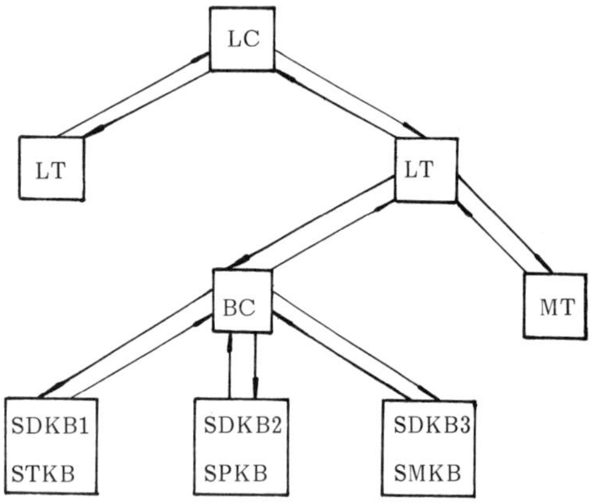

Fig.5. Decomposed knowledge structure

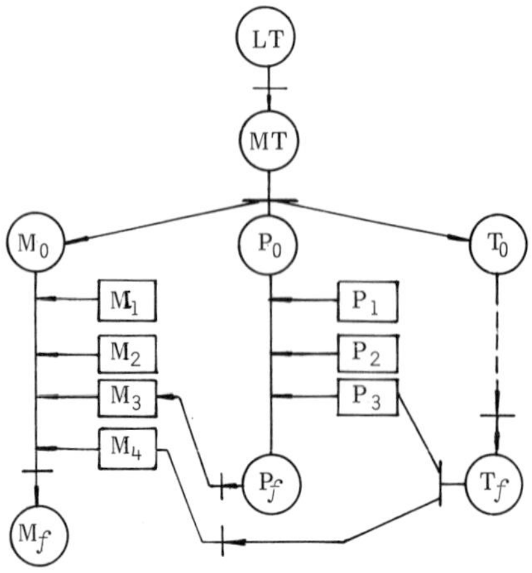

M_1 : Worker man;

M_2 : Machine;

M_3 : Stock information;

M_4 : Machining information;

M_f : Manufacture activity;

P_1 : Raw material supplier;

P_2 : Quality;

P_3 : Technical information

P_f : Raw material procurement activity;

T_f : Technical preparation activity.

Fig.6. Decision structure for ST planning.

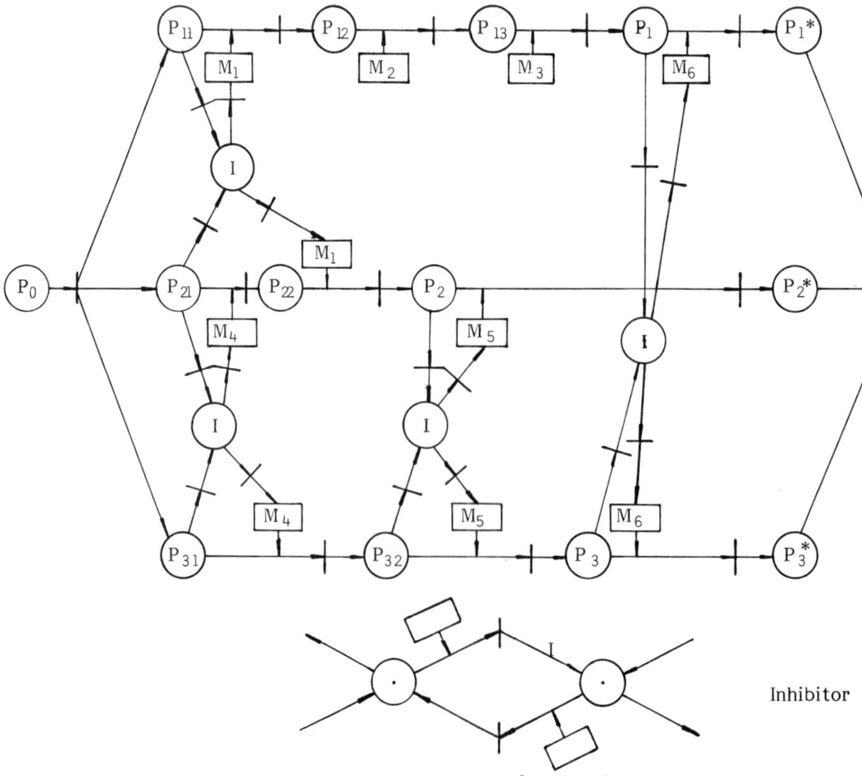

Fig.7. ERAI TMC **net** for manufacture schedule.

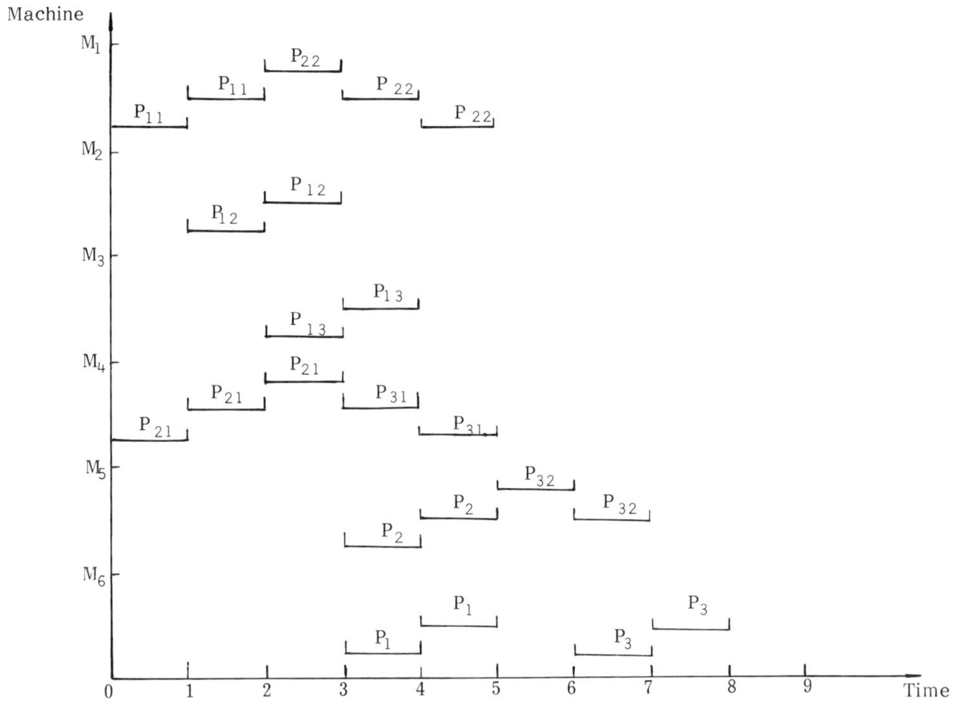

Fig.8. The decision result for manufacture schedule.

EXPERT SYSTEM OF COMPUTER DISPATCH FOR ROAD TRANSPORTATION

Lu Guizhang*, Wang Zhibao*, Tu Fengshen*, Zhang Chaochi*, Wang Xiufeng*, Zhu Yueting*, Lu Kuan and Chen Yan****

*Computer and System Science Department, Nankai University, Tianjin, China
**General Dispatch Office, Bureau of Communication of Tianjin, Tianjin, China

Abstract. This system serves the computer dispatch for road transportation. It is generated on the basis of the necessary and rational knowledge in dispatch operation, i.e. the necessary restrictions encountered in actual work and the requirements of the management. But in some parts of the system like finding the best route or selecting the vehicle team, optimal methods are used with a view to raise the utilization ratio of mileage.

The system has designed a special data structure of the road and created a data base of the road. It can calculate rapidly and correctly the mileage between two arbitrary places.

The system is both effective and feasible. Simulation shows that it can raise the utilization ratio of mileage by 3-5% in a transportation yard.

Keywords. Operations research; transportation; linear programming; optimisation; information processing; data structure of the road; expression of knowledge.

INTRODUCTION

Road transportation dispatch is a typical problem in operations research and can be tackled with linear programming, integer programming, graphic theory and other methods. In real life, however, these methods do not work satisfactorily. To effect an automatic dispatch system, we must work out something new.

This paper deals with the automatic dispatch operation of the comprehensive road transportation in the city with the characteristics of various models of vehicles like dump trucks, trailers, semi-trailers, different types of goods, varied sources of goods, small batches, and both long and short distances. The conditions are so complex that the actual dispatch operation has to be performed under quite a lot of restrictions such as: (a) different types of goods has to be transported by different models of vehicles. For example, the transport of gravels requires dump trucks while that of cement needs trailers, (b) one model of vehicles can only transport cangoes of certain kind. Neither trailers nor dump trucks, for example, can carry reinforced bars or logs, (c) in a return trip, all the goods have to be transported by the same vehicles. Therefore, no return trip should be designated with goods that can't be transported by the same model of vehicles like cement and flour, steel and gravels, and (d) different vehicles and cargoes should be handled with different loading and unloading devices.

Besides, there are also restrictions concerning management such as: (a) restrictions of the work shift, i.e. each vehicle can only work a limited number of hours per day, (b) urgent goods should be given priority in transportation, and (c) in each return trip, the loading and unloading yards must possess the same facilities. Of these restrictions, some are very difficult to be quantized, some, even if quantized, will not be linear restrictions. Even if they can be changed into linear restrictions, the problem will still be too complicated to be solved. Therefore, in overall design, we should not depend solely on the optimal methods of the operational research, but adopt a heuristic method to systemize and regularize the rational and necessary factors of the actual dispatch operation and then turn them into the 'knowledge' necessary for the generating of the system. Of course, optimal methods are still used for some parts of the system. We can, for instance, employ the linear programming or other methods to seek optimum if same model of vehicles to be assigned and the goods can all be transported by same models. Optimal algorithm is also to be processed as a kind of 'knowledge' (programmed knowledge). Using these kinds of knowledge, an expert system for the automatic road transportation dispatch is worked out according to the requirements of the dispatch operation.

To effect a rational dispatch, it is necessary to know the distance between the place of departure from where the goods are delivered and the place of reception for the goods. To use optimal algorithms like linearprogramming, we must also know the distance between any two arbitrary places of departure and destination. These places are not fixed because of the random of customers. Moreover, as the system serves the whole city, the number of places to be encountered in actual transportation could be as great as a few thousands, and it would be impossible to store all the distances of such a large number of places in a micro computer. Therefore, a new method should be worked out to evalute the exact distance between arbitrary places with minimum information that a micro computer with 64k RAM should able to store. For this, two things must be done: first, to determine the form of data, i.e. the way to express the most necessary information for describing a place; then to find the algorithm for calculating the distance. Thise work are both difficult and complex.

In short, to create a system which involves optimisation and at the same time is highly managerial like that of road transportation dispatch, we must retain all the necessary restrictions in actual work while using as many optimal method as possible in some parts of the system. An expert system based on this concept can probably be a effective and feasible means for city road transportation dispatch. This paper describes a system of this kind which has been created in the Fifth Transportation Yard, the First Transportation Company, The Bureau of Communication of Tianjin.

DESING OF THE SYSTEM

1. Expression of Knowledge
Actually all the demands and restrictions on the system present themselves as knowledge, which there are two ways to describe in the computer system.

One is to use the rule 'IF...'. For example (no programming symbols are used here for the sake of clarity): if the surplus of goods exceeds 5% of the total amount or is more than half cargo capacity of a vehicle, then assign one more; if the work shift for the vehicle does not exceed 1.20, then assign; if the goods are not urgent, then leave them to the last to be delivered; if there are more than three departures in a return trip , then break up the trip into small ones. Another way is to use procedure, one kind of 'knowledge' being a subroutine. This knowledge mainly concerns with the optimal algorithms like those to work out the best scheme for assigning empty vehicles, the shortest mileage between the departure and the destination, the best matching for different return trip, etc.

2. System in Blocks
As the system is operated by a microcomputer with 64k RAM which has a limited storage, finally generated will be a system in block, i.e. the whole system is assembled in blocks and each block operates by itself with disk file to carry out the necessary communication between them.

3. The General Block Diagram of the System
The general block diagram of the system is shown in Fig.1. The chief functions of each block are as follows:
A. INPUTS Input the information of the goods. Store in the computer all the information given on the customer's waybill, including the volume of each batch, the place of departure and the destination, types of goods, the models of vehicles that can be used for the task ... as well as all the necessary information on the itinerary like the consigner.

```
    Input the Information of Goods(INPUTS)
            ↓
┌→ Produce Date (INFORM)
│           ↓    <---- Data Base of the Road
│  Calculate Mileage (MILEAG)
│           ↓
│  Calculate the Best Route (OPTIMA)
│           ↓
│  Assign Vehicles (ASSIGN)
│           ↓
│  Print (PRINTS)
no          ↓
└─ See if all the Models are Assigned
            ↓       yes
           End
```

Fig.1 The general block diagram of the system.

B. INFORM Select from each batch of goods those suitable for the same model of vehicles and then turn the information of the localities of the departure and the destination of each selected batch and its volume in a file according to the required data form and then store it in the disk.
C. MILEAG Using the road data base with the information of the localities of the departure and the destination, determine, through programming, the mileage of any departure and destination and turn it into a mileage table according to the required data form and store it in the disk.
D. OPTIMA From the disk read-in the data of the mileage and the volume, and find the shortest mileage of the empty drive and the best assignment scheme for the empty vehicles.
E. ASSIGN Find out the rational route and assign while meeting the requirements of all the practical restrictions.
F. PRINTS Print the final result of the assignment into a table as required and an itinerary (for the driver).
G. The dispatch system operates on the basis of each individual model of vehicles. It assigns cars model by model, and the operation comes to an end when all the models are assigned.

INFORMATION PROCESSING

Information processing consists of two parts: INPUTS and INFORM, whose task is: processing the information of goods sources, creating the data base of the goods according to the customer's waybill, producing the daily file of transportation data from the data base of the goods and, after the transport is finished, draw a monthly production table to indicate what is done and how it is progressing every day.

1. Processing the information of goods sources

This part has the following functions:

A. Creating the data base for the goods sources (JHHW)

Store the information on each waybill: customer; names of goods; the departure and destination and the distance between them; volume of goods; vehicles to be used; types of goods; loading and unloading conditions; accessories to be carried and other necessary signs. The information of each batch of goods is stored in the data base as one record. Divide the sources of goods on the waybills received every day into three groups: the fixed, the semi-fixed, the non-fixed, and then by means of interaction of man and machine, transmit them separately into the data base for goods sources.

B. Creating the data file for goods sources

When all the information for goods sources has entered the data base, start PJHI and select out the goods to be delivered for the day and then create the data file of goods sources for the day.

2. Producing data file INFORM

According to the data file of goods sources of the day, start HCH (for long distance, use CHTU) to returieve information according to each individual model of vehicles and produce a series of data file, including volume of goods, types of goods, numbers of goods, the localities of departure and destination, etc. which will be used for mileage calculation and assign operation.

3. Processing other information

To exercise the management of transportation, it is necessary to transmit timely into the data base the information of the goods already delivered and revise the original data for the next assignment and, in the meantime, complete the report form about the volume of goods delivered. The specific function are as follows:

A. Revision of the data base by deleting the goods already delivered that day from the data base of goods sources. The new data base, then, can receive new information of goods sources. In the meantime the data file for the day's delivery JKLN (K,L,N are for year, month, and day respetively) is created. The file records all the information of each batch of goods delivered that day and can be used for later inquiries.

B. The report form of production

In order to know in time how the delivery is progressing, a report form of monthly production is made, which includes the volume of goods delivered that day, the rotation volume of goods transport and the covered percentage of the goods to be delivered for the month up to that day (calculated separatedly for the delivery volume and the rotation volume), the percentage overfilled (or lost), the daily mileage utilization ratio and the mean mileage utilization ratio.

C. The data contrast table

In the data base JHHW and the data files DRHY and JKLN, all the information appears in codes. To find the original information a data contrast table is needed. Information like addresses, names of goods, models of vehicles, loading and unloading conditions, accessories carried, etc. --- all can exist in telegram codes so that they can be output in Chinese characters.

DATA STRUCTURE OF THE ROAD AND ITS CALCULATION

As we all know, there would be great difficulties if we create the data of the road or calculate its length directly node by node. Therefore, we have created a special data structure and a special algorithm based on that special structure.

1. Four levels of optimization procedure

A. The first level --- Select the optimal node crossing big demarcation with the Hai River as the demarcation line, divide the city of Tianjin into two big areas, the North and South. The bridge is big-demarcation-crossing node i, i=1,2,...,n. If nodes B and E are located in the northern area and the southern area respectively, then i_o exist such that

$$\overline{BE} = \overline{BM_{io}} + \overline{M_{io}E}$$

$$= \min_{1\leqslant i\leqslant n} (\overline{BM_i} + \overline{M_iE})$$

B. The second level --- Select the optimal demarcation

Divide each of the big areas into m small areas and call them in sequence respectively areas 1,2,...,m. The intersection of the kth area and its neighbouring area is the small-demarcation-crossing node $M_{k,i}$ $i=1,2,\ldots,m_k$. If node P and Q are located respectively in the k_1th and k_2 th areas, then there exist M_{k_1}, $M_{k_1}+1$, ..., $M_{k_2}-1$

such that

$$\overline{PQ} = \overline{PM_{k_1}} + \sum_{k=k_1+1}^{x_2-1} \overline{M_{k-1}M_k} + \overline{M_{k_2-1}Q}$$

where

$$\overline{PM_{k_1}} = \min_{1\leqslant i\leqslant m_{k_1}} (\overline{PM_{k_1,i}} + \overline{M_{k_1,i}Q}) - \overline{M_{k_1}Q}$$

For $k=k_1+1, k_1+2, \ldots, k_2-1$

$$\overline{M_{k-1}M_k} = \min_{1\leqslant i\leqslant m_k} (\overline{M_{k-1}M_{k,i}}+\overline{M_{ki}Q})-\overline{M_kQ}$$

C. The third level --- Select a partial reachable node

In a small area, the two ends of each roads are connected respectively with two different nodes, and each node is connected with Roads 1,2,...,l. For two arbitrary nodes in the small area U and V, if the following procedure can be achieved (a) starting from U, choosc a route to reach W, such that

$$\overline{WV} = \min_{1\leqslant i\leqslant l} (\overline{W_iV} \mid W_i \in WI)$$

where WI is the node at one end of all the roads which U is connected with, (b) if W is not V, then assign W to U and turn step (a), (c) if W is V, the procedure ends. The procedure is call the reachable procedure and W is the partial reachable node.

D. The fourth level --- The optimal one-step procedure to deal with stop nodes and dead nodes

In a limited network composed of a limited number of roads and nodes, the procedure can be reachable unless there are no repeated nodes in the procedure.

When the procedure runs from node A to the next node B and comes back from node B to node A, node A is call the stop node. This is the special situation for unreachable nodes. When the procedure is trying to come back from B to A, then close A and force the procedure to pass B. If it is blocked, then node B is called the dead node or 'trap'. Then delete B so that the procedure can get out of the trap and come to life again.

2. Creating and adjusting the special data structure

A. Data file in the small demarcation

For each small area, create a node data file and a road data file, both of which consist of several records. The structural form of the node file records are

$$X\ Y\ W_1\ W_2\ W_3\ W_4\ W_5\ \cdots\ W_l$$

of which X, Y are the abscissa and ordinate of the nodes, W_i (i=1,2,...,l) is the ith road links the nodes (X, Y), the structural form of records of the data file of the road is

$$D\ N_1\ N_2$$

in which D is the distance of the road, N_1 and N_2 are the two nodes at the two ends of the road.

The two files are obviously correlated. They can describe the limited network. It is called reachable network if there have no stop nodes and dead nodes in the procedure and the procedure can be reachable. If stop nodes and dead nodes are in the reachable network, then it is called subreachable network.

The procedure of four-level optimum seeking built on the subreachable network is to be reachable. We hope the road of this reachable procedure is the optimum. Therefore, appropriate adjustment can be made of the coordinate of the nodes in the network.

B. The data file of the big-demarcation-crossing node

Whether the procedure seeking the optimal path crosses big demarcation or small areas, it invariably involves the data of the node. The files created for the data are data files of demarcation-crossing nodes, the structral form of which is

$$X\ \ Y\ \ N_1\ \ N_2$$

in which X and Y are the abscissa and ordinate of the demarcation-crossing nodes, and N_1 and N_2 are two indications of the same node in the two different small areas on the sides of the demarcation line.

ASSIGNMENT OPERATION

This part is the centre of the whole system, whose function is to complete the assignments of vehicles on the basis of the obtained data on mileage, freight volume, etc. This is a subexpert system generated by two kinds of knowledge. The general block diagram is shown in Fig.2.

1. Calculating the minimum mileage of empty drive

After cataloging the sources of goods according to the models of vehicles and the types of goods, the chief task is to assign vehicles according to the optimal routes. This task is based on the calculation of the minimum mileage of empty drive, i.e. the finding out of the optimal dispatch scheme for empty vehicles.

Suppose there are m places of departure and n places of destination, and the matrix of mileage data $X = (x_{ij})$, mxn, the matrix of freight volume $F = (f_{ij})_{mxn}$ if $C = (c_{ij})_{mxn}$ is the mileage of empty drive,

$$Z = \sum_{i=1}^{m} \sum_{j=1}^{n} c_{ij} f_{ij}$$

is the total mileage of empty drive. The problem now is to evalute c_{ij} so that Z = min, the restrictions are

$$\sum_{i=1}^{m} f_{ij} = b_j \qquad j = 1,2,\ldots,n$$

$$\sum_{j=1}^{n} f_{ij} = a_i \qquad i = 1,2,\ldots,m$$

$$f_{ij} \geqslant 0$$

$$c_{ij} \geqslant 0$$

Start TRAFF

↓

Read-in the Data on Mileage and freight volume

↓

Calculate the Minimum Mileage of empty Drive (MINH)

↓

Match the Return Trip (CON)

↓

The Optimal Return Trip Feasible (SECIR)

↓

Assign Vehicles (ASSIGN)

↓

Assign Vehicles II (RANT)

↓

Print the Table of Dispatch

Fig.2 The Block Diagram of Dispatch Operation

This is a typical problem of linear programming and can be solved by any method. In this system, we employ the tabular method. Block MINH will finish the task and the outlet is the optimal scheme for empty drive (the optimal scheme of empty vehicle dispatch).

2. Matching the return trips

According to the data of freight volume and the optimal route of empty drive, match the return trips. The method of matching can vary, but whatever the methods, the total mileage of empty drive will be the same. As there are many other restrictions, the return trip matched on the basis of the optimal scheme of empty drive is not necessary feasible.

3. The optimal return trip feasible
Consider if the return trip matched on the basis of the optimal scheme of empty drive is feasible and then adjust it according to the knowledge obtained in the actual assignment. The knowledge chiefly concerns the following facts: (a) in each return trip the number of places of departure and destination cannot exceed 3; (b) In each return trip the facilities for loading and unloading must be the same; (c) in each return trip the vehicle should not drive longer than the time rated for a work shift, etc.

According to these requirement, adjust the whole return trip with the view to maintain the highest utilization ratio of mileage, and the return trip thus adjust will be the final route for assigning vehicles.

4. Assignment of vehicles
The knowledge on which the assignment is: (a) all the vehicles must be good for the conditions of the roads and the goods yards; (b) select for the assignment the vehicle team that has the shortest mileage of empty drive; (c) If in a return trip all the goods are delivered and the travelling time falls short of the number of hours rated for a work shift, then assign in connection with other assignments, etc.

If there has not enough vehicles, return the remaining goods to the data base of the goods. If there has more vehicles than needed, then apply for new sources of goods.

5. Print the itinerary
After the assignment, print all the information concerning the trip into an itinerary.

The dispatch operation is completed by a combination of optimization and heuristic methods. In seeking the minimum mileage of empty drive, the optimal route feasible, selecting for assignment the vehicle team nearest to the departure and the destination, etc., optimal algorithms are used, but, on the other hand, as these optimal algorithms can only be applied in compliance with the actual conditions, they must be adjusted by heuristic method (the knowledge of 'IF ... THEN'). Only when optimization and heuristic methods are correctly combined can the system raise efficiency and be practical as well.

RESULTS OF EXPERIMENTS

The above system has been tested with the actual data of the goods in the three periods of August, October and December in 1983 (the three months were selected to represent three seasonal of variations of transportation business) and by comparing the results of the experiments with those of the actual assignments in the three months, we can draw the following conclusions.

1. TJNK system is feasible and can be used in the actual assignment of vehicles.

2. TJNK system can increase the utilization ratio of mileage.
Following is a table of comparison

	DISPATCH BY MAN	TJNK
Aug. 83	51.75%	55.36%
Oct. 83	51.04%	55.41%
Dec. 83	51.84%	53.08%

For road transportation in a city, an increase of 1% in the utilization ratio of mileage is a considerable profit.

3. TJNK system combines the best possible use of optimization with its feasibility and, therefore, is more rational than dispatch by man, which generally employs the form of journeying to and fro. But TJNK tries best to match and make multi-node transportation. As a result, the utilization of mileage is increased. Here is an example:
As shown in Fig. 3, The customer consigns two batches of goods for transport: A --- B (steel coils), the distance is 15km; C --- D (billet), the distance is 8km. If dispatch is done by man, vehicles on both routes, will drive to and fro between two places, and the utilization ratio of mileage for either route is less than 0.5, one being 0.48, the other 0.47 (for four return trips). If dispatch is performed by computer, A, B, C, D can be matched in return trip, the weight drive is 23km and the empty drive is 18km, and the utilization ratio is increased.

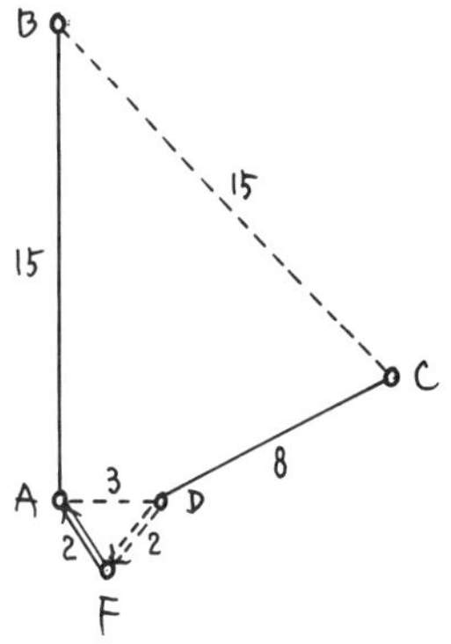

Fig.3 Itinerary of goods transport

Especially for the last trip, the car runs along the A-B route will have to drive 17km to come back from B to F (the garage), and the car running the C-D route will drive 10km to be back from C to F. When they are matched into return trip by computer, it is only 2km from D to F. Thus, for four return trips, the utilization ratio of mileage can reach 0.55, an increase of 7-8%. The dispatcher could hardly think of matching B, at which the car arrives, with C, which is 15km away. But the computer can find the most rational route for the car by seeking the optimum.

4. TJNK system does not depend on the experience of any individual dispatcher. This helps the new dispatcher to learn, and raises the level of dispatch operation.

CONCLUSION

Transportation dispatch is not only a system of optimization, but also highly managerial. Because of the complexity of the actual conditions, it is hardly possible to solve the problems solely by resorting to the operational research.

The expert system discussed in this paper combines optimization and heuristic methods, and proves in practice that it is a good means to effect an automatic dispatch system.

One of the chief difficulties of the automatic dispatch in transportation lies in the automatic calculation of mileage. The road data structure and its calculation put forward in this paper is an algorithm for successful calculation of the mileage of the road on computer. This algorithm, moreover, is universal.

Experiment shows that TJNK can increase the utilization ratio of mileage and has notable economic results. The more extensively it is used, the greater economic results it will yield. To popularize the use of the dispatch expert system will better notably the economic results of the road transportation.

REFERENCES

1. Winston P.H., (1977) Artificial Intelligence, Addison-Wesley Publishing Company.

INTRODUCING FLEXIBLE MANUFACTURING SYSTEMS INTO A DEVELOPING COUNTRY

M. G. Rodd, G. Bloch and S. Meyer

University of the Witwatersrand, Johannesburg, South Africa

Abstract This paper discusses a strategy for the introduction of high-technology manufacturing methods in a country which is characterized by the dual nature of its situation in the international context. On one hand it has an increasing, unsophisticated population desperately searching for work. On the other, it has a highly-sophisticated manufacturing industry which is falling apart because of the adoption by other nations of highly-efficient manufacturing methods. The paper proposes that one of the strategies necessary to reverse this situation involves the rapid adoption of high-technology manufacturing techniques. The paper addresses the issue of education in the broadest sense, ranging from top management, the graduate engineer and the trade-unionist through to the shop floor worker. It is proposed that this education must be accompanied by high-level research and university- and technical-level training. In order to support this, the paper proposes the need to establish a complete full-scale Flexible Manufacturing System which includes all ingredients of modern manufacturing practice, from high-level management tools through to robots on the shop floor. The paper is based on practical experience of a system which is already partly operational and which will be fully operational by the end of 1985.

Keywords Flexible Manufacturing Systems, development technology, control in developing countries, robotics, industrial networks.

INTRODUCTION

Like so many other developing countries, South Africa is caught with one foot in the First world and the other in the Third. For some years now it has had substantial, sophisticated industrial activity aimed primarily at the local market, but to a certain degree supporting exports. This industrial activity is, of course, on a fairly minor scale compared to the vast amounts of minerals, etc. produced and exported from the country. The other side of the coin is that South Africa has a rapidly-expanding, unsophisticated population. Whilst, previously, many of these persons have been absorbed into the industrial sector, an even greater proportion have been involved in the operations necessary to ensure a constant flow of natural resources from the country in the way of gold, coal, iron-ore, etc.. Although the political situation has undoubtedly exacerbated the situation in the past, it is a scenario prevalent in a number of other countries.

International trends towards high-technology manufacturing pioneered in the East, and which are rapidly spreading to North America and Europe, have resulted in a situation in which a country such as South Africa finds itself on the receiving end of what has become known as "exported unemployment". High-quality, low-cost goods produced using modern manufacturing techniques have resulted in locally manufactured goods becoming non-competitive. Previously-exported goods have also a decreasing international market. Generally, only raw materials have continued to be wealth-earners but, from a long-term planning point-of-view, constant exporting of natural resources is highy questionable. It might be a short-term solution, but in the long-term is simply selling one's heritage, and poses a serious problem for many governments. The decreasing local manufacturing industry implies that there are fewer jobs, and the decrease in wealth-earning capability results in less wealth available in the country as a vital resource needed to improve the quality of life. Unemployment figures rise every month and the economy of the country plunges lower and lower. Workers' productivity continues to decrease because of many circumstances, primarily the lack of wealth, which is necessary to train workers and launch new ventures aimed at providing new jobs.

The fundamental premise of this paper is the point-of-view that the only way a developing country can respond to this new scenario (referred to by many as the Second Industrial Revolution) would be to adopt a dual economy. For example, on one hand the wealth-generating areas of the economy, those which can benefit from high-technology manufacturing methods, must be exploited to the full. Imports must decrease and local products must rival (and, in fact, better) those offerings. New export markets must be sought, particularly amongst those countries which are adjacent to the developing nation in question. On the other hand, methods of distributing the wealth must be examined very carefully and labour-intensive industry and other similar endeavours must be established where appropriate. A balanced approach must be adopted. There can be no doubt that the use of high-technology manufacturing methods does cause immediate unemployment - if however, such activities are seen as wealth-generating, then the question becomes a political one relating to the distribution of wealth earned. Claims by the foremost exploiters of high-technology manufacturing that the adoption of such techniques does not cause a reduction in employment in those industries, must be discounted in the face of the real situation. Whilst there may be no immediate increase in unemployment, one cannot deny that there is no additional employment created. The real social result of the technology, however, is only seen in those countries which are on the receiving end of the high-quality, low-cost articles produced and which see their own industries being gutted. The adoption of high-technology methods can be successful from an overall point-of-view only if a totally-integrated strategy is planned.

STRATEGY FOR INTRODUCING HIGH-TECHNOLOGY MANUFACTURING SYSTEMS

In order to ensure the rapid adoption of high-technology manufacturing methods in a developing country, it is vital that an all-pervasive education programme is embarked upon. Clearly, other aspects such as the political philosophy, as well as basic school-level education, must support such technological innovations. This paper focuses on the education required at the tertiary and adult level. There are various aspects to such a large educational exercise. The intended audience must range from politicians to trade-unionists, graduating engineers to senior management, and in order to support such education, fundamental research is also necessary to fully understand the technology in all its aspects. Whilst in the past it has been possible to introduce a new technological innovation, such as a rolling mill, as a turn-key project supplied by a fully-developed nation, in the case of Flexible Manufacturing Systems this approach is not valid. The reason for this is clearly the inherent nature of modern manufacturing methods which are essentially an integration of various technologies - both hardware and software.

To tackle this massive task, the authors and their departments decided on a "total" approach. On one hand, the traditional barriers which exist between the various disciplines of engineering had to be broken down, as well as the barriers which exist between engineer and technician, or engineer and sociologist. Modern manufacturing technologies represent a totally-integrated approach by those who are concerned with the matter. There is no way that this exercise can be undertaken on a theoretical basis only. To ensure a complete approach, it was deemed necessary to create from the outset a full-scale Flexible Manufacturing System, supported by all aspects of computer-assisted engineering appropriate to the modern industrial situation. Whilst it is not proposed to create a full-scale operational factory, the concept is to establish a system which is fully representative of this with full-scale components. Instead of having a large number of separate machine tools, as one would probably find in a typical Flexible Manufacturing System, the approach adopted required the availability of at least one numerically controlled milling machine and lathe. The modern warehouse and its associated technologies has to be an essential ingredient. It is unnecessary to put up a full warehouse; a single stacker crane operating on one line of shelving will suffice - the main point being that the whole structure must be full-scale and must demonstrate all the necessary ingredients.

For the infrastructural support required, it is necessary to have available computer systems which will support actual software being used for such processes as materials requirements, maintenance management, computer-aided design and computer-aided manufacturing. Such systems must be based on available industrial products so that potential users can have a full understanding of their ramifications and, where appropriate, get direct educational experience.

From the point-of-view of research, the facilities described must be configured in such a way that they are flexible and allow more than demonstrations, they must become a base for meaningful applied and, if necessary, fundamental research. Whilst the facility will see a stream of visitors who are really getting only first-level education, the persons who run the system must understand the problems and continually refine the solutions.

There is very little likelihood of a university or a government agency funding such a venture. While it is in the interest of the country to develop high-technology manufacturing, it is equally in the interest of the vendors to display their products in a live situation. Thus, integral to the solution proposed was a close working relationship with various chosen suppliers. Typically, these represented large companies active in various aspects of high-technology manufacturing methods, ranging from robot suppliers to computer vendors and network suppliers. Individual software houses also become involved in the total solution.

Not just a dream, but a reality. The approach discussed has, in a short time, gone very rapidly ahead and is currently within a few months of completion, located off the main university campus where a large space is available, and easy access to visitors ensured. The facility is situated some 20 km from Johannesburg, the main city of South Africa, with equipment valued at approximately $2 ½ million. The UWTec (University of the Witwatersrand Technology Centre) Flexible Manufacturing System, has become the focal point for education and research in the country in modern high-technology manufacturing methods. Visitors to the facility range from politicians to trade-unionists, from school children to artisans. All sectors of the population are seeing the technology in operation and are becoming comfortable with the consequences. Supported by sociologists and economists, the realities of the second industrial revolution can be discussed with potential users and rational decisions made as to where and when these methods should be adopted.

FACILITIES REQUIRED

In order to establish the facility discussed previously it is important to consider the various technologies required to illustrate the total spectrum of activities incorporated into the so-called modern manufacturing system. At the highest level, these will naturally include the various design and planning tools which are now available to the modern industrial, mechanical or electrical engineer. On one hand there is a requirement for true computer-aided design (CAD) facilities which will allow the engineer to exploit and understand fully what capabilities are available. He will also be able to appreciate how these integrate into the whole manufacturing process. The CAD system must be accompanied by computer-aided manufacturing tools, which are increasingly based on the use of Expert systems. Supporting such tools are the fundamental packages which are now only too well-known to the modern industrial engineer. These include support for materials requirement planning (MRP), the implementation of "just-in-time" techniques, simulation of various manufacturing processes, and computer-aided production planning. In addition, these packages need to be supported by large data bases so that real-world situations can be represented. Thus, besides the acquisition of software there are obvious needs for a large amount of computer hardware, which could possibly include networking between the various high-level machines.

Moving closer down to the shop floor, the next aspect of the modern manufacturing system is the real-time computer networking which is so vital to ensure communication between the various hierarchies found within the modern factory. At the higher level, of course, many of the well-known networking techniques are appropriate, but as one moves down the factory, the need for true real-time distributed computer systems becomes vital and these have clearly to be supported by real-time networks. A typical network structure, is demonstrated in Fig. 1, where it will be noticed that this system shows two levels of networking. At the higher level one is concerned only about the transfer of, typically, large amounts of data, while at the lower level, real-time becomes critical in order to synchronize the various shop-floor processes. Many of the techniques of real-time distributed control systems become relevant [Ref. 1]. Moving into the factory area, the Flexible Manufacturing System must start with an Automatic Storage and Retrieval System (AS/RS) i.e. an automated warehouse. This, in turn must consist of stacker cranes controlled through Programmable Logic Controllers (PLCs), through communication processors back into the network. An AS/RS computer on the network will handle the operation of this system, whereas basic higher-level decisions regarding the distribution policies etc. will reside on the upper levels of the control hierarchy.

Goods flowing to and from the warehouse will typically move on either Automatically Guided Vehicles (AGVs) or conveyor systems, be they simple roller conveyors or something more complicated. Thus, the Flexible Manufacturing System described includes large sections of conveyors which provide a means of getting articles from the stacker crane to the robots, etc.. At the sharp end of the production line lies a series of industrial robots, typically configured in a manner flexible enough to allow for the demonstration of various possible features, as well as for experimentation by researchers. Thus the robots range from heavy materials handling robots with capabilities of 100 kgs and up, to smaller high-speed robots used in assembly tasks. As is shown in Fig. 2, the robots are positioned so as to be capable of feeding CNC machines as well as undertaking certain assembly tasks. In the case of assembly, they are accompanied by Computer Vision Systems which are able to communicate, like all the other components, via the real-time network to other parts of the system, and that can ultimately respond to instructions coming down from the top hierarchy. In the case of the assembly stations, it is necessary to have a small local store in the way of a carousel available, so that robots can obtain parts for assembly as fast as possible.

Of particular importance is the physical location of the various aspects of this system. It was initially planned to have the higher-level control computers running the various industrial engineering packages in an office block somewhere away from the workshop floor. It was soon realized, however, that it is vital to demonstrate the totally-integrated approach needed. Thus, whilst one cannot have sophisticated computers running right on the shop floor, it was decided to build a large open-plan computer room immediately above the workshop floor. This means that the computer scientist involved in the design of a new robot language is, in fact, working in close harmony with the object of his research! From the psychological point-of-view this strategy has proved successful in breaking down many of the traditional barriers which exist between persons coming from different technological backgrounds.

INTEGRATING THE PEOPLE

Just as it is important to understand the various technologies which form the chain in modern manufacturing techniques, it is as vital to bring together the various people involved. In order to achieve this, the Flexible Manufacturing System described was integrated into existing research units which had close associations with their parent bodies, i.e. their university departments. The overall control of the manufacturing system and the facilities which housed it were placed under a directorate consisting of the research unit leaders as well as the heads of the home departments. This immediately ensured strong links between the staff and researchers at the home base and those active in the Flexible Manufacturing System. In addition, academics from areas such as industrial sociology and business administration were urged to become involved in the exercise. Strong links were also forged with the suppliers of the equipment used, and wherever possible these people were made to feel part of the overall effort. Being based in a university environment, this is extremely important as it does provide for both continuity and immediate support. However, only a two-way arrangement can work in such a situation, and the many generous suppliers who provided the equipment for the system in terms of initial donations or equipment available on long-term loans, had to see tangible benefits. These benefits result from, on one hand, visitors who become familiar with the equipment as seen in action, and on the other hand, (an even more tangible benefit) the development of local expertise in the product lines being offered. Indeed, one of the spin-offs of the exercise must be to provide an inherent capability within the organization operating the Flexible Manufacturing System. Many of those involved are, naturally, students and researchers who will inevitably move on after having gained considerable experience in the system. This does raise the question of continuity, which in many cases can be handled through a strong relationship with the suppliers. When the persons involved with one particular piece of software leave, the suppliers of the equipment will have developed a large amount of expertise themselves, and should transfer this knowledge to the new researchers.

The aspect of research cannot be over-stressed. It is the belief of UWTec that real understanding of any problem can only be brought about by executing research in that area accompanied by teaching around the subject. Thus, while it is undoubtedly a nuisance from the point-of-view of the researchers involved, the philosophy of the system discussed is that the researchers have a direct responsibility to talk about their area of expertise. No one person, besides possibly the Director, is required to have a global perspective of the approach suggested. A visitor to the facility can get an overview from the Director or his Deputy, after which he will be introduced to the appropriate experts, either research personnel or representatives of the equipment suppliers.

One other aspect of the educational side of the system must relate to the teaching of future engineers and technicians. The facility described is currently used for graduate and undergraduate training. Students in their final year of electrical, mechanical and industrial engineering are able to undertake experimental work in the manufacturing system, or attend formal lecture courses surrounding the technologies. In addition, students from other disciplines (such as sociology) are urged to come and study the technologies so that they in turn can understand some of the fundamental problems which arise from them.

FINANCIAL CONSIDERATIONS

Whilst it has been mentioned that the equipment utilized in the system has been obtained by close collaboration with major suppliers in the area, there are naturally other costs involved in such an infrastructure. Indeed, one of the problems which faced the organizers of the strategy was the fact that the current policy of the government towards universities tended to preclude the activity mentioned. Thus the whole exercise had to become self-funding. The university made land available, but all the infrastructure, buildings, laboratories, etc. had to be paid for out of money generated by the facility. From the point-of-view of an academic institution it is, of course, very difficult to achieve a balancing act between the utilization of the facilities for revenue earning per se, and the need to maintain academic excellence. A compromise is necessary.

In setting up the Flexible Manufacturing System described, it was felt essential to strive for such a balance and to limit contract research to those areas in which the academic merit of the work was deemed sufficient. Likewise, the staff employed in operating the facility should not be seen to be in competition with existing consultants in the technology. Rather, their expertise should come as a result of research, be it self-generated or contract-based, and they should be seen as a resource for consultants in professional practice.

In the hard economic times which face UWTec as it attempts to establish the Flexible Manufacturing System, this balance is not easy to achieve. However, with the generous support of numerous industrial organizations and university benefactors, the system is currently well along the road towards completion. To generate additional income, some of the expertise resulting from the exercise has been used to run fee-paying courses and other Continuing Engineering Education activities. Persons involved in the system are currently offering a wide range of Continuing Engineering Education courses in topics such as robotics, factory automation, image processing, etc.. In addition, some of these courses have extended into highly successful correspondence courses, which provide a regular source of income. Being sporadic in nature, they do not impact too seriously on the day-to-day activities of the researchers involved.

Of course, besides the research persons themselves, a fairly significant infrastructure is necessary to support the overall operation, with very significant cost implications. Clearly, access to workshop facilities (both electrical and mechanical) is vital, as well as the availability of instrumentation and such things as microprocessor development facilities. In addition, facilities for entertaining guests are essential and a fairly significant catering exercise has resulted. In order to draw closer to the industry being served, the whole exercise has expanded, to the point where it offers a complete service to the industry, including the provision of a home for professional bodies, the offering of a variety of courses, etc.. Only with such an integrated approach can such a facility be established.

CONCLUSIONS

This paper has briefly discussed an approach which is being adopted to provide for the upliftment and modernization of the industry of a developing country. Whilst it is admitted that the introduction of high technology manufacturing does have very serious repercussions for a developing country in the sense that it does not directly create jobs, it has to be borne in mind that the destruction of existing jobs is even more serious, especially when accompanied by increasing exports of natural materials and increasing imports of finished articles. Unless methods are found to generate wealth within the developing nations of the world, the world community will be faced with an increasing number of nations which are characterized only by their begging bowls. Countries such as South Africa, which are in the fortunate position of having readily available and extensive sources of metals and energy in the form of coal and hydropower, must be seen to take full advantage of technological innovations in order to exploit their natural resources. In this way the tragic trends which are plaguing the developing countries of the world can be reversed.

REFERENCES

Meyer, S, Macleod, I M, Bloch, G and Rodd, M G, (1985). Real-Time Distributed Computer Control in a Flexible Manufacturing System. Pre-prints, 6th IFAC DCCS Workshop, Monterey, U.S.A.

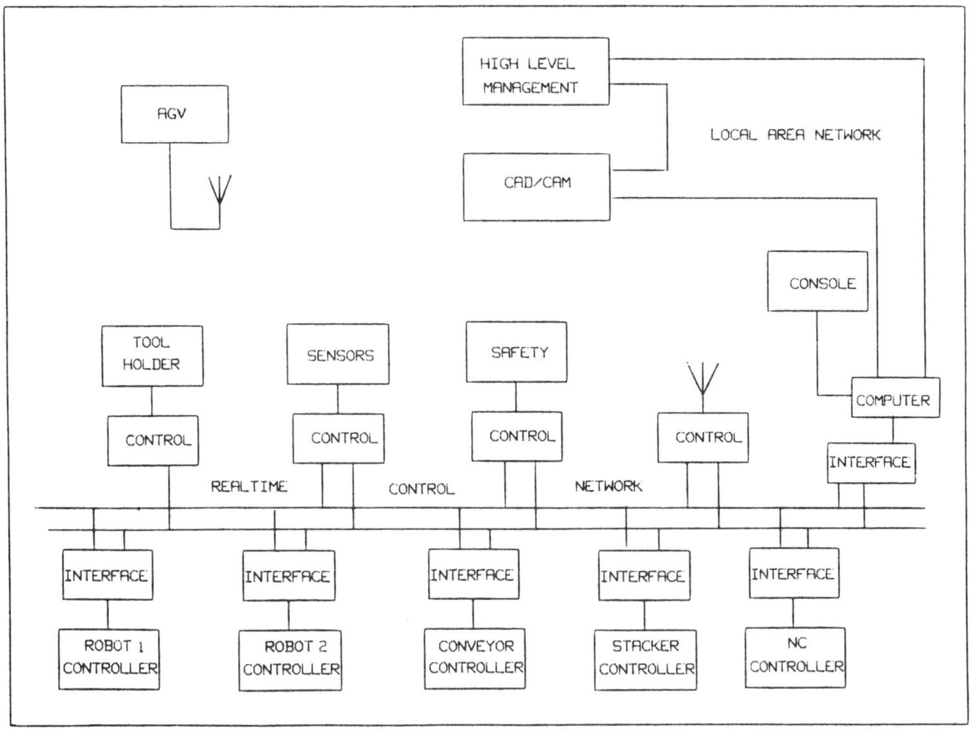

Fig. 1. Flexible Manufacturing System - network structure

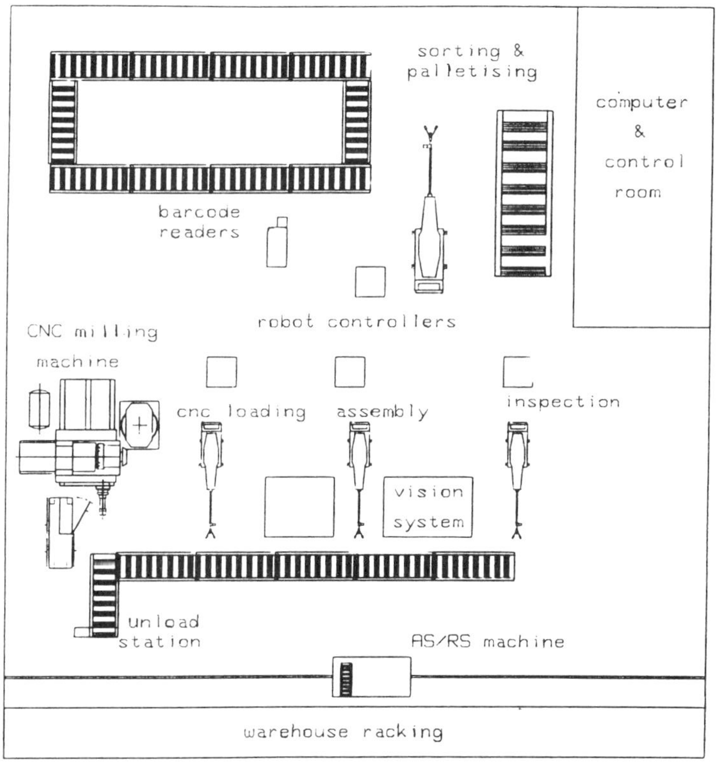

Fig. 2. Experimental Flexible Manufacturing System layout

FUZZY MINIMUM-COST FLOW IN NETWORK AND ITS APPLICATION IN TRANSPORTATION PROBLEMS

Wang Keyi and Wang Zhongtuo

Institute of Systems Engineering, Dalian Institute of Technology, Dalian, China

Abstract. This paper deals with the problem of fuzzy minimum-cost flow in network and it's application in transportation. Two types of problems are tackled. First problem is with fuzzy constraints, second is both with fuzzy constraints and goals. These problems may be treated as fuzzy programming. Methods of solutions are suggested and numerical examples are given.

Keywords. Transportation; optimization; mathematical programming; minimum-cost flow; fuzzy theory.

INTRODUCTION

Minimum-cost flow in network is a vital problem both in theory and practice. In the practical application, there are vague factors which complicate the solution procedure. For example, in the stage of planning of integrated transportation systems of energy, when the network model approach is used, the capacities of railroad can not be estimated accurately. Especially in multi-commodity case, the transportation capacity of each kind of commodity has more uncertainties. If these systems are analyzed with conventional method by using the deterministic models, we must input different data sets and seek a lot of solutions. It is a troublesome task.

In recent years ideas and methods of fuzzy sets had been introduced to network flow analysis. The problems of shortest-path and maximum flow may be treated as fyzzy problems. In this paper, problems of minimum-cost flow in network with vagueness are analyzed. Two types of problems are treated. The first is with fuzzy constraints and second both with fuzzy constraints and goals. These problems may be formulated as non-symmetrical and symmetrical fuzzy programming problems. Methods of solutions are suggested by authors. Numerical examples are given.

FUZZY NETWORK

The general definition of network is $G=\{N,A,W\}$. N is the set of nodes 1,2, ... A is the adjacency matrix of nodes. $W: A \to R^1$.

When the concepts of fuzzy theory is introduced into network, two situations may appear.

One is network with fuzzy structure.

$$\tilde{G}=\{G, \sigma, \mu\}$$

where $G=\{N,A\}$, $\sigma : N \to [0,1]$, $\mu : A \to [0,1]$ and it is true that $\mu(i,j) \leq \sigma(i) \wedge \sigma(j)$, for $\forall$ i,j $\in$ N and $\forall (i,j) \in A$.

The other is network with fuzzy quantities. For example, the lengths of arcs in the network are not exactly determined. There exist a few values, more or less expected of the possible lengths. If the lengths are measured and expressed by integer values (Milan, 1983), then it is possible to describe them by fuzzy subsets of the set I. So, the problem of fuzzy shortest-path appears. Another example is the problem of maximum flow with fuzzy arc capacities in network.

This paper deals with two problems: the first is the minimum-cost flow with fuzzy arc capacities in network (problem I) and the second is the problem of minimum-cost flow both with fuzzy arc capacities and fuzzy cost of transportation (problem II).

THE PROBLEM OF MINIMUM-COST FLOW IN GENERAL NETWORK

It is a problem of mathematical programming.

$$Z=\min \sum_{(i,j)\in A} d(i,j)x(i,j) \qquad (1)$$

subject to:

$$\sum_{(i,j)\in B(j)} x(i,j) - \sum_{k\in A(j)} x(j,k) = \begin{cases} -v & j=1 \\ 0 & j\neq 1,n \\ v & j=n \end{cases} \qquad (2)$$

$$0 \leq x(i,j) \leq c(i,j) \qquad (3)$$

where $d(i,j)$ is the cost of transporting one unit flow on arc (i,j); $c(i,j)$ is the capacity of arc (i,j); $x(i,j)$ denotes the flow value on arc (i,j); v is the value of flow transported from the source 1 to the destination n; B(j) and A(j) are sets of nodes preceding the node j N and following the node j N, respectively.

The problem is to find the flow x(v) which minimize the cost. It can be solved with the minimum-cost flow algorithm of network optimization.

The algorithm is:

Step 1. Initial values bestowed: $x(i,j)=1$; v is the value of flow to be transported; the value of flow transported through the network y=0; the cost of y value flow z=0.

Step 2. Find the shortest path from the source 1 to the destination n with modified Dijkstra's algorithm. The cost of transporting one unit of flow through path π is l. If $l=\infty$, stop.

Step 3. $q=\min\{c(i,j),\ v-y\}$ for $\forall (i,j)\in\pi$.

Step 4. $x(i,j)=x(i,j)+q$ for $\forall (i,j)\in\pi$.

$z=z+q\cdot l$.

$y=y+q$.

If $y=v$, terminate. Otherwise, go to Step 5.

Step 5. Modify the capacity and the cost, for $\forall (i,j)\in\pi$.

$c(i,j)=c(i,j)-q;\ c(j,i)=c(j,i)+q$

$$d(i,j)=\begin{cases} d(i,j) & \text{for } c(i,j)>0 \\ \infty & \text{for } c(i,j)=0 \end{cases}$$

$$d(j,i)=\begin{cases} -d(i,j) & \text{for } c(j,i)>c(i,j) \\ \infty & \text{for } c(j,i)=0 \end{cases}$$

Go to Step 2.

THE PROBLEM I OF FUZZY MINIMUM-COST FLOW

Each capacity constraint $x(i,j)\leq c(i,j)$ can be associated with a fuzzy set $\tilde{C}(i,j)$ of the membership function:

$$\mu_{i,j}(h)=\begin{cases} 1 & \text{for } h\leq \underline{c}(i,j) \\ H(\underline{c}(i,j),\ \bar{c}(i,j);\ h) & \text{for } \underline{c}(i,j)\leq h\leq \bar{c}(i,j) \\ 0 & \text{for } h\geq \bar{c}(i,j) \end{cases} \quad (4)$$

where $\underline{c}(i,j)=\inf \tilde{C}(i,j)$; $\bar{c}(i,j)=\sup \tilde{C}(i,j)$ and H may be a linear or a non-linear function.

Because of the fuzzy capacities of arcs the value of flow to be transported is a fuzzy set $\tilde{V}$ with the member-ship function:

$$\mu_F(v)=\begin{cases} 1 & \text{for } v\geq \bar{v} \\ K(\underline{v},\bar{v};v) & \text{for } \underline{v}\leq v\leq \bar{v} \\ 0 & \text{for } v\leq \underline{v} \end{cases} \quad (5)$$

Where $\underline{v}$ is the maximum value of flow in the network with capacity $\underline{c}(i,j)$ of any arc (i,j); $\bar{v}$ is the maximum value of flow in the network with capacity $\bar{c}(i,j)$ of any arc (i,j) and K may be a linear or a non-linear function.

The problem I of fuzzy minimum-cost flow can be considered as a fuzzy programming with fuzzy constraints.

In fuzzy programming we research how to decide under conditions of fuzzy constraints and fuzzy goals. When fuzzy constraints and fuzzy goals are given in the alternative set we find the decision set first.

Because constraints and goals are defined in different spaces the definition of "f-fuzzy decision" is introduced (ASAI, 1978). X is the alternative set; Z is the goals set and $f: X\to Z$. Fuzzy constraints $\mu_C\in\mathcal{F}(X)$, $C\in\mathcal{P}_i(X)$. Fuzzy goals $\mu_G\in\mathcal{F}(Z)$, $\tilde{G}\in\mathcal{P}_i(Z)$. The "f-fuzzy decision" μ_D^f is the

$$\mu_D^f=\mu_C\wedge\mu_G\circ f \quad (6)$$

$$\mu_D^f(x)=\min[\mu_C(x),\ \mu_G(f(x))] \quad (7)$$

If

$$\mu_D^f(x^*)=\sup_{x\in X}\mu_D^f(x)$$

the x^* is the optimized decision.

In the problem I, the goal is clear. So

$$\mu_G(f(x))=1$$

and

$$\mu_D^f(x)=\mu_C(x)$$

This is a non-symmetrical fuzzy programming. According to the degree of satisfaction with respect to the value v, the decision maker can determine the value of $\mu_F(v)$. $\lambda=\mu_F(v)$, $V\in[\underline{v},\ \bar{v}]$. Then the fuzzy capacity constraints are cut with the value λ.

$$c(i,j)=[\min\{h\mid\mu_{ij}(h)>\lambda\}]$$

where [ ] is the operation of integralization.

Let X(v) be the set satisfying the λ cut set of fuzzy constraints. If X(v) is an empty set $\mu_C(X(v))=0$. Then the above algorithm can be used to find $x(v)\in X(v)$ which minimize the value z of cost. So

$$\mu_D^f(x(v))=\lambda$$

$$x^*(v)=\{x(v)\mid \sup_{x(v)}\mu_D^f(x(v))\}$$

is the result of problem I.

Example 1: See Fig. 1.

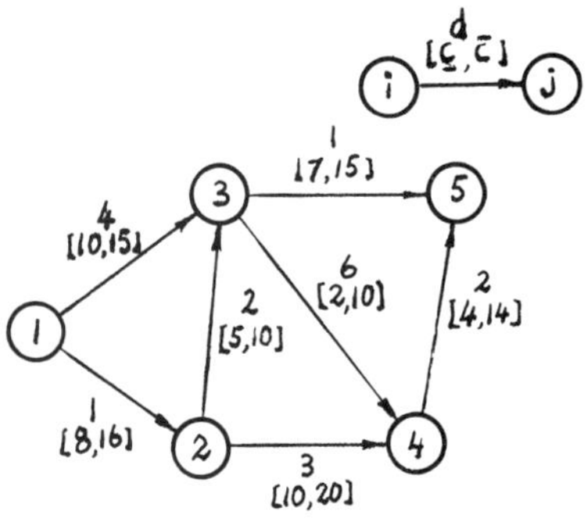

Fig. 1 Example

$\underline{v}=11$, $\bar{v}=29$.

$$H(\underline{c}(i,j),\ \bar{c}(i,j);\ h)=1+(\underline{c}(i,j)-h)/(\bar{c}(i,j)-\underline{c}(i,j))$$

$$\text{for } \underline{c}(i,j) \le h \le \bar{c}(i,j)$$

Table 1 presents the information of calculation and the result.

TABLE 1 The Result of Example 1

v	12	13	14	15	16	17	18*	19
$\mu_F(v)$	0.25	0.30	0.35	0.40	0.45	0.50	0.55	0.60
c(1,2)	14	13	13	12	12	12	11	11
x(1,2)	8	9	10	12	12	12	11*	0
c(1,3)	13	13	13	13	12	12	12	12
x(1,3)	4	4	4	3	4	5	7*	0
c(2,3)	8	8	8	8	7	7	7	7
x(2,3)	8	8	8	8	7	6	3*	0
c(2,4)	17	17	16	16	15	15	14	14
x(2,4)	0	1	2	4	5	6	8*	0
c(3,4)	8	7	7	6	6	6	5	5
x(3,4)	0	0	0	0	0	0	0*	0
c(3,5)	13	12	12	11	11	11	10	10
x(3,5)	12	12	12	11	11	11	10*	0
c(4,5)	11	11	10	10	9	9	8	8
x(4,5)	0	1	2	4	5	6	8*	0
Z	52	58	64	71	78	85	95*	0
$\mu_D(Z)$	0.25	0.30	0.35	0.40	0.45	0.50	0.55	0.00

Z^* and $x^*(i,j)$ are the optimum solutions.

THE PROBLEM II OF FUZZY MINIMUM-COST FLOW

If d(i,j) is a fuzzy set $\tilde{D}(i,j)$ with membership function:

$$\sigma_{i,j}(l)=\begin{cases}0 & \text{for } l \le \underline{d}(i,j)\\ L(\underline{d}(i,j),\ \bar{d}(i,j);l) & \text{for } \underline{d}(i,j) \le l \le \bar{d}(i,j)\\ 1 & \text{for } l \ge \bar{d}(i,j)\end{cases}$$

where $\underline{d}(i,j)=\inf \tilde{D}(i,j)$; $\bar{d}(i,j)=\sup \tilde{D}(i,j)$ and L is a linear or a non-linear function.

When x(i,j) is transported, the cost $\tilde{z}(i,j)=\tilde{D}(i,j)\cdot x(i,j)$ is also a fuzzy subset with membership function:

$$\mu_{i,j}(z)=\sigma_{i,j}(l)$$

Therefore the problem II is a symmotrical fuzzy programming. λ has various values corresponding with v. The fuzzy constraints and fuzzy costs are cut with λ. The symmetrical programming is transformed into the problem of general minimum-cost flow with capacity constraints

$$c(i,j)=\lceil \min\{h \mid \mu_{i,j}(h) \ge \lambda\}\rceil$$

and costs

$$d(i,j)=\min\{l \mid \sigma_{i,j}(l) \ge \lambda\}$$

The solutions are z and x(v).

According to formula (7)

$$\mu_D^f(x(v))=\min\{\mu_C(x),\ \mu_G(f(x))\}$$

$$=\min\{\bigwedge_{(i,j)\in A}\mu_{i,j}(h),\ \bigwedge_{(i,j)\in A}\mu_{i,j}(z)\}=\lambda$$

For $v \in [\underline{v},\ \bar{v}]$

$$x^*(v)=\{x(v) \mid \sup_{x(v)} \mu_D^f(x(v))\}$$

The $x^*(v)$ is the solution of the problem II.

Example 2: it is much same as Example 1.

$$L(\underline{d}(i,j),\ \bar{d}(i,j);l)=1-(\bar{d}(i,j)-l)/(\bar{d}(i,j)-\underline{d}(i,j))$$

$$\text{for } \underline{d}(i,j) \le l \le \bar{d}(i,j)$$

where $\underline{d}(i,j)$ and $\bar{d}(i,j)$ are listed in Table 2.

TABLE 2 $\underline{d}(i,j)$ and $\bar{d}(i,j)$

(i,j)	(1,2)	(1,3)	(2,3)	(2,4)	(3,4)	(3,5)	(4,5)
$\bar{d}(i,j)$	1	4	2	3	6	1	2
$\underline{d}(i,j)$	1	4	1	2	5	1	1

Table 3 presents the information of calculation and the result.

TABLE 3 The Result of Example 2

v		13	14	15	16	17	18*	19
$\mu_F(v)$		0.30	0.35	0.40	0.45	0.50	0.55	0.60
(1,2)	l	1.00	1.00	1.00	1.00	1.00	1.00	1.00
	c	13	13	12	12	12	11	11
	x	13	13	12	12	12	11*	0
(1,3)	l	4.00	4.00	4.00	4.00	4.00	4.00	4.00
	c	13	13	13	12	12	12	12
	x	0	1	3	4	5	7*	0
(2,3)	l	1.30	1.35	1.40	1.45	1.50	1.55	1.60
	c	8	8	8	7	7	7	7
	x	8	8	8	7	6	3*	0
(2,4)	l	2.30	2.35	2.40	2.45	2.50	2.55	2.60
	c	17	16	16	15	15	14	14
	x	5	5	4	5	6	8*	0
(3,4)	l	5.30	5.35	5.40	5.45	5.50	5.55	5.60
	c	7	7	6	6	6	5	5
	x	0	0	0	0	0	0*	0
(3,5)	l	1.00	1.00	1.00	1.00	1.00	1.00	1.00
	c	12	12	11	11	11	10	10
	x	8	9	11	11	11	10*	0

(Continue)

(4,5)	l	1.30	1.35	1.40	1.45	1.50	1.55	1.60
	c	11	10	10	9	9	8	8
	x	5	5	4	5	6	8*	0
Z*		49.40	55.30	61.40	68.65	76.00	86.45*	0
$\mu_D(Z)$		0.30	0.35	0.40	0.45	0.50	0.55	0.00

Z* and x*(i,j) are optimum solutions.

CONCLUSIDN

The concepts and methods of fuzzy programming can be used for solving the problems of fuzzy minimum-cost flow. It can be predicted that fuzzy theory will combine with network flow optimization more and more closely.

REFERENCES

Milan MARES , and Josef HORAK (1983). Fuzzy quantities in network. Fuzzy Sets and Systems, 10, 123-134.
Stefan CHANAS, and Waldemar KOLODZIEJCZYK (1982). Maximum flow in a network with fuzzy arc capacities. Fuzzy Sets and Systems, 8, 165-173.
Stefan CHANAS, and Jerzy KAMBUROWSKI (1981). The use of fuzzy variables in PERT. Fuzzy Sets and Systems, 5, 11-19.
ASAI Kiyoji, and C.V.Negoita (1978). Introduction to Fuzzy System Theory. Ohm, Tokyo.
Jensen P.A., and J.W.Barnes (1980). Network Flow Programming. John Wiley & Sons, New York.

THE INVENTORY MANAGEMENT PROBLEM OF COAL AND THE ESTIMATION OF MONTHLY RECEIPTS BY FILTERING THEORY*

Jing Yuanwei and Zhang Siying

Department of Automatic Control, Northeast Institute of Technology, Shenyang, Liaoning, China

Abstract. In this paper, the inventory management problem of coking coal is considered. Based on the monthly consumption and the inventory of coal in an enterprise, a management scheme is presented. The inventory management is treated as an optimal control problem of a discrete system. In addition, applying Kalman Filtering Theory, the state equation and the equation of output are formulated, and then the monthly receipts are estimated.

Keywords. Optimal control; inventory management; consumption; receipts; discrete system; estimation.

INTRODUCTION

In organizing the production process the inventory management is very necessary for the improvement of economic benefit and the reduction of production cost and spoilage. Taking the inventory and usage problem of coking coal as background, which is one of the major energy resources of an enterprise, we will investigate and discuss the inventory management problem of that enterprise.

Among the purchased energy resources, the coking coal is about 6×10^5 tons, which accounts for 60 per cent of the total amount and it costs ¥39 millions. The coal is used mainly for coking and then the coke is supplied for daily production. In Fig. 1, there are three curves which show the monthly receipts, consumption and inventory through 8 years.

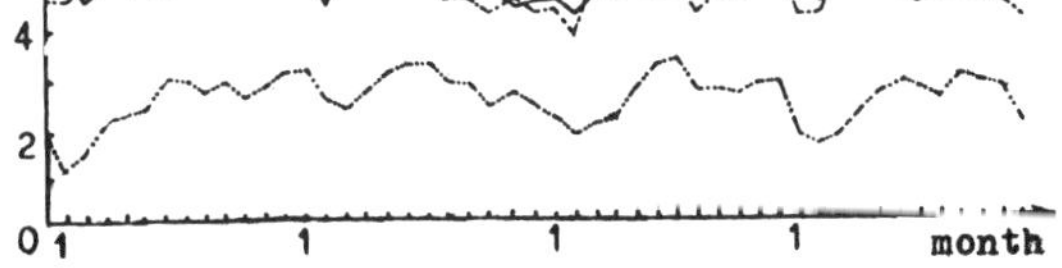

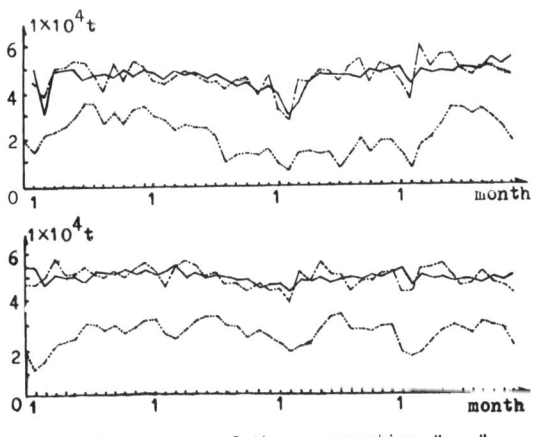

Fig. 1. Curves of the consumption "——", receipts "-·-" and inventory "-..-".

From these data we see that

A. The annual total receipts are about 6×10^5 tons. The consumption of each month is about 5×10^4 tons, and thereby the total amount of consumption in a year is also about 6×10^5 tons.

B. There exist regularities for the monthly receipts. They vary with the seasons evidently.

C. The fluctuation of inventory is rather large.

D. The variation of receipts is in a certain synchronizm with that of consumption.

From the above state it is clearly that, considering both the coke supply and the reduction of inventory, if a reasonable management is possible, then the fluctuation of inventory can be decreased, the stock may be reduced and thereby the amount of circulating fund, the storage costs and the wastage in inventory will be both reduced, the economic benefit will be improved. Based on this point of view and the above analysis, we present a management scheme, in which we treat the inventory management as a optimal control problem of a discrete system and estimate the monthly receipts by Kalman Filtering Theory.

THE PROBLEM

The inventory management may be formulated as a optimal control problem of a discrete system. We use the following symbols:

- $I(k)$ —— The inventory at the end of the kth month or at the beginning of the (k+1)th month.
- $S(k)$ —— The receipts in the (k+1)th month.
- $u(k)$ —— The consumption in the (k+1)th month.
- $I_d(k)$ —— The ideal inventory at the end of the kth month, it is given.
- $u_d(k)$ —— The ideal consumption in the (k+1)th month, it is given.

Generally, the inventory of present month is determined by the receipts and consumption of this month and the inventory of the last month. this can be expressed by

$$I(k+1)=I(k)+S(k)-u(k). \quad k=0,\dots,11. \qquad (2\text{-}1)$$

(2-1) discribes the dynamic process of the variation of inventory in a year. It may be taken as the state equation of our system. In solving the problem, $S(k)$ is treated as a given value. It is the receipts in a month. The initial state $I(0)$

* Project supported by the Science Fund of the Chinese Academy of Sciences.

is given.

$$I(0)=I^{0}. \quad (2\text{-}2)$$

The index performance will be

$$J=\sum_{k=0}^{11}\left\{h\left[I(k)-I_d(k)\right]^2+c\left[u(k)-u_d(k)\right]^2\right\}. \quad (2\text{-}3)$$

where h, c are weighting factors.

Our optimal control problem is that determine the optimal $\bar{u}(k)$, k=0,...,11, to minimize the index performance (2-3). The minimization of the index performance means that: (a) the inventory is stable, (b) the consumption is satisfied, which are the aim of our problem. The weighting factors indicate which term is more important. If the requirement of a stable inventory and a satisfactory consumption are equal, then h, c may have same value. If a stable inventory is more important, then h will be larger than c and vice versa.

THE SOLUTION OF THE PROBLEM

By the "maximum principle" of the discrete system, the Hamiltonian, costate equations and their boundary conditions are as follows:

$$H_k=\lambda(k+1)\left[I(k)+S(k)-u(k)\right]-\left[h(I(k)-I_d(k))^2+c(u(k)-u_d(k))^2\right]. \quad k=0,\dots,11. \quad (3\text{-}1)$$

$$\lambda(k)=\frac{\partial H_k}{\partial I(k)}=\lambda(k+1)-2h\left[I(k)-I_d(k)\right]. \quad k=0,\dots,11. \quad (3\text{-}2)$$

$$\lambda(12)=-\frac{\partial \Phi}{\partial I(12)}=0. \quad (3\text{-}3)$$

The optimal control $\bar{u}(k)$ will be determined by

$$\frac{\partial H_k}{\partial u(k)}=-\lambda(k+1)-2c\left[u(k)-u_d(k)\right]=0. \quad k=0,\dots,11. \quad (3\text{-}4)$$

which leads to

$$\bar{u}(k)=-\frac{1}{2c}\lambda(k+1)+u_d(k). \quad k=0,\dots,11. \quad (3\text{-}5)$$

In order to obtain $\bar{u}(k)$, we have to determine $\lambda(k)$ which are unknown. Substituting (3-5) in (2-1), we obtain:

$$I(k+1)=I(k)+S(k)+\frac{1}{2c}\lambda(k+1)-u_d(k). \quad k=0,\dots,11. \quad (3\text{-}6)$$

Thus, (3-2), (3-3), (3-6) and (2-2) constitute a two-point boundary value problem of difference equations. We may solve this problem as a multivariable linear algebraic equation system of first order. We may write (3-6) as follows.

$$I(k)-I(k+1)+\frac{1}{2c}\lambda(k+1)=u_d(k)-S(k). \quad k=0,\dots,11. \quad (3\text{-}7)$$

or

$$\begin{aligned}&I(0)-I(1)+\tfrac{1}{2c}\lambda(1)=u_d(0)-S(0)\\&I(1)-I(2)+\tfrac{1}{2c}\lambda(2)=u_d(1)-S(1)\\&\cdots\cdots\\&I(11)-I(12)=u_d(11)-S(11).\end{aligned} \quad (3\text{-}8)$$

(3-2) has the form:

$$2hI(k)+\lambda(k)-\lambda(k+1)=2hI_d(k). \quad k=0,\dots,11. \quad (3\text{-}9)$$

or

$$I(k)+\frac{1}{2h}\lambda(k)-\frac{1}{2h}\lambda(k+1)=I_d(k). \quad k=0,\dots,11. \quad (3\text{-}10)$$

By (3-3), we have

$$\begin{aligned}&I(0)+\tfrac{1}{2h}\lambda(0)-\tfrac{1}{2h}\lambda(1)=I_d(0)\\&I(1)+\tfrac{1}{2h}\lambda(1)-\tfrac{1}{2h}\lambda(2)=I_d(1)\\&\cdots\cdots\\&I(11)+\tfrac{1}{2h}\lambda(11)=I_d(11).\end{aligned} \quad (3\text{-}11)$$

I(0) is given. It is on the left side in (3-8) and (3-11). Put it to the right side, we obtain a 24-variale linear algebraic system of first order with I(k), k=1,...,12, (k), k=0,...,11, as unknowns. Its augmented matrix is

$$A_{24\times25}=\begin{bmatrix}A_{11} & A_{12} & B_1\\ A_{21} & A_{22} & B_2\end{bmatrix}. \quad (3\text{-}12)$$

where

$$A_{11}=\begin{bmatrix}-1 & & & \\ 1 & -1 & & \\ & & \ddots & \\ & & 1 & -1\end{bmatrix}_{12\times12}$$

$$A_{21}=\begin{bmatrix}0 & & & \\ 1 & 0 & & \\ & & \ddots & \\ & & 1 & 0\end{bmatrix}_{12\times12}$$

$$A_{12}=\frac{1}{2c}A_{21}^{T}$$

$$A_{22}=-\frac{1}{2h}A_{11}^{T}$$

$$B_1=\begin{bmatrix}u_d(0)-S(0)-I(0)\\ u_d(1)-S(1)\\ \vdots\\ u_d(11)-S(11)\end{bmatrix} \qquad B_2=\begin{bmatrix}I_d(0)-I(0)\\ I_d(1)\\ \vdots\\ I_d(11)\end{bmatrix}$$

Denote

$$\begin{aligned}I_{12}&=\left[I(1),\ I(2),\ \dots,\ I(12)\right]^T\\ \Lambda_{12}&=\left[\lambda(0),\ \lambda(1),\ \dots,\ \lambda(11)\right]^T.\end{aligned} \quad (3\text{-}13)$$

We may write the above system as

$$\begin{bmatrix}A_{11} & A_{12}\\ A_{21} & A_{22}\end{bmatrix}\cdot\begin{bmatrix}I_{12}\\ \Lambda_{12}\end{bmatrix}=\begin{bmatrix}B_1\\ B_2\end{bmatrix}. \quad (3\text{-}14)$$

Solving this system we may obtain I(k), $\lambda(k)$ and then determine $\bar{u}(k)$ by (3-5).

There is already a routine for solving this kind of problem in computer. We may use it with little verification in order to suit our problem. We solved this problem for different h and c.

For S(k) in B_1, we take the average receipts in the last 3 years as the receipts of this year (see TABLE 1).

TABLE 1 Monthly receipts

k	1st year	2nd year	3rd year	average
0	46812	52719	44609	48047
1	50136	45904	38721	44921
2	58019	54026	42630	54892
3	51999	57329	46771	52033
4	51555	55800	56773	54709
5	55009	50253	52910	52724
6	51568	51802	50441	51270
7	50120	47222	43903	47081
8	53341	48109	48600	50044
9	50522	43989	49359	47967
10	53328	47937	52091	51119
11	56820	43744	53412	51326

Let

$I(0)=1.5\times10^4$

$I_d(k)=\text{const}=1.5\times10^4$

$u_d(k)=\text{const}=1.5\times10^4$

For different h and c, we have the following results.

TABLE 2 Results under h=1, c=1

k	(k)	u(k-1)	I(k)
0	3.693×10^3		$1.5\ 10^4$
1	3.693×10^3	4.8181×10^4	1.4866×10^4
2	3.371×10^3	4.8314×10^4	1.1472×10^4
3	-3.684×10^3	5.1842×10^4	1.4522×10^4
4	-4.641×10^3	5.2321×10^4	1.4234×10^4
5	-6.173×10^3	5.3087×10^4	1.5857×10^4
6	-4.460×10^3	5.2230×10^4	1.6531×10^4
7	-1.758×10^3	5.0879×10^4	1.6742×10^4
8	1.726×10^3	4.9137×10^4	1.4686×10^4
9	1.099×10^3	4.9451×10^4	1.5279×10^4
10	1.657×10^3	4.9171×10^4	1.4075×10^4
11	-1.933×10^2	5.0097×10^4	1.5097×10^4
12	0	5.0000×10^4	1.6419×10^4

TABLE 3 Results under h=2, c=1

k	(k)	u(k-1)	I(k)
0	4.240×10^3		1.5×10^4
1	4.240×10^3	4.7880×10^4	1.5167×10^4
2	4.906×10^3	4.7547×10^4	1.2540×10^4
3	-4.933×10^3	5.2467×10^4	1.4965×10^4
4	-5.073×10^3	5.2536×10^4	1.4462×10^4
5	-7.224×10^3	5.3612×10^4	1.5559×10^4
6	-4.987×10^3	5.2493×10^4	1.5790×10^4
7	-1.827×10^3	5.0913×10^4	1.6147×10^4
8	2.761×10^3	4.8620×10^4	1.4606×10^4
9	1.194×10^3	4.9403×10^4	1.5249×10^4
10	2.198×10^3	4.8906×10^4	1.4310×10^4
11	-5.713×10^2	5.0286×10^4	1.5143×10^4
12	0	5.0000×10^4	1.6455×10^4

TABLE 4 Results under h=1, c=2

k	(k)	u($\overline{k}$-1)	I(k)
0	5.285×10^3		1.5×10^4
1	5.285×10^3	4.8679×10^4	1.4368×10^4
2	4.020×10^3	4.8995×10^4	1.0293×10^4
3	-5.393×10^3	5.1348×10^4	1.3837×10^4
4	-7.720×10^3	5.1930×10^4	1.3940×10^4
5	-9.841×10^3	5.2460×10^4	1.6189×10^4
6	-7.463×10^3	5.1866×10^4	1.7047×10^4
7	-3.369×10^3	5.0842×10^4	1.7475×10^4
8	1.581×10^3	4.9605×10^4	1.4952×10^4
9	1.484×10^3	4.9629×10^4	1.5366×10^4
10	2.217×10^3	4.9446×10^4	1.3887×10^4
11	-7.967×10^{-1}	5.0002×10^4	1.5004×10^4
12	0	5.0000×10^4	1.6326×10^4

From these data, we may draw the curves of variation of inventory as Fig. 2.

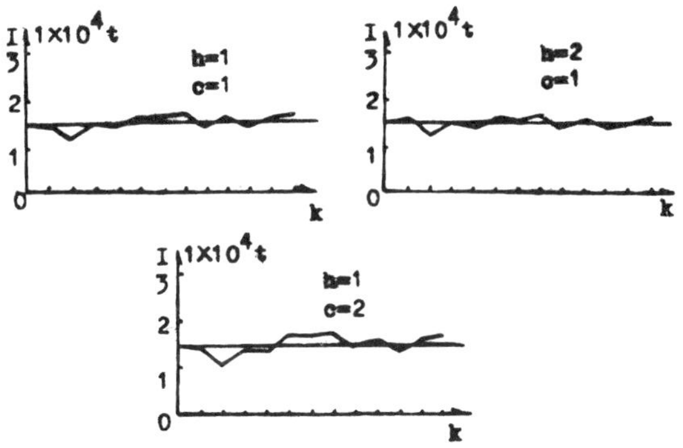

Fig. 2. Inventory curves

There is another method to solve the two-point boundary value problem of (3-2), (3-3), (3-6) and (2-2).

From (3-2) we have

$$I(k)=\frac{1}{2h}\left[\lambda(k+1)-\lambda(k)\right]+I_d(k).\qquad k=0,\ldots,11.\qquad (3\text{-}15)$$

Substituting this expression for I(k) and I(k+1) in (3-6) we obtain

$$\lambda(k+2)-(2+\frac{h}{c})\lambda(k+1)+\lambda(k)=b(k).\qquad k=0,\ldots,10.\qquad (3\text{-}16)$$

where

$$b(k)=2h\left[S(k)-u_d(k)+I_d(k)-I_d(k+1)\right].$$

The characteristic equation of (3-16) is

$$\mu^2-(2+\frac{h}{c})\mu+1=0.\qquad (3\text{-}17)$$

It has two different roots

$$\mu_1,\mu_2=\frac{1}{2}\left[(2+\frac{h}{c})\pm\sqrt{(2+\frac{h}{c})^2-4}\right].\qquad (3\text{-}18)$$

hence the general solution of (3-16) is

$$\lambda(k)=c_1\mu_1^k+c_2\mu_2^k+C(k).\quad k=0,\ldots,12.\qquad (3\text{-}19)$$

where C(k) can be determined by a given initial condition and

$$C(k+2)-(2+\frac{h}{c})C(k+1)+C(k)=b(k). \quad k=0,\ldots,10. \quad (3\text{-}20)$$

Then C_1, C_2 may be determined by I(0) and $\lambda(12)$, and $\lambda(k)$, I(k) and $\bar{u}(k)$ will be determined subsequently.

The calculated results are the same as that obtained by using the first method.

THE ANALYSIS OF THE RESULTS

Comparing the obtained results with the historical data, we see that

A. The total practical consumption in a year is 5.987682×10^5tons, the three calculated results are 6.047091×10^5, 6.046640×10^5, 6.048017×10^5tons respectively. Therefore, in the global account, there is a very little difference between the calculated results and the practical data.

B. In practical case, the fluctuation of inventory is large. In calculated results, the fluctuation is reduced. In the former, I(0) was reduced a maximum amount of 1.3756×10^4tons, in the later, those amounts are 0.35278×10^4, 0.24598×10^4, 0.47607×10^4tons respectively. Thus, for the former it needs more standing inventory than the later.

C. If the inventory is calculated by the practical receipts and the calculated consumption, then the above mentioned amount is 0.97737×10^4tons, which is still less than 1.3756×10^4tons.

Thus, applying our scheme, the monthly coal consumption for production may be guaranteed, the standing inventory may be reduced and the economic benefit may be improved.

FURTHER DISCUSSION: THE ESTIMATION OF S(k) BY KALMAN FILTERING THEORY

In the foregoing paragraphs, under the asumption that S(k) are given, we presented an inventory management scheme (by using the maximum principle). The calculated results are rather good. But, in general, we can not know S(k) in advance. Thus, it gives rise to the problem of estimation and forecasting of the values S(k). We try to solve this problem by Kalman Filtering Theory.

In the Filtering Theory there are two fundamantal equations, i.e. the state equation and the equation of output. In this problem we formulate them as follows.

Since the variation of S(k) is evidently depending on different seasons, this monthly and rather regular variation will provide us a base for constructing the state equation. We take the state equation as the receipts relation between two neighbouring months.

$$S(k)=A(k|k-1)S(k-1)+W(k-1). \quad k=0,\ldots,11. \quad (5\text{-}1)$$

In addition, the receipts will be constrained by the contract. The monthly receipts can not deviate too far from contract. Therefore, we may take the amount of receipts in contract as a quantity of observation and write the output equation as follows.

$$Z(k)=H(k)S(k)+V(k). \quad k=0,\ldots,11. \quad (5\text{-}2)$$

If for S(k) we have m historical data, we may determine A(k|k-1) by minimizing the following equation.

$$M(k)\triangleq\sum_{i=1}^{m}\Big[S_i(k)-A(k|k-1)S_i(k-1)\Big]^2+ + \Big[\sum_{i=1}^{m}(S_i(k)-A(k|k-1)S_i(k-1))\Big]^2. \quad (5\text{-}3)$$

Similarly, we can determine H(k).

In applying the Kalman Filtering Theory, it deals with the variance estimation. Since the quantity of data is rather small, we use the method of "estimation by range" to estimate the variance. We have verified the normal characterestic of S(k), this leads to the application of the estimation by range.

The process of forecasting is as follows.

<u>Known estimation</u>. The estimation $\hat{S}(k-1)$ of state S(k-1) at time k-1 is known. The variance of the error is

$$P_{k-1}\triangleq E\Big[S(k-1)-\hat{S}(k-1)\Big]^2. \quad (5\text{-}4)$$

<u>The first step forecast</u>. Having the estimate for step k-1, we can get the preliminary estimate of S(k) by the state equation (5-1).

$$\hat{S}(k|k-1)\triangleq A(k|k-1)\hat{S}(k-1). \quad (5\text{-}5)$$

The variance of the error is

$$P_{k|k-1}\triangleq E\Big[S(k)-\hat{S}(k|k-1)\Big]^2$$

$$=\Big\{\frac{1}{\alpha}\Big[\max_i(S_i(k)-\hat{S}(k|k-1))-\min_i(S_i(k)-\hat{S}(k|k-1))\Big]\Big\}^2. \quad (5\text{-}6)$$

where α is the coefficient of extremal distance, which may be found in relevant table.

<u>The observational estimate</u>. The ordered quantity in contract is taken as the observational estimate. The variance of the error is

$$R_k\triangleq E\Big[\check{Z}(k)-H(k)S(k)\Big]^2$$

$$=\Big\{\frac{1}{\alpha}\Big[\max_i(\check{Z}(k)-H(k)S_i(k))-\min_i(\check{Z}(k)-H(k)S_i(k))\Big]\Big\}^2. \quad (5\text{-}7)$$

<u>The optimal linear filtering</u>.

$$\hat{S}(k)=\Big[1-B(k)H(k)\Big]\hat{S}(k|k-1)+B(k)\check{Z}(k)$$

$$=\hat{S}(k|k-1)+B(k)\Big[\check{Z}(k)-H(k)\hat{S}(k|k-1)\Big]. \quad (5\text{-}8)$$

Let

$$P_k\triangleq E\Big[S(k)-\hat{S}(k)\Big]^2$$

$$=(1-B(k)H(k))^2P_{k|k-1}+B^2(k)R_k+$$

$$+2\Big[1-B(k)H(k)\Big]B(k)\Big[H(k)Q_k-\check{Z}(k)\bar{S}(k)\Big]. \quad (5\text{-}9)$$

the B(k) that makes the above P_k minimum is

$$\hat{B}(k)=\frac{H(k)P_{k|k-1}-\left[H(k)Q_k-\overset{\circ}{Z}(k)\bar{S}(k)\right]}{H(k)P_{k|k-1}+R_k-2H(k)\left[H(k)Q_k-\overset{\circ}{Z}(k)\bar{S}(k)\right]}. \tag{5-9}$$

where

$$\bar{S}(k)\triangleq ES(k), \qquad Q_k\triangleq ES^2(k).$$

According to the above process, we calculate the intermediate-term (yearly) forecast and the short term (monthly) forecast and obtain the data shown in TABLE 5.

TABLE 5 Monthly and Yearly Forecast

k	Intermediate term	Short-term
1	4.5626×10^4	4.5824×10^4
2	4.1325×10^4	4.0551×10^4
3	5.3606×10^4	5.6061×10^4
4	5.3264×10^4	5.3350×10^4
5	5.5644×10^4	5.6148×10^4
6	5.3117×10^4	5.3574×10^4
7	4.7797×10^4	4.7021×10^4
8	4.3484×10^4	4.3081×10^4
9	4.9824×10^4	5.6562×10^4
10	4.7504×10^4	4.9206×10^4
11	4.9586×10^4	5.0772×10^4
12	4.9288×10^4	4.8395×10^4

From these data we see that the forecasting results are rather close to the practical case. The estimates of intermediate-term, short-term and average value have their own advantages and disadvantages. For the months with large deviation from contract, such as March, the average value is better. For months with unstable receipts, such as September, October and December, the short-term forecast is better. For momths with stable receipts the intermediate-term forecast is better.

CONCLUSION

By the numerical calculations we see that the average inventory may be reduced from 2.5×10^4tons to 1.5×10^4tons. If 1 ton of coal costs ¥65, then the accupied circulating fund, the interest for this fund, the storage cost and the wastage in inventory will be reduced to an amount of ¥6.5×10^5, 4.68×10^4, 8.26×10^3, 1.17×10^4 respectively.

If this scheme is applied and realized for an enterprise, then it will improve the economical benefit.

REFERENCES

A. Bensoussan (1974). Mangement Applications of Modern Control Theory. North-Holland publishing Company.

E. Bryson, Y. C. Ho (1975). Applied Optimal Control. Hemisphere Publishing Corporation.

Edward, A. Silver (1981). Operations research in inventory management: a review and critique. Operations Research, Vol. 29, No. 4, July-August.

R. Suri, Y. C. Ho (1980). Resource management for large system: concepts, algorithms and applications. IEEE, AC. Vol. AC-25, No. 4, August.

SEVERAL WAYS OF FOSTERING STUDENTS OF AUTOMATIC CONTROL SPECIALITY IN COLLEGES OF ENGINEERING IN CHINA

Lei Guo-xiong

Department of Automatic Control and Electronic Engineering, East China Institute of Chemical Technology, Shanghai, China

Abstract. The paper analyses the characteristics in teaching in process control in China and the major goals of reform in teaching under the impact of new technological revolution.

The paper is aimed primarily at the writer's experiences in teaching practice in process control – by organizing Students Papers Small-scale Public Lecture "S.P.S.P.L." setting up Students Science Interest Group "S.S.I.G." and instructing students to complete their theses on practical problems related to the reform of automatic technology on the production site, the students' ability to work independently is extended, the process to educate an automatic control engineer is shortened, and better process control personnel for China's economic construction may be educated.

INTRODUCTION

The automatic control specialization was first established in Chinese technology institutes in 1958.

At present, they can generally be divided into three categories:

1. The automatic control theory and application specialization.

2. The industrial electrical automation specialization.

3. The process control specialization.

There are about twenty chemical technology institutes with the process control specialization in China now, but the number of graduates falls short of demand for automation personnel.

East China Institute of Chemical Technology enrolls nearly 100 process control students each year, and during the past 22 years, over 1000 automatic control graduates from our university have been assigned to work in various design institutions, research institutes, plants and universities.

Large amounts of feedback information showed that the graduates are appreciated for their knowledge, skill and approach to work.

Nevertheless, recently the top authorities in our country pointed out, China's education should "face the modernizations, face the world and face the future".

Upon this instruction, the education of automatic control in China is due to bear new contents and new ideas.

ANALYSIS OF THE CHARACTERISTICS IN TEACHING IN THE AUTOMATIC CONTROL SPECIALIZATION IN CHINA

The teaching aim of the East China Institute of Chemical Technology has been clear-cut since its birth in 1952: "The East China Institute of Chemical Technology is a cradle for chemical engineers".

So an automatic control specialization student has to fulfill a dual task:

-- to learn the related courses and have the training of the basic theories and fundamental skills that an automatic control engineer should master.

-- to develop the ability to work independently after graduation. To realize this, it is necessary to analyze the characteristics in the teaching of automatic control under China's special conditions.

The three characteristics are:

1. An automatic control specialization student should have a wider range of knowledge, moreover, their knowledge should reach a certain depth in the major courses.

In China today, a chemical process control engineer should not only have the ability to design and to deal with problems in the automatic operation of various chemical engineering units, but also have the ability to develop special instruments needed on the production site as well. Therefore during his studies in the institute he should build a knowledge structure with four pillars.

A. Chemical engineering.

B. Mathematics and physics.

C. Electronic technology and computer science.

D. Automatic control theory.

In addition, he has to select other courses concerned. In that case, it is inevitable that there are many courses for a student to learn. Whereas traditional method of teaching usually put the students in a passive position in learning.

2. The demand of new teaching contents and methods.

In recent years, under the impact of the new technological revolution the world over, there have been rapid developments in the field of automatic control. New technologies in measuring and control systems and control strategies have been emerging, and new generations of instruments have been displacing older ones.

Automatic control has risen from a subordinate position of service to the commanding centre in many large and medium-sized plants in China. All these developments greatly stimulate the improvement of the teaching of automatic control. However it brings up new problems to the teaching, such as text-books, instruments for laboratory use, etc. What's more our country is one with vast territory, the economical developments of the various regions are unbalanced and so are the levels of the application of automatic technology.

So it is necessary for the graduates to be able to work at various plants of different levels and foundations in confrontation to the new technical revolution in automations.

"What and how should a teacher teach?" This is a real problem.

3. The demand of the education of ability to work independently and to handle versatile problems on the production site.

Under current conditions, a process control specialization graduate may be assigned to work at medium-sized or small plants where he has to work on automation and on other related posts as well.

Therefore the training of ability to work independently and to handle versatile production problems must be taken into consideration in designing the educational system in colleges and universities.

THE KEY TO THE PROBLEMS

In our opinion, the characteristics in the teaching of the automatic control will be in existence for a long period of time in China. All the problems, however, will give rise so a foundamental teaching reform. Which is characterized by:

-- A correct analysis and study of the current situation and future development in the automatic control field both domestic and abroad.

-- The establishment of a new teaching programme for educating automatic personnel. Clearcutting the aim of each course.

-- Reforming the methods of teaching and study.

-- The reform should serve the national economic construction and be backed up by society.

The two major goals of the reform are:

1. To educate the students to be conscious in their studies, resolute in creative work and strong in mind to challenge difficulties.

2. To extend the student's ability to analyze and solve problems during the whole process of study and training.

This means the goal of teaching is the training of ability rather than the traditional teaching of knowledge.

Under the precondition of extended ability, the students can solve for themselves.

A. The renovation of knowledge.

B. The application of knowledge.

C. The adaptability to work.

D. The creativity in work.

In our opinion, it is much more difficult to extend the students' ability than to make him understand the courses.

The training of ability is itself both a science and an art, and should be dealt with individually according to the different qualities of the students.

PRACTICAL MEASURES TO EXTEND THE STUDENTS' ABILITY

It is easy to see that different countries, different colleges and universities and different teachers have different ideas on this issue. Therefore the exchange of experiences would be of great importance. Upon this, we summed up our experiences in experiments of teaching methods in recent years.

The first. Organizing and holding "S.P.S.P.L."

When the automatic control specialization students have finished their courses of basic technology in automatic control and begun to study specific courses, the teacher should guide the students to extend their ability in self-study and to expand scope of their knowledge to greater widths and depths.

Experiences show that backing the students to hold lecture forums is a good way.

For years we have chosen the course "Controlling instruments" as a break-through.

The concepts and problems to be mastered in teaching can be summed up into several topics, and the students can choose these topics or their own topics and complete their papers within several weeks.

For example, the following topics were chosen on the lecture forum of the 1982 graduates:

A. What is the significance of controlling point deviation and controlling point precision to the engineering? What is its effect to the controlling system?

B. The consequences of windup and the ways to eliminate them.

C. The design principles on double way bumpless transfer in analogue controllers.

D. The combined use of instruments and computers.

E. The evaluation of the reliability of instruments.

Students were enthusiastic in discovering new ideas and eager to compete for honor.

In the end we collected more than 70 papers among the 1982 graduates. The students were urged to form a group to examine the papers, among which eight were chosen. The students organized and presided over the forum.

They invited teachers and engineers to attend the lectures.

From the lecture forums, the students have drawn these useful conclusions:

(a) Never before have they been so absorbed in studying, discussing and reading reference materials.

(b) They have discovered that they really haven't mastered and can not apply most of the knowledge they have learnt before.

(c) They have discovered the close relation between the different subjects.

(d) They are more interested in self-study and in writing articles.

(e) They love the automatic control specialization better.

As to the teachers, the forum is a better way to check their teaching effect and discover the more talented students.

The second. To set up the "S.S.I.G."

This is a good way to develop and train the more talented students. During the past few years, we have organized S.S.I.G. numbering from the 3,4 to 6,7 students in each grade beginning from the senior years. A research problem is assigned to each group and the students select their own group leader. They take an active part in carrying out scientific and technological activities during the summer and winter vacations and other spare time.

For example, during the 1982 summer vacation, a group to make researches on a Wobbe meter calibrating heating values of fuel gases was formed.

First they made investigations and comparisons on the strengths and weaknesses of various Wobbe-index meters and mastered the calibrating methods of heating values accurately.

Then back to the university, they studied papers on the related subjects by themselves, presented a design and made further experiments on the stability of flames and the testing of the interference-proof property of the meter-against the environmental temperature changes. Their work was quite effective and saved several hundred thousand yuan of investment money for the plant.

In our opinion, if favourable conditions can be created and efficient organization work be made, a large number of college students will become a reinforcement in scientific research and their successes in research will greatly encourage the other students to make greater efforts.

We sincerely hope that more successful S.S.I.G.s will emerge in our country, with the patronage and support from the responsible branches concerned.

The third. To instruct the students to complete their theses through practice in the reforms of automatic control technology.

One of the good teaching traditions of the East China Institute of Chemical Technology is to pay great attention to the student's thesis. We think, it is both a general review of all the courses and a rehearsal before taking his post as an automatic control engineer.

How can this last link in teaching be arranged so that the students can gain knowledge and extend their ability to their best?

In our opinion the better way is to let the students contact those problems of automatic technology and management on the production site, and this will speed up the build-up of student's ability.

According to this principle, the following measures have been taken:

A. Select one or two plants that can meet "the required conditions" as the site for the students to write their theses. On the required conditions are as follows:

(1) The plant is determined to make technological reform on automation and there is actually a programme on the agenda.

(2) The director should know what technological progress and benefit automation will bring to the plant and should support the students in taking part in the technological reform.

(3) The plant should have adequate personnel of its own and the required services.

(4) The plant should sign long-term agreements with the institute our university providing technical aid and the plant providing the necessary fund, equipment, instruments, materials and personnel assigned to instruct the students jointly with the teachers.

B. Each year, one or more groups of would be graduates is sent to the plant, and their theses should solve some part of the technological problems listed on the agreement.

For example, in 1980 an agreement on saving energy through automation was signed with an ammonia plant which uses the waste gas from refinery as its raw material and fuel, and its installations needed reform. Since 1980, 4 grades of would be graduates have been sent to the production site, and they completed their assigned task in technological reform and their theses at the same time.

The tasks assigned to them are:

1980 group, in charge of the designs of a N_2:H_2 ratio control system and an H_2O:C ratio control system and the installation of instruments. Another small group took part in the research and manufacture of an online total carbon meter.

1981 group: in charge of the design of automatic control of excess O_2 content in fuel gas in the reforming furnace, including the solution of the continuous measurement of O_2 content and the experiment for improving the total carbon meter.

1982 group: in charge of the design of a control system controlling the heat value of mixed gases in the general network, also took part in the experiment testing the reliability of domestic made heating value index meters.

1983 group: in charge of the design of new types of hydrocarbon total carbon meters and a thorough analysis of error distribution of meters. Nevertheless from our observation in recent years, once the students have been sent to the plant and assigned to fulfill concrete task. The over-reliance on the teacher would hinder the student's academic development and make them incompetent in real work.

In order to get rid of the student's over-reliance on the teacher and to speed up the process to be an competent engineer, we have taken the graduates to the fore front of technological reform.

They have to be responsible for certain projects, even though at first they may not know what to do. We insist that they should make their working plans, present programmes for experiments, make investigations and handle the problems raised during their work.

The teacher should play the role of

-- coordinator

-- encourager

-- guide

-- verifier

In our opinion, courage is an essential quality of the engineer. The students should have courage to hold discussions with the directors and the engineers, to defend their designs on hearings, and to overcome all kinds of difficulties.

The extension of ability is based on courage and will power, therefore the formation of courage should precede training of ability. Thus the students are able to get rid of their over-reliance on the teachers and to work independently sooner.

C. Hard training

To get rid of the over-reliance on the teachers does not mean that the teacher should leave the students alone.

On the contrary, the teacher should concentrate on teaching methods dealing with the students individually according to their ability with far sight.

How to make the best out of the limited time for completing the student's thesis.

The way of hard training in training atheletes may by employed.

The hard training of the students on the production site is to increase their work load, so that their work is busy and in high tension. For example in the winter of 1981 a few students were in charge of the design of a control system controlling the O_2 content in fuel gas, because of the hard training the students completed the design within two weeks and it has been proved to be a good design.

The students' theses were up to standard too. For example the topic of one of the theses was the measureing of total carbon, a special method different from these of using industrial chromatography or P.C. Tranducer (Foxboro patent). The idea of the thesis was highly recommended by the experts of the related departments of the state.

ACKNOWLEDGEMENTS

The writer is most grateful to Professor Jiang Weisun, Associate Professor Zhang Xian-lou, Mr, Yang Jin-ming, Mr. Xu Yi-ming, Mr. Zheng Si-xiong and other colleagues without whose help and constant encouragement, the paper would never have been written.

The writer is much indebted to IPC chairman, Mr. Yan Xiao-jun and Mr. Yu Ren of the East China Institute of Chemical Technology, who read the paper and made valuable suggestions.

The writer also wishes to thank Mr. Zhu Wen-xiong for his help.

A MACROSCOPIC PREDICTIVE MODEL OF TEACHER'S STRUCTURE IN CHINA'S INSTITUTIONS OF HIGHER EDUCATION

Chen Ling and Pan Guozhong

Huazhong University of Science and Technology, Wuhan, China

Abstract. In this paper, the age distribution of university graduates, Masters and doctors as well as the relationship between teacher's promotion probability and age are discussed by using the membership function of fuzzy mathematics. The state transition of teacher with respect to title is regarded as a Markovian process.

Keywords. Fuzzy mathematics; Markovian process; Membership function; Promotion probability; Age distribution; Predictive model; Teacher's structure.

INTRODUCTION

The prediction of teachers' structure is divided into two parts:
1. The number of teachers with different titles must be predicted every year.
2. The number of teachers with different titles at different ages must be predicted every year.

It is easy to predict the number of teachers, but difficult to predict the number of teachers holding the same title but of different ages, which is due to.

(1) Before 1980, replenished teachers in institutions of higher education were mainly selected from among university graduates while nowadays new teachers are replenished from university graduates, Masters and doctors. Owing to the unavailability of statistical data on the age distribution of Masters and doctors, it is necessary to resort to the membership function of fuzzy mathematics.

(2) Because the promotion policy has changed, the relationship between the teacher's promotion probability and age must be found under the new promotion policy. And the relationship must be described by using the membership function of fuzzy mathematics. Because the state transition of teachers with one title to another is a random process and, generally speaking, the random process possesses the Markovian property, the random process is regarded as a Markovian process.

THE SCHEMATIC DIAGRAM OF TEACHERS TITLE STRUCTURE AND THE DEFINITIONS OF SYMBOLS

1. The Schematic Diagram of Teachers' Title Structure

In China's institutions of higher education, the structure of teachers' title is as shown in Fig.1.

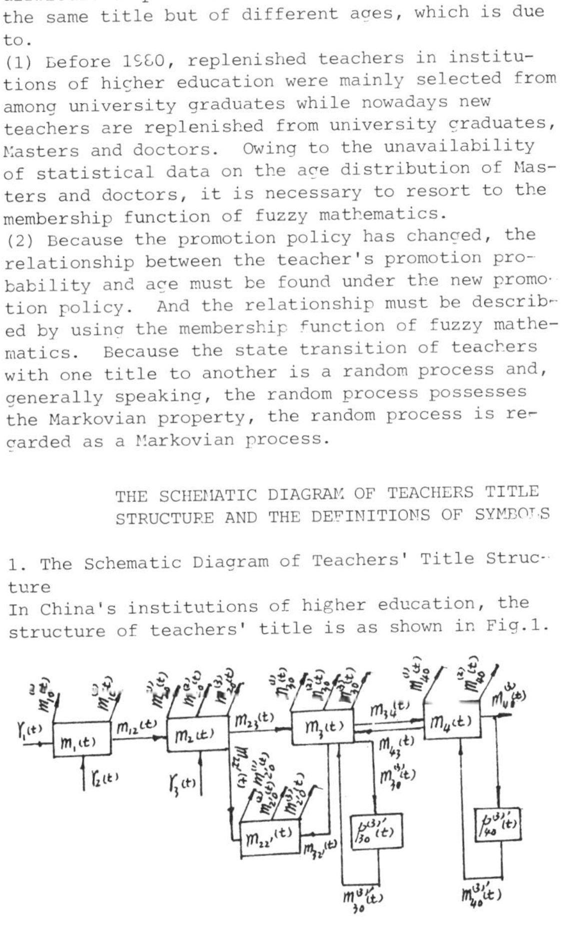

Fig.1. The Schematic Diagraom of Teachers' Title Structure

2. The Definitions of Symbols

(1) $r_k(t)$ is the number of newly replenished teachers in the year t, where k=1,2,3,and they represent university graduates, Masters and doctors,respectively.

(2) $m_i^{(k)}(t)$ is the total number of teachers with i-title and k-educational background in the year t, where i=1,2,2',3,4, and they represent teaching assistants, instructors, senior instructors, associate professors and professors, respectively.

(3) $m_i(t)$ is the total number of teachers with i-title in the year t, where i=1,2,2',3,4.

(4) $m_{io}^{(1)}(t)$ is the number of deceased teachers with i-title in the year t.

(5) $m_{iko}^{(1)}(t)$ is the number of deceased teachers with i-title and k-educational background in the year t.

(6) $m_{io}^{(2)}(t)$ is the number of teachers with i-title flowing from University to other places in the year t.

(7) $m_{iko}^{(2)}(t)$ is the number of teachers with i-title and k-educational background flowing from University to other places in the year t.

(8) $m_{io}^{(3)}(t)$ is the number of teachers with i-title retiring in the year t.

(9) $m_{io}^{(3)}(t)$ is the number of teachers with i-title and k-educational background retiring in the year t.

(10) $m_{io}^{(3)'}(t)$ is the number of retired teachers with i-title and reengaged in the year t.

(11) $m_{iko}^{(3)'}(t)$ is the number of retired teachers with i-title and k-educational background reengaged in the year t.

(12) $p_{io}^{(1)}(t)$ is the death rate of teachers with i-title in the year t.

(13) $p_{iko}^{(1)}(t)$ is the death rate of teachers with i-title and k-educational background in the year t.

(14) $p_{io}^{(2)}(t)$ is the flowing rate of teachers with i-title in the year t.

(15) $p_{iko}^{(2)}(t)$ is the flowing rate of teachers with i-title and k-educational background in the year t.

(16) $p_{io}^{(3)}(t)$ is the retiring rate of teachers with i-title in the year t.

(17) $p_{iko}^{(3)}(t)$ is the retiring rate of teachers with i-title and k-educational background in the year t.

(18) $p_{io}^{(3)'}(t)$ is the engaging rate of teachers with i-title in the year t.

(19) $p_{iko}^{(3)'}(t)$ is the engaging rate of retired teachers with i-title and k-educational background in the year t.

(20) $\bar{p}_{ij}(t)$ is the average state transition probability of teachers with i-title in the year t, where j=1,2,2',3,4.

(21) $\bar{p}_{ij}^{(k)}(t)$ is the average state transition probability of teachers with i-title and k-educational background in the year t.

(22) $m_{ij}(t)$ is the number of teachers with i-title undergoing state transition in the year t.

(23) $m_{ij}^{(k)}(t)$ is the number of teachers with i-title and k-educational background undergoing state transition in the year t.

(24) $m_{i,n}(t)$ is the number of teachers with i-title at age n in the year t, where n=1,2,3,..., 47. When n=1 corresponds to the age of 20, while n=47 corresponds to an age over or equal to 66. And it is stipulated that teachers shall retire at n 42.

(25) $m_{i,n}^{(k)}(t)$ is the number of teachers with i-title and k-educational background at age n in the year t.

THE STRUCTURE VECTOR OF TEACHERS' AGES AND THE CALCULATING FORMULAS OF TEACHERS' NUMBER

1. The death rate and flowing rate of teachers are determined by using the statistical method.
Because the death rate and flowing rate of teachers are very small, they are regarded as of equal-distribution, that is,

$$p_{io}^{(1)}(t)=\sum_{t=-T+1}^{0} m_{io}^{(1)}(t)/\sum_{t=-T+1}^{0} m_i(t-1) \approx 0.2\%, T=1 \sim 4 \quad (1)$$

$$p_{iko}^{(1)}(t)=\sum_{t=-T+1}^{0} m_{iko}^{(1)}(t)/\sum_{t=-T+1}^{0} m_i^{(k)}(t-1) \approx 0.2\%, T=1 \sim 4. \quad (2)$$

$$m_{io}^{(1)}(t)=p_{io}^{(1)}(t)m_i(t-1) \quad (3)$$

$$m_{iko}^{(1)}(t)=p_{iko}^{(1)}(t)m_i^{(k)}(t-1) \quad (4)$$

$$p_{io}^{(2)}(t)=\sum_{t=-T+1}^{0} m_{io}^{(2)}(t)/\sum_{t=-T+1}^{0} m_i(t-1) \approx 2\% \quad (5)$$

$$p_{iko}^{(2)}(t)=\sum_{t=-T+1}^{0} m_{iko}^{(2)}(t)/\sum_{t=-T+1}^{0} m_i^{(k)}(t-1) \approx 2\% \quad (6)$$

$$m_{io}^{(2)}(t)=p_{io}^{(2)}(t)m_i(t-1) \quad (7)$$

$$m_{iko}^{(2)}(t)=p_{iko}^{(2)}(t)m_i^{(k)}(t-1) \quad (8)$$

2. The calculation of teachers' retiring rate
The retiring rate of teachers is determined by age, that is,

$$p_{io}^{(3)}(t)=\sum_{n=42-t+1}^{47} m_{i,n-t+1}^{(t)}(t)/m_i(t-1) \quad (9)$$

$$p_{iko}^{(3)}(t)=\sum_{n=42-t+1}^{47} m_{i,n-t+1}^{(k)}(t)/m_i^{(k)}(t-1) \quad (10)$$

3. The calculation of the engagement rate of teachers
The engagement probability of retired associate professors and professors is a function of age, that is, the older the age is, the smaller the probability of engagement. So the relationship between engagement probability and age is described by the fuzzy membership function of negative exponential distribution, that is,

$$p_{io}^{(3)'}(n,t)=p_{iko}^{(3)'}(n,t)=\begin{cases} \frac{1}{h_i}e^{-R_i(n_i-42)}, & i=3,4, n_i \in n, \ n_i \geq 42; \\ 0, & i=3,4, n_i \in n, \ n_i < 42; \\ 0 & i=1,2,2', \ n_i \in n. \end{cases} \quad (11)$$

Using the Delphi method, we have

$$h_3=2.5, \quad h_4=2.0, \quad R_i=0.2$$

$$p_{io}^{(3)'}(n,t)=p_{iko}^{(3)'}(n,t)=\{p_{iko}^{(3)'}(n,t)\}^T, \quad i=3,4, \ k=1,2,3 \quad (12)$$

4. The distribution function of promotion probability
According to the new promotion policy, the relationship between promotion probability and age can be described by the following fuzzy membership function:

$$p_{12}^{(k)}(n,t)=\frac{1}{\beta(k)}e^{-|n_1(t)-b_2^{(k)}(t)|/\beta^{(k)}},$$
$$n_1(t) \in n, \ n_1(t) \leq b_2^{(k)}, \quad k=1,2, \quad (13)$$

The $\beta^{(k)}$ and $b_2^{(k)}(t)$ values obtained by the Delphi method are listed in Table 1.

Table 1 Values of $\beta^{(k)}$, $b_2^{(k)}(t)$

Time	$\beta^{(1)}$	$\beta^{(2)}$	$b_2^{(1)}(t)$	$b_2^{(2)}(t)$
1985-1990	2.5	2.0	16	21
1991-1995	2.5	2.0	12	15
1996-2000	2.5	2.0	8	10

$$p_{12}^{(k)}(n,t)=\{p_{12}^{(k)}(n,t)\}^T, \quad k=1,2 \quad (14)$$

$$p_{ij}^{(k)}(n,t)=\frac{1}{\sigma_i^{(k)}\sqrt{2\pi}}e^{-[n_i^{(k)}(t)-c_j^{(k)}(t)]^2/2[\sigma_i^{(k)}]^2}$$
$$n_i^{(k)}(t), c_j^{(k)}(t) \in n \quad i=2,3, \ j=i+1, \ k=1,2,3 \quad (15)$$

where the values of $\sigma_i^{(k)}$ are determined by the age set of teachers with i-title and k-educational background. That is, the more age numbers the age set contains, the greater the values of $\sigma_i^{(k)}$. Otherwise, they are small. The values of $C_j^{(k)}(t)$ where determined by the Delphi method. The values of $\sigma_i^{(k)}$ and $c_j^{(k)}(t)$ are listed in Table 2.

Table 2 Values of $\sigma_i^{(k)}$, $c_j^{(k)}(t)$

Time	$\sigma_i^{(1)}$	$\sigma_i^{(2)}$	$\sigma_i^{(3)}$	$c_3^{(1)}(t)$	$c_3^{(2)}(t)$	$c_3^{(3)}(t)$	$c_4^{(1)}(t)$	$c_4^{(2)}(t)$	$c_4^{(3)}(t)$
1985-1990	4.0	3.5	3.0	25	25	25	30	30	30
1991-1995	3.5	3.0	2.5	20	20	20	25	25	25
1996-2000	2.0	2.5	2.0	18	16	14	20	18	16

$$p_{ij}^{(k)}(n,t)=\{p_{ij}^{(k)}(n,t)\}^T, \quad i=2.3, \ j=i+1, \ k=1,2,3 \quad (16)$$

$$p_{22'}^{(k)}(n,t)=\frac{1}{\lambda^{(k)}}e^{-|n_2^{(k)}(t)-d_{2'}^{(k)}(t)|/\lambda^{(k)}}$$
$$n_2^{(k)}(t), \ d_{2'}^{(k)}(t) \in n, \quad k=1,2,3. \quad (17)$$
$$n_2^{(k)}(t) \leq d_2^{(k)}(t) \leq 42$$

The values of $\lambda^{(k)}$, $d_{2'}^{(k)}$ are determined by the Delphi method and listed in Table 3.

$$p_{22'}^{(k)}(n,t)=\{p_{22'}^{(k)}(n,t)\}^T, \quad k=1,2,3 \quad (18)$$

Adopting Delphi method, we have found the rate of demoted assoicate professors and professors.

$$p_{32'}^{(k)}(t)=p_{32'}(t)=p_{43}^{(k)}(t)=0.03 \quad (19)$$

Table 3 Values of $\lambda^{(k)}(t)$, $d_{2'}^{(k)}(t)$, $n_2^{(k)}(t)$

Time	$\lambda^{(k)}$	min $n_2^{(k)}(t)$	$d_{2'}^{(k)}(t)$
1985-1990	5.0	28	42
1991-1995	6.5	24	38

Time	$\lambda^{(k)}$	$\min n_2^{(k)}(t)$	$d_{2'}^{(k)}(t)$
1996-20000	8.0	20	30

5. The age distribution of University graduates, Masters and doctors. Their age distribution can be approximately described by the normal fuzzy membership function, that is,

$$p_r^{(k)}(n,t)=\frac{1}{\sigma^{(k)}\sqrt{2\pi}}e^{-[n_r^{(k)}-e_r^{(k)}]^2/2[\sigma^{(k)}]^2}\ n_r^{(k)},$$

$$e_r^{(k)}\varepsilon n \qquad (20)$$

$$p_r^{(k)}(n,t)=\{p_r^{(k)}(n,t)\}^T \qquad (21)$$

The values of $\sigma^{(k)}$ and $e_r^{(k)}$ are listed in Table 4.

Table 4 Values of $\sigma^{(k)}$, $e_r^{(k)}$

$\sigma^{(1)}$	$\sigma^{(2)}$	$\sigma^{(3)}$	$e_r^{(1)}$	$e_r^{(2)}$	$e_r^{(3)}$
0.7	1.2	2.0	3	7	12

6. The age structure vector of promoted teachers
(1) The age structure vector of teachers promoted to the rank of teaching assistants
All university graduates are promoted to the rank of teaching assistants after one year's work. Hence

$$R_{11}(n,t)=r_1(t-1)p_r^{(1)'}(n,t) \qquad (22)$$

where

$$p_r^{(1)'}(n,t)=\{\frac{1}{\sigma^{(1)}\sqrt{2\pi}}e^{-[n^{(1)}-(e^{(1)}+1)]^2/2[\sigma^{(1)}]^2}\}^T \qquad (23)$$

(2) Masters are teaching assistants and doctors are instructors. Hence

$$R_2(n,t)=r_2(t)p_r^{(2)}(n,t) \qquad (24)$$

$$R_3(n,t)=r_3(t)p_r^{(3)}(n,t) \qquad (25)$$

(3) The age structure vector for the promotion of teachers with other titles.
According to statistical data, the age structure vector of teachers with i-title and k-educational background in the year (t-1) is

$$M^{(k,i,n,\ t-1)}=\{m_{i,n-t+1}^{(k)}(t-1)\}^T \qquad t\geq 1 \qquad (26)$$

Then, in the year t, the age strucutre vector of teachers with i-title and k-educational background that are promoted to j-title (j=i+1) is

$$M^{(k,i.j,n,t)}=\eta_i^{(k)}M^{(k,i,n+1,t-1)'}p_{ij}^{(k)}(n,t) \qquad (27)$$

where

$$M^{(k,i,n+1,t-1)'}=\begin{bmatrix} 0 & & & & 0 \\ m_{i,n-t+1=2}^{(k)(t-1)} & & & & \\ & m_{i,n-t+1=3}^{(k)\ (t-1)} & & & \\ & & \ddots & m_{i,n-1+1=46}^{(k)\ (t-1)} & \\ 0 & & & & m_{i,n-t+1=4}^{(k)\ (t-1)} \end{bmatrix} \qquad (28)$$

$\eta_i^{(k)}$ is equilibrium coefficient, whose values are constant when title i and educational background k are determined, that is

$$m_{ij}^{(k)}(t)=\eta_i^{(k)}[0,M_{i,n-t+1=2}^{(k)\ (t-1)},M_{i,n-t+1=3}^{(k)\ (t-1)},\ldots,$$

$$M_{i,n-t+1=47}^{(k)\ (t-1)}]P_{ij}^{(k)}(n,t),$$

$$\eta_i^{(k)}=m_{ij}^{(k)}(t)/\sum_{n-1}^{47}[m_{i,n-t+1}^{(k)(t-1)}P_{ij}^{(k)}(n,t)] \qquad (29)$$

$m_{ij}^{(k)}(t)$ is determined by means of exchanging information between decision maker and computer.
The total age structure vector of teachers promoted from i-title to j-title is

$$M^{(i,j,n,t)}=\sum_{k=1}^{3}M^{(k,i,j,n,t)} \qquad (30)$$

7. The age structure vector of teachers with i-title in the year t.
(1) Assumed conditions
(a) The death rate of teachers is equi-distributed at each age.
(b) The flow rate of teachers is equi-distributed at each age.
(c) $p_{32'}(t)$ and $p_{43}(t)$ are equi-distributed at each age.
According to the above-mentioned assumed conditions, the age structure vectors of the various titles can be found as follows:
(2) The age structure vector of teaching assistants in the year t.
The age structure vector of teaching assistants who were graduated from university is

$$M^{(1,1,n,t)}=R_{11}(n,t)+[1-p_{1ko}^{(1)}(t)-p_{1ko}^{(2)}(t)]\cdot$$

$$\cdot M^{(1,1,n+1,t-1)}-M^{(1,1,2,n,t)} \qquad (31)$$

The age structure vector of teaching assistants who have won Master's degree is

$$M^{(2,1,n,t)}=R_2(n,t)+[1-p_{1ko}^{(1)}(t)-p_{1ko}^{(2)}(t)]\cdot$$

$$\cdot M^{(2,1,n+1,t-1)}-M^{(2,1,2,n,t)} \qquad (32)$$

The age structure vector of teaching assistants in

$$M^{(1,n,t)}=\sum_{k=1}^{2}M^{(k,1,n,t)} \qquad (33)$$

where the first superscript of M represents the education k, the second represents the title i and the third represents j.
(3) The age structure vector of instructors in the year t.
The age structure vector of instructors in were graduated from university is

$$M^{(1,2,n,t)}=M^{(1,1,2,n,t)}+[1-p_{2ko}^{(1)}(t)-p_{2ko}^{(2)}(t)]\cdot$$

$$\cdot M^{(1,2,n+1,t-1)}-M^{(1,2,3,n,t)}-M^{(1,2,2',n,t)}-$$

$$-M_{2ko}^{(3)}(n,t),\quad k=1 \qquad (34)$$

The age structure vector of instructors who have earned Master's degree is

$$M^{(2,2,n,t)}=M^{(2,1,2,n,t)}+[1-p_{2ko}^{(1)}(t)-p_{2ko}^{(2)}(t)]\cdot$$

$$\cdot M^{(2,2,n+1,t-1)}-M^{(2,2,3,n,t)}-M^{(2,2,2',n,t)}-$$

$$-M_{2ko}^{(3)}(n,t),\qquad k=2 \qquad (35)$$

The age structure vector of instructors who have earned doctor's degree is

$$M^{(3,2,n,t)}=R_3(n,t)+[1-p_{2ko}^{(1)}(t)-p_{2ko}^{(2)}(t)]\cdot$$

$$\cdot M^{(3,2,n+1,t-1)}-M^{(3,2,3',n,t)}-M^{(3,2,2',n,t)}-$$

$$-M_{2ko}^{(3)}(n,t),\qquad k=3 \qquad (36)$$

where

$$M_{2ko}^{(3)}(n,t)=[0,0,\ldots,m_{2,n-t+1=42}^{(k)\ (t)},\ldots,m_{2,n-t+1=47}^{(k)\ (t)}]^T \qquad (37)$$

The age structure vector of instructors is

$$M^{(2,n,t)}=\sum_{k=1}^{3}M^{(k,2,n,t)} \tag{38}$$

(4) The age structure vector of senior instructors in the year t.

The age structure vector of senior instructors with k-educational background is

$$M^{(k,2',n,t)}=M^{(k,2,2',n,t)}+[1-p^{(1)}_{2'ko}(t)-p^{(2)}_{2'ko}(t)]\cdot M^{(k,2' n+1,t-1)}+p_{32'}(t)M^{(k,3,n+1,t-1)}-M^{(3)}_{2'ko}(n,t) \tag{39}$$

where

$$M^{(3)}_{2'ko}(n,t)=[0,0,\ldots,0,\ m^{(k)}_{2'n-t+1=42}(t),\ldots,m^{(k)}_{2'n-t+1=47}(t)]^T \tag{40}$$

$$M^{(2',n,t)}=\sum_{k=1}^{3}M^{(k,2',n,t)} \tag{41}$$

(5) The age structure vector of assoicate professors in the year t.

The age structure vector of associate professors with k-educational background is

$$M^{(k,3,n,t)}=M^{(k,2,3,n,t)}+[1-p^{(1)}_{3ko}(t)-p^{(2)}_{3ko}(t)-p_{32'}(t)]M^{(k,3,n+1,t-1)}-M^{(k,3,4,n,t)}-M^{(3)}_{3ko}(n,t)+M^{(3)'}_{3ko}(n,t)+p_{43}(t)M^{(k,4,n+1,t-1)} \tag{42}$$

where

$$M^{(3)}_{3ko}(n,t)=[0,0,\ldots,m^{(k)}_{3,n-t+1=42}(t),\ldots,m^{(k)}_{3,n-t+1=47}(t)]^T \tag{43}$$

$$M^{(3)'}_{3ko}(n,t)=[M^{(3)}_{3ko}(n,t)]'P^{(3)'}_{30}(n,t) \tag{44}$$

$$[M^{(3)}_{3ko}(n,t)]'=\begin{bmatrix} 0 & & & & \\ & \ddots & & 0 & \\ & & 0 & & \\ & & & m^{(k)}_{3,n-t+1=42}(t) & \\ & & & & \ddots \\ & & & & & m^{(k)}_{3,n-t+1=47}(t) \end{bmatrix} \tag{45}$$

The age structure vector of associate professors is

$$M^{(3,n,t)}=\sum_{k=1}^{3}M^{(k,3,n,t)} \tag{46}$$

(6) The age structure vector of professors in the year t.

The age structure vector of professors with k-educational background is age

$$M^{(k,4,n,t)}=M^{(k,3,4,n,t)}+[1-p^{(1)}_{4ko}(t)-p^{(2)}_{4ko}(t)-p_{43}(t)]\cdot M^{(k,4,n+1,t-1)}-M^{(3)}_{4ko}(n,t)+M^{(3)'}_{4ko}(n,t) \tag{47}$$

where

$$M^{(3)}_{4ko}(n,t)=[0,0,\ldots,m^{(k)}_{4,n-t+1=42}(1),\ldots,m^{(k)}_{4,n-t+1=47}(1)]^T \tag{48}$$

$$M^{(3)'}_{4ko}(n,t)=[M^{(3)}_{4ko}(n,t)]'P^{(3)'}_{40}(n,t) \tag{49}$$

$$[M^{(3)}_{4ko}(n,t)]'=\begin{bmatrix} 0 & & & & \\ & 0 & & & \\ & & \ddots & & \\ & & & 0 & \\ & & & & m^{(k)}_{4,n-t+1=42}(t) & \\ & & & & & \ddots \\ & & & & & & m^{(k)}_{4,n-t+1=47}(t) \end{bmatrix} \tag{50}$$

The age structure vector of professors is

$$M^{(4,n,t)}=\sum_{k=1}^{3}M^{(k,4,n,t)} \tag{51}$$

8. The formulae for calculating the number of teachers with i-title in the year t:

$$m_1^{(1)}(t)=r_1(t-1)+[1-p^{(1)}_{1ko}(t)-p^{(2)}_{1ko}(t)]m_1^{(1)}(t-1-m^{(1)}_{12}(t),\quad k=1; \tag{52}$$

$$m_1^{(2)}(t)=r_2(t)+[1-p^{(1)}_{1ko}(t)-p^{(2)}_{1ko}(t)]m_1^{(2)}(t-1)-m^{(2)}_{12}(t),\quad k=2; \tag{53}$$

$$m_1(t)=\sum_{k=1}^{2}m_1^{(k)}(t) \tag{54}$$

$$m_2^{(1)}(t)=m^{(1)}_{12}(t)+[1-p^{(1)}_{2ko}(t)-p^{(2)}_{2ko}(t)]m_2^{(1)}(t-1)-m^{(1)}_{23}(t)-m^{(1)}_{22'}(t)-m^{(3)}_{2ko}(t)\qquad k=1; \tag{55}$$

$$m_2^{(2)}(t)=m^{(2)}_{12}(t)+[1-p^{(1)}_{2ko}(t)-p^{(2)}_{2ko}(t)]m_2^{(2)}(t-1)-m^{(2)}_{23}(t)-m^{(2)}_{22'}(t)-m^{(3)}_{2ko}(t)\qquad k=2; \tag{56}$$

$$m_2^{(3)}(t)=r_3(t)+[1-p^{(1)}_{2ko}(t)-p^{(2)}_{2ko}(t)]m_2^{(3)}(t-1)-m^{(2)}_{23}(t)-m^{(3)}_{22'}(t)-m^{(3)}_{2ko}(t)\qquad k=3; \tag{57}$$

$$m_2(t)=\sum_{k=1}^{3}m_2^{(k)}(t) \tag{58}$$

$$m_{2'}^{(k)}(t)=m^{(k)}_{22'}(t)+[1-p^{(1)}_{2'ko}(t)-p^{(2)}_{2'ko}(t)]m_{2'}^{(k)}(t-1)+p_{32'}(t)\cdot m_3^{(k)}(t-1)-m^{(3)}_{2'ko}(t) \tag{59}$$

$$m_{2'}(t)=\sum_{k=1}^{3}m_{2'}^{(k)}(t) \tag{60}$$

$$m_3^{(k)}(t)=m^{(k)}_{23}(t)+[1-p^{(1)}_{3ko}(t)-p^{(2)}_{3ko}(t)-p_{32'}(t)]m_3^{(k)}(t-1)-m^{(k)}_{34}(t)-m^{(3)}_{3ko}(t)+m^{(3)'}_{3ko}(t)+p_{43}(t)m_4^{(k)}(t-1) \tag{61}$$

$$m_3(t)=\sum_{k=1}^{3}m_3^{(k)}(t) \tag{62}$$

$$m_4^{(k)}(t)=m^{(k)}_{34}(t)+[1-p^{(1)}_{4ko}(t)-p^{(2)}_{4ko}(t)-p_{43}(t)]m_4^{(k)}(t-1)-m^{(3)}_{4ko}(t)+m^{(3)'}_{4ko}(t) \tag{63}$$

$$m_4(t)=\sum_{k=1}^{3}m_4^{(k)}(t) \tag{64}$$

$$m(t)=\sum_{i=1}^{4}m_i(t),\qquad i=1,2,2',3,4 \tag{65}$$

REFERENCES

1. Ye Pei-hua et al., (1983) Educational Statistics, Chinese People's Education Publishing House.
2. Hao De-Yuan, (1982) Educational and Psychological Statistics, Chinese Education and Science Press.
3. Chen Ling et al., (1984) Promotion Models of Personnel in a Specific Field and an Analysis of the Models' After-Effect, Systems Engineering, pp. 181-190
4. Verhoeven, C.J., (1982) Techniques in Corporate Manpower Planning Method and Applications, Kluwer, Nijhoff Publishing House, Boston/the Hague/London.

REPORT ON PANEL DISCUSSION ON AUTOMATIC CONTROL EDUCATION FOR DEVELOPMENT

Chairman: **W. Schaufelberger,** *Switzerland*
Panelists: **G. Guardabassi,** *Italy*
M. Najim, *MA*
Gu Sheng-Go, *China*

The session was opened by brief statements prepared beforehand by the panellists.

Prof. **Guardabassi** raised the question of the definition of developing countries. In his view, education has three objectives:

- prepare technically updated professionals
- prepare researchers
- contribute in a broad sense to cultural development

One aspect of the complex relationship between control theory and practical applications is the rleative uniformity of the ways the former is taught and developed by researchers all over the world as contrasted with the ample variety of different situations that exist on the professional side, due to historical, cultural, economic and political reasons. On the other hand, if the peculiar features of the practical control problems to be faced in developing countries subtract, in a sense, to their "universal" value, it is equally true that their impact in terms of regional development and technological advancement may be enormous; when imbedded in the appropriate social economic and technological environment, they are then to be seen as important technical problems for which no readily cooked solution is likely to be offered by the however rich practical experience gained in sharply different general settings. In conclusion, a critical mission seen for teachers in developing countries is to recognize and accept this challenge by giving the local "environmental" factors and the role the play in determining specific sets of practical control problems special attention. In particular, this should reflect on the process of shaping the (application oriented) research activities and selecting the specific contents of most (application oriented) courses.

Prof. **Najim** discussed the special needs of developing countries in education. The general availability of computers easies the task of teaching but creates new problems of separation between computers and control education. How can special knowledge for development be created? How can education be tuned to the very different needs of the different developing countries? These are major questions in urgent need of answers. Education needs a minimal infrastructure. To provide this structure is a political question.

Prof. **Gu Sheng-Go** presented some suggestions for education in China. General points are

- the orientation towards the future
- the importance of economic and social development
- the prospects of rapid development are very good.

In his own field of interest (drives), Prof. **Gu Sheng-Go** envisages the following development:

machines → flexible manufacturing systems →
factory automation

Of special importance are the quality of silicon controlled rectifiers and of power supplies.

From these considerations, the following requirements for the education are obtained:

General requirements:
- high level engineering and technical competence
- balanced relation between theory and practice
- basic training must be provided at undergraduate level
- microprocessors and industrial electronics must also be introduced at undergraduate level

After graduation, emphasis must be on:
- design of new products
- availability of electricity
- factory management / technical management
- automatic control systems
- teaching and scientific research

The structure of professional knowledge:
- undergraduates: basic knowledge / economic analysis
- graduates: electrical drives / control systems and theory / applications / interfaces, non-electrical tests / economics / manufacturing processes and technologies

The structure of professional competence:
- undergraduates: analysis tools / reference books / foreign languages / manuals / standards
- graduates: ability in design / computers / interfaces / numerics / research / factory automation

It is vital for China to choose the right technology for the future development. An increasing number of jobs must be provided. Will and political power are available, as the population growth control program shows.

The general discussion was held at different levels. Some of the participants wanted to discuss and solve the problems of the developing countries, while others wanted to restrict the discussion to educational problems.

It was noted that the different developing countries have very differing needs in education and due to the very different political situation also very different possibilities. (The situation in Brazil is very different from the one in China.) It was also suggested to drop "automatic" from the field of "automatic control" due to the possible adverse effects of automation on jobs. Special points raised in addition are: the role of mathematics in engineering and the need for a general education in economics, social sciences and management.

We are in urgent need of a new philosophy.

The IFAC-Proposal of the "Committee for the Support of Control Engineering Education in Developing Countries" was briefly presented at the end of the session. The question why this proposal generated so little interest in IFAC was raised.

REPORT ON PANEL DISCUSSION ON IMPACT OF MICROCOMPUTERS ON INDUSTRY

Chairman: **Yan Xiao-Jun,** *China*
Panelists: **Kazuo Inoue,** *Japan*
Tin-Pui Leung, *Hong Kong*
L. Nemes, *Hungary*

Keywords. Microprocessors; Computer aided design; Industrial control.

This Panel Discussion was held at Beijing, China on August 20, 1985. The number of participants to this Panel Discussion is about 60. After the contributions by the Chairman and the Panelists, it led to a lively flow of questions and discussions. Many questions were raised by participants from developing countries, many are located in the region of Southeast Asia. The contributions from the Chairman and Panelists are as follows:

Prof. Yan Xiao-Jun (PRC):

China is a developing country. The year 1984 can be marked as a year of big impact by microcomputer on Chinese Industry. According to China's official news Agency Xinhua, there are more than 20,000 microcomputers used in Chinese industry at the beginning of 1984. Since then, the number of microcomputers used in China increased much more rapidly. These microcomputers were used successfully in engineering computing and product design, production scheduling and management, data processing and forecasting, supervisoring and monitoring, process control etc. The computers used included both Chinese-made and imported microcomputers. Some of the impacts are cited as follows:

1. Technical transform of traditional industries. The word "traditional industry" usually includes iron and steel, coal, petrochemical, machine-building, textile etc. The application on this field occupies more than 40% of the total number of items of application of microcomputer in China. The automatic control system for transportation and storage of oil products in Taching petro-chemical factory brings an economic benefit of 4,960,000 yuan each year. The automatic control of heating process in the reheating furnace of Wuhan steel company by microcomputer saves energy by 20%. The application of microcomputer to the distillation process of ethyl alcohol brings an economic benefit of 1,670,000 yuan to an installation making 15,000 tons of alcohol each year.

2. Application of microcomputer in machine building industry. There are 116 new products and machine tools exhibited in an Exhibition of Machine-Building Industry in China at the end of 1984. Among those, 54 new products used microcomputers. Two Chinese-made Flexible Manufacturing Cells (FMC) were exhibited. These FMCs have been exported to Japan to be used in a Japanese automated factory.

3. Application of microcomputer to make the instruments " smart ". Some of Chinese made instruments have upgraded by using microcomputer such as geology, hydrology, agricultural, metallurgical, environment protection, medical and electrical instruments.

4. Application of microcomputer in electric power industry. A multi microcomputer distributed system has been implemented in a Chinese hydro-power station. This system increases the generation capacity by 2 %.

5. Application of microcomputer to save energy. The Chungking Industrial Automation Institute has developed a microcom-

puter DDC control system for Chengtu Steel Tube Factory. This system lowers the natural gas consumption for a ring type heating furnace by 29.2%, so that the amount of natural gas saved each year is about 3 million cubic meters. Microcomputer controlled electrolytic tanks for a Chinese Aluminum Factory have saved electric energy by 5 million kilowatt hours each year.
Nevertheless, there are lot of problems needed to be solved in China such as:
1. Many varieties and makers of microcomputers appear in Chinese market, including both Chinese made and those imported from abroad. It brings the difficulties in the standardization and compatibility in both hardwares and softwares.
2. Interface problems for process control systems.
3. Lack of adequate sensors, transducers, actuaters etc. for various industrial control systems.
4. The present achievements are largely limited in smaller scale systems. As to those more complex systems, it is necessary to strengthen the education of modern control theory, optimization theory and systems engineering for the system designers.

Prof. Tin-pui Leung (Hongkong):
A lot of microcomputers are used in Hongkong industries. First, they are used to help in the commercial production and manufacturing systems, e.g. for inventory control, data storage, salary calculation and other simple production control and planning. Secondly, microcomputers are used in automation. Programmable controllers based on microcomputers are widely used in sequential control, mainly for injection moulding machines, electroplating and other machine tools and machining processes. Applications can also be found in air-conditioning systems and building service systems, control of simple robots and manipulators. Microcomputers are used in CAD systems. There are small self-developed CAD programs as well as large imported programs used for specific problems. There is also a CAD system developed by the Hongkong Productivity Center for garment making. This program maximizes the utilization of cloth in garment making. Microcomputers are also used for the control of assembling and testing, especially in electronic factories.

Prof. Kazuo Inoue (Japan):
1. Characteristics of a Microcomputer
   - High computation function
   - Flexibility of function by software
   - High reliability
   - Low cost
2. Utilization of Microcomputer on Industry
   - Flexible Manufacturing Systems
     - Production line
     - Automatic inspection
   - Home electrical equipment
   - Automobile
   - Office Automation
3. Impact by use of microcomputers
   - Decentralization of systems
   - Intellectualization of function
   - Low price of development cost for devices and systems

Dr. L. Nemes (Hungary):
Hungary has established a Committee on Impact of Microelectronics in 1984. They have made a report on the impact of microelectronics on Hungarian industries.

AUTHOR INDEX

SUBJECT INDEX

AKASHI: Control Science and Technology for the Progress of Society, 7 Volumes
ALONSO-CONCHEIRO: Real Time Digital Control Applications
ATHERTON: Multivariable Technological Systems
BABARY & LE LETTY: Control of Distributed Parameter Systems (1982)
BANKS & PRITCHARD: Control of Distributed Parameter Systems (1977)
BARKER & YOUNG: Identification and Systems Parameter Estimation (1985)
BASAR & PAU: Dynamic Modelling and Control of National Economies (1983)
BAYLIS: Safety of Computer Control Systems (1983)
BEKEY & SARIDIS: Identification and System Parameter Estimation (1982)
BINDER & PERRET: Components and Instruments for Distributed Computer Control Systems
BULL: Real Time Programming (1983)
BULL & WILLIAMS: Real Time Programming (1985)
CAMPBELL: Control Aspects of Prosthetics and Orthotics
Van CAUWENBERGHE: Instrumentation and Automation in the Paper, Rubber, Plastics and Polymerisation Industries (1980) (1983)
CHESTNUT, GENSER, KOPACEK & WIERZBICKI: Supplemental Ways for Improving International Stability
CHRETIEN: Automatic Control in Space (1985)
CICHOCKI & STRASZAK: Systems Analysis Applications to Complex Programs
CRONHJORT: Real Time Programming (1978)
CUENOD: Computer Aided Design of Control Systems
DA CUNHA: Planning and Operation of Electric Energy Systems
De GIORGIO & ROVEDA: Criteria for Selecting Appropriate Technologies under Different Cultural, Technical and Social Conditions
DI PILLO: Control Applications of Nonlinear Programming and Optimization
DUBUISSON: Information and Systems
ELLIS: Control Problems and Devices in Manufacturing Technology (1980)
FERRATE & PUENTE: Software for Computer Control (1982)
FLEISSNER: Systems Approach to Appropriate Technology Transfer
GELLIE, FERRATE & BASANEZ: Robot Control "SYROCO '85"
GELLIE & TAVAST: Distributed Computer Control Systems (1982)
GERTLER & KEVICZKY: A Bridge Between Control Science and Technology, 6 Volumes
GHONAIMY: Systems Approach for Development (1977)
HAASE: Real Time Programming (1980)
HAIMES & KINDLER: Water and Related Land Resource Systems
HALME: Modelling and Control of Biotechnical Processes
HARDT: Information Control Problems in Manufacturing Technology (1982)
HARRISON: Distributed Computer Control Systems (1979)
HASEGAWA: Real Time Programming (1981)
HASEGAWA & INOUE: Urban, Regional and National Planning—Environmental Aspects
HERBST: Automatic Control in Power Generation Distribution and Protection
ISERMANN: Identification and System Parameter Estimation (1979)
ISERMANN & KALTENECKER: Digital Computer Applications to Process Control
JANSSEN, PAU & STRASZAK: Dynamic Modelling and Control of National Economies (1980)
JOHANNSEN & RIJNSDORP: Analysis, Design, and Evaluation of Man-Machine Systems
JOHNSON: Modelling and Control of Biotechnological Processes
KLAMT & LAUBER: Control in Transportation Systems
KOTOB: Automatic Control in Petroleum, Petrochemical and Desalination Industries
LANDAU: Adaptive Systems in Control and Signal Processing
LARSEN & HANSEN: Computer Aided Design in Control and Engineering Systems
LAUBER: Safety of Computer Control Systems (1979)
LEININGER: Computer Aided Design of Multivariable Technological Systems
LEONHARD: Control in Power Electronics and Electrical Drives (1977)
LESKIEWICZ & ZAREMBA: Pneumatic and Hydraulic Components and Instruments in Automatic Control
MAFFEZZONI: Modelling and Control of Electric Power Plants
MAHALANABIS: Theory and Application of Digital Control
MANCINI, JOHANNSEN & MARTENSSON: Analysis, Design and Evaluation of Man–Machine Systems (1985)
MARTIN: Design of Work in Automated Manufacturing Systems
MILLER: Distributed Computer Control Systems (1981)
MUNDAY: Automatic Control in Space (1979)
NAJIM & ABDEL-FATTAH: Systems Approach for Development (1980)
NIEMI: A Link Between Science and Applications of Automatic Control, 4 Volumes
NORRIE & TURNER: Automation for Mineral Resource Development
NOVAK: Software for Computer Control (1979)
O'SHEA & POLIS: Automation in Mining, Mineral and Metal Processing (1980)
OSHIMA: Information Control Problems in Manufacturing Technology (1977)
PAUL: Digital Computer Applications to Process Control (1985)
PONOMARYOV: Artificial Intelligence
QUIRK: Safety of Computer Control Systems (1985)
RAUCH: Applications of Nonlinear Programming to Optimization and Control
RAUCH: Control Applications of Nonlinear Programming
REMBOLD: Information Control Problems in Manufacturing Technology (1979)
RIJNSDORP: Case Studies in Automation related to Humanization of Work
RIJNSDORP, PLOMP & MÖLLER: Training for Tomorrow — Educational Aspects of Computerized Automation
RODD: Distributed Computer Control Systems (1983)
SANCHEZ: Fuzzy Information, Knowledge Representation and Decision Analysis
SAWARAGI & AKASHI: Environmental Systems Planning, Design and Control
SINGH & TITLI: Control and Management of Integrated Industrial Complexes
SKELTON & OWENS: Model Error Concepts and Compensation
SMEDEMA: Real Time Programming (1977)
STRASZAK: Large Scale Systems: Theory and Applications (1983)
SUBRAMANYAM: Computer Applications in Large Scale Power Systems
SUSKI: Distributed Computer Control Systems (1985)
TITLI & SINGH: Large Scale Systems: Theory and Applications (1980)
UNBEHAUEN: Adaptive Control of Chemical Processes
VALADARES TAVARES & DA SILVA: Systems Analysis Applied to Water and Related Land Resources
WESTERLUND: Automation in Mining, Mineral and Metal Processing (1983)
van WOERKOM: Automatic Control in Space (1982)
YANG JIACHI: Control Science and Technology for Development
ZWICKY: Control in Power Electronics and Electrical Drives (1983)